Fixed Point Theory in Ordered Sets and Applications

Siegfried Carl • Seppo Heikkilä

Fixed Point Theory in Ordered Sets and Applications

From Differential and Integral Equations
to Game Theory

 Springer

Siegfried Carl
Martin-Luther-Universität
Halle-Wittenberg
Institut für Mathematik
Halle
Germany
siegfried.carl@mathematik.uni-halle.de

Seppo Heikkilä
Department of Mathematical Sciences
University of Oulu
Oulu
Finland
sheikki@sun3.oulu.fi

ISBN 978-1-4899-8178-3   ISBN 978-1-4419-7585-0 (eBook)
DOI 10.1007/978-1-4419-7585-0
Springer New York Dordrecht Heidelberg London

Mathematics Subject Classification Codes (2010): 06Axx, 06Bxx, 03F60, 28B05, 34Axx, 34Bxx, 34Gxx, 34Kxx, 35B51, 35J87, 35K86, 45N05, 46G12, 47H04, 47H10, 47J20, 49J40, 91A10, 91B16, 91B50, 58D25

Printed on acid-free paper

Springer is part of Springer Science+Business Media (www.springer.com)

Dedicated in gratitude and high esteem to

Professor V. Lakshmikantham

Preface

Fixed point theory is one of the most powerful and fruitful tools of modern mathematics and may be considered a core subject of nonlinear analysis. In recent years a number of excellent monographs and surveys by distinguished authors about fixed point theory have appeared such as, e.g., [2, 4, 7, 25, 31, 32, 100, 101, 103, 104, 108, 155, 196]. Most of the books mentioned above deal with fixed point theory related to continuous mappings in topological or even metric spaces (work of Poincaré, Brouwer, Lefschetz–Hopf, Leray–Schauder) and all its modern extensions.

This book focuses on an order-theoretic fixed point theory and its applications to a wide range of diverse fields such as, e.g., (multi-valued) nonlocal and/or discontinuous partial differential equations of elliptic and parabolic type, differential equations and integral equations with discontinuous nonlinearities in general vector-valued normed spaces of non-absolutely integrable functions containing the standard Bochner integrable functions as special case, and mathematical economics and game theory. In all these topics we are faced with the central problem of handling the loss of continuity of mappings and/or missing appropriate geometric and topological structure of their underlying domain of definition. For example, it is noteworthy that, in particular, for proving the existence of certain optimal strategies in game theory, there is a need for purely order-related fixed point results in partially ordered sets that are neither convex nor do they have lattice structure and where the fixed point operator lacks continuity.

The aim of this monograph is to provide a unified and comprehensive exposition of an order-theoretic fixed point theory in partially ordered sets and its various useful interactions with topological structures. A characteristic feature of this fixed point theory, which is developed in detail in Chap. 2, is that it is based on an abstract recursion principle, called the Chain Generating Recursion Principle, which was formulated in [112, 133], and which is the common source of all the order-related fixed point results obtained in this book. In particular, the developed fixed point theory includes the classical order-theoretic fixed point result established by Knaster in [153], which was later extended by Tarski in [215], as well as the fixed point theorems due to Bourbaki and Kneser (cf. [228, Theorem 11.C]) and Amann (cf. [228, Theorem 11.D]). Surprisingly enough, very recently, the classical and seminal Knaster–Tarski fixed point theorem has been applied to computational geometry in [195]. This unexpected application emphasizes even more the importance of an order-theoretic fixed point theory.

Chapter 1 serves as an introduction to the subject and discusses some simple examples of the order-theoretic fixed point results along with simple applications from each of the diverse fields. This will help the reader to get some idea of the theory and its applications before entering the systematic study. Chapter 3 provides preliminary results on multi-valued variational inequalities regarding the topological and order-theoretical structure of solution sets. This chapter, which may be read independently, is of interest on its own and contains new results. Our main emphasis is on Chaps. 4–8 where we demonstrate the power of the developed fixed point theory of Chap. 2, which runs like a thread through the entire book. Attempts have been made to attract a broad audience not only by the diverse fields of applications, but also by emphasizing simple cases and ideas more than complicated refinements. In the treatment of the applications, a wide range of mathematical theories and methods from nonlinear analysis and integration theory are applied; an outline of which has been given in an appendix chapter to make the book self-contained.

This book is an outgrowth of the authors' research on the subject during the past 20 years. However, a great deal of the material presented here has been obtained only in recent years and appears for the first time in book form.

We expect that our book will be accessible and useful to graduate students and researchers in nonlinear analysis, pure and applied mathematics, game theory, and mathematical economics.

We are most grateful to our friends and colleagues who contributed through joint works and papers to the preparation of this book. Rather than inadvertently leaving someone out, we have not listed the names, but we hope our friends and collaborators will be satisfied with our thanks.

Finally, we wish to express our gratitude to the very professional editorial staff of Springer, particularly to Vaishali Damle for her effective and productive collaboration.

Halle   *Siegfried Carl*
Oulu   *Seppo Heikkilä*
September 2010

Contents

1

Introduction

In this introductory chapter we give a short account of the contents of the
book and discuss simple notions and examples of the fixed point theory to be
developed and applied to more involved applications in later chapters. As an
introduction to the fixed point theory and its applications let us recall two
fixed point theorems on a nonempty closed and bounded subset P of $\mathbb{R}^m$, one
purely topological (Brouwer's fixed point theorem) and one order-theoretically
based. A point $x \in P$ is called a fixed point of a function $G : P \to P$ if $x = Gx$.
We assume that $\mathbb{R}^m$ is equipped with Euclidean metric.

Theorem 1.1 (Brouwer's Fixed Point Theorem). *If P is a closed,
bounded, and convex subset of $\mathbb{R}^m$, then every continuous function $G : P \to P$
has a fixed point.*

To formulate the purely order-theoretic fixed point theorem we equip $\mathbb{R}^m$ with
the coordinatewise partial order '$\leq$', i.e., for $x, y \in \mathbb{R}^m$, we define $x \leq y$ if
and only if $x_i \leq y_i$, $i = 1, \ldots, m$. A function $G : P \to P$ is called increasing if
$x \leq y$ implies $Gx \leq Gy$. Further, we will need the notion of a *sup-center* of
the set P, which is defined as follows: A point $c \in P$ is called a sup-center of
P if $\sup\{c, x\} \in P$ for each $x \in P$. The next fixed point theorem is a special
case of Corollary 2.41(a) of Chap. 2.

Theorem 1.2. *If P is a closed and bounded subset of $\mathbb{R}^m$ having a sup-center,
then every increasing function $G : P \to P$ has a fixed point.*

Note that in Theorem 1.2 neither continuity of the fixed point operator nor
convexity of the set P is needed. Let us give two examples of sets P that
have the required properties of Theorem 1.2. The geometrical center $c =
(c_1, \ldots, c_m) \in \mathbb{R}^m$ of every set

$$ P = \{(x_1, \ldots, x_m) \in \mathbb{R}^m : \sum_{i=1}^m |x_i - c_i|^p \leq r^p\}, \ p, r \in (0, \infty), \qquad (1.1) $$

S. Carl and S. Heikkilä, *Fixed Point Theory in Ordered Sets and Applications:
From Differential and Integral Equations to Game Theory*,
DOI 10.1007/978-1-4419-7585-0_1, © Springer Science+Business Media, LLC 2011

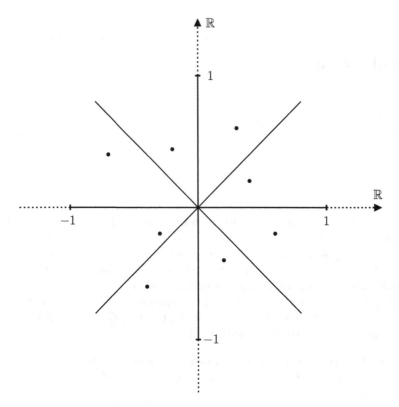

Fig. 1.1. Closed Bounded Set in $\mathbb{R}^2$ with $(0,0)$ as Sup-Center

is a sup-center of P. Because these sets are also closed and bounded, then every increasing mapping $G : P \rightarrow P$ has a fixed point. Notice that P is not convex if $0 < p < 1$, as assumed in Theorem 1.1. If P has the smallest element c, then c is a sup-center of P. If $m = 2$, a necessary and sufficient condition for a point $c = (c_1, c_2)$ of P to be a sup-center of P is that whenever a point $y = (y_1, y_2)$ of P and c are unordered, then $(y_1, c_2) \in P$ if $y_2 < c_2$ and $(c_1, y_2) \in P$ if $y_1 < c_1$. The second example of a set $P \subset \mathbb{R}^2$ is illustrated by Fig. 1.1, where P consists of all the solid lines and the isolated points. One easily verifies that $c = (0,0)$ is a sup-center.

Theorem 1.1 and Theorem 1.2 can be applied, e.g., in the study of the solvability of a finite system of equations. For simplicity consider the system

$$u = u(x,y), \quad v = v(x,y). \tag{1.2}$$

Assume that P is a closed and bounded subset of $\mathbb{R}^2$, and that $G = (u, v)$ maps P into itself. By Theorem 1.1 the system (1.2) has a solution if G is continuous and P is convex. But there is no constructive method to solve system (1.2) under these hypotheses. By Theorem 1.2 the system (1.2) has a solution if G is increasing and P is only assumed to possess a sup-center. As we shall see in Chap. 2 the proof of Theorem 1.2 is constructive. In the special case when strictly monotone sequences of the image $G[P]$ are finite, the following algorithm can be applied to obtain a solution of (1.2) when the sup-center of P is $c = (c_1, c_2)$. Maple commands 'fi;od' in the following program mean 'end if;end do'.

$u := u(x, y) : v := v(x, y) : x := c_1 : y := c_2 :$
for k from 0 while abs$(u - x) +$ abs$(v - y) > 0$ do;
if $(u - x)(v - y) < 0$ then $x := max\{x, u\} : y := max\{y, v\}$
    else $x := u : y := v$:fi;od;
sol $:= (x, y)$;

It is shown in Chap. 2 that the above algorithm can be applied to approximate a solution of (1.2) in the case when G is continuous and increasing, replacing G by its suitable upper and lower estimates.

Consider next generalizations of Theorem 1.1 and Theorem 1.2 to the case when P is a nonempty subset of an infinite-dimensional normed space E. The generalization of Brouwer's fixed point theorem to infinite-dimensional Banach spaces requires the compactness of the fixed point operator. As compact operators play a central role also in later chapters we recall their definition here for convenience, see, e.g., [62, 228].

Definition 1.3. *Let X and Y be normed spaces, and $T : D(T) \subseteq X \to Y$ an operator with domain $D(T)$. The operator T is called* **compact** *iff T is continuous, and T maps bounded sets into relatively compact sets. Compact operators are also called* **completely continuous.**

In Theorem 1.5 we assume that E is ordered by a closed and convex cone E_+ for which $-E_+ \cap E_+ = \{0\}$. A subset A of P is said to have a sup-center in P if there exists a $c \in P$ such that sup$\{c, x\}$ exists in E and belongs to P for every $x \in A$.

Theorem 1.4 (Schauder's Fixed Point Theorem). *Let P be a nonempty, closed, bounded, and convex subset of the Banach space E, and assume that $G : P \to P$ is compact. Then G has a fixed point.*

Theorem 1.5 ([116]). *Let P be a subset of an ordered normed space E, and let $G : P \to P$ be increasing. If the weak closure of $G[P]$ has a sup-center in P, and if monotone sequences of $G[P]$ have weak limits in P, then G has a fixed point.*

If P is, e.g., the closed unit ball in l^2 defined by

$$l^2 = \{x = (x_1, x_2, \dots) : \sum_{i=1}^{\infty} |x_i|^2 < \infty\},$$

then the conclusion of Theorem 1.4 does not hold if $G : P \to P$ is only assumed to be continuous (see Kakutani's counterexample). Thus the result of Theorem 1.4 is not valid if the compactness hypothesis of G is missing. On the other hand, no compactness or continuity is assumed in Theorem 1.5, which is also a consequence of Proposition 2.40(a). The geometrical centers of bounded and closed balls of p-normed spaces l^p, ordered coordinatewise, and $L^p(\Omega)$, $1 \leq p < \infty$, ordered a.e. pointwise, are their sup-centers. This is true also for closed and bounded balls of Sobolev spaces $W^{1,p}(\Omega)$ and $W_0^{1,p}(\Omega)$, $1 < p < \infty$, ordered a.e. pointwise. Moreover, these balls are weakly sequentially closed and their monotone sequences have weak limits. Hence, if P is any of these balls, then every increasing function $G : P \to P$ has a fixed point by Theorem 1.5. To demonstrate the applicability of Theorem 1.4 and Theorem 1.5 let us consider two simple examples of elliptic Dirichlet boundary value problems with homogeneous boundary values.

Example 1.6.

$$-\Delta u(x) = f(x, u(x)) \quad \text{in} \quad \Omega, \quad u = 0 \quad \text{on } \partial\Omega, \tag{1.3}$$

where $\Omega \subset \mathbb{R}^N$ is a bounded domain with Lipschitz boundary $\partial\Omega$. Let us assume that f satisfies the following conditions:

(f1) $f : \Omega \times \mathbb{R} \to \mathbb{R}$ is a Carathéodory function, i.e., $x \mapsto f(x, s)$ is measurable in Ω for all $s \in \mathbb{R}$, and $s \mapsto f(x, s)$ is continuous for almost all (a.a.) $x \in \Omega$.

(f2) The function f fulfills the following growth condition: there is a function $k \in L_+^2(\Omega)$ and a positive constant a such that for a.a. $x \in \Omega$ and for all $s \in \mathbb{R}$ we have

$$|f(x, s)| \leq k(x) + a|s|.$$

By $L_+^2(\Omega)$ we denote the positive cone of all nonnegative functions of $L^2(\Omega)$. Setting $V_0 = W_0^{1,2}(\Omega)$, V_0^* its dual space, $\mathcal{A} = -\Delta$, and defining $\mathcal{A} : V_0 \to V_0^*$ by

$$\langle \mathcal{A}u, \varphi \rangle = \int_\Omega \nabla u \nabla \varphi \, dx, \quad \forall \, \varphi \in V_0,$$

then $\mathcal{A} : V_0 \to V_0^*$ is a strongly monotone, bounded, and continuous operator. Denoting by F the Nemytskij operator associated with f by

$$F(u)(x) = f(x, u(x)),$$

then, in view of (f1)–(f2), $F : L^2(\Omega) \to L^2(\Omega)$ is continuous and bounded. The compact embedding $i : V_0 \hookrightarrow L^2(\Omega)$ readily implies that the operator $\mathcal{F} = i^* \circ F \circ i : V_0 \to V_0^*$ (i^* is the adjoint operator of i) given by

$$\langle \mathcal{F}(u), \varphi \rangle = \int_\Omega F(u) \, \varphi \, dx, \quad \forall \, \varphi \in V_0$$

is completely continuous. With these notations the weak solution of (1.3) can be given the following form: Find $u \in V_0$ such that

$$\mathcal{A}u - \mathcal{F}(u) = 0 \quad \text{in} \quad V_0^*. \tag{1.4}$$

Since $\mathcal{F} : V_0 \to V_0^*$ is completely continuous and bounded, and $\mathcal{A} : V_0 \to V_0^*$ is strongly monotone, continuous, and bounded, it follows that $\mathcal{A} - \mathcal{F} : V_0 \to V_0^*$ is, in particular, continuous, bounded, and pseudomonotone. The classical theory on pseudomonotone operators due to Brezis and Browder (see, e.g., [229]) ensures that if $\mathcal{A} - \mathcal{F} : V_0 \to V_0^*$ is, in addition, coercive, then $\mathcal{A} - \mathcal{F} : V_0 \to V_0^*$ is surjective, which means that (1.4) has a solution, i.e., (1.3) has a weak solution. A sufficient condition to ensure coerciveness of $\mathcal{A} - \mathcal{F}$ is that the positive constant a in (f2) satisfies $a < \lambda_1$, where λ_1 is the first Dirichlet eigenvalue of $\mathcal{A} = -\Delta$, which is known to be positive and simple, see [6]. This can readily be verified by using (f2) and the following variational characterization of the first eigenvalue λ_1 by

$$\lambda_1 = \inf_{0 \neq v \in V_0} \frac{\int_\Omega |\nabla v|^2 \, dx}{\int_\Omega |v|^2 \, dx}.$$

Now we estimate as follows

$$\langle \mathcal{A}u - \mathcal{F}(u), u \rangle \geq \int_\Omega |\nabla u|^2 \, dx - \|k\|_2 \|u\|_2 - a\|u\|_2^2$$
$$\geq \left(1 - \frac{a}{\lambda_1}\right) \|\nabla u\|_2^2 - \frac{\|k\|_2}{\sqrt{\lambda_1}} \|\nabla u\|_2,$$

where $\| \cdot \|_2 = \| \cdot \|_{L^2(\Omega)}$. As $\|u\| = \|\nabla u\|_2$ is an equivalent norm in V_0, we see from the last estimate that

$$\frac{1}{\|\nabla u\|_2} \langle \mathcal{A}u - \mathcal{F}(u), u \rangle \to \infty \quad \text{as} \quad \|\nabla u\|_2 \to \infty,$$

which proves the coercivity, and thus the existence of solutions of (1.4).

An alternative approach to the existence proof for (1.4) that is closely related to the pseudomonotone operator theory is based on Schauder's fixed point theorem (see Theorem 1.4). To this end, problem (1.4) is transformed into a fixed point equation as follows: As $\mathcal{A} = -\Delta : V_0 \to V_0^*$ is a linear, strongly monotone, and bounded operator, it follows that the inverse $\mathcal{A}^{-1} : V_0^* \to V_0$ is linear and bounded, which allows us to rewrite (1.4) in the form: Find $u \in V_0$ such that

$$u = \mathcal{A}^{-1} \circ \mathcal{F}(u) \tag{1.5}$$

holds, i.e., that $u \in V_0$ is fixed point of the operator

$$G = \mathcal{A}^{-1} \circ \mathcal{F}.$$

Since under hypotheses (f1)–(f2), $\mathcal{F} : V_0 \to V_0^*$ is completely continuous, and $\mathcal{A}^{-1} : V_0^* \to V_0$ is linear and bounded, it readily follows that $G : V_0 \to V_0^*$ is continuous and compact. In order to apply Schauder's theorem we are going to verify that under the same assumption on a, i.e., $a < \lambda_1$, G maps a closed ball $B(0, R) \subset V_0$ into itself, which finally allows us to apply Schauder's theorem, and thus the existence of solutions of (1.4). Let $v \in B(0, R)$, and denote $u = Gv$. Then, by definition of the operator G, $u \in V_0$ satisfies

$$\int_\Omega \nabla u \nabla \varphi \, dx = \int_\Omega F(v) \, \varphi \, dx, \quad \forall \, \varphi \in V_0.$$

In particular, the last equation holds for $u = \varphi$, which yields

$$\|\nabla u\|_2^2 = \int_\Omega F(v) \, u \, dx \leq \|F(v)\|_2 \|u\|_2 \leq \|k\|_2 \|u\|_2 + a \|v\|_2 \|u\|_2$$

$$\leq \frac{a}{\lambda_1} \|\nabla v\|_2 \|\nabla u\|_2 + \frac{\|k\|_2}{\sqrt{\lambda_1}} \|\nabla u\|_2,$$

which yields (note $u = Gv$) the norm estimate in V_0

$$\|Gv\|_{V_0} \leq \frac{a}{\lambda_1} \|\nabla v\|_2 + \frac{\|k\|_2}{\sqrt{\lambda_1}}, \quad \forall \, v \in V_0,$$

where $\|u\|_{V_0} := \|\nabla u\|_2$. Thus if $R > 0$ is chosen in such a way that

$$\frac{a}{\lambda_1} R + \frac{\|k\|_2}{\sqrt{\lambda_1}} \leq R,$$

then G provides a mapping of $B(0, R)$ into itself. Such an R always exists, because $\frac{a}{\lambda_1} < 1$. This completes the existence proof via Schauder's fixed point theorem.

Schauder's theorem fails if $\mathcal{F} : V_0 \to V_0^*$ lacks compactness, which may occur, e.g., when in (f2) a critical growth of the form

$$|f(x, s)| \leq k(x) + a|s|^{2^* - 1}$$

is allowed, where 2^* is the critical Sobolev exponent. Lack of compactness occurs also if (1.3) is studied in unbounded domains, or if $s \mapsto f(x, s)$ is no longer continuous. It is Theorem 1.5 that allows us to deal with these kinds of problems provided the fixed point operator G is increasing. In particular, if only continuity of G is violated, then neither monotone operator theory in the sense of Brezis–Browder–Lions–Minty nor fixed point theorems that assume as a least requirement the continuity of the fixed point operator can be applied. To give a simple example, where standard methods fail, consider the next example.

Example 1.7. Let Ω be as in the example before. We study the following discontinuous Dirichlet boundary value problem:

$$-\Delta u(x) = a[u(x)] + k(x) \quad \text{in} \ \Omega, \quad u = 0 \ \text{on} \ \partial\Omega, \tag{1.6}$$

where $a > 0$ is some constant, $k \in L^2(\Omega)$, and $s \mapsto [s]$ stands for the integer function, i.e., $[s]$ denotes the greatest integer with $[s] \leq s$. Apparently, in this case $f(x, s) := a[s] + k(x)$ is discontinuous in $s \in \mathbb{R}$. Set $\tilde{k}(x) = |k(x)| + 1$, then we have $\tilde{k} \in L^2_+(\Omega)$, and the following estimate holds

$$|f(x, s)| \leq \tilde{k}(x) + a|s|.$$

Due to the structure of f the Nemytskij operator $F : L^2(\Omega) \to L^2(\Omega)$ is still well defined and bounded, however, F is no longer continuous. With the same notation as in Example 1.6 we can transform the elliptic problem (1.6) into the fixed point equation in V_0 of the form

$$u = \mathcal{A}^{-1} \circ \mathcal{F}(u). \tag{1.7}$$

The same estimate as in the previous example shows that the fixed point operator $G = \mathcal{A}^{-1} \circ \mathcal{F}$ maps a ball $B(0, \tilde{R}) \subset V_0$ into itself provided $a < \lambda_1$, and $\tilde{R} > 0$ is sufficiently large. Note, however, that the fixed point operator is no longer continuous. Now, we easily observe that $G : V_0 \to V_0$ is increasing with respect to the underlying natural partial order in V_0 defined via the order cone $L^2_+(\Omega)$. The latter is a simple consequence of the fact that $\mathcal{F} : V_0 \to V_0^*$ is increasing, and because of the inverse monotonicity of $\mathcal{A}^{-1}$, which is a consequence of the maximum principle for the Laplacian. Taking into account the comments after Theorem 1.5, we may apply Theorem 1.5 to ensure that G has a fixed point, which proves the existence of weak solutions for (1.6) provided $0 < a < \lambda_1$. It should be noted that the classical fixed point results for increasing self-mappings due to Amann, Tarski, and Bourbaki (see [228]) cannot be applied here.

Further applications of Theorem 1.5 to deal with elliptic problems that lack compactness are demonstrated in [48], where we prove existence results for elliptic problems with critical growth or discontinuity of the data.

The results of Theorem 1.4 and Theorem 1.5 can be extended to set-valued (also called multi-valued) mappings. Let us assume that P is a nonempty subset of a topological space X. In Theorem 1.9 we assume that X is equipped with such a partial ordering that the order intervals $[a, b] = \{x \in X : a \leq x \leq b\}$ are closed. Denote by 2^P the set of all subsets of P. An element x of P is called a *fixed point* of a set-valued mapping $\mathcal{F} : P \to 2^P$ if $x \in \mathcal{F}(x)$. We say that $\mathcal{F}$ is *increasing* if, whenever $x \leq y$ in P, then for every $z \in \mathcal{F}(x)$ there exists a $w \in \mathcal{F}(y)$ such that $z \leq w$, and for every $w \in \mathcal{F}(y)$ there exists a $z \in \mathcal{F}(x)$ such that $z \leq w$.

Theorem 1.8 (Generalized Theorem of Kakutani). *A multi-valued function $\mathcal{F} : P \to 2^P$ has a fixed point if P is a nonempty, compact, and convex set in a locally convex Hausdorff space X, $\mathcal{F} : P \to 2^P$ is upper semi-continuous, and if the set $\mathcal{F}(x)$ is nonempty, closed, and convex for all $x \in P$.*

The following theorem is a consequence of Theorem 2.12, which is proved in Chap. 2.

Theorem 1.9. *A multi-valued function $\mathcal{F} : P \to 2^P$ has a fixed point if $\mathcal{F}$ is increasing, its values $\mathcal{F}(x)$ are nonempty and compact for all $x \in P$, chains of $\mathcal{F}[P]$ have supremums and infimums, and if $\overline{\mathcal{F}[P]}$ has a sup-center in P.*

In particular, if P is any set defined in (1.1), then every increasing mapping $\mathcal{F} : P \to 2^P$ whose values are nonempty closed subsets of $\mathbb{R}^m$ has a fixed point by Theorem 1.9. As a further consequence of Theorem 1.9 one gets the following order-theoretic fixed point result in infinite-dimensional ordered Banach spaces, which is useful in applications to discontinuous differential equations (see Theorem 4.37).

Theorem 1.10. *Let P be a closed and bounded ball in a reflexive lattice-ordered Banach space X, and assume that $\|x^+\| = \|\sup\{0, x\}\| \leq \|x\|$ holds for all $x \in X$. Then every increasing mapping $\mathcal{F} : P \to 2^P$, whose values are nonempty and weakly sequentially closed, has a fixed point.*

To give an idea of how Theorem 1.10 can be applied to differential equations, let us consider a simple example.

Example 1.11. Consider the following slightly extended version of problem (1.6):

$$-\Delta u(x) = a[u(x)] + g(x, u(x)) + k(x) \quad \text{in} \quad \Omega, \quad u = 0 \quad \text{on} \quad \partial\Omega, \qquad (1.8)$$

where $g : \Omega \times \mathbb{R} \to \mathbb{R}$ is a Carathéodory function with the following growth condition

(g) There exist a positive constant b with $b < \lambda_1 - a$, and a $h \in L^2(\Omega)$, such that for a.a. $x \in \Omega$ and for all $s \in \mathbb{R}$

$$|g(x, s)| \leq h(x) + b|s|$$

holds. Here a and λ_1 are as in Example 1.7

If we rewrite the right-hand side of equation (1.8) in the form

$$f(x, s, r) := a[r] + g(x, s) + k(x), \qquad (1.9)$$

then we can distinguish between the continuous and discontinuous dependence of the right-hand side of (1.8). This allows an approach toward the existence of solutions of (1.8) by means of the multi-valued fixed point Theorem 1.9. Note, $s \mapsto f(x, s, r)$ is continuous, and $r \mapsto f(x, s, r)$ is discontinuous and monotone increasing. Let $v \in V_0$ be fixed, and consider the boundary value problem

$$-\Delta u(x) = f(x, u(x), v(x)) \quad \text{in} \quad \Omega, \quad u = 0 \quad \text{on} \quad \partial\Omega. \qquad (1.10)$$

As the function $(x, s) \mapsto f(x, s, v(x))$ with f defined in (1.9) is a Carathéodory function, one can apply the same approach as in Example 1.6 to get the existence of solutions for (1.10). For fixed $v \in V_0$, denote now by Gv the set of all solutions of (1.10). This provides a multi-valued mapping $G : V_0 \to 2^{V_0}$, and certainly any fixed point of G is a solution of the original boundary value problem (1.8), and vice versa. By similar estimates as in Examples 1.6 and 1.7 one can show that under the given assumptions, in particular due to $0 < a + b < \lambda_1$, there is a closed ball $B(0, R) \subset V_0$ such that the multi-valued mapping G maps $B(0, R)$ into itself. As V_0 is a reflexive lattice-ordered Banach space satisfying $\|u^+\| = \|\sup\{0, u\}\| \le \|u\|$ for all $u \in V_0$, for $G : B(0, R) \to 2^{B(0,R)}$ to possess a fixed point it is enough to show that $G : B(0, R) \to 2^{B(0,R)}$ is increasing, and that the images Gv are weakly sequentially closed, see Theorem 4.37. This will be demonstrated for more involved elliptic problems in Chap. 4.

Chapter 3 is devoted to comparison principles for, in general, multi-valued elliptic and parabolic variational inequalities, with an account of the main differences between them. Elliptic multi-valued variational inequalities of the following kind are considered: Let $K \subseteq W^{1,p}(\Omega)$ be a closed convex set. Find $u \in K$, $\eta \in L^q(\Omega)$, and $\xi \in L^q(\partial\Omega)$ satisfying:

$$\eta(x) \in \partial j_1(x, u(x)), \text{ a.e. } x \in \Omega, \quad \xi(x) \in \partial j_2(x, \gamma u(x)), \text{ a.e. } x \in \partial\Omega, \quad (1.11)$$

$$\langle Au - h, v - u \rangle + \int_\Omega \eta\,(v - u)\,dx + \int_{\partial\Omega} \xi\,(\gamma v - \gamma u)\,d\sigma \ge 0, \ \forall\, v \in K, \quad (1.12)$$

where $s \mapsto \partial j_k(x, s)$ are given by Clarke's generalized gradient of locally Lipschitz functions $s \mapsto j_k(x, s)$, $k = 1, 2$, γ is the trace operator, and A is some quasilinear elliptic operator of Leray–Lions type. As for parabolic multi-valued variational inequalities, the underlying solution space is

$$W = \{u \in X : \partial u/\partial t \in X^*\},$$

where $X = L^p(0, \tau; W^{1,p}(\Omega))$, and X^* is its dual space. Consider the time-derivative $L = \frac{\partial}{\partial t} : D(L) \to X^*$ as an operator from the domain $D(L)$ to X^* where $D(L)$ is given by

$$D(L) = \{u \in W : u(0) = 0\},$$

and let $K \subseteq X$ be closed and convex. The following general class of multi-valued parabolic variational inequalities is treated in Chap. 3: Find $u \in K \cap D(L)$, $\eta \in L^q(Q)$, and $\xi \in L^q(\Gamma)$ satisfying:

$$\eta(x, t) \in \partial j_1(x, t, u(x, t)), \text{ for a.e. } (x, t) \in Q, \quad (1.13)$$

$$\xi(x, t) \in \partial j_2(x, t, \gamma u(x, t)), \text{ for a.e. } x \in \Gamma, \text{ and} \quad (1.14)$$

$$\langle Lu + Au - h, v - u \rangle + \int_Q \eta\,(v - u)\,dx dt + \int_\Gamma \xi\,(\gamma v - \gamma u)\,d\Gamma \ge 0, \ \forall\, v \in K, \quad (1.15)$$

where $Q = \Omega \times (0, \tau)$ and $\Gamma = \partial\Omega \times (0, \tau)$. For both problems (1.11)–(1.12) and (1.13)–(1.15) we establish existence and comparison results in terms of appropriately defined sub- and supersolutions, and characterize their solution sets topologically and order-theoretically. We are demonstrating by a number of examples that the variational inequality problems (1.11)–(1.12) and (1.13)–(1.15) include a wide range of specific elliptic and parabolic boundary value problems and variational inequalities. In this sense, Chap. 3 is not only a prerequisite for Chaps. 4 and 5, but it is of interest on its own and can be read independently.

In Chaps. 4 and 5 we apply the fixed point results of Chap. 2 combined with the comparison results of Chap. 3 to deal with discontinuous single and multi-valued elliptic and parabolic problems of different kinds. In particular, we consider nonlocal, discontinuous elliptic and parabolic boundary value problems and multi-valued elliptic problems with discontinuously perturbed Clarke's generalized gradient. In the study of those problems, besides fixed point and comparison results, the existence of *extremal solutions* of certain associated auxiliary problems play an important role. Extremal solution results that are proved in Chap. 3 require rather involved techniques. These results are used to transform a given multi-valued elliptic or parabolic problem into a fixed point equation.

Differential and integral equations treated in Sects. 6.1–6.4 and 7.1–7.2 contain functions that are Henstock–Lebesgue (HL) integrable with respect to the independent variable. A function g from a compact real interval $[a, b]$ to a Banach space E is called HL *integrable* if there is a function $f : [a, b] \to E$, called a *primitive* of g, with the following property: To each $\epsilon > 0$ there corresponds such a function $\delta : [a, b] \to (0, \infty)$ that whenever $[a, b] = \cup_{i=1}^{m}[t_{i-1}, t_i]$ and $\xi_i \in [t_{i-1}, t_i] \subset (\xi_i - \delta(\xi_i), \xi_i + \delta(\xi_i))$ for all $i = 1, \ldots, m$, then

$$\sum_{i=1}^{m} \|f(t_i) - f(t_{i-1}) - g(\xi_i)(t_i - t_{i-1})\| < \epsilon. \tag{1.16}$$

Criteria for HL integrability that are sufficient in most of our applications are given by the following lemma.

Lemma 1.12. *Given a function* $g : [a, b] \to E$, *assume there exists a continuous function* $f : [a, b] \to E$ *and a countable subset* Z *of* $[a, b]$ *such that* $f'(t) = g(t)$ *for all* $t \in [a, b] \setminus Z$. *Then* g *is HL integrable on* $[a, b]$, *and* f *is a primitive of* g.

Proof: Since Z is countable, it can be represented in the form $Z = \{x_j\}_{j \in \mathbb{N}}$. Let $\epsilon > 0$ be given. Since f is continuous, and the values of g have finite norms, then for every $x_j \in Z$ there exists a $\delta(x_j) > 0$ such that $\|f(\bar{t}) - f(t)\| < 2^{-j-1}\epsilon$, and $\|g(x_j)\|(\bar{t} - t) < 2^{-j-1}\epsilon$ whenever $x_j - \delta(x_j) < t \leq x_j \leq \bar{t} < x_j + \delta(x_j)$.

To each $\xi \in [a, b] \setminus Z$ there corresponds, since $f'(\xi)$ exists, such a $\delta(\xi) > 0$ that $\|f(\bar{t}) - f(t) - f'(\xi)(\bar{t} - t)\| < \epsilon(\bar{t} - t)/(b - a)$ whenever $\xi \in [t, \bar{t}] \subset$

$(\xi - \delta(\xi), \xi + \delta(\xi))$. Consequently, if $a = t_0 < t_1 < \cdots < t_m = b$, and if $\xi_i \in [t_{i-1}, t_i] \subset (\xi_i - \delta(\xi_i), \xi_i + \delta(\xi_i))$ for all $i = 1, \ldots, m$, then

$$\sum_{i=1}^{m} \| f(t_i) - f(t_{i-1}) - g(\xi_i)(t_i - t_{i-1}) \|$$

$$\leq \sum_{\xi_i = x_j \in Z} (\| f(t_i) - f(t_{i-1}) \| + \| g(x_j) \| (t_i - t_{i-1}))$$

$$+ \sum_{\xi_i \in [a,b] \backslash Z} \| f(t_i) - f(t_{i-1}) - f'(\xi_i)(t_i - t_{i-1}) \| \leq 2\epsilon.$$

Thus g is HL integrable and f is its primitive. $\qquad\qquad\square$

Remark 1.13. If the set Z in Lemma 1.12 is uncountable, an extra condition, called the Strong Lusin Condition (see Chap. 9), is needed to ensure HL integrability.

Compared with Lebesgue and Bochner integrability, the definition of HL integrability is easier to understand because no measure theory is needed. Moreover, all Bochner integrable (i.e., in real-valued case Lebesgue integrable) functions are HL integrable, but not conversely. For instance, HL integrability encloses improper integrals. Consider the real-valued function f defined on $[0, 1]$ by $f(0) = 0$ and $f(t) = t^2 \cos(1/t^2)$ for $t \in (0, 1]$. This function is differentiable on $[0, 1]$, whence f' is HL integrable by Lemma 1.12. However, f' is not Lebesgue integrable on $[0, 1]$. More generally, let t be called a singular point of the domain interval of a real-valued function that is not Lebesgue integrable on any subinterval that contains t. Then (cf. [167]) there exist "HL integrable functions on an interval that admit a set of singular points with its measure as close as possible but not equal to that of the whole interval."

If g is HL integrable on $[a, b]$, it is HL integrable on every closed subinterval $[c, d]$ of $[a, b]$. The *Henstock–Kurzweil* integral of g over $[c, d]$ is defined by

$$^K\!\!\int_c^d g(s)\, ds := f(d) - f(c), \quad \text{where } f \text{ is a primitive of } g.$$

The main advantage of the Henstock–Kurzweil integral is its applicability for integration of highly oscillatory functions that occur in quantum theory and nonlinear analysis. This integral provides a tool to construct a rigorous mathematical formulation for Feynman's path integral, which plays an important role in quantum physics (see, e.g., [143, 182]).

On the other hand, as stated in [98, p.13], the most important factor preventing a widespread use of the Henstock–Kurzweil integral in engineering, mathematics, and physics has been the lack of a natural Banach space structure for the class of HL integrable functions, even in the case when $E = \mathbb{R}$. However, if E is ordered, the validity of the dominated and monotone convergence theorems, which we prove for order-bounded sequences of HL integrable functions (see Chap. 9), considerably improve the applicability of the

Henstock–Kurzweil integral in Nonlinear Analysis. Combined with fixed point
theorems in ordered normed spaces presented in Chap. 2, these convergence
theorems provide effective tools to solve differential and integral equations
that contain HL integrable valued functions and discontinuous nonlinearities.
All this will be discussed in detail in Chaps. 6 and 7 and shows once more the
importance of the order structure of the underlying function spaces. In particular, the above stated lack of a Banach space structure causes no problems
in our studies. Moreover, as the following simple example shows, the ordering
allows us to determine the smallest and greatest solutions of such equations.

Example 1.14. Determine the smallest and the greatest continuous solutions
of the following Cauchy problem:

$$y'(t) = q(t, y(t), y) \text{ for a.e. } t \in J := [0, 4], \quad y(0) = 0, \qquad (1.17)$$

where

$$\begin{cases} q(t, x, y) = p(t)\,\text{sgn}(x) + h(y)(1 + \cos(t)), \\ p(t) = \left| \cos\left(\frac{1}{t}\right) \right| + \frac{1}{t}\text{sgn}\left(\cos\left(\frac{1}{t}\right)\right) \sin\left(\frac{1}{t}\right), \\ h(y) = \left[2\arctan\left(\int_1^4 y(s)\,ds\right) \right], \quad \text{sgn}(x) = \begin{cases} 1, & x > 0, \\ 0, & x = 0, \\ -1, & x < 0, \end{cases} \\ [x] = \max\{n \in \mathbb{Z} : n \le x\}. \end{cases} \qquad (1.18)$$

Note that the bracket function, called the greatest integer function, occurs in
the function h.

Solution: If $y \in C(J, \mathbb{R})$ and $y(t) > 0$ when $t > 0$, then $\text{sgn}(y(t)) = 1$ when
$t > 0$. Thus

$$q(t, y(t), y) = q_y(t) := p(t) + h(y)(1 + \cos(t)), \quad t \in J.$$

The function $f_y : J \to \mathbb{R}$, defined by

$$f_y(0) = 0, \ f_y(t) = t\left| \cos\left(\frac{1}{t}\right)\right| + h(y)(t + \sin(t)), \quad t \in (0, 4],$$

is continuous, and $f_y'(t) = q_y(t)$ if $t \in (0, 4]$ and $t \ne \frac{1}{(2n+1)\pi}$, $n \in \mathbb{N}_0$. Thus q_y
is HL integrable on J and f_y is its primitive by Lemma 1.12. This result and
the definitions of f_y, q_y and the Henstock–Kurzweil integral imply that

$$Gy(t) := {}^K\!\!\int_0^t q(s, y(s), y)\,ds = f_y(t), \quad t \in J.$$

Moreover, $h(y) = \left[2\arctan\left(\int_1^4 y(s)\,ds\right)\right] \le 3$ for every $y \in C(J, \mathbb{R})$. Thus,
defining

$$y^*(t) = t \left| \cos\left(\frac{1}{t}\right) \right| + 3(t + \sin(t)), \ t \in (0,4], \ y^*(0) = 0,$$

then $f_y(t) \leq y^*(t)$ for all $t \in J$ and $y \in C(J, \mathbb{R})$. On the other hand, it is easy to show that $h(y^*) = \left[2 \arctan\left(\int_1^4 y^*(s) \, ds \right) \right] = 3$. Consequently,

$$y^*(t) = f_{y^*}(t) = Gy^*(t), \quad t \in J.$$

It follows from the above equation by differentiation that

$$(y^*)'(t) = f'_{y^*}(t) = q_{y^*}(t) = q(t, y^*(t), y^*), \quad t \in (0,4], \ t \neq \frac{1}{(2n+1)\pi}, \ n \in \mathbb{N}_0.$$

Moreover $y^*(0) = 0$, so that y^* is a solution of problem (1.17). The above reasoning shows also that if $y \in C(J, \mathbb{R})$ is a solution of problem (1.17), then $y(t) \leq y^*(t)$ for every $t \in J$. Thus y^* is the greatest continuous solution of problem (1.17).

By similar reasoning one can show that the smallest solution of the Cauchy problem (1.17) is

$$y_*(t) = -t \left| \cos\left(\frac{1}{t}\right) \right| - 4(t + \sin(t)), \ t \in (0,4], \ y_*(0) = 0.$$

The function $(t, x, y) \mapsto q(t, x, y)$, defined in (1.18), has the following properties.

- It is HL integrable, but it is neither Lebesgue integrable nor continuous with respect to the independent variable t if $x \neq 0$, because p is not Lebesgue integrable.
- Its dependence on all the variables t, x, and y is discontinuous, since the signum function sgn, the greatest integer function $[\cdot]$, and the function p are discontinuous.
- Its dependence on the unknown function y is nonlocal, since the integral of function y appears in the argument of the arctan-function.
- Its dependence on x is not monotone, since p attains positive and negative values in a infinite number of disjoint sets of positive measure. For instance, $y^*(t) > y_*(t)$ for all $t \in (0,4]$, but the difference function $t \mapsto q(t, y^*(t), y^*) - q(t, y_*(t), y_*) = 2p(t) + 7(t + \sin(t))$ is neither nonnegative-valued nor Lebesgue integrable on J.

The basic theory of Banach-valued HL integrable functions needed in Chaps. 6 and 7 is presented in Chap. 9. However, readers who are interested in Real Analysis may well consider the functions to be real-valued. For those readers who are familiar with Bochner integrability theory, notice that *all the theoretical results of Chaps. 6 and 7 where HL integrability is assumed remain valid if HL integrability is replaced by Bochner integrability. As far as the authors know, even the so obtained special cases are not presented in any other book.*

In Sect. 6.5 we study functional differential equations equipped with a functional initial condition in an ordered Banach space E. There we need fixed point results for an increasing mapping $G : P \to P$, where P is a subset of the Cartesian product of the space $L^1(J, E)$ of Bochner integrable functions from $J := [t_0, t_1]$ to E and the space $C(J_0, E)$ of continuous functions from $J_0 := [t_0 - r, t_0]$ to E. The difficulties one is faced with in the treatment of the considered problems are, first, that only a.e. pointwise weak convergence in $L^1(J, E)$ is available, and second, monotone and bounded sequences of the pointwise ordered space $C(J_0, E)$ need not necessarily have supremums and infimums in $C(J_0, E)$. The following purely order-theoretic fixed point theorem, which is proved in Chap. 2, is the main tool that will allow us to overcome the above described difficulties.

Theorem 1.15. *Let G be an increasing self-mapping of a partially ordered set P such that chains of the range $G[P]$ of G have supremums and infimums in P, and that the set of these supremums and infimums has a sup-center. Then G has minimal and maximal fixed points, as well as fixed points that are increasing with respect to G.*

This fixed point theorem will be applied, in particular, in Sects. 6.5 and 7.3 to prove existence and comparison results for solutions of operator equations in partially ordered sets, integral equations, as well as implicit functional differential problems in ordered Banach spaces. It is noteworthy that the data of the considered problems, i.e., operators and functions involved, are allowed to depend discontinuously on all their arguments. Moreover, we do not suppose the existence of subsolutions and/or supersolutions in the treatment.

The abstract order-theoretic fixed poind theory developed in Chap. 2 has been proved to be an extremely powerful tool in dealing with Nash equilibria for normal form games, which is the subject of Chap. 8.

John Nash invented in [185] an equilibrium concept that now bears his name. Because of its importance in economics, Nash earned the Nobel Prize for Economics in 1994. In [185] Nash described his equilibrium concept in terms of game theory as follows:

"Any N-tuple of strategies, one for each player, may be regarded as a point in the product space obtained by multiplying the N strategy spaces of the players. One such N-tuple counters another if the strategy of each player in countering N-tuple yields the highest possible expectation for its player against the $N - 1$ strategies of the other player in the countered N-tuple. A self-countering N-tuple is called an equilibrium point."

To convert this description into a mathematical concept, we utilize the following notations. Let $\Gamma = \{S_1, \ldots, S_N, u_1, \ldots, u_N\}$ be a *finite normal-form game*, where the *strategy set* S_i for player i is finite, and the *utility function* u_i of player i is real-valued and defined on $S = S_1 \times \cdots \times S_N$. Using the notations

$s_{-i} = (s_1, \ldots, s_{i-1}, s_{i+1}, \ldots, s_N)$ and $u_i(s_1, \ldots, s_N) = u_i(s_i, s_{-i})$, strategies $s_1^*, \ldots, s_N^*$ form a *pure Nash equilibrium* for Γ if and only if

$$u_i(s_i^*, s_{-i}^*) \geq u_i(s_i, s_{-i}^*) \quad \text{for all } s_i \in S_i \text{ and } i = 1, \ldots, N.$$

This definition implies that *the strategies of players form a pure Nash equilibrium if and only if no player can improve his/her utility by changing the strategy when all the other players keep their strategies fixed.*

Besides economics, this equilibrium concept has been applied in other social and behavioral sciences, biology, law, politics, etc., cf. [83, 109, 177, 193, 210, 218, 220, 224]. The Nash equilibrium has found so many applications partly because it can be usefully interpreted in a number of ways (cf. [144]). For instance, in human interaction (social, economic, or political) the utilities of players (individuals, firms, or politicians/parties) are mutually dependent on actions of all players. The Nash equilibrium provides an answer to the question of what is an optimal action for every player. The following simple thought experiment describes the usefulness of the Nash equilibrium concept and its relation to democratic decision procedure.

The traffic board of Sohmu hires a consultant to make a traffic plan for the only crossroad of town having traffic lights. Traffic should be as safe as possible. The consultant seeks Nash safety equilibria and finds two: either every passenger goes toward green light, or, all passengers go toward red light. He suggests to the board that one of these alternatives should be chosen. The state council votes on the choice. Every council member votes according to his/her preferences: Green, Red, or Empty. The result is in Nash equilibrium with respect to the opinions of the council members.

The above thought experiment implies that the concept of Nash equilibrium harmonizes with democratic decision making. It also shows that if actions are in Nash equilibrium, they may give the best result for every participant. It is not a matter of a zero-sum game where someone loses when someone else wins.

Nash used in [186] a version of Theorem 1.8 (see [148]) to prove the existence of Nash equilibrium for a finite normal-form game. Because of finiteness of strategy sets, the application of Kakutani's fixed point theorem was not possible without extensions of S_i to be homeomorphic with convex sets. Thus he extended the strategy sets to contain also strategies that are called *mixed strategies*. This means that players i are allowed to choose independently randomizations of strategies of S_i, that is, each mixed strategy σ_i is a probability measure over S_i. The values of utilities $\mathcal{U}_i$, $i = 1, \ldots, N$, are then the expected values:

$$\mathcal{U}_i(\sigma_1, \ldots, \sigma_N) = \sum_{(s_1, \ldots, s_N) \in S} \sigma_1(\{s_1\}) \cdots \sigma_m(\{s_N\}) u_i(s_1, \ldots, s_N).$$

According to Nash's own interpretation stated above, a *mixed Nash equilibrium* for Γ is a profile of mixed strategies, one for each N players, that has the property that each player's strategy maximizes his/her expected utility against the given strategies of the other players. To put this into a mathematical form, denote $\sigma_{-i} = (\sigma_1, \ldots, \sigma_{i-1}, \sigma_{i+1}, \ldots, \sigma_N)$ and $\mathcal{U}_i(\sigma_1, \ldots, \sigma_N) = \mathcal{U}_i(\sigma_i, \sigma_{-i})$, and let Σ_i denote the set of all mixed strategies of player i. We say that mixed strategies $\sigma_1^*, \ldots, \sigma_N^*$ form a *mixed Nash equilibrium* for Γ if

$$\mathcal{U}_i(\sigma_i^*, \sigma_{-i}^*) = \max_{\sigma_i \in \Sigma_i} \mathcal{U}_i(\sigma_i, \sigma_{-i}^*) \quad \text{for all} \quad i = 1, \ldots, N.$$

As for a variety of areas where the concept of Nash equilibrium is applied, see [144, 183] and the references therein.

In Chap. 8 we present some recent results dealing with Nash equilibria for normal-form games. Our study is focused on games with strategic complementarities, which means roughly speaking that the best response of any player is increasing in actions of the other players. Sections 8.1 and 8.2 are devoted especially to those readers who are interested only in finite games. In section 8.1 we prove the existence for the smallest and greatest pure Nash equilibria of a normal-form game whose strategy spaces S_i are finite sets of real numbers, and the real-valued utility functions u_i possess a finite difference property. If the utilities $u_i(s_1, \ldots, s_N)$ are also increasing (respectively decreasing) in s_j, $j \neq i$, the utilities of the greatest (respectively the smallest) pure Nash equilibrium are shown to majorize the utilities of all pure Nash equilibria. An application to a pricing problem is given.

Our presentation of Sects. 8.2–8.5 has three main purposes.

1. In order to avoid "throwing die" in the search for Nash equilibria, it would be desirable that Γ has a pure Nash equilibrium whose utilities majorize the utilities of all other Nash equilibria for Γ, including mixed Nash equilibria. In such a case it would be of no benefit to seek possible mixed Nash equilibria. If Γ is a finite normal-form game, every player who has at least two pure strategies has uncountably many mixed strategies. On the other hand, the set of its pure strategies, as well as the ranges of its utilities, are finite for each player. Thus one can find pure Nash equilibria in concrete situations by finite methods. In Sect. 8.2 we prove that finite normal-form games, which are supermodular in the sense defined in [218, p. 178], possess the above described desirable properties. The proof is constructive and provides a finite algorithm to determine the most profitable pure Nash equilibrium for Γ. This algorithm and Maple programming is applied to calculate the most profitable pure Nash equilibrium and the corresponding utilities for some concrete pricing games. Proposition 8.61 deals also with finite normal-form games.

2. Theorem 1.9 along with other fixed point theorems presented in Chap. 2 are applied in Sects. 8.3 and 8.4 to derive existence and comparison results for exremal Nash equilibria of normal-form games in more general settings. For instance, the result for finite supermodular games proved in Sect. 8.2 is extended in Sect. 8.4 to the case when the strategy spaces S_i are compact sublattices of complete and separable ordered metric spaces. The easiest case is when strategy spaces are subsets of $\mathbb{R}$. In fact, it has been shown recently (cf.

[86]) that when the strategies of a supermodular normal-form game are real numbers, the mixed extension of that game is supermodular, its equilibria form a non-empty complete lattice, and its extremal equilibria are in pure strategies when mixed strategies are ordered by first order stochastic dominance. A problem that arises when the strategies are not in $\mathbb{R}$ is described in [86, Chap. 3] as follows:

"When the strategy spaces are multidimensional, the set of mixed strategies is not a lattice. This implies that we lack the mathematical structure needed for the theory of complementarities. We need lattice property to make sense of increasing best responses when they are not real-valued. Multiple best responses are always present when dealing with mixed equilibria and there does not seem a simple solution to the requirement that strategy spaces be lattices."

In particular, classical fixed point theorems in complete lattices are not applicable, even for finite normal-form games having multidimensional strategy spaces. Moreover, in such cases the desirable comparison results between pure and mixed strategies cannot be obtained by the methods used, e.g., in [86, 180, 217, 218, 222, 223]. The results of Theorems 2.20 and 2.21 and their duals provide tools to overcome the problem caused by the above stated non-lattice property. Our results imply that the smallest and greatest pure Nash equilibria of supermodular games form lower and upper bounds for all possible mixed Nash equilibria when the set of mixed strategies is ordered by first-order stochastic dominance. These lower and upper bounds have the important property that they are the smallest and greatest rationalizable strategy profile, as shown in [179]. In particular, if these bounds are equal, then there is only one pure Nash equilibrium, which is also a unique rationalizable strategy profile, and no properly mixed Nash equilibria exist. In [218, Sect. 4] the following eight examples of supermodular games are presented: Pricing game with substitute products, production game with complementary products, multimarket oligopoly, arms race game, trading partner search game, optimal consumption game with multiple products, facility location game, and minimum cut game. The first example is studied here more closely. In this example the greatest Nash equilibrium has also the desirable property that the utilities of the greatest Nash equilibrium majorize the utilities of all other Nash equilibria, including mixed Nash equilibria. Concrete examples are solved by using Maple programming.

3. Another property, which restricts the application of the original result of Nash, is that the utility functions are assumed to be real-valued. This requires that differences of values of u_i can be estimated by numbers. The results of Sects. 8.2 and 8.4 are proved in the case when the values of u_i are in ordered vector spaces E_i. Thus we can consider the cases when the values of utilities are random variables by choosing $E_i = L^2(\Omega_i, \mathbb{R})$, ordered a.e. pointwise. If the strategy spaces are finite, the only extra hypothesis is that the ranges of $u_i(\cdot, s_{-i})$ are directed upward. As an application the pricing game is extended

to the form where the values of the utility functions are in $\mathbb{R}^{m_i}$. Concrete examples are solved also in this case.

But in the definition of pure Nash equilibrium a partial ordering of the values of u_i is sufficient. The fixed point theorems presented in Chap. 2 apply to derive existence results for extremal pure Nash equilibria of normal-form games when both the strategy spaces S_i and ranges of utility functions u_i are partially ordered sets. Such results are presented in Sect. 8.3. We also present results dealing with monotone comparative statics, i.e., conditions that ensure the monotone dependence of extremal pure Nash equilibria on a parameter that belongs to a poset. As for applications, see, e.g., [10, 11, 180, 218]. These results can be applied to cases where the utilities of different players are evaluated in different ordinal scales, where all the values of the utility functions need not even be order-related. Thus the way is open for new applications of the theory of Nash equilibrium, for instance, in social and behavioral sciences. In such applications the term 'utility' would approach to one of its meanings: "the greatest happiness of the greatest number."

A necessary condition for the existence of a Nash equilibrium for Γ is that the functions $u_i(\cdot, s_{-i})$ have maximum points. When the ranges of these functions are in partially ordered sets that are not topologized, the classical hypotheses, like upper semicontinuity, are not available. Therefore we define a new concept, called upper closeness, which ensures the existence of required maximum points. Upper semicontinuity implies upper closeness for real-valued functions.

In Sect. 8.5 we study the existence of undominated and weakly dominating strategies of normal-form games when the ranges of the utility functions are in partially ordered sets. A justification for Sects. 8.2–8.5 is a philosophy of economic modeling stated in [13]: "The weakest sufficient conditions for robust conclusions is particularly important to economists." Concrete examples are presented.

The existence of winning strategies for a pursuit and evasion game is proved in Sect. 8.6. As an introduction to the subject consider a finite pursuit and evasion game. Game board P is a nonempty subset of $\mathbb{R}^2$, equipped with coordinatewise partial ordering $\le$. Assume that to every position $x \in P$ of player p (pursuer) there corresponds a nonempty subset $\mathcal{F}(x) \subseteq P$ of possible positions y of player q (quarry). The only rule of the game is:

(R) If (x_n, y_n) and (x_{n+1}, y_{n+1}) are consecutive positions of a play, then $y_n \le y_{n+1}$ whenever $x_n < x_{n+1}$, and $y_{n+1} \le y_n$ whenever $x_{n+1} < x_n$.

We say that a strategy of p is a winning strategy if the use of it yields capturing, i.e., after a finite number of move pairs p and q are in the same position. Player p has a winning strategy if the following conditions hold:

(i)   The set $F[P] = \bigcup \{\mathcal{F}(x) : x \in P\}$ has a sup-center $c \in P$.

(ii) If $x \leq y$ in P, then for every $z \in \mathcal{F}(x)$ for which $z \leq y$ there exists such a $w \in \mathcal{F}(y)$ that $z \leq w$, and for every $w \in \mathcal{F}(y)$ satisfying $x \leq w$ there exists such a $z \in \mathcal{F}(x)$ that $z \leq w$.

(iii) Strictly monotone sequences of $\mathcal{F}[P]$ are finite.

The existence of a winning strategy for p can be justified as follows. Player p starts from $x_0 = c$, and q starts from $y_0 \in \mathcal{F}(x_0)$. If $x_0 = y_0$, then p wins. Otherwise, let x_n and $y_n \in \mathcal{F}(x_n)$ denote positions of p and q after nth move pair of a play. If $x_n \neq y_n$, then p moves to $x_{n+1} = \sup\{c, y_n\}$ if x_n and y_n are unordered, and to $x_{n+1} = y_n$ if x_n and y_n are ordered. If $x_n < x_{n+1}$, then q must obey rule (R) and choose a position y_{n+1} of $\mathcal{F}(x_{n+1})$ such that $y_n \leq y_{n+1}$, which is possible due to condition (ii). If $x_{n+1} < x_n$, then similarly q obeying the rule (R) can choose by condition (ii) $y_{n+1} \in \mathcal{F}(x_{n+1})$ so that $y_{n+1} \leq y_n$. Condition (iii) ensures that every play which follows these rules stops after a finite number of moves to the situation where $x_m = y_m$.

The correspondence $x \mapsto \mathcal{F}(x)$ can be considered also as a set-valued mapping from P to the set $2^P \setminus \emptyset$ of nonempty subsets of P. Since the final positions $x = x_m$ of p and $y = y_m$ of q after a play satisfy $x = y \in \mathcal{F}(x)$, the above reasoning shows that $\mathcal{F}$ has a fixed point under conditions (i)–(iii).

To see that the pursuit–evasion game and the fixed point problem formulated above are different if one of the conditions (i)–(iii) is violated, choose $P = \{a, b\}$, where $a = (0, 1)$ and $b = (1, 0)$. If $\mathcal{F}(a) = \mathcal{F}(b) = P$, then both a and b are fixed points of $\mathcal{F}$. If x_0 is any of the points of P from which p starts, then q can start from the other point of P, and p cannot capture q. In this example conditions (ii) and (iii) hold, but (i) is not valid. This lack can yield also a nonexistence of a fixed point of $\mathcal{F}$ even in the single-valued case, as we see by choosing P as above and defining $\mathcal{F}(a) = \{b\}$ and $\mathcal{F}(b) = \{a\}$. Also in this case conditions (ii) and (iii) are valid. The above results hold true also when P is a partially ordered set (poset), positive and negative directions being determined by a partial ordering of P.

The finite game introduced above is generalized in Sect. 8.6, where we study the existence of winning strategies for pursuit and evasion games that are of ordinal length. The obtained results are then used to study the solvability of equations and inclusions in ordered spaces. Monotonicity hypotheses, like (ii) above, are weaker than those assumed in Chap. 2.

As for the roots of the methods used in the proofs of Theorems 1.2, 1.5, 1.9, 1.15, and related theorems in Chap. 2, and in Sect. 8.6, we refer to Ernst Zermelo's letter to David Hilbert, dated September 24, 1904. This letter contains the first proof that *every set can be well-ordered*, i.e., every set P has such a partial ordering that each nonempty subset A of P has the minimum. The proof in question was published in the same year in *Mathematische Annalen* (see [231]). The influence of that proof is described in [94, p.84] as follows: "The powder keg had been exploded through the match lighted by Zermelo in his first proof of well-ordering theorem." To find out what in that proof was so shocking we give an outline of it. The notations are changed to reveal a recursion principle that is implicitly included in Zermelo's proof. That principle forms a cornerstone to proofs of the main existence and comparison results of this book, including the fixed point results stated above. As for the concepts related to ordering, see Chap. 2.

Let P be a nonempty set, and let f be a *choice function* that selects from the complement $P \setminus U$ of every proper subset U of P an element $f(U)$ $(= \gamma(P \setminus U)$, where γ is "eine Belegung" in Zermelo's proof). We say that a nonempty subset A of P is an f-*set* if A has such an order relation $<$ that the following conditions holds:

(i) $(A, <)$ is *well-ordered*, and if $x \in A$, then $x = f(A^{<x})$,
    where $A^{<x} = \{y \in A : y < x\}$.

Applying a comparison principle for well-ordered sets proved by Georg Cantor in 1897 (see [36]) one can show that *if $A = (A, <)$ and $B = (B, \prec)$ are f-sets and $A \not\subseteq B$, then B is an initial segment of A, and if $x, y \in B$, then $x \prec y$ if and only if $x < y$.* Using these properties it is then elementary to verify that *the union C of f-sets is an f-set, ordered by the union of the orderings of f-sets.*

We have $C = P$, for otherwise $A = C \cup \{f(C)\}$ would be an f-set, ordered by the union of the ordering of C and $\{(y, f(C)) : y \in C\}$, contradicting the fact that C is the union of all f-sets. Thus P is an f-set, and hence well ordered.

The proof is based on three principles. One of them ensures the existence of a choice function f. After his proof Zermelo mentions that "Die Idee, unter Berufung auf dieses Prinzip eine *beliebige* Belegung der Wohlordnung zu grunde zu legen, verdanke ich Herrn Erhard Schmidt." This principle, which is a form of the Axiom of Choice, caused the strongest reactions against Zermelo's proof, because there exists no constructive method to determine f for an arbitrary infinite set P. Another principle used in the proof is Cantor's comparison principle for well-ordered sets. A third principle is hidden in the construction of the union C of f-sets: Because C is an f-set, then $x \in C$ implies $x = f(C^{<x})$. Conversely, if $x = f(C^{<x})$, then $x \in P = C$. Consequently,

(A) $\boxed{x \in C \iff x = f(C^{<x}).}$

In Zermelo's proof f was a choice function. Recently, this special instance is generalized to the following mathematical method, called the **Chain Generating Recursion Principle** (see [112, 133]).

Given any nonempty partially ordered set $P = (P, <)$, a family $\mathcal{D}$ of subsets of P with $\emptyset \in \mathcal{D}$ and a mapping $f : \mathcal{D} \to P$, there is exactly one well-ordered chain C of P such that (A) holds. Moreover, if $C \in \mathcal{D}$, then $f(C)$ is not a strict upper bound of C.

In the proof of this result only elementary properties of set theory are used in [112, 133]. In particular, neither the Axiom of Choice nor Cantor's comparison principle are needed. To get this book more self-contained we give another proof in the Preliminaries, Chap. 2.

To give a simple example, let $\mathcal{D}$ be the family of all finite subsets of the set $P = \mathbb{R}$ of real numbers, and $f(U)$, $U \in \mathcal{D}$, the number of elements of U. By the Chain Generating Recursion Principle there is exactly one subset C of $\mathbb{R}$ that is well-ordered by the natural ordering of $\mathbb{R}$ and satisfies (A). The elements of C are values of f, so that $C \subseteq \mathbb{N}_0 = \{0, 1, \dots\}$. On the other hand, $\mathbb{N}_0$ is a well-ordered subset of $\mathbb{R}$, and $n = f(\mathbb{N}_0^{<n})$, $n \in \mathbb{N}_0$. Thus $\mathbb{N}_0$ is an f-set, whence $\mathbb{N}_0 \subseteq C$. Consequently, $C = \mathbb{N}_0$, so that (A) generates the set of natural numbers.

More generally, given $(P, <, \mathcal{D}, f)$, condition (A) can be considered formally as a 'recursion automate' that generates exactly one well-ordered set C. The amount of admissible quadruples $(P, <, \mathcal{D}, f)$ is so big that no set can accommodate them.

The first elements of C satisfying (A) are

$$x_0 := f(\emptyset), \dots, x_{n+1} := f(\{x_0, \dots, x_n\}), \quad \text{as long as} \quad x_n < f(\{x_0, \dots, x_n\}). \tag{1.19}$$

If $x_{n+1} = x_n$ for some n, then $x_n = \max C$. This property can be used to derive algorithmic methods that apply to determine exact or approximative solutions for many kinds of concrete discontinuous nonlocal problems, as well as to calculate pure Nash equilibria and corresponding utilities for finite normal-form games. The Chain Generating Recursion Principle is applied in this book to introduce generalized iteration methods, which provide the basis for the proofs of our main fixed point theorems including Theorems 1.2, 1.5, 1.9, and 1.15. They are applied to prove existence and comparison results for a number of diverse problems such as, e.g., operator equations and inclusions, partial differential equations and inclusions, ordinary functional differential and integral equations in ordered Banach spaces involving singularities, discontinuities, and also non-absolutely integrable functions. Moreover, these abstract fixed point results are shown to be useful and effective tools to prove existence results for extremal Nash equilibria for normal-form games, and to study the existence of winning strategies for pursuit and evasion games.

2

Fundamental Order-Theoretic Principles

In this chapter we use the Chain Generating Recursion Principle formulated in the Introduction to develop generalized iteration methods and to prove existence and comparison results for operator equations and inclusions in partially ordered sets. Algorithms are designed to solve concrete problems by appropriately constructed Maple programs.

2.1 Recursions and Iterations in Posets

Given a nonempty set P, a relation $x < y$ in $P \times P$ is called a *partial ordering*, if $x < y$ implies $y \not< x$, and if $x < y$ and $y < z$ imply $x < z$. Defining $x \leq y$ if and only if $x < y$ or $x = y$, we say that $P = (P, \leq)$ is a partially ordered set (poset).

An element b of a poset P is called an *upper bound* of a subset A of P if $x \leq b$ for each $x \in A$. If $b \in A$, we say that b is the *greatest element* of A, and denote $b = \max A$. A lower bound of A and the smallest element $\min A$ of A are defined similarly, replacing $x \leq b$ above by $b \leq x$. If the set of all upper bounds of A has the smallest element, we call it a *supremum of A* and denote it by $\sup A$. We say that y is a *maximal element* of A if $y \in A$, and if $z \in A$ and $y \leq z$ imply that $y = z$. An infimum of A, $\inf A$, and a minimal element of A are defined similarly. A poset P is called a *lattice* if $\inf\{x, y\}$ and $\sup\{x, y\}$ exist for all $x, y \in P$. A subset W of P is said to be *upward directed* if for each pair $x, y \in W$ there is a $z \in W$ such that $x \leq z$ and $y \leq z$, and W is *downward directed* if for each pair $x, y \in W$ there is a $w \in W$ such that $w \leq x$ and $w \leq y$. If W is both upward and downward directed it is called *directed*. A set W is said to be a *chain* if $x \leq y$ or $y \leq x$ for all $x, y \in W$. We say that W is *well-ordered* if nonempty subsets of W have smallest elements, and *inversely well-ordered* if nonempty subsets of W have greatest elements. In both cases W is a chain.

A basis to our considerations is the following Chain Generating Recursion Principle (cf. [112, Lemma 1.1], [133, Lemma 1.1.1]).

S. Carl and S. Heikkilä, *Fixed Point Theory in Ordered Sets and Applications:*   23
From Differential and Integral Equations to Game Theory,
DOI 10.1007/978-1-4419-7585-0_2, © Springer Science+Business Media, LLC 2011

Lemma 2.1. *Given a nonempty poset P, a subset $\mathcal{D}$ of $2^P = \{A : A \subseteq P\}$ with $\emptyset \in \mathcal{D}$, where $\emptyset$ denotes the empty set, and a mapping $f : \mathcal{D} \to P$. Then there is a unique well-ordered chain C in P such that*

$$x \in C \quad \text{if and only if } x = f(C^{<x}), \text{ where } \quad C^{<x} = \{y \in C : y < x\}. \quad (2.1)$$

If $C \in \mathcal{D}$, then $f(C)$ is not a strict upper bound of C.

Proof: A nonempty subset A of P is called an f-*set* (with f given in the lemma, and thus the proof is independent on the Axiom of Choice) if it has the following properties.

(i) $(A, <)$ is well-ordered, and if $x \in A$, then $x = f(A^{<x})$, where $A^{<x} = \{y \in A : y < x\}$.

For instance, the singleton $\{f(\emptyset)\}$ is an f-set. These sets possess the following property:

(a) *If A and B are f-sets and $A \nsubseteq B$, then $B = A^{<x}$ for some $x \in A$.*

Namely, according to a comparison principle for well-ordered sets (see [36]) there exists such a bijection $\varphi : B \to A^{<x}$ for some $x \in A$ that $\varphi(u) < \varphi(v)$ if and only if $u < v$ in B. The set $S = \{u \in B : u \neq \varphi(u)\}$ is empty, for otherwise, $y = \min S$ would exist and $B^{<y} = A^{<\varphi(y)}$, which yields a contradiction: $y \neq \varphi(y)$ and $y = f(B^{<y}) = f(A^{<\varphi(y)}) = \varphi(y)$. Thus $B = \varphi[B] = A^{<x}$, which proves (a).

Applying (a) it is then elementary to verify that the union C of all f-sets is an f-set. Hence, $x = f(C^{<x})$ for all $x \in C$. Conversely, if $x \in P$ and $x = f(C^{<x})$, then $C^{<x} \cup \{x\}$ is an f-set, whence $x \in C$. Thus (2.1) holds for C.

To prove uniqueness, let B be a well-ordered subset of P for which $x \in B \Leftrightarrow x = f(B^{<x})$. Since B is an f-set, so $B \subseteq C$. If $B \neq C$, then $B = C^{<x}$ by (a). But then $f(B^{<x}) = f(B) = f(C^{<x}) = x$, and $x \notin B$, which contradicts with $x \in B \Leftrightarrow x = f(B^{<x})$. Thus $B = C$, which proves the uniqueness of C. (the well-ordering condition is needed in this proof, since there may exist other partially ordered sets that satisfy (2.1), cf. [110]).

If $f(C)$ is defined, it cannot be a strict upper bound of C, for otherwise $f(C) \notin C$ and $f(C) = f(C^{<f(C)})$, so that $C \cup f(C)$ would be an f-set, not contained in C, which is the union of all f-sets. This proves the last assertion of the lemma. $\qquad \square$

As a consequence of Lemma 2.1 we get the following result (cf. [116, Lemma 2]).

Lemma 2.2. *Given $G : P \to P$ and $c \in P$, there exists a unique well-ordered chain $C = C(G)$ in P, called a w-o chain of cG-iterations, satisfying*

$$x \in C \quad \text{if and only if } x = \sup\{c, G[C^{<x}]\}. \quad (2.2)$$

Proof: Denote $\mathcal{D} = \{W \subseteq P : W$ is well-ordered and $\sup\{c, G[W]\}$ exists$\}$. Defining $f(W) = \sup\{c, G[W]\}$, $W \in \mathcal{D}$, we get a mapping $f : \mathcal{D} \to P$, and (2.1) is reduced to (2.2). Thus the assertion follows from Lemma 2.1.   □

A subset W of a chain C is called an *initial segment* of C if $x \in W$ and $y < x$ imply $y \in W$. The following application of Lemma 2.2 is used in the sequel.

Lemma 2.3. *Denote by $\mathcal{G}$ the set of all selections from $\mathcal{F} : P \to 2^P \setminus \emptyset$, i.e.,*

$$\mathcal{G} := \{G : P \to P : G(x) \in \mathcal{F}(x) \text{ for all } x \in P\}. \tag{2.3}$$

Given $c \in P$ and $G \in \mathcal{G}$. Let C_G denote the longest initial segment of the w-o chain $C(G)$ of cG-iterations such that the restriction $G|C_G$ of G to C_G is increasing (i.e., $G(x) \leq G(y)$ whenever $x \leq y$ in C_G). Define a partial ordering $\prec$ on $\mathcal{G}$ as follows: Let $F, G \in \mathcal{G}$ then

(O) $F \prec G$ if and only if C_F is a proper initial segment of C_G and $G|C_F = F|C_F$.

Then $(\mathcal{G}, \preceq)$ has a maximal element.

Proof: Let $\mathcal{C}$ be a chain in $\mathcal{G}$. The definition (O) of $\prec$ implies that the sets C_F, $F \in \mathcal{C}$, form a nested family of well-ordered sets of P. Thus the set $C := \cup\{C_F : F \in \mathcal{C}\}$ is well-ordered. Moreover, it follows from (O) that the functions $F|C_F$, $F \in \mathcal{C}$, considered as relations in $P \times P$, are nested. This ensures that $g := \cup\{F|C_F : F \in \mathcal{C}\}$ is a function from C to P. Since each $F \in \mathcal{C}$ is increasing in C_F, then g is increasing, and $g(x) \in \mathcal{F}(x)$ for each $x \in C$. Let G be such a selection from $\mathcal{F}$ that $G|C = g$. Then $G \in \mathcal{G}$, and G is increasing on C. If $x \in C$, then $x \in C_F$ for some $F \in \mathcal{C}$. The definitions of C and the partial ordering $\prec$ imply that C_F is C or its initial segment, whence $C_F^{\leq x} = C^{<x}$. Because $F|C_F = g|C_F = G|C_F$, then

$$x = \sup\{c, F[C_F^{\leq x}]\} = \sup\{c, G[C^{<x}]\}. \tag{2.4}$$

This result implies by (2.2) that C is $C(G)$ or its proper initial segment. Since G is increasing on C, then C is C_G or its proper initial segment. Consequently, G is an upper bound of $\mathcal{C}$ in $\mathcal{G}$. This result implies by Zorn's Lemma that $\mathcal{G}$ has a maximal element.   □

Let $P = (P, \leq)$ be a poset. For $z, w \in P$, we denote

$$[z) = \{x \in P : z \leq x\}, \ (w] = \{x \in P : x \leq w\} \text{ and } [z, w] = [z) \cap (w].$$

A poset X equipped with a topology is called an *ordered topological space* if the order intervals $[z)$ and $(z]$ are closed for each $z \in X$. If the topology of X is induced by a metric, we say that X is an *ordered metric space*. Next we define some concepts for set-valued functions.

Definition 2.4. *Given posets X and P, we say $\mathcal{F} : X \to 2^P \setminus \emptyset$ is **increasing upward** if $x \leq y$ in X and $z \in \mathcal{F}(x)$ imply that $[z) \cap \mathcal{F}(y)$ is nonempty. $\mathcal{F}$ is **increasing downward** if $x \leq y$ in X and $w \in \mathcal{F}(y)$ imply that $(w] \cap \mathcal{F}(x)$ is nonempty. If $\mathcal{F}$ is increasing upward and downward, we say that $\mathcal{F}$ is **increasing**.*

Definition 2.5. *A nonempty subset A of a subset Y of a poset P is called **order compact upward** in Y if for every chain C of Y that has a supremum in P the intersection $\cap\{[y) \cap A : y \in C\}$ is nonempty whenever $[y) \cap A$ is nonempty for every $y \in C$. If for every chain C of Y that has the infimum in P the intersection of all the sets $(y] \cap A$, $y \in C$ is nonempty whenever $(y] \cap A$ is nonempty for every $y \in C$, we say that A is **order compact downward** in Y. If both these properties hold, we say that A is **order compact** in Y. Phrase 'in Y' is omitted if $Y = A$.*

Every poset P is order compact. If a subset A of P has the greatest element (respectively the smallest element), then A is order compact upward (respectively downward) in any subset of P that contains A. Thus an order compact set is not necessarily (topologically) compact, not even closed. On the other hand, every compact subset A of an ordered topological space P is obviously order compact in every subset of P that contains A.

2.2 Fixed Point Results in Posets

In this subsection we prove existence and comparison results for fixed points of set-valued and single-valued functions defined in a poset $P = (P, \leq)$.

Definition 2.6. *Given a poset $P = (P \leq)$ and a set-valued function $\mathcal{F} : P \to 2^P \setminus \emptyset$, denote $Fix(\mathcal{F}) = \{x \in P : x \in \mathcal{F}(x)\}$. Every element of $Fix(\mathcal{F})$ is called a **fixed point** of $\mathcal{F}$. A fixed point of $\mathcal{F}$ is called **minimal, maximal, smallest, or greatest** if it is a minimal, maximal, smallest, or greatest element of $Fix(\mathcal{F})$, respectively. For a single-valued function $G : P \to P$ replace $Fix(\mathcal{F})$ by $Fix(G) = \{x \in P : x = G(x)\}$.*

2.2.1 Fixed Points for Set-Valued Functions

Our first proved fixed point result is an application of Lemma 2.1.

Lemma 2.7. *Assume that $\mathcal{F} : P \to 2^P$ satisfies the following hypothesis.*

(S_+) The set $S_+ = \{x \in P : [x) \cap \mathcal{F}(x) \neq \emptyset\}$ is nonempty, and conditions: C is a nonempty well-ordered chain in S_+, $G : C \to P$ is increasing and $x \leq G(x) \in \mathcal{F}(x)$ for all $x \in C$, imply that $G[C]$ has an upper bound in S_+.

Then $\mathcal{F}$ has a maximal fixed point, which is also a maximal element of S_+.

Proof: Denote

$$\mathcal{D} = \{W \subset S_+ : W \text{ is well-ordered and has a strict upper bound in } S_+\}.$$

Because S_+ is nonempty by the hypothesis (S_+), then $\emptyset \in \mathcal{D}$. Let $f : \mathcal{D} \to P$ be a function that assigns to each $W \in \mathcal{D}$ an element $y = f(W) \in [x) \cap \mathcal{F}(x)$, where x is a fixed strict upper bound of W in S_+. Lemma 2.1 ensures the existence of exactly one well-ordered chain W in P satisfying (2.1). By the above construction and (2.1) each element y of W belongs to $[x) \cap \mathcal{F}(x)$, where x is a fixed strict upper bound of $W^{<y}$ in S_+. It is easy to verify that the set C of these elements x form a well-ordered chain in S_+; that the correspondence $x \mapsto y$ defines an increasing mapping $G : C \to P$; that $x \le G(x) \in \mathcal{F}(x)$ for all $x \in C$; and that $W = G[C]$. It then follows from the hypothesis (S_+) that W has an upper bound $x \in S_+$, which satisfies $x = \max W$. For otherwise $f(W)$ would exist, and as a strict upper bound of W would contradict the last conclusion of Lemma 2.1. By the same reason x is a maximal element of S_+.

Since $x \in S_+$, a $y \in P$ exists such that $x \le y \in \mathcal{F}(x)$. It then follows from the hypothesis (S_+) when $C = \{x\}$ and $G(x) := y$ that $\{y\}$ has an upper bound z in S_+. Because x is a maximal element of S_+, then $z = y = x \in \mathcal{F}(x)$, so that x is a fixed point of $\mathcal{F}$. If z is a fixed point of $\mathcal{F}$ and $x \le z$, then $z \in S_+$, whence $x = z$. Thus x is a maximal fixed point of $\mathcal{F}$.   □

As an application of Lemma 2.7 we obtain the following result.

Proposition 2.8. *Assume that* $\mathcal{F} : P \to 2^P \setminus \emptyset$ *is increasing upward, that the set* $S_+ = \{x \in P : [x) \cap \mathcal{F}(x) \ne \emptyset\}$ *is nonempty, that well-ordered chains of* $\mathcal{F}[S_+]$ *have supremums in* P, *and that the values of* $\mathcal{F}$ *at these supremums are order compact upward in* $\mathcal{F}[S_+]$. *Then* $\mathcal{F}$ *has a maximal fixed point, which is also a maximal element of* S_+.

Proof: It suffices to show that the hypothesis (S_+) of Lemma 2.7 holds. Assume that C is a well-ordered chain in S_+, that $G : C \to P$ is an increasing mapping, and that $x \le G(x) \in \mathcal{F}(x)$ for all $x \in C$. Then $G[C]$ is a well-ordered chain in $\mathcal{F}[S_+]$, so that $y = \sup G[C]$ exists. Since $\mathcal{F}$ is increasing upward, then $[x) \cap \mathcal{F}(y) \ne \emptyset$ for every $x \in G[C]$. Because $\mathcal{F}(y)$ is order compact upward in $\mathcal{F}[S_+]$, then the intersection of the sets $[x) \cap \mathcal{F}(y)$, $x \in G[C]$ contains at least one element w. Thus $G[C]$ has an upper bound w in $\mathcal{F}(y)$. Since $y = \sup G[C]$, then $y \le w$, so that $w \in [y) \cap \mathcal{F}(y)$, i.e., y belongs to S_+.   □

The next result is the dual to Proposition 2.8.

Proposition 2.9. *Assume that* $\mathcal{F} : P \to 2^P \setminus \emptyset$ *is increasing downward, that the set* $S_- = \{x \in P : (x] \cap \mathcal{F}(x) \ne \emptyset\}$ *is nonempty, that inversely well-ordered chains of* $\mathcal{F}[S_-]$ *have infimums in* P, *and that values of* $\mathcal{F}$ *at these infimums are order compact downward in* $\mathcal{F}[S_-]$. *Then* $\mathcal{F}$ *has a minimal fixed point, which is also a minimal element of* S_-.

If the range $\mathcal{F}[P]$ has an upper bound (respectively a lower bound) in P, it belongs to S_- (respectively to S_+). To derive other conditions under which the set S_- or the set S_+ is nonempty, we introduce the following new concepts.

Definition 2.10. *Let A be a nonempty subset of a poset P. The set $ocl(A)$ of all possible supremums and infimums of chains of A is called the* **order closure** *of A. If $A = ocl(A)$, then A is* **order closed**. *We say that a subset A of a poset P has a* **sup-center** *c in P if $c \in P$ and $\sup\{c,x\}$ exists in P for each $x \in A$. If $\inf\{c,x\}$ exists in P for each $x \in A$, we say that c is an* **inf-center** *of A in P. If c has both these properties it is called an* **order center** *of A in P. Phrase "in P" is omitted if $A = P$.*

If P is an ordered topological space, then the order closure $ocl(A)$ of A is contained in the topological closure $\overline{A}$ of A. If c is the greatest element (respectively the smallest element) of P, then c is an inf-center (respectively a sup-center) of P, and trivially c is a sup-center (respectively an inf-center). Therefore, both the greatest and the smallest element of P are order centers. If P is a lattice, then its every point is an order center of P. If P is a subset of $\mathbb{R}^2$, ordered coordinatewise, a necessary and sufficient condition for a point $c = (c_1, c_2)$ of P to be a sup-center of a subset A of P in P is that whenever a point $y = (y_1, y_2)$ of A and c are unordered, then $(y_1, c_2) \in P$ if $y_2 < c_2$ and $(c_1, y_2) \in P$ if $y_1 < c_1$.

The following result is an application of Lemma 2.3.

Proposition 2.11. *Assume that $\mathcal{F} : P \to 2^P \setminus \emptyset$ is increasing upward and that its values are order compact upward in $\mathcal{F}[P]$. If well-ordered chains of $\mathcal{F}[P]$ have supremums, and if the set of these supremums has a sup-center c in P, then the set $S_- = \{x \in P : (x] \cap \mathcal{F}(x) \neq \emptyset\}$ is nonempty.*

Proof: Let $\mathcal{G}$ be defined by (2.3), and let the partial ordering $\prec$ be defined by (O). In view of Lemma 2.3, $(\mathcal{G}, \preceq)$ has a maximal element G. Let $C(G)$ be the w-o chain of cG-iterations, and let $C = C_G$ be the longest initial segment of $C(G)$ on which G is increasing. Thus C is well-ordered and G is an increasing selection from $\mathcal{F}|C$. Since $G[C]$ is a well-ordered chain in $\mathcal{F}[P]$, then $w = \sup G[C]$ exists. Moreover, $\overline{x} = \sup\{c, w\}$ exists in P by the choice of c, and it is easy to see that $\overline{x} = \sup\{c, G[C]\}$. This result and (2.2) imply that for each $x \in C$,

$$x = \sup\{c, G[C^{<x}]\} \leq \sup\{c, G[C]\} = \overline{x}.$$

This proves that $\overline{x}$ is an upper bound of C, and also of $G[C]$. Moreover, $\mathcal{F}$ is increasing upward and $\mathcal{F}(\overline{x})$ is order compact upward in $\mathcal{F}[P]$. Thus the proof of Proposition 2.8 implies that $G[C]$ has an upper bound z in $\mathcal{F}(\overline{x})$, and $w = \sup G[C] \leq z$. To show that $\overline{x} = \max C$, assume on the contrary that $\overline{x}$ is a strict upper bound of C. Let F be a selection from $\mathcal{F}$ whose restriction to $C \cup \{\overline{x}\}$ is $G|C \cup \{(\overline{x}, z)\}$. Since G is increasing on C and $F(x) = G(x) \leq w \leq z = F(\overline{x})$ for each $x \in C$, then F is increasing on $C \cup \{\overline{x}\}$. Moreover,

$$\overline{x} = \sup\{c, G[C]\} = \sup\{c, F[C]\} = \sup\{c, F[\{y \in C \cup \{\overline{x}\} : y < \overline{x}\}]\},$$

whence $C \cup \{\overline{x}\}$ is a subset of the longest initial segment C_F of the w-o chain of cF-iterations where F is increasing. Thus $C = C_G$ is a proper subset of C_F, and $F|C_G = F|C_F$. By (O) this means that $G \prec F$, which, however, is impossible because G is a maximal element of $(\mathcal{G}, \preceq)$. Consequently, $\overline{x} = \max C$. Since G is increasing on C, then $\overline{x} = \sup\{c, G[C]\} = \sup\{c, G(\overline{x})\}$. In particular, $\mathcal{F}(\overline{x}) \ni G(\overline{x}) \leq \overline{x}$, whence $G(\overline{x})$ belongs to the set $(\overline{x}] \cap \mathcal{F}(\overline{x})$. $\square$

As a consequence of Propositions 2.8, 2.9, and 2.11 we obtain the following fixed point result.

Theorem 2.12. *Assume that $\mathcal{F} : P \to 2^P \setminus \emptyset$ is increasing, and that its values are order compact in $\mathcal{F}[P]$. If chains of $\mathcal{F}[P]$ have supremums and infimums, and if $ocl(\mathcal{F}[P])$ has a sup-center or an inf-center in P, then $\mathcal{F}$ has minimal and maximal fixed points.*

Proof: We shall give the proof in the case when $ocl(\mathcal{F}[P])$ has a sup-center in P, as the proof in the case of an inf-center is similar. The hypotheses of Proposition 2.11 are then valid, whence there exists a $\overline{x} \in P$ such that $(\overline{x}] \cap \mathcal{F}(\overline{x}) \neq \emptyset$. Thus the hypotheses of Proposition 2.9 hold, whence $\mathcal{F}$ has by Proposition 2.9 a minimal fixed point x_-. In particular $[x_-) \cap \mathcal{F}(x_-) \neq \emptyset$. The hypotheses of Proposition 2.8 are then valid, whence we can conclude that $\mathcal{F}$ has also a maximal fixed point. $\square$

Example 2.13. Assume that $\mathbb{R}^m$ is ordered as follows. For all $x = (x_1, \ldots, x_m)$, $y = (y_1, \ldots, y_m) \in \mathbb{R}^m$,

$$x \leq y \text{ if and only if } x_i \leq y_i, \ i = 1, \ldots, j, \text{ and } x_i \geq y_i, \ i = j+1, \ldots, m,$$
(2.5)

where $j \in \{0, \ldots, m\}$. Show that if $\mathcal{F} : \mathbb{R}^m \to 2^{\mathbb{R}^m} \setminus \emptyset$ is increasing, and its values are closed subsets of $\mathbb{R}^m$, and if $\mathcal{F}[\mathbb{R}^m]$ is contained in the set

$$B_R^p(c) = \{(x_1, \ldots, x_m) \in \mathbb{R}^m : \sum_{i=1}^m |x_i - c_i|^p \leq R^p\}, \ p, R \in (0, \infty),$$

where $c = (c_1, \ldots, c_m) \in \mathbb{R}^m$, then $\mathcal{F}$ has minimal and maximal fixed points.

Solution: Let $x = (x_1, \ldots, x_m) \in B_R^p(c)$ be given. Since $|\max\{c_i, x_i\} - c_i| \leq |x_i - c_i|$ and $|\min\{c_i, x_i\} - c_i| \leq |x_i - c_i|$ for each $i = 1, \ldots, m$, it follows that $\sup\{c, x\}$ and $\inf\{c, x\}$ belong to $B_R^p(c)$ for all $x \in B_R^p(c)$. Moreover, every $B_R^p(c)$ is a closed and bounded subset of $\mathbb{R}^m$, whence its monotone sequences converge in $B_R^p(c)$ with respect to the Euclidean metric of $\mathbb{R}^m$.

These results, Lemma 2.31 and the given hypotheses imply that chains of $\mathcal{F}[\mathbb{R}^m]$ have supremums and infimums, that c is an order center of $ocl(\mathcal{F}[\mathbb{R}^m])$, and that the values of $\mathcal{F}$ are compact. Thus the hypotheses of Theorem 2.12 are satisfied, whence we conclude that $\mathcal{F}$ has minimal and maximal fixed points.   □

2.2.2 Fixed Points for Single-Valued Functions

Next we present existence and comparison results for fixed points of single-valued functions. The following auxiliary result is a consequence of Proposition 2.11 and its proof. We note that the Axiom of Choice is not needed in the proof.

Proposition 2.14. *Assume that $G : P \to P$ is increasing, that $ocl(G[P])$ has a sup-center c in P, and that $\sup G[C]$ exists whenever C is a nonempty well-ordered chain in P. If C is the w-o chain of cG-iterations, then $\overline{x} = \max C$ exists, $\overline{x} = \sup\{c, G(\overline{x})\} = \sup\{c, G[C]\}$, and*

$$\overline{x} = \min\{z \in P : \sup\{c, G(z)\} \le z\}. \tag{2.6}$$

Moreover, $\overline{x}$ is the smallest solution of the equation $x = \sup\{c, G(x)\}$, and it is increasing with respect to G.

Proof: The mapping $\mathcal{F} := G : P \to 2^P \setminus \emptyset$ is single-valued. Because G is increasing, then C in Lemma 2.3 is the w-o chain of cG-iterations. The hypotheses given for G imply also that c is a sup-center of $ocl(\mathcal{F}[P])$ in P, and that $\sup G[C]$ exists. Since G is single-valued, the values of $\mathcal{F}$ are order compact in $\mathcal{F}[P]$. Thus the proof of Proposition 2.11 implies that $\overline{x} = \max C$ exists, and $\overline{x} = \sup\{c, G(\overline{x})\} = \sup\{c, G[C]\}$. To prove (2.6), let $z \in P$ satisfy $\sup\{c, G(z)\} \le z$. Then $c = \min C \le z$. If $x \in C$ and $\sup\{c, G(y)\} \le z$ for each $y \in C^{<x}$, then $x = \sup\{c, G[C^{<x}]\} \le z$. This implies by transfinite induction that $x \le z$ for each $x \in C$. In particular $\overline{x} = \max C \le z$. From this result and the fact that $\overline{x} = \sup\{c, G(\overline{x})\}$ we infer that $\overline{x} = x$ is the smallest solution of the equation $x = \sup\{c, G(x)\}$, and that (2.6) holds. The last assertion is an immediate consequence of (2.6).   □

The results presented in the next proposition are dual to those of Lemma 2.2 and Proposition 2.14.

Proposition 2.15. *Given $G : P \to P$ and $c \in P$, there exists exactly one inversely well-ordered chain D in P, called an* inversely well-ordered (i.w-o) *chain of cG- iterations, satisfying*

$$x \in D \quad \text{if and only if } x = \inf\{c, G[\{y \in D : x < y\}]\}. \tag{2.7}$$

Assume that G is increasing, that $ocl(G[P])$ has an inf-center c in P, and that $\inf G[D]$ exists whenever D is a nonempty inversely well-ordered chain *in P. If D is the i.w-o chain of cG-iterations, then $\underline{x} = \min D$ exists, $\underline{x} = \inf\{c, G(\underline{x})\} = \inf\{c, G[D]\}$, and*

$$\underline{x} = \max\{z \in P : z \leq \inf\{c, G(z)\}\}. \tag{2.8}$$

Moreover, $\underline{x}$ is the greatest solution of the equation $x = \inf\{c, G(x)\}$, and it is increasing with respect to G.

Our first fixed point result is a consequence of Propositions 2.14 and 2.15.

Theorem 2.16. *Let P be a poset and let $G : P \to P$ be an increasing mapping.*

(a) If $\underline{x} \leq G(\underline{x})$, and if $\sup G[C]$ exists whenever C is a well-ordered chain in $[\underline{x})$ and $x \leq G(x)$ for every $x \in C$, then the w-o chain C of $\underline{x}G$-iterations has a maximum x_* and

$$x_* = \max C = \sup G[C] = \min\{y \in [\underline{x}) : G(y) \leq y\}. \tag{2.9}$$

Moreover, x_* is the smallest fixed point of G in $[\underline{x})$, and x_* is increasing with respect to G.

(b) If $G(\overline{x}) \leq \overline{x}$, and if $\inf G[C]$ exists whenever C is an inversely well-ordered chain $(\overline{x}]$ and $G(x) \leq x$ for every $x \in C$, then the i.w-o chain D of $\overline{x}G$-iterations has a minimum x^* and

$$x^* = \min D = \inf G[D] = \max\{y \in (\overline{x}] : y \leq G(y)\}. \tag{2.10}$$

Moreover, x^* is the greatest fixed point of G in $(\overline{x}]$, and x^* is increasing with respect to G.

Proof: Ad (a) Since G is increasing and $\underline{x} \leq G(\underline{x})$, then $G[[\underline{x})] \subset [\underline{x})$. It is also easy to verify that $x \leq G(x)$ for every element x of the w-o chain C of $\underline{x}G$-iterations. Thus the conclusions of (a) are immediate consequences of the conclusion of Proposition 2.14 when $c = \underline{x}$ and G is replaced by its restriction to $[\underline{x})$.

Ad (b) The proof of (b) is dual to that of (a). □

As an application of Propositions 2.14 and 2.15 and Theorem 2.16 we get the following fixed point results.

Theorem 2.17. *Assume that $G : P \to P$ is increasing, and that $\sup G[C]$ and $\inf G[C]$ exist whenever C is a chain in P.*

(a) If $ocl(G[P])$ has a sup-center or an inf-center in P, then G has minimal and maximal fixed points.

(b) If $ocl(G[P])$ has a sup-center c in P, then G has the greatest fixed point x^* in $(\overline{x}]$, where $\overline{x}$ is the smallest solution of the equation $x = \sup\{c, G(x)\}$. Both $\overline{x}$ and x^* are increasing with respect to G.

(c) If c is an inf-center of $ocl(G[P])$ in P, then G has the smallest fixed point x_* in $[\underline{x})$, where $\underline{x}$ is the greatest solution of the equation $x = \inf\{c, G(x)\}$. Both $\underline{x}$ and x_* are increasing with respect to G.

Theorem 2.16, Proposition 2.8, its proof, and Proposition 2.9 imply the following results.

Proposition 2.18. *Assume that $G : P \to P$ is increasing.*

(a) If the set $S_+ = \{x \in P : x \le G(x)\}$ is nonempty, and if $\sup G[C]$ exists whenever C is a well-ordered chain in S_+, then G has a maximal fixed point. Moreover, G has for every $\underline{x} \in S_+$ the smallest fixed point in $[\underline{x})$, and it is increasing with respect to G.

(b) If the set $S_- = \{x \in P : G(x) \le x\}$ is nonempty, and if $\inf G[D]$ exists whenever D is an inversely well-ordered chain in S_-, then G has a minimal fixed point. Moreover, G has for every $\overline{x} \in S_-$ the greatest fixed point in $(\overline{x}]$, and it is increasing with respect to G.

Example 2.19. Let $\mathbb{R}_+$ be the set of nonnegative reals, and let $\mathbb{R}^m$ be ordered coordinatewise. Assume that $G : \mathbb{R}^m \to \mathbb{R}^m_+$ is increasing and maps increasing sequences of the set $S_+ = \{x \in \mathbb{R}^m_+ : x \le G(x)\}$ to bounded sequences. Show that G has the smallest fixed point and a maximal fixed point.

Solution: The origin is a lower bound of $G[\mathbb{R}^m]$. Let C be a well-ordered chain in S_+. Since G is increasing, then $G[C]$ is a well-ordered chain in $\mathbb{R}^m_+$. If (y_n) is an increasing sequence in $G[C]$, and $x_n = \min\{x \in C : G(x) = y_n\}$, then the sequence (x_n) is increasing and $y_n = G(x_n)$ for every n. Thus (y_n) is bounded by a hypothesis, and hence converges with respect to the Euclidean metric of $\mathbb{R}^m$. This result implies by Lemma 2.31 that $\sup G[C]$ exists. Thus the assertions follow from Proposition 2.18.   □

2.2.3 Comparison and Existence Results

In the next application of Theorem 2.16, fixed points of a set-valued function are bounded from above by a fixed point of a single-valued function.

Theorem 2.20. *Given a poset $X = (X, \le)$, a subset P of X and $\overline{x} \in P$, assume that a function $G : P \to P$ and a set-valued function $\mathcal{F} : X \to 2^X$ have the following properties.*

(Ha) G is increasing, $G(\overline{x}) \le \overline{x}$, and $\inf G[D]$ exists in P whenever D is an inversely well-ordered chain in $(\overline{x}]$.

(Hb) $\overline{x}$ is an upper bound of $\mathcal{F}[X] = \cup_{x \in X} \mathcal{F}(x)$, and if $x \le p$ in X and $p \in P$, then $G(p)$ is an upper bound of $\mathcal{F}(x)$.

Then G has the greatest fixed point x^* in $(\overline{x}]$, and if x is any fixed point of $\mathcal{F}$, then $x \le x^*$.

Proof: Let D be the i.w-o chain of $\overline{x}G$-iterations. Since D is inversely well-ordered, then $x^* = \inf G[D]$ exists and belongs to P by hypothesis (Ha). Moreover, x^* is the greatest fixed point of G in $(\overline{x}]$ by Theorem 2.16 (b). To prove that x^* is an upper bound for fixed points of $\mathcal{F}$, assume on the contrary an existence of a point x of X such that $x \in \mathcal{F}(x)$ and $x \not\leq x^*$. Since $x^* = \min D$ and D is inversely well-ordered, there exists the greatest element p of D such that $x \not\leq p$. Because $x \in \mathcal{F}(x)$, then $x \leq \overline{x} = \max D$ by (Hb), whence $p < \overline{x}$. If $q \in D$ and $p < q$, then $x \leq q$, so that $x \leq G(q)$ by (Hb). Thus x is a lower bound of the set $G[\{q \in D : p < q\}]$. Since $\overline{x}$ is an upper bound of this set, then p is by (2.7) with $c = \overline{x}$ the infimum of $G[\{q \in D : p < q\}]$. But then $x \leq p$, which contradicts with the choice of p. Consequently, $x \leq x^*$ for each fixed point x of $\mathcal{F}$. □

Using the result of Theorem 2.20 we prove the following existence and comparison result for greatest fixed points of set-valued functions.

Theorem 2.21. *Given a nonempty subset P of X and $\mathcal{F} : X \to 2^X$, assume that*

(H0) $\mathcal{F}[X]$ has an upper bound $\overline{x}$ in P.

(H1) If $p \in P$, then $\max \mathcal{F}(p)$ exists, belongs to P, and is an upper bound of $\mathcal{F}[X \cap (p]]$.

(H2) Inversely well-ordered chains of the set $\{\max \mathcal{F}(p) : p \in P\}$ have infimums in P.

Then $\mathcal{F}$ has a greatest fixed point, and it belongs to P. Assume moreover, that $\hat{\mathcal{F}} : X \to 2^X$ is another set-valued function that satisfies the following condition.

(H3) For each $x \in X$ and $y \in \hat{\mathcal{F}}(x)$ there exists a $z \in \mathcal{F}(x)$ such that $y \leq z$.

Then the greatest fixed point of $\mathcal{F}$ is an upper bound for all the fixed points of $\hat{\mathcal{F}}$.

Proof: The hypothesis (H1) ensures that defining

$$G(p) := \max \mathcal{F}(p), \quad p \in P, \tag{2.11}$$

we obtain an increasing mapping $G : P \to P$. Moreover, $G(\overline{x}) \leq \overline{x}$ is by (H0), and the hypothesis (H2) means that every inversely well-ordered chain of $G[P]$ has an infimum in P. Thus the hypothesis (Ha) of Theorem 2.20 holds. The hypothesis (H1) and the definition (2.11) of G imply that also the hypothesis (Hb) of Theorem 2.20 is valid. Thus G has by Theorem 2.20 the greatest fixed point x^*. Because $x^* = G(x^*) = \max \mathcal{F}(x^*) \in \mathcal{F}(x^*)$, then x^* is also a fixed point of $\mathcal{F}$, which is by Theorem 2.20 an upper bound all fixed points of $\mathcal{F}$. Consequently, x^* is the greatest fixed point of $\mathcal{F}$, and $x^* = \max \mathcal{F}(x^*) \in P$ by (H1).

To prove the last assertion, let $\hat{\mathcal{F}} : X \to 2^X$ be such a set-valued function that (H3) holds. The hypotheses (H0) and (H3) imply that $\overline{x}$ is an upper bound of $\hat{\mathcal{F}}[X]$. Moreover, if $x \leq p$ in X and $p \in P$, then for each $y \in \hat{\mathcal{F}}(x)$ there is by (H3) a $z \in \mathcal{F}(x)$ such that $y \leq z$, and $z \leq \max \mathcal{F}(p) = G(p)$ by (H1) and (2.11). Thus the hypotheses of Theorem 2.20 hold when $\mathcal{F}$ is replaced by $\hat{\mathcal{F}}$, whence $x \leq x^*$ for each fixed point x of $\hat{\mathcal{F}}$. $\square$

Remark 2.22. Applying Theorem 2.16 (a) we obtain obvious duals to Theorems 2.20 and 2.21.

2.2.4 Algorithmic Methods

Let P be a poset, and let $G : P \to P$ be increasing. The first elements of the w-o chain C of cG-iterations are: $x_0 = c$, $x_{n+1} = \sup\{c, Gx_n\}$, $n = 0, 1, \ldots$, as long as x_{n+1} exists and $x_n < x_{n+1}$. Assuming that strictly monotone sequences of $G[P]$ are finite, then C is a finite strictly increasing sequence $(x_n)_{n=0}^m$. If $\sup\{c, x\}$ exists for every $x \in G[P]$, then $\overline{x} = \sup\{c, G[C]\} = \max C = x_m$ is the smallest solution of the equation $x = \sup\{c, G(x)\}$ by Proposition 2.14. In particular, $G\overline{x} \leq \overline{x}$. If $G(\overline{x}) < \overline{x}$, then first elements of the i.w-o chain D of $\overline{x}G$-iterations of $\overline{x}$ are $y_0 = \overline{x} = x_m$, $y_{j+1} = Gy_j$, as long as $y_{j+1} < y_j$. Since strictly monotone sequences of $G[P]$ are finite, D is a finite strictly decreasing sequence $(y_j)_{j=0}^k$, and $x^* = \inf G[D] = y_k$ is the greatest fixed point of G in $(\overline{x}]$ by Theorem 2.16.

The above reasoning and its dual imply the following results.

Corollary 2.23. *Conclusions of Theorem 2.17 hold if $G : P \to P$ is increasing and strictly monotone sequences of $G[P]$ are finite, and if $\sup\{c, x\}$ and $\inf\{c, x\}$ exist for every $x \in G[P]$. Moreover, x^* is the last element of the finite sequence determined by the following algorithm:*

(i) $x_0 = c$. For n from 0 while $x_n \neq Gx_n$ do: $x_{n+1} = Gx_n$ if $Gx_n < x_n$ else $x_{n+1} = \sup\{c, Gx_n\}$,

and x_* is the last element of the finite sequence determined by the following algorithm:

(ii) $x_0 = c$. For n from 0 while $x_n \neq Gx_n$ do: $x_{n+1} = Gx_n$ if $Gx_n > x_n$ else $x_{n+1} = \inf\{c, Gx_n\}$.

Let $G : P \to P$ satisfy the hypotheses of Theorem 2.17. The result Corollary 2.23 can be applied to approximate the fixed points x^* and x_* of G introduced in Theorem 2.17 in the following manner. Assume that $\underline{G}, \overline{G} : P \to P$ satisfy the hypotheses given for G in Corollary 2.23, and that

$$\underline{G}(x) \leq G(x) \leq \overline{G}(x) \quad \text{for all } x \in P. \tag{2.12}$$

Since x^* and x_* are increasing with respect to G, it follows from (2.12) that $\underline{x}_* \leq x_* \leq \overline{x}_*$ and $\underline{x}^* \leq x^* \leq \overline{x}^*$, where $\underline{x}^*$ and $\overline{x}^*$ (respectively $\underline{x}_*$ and $\overline{x}_*$) are obtained by algorithm (i) (respectively (ii)) of Corollary 2.23 with G replaced by $\underline{G}$ and $\overline{G}$, respectively.

Since partial ordering is the only structure needed in the proofs, the above results can be applied to problems where only ordinal scales are available. On the other hand, these results have some practical value also in real analysis. We shall demonstrate this by an example where the above described method is applied to a system of the form

$$x_i = G_i(x_1,\ldots,x_m), \quad i = 1,\ldots,m, \tag{2.13}$$

where the functions G_i are real-valued functions of m real variables.

Example 2.24. Approximate a solution $x^* = (x_1, y_1)$ of the system

$$x = G_1(x,y) := \frac{N_1(x,y)}{2 - |N_1(x,y)|}, \quad y = G_2(x,y) := \frac{N_2(x,y)}{3 - |N_2(x,y)|}, \tag{2.14}$$

where

$$N_1(x,y) = \frac{11}{12}x + \frac{12}{13}y + \frac{1}{234} \quad \text{and} \quad N_2(x,y) = \frac{15}{16}x + \frac{14}{15}y - \frac{7}{345}, \tag{2.15}$$

by calculating upper and lower estimates of (x_1, y_1) whose corresponding coordinates differ by less than 10^{-100}.

Solution: The mapping $G = (G_1, G_2)$, defined by (2.14), (2.15) maps the set $P = \{(x,y) \in \mathbb{R}^2 : |x| + |y| \le \frac{1}{2}\}$ into P, and is increasing on P. It follows from Example 2.1 that $c = (0,0)$ is an order center of P, and that chains of P have supremums and infimums. Thus the results of Theorem 2.17 are valid.

Upper and lower estimates to the fixed point $x^* = (x_1, y_1)$ of G, and hence to a solution (x_1, y_1) of system (2.14), (2.15), can be obtained by applying the algorithm (i) given in Corollary 2.23 to operators $\overline{G}$ and $\underline{G}$, defined by

$$\begin{cases} \overline{G}(x,y) = (10^{-101}\text{ceil}(10^{101}G_1(x,y)), 10^{-101}\text{ceil}(10^{101}G_2(x,y)), \\ \underline{G}(x,y) = (10^{-101}\text{floor}(10^{101}G_1(x,y)), 10^{-101}\text{floor}(10^{101}G_2(x,y)), \end{cases} \tag{2.16}$$

where ceil(x) is the smallest integer $\ge x$ and floor(x) is the greatest integer $\le x$. The so defined operators $\underline{G}, \overline{G}$ are increasing and map the set $P = \{(x,y) \in \mathbb{R}^2 : |x| + |y| \le \frac{1}{2}\}$ into finite subsets of P, and (2.12) holds. We are going to show that the required upper and lower estimates are obtained by algorithm (i) of Corollary 2.23 with G replaced by $\underline{G}$ and $\overline{G}$, respectively. The following Maple program is used in calculations of the upper estimate $\overline{x}^* = (x1, y1)$.

$(N1, N2) := (11/12 * x + 12/13 * y + 1/234, 15/16 * x + 14/15 * y - 7/345):$

$(z, w) := (N1/(2 - abs(N1)), N2/(3 - abs(N2)))$:
$(G1, G2) := (ceil(10^{101}z)/10^{101}, ceil(10^{101}w)/10^{101})$:
$(x0, y0) := (0, 0); x := x0 : y := y0 : u := G1 : v := G2 : b[0] := [x, y]$:
for k from 1 while $abs(u - x) + abs(v - y) > 0$ do :
if $u <= x$ and $v <= y$ then $(x, y) := (u, v)$
else $(x, y) := (\max\{x, u\}, \max\{y, v\}) : fi$:
$u := G1 : v := G2 : b[k] := [x, y] : od : n := k - 1 : x1 := x; y1 = y;$

The above program yields the following results (n=1246).

$$\begin{aligned} x1 =& -0.007753186849780811654910693041037019619471431 38\\ &774717254950456999535626408273278584836718225237250043,\\ y1 =& -0.0135996154246109014898367199131292800245242544 01\\ &2899273758805991617838548683927620135569441397855721 \end{aligned}$$

In particular, $(x1, y1)$ is the fixed point $\overline{x}^*$ of $\overline{G}$.

Replacing 'ceil' by 'floor' in the above program, we obtain components of the fixed point $\underline{x}^* = (x2, y2)$ of $\underline{G}$ (n:=1248).

$$\begin{aligned} x2 =& -0.0077531868497808116549106930410370196194714313 87\\ &7471725495045699953562640827327858483671822523725005,\\ y2 =& -0.0135996154246109014898367199131292800245242544012\\ &8992737588059916178385486839276201355694413978557215 \end{aligned}$$

The above calculated components of $\overline{x}^*$ and $\underline{x}^*$ are exact, and their differences are $< 10^{-100}$. According to the above reasoning the exact fixed point x^* of G belongs to order interval $[\underline{x}^*, \overline{x}^*]$. In particular, both $(x1, y1)$ and $(x2, y2)$ approximate an exact solution (x_1, y_1) of system (2.14), (2.15) with the required precision. Moreover, $x1 \leq x_1 \leq y1$ and $x2 \leq y_1 \leq y2$. □

2.3 Solvability of Operator Equations and Inclusions

In this section we apply the results of Sect. 2.2 to study the solvability of operator equations in the form $Lu = Nu$, where L and N are single-valued mappings from a poset $V = (V, \leq)$ to another poset $P = (P, \leq)$. The solvability of the corresponding inclusions $Lu \in \mathcal{N}u$, where $\mathcal{N} : V \to 2^P \setminus \emptyset$, is studied as well. In order to obtain solvability results applicable to implicit equations and inclusions, we make use of the so-called *graph ordering* of V, defined by

$$u \preceq v \ \text{ if and only if } u \leq v \text{ and } \ Lu \leq Lv. \tag{2.17}$$

2.3.1 Inclusion Problems

As an application of Theorem 2.12 and Propositions 2.8 and 2.9 we prove the following existence result for the inclusion problem $Lu \in \mathcal{N}u$.

Theorem 2.25. *Let $L : V \to P$ and $\mathcal{N} : V \to 2^P \setminus \emptyset$ satisfy the following hypotheses.*

(L)   The equation $Lu = x$ has for each $x \in P$ smallest and greatest solutions, and they are increasing in x.

($\mathcal{N}$1) Chains of $\mathcal{N}[V]$ have supremums and infimums in P, and $ocl(\mathcal{N}[V])$ has a sup-center or an inf-center in P.

($\mathcal{N}$2) $\mathcal{N}$ is increasing in $(V, \preceq)$ or in $(V, \leq)$, and its values are order compact in $\mathcal{N}[V]$.

Then $Lu \in \mathcal{N}u$ has minimal and maximal solutions in $(V, \preceq)$.

Proof: Denote $V_- = \{\min L^{-1}\{x\} : x \in P\}$ and $L_- = L|V_-$. Define a mapping $\mathcal{F} : P \to 2^P \setminus \emptyset$ by

$$\mathcal{F}(x) := \mathcal{N}(L_-^{-1}x), \quad x \in P. \tag{2.18}$$

Assume first that $\mathcal{N}$ is increasing in $(V, \preceq)$, and that its values are order compact in $\mathcal{N}[V]$. To show that $\mathcal{F}$ is increasing, assume that $x \leq y$ in P. Then $u := \min L^{-1}\{x\} \leq v := \min L^{-1}\{y\}$ by condition (L), and $Lu = x \leq y = Lv$, whence $u \preceq v$. Since $\mathcal{N}$ is increasing in $(V, \preceq)$, thus $[z) \cap \mathcal{F}(y) = [z) \cap \mathcal{N}v \neq \emptyset$ for each $z \in \mathcal{N}u = \mathcal{F}(x)$ and $(w] \cap \mathcal{F}(x) = (w] \cap \mathcal{N}u \neq \emptyset$ for each $w \in \mathcal{N}v = \mathcal{F}(y)$. This proves that $\mathcal{F}$ is increasing.

As $\mathcal{F}[P]$ is contained in $\mathcal{N}[V]$ by (2.18), the chains of $\mathcal{F}[P]$ have supremums and infimums in P by condition ($\mathcal{N}$), and the values of $\mathcal{F}$ are order compact in $\mathcal{F}[P]$. Moreover, $ocl(\mathcal{F}[P])$ has a sup-center or an inf-center by hypothesis.

The above proof shows that $\mathcal{F}$ satisfies the hypotheses of Theorem 2.12, whence we conclude that it has minimal and maximal fixed points. If x is any fixed point of $\mathcal{F}$, and $u = L_-^{-1}x$, then $Lu = x \in \mathcal{F}(x) = \mathcal{N}u$, which shows that u is a solution of the inclusion problem $Lu \in \mathcal{N}u$.

To prove the existence of a minimal solution of $Lu \in \mathcal{N}u$, let x_- be a minimal fixed point of $\mathcal{F}$. Then $u_- = L_-^{-1}x_-$ is a solution of the inclusion problem $Lu \in \mathcal{N}u$. Let $v \in V$ satisfy $Lv \in \mathcal{N}v$, $v \leq u_-$ and $Lv \leq Lu_-$. Denoting $y = Lv$ and $u = L_-^{-1}y$, then $u \leq v$ and $Lu = Lv$, that is $u \preceq v$. Then $(y] \cap \mathcal{N}u \neq \emptyset$ because $\mathcal{N}$ is increasing. Since $\mathcal{N}u = \mathcal{N}L_-^{-1}y = \mathcal{F}(y)$, we have $(y] \cap \mathcal{F}(y) \neq \emptyset$. Thus y belongs to the set $S_- = \{x \in P : (x] \cap \mathcal{F}(x) \neq \emptyset\}$. Because $y = Lv \leq Lu_- = x_-$ and x_- is, by Proposition 2.9, a minimal element of S_-, then $y = x_-$. Hence it follows $u = L_-^{-1}y = L_-^{-1}x_- = u_-$. Moreover, $u \leq v \leq u_-$, whence $v = u_-$, which proves that u_- is a minimal solution of $Lu = \mathcal{N}u$ with respect to the graph ordering of V.

Denoting $V_+ = \{\max L^{-1}\{x\} : x \in P\}$ and $L_+ = L|V_+$, and replacing L_- in (2.18) by L_+ we obtain another mapping $\mathcal{F} : P \to 2^P \setminus \emptyset$, which satisfies the hypotheses of Theorem 2.12. Thus $\mathcal{F}$ has minimal and maximal fixed points, and to each fixed point x of $\mathcal{F}$ there corresponds a solution $u = \max L^{-1}\{x\}$ of the inclusion problem $Lu \in \mathcal{N}u$. Moreover, if x_+ is a maximal fixed point of $\mathcal{F}$, then by applying Proposition 2.8 one can show that $u_+ = L_+^{-1}x_+$ is a maximal solution of $Lu \in \mathcal{N}u$ in $(V, \preceq)$.

If $\mathcal{N}$ is increasing in $(V, \leq)$, it is increasing also in $(V, \preceq)$. Thus $Lu \in \mathcal{N}u$ has, by the above proof, minimal and maximal solutions in $(V, \preceq)$.   □

2.3.2 Single-Valued Problems

Consider next the single-valued case. As an application of Theorems 2.16 and 2.25 and Propositions 2.14 and 2.15 we obtain the following existence and comparison results for the equation $Lu = Nu$.

Theorem 2.26. *Given posets V and P, mappings $L, N : V \to P$, assume that L satisfies the hypothesis (L), that N is increasing in $(V, \preceq)$ or in $(V, \leq)$. If $ocl(N[V])$ has an order center c in P, and if chains of $N[V]$ have supremums and infimums in P, then the following results hold.*

(a) *The equation $Lu = \sup\{c, Nu\}$ has the smallest solution $\overline{u}$ in $(V_+, \leq)$, where $V_+ = \{\max L^{-1}\{x\} : x \in P\}$, and the equation $Lu = \inf\{c, Nu\}$ has the greatest solution $\underline{u}$ in $(V_-, \leq)$, where $V_- = \{\min L^{-1}\{x\} : x \in P\}$.*
(b) *If N is increasing in $(V, \preceq)$, then the equation $Lu = Nu$ has smallest and greatest solutions in the order interval $[\underline{u}, \overline{u}]$ of $(V, \preceq)$, and they are increasing in $(V, \leq)$ with respect to N.*
(c) *If N is increasing in $(V, \leq)$, then the equation $Lu = Nu$ has smallest and greatest solutions in the order interval $[\underline{u}, \overline{u}]$ of $(V, \leq)$, and they are increasing in $(V, \leq)$ with respect to N.*
(d) *The equation $Lu = Nu$ has minimal and maximal solutions in $(V, \preceq)$.*

Proof: Ad (a) The given hypotheses ensure that relation

$$G(x) := NL_+^{-1}x, \quad x \in P, \text{ where } L_+ = L|V_+, \tag{2.19}$$

defines an increasing mapping $G : P \to P$. Moreover, $N[V_+] = G[P]$, which in view of the hypotheses implies that chains of $G[P]$ have supremums and infimums in P. Moreover, c is a sup-center of $ocl(N[V_+]) = ocl(G[P])$ in P. Thus, by Proposition 2.14, the equation $x = \sup\{c, G(x)\}$ has the smallest solution $\overline{x}$. Denoting by $\overline{u} = L_+^{-1}\overline{x}$, then $G(\overline{x}) = N\overline{u}$, whence $L\overline{u} = \overline{x} = \sup\{c, G(\overline{x})\} = \sup\{c, N\overline{u}\}$. Thus $\overline{u}$ is a solution of the equation $Lu = \sup\{c, Nu\}$.

Assume that $v = L_+^{-1}x$, $x \in P$, and $Lv = \sup\{c, Nv\}$. Then

$$x = Lv = \sup\{c, Nv\} = \sup\{c, NL_+^{-1}x\} = \sup\{c, G(x)\}.$$

Since $\overline{x}$ is the smallest solution of the equation $x = \sup\{c, G(x)\}$, we get $\overline{x} \leq x$. This implies by condition (L) that $\overline{u} = L_+^{-1}\overline{x} \leq L_+^{-1}x = v$. Thus $\overline{u}$ is the smallest solution of $Lu = \sup\{c, Nu\}$ in $(V_+, \leq)$.

Ad (b) Assume that N is increasing in $(V, \preceq)$. By the proof of (a) $\bar{u} = L_+^{-1}\bar{x}$, where $\bar{x}$ is the smallest solution of the equation $x = \sup\{c, G(x)\}$. Then $G(\bar{x}) \leq \bar{x}$, whence G has the greatest fixed point x^* in $(\bar{x}]$ by Theorem 2.16. Denoting $u^* = L_+^{-1}x^*$, we see that u^* is a solution of the equation $Lu = Nu$ and $u^* \leq \bar{u}$ due to (L).

Let $u \in V$ be a solution of $Lu = Nu$ with $u \preceq \bar{u}$. Denoting $x = Lu$ and $v = L_+^{-1}x$, we infer $u \leq v$ and $Lu = Lv$, whence $u \preceq v$. Thus $x = Lv = Lu \leq L\bar{u} = \bar{x}$ and $x = Nu \leq Nv = NL_+^{-1}x = G(x)$. Since x^* is the greatest fixed point of G in $(\bar{x}]$, it follows that $Lv = x \leq x^* = Lu^*$ by (2.10). This implies by condition (L) that $v = L_+^{-1}x \leq L_+^{-1}x^* = u^*$, whence $v \preceq u^*$. Since $u \preceq v$, then $u \preceq u^*$, and thus u^* is the greatest solution of $Lu = Nu$ within the order interval $(\bar{u}]$ of $(V, \preceq)$.

To prove that u^* is increasing with respect to N, assume that the hypotheses imposed on N remain valid when N is replaced by $\hat{N} : V \to P$, and that

$$\hat{N}u \leq Nu \quad \text{for all } u \in V. \tag{2.20}$$

The above proof shows that the equation $Lu = \hat{N}u$ has the greatest solution v of the form $v = L_+^{-1}y$. Applying (2.19) and (2.20) we see that $y = \hat{N}v \leq Nv = G(y)$. This result and (2.10) imply that $y \leq x^*$, which results in $v \leq u^*$ by condition (L). This shows that u^* is increasing with respect to N.

By dual reasoning one can show that the equation $Lu = Nu$ has the smallest solution u_* within the order interval $[\underline{u})$ of $(V, \preceq)$, and that u_* is increasing with respect to N in $(V, \leq)$. In particular, u_* and u^* are smallest and greatest solutions in the order interval $[\underline{u}, \bar{u}]$ of $(V, \preceq)$.

Ad (c) Assume that N is increasing in $(V, \leq)$. Then N is increasing in $(V, \preceq)$. Let u^* be the solution constructed in the proof of (a). By the above proof we have $u^* \leq \bar{u}$. Let $u \in V$ be a solution of $Lu = Nu$ that satisfies $u \leq \bar{u}$. Denoting $x = Lu$ and $v = L_+^{-1}x$, then $u \leq v$ and $x = Lv = Nu \leq N\bar{u} \leq L\bar{u} = \bar{x}$, and $x = Lv \leq Nv = NL_+^{-1}x = G(x)$. Since x^* is the greatest fixed point of G in $(\bar{x}]$, it follows that $x \leq x^*$ by (2.10). This implies by condition (L) that $v = L_+^{-1}x \leq L_+^{-1}x^* = u^*$. Because $u \leq v$, we obtain $u \leq u^*$. Thus u^* is the greatest solution of $Lu = Nu$ within the order interval $(\bar{u}]$ of $(V, \leq)$. The proof that $Lu = Nu$ has the smallest solution u_* within the order interval $[\underline{u})$ of $(V, \leq)$ is done in similar way. In particular, u_* and u^* are smallest and greatest solutions within the order interval $[\underline{u}, \bar{u}]$ of $(V, \leq)$.

Ad (d) The hypotheses of Theorem 2.25 are valid, whence the equation $Lu = Nu$ has minimal and maximal solutions in $(V, \preceq)$. $\qquad\square$

The hypotheses of Theorem 2.26 hold true if $L, N : V \to P$ and $c \in P$ fulfil the following conditions:

(L1) L is a bijection and L^{-1} is increasing.
(N1) N is increasing, and strictly monotone sequences of $N[V]$ are finite.
(N2) $\sup\{c, x\}$ and $\inf\{c, x\}$ exist for every $x \in N[V]$.

Applying the algorithms (i) and (ii) of Corollary 2.23 to $G = N \circ L^{-1}$, we obtain the following result.

Corollary 2.27. *Let the hypotheses (L1), (N1), and (N2) hold for* $L, N :$ $V \to P$ *and* $c \in P$. *Then the solutions* u^* *and* u_* *of the equation* $Lu = Nu$ *introduced in Theorem 2.26(b) are the last elements of the sequences determined by the following algorithms.*

(iii) $Lu_0 = c$. *For* n *from* 0 *while* $Lu_n \neq Nu_n$ *do:* $Lu_{n+1} = Nu_n$ *if* $Nu_n <$ Lu_n *else* $Lu_{n+1} = \sup\{c, Nu_n\}$.
(iv) $Lu_0 = c$. *For* n *from* 0 *while* $Lu_n \neq Nu_n$ *do:* $Lu_{n+1} = Nu_n$ *if* $Nu_n >$ Lu_n *else* $Lu_{n+1} = \inf\{c, Nu_n\}$.

The algorithms (iii) and (iv) can be used, for instance, to calculate exact or approximative solutions for the equations $Lu = Nu$ in $\mathbb{R}^m$, and hence also for systems of the form

$$L_i(u_1, \ldots, u_m) = N_i(u_1, \ldots, u_m), \quad i = 1, \ldots, m, \tag{2.21}$$

where L_i and N_i are real-valued functions of m real variables. Algorithms (iii) and (iv) are applied in Sect. 6.5 to calculate exact solutions of an implicit functional initial function problem.

In the case when the range of N has an upper bound or a lower bound we have the following result.

Proposition 2.28. *Given posets* V *and* P *and mappings* $L, N : V \to P$, *assume that* L *satisfies the hypothesis (L), and that* N *is increasing in* $(V, \leq)$ *or in* $(V, \preceq)$.

(a) If $N[V]$ *has an upper bound in* P, *and if chains of* $N[V]$ *have infimums, then the equation* $Lu = Nu$ *has a minimal solution in* $(V, \preceq)$. *The equation has in* $(V, \leq)$ *the greatest solution, which is increasing with respect to* N.
(b) If $N[V]$ *has a lower bound in* P, *and if chains of* $N[V]$ *have supremums, then the equation* $Lu = Nu$ *has a maximal solution in* $(V, \preceq)$. *It has in* $(V, \leq)$ *the smallest solution, which is increasing with respect to* N.

Proof: Ad (a) Assume that $N[V] \subseteq (\overline{x}]$ for some $\overline{x} \in P$. Then $\overline{x}$ is an inf-center of $\mathrm{ocl}(N[V])$, so that the hypotheses of Theorem 2.25 are valid. Thus the equation $Lu = Nu$ has a minimal solution in $(V, \preceq)$.

Next we prove the existence of the greatest solution. The given hypotheses ensure that relation (2.19) defines an increasing mapping $G : P \to (\overline{x}]$. Moreover, the chains of $G[P] \subseteq N[V]$ have infimums in P. Thus G has by Proposition 2.18 the greatest fixed point x^* in $(\overline{x}]$. Denoting $u^* = L_+^{-1}x^*$, it follows that u^* is a solution of the equation $Lu = Nu$.

Let $u \in V$ be a solution of $Lu = Nu$. Denoting $x = Lu$ and $v = L_+^{-1}x$, then $u \leq v$ and $Lu = Lv$, whence $u \preceq v$. Thus $x = Nu \leq Nv = NL_+^{-1}x = G(x) \leq \overline{x}$. Since x^* is the greatest fixed point of G in $(\overline{x}]$, we have $x \leq x^*$ by

(2.10). This implies by condition (L) that $v = L_+^{-1}x \leq L_+^{-1}x^* = u^*$. As $u \leq v$, we conclude $u \leq u^*$. Thus u^* is the greatest solution of $Lu = Nu$ in $(V, \leq)$. The monotone dependence of u^* with respect to N can be shown in just the same way as in the proof of Theorem 2.26.

   Ad (b) The proof of (b) is dual to the above proof.   □

Remark 2.29. (i) If $Q : V \times P \to P$ is increasing with respect to the product ordering of $(V, \leq)$ and $(P, \leq)$, then $N := u \mapsto Q(u, Lu)$ is increasing with respect to the graph ordering $\preceq$ of V, defined by (2.17). Thus the result of Theorem 2.26(b) can be applied to the implicit problem $Lu = Q(u, Lu)$. Similarly, the result of Theorem 2.25 is applicable to the implicit inclusion problem $Lu \in \mathcal{Q}(u, Lu)$, where $\mathcal{Q} : V \times P \to 2^P \setminus \emptyset$.

   (ii) In Sect. 8.6 we present results in the case when the functions $\mathcal{F}$, G, N, and $\mathcal{N}$ satisfy weaker monotonicity conditions as assumed above. The case when V is not ordered is studied as well.

2.4 Special Cases

In this section we first formulate some fixed point results in ordered topological spaces derived in Sect. 2.2. Second, we present existence and comparison results for equations and inclusions in ordered normed spaces.

2.4.1 Fixed Point Results in Ordered Topological Spaces

Let $P = (P, \leq)$ be an ordered topological space, i.e., for each $a \in P$ the order intervals $[a) = \{x \in P : a \leq x\}$ and $(a] = \{x \in P : x \leq a\}$ are closed in the topology of P.

Definition 2.30. *A sequence* $(z_n)_{n=0}^\infty$ *of a poset is called* **increasing** *if* $z_n \leq z_m$ *whenever* $n \leq m$, **decreasing** *if* $z_m \leq z_n$ *whenever* $n \leq m$, *and* **monotone** *if it is increasing or decreasing. If the above inequalities are strict, the sequence* $(z_n)_{n=0}^\infty$ *is called strictly increasing, strictly decreasing, or strictly monotone, respectively.*

   In what follows, we assume that P has the following property:

(C) Each well-ordered chain C of P whose increasing sequences have limits in P contains an increasing sequence that converges to $\sup C$, and each inversely well-ordered chain C of P whose decreasing sequences have limits in P contains a decreasing sequence that converges to $\inf C$.

Lemma 2.31. *A second countable or metrizable ordered topological space has property (C).*

Proof: If P is an ordered topological space that satisfies the second countability axiom, then each chain of P is separable, whence P has property (C) by the result of [133, Lemma 1.1.7] and its dual. If P is metrizable, the assertion follows from [133, Proposition 1.1.5], and from its dual. □

The following result is a consequence of Proposition 2.18.

Proposition 2.32. *Given an ordered topological space P with property (C), assume that $G : P \to P$ is an increasing function.*

(a) *If the set $S_+ = \{x \in P : x \leq G(x)\}$ is nonempty, and if G maps increasing sequences of S_+ to convergent sequences, then G has a maximal fixed point. Moreover, G has for every $\underline{x} \in S_+$ the smallest fixed point in $[\underline{x})$, and it is increasing with respect to G.*

(b) *If the set $S_- = \{x \in P : G(x) \leq x\}$ is nonempty, and if G maps decreasing sequences of S_- to convergent sequences, then G has a minimal fixed point. Moreover, G has for every $\overline{x} \in S_-$ the greatest fixed point in $(\overline{x}]$, and it is increasing with respect to G.*

Proof: Ad (a) Let C be a well-ordered chain in S_+. Since G is increasing, then $G[C]$ is well-ordered. Every increasing sequence of $G[C]$ is of the form $(G(x_n))$, where (x_n) is an increasing sequence in C. Thus the hypotheses of (a) and property (C) imply that $\sup G[C]$ exists in P, and, therefore, the conclusions of (a) follows from Proposition 2.18(a).

Ad (b) The conclusions of (b) are similar consequences of Proposition 2.18(b). □

The next result is a consequence of Theorem 2.17.

Theorem 2.33. *Given an ordered topological space P with property (C), assume that $G : P \to P$ is increasing and maps monotone sequences of P to convergent sequences.*

(a) *If c is a sup-center of $\overline{G[P]}$ in P, then G has minimal and maximal fixed points. Moreover, G has the greatest fixed point x^* in $(\overline{x}]$, where $\overline{x}$ is the smallest solution of the equation $x = \sup\{c, G(x)\}$. Both $\overline{x}$ and x^* are increasing with respect to G.*

(b) *If c is an inf-center of $\overline{G[P]}$ in P, then G has minimal and maximal fixed points. Moreover, G has the smallest fixed point x_* in $[\underline{x})$, where $\underline{x}$ is the greatest solution of the equation $x = \inf\{c, G(x)\}$. Both $\underline{x}$ and x_* are increasing with respect to G.*

As a consequence of Propositions 2.8 and 2.9 and Theorem 2.12 we obtain the following proposition.

Proposition 2.34. *Let P be an ordered topological space with property (C), and let the values of $\mathcal{F} : P \to 2^P \setminus \emptyset$ be compact.*

(a) If $\mathcal{F}$ is increasing upward, if the set $S_+ = \{x \in P : [x) \cap \mathcal{F}(x) \neq \emptyset\}$ is nonempty, and if (y_n) converges whenever it is increasing and $y_n \in \mathcal{F}(x_n)$, for every n, where (x_n) is an increasing sequence of S_+, then $\mathcal{F}$ has a maximal fixed point.

(b) If $\mathcal{F}$ is increasing downward, if the set $S_- = \{x \in P : (x] \cap \mathcal{F}(x) \neq \emptyset\}$ is nonempty, and if (y_n) converges whenever it is decreasing and $y_n \in \mathcal{F}(x_n)$, for every n, where (x_n) is a decreasing sequence of S_-, then $\mathcal{F}$ has a minimal fixed point.

(c) If $\mathcal{F}$ is increasing, if $\overline{\mathcal{F}[P]}$ has a sup-center or an inf-center in P, and if (y_n) converges whenever $y_n \in \mathcal{F}(x_n)$, for every n, and both (x_n) and (y_n) are either increasing or decreasing sequences of P, then $\mathcal{F}$ has minimal and maximal fixed points.

The next theorem is a special case of Theorem 2.20.

Theorem 2.35. *Given an ordered topological space X with property (C) and a subset P of X, assume that a function $G : P \to P$ and a multifunction $\mathcal{F} : X \to 2^X$ have the following properties:*

(ha) *G is increasing, $G[P]$ has an upper bound $\bar{x}$ in P, and G maps every decreasing sequence (x_n) of P to a convergent sequence whose limit is in P.*

(hb) *$\bar{x}$ is an upper bound of $\mathcal{F}[X]$, and if $x \leq p$ in X and $p \in P$, then $G(p)$ is an upper bound of $\mathcal{F}(x)$.*

Then G has the greatest fixed point x^*, and if x is any fixed point of $\mathcal{F}$, then $x \leq x^*$.

As a special case of Theorem 2.21 we formulate the following existence and comparison result for greatest fixed points of multifunctions.

Theorem 2.36. *Given an ordered topological space X with property (C) and a subset P of X. Assume that a multifunction $\mathcal{F} : X \to 2^X$ has the following properties:*

(h0) *$\mathcal{F}[X]$ has an upper bound in P.*

(h1) *If $p \in P$, then $\max \mathcal{F}(p)$ exists, belongs to P and is an upper bound of $\mathcal{F}[X \cap (p]]$.*

(h2) *Decreasing sequences of $\{\max \mathcal{F}(p) : p \in P\}$ have limits in X and they belong to P.*

Then $\mathcal{F}$ has the greatest fixed point, and it belongs to P. Assume moreover, that $\hat{\mathcal{F}} : X \to 2^X$ is another multifunction that satisfies the following condition:

(h3) *For each $x \in X$ and $y \in \hat{\mathcal{F}}(x)$ there exists a $z \in \mathcal{F}(x)$ such that $y \leq z$.*

Then the greatest fixed point of $\mathcal{F}$ majorizes all the fixed points of $\hat{\mathcal{F}}$.

2.4.2 Equations and Inclusions in Ordered Normed Spaces

Definition 2.37. *A closed subset E_+ of a normed space E is called an* **order cone** *if $E_+ + E_+ \subseteq E_+$, $E_+ \cap (-E_+) = \{0\}$, and $cE_+ \subseteq E_+$ for each $c \geq 0$. The space E, equipped with an order relation '$\leq$', defined by*

$$x \leq y \quad \text{if and only if} \quad y - x \in E_+,$$

is called an **ordered normed space**.

It is easy to see that the above defined order relation $\leq$ is a partial ordering in E.

Lemma 2.38. *An ordered normed space E is an ordered topological space with respect to the weak and the norm topologies. Moreover, property (C) holds in both cases.*

Proof: It is easy to verify that the first assertion holds. To prove the second assertion, let C be a well-ordered chain in E. If all increasing sequences of C have weak limits, there is, by [44, Lemma A.3.1], an increasing sequence (x_n) in C that converges weakly to $x = \sup C$. If C is inversely well-ordered and its decreasing sequences have weak limits, then $-C$ is a well-ordered chain whose increasing sequences have weak limits. Thus there exists an increasing sequence (x_n) of $-C$ that converges weakly to $\sup(-C) = -\inf C$. Denoting $y_n = -x_n$, we obtain a decreasing sequence (y_n) of C, which converges weakly to $\inf C$. If E is equipped with norm topology, it is an ordered metric space, whence the conclusion follows from Lemma 2.31.   □

The next fixed point result is a consequence of Proposition 2.32 and Lemma 2.38.

Proposition 2.39. *Let P be a subset of an ordered normed space, and let $G : P \to P$ be increasing.*

(a) *If the set $S_+ = \{x \in P : x \leq G(x)\}$ is nonempty, and if G maps increasing sequences of S_+ to sequences that have weak or strong limits in P, then G has a maximal fixed point. Moreover, G has for every $\underline{x} \in S_+$ the smallest fixed point in $[\underline{x})$, and it is increasing with respect to G.*

(b) *If the set $S_- = \{x \in P : G(x) \leq x\}$ is nonempty, and if G maps decreasing sequences of S_- to sequences that have weak or strong limits in P, then G has a minimal fixed point. Moreover, G has for every $\overline{x} \in S_-$ the greatest fixed point in $(\overline{x}]$, and it is increasing with respect to G.*

As a special case of Theorem 2.33 we obtain the following proposition.

Proposition 2.40. *Given a subset P of an ordered normed space E, assume that $G : P \to P$ is increasing, and that monotone sequences of $G[P]$ have weak limits in P.*

(a) *If the weak closure of $G[P]$ has a sup-center c in P, then G has minimal and maximal fixed points. Moreover, G has the greatest fixed point x^* in $(\overline{x}]$, where $\overline{x}$ is the smallest solution of the equation $x = \sup\{c, G(x)\}$. Both $\overline{x}$ and x^* are increasing with respect to G.*

(b) *If the weak closure of $G[P]$ has an inf-center c in P, then G has minimal and maximal fixed points. Moreover, G has the smallest fixed point x_* in $[\underline{x})$, where $\underline{x}$ is the greatest solution of the equation $x = \inf\{c, G(x)\}$. Both $\underline{x}$ and x_* are increasing with respect to G.*

In case that E is $\mathbb{R}^m$ equipped with Euclidean norm and ordered coordinatewise, we obtain the following consequence of Proposition 2.40.

Corollary 2.41. *Let P be a closed and bounded subset of $\mathbb{R}^m$, and assume that $G : P \to P$ is increasing.*

(a) *If P has a sup-center c, then G has minimal and maximal fixed points. Moreover, G has the greatest fixed point x^* in $(\overline{x}]$, where $\overline{x}$ is the smallest solution of the equation $x = \sup\{c, G(x)\}$. Both $\overline{x}$ and x^* are increasing with respect to G.*

(b) *If P has an inf-center c, then G has minimal and maximal fixed points. Moreover, G has the smallest fixed point x_* in $[\underline{x})$, where $\underline{x}$ is the greatest solution of the equation $x = \inf\{c, G(x)\}$. Both $\underline{x}$ and x_* are increasing with respect to G.*

As a consequence of Proposition 2.28 and Lemma 2.38 we obtain the following proposition.

Proposition 2.42. *Given a poset V and a subset P of an ordered normed space, assume that mappings L, $N : V \to P$ satisfy the following hypotheses.*

(L) *The equation $Lu = x$ has for each $x \in P$ smallest and greatest solutions, and they are increasing with respect to x.*

(N) *N is increasing in $(V, \le)$ or in $(V, \preceq)$.*

Then the following assertions hold.

(a) *If $N[V]$ has an upper bound in P, and if decreasing sequences of $N[V]$ have weak or strong limits in P, then the equation $Lu = Nu$ has a minimal solution in $(V, \preceq)$. It has in $(V, \le)$ the greatest solution, which is increasing with respect to N.*

(b) *If $N[V]$ has a lower bound in P, and if increasing sequences of $N[V]$ have weak or strong limits in P, then the equation $Lu = Nu$ has a maximal solution in $(V, \preceq)$. It has in $(V, \le)$ the smallest solution, which is increasing with respect to N.*

In what follows, E is an ordered normed space having the following properties.

(E0) Bounded and monotone sequences of E have weak or strong limits.

(E1) $x^+ = \sup\{0, x\}$ exists, and $\|x^+\| \le \|x\|$ for every $x \in E$.

When $c \in E$ and $R \in [0, \infty)$, denote $B_R(c) := \{x \in E : \|x - c\| \le R\}$. Recall (cf., e.g., [227]) that if a sequence (x_n) of a normed space E converges weakly to x, then (x_n) is bounded, i.e., $\sup_n \|x_n\| < \infty$, and

$$\|x\| \le \liminf_{n \to \infty} \|x_n\|. \tag{2.22}$$

The next auxiliary result is needed in the sequel.

Lemma 2.43. *If $c \in E$ and $R \in (0, \infty)$, then c is an order center of $B_R(c)$, and for every chain C of $B_R(c)$ both $\sup C$ and $\inf C$ exist and belong to $B_R(c)$.*

Proof: Since

$$\sup\{c, x\} = (x - c)^+ - c \quad \text{and} \quad \inf\{c, x\} = c - (c - x)^+, \quad \text{for all} \quad x \in E, \tag{2.23}$$

(E1) and (2.23) imply that

$$\| \sup\{c, x\} - c \| = \| \inf\{c, x\} - c \| = \|(x - c)^+\| \le \|x - c\| \le R$$

for every $x \in B_R(c)$. Thus both $\sup\{c, x\}$ and $\inf\{c, x\}$ belong to $B_R(c)$. Let C be a chain in $B_R(c)$. Since C is bounded, there is an increasing sequence (x_n) in C that converges weakly to $x = \sup C$ due to (E0) and Lemma 2.38. Since $\|x_n - c\| \le R$ for each n, it follows from (2.22) that

$$\|x - c\| \le \liminf_{n \to \infty} \|x_n - c\| \le R.$$

Thus $x = \sup C$ exists and belongs to $B_R(c)$. Similarly one can show that $\inf G[C]$ exists in E and belongs to $B_R(c)$. □

Applying Theorem 2.17 and Lemma 2.43 we obtain the following fixed point results.

Theorem 2.44. *Given a subset P of E, assume that $G : P \to P$ is increasing, and that $G[P] \subseteq B_R(c) \subseteq P$ for some $c \in P$ and $R \in (0, \infty)$. Then G has*

(a) minimal and maximal fixed points;
(b) smallest and greatest fixed points x_* and x^* in the order interval $[\underline{x}, \overline{x}]$ of P, where $\underline{x}$ is the greatest solution of $x = \inf\{c, G(x)\}$, and $\overline{x}$ is the smallest solution of $x = \sup\{c, G(x)\}$.

Moreover, x^*, x_*, $\underline{x}$ and $\overline{x}$ are all increasing with respect to G.

Proof: Let C be a chain in P. Since $G[C]$ is a chain in $B_R(c)$, both $\sup G[C]$ and $\inf G[C]$ exist in E and belong to $B_R(c) \subseteq P$ by Lemma 2.43. Because c is an order center of $B_R(c)$ and $ocl(G[P]) \subseteq \overline{G[P]} \subseteq B_R(c) \subseteq P$, then c is an order center of $ocl(G[P])$ in P. Thus the hypotheses of Theorem 2.17 are valid. □

Next we assume also that $V = (V, \le)$ is a poset, and that $P \subseteq E$. By means of Theorem 2.26 we obtain the following results.

Theorem 2.45. *Assume that the hypothesis (L) holds for $L : V \to P$, that $N : V \to P$ is increasing in $(V, \leq)$ (respectively in $(V, \preceq)$), and that $N[V] \subseteq B_R(c) \subseteq P$ for some $c \in E$ and $R \in (0, \infty)$. Then the equation $Lu = Nu$ has*

(a) minimal and maximal solutions in $(V, \preceq)$;

(b) smallest and greatest solutions u_*, u^* within the order interval $[\underline{u}, \overline{u}]$ of $(V, \leq)$ (respectively $(V, \preceq)$), where $\underline{u}$ is the greatest solution of $Lu = \inf\{c, Nu\}$ in $V_- = \{\min L^{-1}\{x\} : x \in P\}$, and $\overline{u}$ is the smallest solution of $Lu = \sup\{c, Nu\}$ in $V_+ = \{\max L^{-1}\{x\} : x \in P\}$.

Moreover, u_*, u^*, $\underline{u}$ and $\overline{u}$ are all increasing with respect to N in $(V, \leq)$.

Proof: The given hypotheses imply by Lemma 2.43 (cf. the proof of Theorem 2.44) that the hypotheses of Theorem 2.26 hold. The assertions follow then from Theorem 2.26. □

Noticing that $\inf\{0, v\} = -(-v)^+$, the next result is a consequence of Theorem 2.45.

Corollary 2.46. *Given a poset V and $R \in (0, \infty)$, assume that $L, N : V \to B_R(0)$ satisfy the following hypotheses.*

(LN) L is a bijection, and both L^{-1} and N are increasing.

Then the equation $Lu = Nu$ has

(a) minimal and maximal solutions in $(V, \preceq)$;

(b) smallest and greatest solutions u_*, u^* within the order interval $[\underline{u}, \overline{u}]$ of $(V, \leq)$, where $\underline{u}$ is the greatest solution of $Lu = -(-Nu)^+$ and $\overline{u}$ is the smallest solution of $Lu = (Nu)^+$ in $(V, \leq)$.

Moreover, u_*, u^*, $\underline{u}$ and $\overline{u}$ are all increasing with respect to N in $(V, \leq)$.

In the set-valued case we have the following consequences of Theorems 2.12 and 2.25.

Theorem 2.47. *Assume that P is a subset of E which contains $B_R(c)$ for some $c \in E$ and $R \in (0, \infty)$. Then the following holds.*

(a) Let $\mathcal{F} : P \to 2^P \setminus \emptyset$ be an increasing mapping whose values are weakly compact in E, and whose range $\mathcal{F}[P]$ is contained in $B_R(c)$. Then $\mathcal{F}$ has minimal and maximal fixed points.

(b) Assume that $\mathcal{N} : V \to 2^P \setminus \emptyset$ is increasing in $(V, \leq)$ or in $(V, \preceq)$, that its values are weakly compact in E, and that $\mathcal{N}[V] \subseteq B_R(c)$. If $L : V \to P$ satisfies the hypothesis (L), then the inclusion problem $Lu \in \mathcal{N}u$ has minimal and maximal solutions in $(V, \preceq)$.

The next result is also a consequence of Theorem 2.45.

Theorem 2.48. *Given a lattice-ordered Banach space E with properties (E0) and (E1) and a poset W, assume that mappings $\Lambda, F : W \to E$ satisfy the following hypotheses:*

(Λ)   The equation $\Lambda u = v$ has for each $v \in E$ smallest and greatest solutions, and they are increasing with respect to v.
(F) F is increasing.
(ΛF) $\|F(u)\| \leq q(\|\Lambda u\|)$ for all $u \in W$, where $q : \mathbb{R}_+ \to \mathbb{R}_+$ is increasing, and there exists a $R > 0$ such that $R = q(R)$, and, moreover, if $s \leq q(s)$, then $s \leq R$.

Then the equation

$$\Lambda u = Fu \tag{2.24}$$

has minimal and maximal solutions.

Proof: Let $R > 0$ be the constant in the hypothesis (ΛF). Define

$$P := B_R(0), \quad \text{and} \quad V := \{u \in W : \|\Lambda u\| \leq R\}. \tag{2.25}$$

The growth condition (ΛF) implies that for each $u \in V$,

$$\|Fu\| \leq q(\|\Lambda u\|) \leq q(R) = R,$$

so that $F[V] \subseteq P$ by (2.25).

Defining $L = \Lambda_{|V}$ and $N = F_{|V}$, then $L, N : V \to P$, L satisfies the hypothesis (L) and N is increasing. Therefore, the results of Theorem 2.45 can be applied. In particular, it follows that the equation $Lu = Nu$ has minimal and maximal solutions. If $u \in W$ and $\Lambda u = Fu$, then

$$\|\Lambda u\| = \|Fu\| \leq q(\|\Lambda u\|)$$

by (ΛF), which implies that $\|\Lambda u\| \leq R$, i.e., $u \in V$. Thus all the solutions of (2.24) are contained in V, whence we conclude that (2.24) and the equation $Lu = Nu$ have the same solutions. In particular, (2.24) has minimal and maximal solutions. □

The next result is a direct consequence of Theorem 2.48

Corollary 2.49. *Let E be a lattice-ordered Banach space E with properties (E0) and (E1), let W be a poset, and assume that $\Lambda : W \to E$ satisfies the hypothesis (Λ), and that $F : W \to E$ is increasing and bounded. Then the equation (2.24) has minimal and maximal solutions.*

Proof: Defining $q(s) \equiv R > \sup\{\|Fu\| : u \in W\}$, we see that the hypothesis (ΛF) holds. □

Remark 2.50. As an application of Theorem 2.45(b) one can formulate other existence results for equation (2.24). In particular, one can construct such solutions of (2.24) that are increasing with respect to F.

According to the hypothesis (F) the operator F may be discontinuous and noncompact. Moreover, the growth condition of (ΛF) does not provide means to construct a priori upper and/or lower solutions for equation (2.24).

Thus the standard theories such as the theory of monotone operators due to Brezis and Browder, Schauder's fixed point theorem, or the method of sub- and supersolutions are, in general, not applicable under the hypotheses given above to solve (2.24).

Each of the following spaces has properties (E0) and (E1) (as for the proofs, see, e.g., [22, 44, 48, 118, 133, 152, 170]):

(a) A Sobolev space $W^{1,p}(\Omega)$ or $W_0^{1,p}(\Omega)$, $1 < p < \infty$, ordered a.e. pointwise, where Ω is a bounded domain in $\mathbb{R}^N$.
(b) A finite-dimensional normed space ordered by a cone generated by a basis.
(c) l^p, $1 \leq p < \infty$, normed by p-norm and ordered coordinatewise.
(d) $L^p(\Omega)$, $1 \leq p < \infty$, normed by p-norm and ordered a.e. pointwise, where Ω is a σ-finite measure space.
(e) A separable Hilbert space whose order cone is generated by an orthonormal basis.
(f) A weakly complete Banach lattice or a UMB-lattice (cf. [22]).
(g) $L^p(\Omega, Y)$, $1 \leq p < \infty$, normed by p-norm and ordered a.e. pointwise, where Ω is a σ-finite measure space and Y is any of the spaces (b)–(f).
(h) Newtonian spaces $N^{1,p}(Y)$, $1 < p < \infty$, ordered a.e. pointwise, where Y is a metric measure space.

Thus the results of Theorems 2.44–2.48 hold if E is any of the spaces listed in (a)–(g).

2.5 Fixed Point Results for Maximalizing Functions

In this section we prove fixed point results for a self-mapping G of a poset P by assuming that G is **maximalizing**, i.e., $G(x)$ is a maximal element of $\{x, G(x)\}$ for all $x \in P$. Concrete examples of maximalizing functions that have or don't have fixed points are presented. The generalized iteration method introduced in Lemma 2.2 is used in the proofs.

2.5.1 Preliminaries

The following result helps to analyze the w-o chain of cG-iterations defined in (2.2).

Lemma 2.51. *Let A and B be nonempty subsets of P. If $\sup A$ and $\sup B$ exists, then the equation*

$$\sup(A \cup B) = \sup\{\sup A, \sup B\} \tag{2.26}$$

is valid whenever either of its sides is defined.

Proof: The sets $A \cup B$ and $\{\sup A, \sup B\}$ have the same upper bounds, which implies the assertion. $\square$

Recall that a subset W of a chain C is called an *initial segment* of C if $x \in W$, $y \in C$, and $y < x$ imply $y \in W$. If W is well-ordered, then every element x of W that is not the possible maximum of W has a **successor**: $Sx = \min\{y \in W : x < y\}$, in W.

A characterization of elements of the w-o chain of cG-iterations, defined by (2.2), is provided by the following lemma.

Lemma 2.52. *Given $G : P \to P$ and $c \in P$, and let C be the w-o chain of cG-iterations. Then the elements of C have the following properties:*

(a) $\min C = c$.
(b) An element x of C has a successor in C if and only if $\sup\{x, G(x)\}$ exists and $x < \sup\{x, G(x)\}$, and then $Sx = \sup\{x, G(x)\}$.
(c) If W is an initial segment of C and $y = \sup W$ exists, then $y \in C$.
(d) If $c < y \in C$ and y is not a successor, then $y = \sup C^{<y}$.
(e) If $y = \sup C$ exists, then $y = \max C$.
(f) If $x_* = \sup\{c, G[C]\}$ exists in P, then $x_* = \max C$, and $G(x_*) \le x_*$.

Proof: Ad (a) $\min C = \sup\{c, G[C^{<\min C}]\} = \sup\{c, G[\emptyset]\} = \sup\{c, \emptyset\} = c$.

Ad (b) Assume first that $x \in C$, and that Sx exists in C. Applying (2.2), Lemma 2.51, and the definition of Sx we obtain

$$Sx = \sup\{c, G[C^{<Sx}]\} = \sup\{c, G[C^{<x}] \cup \{G(x)\}\} = \sup\{x, G(x)\}.$$

Moreover, $x < Sx$, by definition, whence $x < \sup\{x, G(x)\}$.

Assume next that $x \in C$, that $y = \sup\{x, G(x)\}$ exists, and that $x < \sup\{x, G(x)\}$. The above proof implies that

(i) there is no element $w \in C$ that satisfies $x < w < \sup\{x, G(x)\}$.

Then $\{z \in C : z \le x\} = C^{<y}$, so that

$$\begin{aligned} x < \sup\{x, G(x)\} &= \sup\{\sup\{c, G[C^{<x}]\}, G(x)\} \\ &= \sup\{\{c\} \cup G[C^{<x}] \cup \{G(x)\}\} = \sup\{c, G[\{z \in C : z \le x\}]\} \\ &= \sup\{c, G[C^{<y}]\}. \end{aligned}$$

Thus $y = \sup\{x, G(x)\} \in C$ by (2.2). This result and (i) imply that $y = \sup\{x, G(x)\} = \min\{z \in C : x < z\} = Sx$.

Ad (c) Assume that W is an initial segment of C, and that $y = \sup W$ exists. If there is $x \in W$ that does not have the successor, then $x = \max W = y$, so that $y \in C$. Assume next that every element x of W has the successor Sx in W. Since $Sx = \sup\{x, G(x)\}$ by (b), then $G(x) \le Sx < y$. This holds for all $x \in W$. Since $c = \min C = \min W < y$, then y is an upper bound of $\{c\} \cup G[W]$. If z is an upper bound of $\{c\} \cup G[W]$, then $x = \sup\{c, G[C^{<x}]\} = \sup\{c, G[W^{<x}]\} \le z$ for every $x \in W$. Thus z is an upper bound of W, whence

$y = \sup W \leq z$. But then $y = \sup\{c, G[W]\} = \sup\{c, G[C^{<y}]\}$, so that $y \in C$ by (2.2).

Ad (d) Assume that $c < y \in C$, and that y is not a successor of any element of C. Obviously, y is an upper bound of $C^{<y}$. Let z be an upper bound of $C^{<y}$. If $x \in C^{<y}$, then also $Sx \in C^{<y}$ since y is not a successor. Because $Sx = \sup\{x, G(x)\}$ by (b), then $G(x) \leq Sx \in C^{<y}$. This holds for every $x \in C^{<y}$. Since also $c \in C^{<y}$, then z is an upper bound of $\{c\} \cup G[C^{<y}]$. Thus $y = \sup\{c, G[C^{<y}]\} \leq z$. This holds for every upper bound z of $C^{<y}$, whence $y = \sup C^{<y}$.

Ad (e) If $y = \sup C$ exists, then $y \in C$ by (c) when $W = C$, whence $y = \max C$.

Ad (f) Assume that $x_* = \sup\{c, G[C]\}$ exists. If $x \in C$, then $x = \sup\{c, G[C^{<x}] \leq \sup\{c, G[C]\} = x_*$, whence x_* is an upper bound of C. If x_* is a strict upper bound of C, then $C = C^{<x_*}$, so that $x_* = \sup\{c, G[C^{<x_*}]\}$. But then $x_* \in C$ by (2.2), and x_* is a strict upper bound of C, a contradiction. Thus $x_* = \max C$. In particular, $G(x_*) \in G[C]$, whence $G(x_*) \leq \sup\{c, G[C]\} = x_*$. □

2.5.2 Main Results

Let $P = (P, \leq)$ be a nonempty poset. As an application of Lemma 2.52(f) we shall prove our first existence result.

Theorem 2.53. *A function $G : P \to P$ has a fixed point if G is* maximalizing, *i.e., $G(x)$ is a maximal element of $\{x, G(x)\}$ for all $x \in P$, and if $x_* = \sup\{c, G[C]\}$ exists in P for some $c \in P$ where C is the w-o chain of cG-iterations.*

Proof: If C is the w-o chain of cG-iterations, and if $x_* = \sup\{c, G[C]\}$ exists in P, then $x_* = \max C$ and $G(x_*) \leq x_*$ by Lemma 2.52(f). Since G is maximalizing, then $G(x_*) = x_*$, i.e., x_* is a fixed point of G. □

The following proposition is a consequence of Theorem 2.53 and Lemma 2.52(b),(e).

Proposition 2.54. *Assume that $G : P \to P$ is maximalizing. Given $c \in P$, let C be a w-o chain of cG-iterations. If $z = \sup C$ exists, then z is a fixed point of G if and only if $x_* = \sup\{z, G(z)\}$ exists.*

Proof: Assume that $z = \sup C$ exists. It follows from Lemma 2.52(e) that $z = \max C$. If z is a fixed point of G, i.e., $z = G(z)$, then $x_* = \sup\{z, G(z)\} = z$.

Assume next that $\sup\{z, G(z)\}$ exists. Since Sz (successor) does not exist, it follows from Lemma 2.52(b) that $z \not< \sup\{z, G(z)\}$. Thus $x_* = \sup\{z, G(z)\} = z$, so that $G(z) \leq z$. Moreover, $G(z) \not< z$ because G is maximalizing, whence $G(z) = z$. □

As a consequence of Proposition 2.54 we obtain the following corollary.

Corollary 2.55. *Assume that all nonempty chains of P have supremums in P. If $G : P \to P$ is maximizing, and if $\sup\{x, G(x)\}$ exists for all $x \in P$, then for each $c \in P$ the maximum of the w-o chain of cG-iterations exists and is a fixed point of G.*

Proof: Let C be the w-o chain of cG-iterations. The given hypotheses imply that both $z = \sup C$ and $x_* = \sup\{z, G(z)\}$ exist. Thus the hypotheses of Proposition 2.54 are valid.   □

For completeness we formulate the obvious duals of the above results.

Theorem 2.56. *A function $G : P \to P$ has a fixed point if G is minimizing, i.e., $G(x)$ is a minimal element of $\{x, G(x)\}$ for all $x \in P$, and if $\inf\{c, G[W]\}$ exists in P for some $c \in P$ whenever W is a non-empty chain in P.*

Proposition 2.57. *A minimalizing function $G : P \to P$ has a fixed point if every nonempty chain P has the infimum in P, and if $\inf\{x, G(x)\}$ exists for all $x \in P$.*

Remark 2.58. The hypothesis that $G : X \to X$ is maximizing can be weakened in Theorem 2.53 and in Proposition 2.54 to the form: $G|\{x_*\}$ is maximizing, i.e., $G(x_*)$ is a maximal element of $\{x_*, G(x_*)\}$.

2.5.3 Examples and Remarks

We shall first present an example of a maximizing mapping whose fixed point is obtained as the maximum of the w-o chain of cG-iterations.

Example 2.59. Let P be a closed disc $P = \{(u, v) \in \mathbb{R}^2 : u^2 + v^2 \le 2\}$, ordered coordinatewise. Let $[u]$ denote the greatest integer $\le u$ when $u \in \mathbb{R}$. Define a function $G : P \to \mathbb{R}^2$ by

$$G(u, v) = \left(\min\{1, 1 - [u] + [v]\}, \frac{1}{2}([u] + v^2) \right), \quad (u, v) \in P. \qquad (2.27)$$

It is easy to verify that $G[P] \subset P$, and that G is maximizing. To find a fixed point of G, choose $c = (1, 0)$. It follows from Lemma 2.52 (a) and (b) that the first elements of the w-o chain of cG-iterations are successive approximations

$$x_0 = c, \quad x_{n+1} = Sx_n = \sup\{x_n, G(x_n)\}, \quad n = 0, 1, \ldots, \qquad (2.28)$$

as long as Sx_n is defined. Denoting $x_n = (u_n, v_n)$, these successive approximations can be rewritten in the form

$$u_0 = 1, \quad u_{n+1} = \max\{u_n, \min\{1, 1 - [u_n] + [v_n]\}\},$$
$$v_0 = 0, \quad v_{n+1} = \max\{v_n, \frac{1}{2}([u_n] + v_n^2)\}, \quad n = 0, 1, \ldots, \qquad (2.29)$$

as long as $u_n \leq u_{n+1}$ and $v_n \leq v_{n+1}$, and at least one of these inequalities is strict. Elementary calculations show that $u_n = 1$, for every $n \in \mathbb{N}_0$. Thus (2.29) can be rewritten as

$$u_n = 1, \quad v_0 = 0, \quad v_{n+1} = \max\{v_n, \frac{1}{2}(1 + v_n^2)\}, \quad n = 0, 1, \ldots. \tag{2.30}$$

Since the function $g(v) = \frac{1}{2}(1 + v^2)$ is increasing in $\mathbb{R}_+$, then $v_n < g(v_n)$ for every $n = 0, 1, \ldots$. Thus (2.30) can be reduced to the form

$$u_n = 1, \quad v_0 = 0, \quad v_{n+1} = g(v_n) = \frac{1}{2}(1 + v_n^2), \quad n = 0, 1, \ldots. \tag{2.31}$$

The sequence $(g(v_n))_{n=0}^\infty$ is strictly increasing, whence also $(v_n)_{n=0}^\infty$ is strictly increasing by (2.31). Thus the set $W = \{(1, g(v_n))\}_{n \in \mathbb{N}_0}$ is an initial segment of C. Moreover, $v_0 = 0 < 1$, and if $0 \leq v_n < 1$, then $0 < g(v_n) < 1$. Since $(g(v_n))_{n=0}^\infty$ is bounded above by 1, then $v_* = \lim_n g(v_n)$ exists, and $0 < v_* \leq 1$. Thus $(1, v_*) = \sup W$, and it belongs to X, whence $(1, v_*) \in C$ by Lemma 2.52(c). To determine v_*, notice that $v_{n+1} \to v_*$ by (2.31). Thus $v_* = g(v_*)$, or equivalently, $v_*^2 - 2v_* + 1 = 0$, so that $v_* = 1$. Since $\sup W = (1, v_*) = (1, 1)$, then $(1, 1) \in C$ by Lemma 2.52(c). Because $(1, 1)$ is a maximal element of X, then $(1, 1) = \max C$. Moreover, $G(1, 1) = (1, 1)$, so that $(1, 1)$ is a fixed point of G.

The first $m + 1$ elements of the w-o chain C of cG-iterations can be estimated by the following Maple program (floor($\cdot$)=[$\cdot$]):

$x := min\{1, 1 - \text{floor}(u) + \text{floor}(v)\} : y := (\text{floor}(u) + v^2)/2 :$

$(u, v) := (1, 0) : c[0] := (u, v) :$

for n to m do $(u, v) := (\max\{x, u\}, \text{evalf}(\max\{y, v\})); c[n] := (u, v)$ end do;

For instance, c[100000]=(1,0.99998).

The verification of the following properties are left to the reader:

- If $c = (u, v) \in X$, $u < 1$ and $v < 1$, then the elements of w-o chain C of cG-iterations, after two first terms if $u < 1$, are of the form $(1, w_n)$, $n = 0, 1, \ldots$, where $(w_n)_{n=0}^\infty$ is increasing and converges to 1. Thus $(1, 1)$ is the maximum of C and a fixed point of G.
- If $c = (u, 1)$, $u < 1$, or $c = (1, -1)$, then $C = \{c, (1, 1)\}$.
- If $= (1, 0)$, then $G^{2k}c = (1, z_k)$ and $G^{2k+1}c = (0, y_k)$, $k \in \mathbb{N}_0$, where the sequences (z_k) and (y_k) are bounded and increasing. The limit z of (z_k) is the smaller real root of $z^4 - 8z + 4 = 0$; $z \approx 0.50834742498666121699$, and the limit y of (y_k) is $y = \frac{1}{2}z^2 \approx 0.12920855224528457650$. Moreover $G(1, y) = (0, z)$ and $G(0, z) = (1, y)$, whence no subsequence of the iteration $(G^n c)$ converges to a fixed point of G.
- For any choice of $c = (u, v) \in P \setminus \{(1, 1)\}$ the iterations $G^n c$ and $G^{n+1} c$ are not order related when $n \geq 2$. The sequence $(G^n c)$ does not converge, and no subsequence of it converges to a fixed point of G.

- Denote $Y = \{(u,v) \in \mathbb{R}_+^2 : u^2 + v^2 \le 2, v > 0\} \cup \{(1,0)\}$. The function G, defined by (2.27) satisfies $G[Y] \subset Y$, and is maximalizing. The maximum of the w-o chain of cG-iterations with $c = (1,0)$ is $x_* = (1,1)$, and x_* is a fixed point of G. If $x \in Y \setminus \{x_*\}$, then x and $G(x)$ are not comparable.

The following example shows that G need not have a fixed point if either of the hypothesis of Theorem 2.53 is not valid.

Example 2.60. Denote $a = (1,y)$ and $b = (0,z)$, where y and z are as in Example 2.59. Choose $X = \{a,b\}$, and let $G : X \to X$ be defined by (2.27). G is maximalizing, but G has no fixed points, since $G(a) = b$ and $G(b) = a$. The last hypothesis of Theorem 2.53 is not satisfied.

Denoting $c = (1,z)$, then the set $X = \{a,b,c\}$ is a complete join lattice, i.e., every nonempty subset of X has the supremum in X. Let $G : X \to X$ satisfy $G(a) = b$ and $G(b) = G(c) = a$. G has no fixed points, but G is not maximalizing, since $G(c) < c$.

Example 2.61. The components: $u = 1$, $v = 1$ of the fixed point of G in Example 2.59 form also a solution of the system

$$u = \min\{1, 1 - [u] + [v]\}, \qquad v = \frac{[u] + v^2}{2}.$$

Moreover, a Maple program introduced in Example 2.59 serves a method to estimate this solution. When $m = 100000$, the estimate is: $u = 1$, $v = .99998$.

Remark 2.62. (i) The standard 'solve' and 'fsolve' commands of Maple 13 don't give a solution or its approximation for the system of Example 2.61.

(ii) In Example 2.59 the mapping G is non-increasing, non-extensive, non-ascending, not semi-increasing upward, and non-continuous.

(iii) The generalized iteration method presented in Lemma 2.2 is based on the w-o chain of cG-iterations, defined by (2.2), where G is a self-mapping of a poset P and $c \in P$. In the case when $c \le G(c)$, this chain equals to the w-o chain $C = C(c)$ of G-iterations of c, defined by

$$c = \min C, \text{ and } x \in C \setminus \{c\} \text{ if and only if } x = \sup G[C^{<x}]. \tag{2.32}$$

(iv) As for the use of $C(c)$ in fixed point theory and in the theory of discontinuous differential and integral equations, see, e.g., [44, 133] and the references therein.

(v) Chain $C(c)$ is compared in [175] with three other chains that generalize the sequence of ordinary iterations $(G^n(c))_{n=0}^\infty$, and which are used to prove fixed point results for G. These chains are: the generalized orbit $O(c)$ defined in [175] (being identical with the set $W(c)$ defined in [1]), the smallest admissible set $I(c)$ containing c (cf. [33, 34, 115]), and the smallest complete G-chain $B(c)$

containing c (cf. [96, 175]). If G is extensive, and if nonempty chains of X have supremums, then $C(c) = O(c) = I(c)$, and $B(c)$ is their cofinal subchain (cf. [175], Corollary 7). The common maximum x_* of these four chains is a fixed point of G. This result implies Bourbaki's Fixed Point Theorem, cf. [28, p. 37].

(vi) On the other hand, if the hypotheses of Theorem 2.56 hold and $x \in C(c) \setminus \{c, x_*\}$, then x and $G(x)$ are not necessarily comparable. The successor of such an x in $C(c)$ is $\sup\{x, G(x)\}$ by [115, Prop. 5]. In such a case the chains $O(c)$, $I(c)$, and $B(c)$ attain neither x nor any fixed point of G. For instance when $c = (0,0)$ in Example 2.59, then $C(c) = \{(0,0)\} \cup C$, where C is the w-o chain of $(1,0)G$-iterations. Since $(G^n(0,0))_{n=0}^{\infty} = \{(0,0)\} \cup (G^n(1,0))_{n=0}^{\infty}$, then $B(c)$ does not exist, $O(c) = I(c) = \{(0,0), (1,0)\}$ (see [175]). Thus only $C(c)$ attains a fixed point of G as its maximum. As shown in Example 2.59, the consecutive elements of the iteration sequence $(G^n(1,0))_{n=0}^{\infty}$ are unordered, and their limits are not fixed points of G. Hence, in these examples also finite combinations of chains $W(c_i)$ used in [15, Theorem 4.2] to prove a fixed point result are insufficient to attain a fixed point of G. As for other examples of such cases, see [110, Example 3], and [115, Example 16].

(vii) Neither the above mentioned four chains nor their duals are available to find fixed points of G if a and $G(a)$ are not ordered. For instance, they cannot be applied to prove Theorems 2.53 and 2.56 or Propositions 2.54 and 2.57.

2.6 Notes and Comments

In Sect. 2.1, the Chain Generating Recursion Principle is established and applied to derive generalized iteration methods. As noticed in the Introduction, the argumentation in the proof of the Chain Generating Recursion Principle is similar to that used in [231] to prove Zermelo's first well-ordering theorem. The importance of such an argumentation to set theory, to fixed point theory in posets, to theories of inductive definitions, and to computer science is studied in [147].

The recursion and iteration methods presented in Sect. 2.1 are applied to prove fixed point results and existence and comparison results for solutions of operator equations and inclusions in Sects. 2.2–2.5. The material of these sections is based on papers [52, 117, 119, 121, 125, 127].

3

Multi-Valued Variational Inequalities

In this chapter we provide existence, comparison, and extremality results for multi-valued elliptic and parabolic variational inequalities that will be used in subsequent chapters about discontinuously perturbed problems of this kind. The subject of this chapter is not only a prerequisite for the following chapters, but also is of independent interest. Our presentation is based on and includes results recently obtained in [38, 39, 66, 69, 70, 72], which partly generalize related results of [62] on this topic. Moreover, the theory about multi-valued elliptic and parabolic variational inequalities allows us to treat a wide range of nonsmooth elliptic and parabolic problems in a unified way.

3.1 Introductory Example

To motivate our study of multi-valued variational problems let us consider first the following simple Dirichlet boundary value problem (BVP for short):

$$-\Delta_p u + g(u) = h \quad \text{in} \quad \Omega, \quad u = 0 \quad \text{on} \quad \partial\Omega, \tag{3.1}$$

where $\Omega \subset \mathbb{R}^N$ is a bounded domain with Lipschitz boundary $\partial\Omega$, and $\Delta_p u = \text{div}(|\nabla u|^{p-2}\nabla u)$ is the p-Laplacian with $1 < p < \infty$. The nonlinearity $g : \mathbb{R} \to \mathbb{R}$ is assumed to be the Heaviside step function, i.e.,

$$g(s) = \begin{cases} 0 & \text{if } s \le 0, \\ 1 & \text{if } s > 0, \end{cases} \tag{3.2}$$

and $h(x) \equiv 1$ is taken for simplicity. We are looking for weak solutions u of (3.1) from the Sobolev space $V_0 = W_0^{1,p}(\Omega)$, which means that $g(u) \in L^q(\Omega)$ $(1/p + 1/q = 1)$, and u satisfies

$$u \in V_0 : \quad \int_\Omega \left(|\nabla u|^{p-2}\nabla u \nabla\varphi + g(u)\varphi \right) dx = \int_\Omega h\varphi \, dx \quad \text{for all} \quad \varphi \in V_0. \tag{3.3}$$

S. Carl and S. Heikkilä, *Fixed Point Theory in Ordered Sets and Applications:* *From Differential and Integral Equations to Game Theory,* DOI 10.1007/978-1-4419-7585-0_3, © Springer Science+Business Media, LLC 2011

Due to the discontinuous function g, the well-known method of sub-superso-lution developed for nonlinear elliptic (and parabolic) BVPs (see, e.g., [62]) does not apply to the simple BVP (3.1) as will be seen next. Recall that a function $\overline{u} \in V = W^{1,p}(\Omega)$ is called a supersolution of (3.1) if $g(\overline{u}) \in L^q(\Omega)$, $\overline{u} \geq 0$ on $\partial\Omega$, and $\overline{u}$ satisfies the inequality

$$\int_{\Omega} \left(|\nabla \overline{u}|^{p-2} \nabla \overline{u} \nabla \varphi + g(\overline{u})\varphi \right) dx \geq \int_{\Omega} h\varphi \, dx \text{ for all } \varphi \in V_0 \cap L^p_+(\Omega), \quad (3.4)$$

where $L^p_+(\Omega)$ is the positive cone of all nonnegative function of $L^p(\Omega)$. In an obvious similar way a subsolution $\underline{u} \in V$ of (3.1) is defined by replacing $\overline{u}$ by $\underline{u}$, and reversing the inequality sign in the above inequalities.

One readily verifies that for any constant $c > 0$, the constant functions $\underline{u}(x) \equiv -c$ and $\overline{u}(x) \equiv c$ form an ordered pair of sub- and supersolutions of (3.1). However, the BVP (3.1) has no solutions within the interval $[-c, c]$, and even more, there are no solutions at all. To see this we argue by contradiction. Suppose u was a solution of (3.1), i.e., $u \in V_0$ satisfies (3.3), which yields by using $h(x) \equiv 1$ the equality

$$\int_{\Omega} |\nabla u|^{p-2} \nabla u \nabla \varphi \, dx = \int_{\Omega} (1 - g(u))\varphi \, dx \text{ for all } \varphi \in V_0. \quad (3.5)$$

In particular, (3.5) holds for $\varphi = u$, which results in

$$\int_{\Omega} |\nabla u|^p \, dx = \int_{\Omega} (1 - g(u))u \, dx. \quad (3.6)$$

Applying the definition of the nonlinearity g to the right-hand side of (3.6) we get

$$\int_{\Omega} (1 - g(u))u \, dx = \int_{\{x \in \Omega : u \leq 0\}} (1 - g(u))u \, dx + \int_{\{x \in \Omega : u > 0\}} (1 - g(u))u \, dx \leq 0,$$

and thus we obtain from (3.6)

$$\int_{\Omega} |\nabla u|^p \, dx = 0,$$

which implies $u = 0$. However, the latter is a contradiction, because $u = 0$ is apparently not a solution of the BVP (3.1) (note $h(x) \equiv 1$).

Problem (3.1) with the nonlinearity g given by the Heaviside function is now embedded into a multi-valued setting replacing g by the following multi-function $\theta : \mathbb{R} \to 2^{\mathbb{R}} \setminus \emptyset$ that arises from g by filling in the gap at the point of discontinuity, i.e.,

$$\theta(s) = \begin{cases} 0 & \text{if } s < 0, \\ [0, 1] & \text{if } s = 0, \\ 1 & \text{if } s > 0. \end{cases} \quad (3.7)$$

Let $j : \mathbb{R} \to \mathbb{R}$ be the primitive of g, i.e.,

$$j(s) = \int_0^s g(t)\,dt.$$

By elementary calculation one easily verifies that $j(s) = s^+$ where $s^+ = \max\{s,0\}$, which is a convex function, and $\theta(s) = \partial j(s)$, where ∂j is the subdifferential of j. The relaxed multi-valued BVP associated with (3.1) reads now as follows:

$$-\Delta_p u + \partial j(u) \ni h \quad \text{in } \Omega, \quad u = 0 \quad \text{on } \partial\Omega, \tag{3.8}$$

and we call a function $u \in V_0$ a solution of (3.8), if there is an $\eta \in L^q(\Omega)$ such that $\eta(x) \in \partial j(u(x))$ for a.e. $x \in \Omega$ and the equality

$$\int_\Omega \left(|\nabla u|^{p-2}\nabla u \nabla\varphi + \eta\varphi \right) dx = \int_\Omega h\varphi\,dx$$

holds for all $\varphi \in V_0$. As a natural generalization of the notion of sub-supersolution for the multi-valued problem (3.8), we say $\overline{u}$ is a supersolution if $\overline{u} \geq 0$ on $\partial\Omega$, and if there is an $\overline{\eta} \in L^q(\Omega)$ such that $\overline{\eta}(x) \in \partial j(\overline{u}(x))$ for a.e. $x \in \Omega$ and the following inequality is satisfied:

$$\int_\Omega \left(|\nabla\overline{u}|^{p-2}\nabla\overline{u}\nabla\varphi + \overline{\eta}\varphi \right) dx \geq \int_\Omega h\varphi\,dx \quad \text{for all } \varphi \in V_0 \cap L^p_+(\Omega). \tag{3.9}$$

In an obvious similar way we define a subsolution $\underline{u}$. One readily verifies that the constant functions $\underline{u}(x) \equiv -c$ and $\overline{u} \equiv c$ form an ordered pair of sub-supersolutions for the multi-valued BVP (3.8) for any constant $c > 0$. As the method of sub-supersolution for the multi-valued BVP (3.8) holds true (see [62]), $u = 0$ must be a solution of (3.8), which can readily be verified. Moreover, $u = 0$ is the unique solution of (3.8), because $\partial j(u) : \mathbb{R} \to 2^\mathbb{R} \setminus \emptyset$ is a maximal monotone graph in $\mathbb{R}^2$. The latter can be justified also in the following alternative way. Consider the functional $E : V_0 \to \mathbb{R}$ defined by

$$E(u) = \frac{1}{p}\int_\Omega |\nabla u|^p\,dx + \int_\Omega (j(u) - hu)\,dx,$$

which is a strictly convex, continuous, and coercive functional, and thus the optimization problem

$$E(u) = \inf_{v \in V_0} E(v) \tag{3.10}$$

has a unique solution, which necessarily satisfies $0 \in \partial E(u)$, where $\partial E(u)$ is the subdifferential of E at u. In this case the inclusion condition is also sufficient. The inclusion $0 \in \partial E(u)$ may be considered as a kind of Euler equation for the minimum problem (3.10) and is nothing but the multi-valued BVP (3.8).

Let us consider the problem of minimization of the functional E under some constraint given in terms of a closed and convex subset $K \neq \emptyset$ of V_0, i.e.,

$$u \in K: \quad E(u) = \inf_{v \in K} E(v). \tag{3.11}$$

Applying the main theorem of convex optimization (see [230, Theorem 47.C.], a necessary and sufficient condition for $u \in K$ to be a solution of (3.11) is that the directional derivative $E'(u; \varphi - u)$ of E at u in any direction $\varphi - u$ defined by

$$E'(u; \varphi - u) = \lim_{t \downarrow 0} \frac{E(u + t(\varphi - u)) - E(u)}{t},$$

fulfills

$$u \in K: \quad E'(u; \varphi - u) \geq 0, \quad \forall \, \varphi \in K. \tag{3.12}$$

Evaluating the left-hand side of the last inequality we get the following *variational-hemivariational inequality*

$$\int_\Omega |\nabla u|^{p-2} \nabla u \nabla (\varphi - u) \, dx + \int_\Omega j'(u; \varphi - u) \, dx - \int_\Omega h(\varphi - u) \, dx \geq 0, \; \forall \varphi \in K, \tag{3.13}$$

where $j'(s; r)$ is the directional derivative of the convex function $j : \mathbb{R} \to \mathbb{R}$ at s in the direction r. As the functional E is strictly convex on $K \subseteq V_0$ we infer the existence of a unique solution of (3.11), which implies the existence of a unique solution of (3.13).

Let us associate to (3.13) the following *multi-valued variational inequality*: Find $u \in K$ such that there is an $\eta \in L^q(\Omega)$ satisfying $\eta(x) \in \partial j(u(x))$ and the inequality

$$\int_\Omega |\nabla u|^{p-2} \nabla u \nabla (\varphi - u) \, dx + \int_\Omega \eta \, (\varphi - u) \, dx - \int_\Omega h(\varphi - u) \, dx \geq 0, \; \forall \varphi \in K. \tag{3.14}$$

Since $s \mapsto \partial j(s)$ is maximal monotone, the variational inequality (3.14) has at most one solution. To verify this, let u_1 and u_2 be solutions of (3.14). Then we obtain from (3.14) with $\varphi = u_2$ and $\varphi = u_1$ for the solution u_1 and u_2, respectively, the inequalities

$$\int_\Omega |\nabla u_1|^{p-2} \nabla u_1 \nabla (u_2 - u_1) \, dx + \int_\Omega \eta_1 \, (u_2 - u_1) \, dx - \int_\Omega h(u_2 - u_1) \, dx \geq 0,$$

$$\int_\Omega |\nabla u_2|^{p-2} \nabla u_2 \nabla (u_1 - u_2) \, dx + \int_\Omega \eta_2 \, (u_1 - u_2) \, dx - \int_\Omega h(u_1 - u_2) \, dx \geq 0,$$

where $\eta_k(x) \in \partial j(u_k(x))$ for a.e. $x \in \Omega$, $k = 1, 2$. Adding the last two inequalities yields

$$0 \leq \int_\Omega \left(|\nabla u_2|^{p-2} \nabla u_2 - |\nabla u_1|^{p-2} \nabla u_1 \right) \nabla (u_2 - u_1) \, dx \leq \int_\Omega (\eta_2 - \eta_1)(u_1 - u_2) \, dx.$$

Because $s \mapsto \partial j(s)$ is maximal monotone, we get

$$\int_\Omega (\eta_2 - \eta_1)(u_1 - u_2) \, dx \leq 0,$$

which implies $\nabla(u_2 - u_1) = 0$ a.e. in Ω, and thus $u_2 - u_1 = 0$ (note $u_1, u_2 \in K \subseteq V_0$).

If we denote by $\partial_c j$ Clarke's generalized gradient of j and by $j^o(s; r)$ the generalized directional derivative due to Clarke of j at s in direction r, then for the convex function $j : \mathbb{R} \to \mathbb{R}$ we have

$$\partial_c j(s) = \partial j(s), \quad \forall \, s \in \mathbb{R}, \quad j^o(s;r) = j'(s;r) \ \forall \, s,r \in \mathbb{R},$$

see [80, Proposition 2.2.7]. Therefore, in what follows we can use the notation ∂j instead of $\partial_c j$ for Clarke's generalized gradient. It is an immediate consequence of the definition of Clarke's gradient that any solution of (3.14) is a solution of (3.13) as well, which again verifies the unique solvability of (3.14), because (3.13) has a unique solution. In fact, one can show that the two variational inequalities (3.13) and (3.14) are equivalent under the specific conditions of this section. To justify this equivalence we note that the optimization problem (3.11) is equivalent to

$$u \in V_0 : \quad E(u) = \inf_{v \in V_0} E(v) + I_K(v), \tag{3.15}$$

where $I_K : V_0 \to \mathbb{R} \cup \{+\infty\}$ denotes the indicator function related to K defined by

$$I_K(v) = \begin{cases} 0 & \text{if } v \in K, \\ +\infty & \text{if } v \notin K. \end{cases} \tag{3.16}$$

A necessary and sufficient condition for u to be a solution of (3.15) is

$$0 \in \partial(E + I_K)(u).$$

Applying the sum rule for the subgradient of the sum of convex functionals (see [230, Theorem 47.B]) we get as necessary and sufficient condition

$$0 \in \partial(E + I_K)(u) = \partial E(u) + \partial I_K(u), \tag{3.17}$$

which is equivalent to (3.14).

Remark 3.1. The considerations above deal with the simple example that $j(s) = s^+$ and $h(x) \equiv 1$. We should mention that all the results concerning the solvability of the variational-hemivariational inequality (3.13) and the multi-valued variational inequality (3.14) and their interrelation remain true also in case that $h \in V_0^*$, and that $j : \mathbb{R} \to \mathbb{R}$ is any convex function whose subdifferential $\partial j : \mathbb{R} \to 2^{\mathbb{R}}$ satisfies the following growth condition:

$$|\eta| \le c(1 + |s|^{p-1}, \quad \forall \, \eta \in \partial j(s), \ \forall \, s \in \mathbb{R}. \tag{3.18}$$

As can be seen from the above treatment, the arguments only rely on basic tools from Convex Analysis.

The main goal of the subsequent section is to establish the method of sub-supersolutions for multi-valued quasilinear elliptic variational inequalities in a much more general framework that allows us to include nonlinear elliptic operators of non-potential type as well as nonconvex functions j. Comparison results based on the sub-supersolution method will play an important role in the study of related discontinuous multi-valued variational problems.

3.2 Multi-Valued Elliptic Variational Inequalities

Throughout this section let $V = W^{1,p}(\Omega)$ and $V_0 = W_0^{1,p}(\Omega)$, with $1 < p < \infty$, denote the usual Sobolev spaces, where $\Omega \subset \mathbb{R}^N$, $N \geq 1$, is a bounded domain with Lipschitz boundary $\partial\Omega$. We denote by V^* and V_0^* the dual spaces corresponding to V and V_0, respectively, and by $\langle \cdot, \cdot \rangle$ the duality pairing. Let K be a closed, convex subset of V, and let $j_1 : \Omega \times \mathbb{R} \to \mathbb{R}$ and $j_2 : \partial\Omega \times \mathbb{R} \to \mathbb{R}$ be functions that are only supposed to be measurable in their first and locally Lipschitz continuous with respect to their second argument. The main goal of this section is to extend the idea of the sub-supersolution method in a natural and systematic way to quasilinear multi-valued elliptic variational inequalities. Let q denote the Hölder conjugate to p, i.e., q satisfies $1/p + 1/q = 1$. We deal with the following problem.

Find $u \in K$, $\eta \in L^q(\Omega)$, and $\xi \in L^q(\partial\Omega)$ satisfying:

$$\eta(x) \in \partial j_1(x, u(x)), \text{ a.e. } x \in \Omega, \quad \xi(x) \in \partial j_2(x, \gamma u(x)), \text{ a.e. } x \in \partial\Omega, \quad (3.19)$$

$$\langle Au - h, v - u \rangle + \int_\Omega \eta\,(v - u)\,dx + \int_{\partial\Omega} \xi\,(\gamma v - \gamma u)\,d\sigma \geq 0, \ \forall\, v \in K, \quad (3.20)$$

where

- the multi-valued functions $s \mapsto \partial j_k(x, s)$ are given by Clarke's generalized gradient of the locally Lipschitz functions $s \mapsto j_k(x, s)$, $k = 1, 2$, defined by

$$\partial j_k(x, s) := \{\zeta \in \mathbb{R} : j_k^o(x, s; r) \geq \zeta r, \ \forall\, r \in \mathbb{R}\}$$

for a.a. $x \in \Omega$ in case of $k = 1$, and for a.a. $x \in \partial\Omega$ in case of $k = 2$, with $j_k^o(x, s; r)$ denoting the generalized directional derivative of $s \mapsto j_k(x, s)$ at s in the direction r given by

$$j_k^o(x, s; r) = \limsup_{y \to s,\ t\downarrow 0} \frac{j_k(x, y + t\,r) - j_k(x, y)}{t},$$

(cf., e.g., [80, Chap. 2]),
- A is a second-order quasilinear elliptic differential operator of the form

$$Au(x) = -\sum_{i=1}^N \frac{\partial}{\partial x_i} a_i(x, \nabla u(x)), \quad \text{with } \nabla u = \left(\frac{\partial u}{\partial x_1}, \ldots, \frac{\partial u}{\partial x_N} \right),$$

- $\gamma : V \to L^p(\partial\Omega)$ denotes the trace operator, which is known to be linear and compact from V into $L^p(\partial\Omega)$,
- $h \in V^*$.

Before we discuss various special cases of the general setting of the multi-valued variational inequality (3.19)–(3.20), let us formulate the assumptions on the operator A. We assume the following hypotheses of Leray–Lions type on the coefficient functions a_i, $i = 1, ..., N$, of the operator A:

(A1) Each $a_i : \Omega \times \mathbb{R}^N \to \mathbb{R}$ satisfies the Carathéodory conditions, i.e., $a_i(x, \zeta)$ is measurable in $x \in \Omega$ for all $\zeta \in \mathbb{R}^N$, and continuous in ζ for a.a. $x \in \Omega$. There exist a constant $c_0 > 0$ and a function $k_0 \in L^q(\Omega)$ such that

$$|a_i(x, \zeta)| \le k_0(x) + c_0 |\zeta|^{p-1},$$

for a.a. $x \in \Omega$ and for all $\zeta \in \mathbb{R}^N$.

(A2) For a.a. $x \in \Omega$, and for all $\zeta, \zeta' \in \mathbb{R}^N$ with $\zeta \ne \zeta'$ the following monotonicity holds:

$$\sum_{i=1}^{N}(a_i(x, \zeta) - a_i(x, \zeta'))(\zeta_i - \zeta_i') > 0.$$

(A3) There is some constant $\nu > 0$ such that for a.a. $x \in \Omega$ and for all $\zeta \in \mathbb{R}^N$ the inequality

$$\sum_{i=1}^{N} a_i(x, \zeta)\zeta_i \ge \nu|\zeta|^p - k_1(x)$$

is satisfied for some function $k_1 \in L^1(\Omega)$.

A particular case of the operator A satisfying (A1)–(A3) is the negative p-Laplacian, i.e., $A = -\Delta_p$, which is obtained if

$$a_i(x, \zeta) = |\zeta|^{p-2}\zeta_i, \quad i = 1, ..., N.$$

In view of (A1), (A2), the operator A defined by

$$\langle Au, \varphi \rangle := \int_\Omega \sum_{i=1}^{N} a_i(x, \nabla u)\frac{\partial\varphi}{\partial x_i} \, dx, \quad \forall \, \varphi \in V_0$$

is known to provide a continuous, bounded, and monotone (resp. strictly monotone) mapping from V (resp. V_0) into V_0^*.

To demonstrate the general framework provided by (3.19) let us consider a few important special cases.

Example 3.2. Let $f : \Omega \times \mathbb{R} \to \mathbb{R}$ be a Carathéodory function. Consider its primitive given by

$$j(x, s) := \int_0^s f(x, t) \, dt.$$

Then the function $s \mapsto j(x, s)$ is continuously differentiable, and thus Clarke's gradient reduces to a singleton, i.e.,

$$\partial j(x, s) = \{\partial j(x, s)/\partial s\} = \{f(x, s)\}.$$

If we set $\hat{j}_1(x, s) = j(x, s) + j_1(x, s)$ with j_1 given above, then $\hat{j}_1$ satisfies the same qualitative properties as j_1, i.e., $x \mapsto \hat{j}_1(x, s)$ is measurable, and $s \mapsto \hat{j}_1(x, s)$ is locally Lipschitz continuous. Therefore we can replace j_1 in (3.19) by $\hat{j}_1$, which, due to the calculus of Clarke's gradient (see Chap. 9)

$$\partial \hat{j}_1(x, s) = f(x, s) + \partial j_1(x, s)$$

yields

$$\langle Au + F(u) - h, v - u \rangle + \int_{\Omega} \eta (v - u) \, dx + \int_{\partial \Omega} \xi (\gamma v - \gamma u) \, d\sigma \geq 0, \quad \forall \, v \in K,$$
(3.21)

where F is the Nemytskij operator generated by f through $F(u)(x) = f(x, u(x))$. Thus problems with elliptic operators $\hat{A}u = Au + F(u)$ involving additional lower order terms of the form $a_0(x, u) = f(x, u)$ are already included in (3.19)–(3.20).

Example 3.3. Let $f_1 : \Omega \times \mathbb{R} \to \mathbb{R}$ and $f_2 : \partial \Omega \times \mathbb{R} \to \mathbb{R}$ be measurable functions, which are locally bounded with respect to their second argument. If j_k are the primitives of f_k, i.e.,

$$j_k(x, s) := \int_0^s f_k(x, t) \, dt,$$
(3.22)

then $s \mapsto j_k(x, s)$ are locally Lipschitz and their generalized Clarke's gradients are given by

$$\partial j_k(x, s) = [\underline{f_k}(x, s), \overline{f_k}(x, s)],$$
(3.23)

where

$$\underline{f_k}(x, t) := \lim_{\delta \to 0^+} \operatorname*{ess\,inf}_{|\tau - t| < \delta} f(x, \tau), \quad \overline{f_k}(x, t) := \lim_{\delta \to 0^+} \operatorname*{ess\,sup}_{|\tau - t| < \delta} f(x, \tau).$$

Example 3.4. If $K = V$, $f_1 : \Omega \times \mathbb{R} \to \mathbb{R}$, and $f_2 : \partial \Omega \times \mathbb{R} \to \mathbb{R}$ are Carathéodory functions, and if j_k, $k = 1, 2$, are the primitives of f_k given by (3.22), then (3.19)–(3.20) reduces to the following quasilinear elliptic BVP

$$\langle Au - h, v \rangle + \int_{\Omega} f_1(x, u) \, v \, dx + \int_{\partial \Omega} f_2(x, \gamma u) \, \gamma v \, d\sigma = 0, \quad \forall \, v \in V, \quad (3.24)$$

which is the formulation of the weak solution of the BVP

$$Au + f_1(x, u) = h \text{ in } \Omega, \quad \frac{\partial u}{\partial \nu} + f_2(x, u) = 0 \text{ on } \partial \Omega,$$
(3.25)

where $\partial/\partial \nu$ denotes the outward pointing conormal derivative associated with A.

Example 3.5. If $K = V_0$, and j_k as in Example 3.4, then (3.19)–(3.20) is equivalent to

$$u \in V_0 : \quad \langle Au - h, v \rangle + \int_\Omega f_1(x, u) \, v \, dx = 0, \quad \forall \, v \in V_0, \tag{3.26}$$

which is nothing but the weak formulation of the homogeneous Dirichlet problem

$$Au + f_1(x, u) = h \quad \text{in} \quad \Omega, \quad u = 0 \quad \text{on} \quad \partial\Omega. \tag{3.27}$$

Example 3.6. If $K = V_0$ or $K = V$, then (3.19)–(3.20) reduces to elliptic inclusion problems, which for $K = V_0$ yields the following Dirichlet problem

$$Au + \partial j_1(x, u) \ni h \quad \text{in} \quad \Omega, \quad u = 0 \quad \text{on} \quad \partial\Omega, \tag{3.28}$$

and for $K = V$ the elliptic inclusion problem

$$Au + \partial j_1(x, u) \ni h \quad \text{in} \quad \Omega, \quad \frac{\partial u}{\partial \nu} + \partial j_2(x, u) \ni 0 \quad \text{on} \quad \partial\Omega. \tag{3.29}$$

We note that nonhomogeneous Dirichlet conditions that are given by the trace $\gamma\varphi$ with $\varphi \in V$ can always be transformed to a homogeneous Dirichlet problem. Moreover, by an appropriate choice of K also BVP with mixed boundary conditions can be seen to be included as special case of the general formulation (3.19)–(3.20). To this end let Γ_1 and Γ_2 be relatively open subsets of $\partial\Omega$ satisfying $\overline{\Gamma_1} \cup \overline{\Gamma_2} = \partial\Omega$ and $\Gamma_1 \cap \Gamma_2 = \emptyset$. Then we obtain the following special case of (3.19)–(3.20).

Example 3.7. If $K \subseteq V$ is the closed subspace given by

$$K = \{v \in V : \gamma v = 0 \quad \text{on} \quad \Gamma_1\},$$

then (3.19)–(3.20) reduces to

$$Au + \partial j_1(x, u) \ni h \quad \text{in} \quad \Omega, \quad \frac{\partial u}{\partial \nu} + \partial j_2(x, u) \ni 0 \quad \text{on} \quad \Gamma_2, \quad u = 0 \quad \text{on} \quad \Gamma_1. \tag{3.30}$$

Example 3.8. If $K \subseteq V$, and $j_k = 0$, then (3.19)–(3.20) is equivalent to the usual variational inequality of the form

$$u \in K : \quad \langle Au - h, v - u \rangle \geq 0, \quad \forall \, v \in K.$$

Remark 3.9. Applying the definition of Clarke's generalized gradient of j_k, $k = 1, 2$, and assuming standard growth conditions for $s \mapsto \partial j_k(x, s)$, one readily verifies that any solution of the multi-valued variational inequality (3.19)–(3.20) must be a solution of the following variational-hemivariational inequality: Find $u \in K$ such that

$$\langle Au - h, v - u \rangle + \int_\Omega j_1^o(x, u; v - u) \, dx + \int_{\partial\Omega} j_2^o(x, \gamma u; \gamma v - \gamma u) \, d\sigma \geq 0, \ \forall \, v \in K. \tag{3.31}$$

The reverse is well known to be true if the locally Lipschitz functions $s \mapsto j_k(x, s)$ are assumed to be regular in the sense of Clarke, see [80, Chap. 2.3]. For example, if the functions $s \mapsto j_k(x, s)$ are smooth (i.e., differentiable) or convex then they are regular in the sense of Clarke.

In Sect. 3.2.4 we are going to show that the reverse still holds true without imposing any additional regularity on the functions $s \mapsto j_k(x, s)$ provided the lattice condition (3.32) on K is fulfilled. This result on the equivalence of problems (3.19)–(3.20) and (3.31) (see [41]) fills a gap in the literature where both problems are treated separately. The main tools in the proof are existence and comparison results based on an appropriate sub-supersolution method that will be developed in what follows.

3.2.1 The Sub-Supersolution Method

The main goal of this subsection is to establish the method of sub-supersolution for the multi-valued variational inequality (3.19)–(3.20). Our presentation is motivated by and relies on results recently obtained in [38, 39, 62, 66, 69, 70]. It should be noted that the results we are going to present here include most of the results in the above works, which is mainly due to the weakened assumptions on the nonlinearities j_k, $k = 1, 2$, whose generalized gradients $s \mapsto \partial j_k(x, s)$ need only satisfy a local growth condition with respect to an ordered pair of sub-supersolutions whose definition will be given next.

For functions w, z and sets W and Z of functions defined on Ω or $\partial\Omega$ we use the notations: $w \wedge z = \min\{w, z\}$, $w \vee z = \max\{w, z\}$, $W \wedge Z = \{w \wedge z : w \in W, z \in Z\}$, $W \vee Z = \{w \vee z : w \in W, z \in Z\}$, and $w \wedge Z = \{w\} \wedge Z$, $w \vee Z = \{w\} \vee Z$. The following lattice property of the closed convex subset $K \subseteq V$:

$$K \wedge K \subseteq K, \quad K \vee K \subseteq K \tag{3.32}$$

will play a crucial role in the proof of the existence of extremal solutions.

Our basic notion of sub- and supersolution of the multi-valued variational inequality (3.19)–(3.20) reads as follows.

Definition 3.10. *A function $\underline{u} \in V$ is called a **subsolution** of (3.19)–(3.20) if there is an $\underline{\eta} \in L^q(\Omega)$ and a $\underline{\xi} \in L^q(\partial\Omega)$ satisfying*

(i) $\underline{u} \vee K \subseteq K$,

(ii) $\underline{\eta}(x) \in \partial j_1(x, \underline{u}(x))$, *a.e.* $x \in \Omega$, $\underline{\xi}(x) \in \partial j_2(x, \gamma\underline{u}(x))$, *a.e.* $x \in \partial\Omega$,

(iii) $\langle A\underline{u} - h, v - \underline{u}\rangle + \displaystyle\int_\Omega \underline{\eta}\,(v - \underline{u})\,dx + \int_{\partial\Omega} \underline{\xi}\,(\gamma v - \gamma\underline{u})\,d\sigma \geq 0$,

    for all $v \in \underline{u} \wedge K$.

Definition 3.11. *A function $\overline{u} \in V$ is called a **supersolution** of (3.19)–(3.20) if there is an $\overline{\eta} \in L^q(\Omega)$ and a $\overline{\xi} \in L^q(\partial\Omega)$ satisfying*

(i) $\overline{u} \wedge K \subseteq K$,

(ii) $\overline{\eta}(x) \in \partial j_1(x, \overline{u}(x))$, a.e. $x \in \Omega$, $\overline{\xi}(x) \in \partial j_2(x, \gamma \overline{u}(x))$, a.e. $x \in \partial \Omega$,

(iii) $\langle A\overline{u} - h, v - \overline{u}\rangle + \int_{\Omega} \overline{\eta}\,(v - \overline{u})\,dx + \int_{\partial\Omega} \overline{\xi}\,(\gamma v - \gamma\overline{u})\,d\sigma \geq 0$,

     for all $v \in \overline{u} \vee K$.

Remark 3.12. Note that the notions for sub- and supersolution defined in Definition 3.10 and Definition 3.11 have a symmetric structure, i.e., one obtains the definition for the supersolution $\overline{u}$ from the definition of the subsolution by replacing $\underline{u}$ in Definition 3.10 by $\overline{u}$, and interchanging $\vee$ by $\wedge$. Furthermore, the lattice condition (3.32) readily implies that any solution of the multi-valued variational inequality (3.19)–(3.20) is both a subsolution and a supersolution for (3.19)–(3.20).

To justify that Definitions 3.10 and 3.11 are in fact natural extensions of the usual notions of sub-supersolutions for elliptic boundary value problems, let us discuss several special cases.

Example 3.13. Consider Example 3.4, i.e., $K = V$, $f_1 : \Omega \times \mathbb{R} \to \mathbb{R}$ and $f_2 : \partial\Omega \times \mathbb{R} \to \mathbb{R}$ are Carathéodory functions, and j_k, $k = 1, 2$, are the primitives of f_k given by (3.22), then Clarke's generalized gradient ∂j_k reduces to a singleton, i.e.,

$$\partial j_k(x, s) = \{f_k(x, s)\},$$

and (3.19)–(3.20) reduces to the quasilinear elliptic BVP (3.25). If $\underline{u} \in V$ is a subsolution according to Definition 3.10, then the first condition (i) is trivially satisfied. The second condition (ii) of Definition 3.10 means that

$$\underline{\eta}(x) = f_1(x, \underline{u}(x)), \text{ a.e. } x \in \Omega, \ \underline{\xi}(x) = f_2(x, \gamma\underline{u}(x)), \text{ a.e. } x \in \partial\Omega.$$

Since $K = V$, any $v \in \underline{u} \wedge V$ has the form $v = \underline{u} \wedge \varphi = \underline{u} - (\underline{u} - \varphi)^+$ with $\varphi \in V$, where $w^+ = \max\{w, 0\}$, condition (iii) becomes

$$\langle A\underline{u} - h, -(\underline{u} - \varphi)^+\rangle + \int_{\Omega} f_1(\cdot, \underline{u})\,(-(\underline{u} - \varphi)^+)\,dx$$
$$+ \int_{\partial\Omega} f_2(\cdot, \gamma\underline{u})\,(-(\gamma\underline{u} - \gamma\varphi)^+)\,d\sigma \geq 0, \quad (3.33)$$

for all $\varphi \in V$. Since $\underline{u} \in V$, we have

$$M = \{(\underline{u} - \varphi)^+ : \varphi \in V\} = V \cap L_+^p(\Omega),$$

where $L_+^p(\Omega)$ is the positive cone of $L^p(\Omega)$, and thus we obtain from inequality (3.33)

$$\langle A\underline{u} - h, \chi\rangle + \int_{\Omega} f_1(x, \underline{u})\,\chi\,dx + \int_{\partial\Omega} f_2(x, \gamma\underline{u})\,\gamma\chi\,d\sigma \leq 0, \quad \forall \chi \in V \cap L_+^p(\Omega,$$

$$(3.34)$$

which is nothing but the usual notion of a (weak) subsolution for the BVP (3.25). Similarly, one verifies that $\overline{u} \in V$, which is a supersolution according to Definition 3.11, is equivalent with the usual supersolution of the BVP (3.25).

Example 3.14. In case that $K = V_0$, and j_k as in Example 3.13, then (3.19)–(3.20) is equivalent to the BVP (3.26) (resp. (3.27)). Let us consider the notion of subsolution in this case given via Definition 3.10. For $\underline{u} \in V$ condition (i) means $\underline{u} \vee V_0 \subseteq V_0$. This last condition is satisfied if and only if

$$\gamma \underline{u} \leq 0 \quad \text{i.e.,} \quad \underline{u} \leq 0 \ \text{ on } \ \partial\Omega, \tag{3.35}$$

and condition (ii) means, as above,

$$\underline{\eta}(x) = f_1(x, \underline{u}(x)), \ \text{a.e. } x \in \Omega.$$

(Note the boundary integral vanishes since $\gamma v = 0$ for $v \in V_0$.) Since any $v \in \underline{u} \wedge V_0$ can be represented in the form $v = \underline{u} - (\underline{u} - \varphi)^+$ with $\varphi \in V_0$, from (iii) of Definition 3.10 we obtain

$$\langle A\underline{u} - h, -(\underline{u} - \varphi)^+ \rangle + \int_{\Omega} f_1(\cdot, \underline{u}) \left(-(\underline{u} - \varphi)^+ \right) dx \geq 0, \ \forall \ \varphi \in V_0. \tag{3.36}$$

Set $\chi = (\underline{u} - \varphi)^+$, then (3.36) results in

$$\langle A\underline{u} - h, \chi \rangle + \int_{\Omega} f_1(\cdot, \underline{u}) \chi \, dx \leq 0, \ \forall \ \chi \in M_0, \tag{3.37}$$

where $M_0 := \{\chi \in V : \chi = (\underline{u} - \varphi)^+, \ \varphi \in V_0\} \subseteq V_0 \cap L_+^p(\Omega)$. One can prove that the set M_0 is a dense subset of $V_0 \cap L_+^p(\Omega)$ (cf. [62]), which shows that (3.37) together with (3.35) is nothing but the weak formulation for the subsolution of the Dirichlet problem (3.26) (resp. (3.27)). Similarly, $\overline{u} \in V$ given by Definition 3.11 is shown to be a supersolution of the Dirichlet problem (3.26) (resp. (3.27)).

In the same way one can verify that the notion of sub- and supersolution defined via Definition 3.10 and Definition 3.11, respectively, turns out to be equivalent with the well established notion of sub-supersolution for all other special cases as well.

Assume the existence of an ordered pair $(\underline{u}, \overline{u})$ of sub-supersolutions of the multi-valued variational inequality (3.19)–(3.20) satisfying $\underline{u} \leq \overline{u}$. With respect to this ordered pair we impose the following hypotheses on the non-linearities j_k, $k = 1, 2$.

(E-j1) $j_1 : \Omega \times \mathbb{R} \to \mathbb{R}$ satisfies
        (i) $x \mapsto j_1(x, s)$ is measurable in Ω for all $s \in \mathbb{R}$, and $s \mapsto j_1(x, s)$ is locally Lipschitz continuous in $\mathbb{R}$ for a.e. $x \in \Omega$.

(ii) There exists a function $k_\Omega \in L^q_+(\Omega)$ such that for a.e. $x \in \Omega$ and for all $s \in [\underline{u}(x), \overline{u}(x)]$ the growth condition

$$|\eta| \leq k_\Omega(x), \quad \forall \, \eta \in \partial j_1(x, s)$$

is fulfilled.

(E-j2) $j_2 : \partial\Omega \times \mathbb{R} \to \mathbb{R}$ satisfies

(i) $x \mapsto j_2(x, s)$ is measurable in $\partial\Omega$ for all $s \in \mathbb{R}$, and $s \mapsto j_2(x, s)$ is locally Lipschitz continuous in $\mathbb{R}$ for a.e. $x \in \partial\Omega$.

(ii) There exists a function $k_{\partial\Omega} \in L^q_+(\partial\Omega)$ such that for a.e. $x \in \partial\Omega$ and for all $s \in [\gamma\underline{u}(x), \gamma\overline{u}(x)]$ the growth condition

$$|\xi| \leq k_{\partial\Omega}(x), \quad \forall \, \xi \in \partial j_2(x, s)$$

is fulfilled.

Remark 3.15. We note that by the growth condition (ii) of (E-j1) and (E-j2) only a local L^q-boundedness condition on Clarke's generalized gradient ∂j_k is assumed, which is trivially satisfied, in particular, if we suppose the following natural growth condition on ∂j_k: There exist $c > 0$, $k_\Omega \in L^q(\Omega)$ and $k_{\partial\Omega} \in L^q_+(\partial\Omega)$ such that

$$|\eta| \leq k_\Omega(x) + c|s|^{p-1}, \quad \forall \, \eta \in \partial j_1(x, s),$$

for a.a. $x \in \Omega$ and for all $s \in \mathbb{R}$, and

$$|\xi| \leq k_{\partial\Omega}(x) + c|s|^{p-1}, \quad \forall \, \xi \in \partial j_2(x, s),$$

for a.e. $x \in \partial\Omega$ and for all $s \in \mathbb{R}$.

Since we are going to prove the existence of solutions of the multi-valued variational inequality (3.19)–(3.20) within the ordered interval $[\underline{u}, \overline{u}]$ of the given sub- and supersolution of (3.19)–(3.20), truncation and comparison techniques will play an important role. Therefore, we next provide some preliminaries in this direction.

Preliminaries

Let us first briefly recall the definition of pseudomonotonicity for a multivalued operator $\mathcal{A} : X \to 2^{X^*}$ defined on a real reflexive Banach space X, see, e.g., [184], and Chap. 9, Definition 9.88.

Definition 3.16. *Let X be a real reflexive Banach space with dual space X^*. The operator $\mathcal{A} : X \to 2^{X^*}$ is called* **pseudomonotone** *if the following conditions hold:*

(i)   The set $\mathcal{A}(u)$ is nonempty, bounded, closed and convex for all $u \in X$;

(ii) $\mathcal{A}$ is upper semicontinuous from each finite dimensional subspace of X to X^* equipped with the weak topology;

(iii) If $(u_n) \subset X$ with $u_n \rightharpoonup u$, and if $u_n^* \in \mathcal{A}(u_n)$ is such that $\limsup\langle u_n^*, u_n - u\rangle \leq 0$, then to each element $v \in X$ there exists $u^*(v) \in \mathcal{A}(u)$ with

$$\liminf\langle u_n^*, u_n - v\rangle \geq \langle u^*(v), u - v\rangle.$$

The following proposition provides sufficient conditions for an operator $\mathcal{A} : X \to 2^{X^*}$ to be pseudomonotone.

Proposition 3.17. *Let X be a real reflexive Banach space, and assume that $\mathcal{A} : X \to 2^{X^*}$ satisfies the following conditions:*

(i)   For each $u \in X$ we have that $\mathcal{A}(u)$ is a nonempty, closed, and convex subset of X^*;
(ii) $\mathcal{A} : X \to 2^{X^*}$ is bounded;
(iii) If $u_n \rightharpoonup u$ in X and $u_n^* \rightharpoonup u^*$ in X^* provided one has that $u_n^* \in \mathcal{A}(u_n)$ and $\limsup\langle u_n^*, u_n - u\rangle \leq 0$, then $u^* \in \mathcal{A}(u)$ and $\langle u_n^*, u_n\rangle \to \langle u^*, u\rangle$.

Then the operator $\mathcal{A} : X \to 2^{X^*}$ is pseudomonotone.

As for the proof of Proposition 3.17 we refer, e.g., to [184, Chap. 2]. In the proof of our main result we make use of the following surjectivity result for multi-valued pseudomonotone mappings perturbed by maximal monotone operators in reflexive Banach spaces (cf., e.g., [184, Theorem 2.12]).

Theorem 3.18. *Let X be a real reflexive Banach space with dual space X^*, $\Phi : X \to 2^{X^*}$ a maximal monotone operator, and $v_0 \in dom(\Phi)$. Let $\mathcal{A} : X \to 2^{X^*}$ be a pseudomonotone operator, and assume that either $\mathcal{A}_{v_0}$ is quasi-bounded or Φ_{v_0} is strongly quasi-bounded. Assume further that $\mathcal{A} : X \to 2^{X^*}$ is v_0-coercive, i.e., there exists a real-valued function $a : \mathbb{R}_+ \to \mathbb{R}$ with $a(r) \to +\infty$ as $r \to +\infty$ such that for all $(u, u^*) \in graph(\mathcal{A})$ one has $\langle u^*, u - v_0\rangle \geq a(\|u\|_X)\|u\|_X$. Then $\mathcal{A} + \Phi$ is surjective, i.e., $range(\mathcal{A} + \Phi) = X^*$.*

Remark 3.19. The operators $\mathcal{A}_{v_0}$ and Φ_{v_0} that appear in the theorem above are defined by $\mathcal{A}_{v_0}(v) := \mathcal{A}(v_0 + v)$ and similarly for Φ_{v_0}. As for the notion of *quasi-bounded* and *strongly quasi-bounded* we refer to [184, p.51]. In particular, any bounded operator is quasi-bounded and strongly quasi-bounded as well.

The proof of our main existence and comparison result strongly relies on an appropriate modification of the functions j_k outside the interval $[\underline{u}, \overline{u}]$ formed by the given sub- and supersolutions. Let $(\underline{u}, \underline{\eta}, \underline{\xi}) \in V \times L^q(\Omega) \times L^q(\partial\Omega)$ and $(\overline{u}, \overline{\eta}, \overline{\xi}) \in V \times L^q(\Omega) \times L^q(\partial\Omega)$ satisfy the conditions of Definition 3.10 and Definition 3.11, respectively, with $\underline{u} \leq \overline{u}$. Then we define the following modifications $\tilde{j}_k$ of the given j_k:

$$\tilde{j}_1(x, s) = \begin{cases} j_1(x, \underline{u}(x)) + \underline{\eta}(x)(s - \underline{u}(x)) & \text{if } s < \underline{u}(x), \\ j_1(x, s) & \text{if } \underline{u}(x) \leq s \leq \overline{u}(x), \\ j_1(x, \overline{u}(x)) + \overline{\eta}(x)(s - \overline{u}(x)) & \text{if } s > \overline{u}(x), \end{cases} \qquad (3.38)$$

and similarly

$$\tilde{j}_2(x,s) = \begin{cases} j_2(x,\gamma\underline{u}(x)) + \underline{\xi}(x)(s - \gamma\underline{u}(x)) & \text{if } s < \gamma\underline{u}(x), \\ j_2(x,s) & \text{if } \gamma\underline{u}(x) \leq s \leq \gamma\overline{u}(x), \\ j_2(x,\gamma\overline{u}(x)) + \overline{\xi}(x)(s - \gamma\overline{u}(x)) & \text{if } s > \gamma\overline{u}(x). \end{cases} \quad (3.39)$$

The following two lemmas list essential properties of the modified functions $\tilde{j}_k$.

Lemma 3.20. *Let hypotheses (E-j1) be satisfied. Then the function $\tilde{j}_1$ has the following properties:*

(P1) $\tilde{j}_1 : \Omega \times \mathbb{R} \to \mathbb{R}$ satisfies

    (i) $x \mapsto \tilde{j}_1(x,s)$ is measurable in Ω for all $s \in \mathbb{R}$, and $s \mapsto \tilde{j}_1(x,s)$ is Lipschitz continuous in $\mathbb{R}$ for a.e. $x \in \Omega$.

    (ii) Let $\partial\tilde{j}_1$ denote Clarke's generalized gradient of $s \mapsto \tilde{j}_1(x,s)$, then for a.e. $x \in \Omega$ and for all $s \in \mathbb{R}$ the growth

$$|\eta| \leq k_\Omega(x), \quad \forall \, \eta \in \partial\tilde{j}_1(x,s)$$

    is fulfilled.

    (iii) Clarke's generalized gradient of $s \mapsto \tilde{j}_1(x,s)$ is given by

$$\partial\tilde{j}_1(x,s) = \begin{cases} \underline{\eta}(x) & \text{if } s < \underline{u}(x), \\ \partial\tilde{j}_1(x,\underline{u}(x)) & \text{if } s = \underline{u}(x), \\ \partial j_1(x,s) & \text{if } \underline{u}(x) < s < \overline{u}(x), \\ \partial\tilde{j}_1(x,\overline{u}(x)) & \text{if } s = \overline{u}(x), \\ \overline{\eta}(x) & \text{if } s > \overline{u}(x), \end{cases} \quad (3.40)$$

    and the inclusions $\partial\tilde{j}_1(x,\underline{u}(x)) \subseteq \partial j_1(x,\underline{u}(x))$ and $\partial\tilde{j}_1(x,\overline{u}(x)) \subseteq \partial j_1(x,\overline{u}(x))$ hold true.

Proof: The proof follows immediately from the definition (3.38) of $\tilde{j}_1$, and using the assumptions (E-j1) on j_1 as well as from the fact that Clarke's generalized gradient $\partial j_1(x,s)$ is a convex set. $\qquad\square$

Lemma 3.21. *Let hypotheses (E-j2) be satisfied. Then the function $\tilde{j}_2$ has the following properties:*

(P2) $\tilde{j}_2 : \partial\Omega \times \mathbb{R} \to \mathbb{R}$ satisfies

    (i) $x \mapsto \tilde{j}_2(x,s)$ is measurable in $\partial\Omega$ for all $s \in \mathbb{R}$, and $s \mapsto \tilde{j}_2(x,s)$ is Lipschitz continuous in $\mathbb{R}$ for a.e. $x \in \partial\Omega$.

    (ii) Let $\partial\tilde{j}_2$ denote Clarke's generalized gradient of $s \mapsto \tilde{j}_2(x,s)$, then for a.e. $x \in \partial\Omega$ and for all $s \in \mathbb{R}$ the growth

$$|\xi| \leq k_{\partial\Omega}(x), \quad \forall \, \xi \in \partial\tilde{j}_2(x,s)$$

    is fulfilled.

(iii)Clarke's generalized gradient of $s \mapsto \tilde{j}_2(x, s)$ is given by

$$\partial \tilde{j}_2(x, s) = \begin{cases} \xi(x) & \text{if } s < \gamma \underline{u}(x), \\ \partial \tilde{j}_2(x, \gamma \underline{u}(x)) & \text{if } s = \gamma \underline{u}(x), \\ \partial j_2(x, s) & \text{if } \gamma \underline{u}(x) < s < \gamma \overline{u}(x), \\ \partial \tilde{j}_2(x, \gamma \overline{u}(x)) & \text{if } s = \gamma \overline{u}(x), \\ \overline{\xi}(x) & \text{if } s > \gamma \overline{u}(x), \end{cases} \qquad (3.41)$$

and the inclusions $\partial \tilde{j}_2(x, \gamma \underline{u}(x)) \subseteq \partial j_2(x, \gamma \underline{u}(x))$ and $\partial \tilde{j}_2(x, \gamma \overline{u}(x)) \subseteq \partial j_2(x, \gamma \overline{u}(x))$ hold true.

Proof: The proof is similar as for Lemma 3.20.   □

By means of $\tilde{j}_1$ and $\tilde{j}_2$ we introduce integral functionals $\tilde{J}_1$ and $\tilde{J}_2$ defined on $L^p(\Omega)$ and $L^p(\partial\Omega)$, respectively, and given by

$$\tilde{J}_1(u) = \int_{\Omega} \tilde{j}_1(x, u(x)) \, dx, \quad u \in L^p(\Omega),$$

$$\tilde{J}_2(v) = \int_{\partial\Omega} \tilde{j}_2(x, v(x)) \, d\sigma, \quad v \in L^p(\partial\Omega).$$

Due to (P1)(ii) and (P2) (ii) and applying Lebourg's mean value theorem (see [62, Theorem 2.177]), the functionals $\tilde{J}_1 : L^p(\Omega) \to \mathbb{R}$ and $\tilde{J}_2 : L^p(\partial\Omega) \to \mathbb{R}$ are well-defined and Lipschitz continuous, so that Clarke's generalized gradients $\partial \tilde{J}_1 : L^p(\Omega) \to 2^{(L^p(\Omega))^*}$ and $\partial \tilde{J}_2 : L^p(\partial\Omega) \to 2^{(L^p(\partial\Omega))^*}$ are well-defined, too. Moreover, Aubin–Clarke theorem (cf. [80, p. 83]) provides the following characterization of the generalized gradients. For $u \in L^p(\Omega)$ we have

$$\tilde{\eta} \in \partial \tilde{J}_1(u) \Longrightarrow \tilde{\eta} \in L^q(\Omega) \text{ with } \tilde{\eta}(x) \in \partial \tilde{j}_1(x, u(x)) \text{ for a.e. } x \in \Omega, \quad (3.42)$$

and similarly for $v \in L^p(\partial\Omega)$

$$\tilde{\xi} \in \partial \tilde{J}_2(v) \Longrightarrow \tilde{\xi} \in L^q(\partial\Omega) \text{ with } \tilde{\xi}(x) \in \partial \tilde{j}_2(x, v(x)) \text{ for a.e. } x \in \partial\Omega. \quad (3.43)$$

By means of Clarke's generalized gradient $\partial \tilde{J}_k$ we introduce the following multi-valued operators:

$$\Phi_1(u) := (i^* \circ \partial \tilde{J}_1 \circ i)(u), \quad \Phi_2(u) := (\gamma^* \circ \partial \tilde{J}_2 \circ \gamma)(u), \quad u \in V, \quad (3.44)$$

where $i^* : L^q(\Omega) \to V^*$ and $\gamma^* : L^q(\partial\Omega) \to V^*$ denote the adjoint operators of the embedding $i : V \hookrightarrow L^p(\Omega)$ and the trace operator $\gamma : V \to L^p(\partial\Omega)$, respectively, defined by: If $\eta \in L^q(\Omega)$ and $\xi \in L^q(\partial\Omega)$ then

$$\langle i^*\eta, \varphi \rangle = \int_{\Omega} \eta \varphi \, dx, \quad \langle \gamma^*\xi, \varphi \rangle = \int_{\partial\Omega} \xi \gamma\varphi \, d\sigma, \quad \forall \varphi \in V. \quad (3.45)$$

The operators Φ_k, $k = 1, 2$, possess the following properties.

Lemma 3.22. *The operators $\Phi_k : V \to 2^{V^*}$, $k = 1, 2$, are bounded and pseudomonotone.*

Proof: As for $\Phi_1 : V \to 2^{V^*}$ we refer to [38, Lemma 3.1], and for $\Phi_1 : V \to 2^{V^*}$ the proof is given in [38, Lemma 3.2]   □

Let b be the cut-off function defined as follows

$$b(x, s) = \begin{cases} (s - \overline{u}(x))^{p-1} & \text{if } s > \overline{u}(x) \\ 0 & \text{if } \underline{u}(x) \leq s \leq \overline{u}(x) \\ -(\underline{u}(x) - s)^{p-1} & \text{if } s < \underline{u}(x). \end{cases}$$

Apparently, $b : \Omega \times \mathbb{R} \to \mathbb{R}$ is a Carathéodory function satisfying the growth condition

$$|b(x, s)| \leq k_2(x) + c_1|s|^{p-1} \tag{3.46}$$

for a.e. $x \in \Omega$ and for all $s \in \mathbb{R}$, where $c_1 > 0$ is a constant and $k_2 \in L^q(\Omega)$. Moreover, one has the following estimate

$$\int_\Omega b(x, u(x))\, u(x)\, dx \geq c_2\|u\|_{L^p(\Omega)}^p - c_3, \quad \forall u \in L^p(\Omega), \tag{3.47}$$

for some constants $c_2 > 0$ and $c_3 > 0$. Let B denote the Nemytskij operator associated with b, i.e.,

$$B(u)(x) = b(x, u(x)).$$

In view of (3.46) the Nemytskij operator $B : L^p(\Omega) \to L^q(\Omega)$ is continuous and bounded, and thus due to the compact embedding $V \hookrightarrow L^p(\Omega)$, it follows that the composed operator $\tilde{B} := i^* \circ B \circ i : V \to V^*$ is completely continuous.

Consider the following multi-valued operator $\mathcal{A}$ defined by

$$\mathcal{A}(u) = Au + \tilde{B}(u) + \Phi_1(u) + \Phi_2(u). \tag{3.48}$$

Lemma 3.23. *The operator* $\mathcal{A} : V \to 2^{V^*}$ *is bounded, pseudomonotone, and* v_0-*coercive for* $v_0 \in K$.

Proof: Hypotheses (A1)–(A2) imply that $A : V \to V^*$ is continuous, bounded, and monotone, and thus, in particular, pseudomonotone, see, e.g., [62, Theorem 2.109]. The cut-off operator $\tilde{B} : V \to V^*$ is bounded and completely continuous. Therefore, the (single-valued) operator $A + \tilde{B} : V \to V^*$ is continuous, bounded and pseudomonotone. Due to Lemma 3.22, the multi-valued operators $\Phi_k : V \to 2^{V^*}$, $k = 1, 2$, are bounded and pseudomonotone. Since pseudomonotonicity is invariant under addition (see [184, Chap. 2]), it follows that $\mathcal{A} : V \to 2^{V^*}$ is bounded and pseudomonotone, and so it remains to show that $\mathcal{A}$ is v_0-coercive for $v_0 \in K$, i.e., we need to verify that there exists a real-valued function $a : \mathbb{R}_+ \to \mathbb{R}$ with $a(r) \to +\infty$ as $r \to +\infty$ such that for all $(u, u^*) \in \text{graph}\,(\mathcal{A})$ one has

$$\langle u^*, u - v_0 \rangle \geq a(\|u\|_V)\|u\|_V. \tag{3.49}$$

To check (3.49) let $u^* \in \mathcal{A}(u)$, i.e., u^* has the form

$$u^* = Au + \tilde{B}(u) + i^*\tilde{\eta} + \gamma^*\tilde{\xi},$$

where $\tilde{\eta} \in L^q(\Omega)$ with $\tilde{\eta}(x) \in \partial \tilde{j}_1(x, u(x))$, for a.e. $x \in \Omega$, and $\tilde{\xi} \in L^q(\partial\Omega)$ with $\tilde{\xi}(x) \in \partial \tilde{j}_2(x, \gamma u(x))$, for a.e. $x \in \partial\Omega$. Hypotheses (A1), (A2), and the estimates (3.46) and (3.47), as well as the uniform boundedness of $\partial \tilde{j}_1$ and $\partial \tilde{j}_2$ in view of (P1) (ii) of Lemma 3.20 and (P2) (ii) of Lemma 3.21, respectively, allows us to estimate as follows:

$$|\langle Au + \tilde{B}(u) + i^*\tilde{\eta} + \gamma^*\tilde{\xi}, v_0\rangle| \leq c(1 + \|u\|_V^{p-1}), \quad \forall\, u \in V, \tag{3.50}$$

for some constant $c > 0$, and

$$\begin{aligned}
\langle u^*, u\rangle &= \langle Au + \tilde{B}(u) + i^*\tilde{\eta} + \gamma^*\tilde{\xi}, u\rangle \\
&\geq \nu\|\nabla u\|_{L^p(\Omega)}^p - c_4 + c_2\|u\|_{L^p(\Omega)}^p - c_3 - c_5\|u\|_{L^p(\Omega)} - c_6\|\gamma u\|_{L^p(\partial\Omega)} \\
&\geq c_7\|u\|_V^p - c_8\|u\|_V - c_9.
\end{aligned} \tag{3.51}$$

From (3.50) and (3.51) we get the estimate

$$\langle Au + \tilde{B}(u) + i^*\tilde{\eta} + \gamma^*\tilde{\xi}, u - v_0\rangle \geq c_7\|u\|_V^p - c\|u\|_V^{p-1} - c_8\|u\|_V - c_{10},$$

for some positive constants c_i, which proves the v_0-coercivity. $\square$

Existence and Comparison Result, Sub-Supersolution Method

We are now in a position to prove the main existence and comparison result for the multi-valued variational inequality (3.19)–(3.20).

Theorem 3.24 (Sub-Supersolution Method). *Let (A1)–(A3) be satisfied. Assume the existence of sub- and supersolutions $\underline{u}$ and $\overline{u}$, respectively, of the multi-valued variational inequality (3.19)–(3.20) with $\underline{u} \leq \overline{u}$ such that (E-j1)–(E-j2) is fulfilled. Then there exist solutions of (3.19)–(3.20) within the ordered interval $[\underline{u}, \overline{u}]$.*

Proof: Let $I_K : V \to \mathbb{R} \cup \{+\infty\}$ be the indicator function related to the given closed convex set $K \neq \emptyset$, i.e.,

$$I_K(u) = \begin{cases} 0 & \text{if } u \in K, \\ +\infty & \text{if } u \notin K, \end{cases}$$

which is known to be proper, convex, and lower semicontinuous. Consider the following multi-valued operator

$$\mathcal{A} + \partial I_K : V \to 2^{V^*}, \tag{3.52}$$

where $\mathcal{A}$ is given by (3.48), and ∂I_K is the usual subdifferential of I_K, which is known to be a maximal monotone operator, cf., e.g., [229]. Taking into account Lemma 3.23, we may apply the surjectivity result of Theorem 3.18, which implies the existence of $u \in K$ such that $h \in \mathcal{A}(u) + \partial I_K(u)$. By definition of $\mathcal{A}$ and ∂I_K the latter inclusion implies the existence of $\eta^* \in \Phi_1(u)$, $\xi^* \in \Phi_2(u)$, and $\theta^* \in \partial I_K(u)$ with $\eta^* = i^* \tilde{\eta}$ and $\xi^* = \gamma^* \tilde{\xi}$ such that equation

$$Au + \tilde{B}(u) + i^* \tilde{\eta} + \gamma^* \tilde{\xi} + \theta^* = h, \quad \text{in } V^* \tag{3.53}$$

holds, where in view of (3.42)–(3.43) we have $\tilde{\eta} \in L^q(\Omega)$ with $\tilde{\eta}(x) \in \partial \tilde{j}_1(x, u(x))$, for a.e. $x \in \Omega$, and $\tilde{\xi} \in L^q(\partial\Omega)$ with $\tilde{\xi}(x) \in \partial \tilde{j}_2(x, \gamma u(x))$, for a.e. $x \in \partial\Omega$. By using the definition of $\partial I_K(u)$, the solution u of equation (3.53) is seen to be a solution of the following multi-valued variational inequality

$$u \in K: \quad \langle Au - h + \tilde{B}(u) + i^* \tilde{\eta} + \gamma^* \tilde{\xi}, v - u \rangle \geq 0, \ \forall \, v \in K, \tag{3.54}$$

which is equivalent to

$$u \in K: \quad \langle Au - h, v - u \rangle + \int_\Omega b(x, u) \, (v - u) \, dx + \int_\Omega \tilde{\eta} \, (v - u) \, dx$$
$$+ \int_{\partial\Omega} \tilde{\xi} \, (\gamma v - \gamma u) \, d\sigma \geq 0, \ \forall \, v \in K. \tag{3.55}$$

We are going to show that any solution u of (3.55) is a solution of the multi-valued variational inequality (3.19)–(3.20) satisfying $u \in [\underline{u}, \overline{u}]$. To this end we first show that u indeed satisfies $\underline{u} \leq u \leq \overline{u}$. To prove the inequality $u \leq \overline{u}$ we apply the special test function $v = \overline{u} \vee u = \overline{u} + (u - \overline{u})^+$ in Definition 3.11 (iii), and $v = \overline{u} \wedge u = u - (u - \overline{u})^+ \in K$ in (3.55), and get by adding the resulting inequalities the following:

$$\langle A\overline{u} - Au, (u - \overline{u})^+ \rangle - \int_\Omega b(x, u) \, (u - \overline{u})^+ \, dx + \int_\Omega (\overline{\eta} - \tilde{\eta}) \, (u - \overline{u})^+ \, dx$$
$$+ \int_{\partial\Omega} (\overline{\xi} - \tilde{\xi}) \, (\gamma u - \gamma \overline{u})^+ \, d\sigma \geq 0. \tag{3.56}$$

With the notation $\{u > \overline{u}\} = \{x \in \Omega : u(x) > \overline{u}(x)\}$ and $\{\gamma u > \gamma \overline{u}\} = \{x \in \partial\Omega : \gamma u(x) > \gamma \overline{u}(x)\}$, and by applying the results (P1)(iii) of Lemma 3.20 and (P2)(iii) of Lemma 3.21 we obtain

$$\int_\Omega (\overline{\eta} - \tilde{\eta}) \, (u - \overline{u})^+ \, dx = \int_{\{u > \overline{u}\}} (\overline{\eta} - \tilde{\eta}) \, (u - \overline{u}) \, dx = 0, \tag{3.57}$$

because $\tilde{\eta}(x) = \overline{\eta}(x)$ for $x \in \{u > \overline{u}\}$, and

$$\int_{\partial\Omega} (\overline{\xi} - \tilde{\xi}) \, (\gamma u - \gamma \overline{u})^+ \, d\sigma = \int_{\{\gamma u > \gamma \overline{u}\}} (\overline{\xi} - \tilde{\xi}) \, (\gamma u - \gamma \overline{u}) \, d\sigma = 0, \tag{3.58}$$

because $\tilde{\xi}(x) = \overline{\xi}(x)$ for $x \in \{\gamma u > \gamma \overline{u}\}$. Taking the definition of the cut-off function b into account we get

$$\int_{\Omega} b(x, u)\,(u - \overline{u})^+\,dx = \int_{\Omega}\left((u - \overline{u})^+\right)^p dx, \tag{3.59}$$

and by means of (A2) we obtain the estimate

$$\langle A\overline{u} - Au, (u - \overline{u})^+\rangle = -\langle Au - A\overline{u}, (u - \overline{u})^+\rangle \le 0. \tag{3.60}$$

Applying the results (3.57)–(3.60) to (3.56) finally yields

$$\int_{\Omega}\left((u - \overline{u})^+\right)^p dx = 0,$$

which implies $(u - \overline{u})^+ = 0$, and thus $u \le \overline{u}$. The proof for $\underline{u} \le u$ can be done in a similar way. So far we have shown that any solution u of the auxiliary multi-valued variational inequality (3.55) belongs to the interval $[\underline{u}, \overline{u}]$, and thus satisfies: $u \in K$, $b(x, u(x)) = 0$, $\tilde{\eta} \in L^q(\Omega)$, $\tilde{\xi} \in L^q(\partial\Omega)$ and

$$\tilde{\eta}(x) \in \partial\tilde{j}_1(x, u(x)), \text{ a.e. } x \in \Omega, \quad \tilde{\xi}(x) \in \partial\tilde{j}_2(x, \gamma u(x)), \text{ a.e. } x \in \partial\Omega, \tag{3.61}$$

$$\langle Au - h, v - u\rangle + \int_{\Omega} \tilde{\eta}\,(v - u)\,dx + \int_{\partial\Omega} \tilde{\xi}\,(\gamma v - \gamma u)\,d\sigma \ge 0, \ \forall\, v \in K. \tag{3.62}$$

From (P1)(iii) of Lemma 3.20 we see that $\partial\tilde{j}_1(x, u(x)) \subseteq \partial j_1(x, u(x))$ for any $u \in [\underline{u}, \overline{u}]$, and from (P2)(iii) of Lemma 3.21 we see that $\partial\tilde{j}_2(x, \gamma u(x)) \subseteq \partial j_2(x, \gamma u(x))$ for $\gamma u \in [\gamma\underline{u}, \gamma\overline{u}]$, and therefore we also have

$$\tilde{\eta}(x) \in \partial j_1(x, u(x)), \text{ a.e. } x \in \Omega, \quad \tilde{\xi}(x) \in \partial j_2(x, \gamma u(x)), \text{ a.e. } x \in \partial\Omega,$$

which shows that the solution $u \in [\underline{u}, \overline{u}]$ of the auxiliary problem is in fact a solution of the original multi-valued inequality (3.19)–(3.20). This completes the proof. ☐

Remark 3.25. It should be mentioned that the existence and comparison result of Theorem 3.24 can be extended to more general elliptic operators A of Leray–Lions type such as

$$Au(x) = -\sum_{i=1}^{N} \frac{\partial}{\partial x_i} a_i(x, u, \nabla u(x)) + a_0(x, u, \nabla u(x)),$$

where the coefficients $a_i : \Omega \times \mathbb{R} \times \mathbb{R}^N \to \mathbb{R}$, $i = 0, 1, \ldots, N$ satisfy the following conditions:

(A1') Each $a_i(x, s, \zeta)$ satisfies Carathéodory conditions, i.e., is measurable in $x \in \Omega$ for all $(s, \zeta) \in \mathbb{R} \times \mathbb{R}^N$ and continuous in (s, ζ) for a.e. $x \in \Omega$. There exist a constant $c_0 > 0$ and a function $k_0 \in L^q(\Omega)$ so that

$$|a_i(x, s, \zeta)| \le k_0(x) + c_0(|s|^{p-1} + |\zeta|^{p-1})$$

for a.e. $x \in \Omega$ and for all $(s, \zeta) \in \mathbb{R} \times \mathbb{R}^N$, with $|\zeta|$ denoting the Euclidean norm of the vector ζ.

(A2') The coefficients a_i satisfy a monotonicity condition with respect to ζ in the form

$$\sum_{i=1}^{N}(a_i(x,s,\zeta) - a_i(x,s,\zeta'))(\zeta_i - \zeta_i') > 0$$

for a.e. $x \in \Omega$, for all $s \in \mathbb{R}$, and for all $\zeta, \zeta' \in \mathbb{R}^N$ with $\zeta \neq \zeta'$.

(A3')

$$\sum_{i=1}^{N} a_i(x,s,\zeta)\zeta_i \geq \nu|\zeta|^p - k(x)$$

for a.e. $x \in \Omega$, for all $s \in \mathbb{R}$, and for all $\zeta \in \mathbb{R}^N$ with some constant $\nu > 0$ and some function $k \in L^1(\Omega)$.

The proof of Theorem 3.24 under these more general assumptions can be done by appropriately modifying the multi-valued operator $\mathcal{A}$ defined by (3.48) in the following way:

$$\mathcal{A}(u) = A_T u + \lambda \tilde{B}(u) + \Phi_1(u) + \Phi_2(u), \tag{3.63}$$

where A_T is given by

$$A_T u(x) = -\sum_{i=1}^{N} \frac{\partial}{\partial x_i} a_i(x, Tu, \nabla u(x)) + a_0(x, Tu, \nabla Tu(x)),$$

and T is the following truncation mapping

$$(Tu)(x) = \begin{cases} \overline{u}(x) & \text{if } u(x) > \overline{u}(x) \\ u(x) & \text{if } \underline{u}(x) \leq u(x) \leq \overline{u}(x) \\ \underline{u}(x) & \text{if } u(x) < \underline{u}(x). \end{cases}$$

Then for $\lambda > 0$ sufficiently large one can prove that the modified operator $\mathcal{A} : V \rightarrow 2^{V^*}$ is bounded, pseudomonotone, and v_0-coercive for $v_0 \in K$.

3.2.2 Directedness of Solution Set

Let $\mathcal{S}$ denote the set of all solutions of the multi-valued variational inequality (3.19)–(3.20) that belong to the ordered interval $[\underline{u}, \overline{u}]$ formed by the given sub- and supersolution. In view of Theorem 3.24, we have that $\mathcal{S} \neq \emptyset$. The main goal of this subsection is to show that $\mathcal{S}$ is a *directed set*.

Definition 3.26 (directed set). *Let $(\mathcal{P}, \leq)$ be a partially ordered set. A subset $\mathcal{C}$ of $\mathcal{P}$ is said to be* **upward directed** *if for each pair $x, y \in \mathcal{C}$ there is a $z \in \mathcal{C}$ such that $x \leq z$ and $y \leq z$. Similarly, $\mathcal{C}$ is* **downward directed** *if for each pair $x, y \in \mathcal{C}$ there is a $w \in \mathcal{C}$ such that $w \leq x$ and $w \leq y$. If $\mathcal{C}$ is both upward and downward directed it is called* **directed.**

Our main result is given by the following theorem.

Theorem 3.27. *Let the hypotheses of Theorem 3.24 and the lattice condition (3.32) be satisfied. Then the solution set S of all solutions of (3.19)–(3.20) within the interval $[\underline{u}, \overline{u}]$ equipped with the natural partial ordering of functions introduced by the order cone $L^p_+(\Omega)$ is a directed set.*

Proof: We are going to show that S is upward directed only, since the proof for S being downward directed can be done in a similar way. To this end let $u_1, u_2 \in S$. Our goal is to prove the existence of an element $u \in S$ such that $u \geq u_k$, $k = 1, 2$. The proof will be done in several steps and crucially relies on an appropriately designed auxiliary problem.

Step 1: Auxiliary Problem

Let $u_k \in S$, $k = 1, 2$, i.e., $u_k \in K$, and there exist $\eta_k \in L^q(\Omega)$, $\xi_k \in L^q(\partial\Omega)$ such that $\eta_k(x) \in \partial j_1(x, u_k(x))$ for a.e. $x \in \Omega$, $\xi_k(x) \in \partial j_2(x, \gamma u_k(x))$ for a.e. $x \in \partial\Omega$, and the following variational inequality is satisfied:

$$\langle Au_k - h, v - u_k \rangle + \int_\Omega \eta_k\,(v - u_k)\,dx + \int_{\partial\Omega} \xi_k\,(\gamma v - \gamma u_k)\,d\sigma \geq 0, \quad \forall\, v \in K. \tag{3.64}$$

Let us define $u_0 := \max\{u_1, u_2\}$, and η_0 as follows

$$\eta_0(x) = \begin{cases} \eta_1(x) & \text{if } x \in \{u_1 \geq u_2\}, \\ \eta_2(x) & \text{if } x \in \{u_2 > u_1\}, \end{cases}$$

as well as define ξ_0 by

$$\xi_0(x) = \begin{cases} \xi_1(x) & \text{if } x \in \{\gamma u_1 \geq \gamma u_2\}, \\ \xi_2(x) & \text{if } x \in \{\gamma u_2 > \gamma u_1\}. \end{cases}$$

By the definition of η_0, and ξ_0 we readily see that $\eta_0 \in L^q(\Omega)$, and $\xi_0 \in L^q(\partial\Omega)$, and

$$\eta_0(x) \in \partial j_1(x, u_0(x)) \text{ for a.e. } x \in \Omega, \xi_0(x) \in \partial j_2(x, \gamma u_0(x)) \text{ for a.e. } x \in \partial\Omega. \tag{3.65}$$

By means of η_0, ξ_0, u_0, and $\overline{\eta}$, $\overline{\xi}$, $\overline{u}$ of Definition 3.11 we introduce the following modifications $\tilde{j}_1 : \Omega \times \mathbb{R} \to \mathbb{R}$ and $\tilde{j}_2 : \partial\Omega \times \mathbb{R} \to \mathbb{R}$ of the given j_1 and j_2, respectively:

$$\tilde{j}_1(x, s) = \begin{cases} j_1(x, u_0(x)) + \eta_0(x)(s - u_0(x)) & \text{if } s < u_0(x), \\ j_1(x, s) & \text{if } u_0(x) \leq s \leq \overline{u}(x), \\ j_1(x, \overline{u}(x)) + \overline{\eta}(x)(s - \overline{u}(x)) & \text{if } s > \overline{u}(x), \end{cases} \tag{3.66}$$

and similarly

$$\tilde{j}_2(x, s) = \begin{cases} j_2(x, \gamma u_0(x)) + \xi_0(x)(s - \gamma u_0(x)) & \text{if } s < \gamma u_0(x), \\ j_2(x, s) & \text{if } \gamma u_0(x) \leq s \leq \gamma\overline{u}(x), \\ j_2(x, \gamma\overline{u}(x)) + \overline{\xi}(x)(s - \gamma\overline{u}(x)) & \text{if } s > \gamma\overline{u}(x). \end{cases} \tag{3.67}$$

In view of hypotheses (E-j1) and (E-j2) the modified function $\tilde{j}_1$ and $\tilde{j}_2$ given by (3.66) and (3.67) enjoy the properties of Lemma 3.20 and Lemma 3.21, respectively, with the only difference that Clarke's gradient of $\tilde{j}_1$ and $\tilde{j}_2$ is now given by:

$$\partial \tilde{j}_1(x,s) = \begin{cases} \eta_0(x) & \text{if } s < u_0(x), \\ \partial \tilde{j}_1(x,u_0(x)) & \text{if } s = u_0(x), \\ \partial j_1(x,s) & \text{if } u_0(x) < s < \overline{u}(x), \\ \partial \tilde{j}_1(x,\overline{u}(x)) & \text{if } s = \overline{u}(x), \\ \overline{\eta}(x) & \text{if } s > \overline{u}(x), \end{cases} \qquad (3.68)$$

with $\partial \tilde{j}_1(x,u_0(x)) \subseteq \partial j_1(x,u_0(x))$ and $\partial \tilde{j}_1(x,\overline{u}(x)) \subseteq \partial j_1(x,\overline{u}(x))$ being satisfied, and similarly,

$$\partial \tilde{j}_2(x,s) = \begin{cases} \xi_0(x) & \text{if } s < \gamma u_0(x), \\ \partial \tilde{j}_2(x,\gamma u_0(x)) & \text{if } s = \gamma u_0(x), \\ \partial j_2(x,s) & \text{if } \gamma u_0(x) < s < \gamma \overline{u}(x), \\ \partial \tilde{j}_2(x,\gamma \overline{u}(x)) & \text{if } s = \gamma \overline{u}(x), \\ \overline{\xi}(x) & \text{if } s > \gamma \overline{u}(x), \end{cases} \qquad (3.69)$$

with the inclusions $\partial \tilde{j}_2(x,\gamma u_0(x)) \subseteq \partial j_2(x,\gamma u_0(x))$ and $\partial \tilde{j}_2(x,\gamma \overline{u}(x)) \subseteq \partial j_2(x,\gamma \overline{u}(x))$. Furthermore, we introduce functions $g_{1,i} : \Omega \times \mathbb{R} \to \mathbb{R}$, $g_{2,i} : \partial\Omega \times \mathbb{R} \to \mathbb{R}$ related to η_i, ξ_i and u_i, $i = 0,1,2$, and defined by:

$$g_{1,0}(x,s) = \begin{cases} \eta_0(x) & \text{if } s \le u_0(x), \\ \eta_0(x) + \frac{\overline{\eta}(x)-\eta_0(x)}{\overline{u}(x)-u_0(x)}(s-u_0(x)) & \text{if } u_0(x) < s < \overline{u}(x), \\ \overline{\eta}(x) & \text{if } s \ge \overline{u}(x), \end{cases} \qquad (3.70)$$

and for $k = 1,2$

$$g_{1,k}(x,s) = \begin{cases} \eta_k(x) & \text{if } s \le u_k(x), \\ \eta_k(x) + \frac{\eta_0(x)-\eta_k(x)}{u_0(x)-u_k(x)}(s-u_k(x)) & \text{if } u_k(x) < s < u_0(x), \\ g_{1,0}(x,s) & \text{if } s \ge u_0(x), \end{cases} \qquad (3.71)$$

as well as

$$g_{2,0}(x,s) = \begin{cases} \xi_0(x) & \text{if } s \le \gamma u_0(x), \\ \xi_0(x) + \frac{\overline{\xi}(x)-\xi_0(x)}{\gamma\overline{u}(x)-\gamma u_0(x)}(s-\gamma u_0(x)) & \text{if } \gamma u_0(x) < s < \gamma\overline{u}(x), \\ \overline{\xi}(x) & \text{if } s \ge \gamma\overline{u}(x), \end{cases}$$

$$\qquad (3.72)$$

and for $k = 1,2$

$$g_{2,k}(x,s) = \begin{cases} \xi_k(x) & \text{if } s \le \gamma u_k(x), \\ \xi_k(x) + \frac{\xi_0(x)-\xi_k(x)}{\gamma u_0(x)-\gamma u_k(x)}(s-\gamma u_k(x)) & \text{if } \gamma u_k(x) < s < \gamma u_0(x), \\ g_{2,0}(x,s) & \text{if } s \ge \gamma u_0(x). \end{cases}$$

$$\qquad (3.73)$$

Apparently, the functions $g_{1,i} : \Omega \times \mathbb{R} \to \mathbb{R}$, $g_{2,i} : \partial\Omega \times \mathbb{R} \to \mathbb{R}$, $i = 0, 1, 2$, are Carathéodory functions that are piecewise linear with respect to s and uniformly $L^q(\Omega)$ and $L^q(\partial\Omega)$-bounded, respectively. Finally, define the cut-off function $\hat{b}$ related to the pair u_0 and $\bar{u}$ by:

$$\hat{b}(x, s) = \begin{cases} (s - \bar{u}(x))^{p-1} & \text{if } s > \bar{u}(x) \\ 0 & \text{if } u_0(x) \leq s \leq \bar{u}(x) \\ -(u_0(x) - s)^{p-1} & \text{if } s < u_0(x), \end{cases} \tag{3.74}$$

which qualitatively satisfies the same estimates as b in (3.46) and (3.47). The function $\hat{b} : \Omega \times \mathbb{R} \to \mathbb{R}$ is a Carathéodory function, and its associated Nemytskij operator $\hat{B} : L^p(Q) \to L^q(Q)$ defined by $\hat{B}u(x,t) = \hat{b}(x, t, u(x,t))$ is continuous and bounded. With $i^* : L^q(\Omega) \hookrightarrow V^*$ being the adjoint operator of the embedding $i : V \hookrightarrow L^p(\Omega)$, the operator $\hat{\mathcal{B}} = i^* \circ \hat{B} \circ i : V \to V^*$ is completely continuous and bounded, due to the compact embedding $V \hookrightarrow L^p(\Omega)$. By means of the Carathéodory functions $g_{k,i}$ introduced in (3.70)–(3.73) we define functions $g_1 : \Omega \times \mathbb{R} \to \mathbb{R}$ and $g_2 : \partial\Omega \times \mathbb{R} \to \mathbb{R}$ as follows:

$$g_j(x, s) = \sum_{l=1}^{2} |g_{j,l}(x, s) - g_{j,0}(x, s)|, \quad j = 1, 2, \tag{3.75}$$

which are Carathéodory functions in their respective domains of definition, and which are uniformly $L^q(\Omega)$ and $L^q(\partial\Omega)$-bounded, respectively. Thus, the associated Nemytskij operators $G_1 : L^p(\Omega) \to L^q(\Omega)$ and $G_2 : L^p(\partial\Omega) \to L^q(\partial\Omega)$ are continuous and bounded, which due to the compact embedding $V \hookrightarrow L^p(\Omega)$ and the compactness of the trace operator γ implies that

$$\mathcal{G}_1 = i^* \circ G_1 \circ i : V \to V^*, \quad \mathcal{G}_2 = \gamma^* \circ G_2 \circ \gamma : V \to V^* \tag{3.76}$$

is bounded and completely continuous. Further, in a similar way as in Sect. 3.2.1, we let $\tilde{J}_1 : L^p(\Omega) \to \mathbb{R}$ and $\tilde{J}_2 : L^p(\partial\Omega) \to \mathbb{R}$ be the integral functionals, here associated to $\tilde{j}_1$ and $\tilde{j}_2$ defined by (3.66) and (3.67), respectively. With Clarke's generalized gradients $\partial\tilde{J}_k$ of $\tilde{J}_k$ we define the multi-valued functions Φ_k as in (3.44), i.e.,

$$\Phi_1(u) := (i^* \circ \partial\tilde{J}_1 \circ i)(u), \quad \Phi_2(u) := (\gamma^* \circ \partial\tilde{J}_2 \circ \gamma)(u), \quad u \in V.$$

Further, define the multi-valued operator $\mathcal{A}$ as

$$\mathcal{A}(u) = Au + \hat{\mathcal{B}}(u) + \Phi_1(u) + \Phi_2(u) - \mathcal{G}_1(u) - \mathcal{G}_2(u). \tag{3.77}$$

Since $\mathcal{A}$ given by (3.77) differs qualitatively from (3.48) basically only by uniformly bounded and completely continuous perturbations $\mathcal{G}_k$, we readily observe that $\mathcal{A}$ given by (3.77) fulfills Lemma 3.23, i.e., $\mathcal{A} : V \to 2^{V^*}$ is bounded, pseudomonotone, and v_0-coercive for $v_0 \in K$. Therefore, we may apply the surjectivity result of Theorem 3.18 to the multi-valued operator

$$A + \partial I_K : V \to 2^{V^*},$$

where I_K is the indicator function related to K. As a consequence of Theorem 3.18 we obtain the existence of a $u \in K$ such that $h \in \mathcal{A}(u) + \partial I_K(u)$, which by definition (3.77) of $\mathcal{A}$ and ∂I_K results in the existence of $\eta^* \in \Phi_1(u)$, $\xi^* \in \Phi_2(u)$, and $\theta^* \in \partial I_K(u)$ with $\eta^* = i^* \tilde{\eta}$, $\xi^* = \gamma^* \tilde{\xi}$ such that equation

$$Au + \hat{\mathcal{B}}(u) + i^* \tilde{\eta} + \gamma^* \tilde{\xi} - \mathcal{G}_1(u) - \mathcal{G}_2(u) + \theta^* = h, \quad \text{in } V^* \tag{3.78}$$

holds, where $\tilde{\eta} \in L^q(\Omega)$ with $\tilde{\eta}(x) \in \partial \tilde{j}_1(x, u(x))$, for a.e. $x \in \Omega$, and $\tilde{\xi} \in L^q(\partial\Omega)$ with $\tilde{\xi}(x) \in \partial \tilde{j}_2(x, \gamma u(x))$, for a.e. $x \in \partial\Omega$. By using the definition of $\partial I_K(u)$, the solution u of equation (3.78) is seen to solve the following auxiliary multi-valued variational inequality

$$u \in K : \langle Au - h + \hat{\mathcal{B}}(u) + i^* \tilde{\eta} + \gamma^* \tilde{\xi} - \mathcal{G}_1(u) - \mathcal{G}_2(u), v - u \rangle \geq 0, \ \forall v \in K, \tag{3.79}$$

which is equivalent to

$$u \in K : \ \langle Au - h, v - u \rangle + \int_\Omega \hat{b}(x, u) (v - u) \, dx + \int_\Omega (\tilde{\eta} - g_1(x, u)) (v - u) \, dx$$

$$+ \int_{\partial\Omega} (\tilde{\xi} - g_2(x, \gamma u)) (\gamma v - \gamma u) \, d\sigma \geq 0, \ \forall \, v \in K. \tag{3.80}$$

Step 2: Comparison

Here we are going to show that any solution u of the auxiliary multi-valued variational inequality (3.80) satisfies $u_0 \leq u \leq \overline{u}$.

Let us first verify $u \leq \overline{u}$. To this end we recall Definition 3.11, and take as special test function in Definition 3.11 (iii) $v = \overline{u} \vee u = \overline{u} + (u - \overline{u})^+$, and in (3.80) the special test function $v = \overline{u} \wedge u = u - (u - \overline{u})^+ \in K$ is applied. By adding the resulting inequalities we obtain the following:

$$\langle A\overline{u} - Au, (u - \overline{u})^+ \rangle - \int_\Omega \hat{b}(x, u) (u - \overline{u})^+ \, dx$$

$$+ \int_\Omega (\overline{\eta} - \tilde{\eta} + g_1(x, u)) (u - \overline{u})^+ \, dx$$

$$+ \int_{\partial\Omega} (\overline{\xi} - \tilde{\xi} + g_2(x, \gamma u)) (\gamma u - \gamma \overline{u})^+ \, d\sigma \geq 0. \tag{3.81}$$

We recall that $\tilde{\eta}(x) \in \partial \tilde{j}_1(x, u(x))$ and $\tilde{\xi}(x) \in \partial \tilde{j}_2(x, \gamma u(x))$ with $\partial \tilde{j}_1$ and $\partial \tilde{j}_2$ given by (3.68) and (3.69), respectively. Therefore, if $x \in \{u > \overline{u}\}$ then $\overline{\eta}(x) = \tilde{\eta}(x)$, and due to (3.71) $g_{1,l}(x, u(x)) = g_{1,0}(x, u(x))$, for $l = 1, 2$, which results in $g_1(x, u(x)) = 0$, and thus

$$\int_\Omega (\overline{\eta} - \tilde{\eta} + g_1(x, u)) (u - \overline{u})^+ \, dx = \int_{\{u > \overline{u}\}} (\overline{\eta} - \tilde{\eta} + g_1(x, u)) (u - \overline{u}) \, dx = 0. \tag{3.82}$$

If $x \in \{\gamma u > \gamma \bar{u}\}$ then $\overline{\xi}(x) = \tilde{\xi}(x)$, and due to (3.73), $g_{2,l}(x, \gamma u(x)) = g_{2,0}(x, \gamma u(x))$, for $l = 1, 2$, which results in $g_2(x, \gamma u(x)) = 0$, and thus

$$\int_{\partial\Omega} (\overline{\xi} - \tilde{\xi} + g_2(x, \gamma u))(\gamma u - \gamma \bar{u})^+ \, d\sigma = 0. \tag{3.83}$$

Taking the definition of the cut-off function $\hat{b}$ into account we get

$$\int_\Omega \hat{b}(x, u)(u - \bar{u})^+ \, dx = \int_\Omega ((u - \bar{u})^+)^p \, dx, \tag{3.84}$$

and by means of (A2) we obtain the estimate

$$\langle A\bar{u} - Au, (u - \bar{u})^+ \rangle = -\langle Au - A\bar{u}, (u - \bar{u})^+ \rangle \leq 0. \tag{3.85}$$

Applying the results (3.82)–(3.85) to (3.81) finally yields

$$\int_\Omega ((u - \bar{u})^+)^p \, dx = 0,$$

which implies $(u - \bar{u})^+ = 0$, and thus $u \leq \bar{u}$.

In order to show that $u \geq u_0 = \max\{u_1, u_2\}$, we show that $u \geq u_k$, $k = 1, 2$. We recall that $u_k \in S$ means that u_k is a solution of (3.19)–(3.20) within the interval $[\underline{u}, \bar{u}]$, i.e., there exist $\eta_k \in L^q(\Omega)$, $\xi_k \in L^q(\partial\Omega)$ such that $\eta_k(x) \in \partial j_1(x, u_k(x))$, for a.e. $x \in \Omega$, $\xi_k(x) \in \partial j_2(x, \gamma u_k(x))$ for a.e. $x \in \partial\Omega$, and the variational inequality (3.64) holds. If we test (3.64) with $v = u_k \wedge u = u_k - (u_k - u)^+ \in K$, and the auxiliary multi-valued variational inequality (3.80) with $v = u \vee u_k = u + (u_k - u)^+ \in K$ we get by adding the resulting inequalities the following estimate:

$$\langle Au - Au_k, (u_k - u)^+ \rangle + \int_\Omega \hat{b}(x, u)(u_k - u)^+ \, dx$$

$$+ \int_\Omega (\tilde{\eta} - g_1(x, u) - \eta_k)(u_k - u)^+ \, dx$$

$$+ \int_{\partial\Omega} (\tilde{\xi} - g_2(x, \gamma u) - \xi_k)(\gamma u_k - \gamma u)^+ \, d\sigma \geq 0. \tag{3.86}$$

We estimate the terms on the left-hand side of the last inequality individually. By means of (A2) we get

$$\langle Au - Au_k, (u_k - u)^+ \rangle = -\langle Au_k - Au, (u_k - u)^+ \rangle \leq 0. \tag{3.87}$$

For $x \in \{u_k > u\}$, from (3.70) it follows $g_{1,0}(x, u(x)) = \eta_0(x)$, and (3.71) yields $g_{1,k}(x, u(x)) = \eta_k(x)$, hence we obtain by using the definition of g_1 given in (3.75) the equation (with $l \neq k$, and $l, k \in \{1, 2\}$)

$$g_1(x, u(x)) = |\eta_k(x) - \eta_0(x)| + |g_{1,l}(x, u(x)) - g_{1,0}(x, u(x))|, \quad x \in \{u_k > u\}. \tag{3.88}$$

Further, for $x \in \{u_k > u\} \subseteq \{u_0 > u\}$ from (3.68) it follows $\tilde{\eta}(x) \in \partial \tilde{j}_1(x, u(x)) = \eta_0(x)$, which in conjunction with (3.88) results in the following estimate of the third term on the left-hand side of (3.86): $l \neq k$

$$\int_\Omega (\tilde{\eta} - g_1(x, u) - \eta_k) (u_k - u)^+ \, dx = \int_{\{u_k > u\}} (\tilde{\eta} - g_1(x, u) - \eta_k) (u_k - u) \, dx$$

$$= \int_{\{u_k > u\}} \left(\eta_0 - \eta_k - |\eta_k - \eta_0| - |g_{1,l}(x, u) - g_{1,0}(x, u)| \right) (u_k - u) \, dx$$

$$\leq 0. \tag{3.89}$$

Similarly, for $x \in \{\gamma u_k > \gamma u\}$, from (3.72) it follows $g_{2,0}(x, \gamma u(x)) = \xi_0(x)$, and (3.73) yields $g_{2,k}(x, \gamma u(x)) = \xi_k(x)$, hence we obtain by using the definition of g_2 given in (3.75) the following equation (with $l \neq k$, and $l, k \in \{1, 2\}$)

$$g_2(x, \gamma u(x)) = |\xi_k(x) - \xi_0(x)| + |g_{2,l}(x, \gamma u(x)) - g_{2,0}(x, \gamma u(x))| \tag{3.90}$$

for $x \in \{\gamma u_k > \gamma u\}$. Further, for $x \in \{\gamma u_k > \gamma u\} \subseteq \{\gamma u_0 > \gamma u\}$ from (3.69) it follows $\tilde{\xi}(x) \in \partial \tilde{j}_2(x, \gamma u(x)) = \xi_0(x)$, which in conjunction with (3.90) implies the following estimate of the 4th term on the left-hand side of (3.86): $l \neq k$

$$\int_{\partial\Omega} (\tilde{\xi} - g_2(x, \gamma u) - \xi_k) (\gamma u_k - \gamma u)^+ \, d\sigma$$

$$= \int_{\{\gamma u_k > \gamma u\}} (\tilde{\xi} - g_2(x, \gamma u) - \xi_k) (\gamma u_k - \gamma u) \, d\sigma$$

$$= \int_{\{\gamma u_k > \gamma u\}} \left(\xi_0 - \xi_k - |\xi_k - \xi_0| - |g_{2,l}(x, \gamma u) - g_{2,0}(x, \gamma u)| \right) (\gamma u_k - \gamma u) \, d\sigma$$

$$\leq 0. \tag{3.91}$$

Using (3.87), (3.89), and (3.91), we get from (3.86) the estimate

$$\int_\Omega \hat{b}(x, u) (u_k - u)^+ \, dx \geq 0,$$

which by applying the definition of $\hat{b}$ finally results in

$$0 \leq \int_\Omega \hat{b}(x, u) (u_k - u)^+ \, dx = \int_{\{u_k > u\}} \hat{b}(x, u) (u_k - u) \, dx$$

$$= -\int_{\{u_k > u\}} (u_0 - u)^{p-1} (u_k - u) \, dx \leq -\int_{\{u_k > u\}} (u_k - u)^p \, dx$$

$$\leq -\int_\Omega ((u_k - u)^+)^p \, dx \leq 0.$$

The last inequality implies

$$\int_\Omega ((u_k - u)^+)^p \, dx = 0,$$

and hence it follows $(u_k - u)^+ = 0$, i.e., $u_k \leq u$, $k = 1, 2$, which completes the comparison.

Step 3: $\mathcal{S}$ Is Upward Directed

In this final step we complete the proof of Theorem 3.27 showing that $\mathcal{S}$ is indeed upward directed. To this end we only need to show that a solution u of the auxiliary multi-valued variational inequality (3.80), which due to Step 2 above satisfies $u_0 \leq u \leq \overline{u}$, belongs to $\mathcal{S}$. As $u \in [u_0, \overline{u}]$, it follows that $\hat{b}(x, u) = 0$, and $\gamma u \in [\gamma u_0, \gamma \overline{u}]$. Moreover, (3.70)–(3.73) imply $g_1(x, u) = 0$ and $g_2(x, \gamma u) = 0$, and hence the solution u of (3.80) satisfies: $u \in K$ and

$$\langle Au - h, v - u \rangle + \int_\Omega \tilde{\eta}\,(v - u)\,dx + \int_{\partial\Omega} \tilde{\xi}\,(\gamma v - \gamma u)\,d\sigma \geq 0, \ \forall\, v \in K \quad (3.92)$$

where $\tilde{\eta}(x) \in \partial \tilde{j}_1(x, u(x))$, and $\tilde{\xi} \in \partial \tilde{j}_2(x, \gamma u(x))$. Because $u \in [u_0, \overline{u}]$, from (3.68) and (3.69) it follows that $\partial \tilde{j}_1(x, u(x)) \subseteq \partial j_1(x, u(x))$ as well as $\partial \tilde{j}_2(x, \gamma u(x)) \subseteq \partial j_2(x, \gamma u(x))$, which proves that u is in fact a solution of our original multi-valued variational inequality (3.19)–(3.20). This completes the proof. $\qquad\qquad\qquad\qquad\qquad\qquad\qquad\qquad\qquad\qquad\qquad\qquad\qquad\quad$ $\square$

Remark 3.28. If $K = V$, or $K = V_0$, or K is the the closed subspace of V given in Example 3.7, then the multi-valued variational inequality (3.19)–(3.20) reduces to a multi-valued variational equation. In this case Theorem 3.27 on the directedness of $\mathcal{S}$ remains true also if the elliptic operator A is replaced by the more general Leray–Lions operator as in Remark 3.25, i.e.,

$$Au(x) = -\sum_{i=1}^{N} \frac{\partial}{\partial x_i} a_i(x, u, \nabla u(x)) + a_0(x, u, \nabla u(x)),$$

where the coefficients $a_i : \Omega \times \mathbb{R} \times \mathbb{R}^N \to \mathbb{R}$, $i = 0, 1, \ldots, N$ satisfy assumptions (A1')–(A3') as in Remark 3.25, and the coefficients a_i with $i = 1, \ldots, N$ satisfy, in addition, the following hypothesis (A4'):

(A4') There is a function $k_2 \in L_+^q(\Omega)$ and a continuous function $\omega : \mathbb{R}_+ \to \mathbb{R}_+$ such that

$$|a_i(x, s, \zeta) - a_i(x, s', \zeta)| \leq [k_2(x) + |s|^{p-1} + |s'|^{p-1} + |\zeta|^{p-1}]\omega(|s - s'|),$$

holds for a.a. $x \in \Omega$, for all $s, s' \in \mathbb{R}$, and for all $\zeta \in \mathbb{R}^N$, where ω satisfies, for each $\varepsilon > 0$,

$$\int_0^\varepsilon \frac{dr}{\omega(r)} = +\infty. \quad (3.93)$$

Relation (3.93) means that the integral diverges near zero. Hypothesis (A4') includes, for example, $\omega(r) = cr$, for $c > 0$, and for all $r \geq 0$, which means that the coefficients $a_i(x, s, \zeta)$ are Lipschitz continuous with respect to s. The proof of the directedness makes use of a special test function technique analogous to that for evolution variational equations as will be demonstrated in Sect. 3.3.

3.2.3 Extremal Solutions

In this subsection, the solution set $\mathcal{S}$ of all solutions of the multi-valued variational inequality (3.19)–(3.20) within the ordered interval $[\underline{u}, \overline{u}]$ is shown to possess *extremal solutions*, i.e., $\mathcal{S}$ has a *smallest solution* u_* and a *greatest solution* u^* with respect to the natural underlying partial ordering induced by the order cone $L_+^p(\Omega)$. A first step toward this goal is the following topological characterization of $\mathcal{S}$.

Lemma 3.29. *The solution set $\mathcal{S}$ is a compact set in $K \subseteq V$.*

Proof: We first prove that $\mathcal{S}$ is bounded in V. Let $c > 0$ be a generic constant that may change size, but which is independent of u. As $\mathcal{S} \subseteq [\underline{u}, \overline{u}]$, for any $u \in \mathcal{S}$ one has $\gamma u \in [\gamma\underline{u}, \gamma\overline{u}]$, and thus

$$\|u\|_{L^p(\Omega)} \le c, \quad \|\gamma u\|_{L^p(\partial\Omega)} \le c, \quad \|u\|_V \le c(1 + \|\nabla u\|_{L^p(\Omega)}), \quad \forall\, u \in \mathcal{S}. \tag{3.94}$$

Therefore, the boundedness of $\mathcal{S}$ in V is proved provided that $\|\nabla u\|_{L^p(\Omega)}$ is bounded for all $u \in \mathcal{S}$. Let $v_0 \in K$ be fixed, and $u \in \mathcal{S}$. Then u, in particular, satisfies: $u \in K$ and

$$\langle Au - h, v_0 - u\rangle + \int_\Omega \eta\,(v_0 - u)\,dx + \int_{\partial\Omega} \xi\,(\gamma v_0 - \gamma u)\,d\sigma \ge 0, \tag{3.95}$$

where $\eta(x) \in \partial j_1(x, u(x))$, for a.e. $x \in \Omega$, and $\xi(x) \in \partial j_2(x, \gamma u(x))$, for a.e. $x \in \partial\Omega$. By using (E-j1) and (E-j2) we immediately get

$$\left|\int_\Omega \eta\,(v_0 - u)\,dx\right| \le c, \quad \left|\int_{\partial\Omega} \xi\,(\gamma v_0 - \gamma u)\,d\sigma\right| \le c, \quad \forall\, u \in \mathcal{S}. \tag{3.96}$$

Assumption (A1) implies

$$|\langle Au - h, v_0\rangle| \le c(1 + \|\nabla u\|_{L^p(\Omega)}^{p-1}), \quad \forall\, u \in \mathcal{S}. \tag{3.97}$$

Further we have

$$|\langle h, u\rangle| \le \|h\|_{V^*}\|u\|_V \le c(1 + \|\nabla u\|_{L^p(\Omega)}) \quad \forall\, u \in \mathcal{S}. \tag{3.98}$$

From (3.95), we obtain by using (3.96)–(3.98), as well as (A3), the following estimate

$$\nu\,\|\nabla u\|_{L^p(\Omega)}^p \le c\,(1 + \|\nabla u\|_{L^p(\Omega)} + \|\nabla u\|_{L^p(\Omega)}^{p-1}), \quad \forall\, u \in \mathcal{S}, \tag{3.99}$$

which proves $\|\nabla u\|_{L^p(\Omega)} \le c$, and thus

$$\|u\|_V \le c, \quad \forall\, u \in \mathcal{S}. \tag{3.100}$$

Let $(u_n) \subseteq \mathcal{S}$. Since V is reflexive, there exists a weakly convergent subsequence (again denoted by (u_n)), which due to the compact embedding

$V \hookrightarrow L^p(\Omega)$ and the compactness of the trace operator $\gamma : V \to L^p(\partial\Omega)$, has the following convergence properties

$$u_n \rightharpoonup u \quad \text{in } V,$$
$$u_n \to u \quad \text{in } L^p(\Omega) \text{ and a.e. in } \Omega, \qquad (3.101)$$
$$\gamma u_n \to \gamma u \quad \text{in } L^p(\partial\Omega) \text{ and a.e. in } \partial\Omega.$$

Since $(u_n) \subset K$, and K is closed in V and convex, so K is weakly closed, and thus $u \in K$. We are going to show that $u_n \to u$ (strongly) in V and that $u \in S$, which completes the proof of the lemma. Testing the variational inequality (3.20) with $v = u$, then each u_n satisfies

$$\langle Au_n - h, u - u_n \rangle + \int_\Omega \eta_n \, (u - u_n) \, dx + \int_{\partial\Omega} \xi_n \, (\gamma u - \gamma u_n) \, d\sigma \geq 0, \quad (3.102)$$

where $\eta_n(x) \in \partial j_1(x, u_n(x))$, for a.e. $x \in \Omega$, and $\xi_n(x) \in \partial j_2(x, \gamma u_n(x))$, for a.e. $x \in \partial\Omega$. Note $u_n \in S$, which, in particular, implies $u_n \in [\underline{u}, \overline{u}]$, and therefore, by (E-j1) and (E-j2) the sequences (η_n) and (ξ_n) are uniformly bounded in $L^q(\Omega)$ and $L^q(\partial\Omega)$, respectively. The latter in conjunction with (3.102) and the convergence properties (3.101) result in

$$\limsup_{n\to\infty} \langle Au_n, u_n - u \rangle \leq 0. \qquad (3.103)$$

Taking into account that the operator A enjoys the (S_+)-property (see Chap. 9), $u_n \rightharpoonup u$ and (3.103) imply the strong convergence $u_n \to u$ in V. Due to $\|\eta_n\|_{L^q(\Omega)} \leq c$ and $\|\xi_n\|_{L^q(\partial\Omega)} \leq c$, we get by passing to a further subsequence if necessary (again denoted by (η_n) and (ξ_n))

$$\eta_n \rightharpoonup \eta \text{ in } L^q(\Omega), \quad \xi_n \rightharpoonup \xi \text{ in } L^q(\partial\Omega). \qquad (3.104)$$

The strong convergence $u_n \to u$ in V and (3.104) allow to pass to the limit in

$$\langle Au_n - h, v - u_n \rangle + \int_\Omega \eta_n \, (v - u_n) \, dx + \int_{\partial\Omega} \xi_n \, (\gamma v - \gamma u_n) \, d\sigma \geq 0, \quad v \in K$$

which yields for the strong limit $u \in K$

$$\langle Au - h, v - u \rangle + \int_\Omega \eta \, (v - u) \, dx + \int_{\partial\Omega} \xi \, (\gamma v - \gamma u) \, d\sigma \geq 0, \quad v \in K. \quad (3.105)$$

Apparently $u \in [\underline{u}, \overline{u}]$. To complete the proof we need to verify that $\eta(x) \in \partial j_1(x, u(x))$, for a.e. $x \in \Omega$, and $\xi(x) \in \partial j_2(x, \gamma u(x))$, for a.e. $x \in \partial\Omega$, which together with (3.105) proves that $u \in S$. Let us check $\eta(x) \in \partial j_1(x, u(x))$. For this purpose, one only needs to prove that

$$\eta(x) r \leq j_1^o(x, u(x); r) \qquad (3.106)$$

for all $r \in \mathbb{R}$ and a.e. $x \in \Omega$. Let r be a fixed real number and E be any measurable subset of Ω. From the upper semicontinuity of $(s, r) \mapsto j_1^o(x, s; r)$ (cf., e.g., [80]) and $u_n(x) \to u(x)$ for a.e. $x \in \Omega$ we get

$$j_1^o(x, u(x); r) \geq \limsup j_1^o(x, u_n(x); r),$$

for a.e. $x \in \Omega$, and thus

$$\int_E j_1^o(x, u(x); r) dx \geq \int_E \limsup j_1^o(x, u_n(x); r) dx. \tag{3.107}$$

On the other hand, since

$$j_1^o(x, s; r) = \max\{\zeta r : \zeta \in \partial j_1(x, s)\},$$

it follows from (E-j1) that

$$|j_1^o(x, s; r)| \leq |r| \, k_\Omega(x), \ \forall s \in [\underline{u}(x), \overline{u}(x)]. \tag{3.108}$$

Together with $u_n \in [\underline{u}, \overline{u}]$, this implies that

$$|j_1^o(x, u_n(x); r)| \leq |r| \, k_\Omega(x) \tag{3.109}$$

for a.e. $x \in \Omega$. Since the right-hand side is, in particular, in $L^1(\Omega)$, one can apply Fatou's lemma to get

$$\int_E \limsup j_1^o(x, u_n(x); r) dx \geq \limsup \int_E j_1^o(x, u_n(x); r) dx.$$

This estimate, together with (3.107), yields

$$\int_E j_1^o(x, u(x); r) dx \geq \limsup \int_E j_1^o(x, u_n(x); r) dx. \tag{3.110}$$

From the weak convergence (3.104) of η_n to η it follows

$$\int_E \eta_n(x) r dx \to \int_E \eta(x) r dx \ \text{ as } \ n \to \infty. \tag{3.111}$$

For each n, we have $\eta_n(x) \in \partial j_1(x, u_n(x))$ and consequently $j_1^o(x, u_n(x); r) \geq \eta_n(x) r$ for a.e. $x \in \Omega$. Hence,

$$\int_E j_1^o(x, u_n(x); r) dx \geq \int_E \eta_n(x) r dx, \ \forall \ n. \tag{3.112}$$

Letting $n \to \infty$ in (3.112) and making use of (3.111) and (3.110), we obtain

$$\int_E j_1^o(x, u(x); r) dx \geq \int_E \eta(x) r dx.$$

Since this inequality holds for all measurable subsets E of Ω, we must have (3.106) for a.e. $x \in \Omega$. The proof of $\xi(x) \in \partial j_2(x, \gamma u(x))$ can be done in an obvious similar way, which completes our compactness proof.   □

Theorem 3.27 on the directedness of $\mathcal{S}$ and Lemma 3.29 on the compactness of $\mathcal{S}$ allow us to prove the following extremality property of $\mathcal{S}$.

Theorem 3.30. *Let the hypotheses of Theorem 3.27 be satisfied. Then the solution set S has the greatest element u^* and the smallest element u_*, i.e., there exist greatest and smallest solutions u^* and u_*, respectively, of the multi-valued variational inequality (3.19)–(3.20) within the ordered interval $[\underline{u}, \overline{u}]$ in the sense that if u is any solution of (3.19)–(3.20) in $[\underline{u}, \overline{u}]$ then it satisfies $u_* \leq u \leq u^*$.*

Proof: Let us focus on the existence of the greatest element of S. Since V is separable, it follows that $S \subset V$ is separable, too, so there exists a countable, dense subset $Z = \{z_n : n \in \mathbb{N}\}$ of S. Since S is, in particular, upward directed, we can construct an increasing sequence $(u_n) \subseteq S$ as follows. Let $u_1 = z_1$. Select $u_{n+1} \in S$ such that

$$\max\{z_n, u_n\} \leq u_{n+1} \leq \overline{u}.$$

The existence of u_{n+1} is due to the directedness of S by Theorem 3.27. From the compactness of S according to Lemma 3.29, there exists a subsequence of (u_n), denoted again (u_n), and an element $u \in S$ such that $u_n \to u$ in V, and $u_n(x) \to u(x)$ a.e. in Ω. This last property of (u_n) combined with its increasing monotonicity implies that the entire sequence is convergent in V and, moreover, $u = \sup_n u_n$. By construction, we see that

$$\max\{z_1, z_2, \ldots, z_n\} \leq u_{n+1} \leq u, \quad \text{for all } n \in \mathbb{N},$$

thus $Z \subseteq [\underline{u}, u]$. Since the interval $[\underline{u}, u]$ is closed in V, we infer

$$S \subseteq \overline{Z} \subseteq \overline{[\underline{u}, u]} = [\underline{u}, u],$$

which in conjunction with $u \in S$ ensures that $u = u^*$ is the greatest solution. The existence of the smallest solution u_* can be proved in a similar way using the fact that S is downward directed.   □

3.2.4 Equivalence to Variational-Hemivariational Inequality

Here we are going to verify the assertion on the equivalence of problems (3.19)–(3.20) and (3.31) that was already anticipated in Remark 3.9. The standard growth conditions on the functions $j_1 : \Omega \times \mathbb{R} \to \mathbb{R}$ and $j_2 : \partial\Omega \times \mathbb{R} \to \mathbb{R}$ mentioned in Remark 3.9 are as follows:

(G-j1) (i) $x \mapsto j_1(x, s)$ is measurable in Ω for all $s \in \mathbb{R}$, and $s \mapsto j_1(x, s)$ is locally Lipschitz continuous in $\mathbb{R}$ for a.e. $x \in \Omega$.

(ii) There exist a constant $c > 0$ and a function $k_\Omega \in L^q_+(\Omega)$ such that for a.e. $x \in \Omega$ and for all $s \in \mathbb{R}$ the growth condition

$$|\eta| \leq k_\Omega(x) + c|s|^{p-1}, \quad \forall \, \eta \in \partial j_1(x, s)$$

is fulfilled.

(G-j2) (i) $x \mapsto j_2(x,s)$ is measurable in $\partial\Omega$ for all $s \in \mathbb{R}$, and $s \mapsto j_2(x,s)$ is locally Lipschitz continuous in $\mathbb{R}$ for a.e. $x \in \partial\Omega$.

(ii) There exist a constant $c > 0$ and a function $k_{\partial\Omega} \in L_+^q(\partial\Omega)$ such that for a.e. $x \in \partial\Omega$ and for all $s \in \mathbb{R}$ the growth condition

$$|\xi| \leq k_{\partial\Omega}(x) + c|s|^{p-1}, \quad \forall\, \xi \in \partial j_2(x,s)$$

is fulfilled.

Before proving the equivalence result, let us first provide a short account of the sub-supersolution method for the variational-hemivariational inequality (3.31) that has been established in [39, 73]. Consider (3.31), which is: Find $u \in K$ such that

$$\langle Au - h, v - u \rangle + \int_\Omega j_1^o(x,u;v-u)\,dx + \int_{\partial\Omega} j_2^o(x,\gamma u;\gamma v - \gamma u)\,d\sigma \geq 0, \ \forall\, v \in K.$$

By specifying the closed convex set $K \subseteq V$, one can see in a similar way as in Sect. 3.2.1 that the variational-hemivariational (3.31) includes various elliptic boundary value problems as special cases.

We introduce the following notion of sub- and supersolution, see [39, 62].

Definition 3.31. *A function $\underline{u} \in V$ is called a **subsolution** of (3.31) if the following holds:*

(i) $\underline{u} \vee K \subseteq K$,

(ii) $\langle A\underline{u} - h, v - \underline{u} \rangle + \displaystyle\int_\Omega j_1^o(x,\underline{u};v-\underline{u})\,dx + \int_{\partial\Omega} j_2^o(x,\gamma\underline{u};\gamma v - \gamma\underline{u})\,d\sigma \geq 0$

for all $v \in \underline{u} \wedge K$.

Definition 3.32. *A function $\overline{u} \in V$ is called a **supersolution** of (3.31) if the following holds:*

(i) $\overline{u} \wedge K \subseteq K$,

(ii) $\langle A\overline{u} - h, v - \overline{u} \rangle + \displaystyle\int_\Omega j_1^o(x,\overline{u};v-\overline{u})\,dx + \int_{\partial\Omega} j_2^o(x,\gamma\overline{u};\gamma v - \gamma\overline{u})\,d\sigma \geq 0$

for all $v \in \overline{u} \vee K$.

Remark 3.33. Note again that the notion for sub- and supersolution defined in Definition 3.31 and Definition 3.32 have a symmetric structure, i.e., one obtains the definition for the supersolution $\overline{u}$ from the definition of the subsolution by replacing $\underline{u}$ in Definition 3.31 by $\overline{u}$, and interchanging $\vee$ by $\wedge$. If one applies Definitions 3.31 and 3.32 to specific K such as, e.g., K being some subspace of V, then in a similar way as in Sect. 3.2.1 one can show that the above definitions are in fact natural extensions of the usual notions of sub-supersolutions for elliptic boundary value problems.

As for the following existence and comparison result we refer to [39, 73].

Theorem 3.34. *Let (A1)–(A3) be satisfied. Assume the existence of sub- and supersolutions $\underline{u}$ and $\overline{u}$, respectively, of the variational-hemivariational inequality (3.31) with $\underline{u} \leq \overline{u}$ such that (E-j1)–(E-j2) are fulfilled. Then there exist solutions of (3.31)) within the ordered interval $[\underline{u}, \overline{u}]$.*

Note that by Remark 3.15 the conditions (E-j1)–(E-j2) are trivially satisfied if (G-j1) and (G-j2) are fulfilled.

The main result of this subsection is the following theorem.

Theorem 3.35. *Let hypotheses (A1)–(A3) and (G-j1)–(G-j2) be satisfied, and assume the lattice condition (3.32) of the closed convex subset $K \subseteq V$, i.e.,*

$$K \wedge K \subseteq K, \quad K \vee K \subseteq K.$$

Then u is a solution of the multi-valued variational inequality (3.19)–(3.20) if and only if u is a solution of the variational-hemivariational inequality (3.31).

Proof: Our aim is to show that any solution of the multi-valued variational inequality (3.19)–(3.20) is a solution of the variational-hemivariational inequality (3.31), and vice versa. The basic tools to achieve our goal are Theorem 3.24 and Theorem 3.34 on the sub-supersolution method for problems (3.19)–(3.20) and (3.31), respectively.

Let $u \in K$ be a solution of (3.19)–(3.20), i.e., there is an $\eta \in L^q(\Omega)$, and a $\xi \in L^q(\partial\Omega)$ such that $\eta(x) \in \partial j_1(x, u(x))$, for a.e. $x \in \Omega$, $\xi(x) \in \partial j_2(x, \gamma u(x))$, for a.e. $x \in \partial\Omega$, and the following variational inequality is satisfied:

$$\langle Au - h, v - u \rangle + \int_\Omega \eta \, (v - u) \, dx + \int_{\partial\Omega} \xi \, (\gamma v - \gamma u) \, d\sigma \geq 0, \ \forall \, v \in K. \ (3.113)$$

By the definition of Clarke's generalized gradient we readily obtain for any $v \in K$:

$$j_1^o(x, u(x); v(x) - u(x)) \geq \eta(x) \, (v(x) - u(x)), \qquad (3.114)$$
$$j_2^o(x, \gamma u(x); \gamma v(x) - \gamma u(x)) \geq \xi(x) \, (\gamma v(x) - \gamma u(x)), \qquad (3.115)$$

for a.e. $x \in \Omega$ and for a.e. $x \in \partial\Omega$ in (3.114) and (3.115), respectively. By (G-j1)(ii) and (G-j2)(ii), the absolute value of the left-hand sides of (3.114) and (3.115) can be bounded by $L^1(\Omega)$ and $L^1(\partial\Omega)$ functions, respectively. Since the functions $(s, r) \mapsto j_k^o(x, s; r)$, $k = 1, 2$, are superpositionally measurable, from (3.114) and (3.115) we obtain

$$\int_\Omega j_1^o(x, u(x); v(x) - u(x)) \, dx \geq \int_\Omega \eta(x) \, (v(x) - u(x)) \, dx,$$

$$\int_{\partial\Omega} j_2^o(x, \gamma u(x); \gamma v(x) - \gamma u(x)) \, dx \geq \int_{\partial\Omega} \xi(x) \, (\gamma v(x) - \gamma u(x)) \, d\sigma. \qquad (3.116)$$

Thus, the variational inequality (3.113) along with (3.116) implies that u is a solution of variational-hemivariational inequality (3.31). One readily observes that this direction of the proof basically follows from the definition of Clarke's generalized gradient.

To prove the reverse, let u be any solution of (3.31), i.e., $u \in K$ and u satisfies (3.31), which is

$$\langle Au - h, v - u \rangle + \int_\Omega j_1^o(x, u; v - u)\, dx + \int_{\partial\Omega} j_2^o(x, \gamma u; \gamma v - \gamma u)\, d\sigma \geq 0, \ \forall\, v \in K.$$
(3.117)

The lattice condition (3.32) implies that u is both a subsolution and a supersolution for (3.31), i.e., of (3.117). Next we are going to show that u must be both a subsolution and a supersolution for (3.19)–(3.20). Let us show first that u is a subsolution of (3.19)–(3.20). Since u is a subsolution of (3.31), the inequality (3.117) is satisfied, in particular, for all $v \in u \wedge K$, i.e., $v = u - (u - \varphi)^+$ with $\varphi \in K$, which yields

$$\langle Au - h, -(u - \varphi)^+ \rangle + \int_\Omega j_1^o(x, u; -(u - \varphi)^+)\, dx$$
$$+ \int_{\partial\Omega} j_2^o(x, \gamma u; -(\gamma u - \gamma\varphi)^+)\, d\sigma \geq 0, \ \forall\, \varphi \in K.$$

Because $r \mapsto j_k^o(\cdot, s; r)$ is positively homogeneous, the last inequality is equivalent to

$$\langle Au - h, -(u - \varphi)^+ \rangle + \int_\Omega j_1^o(x, u; -1)(u - \varphi)^+\, dx$$
$$+ \int_{\partial\Omega} j_2^o(x, \gamma u; -1)(\gamma u - \gamma\varphi)^+\, d\sigma \geq 0, \ \forall\, \varphi \in K.$$
(3.118)

Using again for any $v \in u \wedge K$ its representation in the form $v = u - (u - \varphi)^+$ with $\varphi \in K$, (3.118) is equivalent to

$$\langle Au - h, v - u \rangle - \int_\Omega j_1^o(x, u; -1)(v - u)\, dx$$
$$- \int_{\partial\Omega} j_2^o(x, \gamma u; -1)(\gamma v - \gamma u)\, d\sigma \geq 0, \ \forall\, v \in u \wedge K.$$
(3.119)

By [80, Proposition 2.1.2] we have

$$j_1^o(x, u(x); -1) = \max\{-\theta(x) : \theta(x) \in \partial j_1(x, u(x))\}$$
$$= -\min\{\theta(x) : \theta(x) \in \partial j_1(x, u(x))\} = -\underline{\eta}(x), \quad (3.120)$$

where

$$\underline{\eta}(x) = \min\{\theta(x) : \theta(x) \in \partial j_1(x, u(x))\} \in \partial j_1(x, u(x)), \quad \text{for a.e. } x \in \Omega.$$
(3.121)

Since $x \mapsto j_1^o(x, u(x); -1)$ is a measurable function, it follows that $x \mapsto \underline{\eta}(x)$ is measurable in Ω, too, and in view of the growth condition (G-j1)(ii), we infer $\underline{\eta} \in L^q(\Omega)$.

In a similar way one can show that there is a $\underline{\xi} \in L^q(\partial\Omega)$ with

$$\underline{\xi}(x) = \min\{\theta(x) : \theta(x) \in \partial j_2(x, \gamma u(x))\} \in \partial j_2(x, \gamma u(x)), \quad \text{for a.e. } x \in \partial\Omega, \tag{3.122}$$

such that

$$j_2^o(x, \gamma u(x); -1) = -\underline{\xi}(x), \quad \text{for a.e. } x \in \partial\Omega. \tag{3.123}$$

Taking (3.120)–(3.123) into account, (3.119) yields

$$\langle Au - h, v - u \rangle + \int_\Omega \underline{\eta} (v - u) \, dx + \int_{\partial\Omega} \underline{\xi} (\gamma v - \gamma u) \, d\sigma \geq 0, \ \forall \, v \in u \wedge K, \tag{3.124}$$

which proves that u is a subsolution of (3.19)–(3.20). By applying similar arguments, one shows that u is also supersolution of (3.19)–(3.20), i.e., there is an $\overline{\eta} \in L^q(\Omega)$, and a $\overline{\xi} \in L^q(\partial\Omega)$ such that $\overline{\eta}(x) \in \partial j_1(x, u(x))$, for a.e. $x \in \Omega$, $\overline{\xi}(x) \in \partial j_2(x, \gamma u(x))$, for a.e. $x \in \partial\Omega$, and the following inequality is satisfied:

$$\langle Au - h, v - u \rangle + \int_\Omega \overline{\eta} (v - u) \, dx + \int_{\partial\Omega} \overline{\xi} (\gamma v - \gamma u) \, d\sigma \geq 0, \ \forall \, v \in u \vee K. \tag{3.125}$$

So far we have shown that any solution u of the variational-hemivariational inequality (3.31) is both a subsolution and a supersolution of the multi-valued variational inequality (3.19)–(3.20). Therefore, Theorem 3.24 ensures the existence of a solution $\tilde{u}$ of (3.19)–(3.20) within the interval $[u, u] = \{u\}$, which implies $u = \tilde{u}$. This proves that any solution u of (3.31) must be a solution of (3.19)–(3.20), which completes the proof. □

Remark 3.36. The lattice condition (3.32) on the closed, convex subset $K \subseteq V$ is fulfilled for a number of important models in applied sciences, see, e.g., [62, p. 216]. Moreover, Theorem 3.35 remains true also in case that the operator A is replaced by a more general Leray–Lions operator in the form:

$$Au(x) = -\sum_{i=1}^N \frac{\partial}{\partial x_i} a_i(x, u, \nabla u(x)) + a_0(x, u, \nabla u(x)).$$

3.3 Multi-Valued Parabolic Variational Inequalities

In this section we are going to establish existence and comparison principles for the parabolic counterpart to the multi-valued elliptic variational inequality (3.19)–(3.20) of Sect. 3.2. To formulate the problem to be considered here, let us introduce some notations, cf. Sect. 9.5. As in Sect. 3.2 let $\Omega \subset \mathbb{R}^N$, $N \geq 1$, be a bounded domain with Lipschitz boundary $\partial\Omega$, and let $V = W^{1,p}(\Omega)$

and $V_0 = W_0^{1,p}(\Omega)$ denote the usual Sobolev spaces with their dual spaces V^* and V_0^*, respectively. Let $Q = \Omega \times (0,\tau)$, be a cylindrical domain, and denote by $\Gamma = \partial\Omega \times (0,\tau)$, its lateral boundary, with $\tau > 0$. For the sake of simplicity we assume throughout this section $2 \leq p < \infty$ with q being its Hölder conjugate, i.e., $1/p + 1/q = 1$. Then $W^{1,p}(\Omega) \hookrightarrow L^2(\Omega) \hookrightarrow (W^{1,p}(\Omega))^*$ (resp. $W_0^{1,p}(\Omega) \hookrightarrow L^2(\Omega) \hookrightarrow (W_0^{1,p}(\Omega))^*$) forms an evolution triple with all the embeddings being continuous, dense, and compact, cf. [229]. We set $X = L^p(0,\tau;W^{1,p}(\Omega))$, with its dual space X^*, and denote the norms in X and X^* by $\|\cdot\|_X$ and $\|\cdot\|_{X^*}$, respectively, which are given by

$$\|u\|_X = \left(\int_0^\tau \|u(t)\|_V^p \, dt \right)^{1/p}, \quad \|u\|_{X^*} = \left(\int_0^\tau \|u(t)\|_{V^*}^q \, dt \right)^{1/q}.$$

As the derivative $\partial u/\partial t$ explicitly appears in the multi-valued problem to be considered (see below), a natural underlying solution space is the function space W defined by

$$W = \{u \in X : \partial u/\partial t \in X^*\},$$

where the derivative $\partial u/\partial t$ is understood in the sense of vector-valued distributions, cf. Sect. 9.5. The space W endowed with the graph norm of the operator $\partial/\partial t$

$$\|u\|_W = \|u\|_X + \|\partial u/\partial t\|_{X^*}$$

is a Banach space that is separable and reflexive due to the separability and reflexivity of X and X^*, respectively. Furthermore, it is well known that the embedding $W \hookrightarrow C([0,\tau], L^2(\Omega))$ is continuous, cf. [229], and because $W^{1,p}(\Omega)$ is compactly embedded in $L^p(\Omega)$, we have by Aubin's lemma that the $W \hookrightarrow L^p(Q)$ is compactly embedded, cf. Theorem 9.98. Similarly, we set $X_0 = L^p(0,\tau;W_0^{1,p}(\Omega))$, whose dual space is given by $X_0^* = L^q(0,\tau;V^*) = L^q(0,\tau;W^{-1,q}(\Omega))$, and introduce W_0 defined by

$$W_0 = \{u \in X_0 : \partial u/\partial t \in X_0^*\}.$$

We use the notation $\langle\cdot,\cdot\rangle$ for any of the dual pairings between X and X^*, X_0 and X_0^*, $W^{1,p}(\Omega)$ and $(W^{1,p}(\Omega))^*$, and $W_0^{1,p}(\Omega)$ and $W^{-1,q}(\Omega)$. For example, with $f \in X^*, u \in X$,

$$\langle f, u \rangle = \int_0^\tau \langle f(t), u(t) \rangle \, dt.$$

Let $L := \partial/\partial t$, and its domain of definition $D(L)$ be given by

$$D(L) = \{u \in W : u(0) = 0\}.$$

The linear operator $L : D(L) \subset X \to X^*$ is closed, densely defined, and maximal monotone, e.g., cf. [229, Chap. 32], and Lemma 9.106.

Let K be a closed, convex subset of X, and let $j_1 : Q \times \mathbb{R} \to \mathbb{R}$ and $j_2 : \Gamma \times \mathbb{R} \to \mathbb{R}$ be functions that are only supposed to be measurable in their first and locally Lipschitz continuous with respect to their second argument. The main goal of this section is to extend the sub-supersolution method to quasilinear multi-valued parabolic variational inequalities of the following form:

Find $u \in K \cap D(L)$, $\eta \in L^q(Q)$, and $\xi \in L^q(\Gamma)$ satisfying:

$$\eta(x,t) \in \partial j_1(x,t,u(x,t)), \text{ for a.e. } (x,t) \in Q, \tag{3.126}$$

$$\xi(x,t) \in \partial j_2(x,t,\gamma u(x,t)), \text{ for a.e. } (x,t) \in \Gamma, \text{ and} \tag{3.127}$$

$$\langle Lu + Au - h, v - u \rangle + \int_Q \eta \, (v - u) \, dxdt + \int_\Gamma \xi \, (\gamma v - \gamma u) \, d\Gamma \geq 0, \ \forall \, v \in K, \tag{3.128}$$

where

- the multi-valued functions $s \mapsto \partial j_k(x,t,s)$ are given by Clarke's generalized gradient of the locally Lipschitz functions $s \mapsto j_k(x,t,s)$, $k = 1,2$, defined by

$$\partial j_k(x,t,s) := \{\zeta \in \mathbb{R} : j_k^o(x,t,s;r) \geq \zeta r, \ \forall \, r \in \mathbb{R}\}$$

for a.a. $(x,t) \in Q$ in case $k = 1$, and for a.a. $(x,t) \in \Gamma$ in case $k = 2$, with $j_k^o(x,t,s;r)$ denoting the generalized directional derivative of $s \mapsto j_k(x,t,s)$ at s in the direction r given by

$$j_k^o(x,t,s;r) = \limsup_{y \to s, \ \varepsilon \downarrow 0} \frac{j_k(x,t,y+\varepsilon r) - j_k(x,t,y)}{\varepsilon}.$$

- A is a second-order quasilinear elliptic differential operator of the form

$$Au(x,t) = -\sum_{i=1}^{N} \frac{\partial}{\partial x_i} a_i(x,t,u(x,t),\nabla u(x,t)),$$

with $\nabla u = (\partial u/\partial x_1, \ldots, \partial u/\partial x_N)$.
- $\gamma : X \to L^p(\Gamma)$ denotes the trace operator, which is linear and continuous, and $\gamma : W \to L^p(\Gamma)$ is linear and compact, see [42, Lemma 3.1], or Proposition 9.100.
- $h \in X^*$.

A partial ordering in $L^p(Q)$ is defined by $u \leq w$ if and only if $w - u$ belongs to the positive cone $L^p_+(Q)$ of all nonnegative elements of $L^p(Q)$. This induces corresponding partial orderings in the subspaces W and X of $L^p(Q)$. The partial ordering on X implies a corresponding partial ordering for the traces, namely, if $u, w \in X$ and $u \leq w$ then $\gamma u \leq \gamma w$ in $L^p(\Gamma)$. If $u, w \in W$ with $u \leq w$, the order interval formed by u and w is the set

$$[u,w] = \{v \in W : u \leq v \leq w\}.$$

As in Sect. 3.2, for $u, v \in X$, and $U_1, U_2 \subseteq X$, we use the notation $u \wedge v = \min\{u, v\}$, $u \vee v = \max\{u, v\}$, $U_1 * U_2 = \{u * v : u \in U_1, v \in U_2\}$, $u * U_1 = \{u\} * U_1$ with $* \in \{\wedge, \vee\}$.

We assume the following Leray–Lions conditions on the coefficient functions a_i, $i = 1, \ldots, N$, entering the definition of the operator A.

(AP1) $a_i : Q \times \mathbb{R} \times \mathbb{R}^N \to \mathbb{R}$ are Carathéodory functions, i.e., $a_i(\cdot, \cdot, s, \zeta) : Q \to \mathbb{R}$ is measurable for all $(s, \zeta) \in \mathbb{R} \times \mathbb{R}^N$ and $a_i(x, t, \cdot, \cdot) : \mathbb{R} \times \mathbb{R}^N \to \mathbb{R}$ is continuous for a.a. $(x, t) \in Q$. In addition, the following growth condition holds:

$$|a_i(x, t, s, \zeta)| \le k_0(x, t) + c_0 \left(|s|^{p-1} + |\zeta|^{p-1}\right)$$

for a.a. $(x, t) \in Q$ and for all $(s, \zeta) \in \mathbb{R} \times \mathbb{R}^N$, for some constant $c_0 > 0$ and some function $k_0 \in L^q(Q)$.

(AP2) For a.a. $(x, t) \in Q$, for all $s \in \mathbb{R}$ and for all $\zeta, \zeta' \in \mathbb{R}^N$ with $\zeta \ne \zeta'$ the following monotonicity in ζ holds:

$$\sum_{i=1}^N (a_i(x, t, s, \zeta) - a_i(x, t, s, \zeta'))(\zeta_i - \zeta_i') > 0.$$

(AP3) There is some constant $\nu > 0$ such that for a.a. $(x, t) \in Q$ and for all $(s, \zeta) \in \mathbb{R} \times \mathbb{R}^N$ the inequality

$$\sum_{i=1}^N a_i(x, t, s, \zeta)\zeta_i \ge \nu|\zeta|^p - k_1(x, t)$$

is satisfied for some function $k_1 \in L^1(Q)$.

(AP4) A function $k_2 \in L_+^q(Q)$ and a continuous function $\omega : \mathbb{R}_+ \to \mathbb{R}_+$ exist such that

$$|a_i(x, t, s, \zeta) - a_i(x, t, s', \zeta)| \le [k_2(x, t) + |s|^{p-1} + |s'|^{p-1} + |\zeta|^{p-1}]\omega(|s - s'|),$$

holds for a.a. $(x, t) \in Q$, for all $s, s' \in \mathbb{R}$, and for all $\zeta \in \mathbb{R}^N$, where ω satisfies, for each $\varepsilon > 0$,

$$\int_0^\varepsilon \frac{dr}{\omega(r)} = +\infty. \tag{3.129}$$

Remark 3.37. Relation (3.129) means that the integral diverges near zero. Hypothesis (AP4) includes, for example, $\omega(r) = cr$, for $c > 0$, and for all $r \ge 0$, which means that the coefficients $a_i(x, t, s, \xi)$ are Lipschitz continuous with respect to s.

In view of (AP1), the operator A defined by

$$\langle Au, \varphi \rangle := \int_Q \sum_{i=1}^N a_i(x, t, u, \nabla u) \frac{\partial \varphi}{\partial x_i} \, dx \, dt,$$

is continuous and bounded from X (resp. X_0) into its dual space.

Before we define our basic notion of sub-supersolutions for (3.126)–(3.128), let us consider important special cases that arise from (3.126)–(3.128) by specifying K.

Example 3.38. If $K = X$, and $s \mapsto j_k(x, t, s)$ are given in terms of the primitives of Carathéodory functions f_k, i.e.,

$$j_k(x, t, s) = \int_0^s f_k(x, t, r) \, dr,$$

then $\partial j_k(x, t, s) = \{f_k(x, t, s)\}$ is a singleton, and the multi-valued parabolic variational inequality (3.126)–(3.128) reduces to the following (single-valued) parabolic initial boundary value problem: Find $u \in D(L)$ such that

$$\langle Lu + Au - h, v \rangle + \int_Q f_1(x, t, u) \, v \, dx dt + \int_\Gamma f_2(x, t, \gamma u) \, \gamma v \, d\Gamma = 0, \quad \forall \, v \in X,$$
(3.130)

which is the formulation of the weak solution of the initial boundary value problem

$$u_t + Au + f_1(x, t, u) = h \ \text{ in } \ Q, \ \ u(\cdot, 0) = 0 \ \text{ in } \ \Omega, \ \ \frac{\partial u}{\partial \nu} + f_2(x, t, u) = 0 \ \text{ on } \ \Gamma,$$
(3.131)

where $\partial / \partial \nu$ denotes the outward pointing conormal derivative associated with A.

Example 3.39. If $K = X_0$, and j_k as in Example 3.38, then (3.126)–(3.128) is equivalent to: Find $u \in D(L) \cap X_0$ such that

$$\langle Lu + Au - h, v \rangle + \int_Q f_1(x, t, u) \, v \, dx dt = 0, \quad \forall \, v \in X_0,$$
(3.132)

which is nothing but the weak formulation of the homogeneous initial Dirichlet boundary value problem

$$u_t + Au + f_1(x, t, u) = h \ \text{ in } \ Q, \ \ u(\cdot, 0) = 0 \ \text{ in } \ \Omega, \ \ u = 0 \ \text{ on } \ \Gamma. \quad (3.133)$$

Example 3.40. If $K = X_0$ or $K = X$, and $s \mapsto j_k(x, t, s)$ not necessarily differentiable, then (3.126)–(3.128) reduces to parabolic inclusion problems of hemivariational type, which for $K = X_0$ yields the following multi-valued initial Dirichlet boundary value problem

$$u_t + Au + \partial j_1(x, t, u) \ni h \ \text{ in } \ Q, \ \ u(\cdot, 0) = 0 \ \text{ in } \ \Omega, \ \ u = 0 \ \text{ on } \ \Gamma, \quad (3.134)$$

and for $K = X$ the parabolic inclusion problem

$$u_t + Au + \partial j_1(x, t, u) \ni h \ \text{ in } \ Q, \ \ u(\cdot, 0) = 0 \text{ in } \Omega, \ \ \frac{\partial u}{\partial \nu} + \partial j_2(x, t, u) \ni 0 \ \text{ on } \Gamma.$$
(3.135)

Let us remark that nonhomogeneous initial and Dirichlet conditions can always be transformed to homogeneous ones by simple translation provided the initial and boundary values arise as traces of some function from W. By specifying K appropriately, also mixed boundary conditions can be seen to be a special case of the general formulation (3.126)–(3.128). To this end let $\Gamma_1 = S_1 \times (0,\tau)$ and $\Gamma_2 = S_2 \times (0,\tau)$, where S_1 and S_2 are relatively open subsets of $\partial\Omega$ satisfying $\overline{S_1} \cup \overline{S_2} = \partial\Omega$ and $S_1 \cap S_2 = \emptyset$.

Example 3.41. If $K \subseteq X$ is the following closed subspace

$$K = \{v \in X : \gamma v = 0 \text{ on } \Gamma_1\},$$

then (3.126)–(3.128) reduces to

$$u_t + Au + \partial j_1(x,t,u) \ni h \text{ in } Q,$$

$$u(x,0) = 0 \text{ in } \Omega, \quad \frac{\partial u}{\partial \nu} + \partial j_2(x,t,u) \ni 0 \text{ on } \Gamma_2, \quad u = 0 \text{ on } \Gamma_1.$$

Example 3.42. If $K \subseteq X$, and $j_k = 0$, then (3.126)–(3.128) is equivalent to the parabolic variational inequality of the form

$$u \in K \cap D(L): \quad \langle Lu + Au - h, v - u \rangle \geq 0, \quad \forall\, v \in K, \tag{3.136}$$

which has been treated, e.g., in [61, 62, 63].

Remark 3.43. Applying the definition of Clarke's generalized gradient of j_k, $k = 1,2$, and assuming standard growth conditions for $s \mapsto \partial j_k(x,t,s)$, one readily verifies that any solution of the multi-valued variational inequality (3.126)–(3.128) must be a solution of the following parabolic variational-hemivariational inequality: Find $u \in K \cap D(L)$ such that

$$\langle Lu + Au - h, v - u \rangle + \int_Q j_1^o(\cdot,\cdot,u;v-u)\,dxdt$$

$$+ \int_\Gamma j_2^o(\cdot,\cdot,\gamma u;\gamma v - \gamma u)\,d\Gamma \geq 0, \;\forall\, v \in K. \tag{3.137}$$

The reverse holds true if K satisfies the lattice condition

$$K \wedge K \subseteq K, \quad K \vee K \subseteq K,$$

and provided that to the multi-valued variational inequality (3.126)–(3.128) the method of sub-supersolution applies. Under these assumptions, the proof for the latter follows basically the arguments of Sect. 3.2.4. For periodic-Dirichlet problems, the equivalence of quasilinear parabolic inclusions and the associated parabolic hemivariational inequalities has been proved among others in [40].

3.3.1 Notion of Sub-Supersolution

We first introduce our basic notion of sub-supersolution for the multi-valued parabolic variational inequality (3.126)–(3.128), which, by specifying the closed convex set K, will be seen as a natural extension of the well-known notion of sub-supersolutions for single- and multi-valued initial and boundary value problems such as, e.g., those of Examples 3.38–3.40.

Definition 3.44. *A function* $\underline{u} \in W$ *is called a* **subsolution** *of (3.126)–(3.128), if there is an* $\underline{\eta} \in L^q(Q)$ *and a* $\underline{\xi} \in L^q(\Gamma)$ *such that the following holds:*

    (i) $\underline{u} \vee K \subseteq K$, $\underline{u}(\cdot, 0) \le 0$ *in* Ω,

    (ii) $\underline{\eta} \in \partial j_1(\cdot, \cdot, \underline{u})$, $\underline{\xi} \in \partial j_2(\cdot, \cdot, \gamma \underline{u})$,

    (iii) $\langle \underline{u}_t + A\underline{u} - h, v - \underline{u} \rangle + \displaystyle\int_Q \underline{\eta}\,(v - \underline{u})\,dxdt + \int_\Gamma \underline{\xi}\,(\gamma v - \gamma \underline{u})\,d\Gamma \ge 0$

    for all $v \in \underline{u} \wedge K$.

Definition 3.45. *A function* $\overline{u} \in W$ *is called a* **supersolution** *of (3.126)–(3.128), if there is an* $\overline{\eta} \in L^q(Q)$ *and a* $\overline{\xi} \in L^q(\Gamma)$ *such that the following holds:*

    (i) $\overline{u} \wedge K \subseteq K$, $\overline{u}(\cdot, 0) \ge 0$ *in* Ω,

    (ii) $\overline{\eta} \in \partial j_1(\cdot, \cdot, \overline{u})$, $\overline{\xi} \in \partial j_2(\cdot, \cdot, \gamma \overline{u})$,

    (iii) $\langle \overline{u}_t + A\overline{u} - h, v - \overline{u} \rangle + \displaystyle\int_Q \overline{\eta}\,(v - \overline{u})\,dxdt + \int_\Gamma \overline{\xi}\,(\gamma v - \gamma \overline{u})\,d\Gamma \ge 0$

    for all $v \in \overline{u} \vee K$.

Remark 3.46. Note that we use the notations u_t or $\partial u / \partial t$ if u does not necessarily belong to $D(L)$, as it is the case, e.g., for the sub- and supersolution.

Let us next consider Definition 3.44 for specific K.

Example 3.47. Consider the Example 3.38, i.e., $K = X$, and $s \mapsto j_k(x, t, s)$ are given in terms of the primitives of Carathéodory functions f_k, so the multi-valued parabolic variational inequality (3.126)–(3.128) reduces to (3.130). If $\underline{u} \in W$ is a subsolution according to Definition 3.44, then the first condition in (i) is trivially satisfied, because X possesses lattice structure, and (i) becomes

$$(i') \; \underline{u}(\cdot, 0) \le 0 \; \text{in} \; \Omega.$$

As $\partial j_k(\cdot, \cdot, s)$ is a singleton, (ii) of Definition 3.44 becomes

    (ii') $\underline{\eta}(x, t) = f_1(x, t, \underline{u}(x, t))$, a.e. $(x, t) \in Q$,

           $\underline{\xi}(x, t) = f_2(x, t, \gamma \underline{u}(x, t))$, a.e. $(x, t) \in \Gamma$.

Since $K = X$, any $v \in \underline{u} \wedge X$ has the form $v = \underline{u} \wedge \varphi = \underline{u} - (\underline{u} - \varphi)^+$ with $\varphi \in X$, and condition (iii) becomes

$$(iii')\quad \langle \underline{u}_t + A\underline{u} - h, -(\underline{u} - \varphi)^+ \rangle + \int_Q f_1(\cdot, \cdot, \underline{u})\,(-(\underline{u} - \varphi)^+)\,dxdt$$

$$+ \int_\Gamma f_2(\cdot, \cdot, \gamma\underline{u})\,(-(\gamma\underline{u} - \gamma\varphi)^+)\,d\Gamma \geq 0,$$

for all $\varphi \in X$. Because $\underline{u} \in X$, we have

$$M = \{(\underline{u} - \varphi)^+ : \varphi \in X\} = X \cap L_+^p(Q),$$

where $L_+^p(Q)$ is the positive cone of $L^p(Q)$, and thus (iii') is equivalent to

$$(iii')\quad \langle \underline{u}_t + A\underline{u} - h, \chi \rangle + \int_Q f_1(\cdot, \cdot, \underline{u})\,\chi\,dxdt$$

$$+ \int_\Gamma f_2(\cdot, \cdot, \gamma\underline{u})\,\gamma\chi\,d\Gamma \leq 0$$

for all $\chi \in X \cap L_+^p(Q)$, which shows that the notion of subsolution according to Definition 3.44 of the special case given in Example 3.38 reduces to (i')–(iii'). The latter is nothing but the usual notion of a (weak) subsolution for the initial boundary value problem (3.130) or equivalently of (3.131). Analogous arguments apply for the supersolution.

Example 3.48. Let $K = X_0$, and let j_k as before in Example 3.47, then (3.126)–(3.128) is equivalent to the initial Dirichlet boundary value problem (3.132): Find $u \in D(L) \cap X_0$ such that

$$\langle Lu + Au - h, v \rangle + \int_Q f_1(x, t, u)\,v\,dxdt = 0, \quad \forall\, v \in X_0.$$

We consider the notion of subsolution in this case given via Definition 3.44. For $\underline{u} \in W$, condition (i) means $\underline{u}(\cdot, 0) \leq 0$ in Ω, and $\underline{u} \vee X_0 \subseteq X_0$. This last condition is satisfied if and only if

$$\gamma\underline{u} \leq 0, \quad \text{i.e.,} \quad \underline{u} \leq 0 \ \text{ on } \Gamma, \tag{3.138}$$

and condition (ii) means as above

$$\underline{\eta}(x, t) = f_1(x, t, \underline{u}(x, t)), \quad \text{a.e. } (x, t) \in Q.$$

(Again, the boundary integral vanishes since $\gamma v = 0$ for $v \in X_0$.) Since any $v \in \underline{u} \wedge X_0$ can be represented in the form $v = \underline{u} - (\underline{u} - \varphi)^+$ with $\varphi \in X_0$, from (iii) of Definition 3.44 we obtain

$$\langle \underline{u}_t + A\underline{u} - h, -(\underline{u} - \varphi)^+ \rangle + \int_Q f_1(\cdot, \cdot, \underline{u})\,(-(\underline{u} - \varphi)^+)\,dxdt \geq 0, \ \forall\, \varphi \in X_0.$$

$$\tag{3.139}$$

Set $\chi = (\underline{u} - \varphi)^+$, then (3.139) results in

$$\langle \underline{u}_t + A\underline{u} - h, \chi \rangle + \int_Q f_1(\cdot, \cdot, \underline{u}) \, \chi \, dx dt \leq 0, \ \forall \, \chi \in M_0, \qquad (3.140)$$

where $M_0 := \{\chi \in X_0 : \chi = (\underline{u} - \varphi)^+, \ \varphi \in X_0\} \subseteq X_0 \cap L_+^p(\Omega)$. In [62, Lemma 5.36] it has been shown that the set M_0 is a dense subset of $X_0 \cap L_+^p(\Omega)$, which shows that (3.140), (3.138) together with $\underline{u}(\cdot, 0) \leq 0$ in Ω, is nothing but the weak formulation for the subsolution of the initial Dirichlet problem (3.132) (resp. (3.133)). Similarly, $\overline{u} \in W$ given by Definition 3.45 is shown to be a supersolution of the initial Dirichlet problem (3.132) (resp. (3.133)).

Also in case of other parabolic initial boundary value problems one can check that the notion of sub- and supersolution defined via Definition 3.44 and Definition 3.45 provides an appropriate general framework. As for the hypotheses on j_k, in what follows we assume the existence of an ordered pair $(\underline{u}, \overline{u})$ of sub-supersolutions of the multi-valued parabolic variational inequality (3.126)–(3.128) satisfying $\underline{u} \leq \overline{u}$. With respect to this ordered pair we impose the following hypotheses on the nonlinearities j_k, $k = 1, 2$.

(P-j1) $j_1 : Q \times \mathbb{R} \to \mathbb{R}$ satisfies
   (i) $(x, t) \mapsto j_1(x, t, s)$ is measurable in Q for all $s \in \mathbb{R}$, and $s \mapsto j_1(x, t, s)$ is locally Lipschitz continuous in $\mathbb{R}$ for a.e. $(x, t) \in Q$.
   (ii) There exists a function $k_Q \in L_+^q(Q)$ such that for a.e. $(x, t) \in Q$ and for all $s \in [\underline{u}(x, t), \overline{u}(x, t)]$ the growth condition

$$|\eta| \leq k_Q(x, t), \quad \forall \, \eta \in \partial j_1(x, t, s)$$

   is fulfilled.
(P-j2) $j_2 : \Gamma \times \mathbb{R} \to \mathbb{R}$ satisfies
   (i) $(x, t) \mapsto j_2(x, t, s)$ is measurable in Γ for all $s \in \mathbb{R}$, and $s \mapsto j_2(x, t, s)$ is locally Lipschitz continuous in $\mathbb{R}$ for a.e. $(x, t) \in \Gamma$.
   (ii) There exists a function $k_\Gamma \in L_+^q(\Gamma)$ such that for a.e. $(x, t) \in \Gamma$ and for all $s \in [\gamma\underline{u}(x, t), \gamma\overline{u}(x, t)]$ the growth condition

$$|\xi| \leq k_\Gamma(x, t), \quad \forall \, \xi \in \partial j_2(x, t, s)$$

   is fulfilled.

Remark 3.49. A similar remark as for (E-j1) and (E-j2) can be made here. We note that by the growth condition (ii) of (P-j1) and (P-j2) only a local L^q-boundedness condition on Clarke's generalized gradient ∂j_k is required, which is trivially satisfied, in particular, if we assume the following natural growth condition on ∂j_k: There exist $c > 0$, $k_Q \in L_+^q(Q)$ and $k_\Gamma \in L_+^q(\Gamma)$ such that

$$|\eta| \leq k_Q(x, t) + c|s|^{p-1}, \quad \forall \, \eta \in \partial j_1(x, t, s),$$

for a.a. $(x, t) \in Q$ and for all $s \in \mathbb{R}$, and

$$|\xi| \leq k_\Gamma(x, t) + c|s|^{p-1}, \quad \forall \, \xi \in \partial j_2(x, t, s),$$

for a.e. $(x, t) \in \Gamma$ and for all $s \in \mathbb{R}$.

3.3.2 Multi-Valued Parabolic Equation

We consider in this subsection the case when the closed, convex subset K of X is a closed subspace of X. As a prototype we assume $K = X_0$. As seen in Sect. 3.3.1, in this case the multi-valued parabolic variational inequality (3.126)–(3.128) reduces to the following multi-valued parabolic variational equation: Find $u \in X_0 \cap D(L)$ such that there is an $\eta \in L^q(Q)$ satisfying

$$(i) \; \eta(x,t) \in \partial j_1(x,t,u(x,t)) \;\text{ for a.a. } (x,t) \in Q, \tag{3.141}$$

$$(ii) \; \langle Lu + Au - h, \varphi \rangle + \int_Q \eta \varphi \, dx dt = 0, \quad \forall \, \varphi \in X_0, \tag{3.142}$$

which is equivalent to the multi-valued initial Dirichlet boundary value problem (Example 3.40):

$$u_t + Au + \partial j_1(\cdot, \cdot, u) \ni h \;\text{ in } Q,$$
$$u(\cdot, 0) = 0 \;\text{ in } \Omega, \quad u = 0 \;\text{ on } \Gamma.$$

Remark 3.50. Let $D_0(L)$ denote the domain of the operator $L = \partial/\partial t$ given by

$$D_0(L) = \{u \in W_0 : u(\cdot, 0) = 0\}.$$

By Hahn–Banach's theorem we have $D_0(L) \subseteq X_0 \cap D(L)$. Therefore, any solution $u \in D_0(L)$ of (3.141)–(3.142) belongs to $X_0 \cap D(L)$. Due to $X^* \hookrightarrow X_0^*$, any solution $u \in X_0 \cap D(L)$ of (3.141)–(3.142) may be identified with a solution $u \in D_0(L)$.

Similar to Example 3.48 we deduce from Definition 3.44 and Definition 3.45 the following notion of sub- and supersolution for (3.141)–(3.142).

Definition 3.51. *A function $\underline{u} \in W$ is called a* **subsolution** *of (3.141)–(3.142) if there is an $\underline{\eta} \in L^q(Q)$ satisfying $\underline{\eta}(x,t) \in \partial j_1(x,t,\underline{u}(x,t))$ for a.a. $(x,t) \in Q$ such that $\underline{u}(x,0) \leq 0$ for a.a. $x \in \Omega$, $\underline{u}|_\Gamma \leq 0$, and*

$$\langle \underline{u}_t + A\underline{u}, \varphi \rangle + \int_Q \underline{\eta} \varphi \, dx \, dt \leq \langle h, \varphi \rangle, , \quad \forall \, \varphi \in X_0 \cap L_+^p(Q). \tag{3.143}$$

Definition 3.52. *A function $\overline{u} \in W$ is called a* **supersolution** *of (3.141)–(3.142) if there is an $\overline{\eta} \in L^q(Q)$ satisfying $\overline{\eta}(x,t) \in \partial j_1(x,t,\overline{u}(x,t))$ for a.a. $(x,t) \in Q$ such that $\overline{u}(x,0) \geq 0$ for a.a. $x \in \Omega$, $\overline{u}|_\Gamma \geq 0$, and*

$$\langle \overline{u}_t + A\overline{u}, \varphi \rangle + \int_Q \overline{\eta} \varphi \, dx \, dt \geq \langle h, \varphi \rangle, \quad \forall \, \varphi \in X_0 \cap L_+^p(Q). \tag{3.144}$$

Our goal is to establish the method of sub- and supersolution for (3.141)–(3.142), i.e., we are going to show the existence of solutions within the ordered interval $[\underline{u}, \overline{u}]$ of the given sub- and supersolution. Moreover, the existence of extremal solutions within $[\underline{u}, \overline{u}]$ will be proved as well. In view of Remark 3.50 we may use the following equivalent notion of solution for (3.141)–(3.142).

Definition 3.53. *A function* $u \in D_0(L) \subset W_0$ *is called a* **solution** *of* *(3.141)–(3.142) if there is an* $\eta \in L^q(Q)$ *with* $\eta(x,t) \in \partial j_1(x,t,u(x,t))$ *for a.a.* $(x,t) \in Q$ *such that*

$$\langle Lu + Au, \varphi \rangle + \int_Q \eta \, \varphi \, dx \, dt = \langle h, \varphi \rangle, \quad \forall \, \varphi \in X_0. \tag{3.145}$$

Preliminaries

We briefly recall a general surjectivity result for multi-valued operators in a real reflexive Banach space Y, which will be used later. To this end we introduce the notion of multi-valued pseudomonotone operators with respect to the graph norm topology of the domain $D(\hat{L})$ (w.r.t. $D(\hat{L})$ for short) of some linear, closed, densely defined, and maximal monotone operator $\hat{L}$: $D(\hat{L}) \subseteq Y \to Y^*$.

Definition 3.54. *Let* $\hat{L} : D(\hat{L}) \subseteq Y \to Y^*$ *be a linear, closed, densely defined, and maximal monotone operator. The operator* $\mathcal{A} : Y \to 2^{Y^*}$ *is called* **pseudomonotone w.r.t.** $\mathbf{D}(\hat{\mathbf{L}})$ *if the following conditions are satisfied:*

(i)   *The set* $\mathcal{A}(u)$ *is nonempty, bounded, closed, and convex for all* $u \in Y$.
(ii) $\mathcal{A}$ *is upper semicontinuous from each finite dimensional subspace of* Y *to* Y^* *equipped with the weak topology.*
(iii) *If* $(u_n) \subset D(\hat{L})$ *with* $u_n \rightharpoonup u$ *in* Y, $\hat{L}u_n \rightharpoonup \hat{L}u$ *in* Y^*, $u_n^* \in \mathcal{A}(u_n)$ *with* $u_n^* \rightharpoonup u^*$ *in* Y^* *and* $\limsup \langle u_n^*, u_n - u \rangle \leq 0$, *then* $u^* \in \mathcal{A}(u)$ *and* $\langle u_n^*, u_n \rangle \to \langle u^*, u \rangle$.

Definition 3.55. *The operator* $\mathcal{A} : Y \to 2^{Y^*}$ *is called* **coercive** *iff either the domain of* $\mathcal{A}$ *denoted by* $D(\mathcal{A})$ *is bounded or* $D(\mathcal{A})$ *is unbounded and*

$$\frac{\inf\{\langle v^*, v \rangle : v^* \in \mathcal{A}(v)\}}{\|v\|_Y} \to +\infty \quad as \quad \|v\|_Y \to \infty, \ v \in D(\mathcal{A}).$$

The following surjectivity result, which will be used later, can be found, e.g., in [82, Theorem 1.3.73, p. 62].

Theorem 3.56. *Let* Y *be a real reflexive, strictly convex Banach space with dual space* Y^*, *and let* $\hat{L} : D(\hat{L}) \subseteq Y \to Y^*$ *be a closed, densely defined, and maximal monotone operator. If the multi-valued operator* $\mathcal{A} : Y \to 2^{Y^*}$ *is pseudomonotone w.r.t.* $D(\hat{L})$, *bounded and coercive, then* $\hat{L} + \mathcal{A}$ *is surjective, i.e.,* range $(\hat{L} + \mathcal{A}) = Y^*$.

The next abstract theorem represents a version of the well-known result concerning the sum of pseudomonotone operators.

Theorem 3.57. *Let* Y *be a real reflexive Banach space with dual space* Y^*, *let* $\hat{L} : D(\hat{L}) \subseteq Y \to Y^*$ *be a linear, closed, densely defined, maximal monotone operator, let* $A_1 : Y \to 2^{Y^*}$ *be a multi-valued operator which is bounded and pseudomonotone w.r.t.* $D(\hat{L})$, *and let* $A_2 : Y \to Y^*$ *be a bounded and pseudomonotone operator. Then the multivalued operator* $A_1 + A_2 : Y \to 2^{Y^*}$ *is pseudomonotone w.r.t.* $D(\hat{L})$.

Proof: It suffices to check condition (iii) of Definition 3.54 for $A = A_1 + A_2$. To this end let sequences $(u_n) \subset D(\hat{L})$ and $(u_n^*) \subset Y^*$ with $u_n \rightharpoonup u$ in Y, $\hat{L}u_n \rightharpoonup \hat{L}u$ in Y^*, $u_n^* \in A_1(u_n)$ with $u_n^* + A_2(u_n) \rightharpoonup u^*$ in Y^*, and

$$\limsup_n \langle u_n^* + A_2(u_n), u_n - u \rangle \leq 0. \tag{3.146}$$

We claim that

$$\limsup_n \langle u_n^*, u_n - u \rangle \leq 0. \tag{3.147}$$

Arguing by contradiction, we find subsequences $(u_{n_k}) \subset D(\hat{L})$ and $(u_{n_k}^*) \subset Y^*$ such that

$$\limsup_n \langle u_n^*, u_n - u \rangle = \lim_k \langle u_{n_k}^*, u_{n_k} - u \rangle > 0.$$

We derive from (3.146) that

$$\limsup_k \langle A_2(u_{n_k}), u_{n_k} - u \rangle \leq - \lim_k \langle u_{n_k}^*, u_{n_k} - u \rangle < 0. \tag{3.148}$$

The pseudomonotonicity of A_2 and (3.148) guarantee that for every $w \in Y$ we have

$$\langle A_2(u), u - w \rangle \leq \liminf_k \langle A_2(u_{n_k}), u_{n_k} - w \rangle.$$

Setting $w = u$ and using (3.148) we reach a contradiction, which establishes (3.147).

Now we claim that

$$\limsup_n \langle A_2(u_n), u_n - u \rangle \leq 0. \tag{3.149}$$

On the contrary, we would find a subsequence (u_{n_k}) such that

$$\limsup_n \langle A_2(u_n), u_n - u \rangle = \lim_k \langle A_2(u_{n_k}), u_{n_k} - u \rangle > 0.$$

Taking into account the boundedness of A_1, there exists a subsequence $(u_{n_k}^*) \subset Y^*$ satisfying $u_{n_k}^* \rightharpoonup \xi$ in Y^*, for some $\xi \in Y^*$. By (3.146) it follows that

$$\limsup_n \langle u_{n_k}^*, u_{n_k} - u \rangle \leq - \lim_k \langle A_2(u_{n_k}), u_{n_k} - u \rangle < 0.$$

Since A_1 is pseudomonotone w.r.t. $D(\hat{L})$, it turns out that $\xi \in A_1(u)$ and $\langle u_{n_k}^*, u_{n_k} \rangle \to \langle \xi, u \rangle$, which results in

$$\limsup_n \langle u_{n_k}^* + A_2 u_{n_k}, u_{n_k} - u \rangle = \lim_k \langle A_2(u_{n_k}), u_{n_k} - u \rangle > 0.$$

This contradicts (3.146), so (3.149) holds true.

The boudedness of A_2 ensures that there is a subsequence (u_{n_k}) with $A_2(u_{n_k}) \rightharpoonup \eta$ in Y^*, for some $\eta \in Y^*$. The property of A_2 to be pseudomonotone and (3.149) imply that $\eta = A_2(u)$, thus $A_2(u_n) \rightharpoonup A_2(u)$ in Y^*, which reads as $u_n^* \rightharpoonup u^* - A_2(u)$ in Y^*, and

$$\langle A_2(u_n), u_n \rangle \rightarrow \langle A_2(u), u \rangle. \tag{3.150}$$

In addition, by (3.147) and the pseudomonotonicity of A_1 w.r.t. $D(\hat{L})$, it is seen that $u^* - A_2(u) \in A_1(u)$ and $\langle u_n^*, u_n \rangle \rightarrow \langle u^* - A_2(u), u \rangle$. Due to (3.150), the proof is complete. $\qquad\square$

The Sub-Supersolution Method

Let $\underline{u} \leq \overline{u}$ be an ordered pair of sub-supersolution of (3.141)–(3.142), i.e., $(\underline{u}, \underline{\eta}) \in W \times L^q(Q)$ and $(\overline{u}, \overline{\eta}) \in W \times L^q(Q)$ satisfy the conditions in Definition 3.51 and Definition 3.52, respectively, with $\underline{\eta}(x,t) \in \partial j_1(x, t, \underline{u}(x,t))$ and $\overline{\eta}(x,t) \in \partial j_1(x, t, \overline{u}(x,t))$. We define the following modification $\tilde{j}_1 : Q \times \mathbb{R} \rightarrow \mathbb{R}$ of the given j_1:

$$\tilde{j}_1(x,t,s) = \begin{cases} j_1(x,t,\underline{u}(x,t)) + \underline{\eta}(x,t)(s - \underline{u}(x,t)) & \text{if } s < \underline{u}(x,t), \\ j_1(x,t,s) & \text{if } \underline{u}(x,t) \leq s \leq \overline{u}(x,t), \\ j_1(x,t,\overline{u}(x,t)) + \overline{\eta}(x,t)(s - \overline{u}(x,t)) & \text{if } s > \overline{u}(x,t). \end{cases} \tag{3.151}$$

Lemma 3.58. *Let j_1 satisfy (P-j1). Then the modified function $\tilde{j}_1 : Q \times \mathbb{R} \rightarrow \mathbb{R}$ has the following properties:*

(i) $(x,t) \mapsto \tilde{j}_1(x,t,s)$ *is measurable in Q for all $s \in \mathbb{R}$, and $s \mapsto \tilde{j}_1(x,t,s)$ is Lipschitz continuous in $\mathbb{R}$ for a.a. $(x,t) \in Q$.*

(ii) Clarke's generalized gradient $\partial \tilde{j}_1(x,t,s)$ of $s \mapsto \tilde{j}_1(x,t,s)$ satisfies the estimate

$$|\eta| \leq k_Q(x,t), \quad \forall\, \eta \in \partial \tilde{j}_1(x,t,s)$$

for a.a. $(x,t) \in Q$ and for all $s \in \mathbb{R}$.

(iii) Clarke's generalized gradient of $s \mapsto \partial \tilde{j}_1(x,t,s)$ is given by

$$\partial \tilde{j}_1(x,t,s) = \begin{cases} \underline{\eta}(x,t) & \text{if } s < \underline{u}(x,t), \\ \partial \tilde{j}_1(x,t,\underline{u}(x,t)) & \text{if } s = \underline{u}(x,t), \\ \partial j_1(x,t,s) & \text{if } \underline{u}(x,t) < s < \overline{u}(x,t), \\ \partial \tilde{j}_1(x,t,\overline{u}(x,t)) & \text{if } s = \overline{u}(x,t), \\ \overline{\eta}(x,t) & \text{if } s > \overline{u}(x,t), \end{cases} \tag{3.152}$$

and the inclusions $\partial \tilde{j}_1(x,t,\underline{u}(x,t)) \subseteq \partial j_1(x,t,\underline{u}(x,t))$ as well as $\partial \tilde{j}_1(x,t,\overline{u}(x,t)) \subseteq \partial j(x,t,\overline{u}(x,t))$ hold true.

Proof: The proof follows immediately from the definition (3.152) of $\tilde{j}$, and using the assumptions (P-j1) on j as well as from the fact that Clarke's generalized gradient $\partial j(x,t,s)$ is a convex set. $\qquad\square$

By means of $\tilde{j}_1$ we introduce the integral functional $\tilde{J}_1$ given by

$$\tilde{J}_1(v) = \int_Q \tilde{j}_1(x,t,v(x,t))\,dx dt, \quad v \in L^p(Q).$$

Due to hypotheses (P-j1), and applying Lebourg's mean value theorem (see [62, Theorem 2.177]), the functional $\tilde{J}_1 : L^p(Q) \to \mathbb{R}$ is well defined and Lipschitz continuous, so that Clarke's generalized gradient $\partial \tilde{J}_1 : L^p(Q) \to 2^{L^q(Q)}$ is well defined, too. Moreover, Aubin–Clarke theorem (cf. [80, p. 83]) provides the following characterization of the generalized gradient. For $u \in L^p(Q)$ we have

$$\tilde{\eta} \in \partial \tilde{J}_1(u) \Longrightarrow \tilde{\eta} \in L^q(Q) \text{ with } \tilde{\eta}(x,t) \in \partial \tilde{j}_1(x,t,u(x,t)) \text{ for a.a. } (x,t) \in Q. \tag{3.153}$$

By means of Clarke's generalized gradient $\partial \tilde{J}_1$ we further introduce the following multi-valued operator:

$$\Psi(u) := (i^* \circ \partial \tilde{J}_1 \circ i)(u), \quad u \in X_0,$$

where $i^* : L^q(Q) \hookrightarrow X_0^*$ is the adjoint operator of the embedding $i : X_0 \hookrightarrow L^p(Q)$. As the operator Ψ is easily seen to be bounded due to hypothesis (P-j1), we obtain in conjunction with [62, Lemma 6.22] the following result.

Lemma 3.59. *Let hypothesis (P-j1) be fulfilled. Then the operator* $\Psi : X_0 \to 2^{X_0^*}$ *is bounded and pseudomonotone w.r.t.* $D_0(L)$.

The sub-supersolution method for (3.141)–(3.142) is established by the following theorem.

Theorem 3.60. *Let hypotheses (AP1)–(AP3) and (P-j1) be satisfied, and let* $\underline{u}$ *and* $\overline{u}$ *be the given sub- and supersolution, respectively, satisfying* $\underline{u} \le \overline{u}$. *Then there exist solutions of (3.141)–(3.142) within the ordered interval* $[\underline{u}, \overline{u}]$.

Proof: The following auxiliary, multi-valued, initial-boundary value problem plays a crucial role in the proof of the theorem:

$$u_t + A_T u + \hat{B}(u) + \partial \tilde{j}_1(\cdot, \cdot, u) \ni h \quad \text{in } Q, \tag{3.154}$$

$$u(\cdot, 0) = 0 \text{ in } \Omega, \quad u = 0 \quad \text{on } \Gamma, \tag{3.155}$$

where $\tilde{j}_1$ is defined in (3.151), and A_T is given by

$$\langle A_T u, \varphi \rangle := \int_Q \sum_{i=1}^N a_i(x,t,Tu,\nabla u) \frac{\partial \varphi}{\partial x_i} \, dx \, dt, \quad \forall \varphi \in X_0,$$

with T being the following truncation operator

$$(Tu)(x,t) = \begin{cases} \overline{u}(x,t) & \text{if } u(x,t) > \overline{u}(x,t) \\ u(x,t) & \text{if } \underline{u}(x,t) \le u(x,t) \le \overline{u}(x,t) \\ \underline{u}(x,t) & \text{if } u(x,t) < \underline{u}(x,t). \end{cases}$$

Let B be the Nemytskij operator associated with the cut-off function $b : Q \times \mathbb{R} \to \mathbb{R}$ given by

$$b(x,t,s) = \begin{cases} (s - \overline{u}(x,t))^{p-1} & \text{if } s > \overline{u}(x,t) \\ 0 & \text{if } \underline{u}(x,t) \leq s \leq \overline{u}(x,t) \\ -(\underline{u}(x,t) - s)^{p-1} & \text{if } s < \underline{u}(x,t). \end{cases}$$

Then $b : Q \times \mathbb{R} \to \mathbb{R}$ is a Carathéodory function satisfying the growth condition

$$|b(x,t,s)| \leq k_2(x,t) + c_1|s|^{p-1} \tag{3.156}$$

for a.e. $(x,t) \in Q$ and for all $s \in \mathbb{R}$, where $c_1 > 0$ is a constant and $k_2 \in L^q(Q)$. Moreover, one has the following estimate

$$\int_Q b(x,t,u(x,t))\, u(x,t)\, dx\, dt \geq c_2\|u\|^p_{L^p(Q)} - c_3, \quad \forall u \in L^p(Q), \tag{3.157}$$

for some constants $c_2 > 0$ and $c_3 > 0$. Thus in view of (3.156), the Nemytski operator $B : L^p(Q) \to L^q(Q)$ defined by $Bu(x,t) = b(x,t,u(x,t))$ is continuous and bounded. Therefore, the operator $\hat{B} = i^* \circ B \circ i : X_0 \to X_0^*$ is continuous, bounded, and pseudomonotone w.r.t. $D_0(L)$ due to the compact embedding $W_0 \hookrightarrow L^p(Q)$. The weak formulation of problem (3.154)–(3.155) is as follows: Find $u \in D_0(L)$ such that

$$\langle Lu + A_T u + \hat{B}(u), \varphi \rangle + \int_Q \tilde{\eta}\, \varphi\, dx\, dt = \langle h, \varphi \rangle, \ \forall\, \varphi \in X_0, \tag{3.158}$$

where $\tilde{\eta} \in L^q(Q)$ and $\tilde{\eta}(x,t) \in \partial \tilde{j}_1(x,t,u(x,t))$ a.e. in Q. The proof of the theorem is accomplished in two steps. First, we are going to show the existence of solutions of (3.158). Secondly, we are going to verify that any solution of (3.158) belongs to the interval $[\underline{u}, \overline{u}]$, which completes the proof, because then $A_T u = Au$, $\hat{B}(u) = 0$, and $\tilde{\eta}(x,t) \in \partial \tilde{j}_1(x,t,u(x,t)) \subseteq \partial j_1(x,t,u(x,t))$. The latter is due to Lemma 3.58 (iii).

Step 1: Existence for (3.158)

As for the existence of solutions of (3.158) let us consider the multi-valued mapping

$$\mathcal{A}(u) := A_T u + \hat{B}(u) + \Psi(u) : X_0 \to 2^{X_0^*}. \tag{3.159}$$

Due to (A1)–(A3) and since $T : X_0 \to X_0$ is continuous and bounded, the operator $A_T : X_0 \to X_0^*$ is bounded, continuous, and pseudomonotone w.r.t. $D_0(L)$, cf. Theorem 9.109. By Lemma 3.59 the multi-valued operator Ψ is pseudomonotone w.r.t. $D_0(L)$, and in view of Lemma 3.58 even uniformly bounded. As $\hat{B} : X_0 \to X_0^*$ is continuous, bounded, and pseudomonotone w.r.t. $D_0(L)$, we see that $\mathcal{A} : X_0 \to 2^{X_0^*}$ is bounded and pseudomonotone w.r.t. $D_0(L)$ in the sense of Definition 3.54, and thanks to (A3) and (3.157) it is also coercive. Thus, by Theorem 3.56 the operator $L + \mathcal{A} : X_0 \to 2^{X_0^*}$ is surjective, which implies the existence of $u \in D_0(L)$ such that

$$h \in Lu + \mathcal{A}(u) \quad \text{in } X_0^*.$$

The latter means that there is an $\tilde{\eta} \in \partial \tilde{J}_1(i(u)) = \partial \tilde{J}_1(u)$ such that

$$Lu + A_T u + \hat{B}(u) + i^* \tilde{\eta} = h \quad \text{in } X_0^*. \tag{3.160}$$

From (3.153) and (3.160) we see that the solution u of (3.160) is in fact a solution of the auxiliary problem (3.158). To complete the proof it remains to verify that any solution of (3.158) belongs to the interval $[\underline{u}, \overline{u}]$.

Step 2: Comparison

Let us prove that $u \leq \overline{u}$ holds, where $\overline{u}$ is the given supersolution. Subtracting (3.144) from (3.160) we obtain the following inequality:

$$\langle u_t - \overline{u}_t + A_T u - A\overline{u} + \hat{B}(u) + i^* \tilde{\eta} - i^* \overline{\eta}, \varphi \rangle \leq 0, \ \forall \, \varphi \in X_0 \cap L^p_+(Q), \tag{3.161}$$

where $\tilde{\eta}(x, t) \in \partial \tilde{j}_1(x, t, u(x, t))$ and $\overline{\eta}(x, t) \in \partial j_1(x, t, \overline{u}(x, t))$ for a.a. $(x, t) \in Q$. With the special test function $\varphi = (u - \overline{u})^+ \in X_0 \cap L^p_+(Q)$ in (3.161) we get for the individual terms on the left-hand side of inequality (3.161) the following estimates:

$$\langle u_t - \overline{u}_t, (u - \overline{u})^+ \rangle \geq 0, \quad \text{(cf. 9.78)} \tag{3.162}$$

$$\langle A_T u - A\overline{u}, (u - \overline{u})^+ \rangle$$

$$= \int_{\{u > \overline{u}\}} \sum_{i=1}^{N} \left(a_i(x, t, \overline{u}, \nabla u) - a_i(x, t, \overline{u}, \nabla \overline{u}) \right) \frac{\partial(u - \overline{u})}{\partial x_i} \, dx \, dt \geq 0, \tag{3.163}$$

where $\{u > \overline{u}\} = \{(x, t) \in Q : u(x, t) > \overline{u}(x, t)\}$, and

$$\langle i^* \tilde{\eta} - i^* \overline{\eta}, (u - \overline{u})^+ \rangle = \int_{\{u > \overline{u}\}} (\tilde{\eta} - \overline{\eta})(u - \overline{u}) \, dx \, dt = 0, \tag{3.164}$$

because for $(x, t) \in \{u > \overline{u}\}$ we get by Lemma 3.58 (iii) that $\tilde{\eta}(x, t) = \overline{\eta}(x, t)$. In view of (3.162)–(3.164) we obtain from (3.161) the inequality

$$\langle \hat{B}(u), (u - \overline{u})^+ \rangle = \int_Q b(x, t, u)(u - \overline{u})^+ \, dx \, dt \leq 0,$$

which by using the definition of b implies

$$0 \leq \int_Q [(u - \overline{u})^+]^p \, dx \, dt \leq 0, \tag{3.165}$$

and thus it follows $(u - \overline{u})^+ = 0$, i.e., $u \leq \overline{u}$. The inequality $\underline{u} \leq u$ can be shown analogously, which completes the proof of the theorem. $\quad \square$

Directedness of the Solution Set

Let S denote the set of all solutions of (3.141)–(3.142) that belong to the ordered interval $[\underline{u}, \overline{u}]$ formed by the given sub- and supersolution. In view of Theorem 3.60, we have that $S \neq \emptyset$.

Theorem 3.61. *Let hypotheses (AP1)–(AP4) and (P-j1) be satisfied. Then the solution set $\mathcal{S}$ equipped with the natural partial ordering of functions introduced by the order cone $L^p_+(Q)$ is a directed set.*

Proof: We are going to show that $\mathcal{S}$ is upward directed only, since the proof for $\mathcal{S}$ being downward directed can be done in a similar way. To this end let $u_1, u_2 \in \mathcal{S}$. Our goal is to prove the existence of an element $u \in \mathcal{S}$ such that $u \geq u_k$, $k = 1, 2$. The proof will be done in 4 steps and crucially relies on an appropriately designed auxiliary problem.

Step 1: Auxiliary Problem

Let $u_k \in \mathcal{S}$, $k = 1, 2$, i.e., $u_k \in D_0(L) \subset W_0$, and there are $\eta_k \in L^q(Q)$ with $\eta_k(x,t) \in \partial j_1(x,t,u_k(x,t))$ for a.e. $(x,t) \in Q$ such that

$$\langle Lu_k + Au_k, \varphi \rangle + \int_Q \eta_k \varphi \, dx \, dt = \langle h, \varphi \rangle, \quad \forall\, \varphi \in X_0. \tag{3.166}$$

Let us define $u_0 := \max\{u_1, u_2\}$, and η_0 by

$$\eta_0(x,t) = \begin{cases} \eta_1(x,t) & \text{if } (x,t) \in \{u_1 \geq u_2\}, \\ \eta_2(x,t) & \text{if } (x,t) \in \{u_2 > u_1\}, \end{cases}$$

where $\{u_j \geq (>) u_k\}$ stands for $\{(x,t) \in Q : u_j(x,t) \geq (>) u_k(x,t)\}$. By the definition of η_0 we readily see that $\eta_0 \in L^q(Q)$, and

$$\eta_0(x,t) \in \partial j(x,t,u_0(x,t)) \text{ for a.e. } (x,t) \in Q. \tag{3.167}$$

By means of η_0, u_0, and $\overline{\eta}$, $\overline{u}$ of Definition 3.52 we introduce the following modification $\tilde{j}_1 : Q \times \mathbb{R} \to \mathbb{R}$ of the given j:

$$\tilde{j}_1(x,t,s) = \begin{cases} j_1(x,t,u_0(x,t)) + \eta_0(x,t)(s - u_0(x,t)) & \text{if } s < u_0(x,t), \\ j_1(x,t,s) & \text{if } u_0(x,t) \leq s \leq \overline{u}(x,t), \\ j_1(x,t,\overline{u}(x,t)) + \overline{\eta}(x,t)(s - \overline{u}(x,t)) & \text{if } s > \overline{u}(x,t). \end{cases} \tag{3.168}$$

In view of hypothesis (P-j1) the modified function $\tilde{j}_1 : Q \times \mathbb{R} \to \mathbb{R}$ has the following properties:

(a) $(x,t) \mapsto \tilde{j}_1(x,t,s)$ is measurable in Q for all $s \in \mathbb{R}$, and $s \mapsto \tilde{j}_1(x,t,s)$ is Lipschitz continuous in $\mathbb{R}$ for a.a. $(x,t) \in Q$.
(b) Clarke's generalized gradient $s \mapsto \partial \tilde{j}_1(x,t,s)$ is uniformly bounded, i.e.,

$$|\eta| \leq k_Q(x,t), \quad \forall\, \eta \in \partial \tilde{j}_1(x,t,s) \tag{3.169}$$

for a.a. $(x,t) \in Q$ and for all $s \in \mathbb{R}$.
(c) Clarke's generalized gradient of $s \mapsto \tilde{j}_1(x,t,s)$ is given by

$$\partial \tilde{j}_1(x,t,s) = \begin{cases} \eta_0(x,t) & \text{if } s < u_0(x,t), \\ \partial \tilde{j}_1(x,t,u_0(x,t)) & \text{if } s = u_0(x,t), \\ \partial j_1(x,t,s) & \text{if } u_0(x,t) < s < \overline{u}(x,t), \\ \partial \tilde{j}_1(x,t,\overline{u}(x,t)) & \text{if } s = \overline{u}(x,t), \\ \overline{\eta}(x,t) & \text{if } s > \overline{u}(x,t). \end{cases} \tag{3.170}$$

Moreover, the inclusions

$$\partial \tilde{j}_1(x,t,u_0(x,t)) \subseteq \partial j_1(x,t,u_0(x,t)), \quad \partial \tilde{j}_1(x,t,\overline{u}(x,t)) \subseteq \partial j_1(x,t,\overline{u}(x,t))$$

(3.171)

hold true due to the fact that Clarke's generalized gradient $\partial j(x,t,s)$ is a convex set.

Furthermore, we introduce the functions $g_i : Q \times \mathbb{R} \to \mathbb{R}$ related with η_i and u_i, for $i = 0, 1, 2$, and defined by:

$$g_0(x,t,s) = \begin{cases} \eta_0(x,t) & \text{if } s \leq u_0(x,t), \\ \eta_0(x,t) + \frac{\overline{\eta}(x,t) - \eta_0(x,t)}{\overline{u}(x,t) - u_0(x,t)}(s - u_0(x,t)) & \text{if } u_0(x,t) < s < \overline{u}(x,t), \\ \overline{\eta}(x,t) & \text{if } s \geq \overline{u}(x,t), \end{cases}$$

(3.172)

and for $k = 1, 2$

$$g_k(x,t,s) = \begin{cases} \eta_k(x,t) & \text{if } s \leq u_k(x,t), \\ \eta_k(x,t) + \frac{\eta_0(x,t) - \eta_k(x,t)}{u_0(x,t) - u_k(x,t)}(s - u_k(x,t)) & \text{if } u_k(x,t) < s < u_0(x,t), \\ g_0(x,t,s) & \text{if } s \geq u_0(x,t). \end{cases}$$

(3.173)

Apparently, the functions $(x,t,s) \mapsto g_i(x,t,s)$ are Carathéodory functions that are piecewise linear with respect to s. Finally, define the cut-off function $\hat{b}$ related to the pair u_0 and $\overline{u}$ by:

$$\hat{b}(x,t,s) = \begin{cases} (s - \overline{u}(x,t))^{p-1} & \text{if } s > \overline{u}(x,t) \\ 0 & \text{if } u_0(x,t) \leq s \leq \overline{u}(x,t) \\ -(u_0(x,t) - s)^{p-1} & \text{if } s < u_0(x,t). \end{cases}$$

(3.174)

Apparently $\hat{b} : Q \times \mathbb{R} \to \mathbb{R}$ is a Carathéodory function satisfying the growth condition

$$|\hat{b}(x,t,s)| \leq k_3(x,t) + c_1|s|^{p-1}$$

(3.175)

for a.a. $(x,t) \in Q$ and for all $s \in \mathbb{R}$, where $c_1 > 0$ is a constant and $k_3 \in L^q(Q)$. Moreover, one has the following estimate

$$\int_Q \hat{b}(x,t,u(x,t))\,u(x,t)\,dx\,dt \geq c_2\|u\|_{L^p(Q)}^p - c_3, \quad \forall u \in L^p(Q),$$

(3.176)

for some constants $c_2 > 0$ and $c_3 > 0$. By (3.175), the associated Nemytskij operator $\hat{B} : L^p(Q) \to L^q(Q)$ defined by $\hat{B}u(x,t) = \hat{b}(x,t,u(x,t))$ is continuous and bounded. If $i^* : L^q(Q) \hookrightarrow X_0^*$ is the adjoint operator of the embedding $i : X_0 \hookrightarrow L^p(Q)$, then the operator $\hat{\mathcal{B}} = i^* \circ \hat{B} \circ i : X_0 \to X_0^*$ is continuous, bounded, and pseudomonotone w.r.t. $D_0(L)$ due to the compact embedding $W_0 \hookrightarrow L^p(Q)$. Now we are ready to introduce the following auxiliary problem, which is crucial for the rest of the proof: Let $\tilde{j}_1$, g_i, and $\hat{\mathcal{B}}$ be given by (3.168), (3.172)–(3.174), we consider

$$u_t + A_T u + \hat{B}(u) + \partial \tilde{j}(x, t, u) - \sum_{l=1}^{2} |g_l(x, t, u) - g_0(x, t, u)| \ni h \quad \text{in } Q,$$

(3.177)

$$u(\cdot, 0) = 0 \quad \text{in } \Omega, \quad u = 0 \quad \text{on } \Gamma,$$

(3.178)

with the operator A_T defined by

$$\langle A_T u, \varphi \rangle := \int_Q \sum_{i=1}^{N} a_i(x, t, Tu, \nabla u) \frac{\partial \varphi}{\partial x_i} \, dx \, dt, \quad \forall \varphi \in X_0,$$

where here T denotes the truncation operator given by

$$(Tu)(x, t) = \begin{cases} \overline{u}(x, t) & \text{if } u(x, t) > \overline{u}(x, t) \\ u(x, t) & \text{if } u(x, t) \leq \overline{u}(x, t), \end{cases}$$

which is known to be continuous from X into X. In the steps to follow we are going to show that the auxiliary Problem (3.177)–(3.178) enjoys the following properties:

(i)   Existence: There exist solutions of Problem (3.177)–(3.178).
(ii) Comparison: Any solution u of (3.177)–(3.178) satisfies $u_k \leq u \leq \overline{u}$, $k = 1, 2$.
(iii) Upward Directedness: Any solution u of (3.177)–(3.178) is a solution of the original problem (3.141)–(3.142), which due to (ii) exceeds u_k, and thus S is upward directed.

Step 2: Existence of Solutions of the Auxiliary Problem

To prove the existence of solutions of (3.177)–(3.178) let

$$g(x, t, u) := \sum_{l=1}^{2} |g_l(x, t, u) - g_0(x, t, u)|.$$

(3.179)

By definitions (3.172)–(3.173), the functions $g_i : Q \times \mathbb{R} \to \mathbb{R}$ are Carathéodory functions that are uniformly $L^q(Q)$-bounded. Therefore, the associated Nemytskij operator $G : L^p(Q) \to L^q(Q)$ is continuous and bounded, and $\mathcal{G} = i^* \circ G \circ i : X_0 \to X_0^*$ is continuous, bounded, and pseudomonotone w.r.t. $D_0(L)$ due to the compact embedding $W_0 \hookrightarrow L^p(Q)$. Thus the weak formulation of problem (3.177)–(3.178) is as follows: Find $u \in D_0(L)$ such that

$$\langle Lu + A_T u + \hat{B}(u) - \mathcal{G}(u), \varphi \rangle + \int_Q \tilde{\eta} \varphi \, dx \, dt = \langle h, \varphi \rangle, \ \forall \, \varphi \in X_0,$$

(3.180)

where $\tilde{\eta} \in L^q(Q)$ and $\tilde{\eta}(x, t) \in \partial \tilde{j}_1(x, t, u(x, t))$ a.e. on Q. By means of $\tilde{j}_1$ we introduce the integral functional $\tilde{J}_1$ given by

$$\tilde{J}_1(v) = \int_Q \tilde{j}_1(x,t,v(x,t))\,dxdt, \quad v \in L^p(Q).$$

Hypotheses (P-j1) together with Lebourg's mean value theorem imply that the functional $\tilde{J}_1 : L^p(Q) \to \mathbb{R}$ is well defined and even Lipschitz continuous, so that Clarke's generalized gradient $\partial\tilde{J}_1 : L^p(Q) \to 2^{L^q(Q)}$ is well defined, too. Moreover, Aubin–Clarke's theorem (cf. [80, p. 83]) provides the following characterization of the generalized gradient. For $u \in L^p(Q)$ we have

$$\tilde{\eta} \in \partial\tilde{J}_1(u) \implies \tilde{\eta} \in L^q(Q) \text{ with } \tilde{\eta}(x,t) \in \partial\tilde{j}_1(x,t,u(x,t)) \text{ for a.a. } (x,t) \in Q. \tag{3.181}$$

By means of Clarke's generalized gradient $\partial\tilde{J}_1$ we further introduce the following multi-valued operator:

$$\tilde{\Psi}(u) := (i^* \circ \partial\tilde{J}_1 \circ i)(u), \quad u \in X_0.$$

As the operator $\tilde{\Psi}$ is bounded due to hypothesis (P-j1), we obtain, in conjunction with [62, Lemma 6.22], that the multi-valued operator $\tilde{\Psi} : X_0 \to 2^{X_0^*}$ is bounded and pseudomonotone w.r.t. $D_0(L)$. To show the existence of solutions of (3.180) let us consider the following multi-valued mapping

$$\mathcal{A}(u) := A_T u + \hat{\mathcal{B}}(u) - \mathcal{G}(u) + \tilde{\Psi}(u). \tag{3.182}$$

As pseudomonotonicity w.r.t. $D_0(L)$ is invariant under addition, the operator $\mathcal{A} : X_0 \to 2^{X_0^*}$ of (3.182) is bounded and pseudomonotone w.r.t. $D_0(L)$ in the sense of Definition 3.54. By the definitions of the g_i we deduce from (3.172)–(3.173) that $\mathcal{G} : X_0 \to X_0^*$ is globally bounded. From (3.169) it follows that also $\tilde{\Psi} : X_0 \to 2^{X_0^*}$ is globally bounded. Since $\hat{b}$ defining the operator $\hat{\mathcal{B}}$ satisfies an estimate of the form (3.176), we infer from hypothesis (A3) that $\mathcal{A} : X_0 \to 2^{X_0^*}$ is coercive. This allows us to apply Theorem 3.56. Hence it follows that the operator $L + \mathcal{A} : X_0 \to 2^{X_0^*}$ is surjective, which implies the existence of $u \in D_0(L)$ such that

$$h \in Lu + \mathcal{A}(u) \quad \text{in } X_0^*.$$

The latter means that there is an $\tilde{\eta} \in \partial\tilde{J}_1(i(u)) = \partial\tilde{J}_1(u)$ such that

$$Lu + A_T u + \hat{\mathcal{B}}(u) - \mathcal{G}(u) + i^*\tilde{\eta} = h \quad \text{in } X_0^*, \tag{3.183}$$

which in view of (3.181) completes the existence proof of the auxiliary problem (3.177)–(3.178) in its weak form (3.180).

Step 3: Comparison

Next we show that any solution u of the auxiliary problem (3.177)–(3.178) satisfies $u_k \le u \le \bar{u}$. Let us prove first $u \le \bar{u}$. Subtracting the inequality (3.144) satisfied by the supersolution $\bar{u}$ from the equation (3.183) and testing the resulting inequality by $(u - \bar{u})^+ \in X_0 \cap L_+^p(Q)$, we obtain:

$$\langle u_t - \overline{u}_t + A_T u - A\overline{u} + \hat{\mathcal{B}}(u) - \mathcal{G}(u) + i^*\tilde{\eta} - i^*\overline{\eta}, (u - \overline{u})^+ \rangle \leq 0. \qquad (3.184)$$

We estimate the individual terms on the left-hand side of (3.184) as follows:

$$\langle u_t - \overline{u}_t, (u - \overline{u})^+ \rangle \geq 0, \qquad (3.185)$$

due to (9.78), and because $(u - \overline{u})^+(\cdot, 0) = 0$. In view of (A2) and due to the definition of A_T we get

$$\langle A_T u - A\overline{u}, (u - \overline{u})^+ \rangle$$

$$= \int_{\{u > \overline{u}\}} \sum_{i=1}^{N} \Big(a_i(x, t, \overline{u}, \nabla u) - a_i(x, t, \overline{u}, \nabla \overline{u}) \Big) \frac{\partial(u - \overline{u})}{\partial x_i} \, dx \, dt \geq 0, \quad (3.186)$$

and by (3.172), (3.173)

$$\langle -\mathcal{G}(u), (u - \overline{u})^+ \rangle = \int_{\{u > \overline{u}\}} -G(u)\,(u - \overline{u})\, dx\, dt = 0. \qquad (3.187)$$

From (3.170) we see that

$$\langle i^*\tilde{\eta} - i^*\overline{\eta}, (u - \overline{u})^+ \rangle = \int_{\{u > \overline{u}\}} (\tilde{\eta} - \overline{\eta})\,(u - \overline{u})\, dx\, dt = 0. \qquad (3.188)$$

Taking into account (3.185)–(3.188), and (3.174), from (3.184) we finally get

$$\langle \hat{\mathcal{B}}(u), (u - \overline{u})^+ \rangle = \int_{\{u > \overline{u}\}} \hat{b}(x, t, u)\,(u - \overline{u})\, dx\, dt = \int_{\{u > \overline{u}\}} (u - \overline{u})^p\, dx\, dt \leq 0,$$

and thus $(u - \overline{u})^+ = 0$, i.e., $u \leq \overline{u}$.

The proof of $u_k \leq u$ for $k = 1, 2$ requires more involved tools. The solutions $u_k \in \mathcal{S}$ satisfy (3.166). Subtracting the equation (3.180) (or equivalently (3.183)) satisfied by a solution u of the auxiliary problem from (3.166), and noting that $Tu = u$ by the previous comparison, yields

$$\langle (u_k - u)_t + Au_k - Au - \hat{\mathcal{B}}(u) + \mathcal{G}(u) + i^*\eta_k - i^*\tilde{\eta}, \varphi \rangle = 0, \quad \forall\, \varphi \in X_0. \quad (3.189)$$

Unlike in the previous comparison procedure, a more subtle test function technique is required here. This is because u has to be compared with both u_1 and u_2, and the operator A is not specified for one particular u_k. It is here where the assumption (A4) comes into play. For this purpose we construct a special test function φ for (3.189). By (A4), for any fixed $\varepsilon > 0$ there exists $\delta(\varepsilon) \in (0, \varepsilon)$ such that

$$\int_{\delta(\varepsilon)}^{\varepsilon} \frac{1}{\omega(r)}\, dr = 1.$$

We define the function $\theta_\varepsilon : \mathbb{R} \to \mathbb{R}_+$ as follows:

$$\theta_\varepsilon(s) = \begin{cases} 0 & \text{if } s < \delta(\varepsilon) \\ \displaystyle\int_{\delta(\varepsilon)}^{s} \frac{1}{\omega(r)}\, dr & \text{if } \delta(\varepsilon) \le s \le \varepsilon \\ 1 & \text{if } s > \varepsilon. \end{cases}$$

One readily verifies that, for each $\varepsilon > 0$, the function θ_ε is continuous, piecewise differentiable, and the derivative is nonnegative and bounded. Therefore the function θ_ε is Lipschitz continuous and nondecreasing. In addition, it satisfies

$$\theta_\varepsilon \to \chi_{\{s>0\}} \quad \text{as } \varepsilon \to 0, \tag{3.190}$$

where $\chi_{\{s>0\}}$ is the characteristic function of the real half line $\{s \in \mathbb{R} : s > 0\}$, and one has

$$\theta'_\varepsilon(s) = \begin{cases} 1/\omega(s) & \text{if } \delta(\varepsilon) < s < \varepsilon \\ 0 & \text{if } s \notin [\delta(\varepsilon), \varepsilon]. \end{cases}$$

As our special test function we choose the composition of θ_ε with $(u_k - u)$, i.e.,

$$\varphi = \theta_\varepsilon(u_k - u) \in X_0 \cap L^p_+(Q), \tag{3.191}$$

and note that

$$\frac{\partial}{\partial x_i}\theta_\varepsilon(u_k - u) = \theta'_\varepsilon(u_k - u)\frac{\partial(u_k - u)}{\partial x_i}.$$

We use (3.191) to test (3.189), and estimate the terms on the left-hand side individually. Let Θ_ε be the primitive of the function θ_ε defined by

$$\Theta_\varepsilon(s) = \int_0^s \theta_\varepsilon(r)\, dr.$$

We obtain for the first term on the left-hand side of (3.189) by using Lemma 9.103

$$\left\langle \frac{\partial(u_k - u)}{\partial t}, \theta_\varepsilon(u_k - u) \right\rangle = \int_\Omega \Theta_\varepsilon(u_k - u)(x, \tau)\, dx \ge 0. \tag{3.192}$$

Making use of (A2) and (A4), the term $\langle Au_k - Au, \theta_\varepsilon(u_k - u) \rangle$ can be estimated as follows:

$$\langle Au_k - Au, \theta_\varepsilon(u_k - u) \rangle$$

$$= \int_Q \sum_{i=1}^N \Big(a_i(x, t, u_k, \nabla u_k) - a_i(x, t, u, \nabla u) \Big) \frac{\partial}{\partial x_i}\theta_\varepsilon(u_k - u)\, dx\, dt$$

$$= \int_Q \sum_{i=1}^N \Big(a_i(x, t, u_k, \nabla u_k) - a_i(x, t, u_k, \nabla u) \Big) \theta'_\varepsilon(u_k - u)\frac{\partial(u_k - u)}{\partial x_i}\, dx\, dt$$

$$+ \int_Q \sum_{i=1}^N \Big(a_i(x, t, u_k, \nabla u) - a_i(x, t, u, \nabla u) \Big) \theta'_\varepsilon(u_k - u)\frac{\partial(u_k - u)}{\partial x_i}\, dx\, dt,$$

which yields the estimate

$$\langle Au_k - Au, \theta_\varepsilon(u_k - u)\rangle$$

$$\geq \int_Q \sum_{i=1}^N \Big(a_i(x,t,u_k,\nabla u) - a_i(x,t,u,\nabla u)\Big)\theta'_\varepsilon(u_k - u)\frac{\partial(u_k - u)}{\partial x_i}\, dx\, dt$$

$$\geq -N\int_Q (k_2 + |u|^{p-1} + |u_k|^{p-1} + |\nabla u|^{p-1})\times$$

$$\times\, \omega(|u_k - u|)\theta'_\varepsilon(u_k - u)|\,\nabla(u_k - u)|\, dx\, dt$$

$$\geq -N\int_{\{\delta(\varepsilon)<u_k-u<\varepsilon\}} \varrho\,|\nabla(u_k - u)|\, dx\, dt,$$

$$(3.193)$$

where $\varrho = k_2 + |u|^{p-1} + |u_k|^{p-1} + |\nabla u|^{p-1} \in L^q(Q)$. The term on the right-hand side of (3.193) tends to zero as $\varepsilon \to 0$. Next we estimate the term

$$\langle \mathcal{G}(u) + i^*\eta_k - i^*\tilde{\eta}, \theta_\varepsilon(u_k - u)\rangle$$

$$= \int_Q \Big(\sum_{l=1}^2 |g_l(x,t,u) - g_0(x,t,u)| + \eta_k - \tilde{\eta}\Big)\theta_\varepsilon(u_k - u)\, dx\, dt.$$

$$(3.194)$$

By using (3.190) and applying Lebesgue's dominated convergence theorem, and taking into account the definitions of g_l $(l = 0,1,2)$ according to (3.172) and (3.173), from (3.190) we obtain

$$\lim_{\varepsilon\to 0}\langle \mathcal{G}(u) + i^*\eta_k - i^*\tilde{\eta}, \theta_\varepsilon(u_k - u)\rangle$$

$$= \int_Q \Big(\sum_{l=1}^2 |g_l(x,t,u) - g_0(x,t,u)| + \eta_k - \tilde{\eta}\Big)\chi_{\{u_k-u>0\}}\, dx\, dt.$$

$$(3.195)$$

If $u_k - u > 0$, then $u < u_0$, and thus for $(x,t) \in \{u_k - u > 0\}$ we get $\tilde{\eta}(x,t) = \eta_0(x,t)$, $g_0(x,t,u) = \eta_0(x,t)$, and $g_k(x,t,u) = \eta_k(x,t)$, which implies the following estimate of the right-hand side of (3.195) (with $l \neq k$, and $l,k \in \{1,2\}$)

$$\int_Q \Big(\sum_{l=1}^2 |g_l(x,t,u) - g_0(x,t,u)| + \eta_k - \tilde{\eta}\Big)\chi_{\{u_k-u>0\}}\, dx\, dt$$

$$= \int_Q \Big(|g_k(x,t,u) - g_0(x,t,u)| + \eta_k - \tilde{\eta}\Big)\chi_{\{u_k-u>0\}}\, dx\, dt$$

$$+ \int_Q |g_l(x,t,u) - g_0(x,t,u)|\,\chi_{\{u_k-u>0\}}\, dx\, dt$$

$$\geq \int_{\{u_k-u>0\}} \Big(|\eta_k - \eta_0| + \eta_k - \eta_0\Big)dx\, dt \geq 0.$$

$$(3.196)$$

Using (3.192)–(3.196) we finally obtain from (3.189) as $\varepsilon \to 0$ the following inequality:

$$\lim_{\varepsilon \to 0} \langle -\hat{\mathcal{B}}(u), \theta_\varepsilon(u_k - u) \rangle = -\int_Q \hat{b}(x, t, u) \chi_{\{u_k - u > 0\}} \, dx \, dt \leq 0. \qquad (3.197)$$

As $u_k \leq u_0$, and for $(x,t) \in \{u_k - u > 0\}$ the cut-off function becomes $\hat{b}(x, t, u) = -(u_0(x, t) - u(x, t))^{p-1}$, inequality (3.197) yields the estimate

$$0 \leq \int_{\{u_k - u > 0\}} (u_k(x, t) - u(x, t))^{p-1} \, dx \, dt$$

$$\leq \int_{\{u_k - u > 0\}} (u_0(x, t) - u(x, t))^{p-1} \, dx \, dt \leq 0,$$

which implies

$$\int_Q \left((u_k(x, t) - u(x, t))^+ \right)^{p-1} dx \, dt = 0,$$

and thus $(u_k - u)^+ = 0$, i.e., $u_k \leq u$ in Q, for $k = 1, 2$.

Step 4: S Is Upward Directed

From Step 1–Step 3 we know that any solution u of the auxiliary problem (3.177)–(3.178) (or equivalently (3.180)) satisfies $u_k \leq u \leq \overline{u}$, where $u_k \in S$, $k = 1, 2$, and therefore $u \geq u_0$. The proof of the upward directedness is complete, if the solution u of the auxiliary problem in fact turns out to be a solution of the original problem (3.141)–(3.142). Because $u_k \leq u \leq \overline{u}$ it is clear that $u \in [\underline{u}, \overline{u}]$. Due to $u_0 \leq u \leq \overline{u}$, it follows that $A_T u = Au$, $\hat{B}(u) = 0$, and $G(u) = 0$, because then $g_0(x, t, u) = g_l(x, t, u)$, $l = 1, 2$, see (3.173). Hence, the solution of the auxiliary problem in fact fulfills: $u \in D_0(L)$ such that

$$\langle Lu + Au, \varphi \rangle + \int_Q \tilde{\eta} \varphi \, dx \, dt = \langle h, \varphi \rangle, \ \forall \, \varphi \in X_0, \qquad (3.198)$$

where $\tilde{\eta} \in L^q(Q)$ and $\tilde{\eta}(x, t) \in \partial \tilde{j}_1(x, t, u(x, t))$ a.e. on Q. In view of (3.170)–(3.171) we see that $\partial \tilde{j}_1(x, t, u(x, t)) \subseteq \partial j_1(x, t, u(x, t))$, which shows that u is in fact a solution of the original problem (3.141)–(3.142).   □

Remark 3.62. The directedness of the solution set S ensured by Theorem 3.61 is the crucial step toward the existence of the smallest and greatest element of S, which are the smallest and greatest solution of (3.141)–(3.142) within the interval $[\underline{u}, \overline{u}]$ (also called the *extremal solutions*). The proof of the extremal solutions will be done in the next paragraph.

Extremal Solutions

Let S denote the set of all solutions of (3.141)–(3.142) within the interval $[\underline{u}, \overline{u}]$. A first step toward the existence of extremal elements of S, i.e., the extremal solutions of (3.141)–(3.142) within $[\underline{u}, \overline{u}]$, is the following topological characterization of S.

Lemma 3.63. *The solution set* $\mathcal{S}$ *is compact in* X_0, *and weakly sequentially compact in* W_0.

Proof: Due to $\mathcal{S} \subseteq [\underline{u}, \overline{u}]$, $\mathcal{S}$ is bounded in $L^p(Q)$. Let $u \in \mathcal{S}$ be given. Then with the special test function $\varphi = u$ in (3.145) one gets

$$\langle Lu + Au, u \rangle + \int_Q \eta\, u\, dxdt = \langle h, u \rangle \qquad (3.199)$$

with $\eta \in \partial j_1(\cdot, \cdot, u)$. As $u \in D_0(L)$, it follows

$$\langle Lu, u \rangle = \langle u_t, u \rangle = \frac{1}{2} \|u(\cdot, \tau)\|_{L^2(\Omega)}^2 \geq 0.$$

By means of (A3) and the uniform boundedness of $\partial j_1(\cdot, \cdot, u)$ due to (P-j1) and using the last inequality, we get from (3.199) a uniform bound for the gradient $\|\nabla u\|_{L^p(Q)}$, which together with the $L^p(Q)$-boundedness of $\mathcal{S}$ results in

$$\|u\|_{X_0} \leq c, \quad \forall\, u \in \mathcal{S}. \qquad (3.200)$$

Any solution $u \in \mathcal{S}$ satisfies (3.145), i.e.,

$$Lu + Au + i^*\eta = h \quad \text{in } X_0^*.$$

From the last equation, and (3.200) along with the boundedness of $A : X_0 \to X_0^*$, we readily infer

$$\|u_t\|_{X_0^*} \leq c, \quad \forall\, u \in \mathcal{S}, \qquad (3.201)$$

and thus, from (3.200) and (3.201) we obtain

$$\|u\|_{W_0} \leq c, \quad \forall\, u \in \mathcal{S}. \qquad (3.202)$$

Let $(u_n) \subseteq \mathcal{S}$ be any sequence. Then, due to (3.202), a weakly convergent subsequence (u_k) exists with

$$u_k \rightharpoonup u \quad \text{in } W_0.$$

As u_k are solutions of (3.145), we get by testing (3.145) with $\varphi = u_k - u$ the estimate

$$\langle Au_k, u_k - u \rangle = \langle Lu_k + i^*\eta_k - h, u - u_k \rangle \leq \langle Lu + i^*\eta_k - h, u - u_k \rangle \to 0,$$

which yields

$$\limsup_k \langle Au_k, u_k - u \rangle \leq 0. \qquad (3.203)$$

As $A : X_0 \to X_0^*$ is pseudomonotone w.r.t. $D_0(L)$, from (3.203) it follows that

$$Au_k \rightharpoonup Au, \quad \langle Au_k, u_k \rangle \to \langle Au, u \rangle,$$

and because A is of class (S_+) w.r.t. to $D_0(L)$ (cf., e.g., [62, Theorem 2.153]), the strong convergence $u_k \to u$ in X_0 holds. Thus, the proof is complete provided the limit u belongs to $\mathcal{S}$. As u apparently belongs to the interval $[\underline{u}, \overline{u}]$, we only need to justify that u is a solution of (3.145). The u_k fulfill (3.145), i.e., we have

$$\langle Lu_k + Au_k, \varphi \rangle + \int_Q \eta_k \, \varphi \, dxdt = \langle h, \varphi \rangle, \tag{3.204}$$

where $\eta_k \in \partial j_1(\cdot, \cdot, u_k)$. By (P-j1) (η_k) is bounded in $L^q(Q)$, and thus passing to a subsequence if necessary, again denoted by (η_k)), we get

$$\eta_k \rightharpoonup \eta \text{ in } L^q(Q) \text{ as } k \to \infty.$$

In just the same way as in Sect. 3.2.3 we can prove that $\eta \in \partial j_1(\cdot, \cdot, u)$ holds true. Using the convergence properties of (u_k) derived above, and passing to the limit in (3.204) as $k \to \infty$, completes the proof. □

Theorem 3.61 and Lemma 3.63 are the main ingredients for proving the following theorem on the existence of extremal solutions. The proof follows similar arguments as in the proof of Theorem 3.30, and therefore will be omitted.

Theorem 3.64. *Let hypotheses (AP1)–(AP4) and (P-j1) be satisfied. Then the solution set $\mathcal{S}$ has the greatest element u^* and the smallest element u_*.*

Remark 3.65. If $K = X$, then (see Example 3.40) (3.126)–(3.128) reduces to the multi-valued parabolic problem (3.135) including multi-valued Robin-type boundary conditions. Similarly as in Example 3.47, we deduce the corresponding notion of sub-supersolution from the general notions given by Definition 3.44 and Definition 3.45. We note that under (AP1)–(AP4), and (P-j1)–(P-j2), all the results obtained above for the case $K = X_0$ can be extended in a straightforward way to the case $K = X$.

3.3.3 Parabolic Variational Inequality

In this subsection the closed, convex subset K is not necessarily a closed subspace of X. Unlike in the elliptic case, in the treatment of parabolic variational inequalities we are faced with additional difficulties. In order to demonstrate of what kind these difficulties are, and how to overcome them, we consider some special case of the general variational inequality formulated in (3.126)–(3.128). That is, throughout this subsection we assume $j_1 = 0$, and assume a_i to be independent of u, i.e., the operator A is of the form

$$Au(x,t) = -\sum_{i=1}^{N} \frac{\partial}{\partial x_i} a_i(x, t, \nabla u(x,t)),$$

with coefficients a_i satisfying (AP1)–(AP3). Note, in this case (AP4) is trivially fulfilled. The special multi-valued variational inequality we are going to study here is as follows: Find $u \in K \cap D(L)$, and $\xi \in L^q(\Gamma)$ such that

$$\langle Lu + Au - h, v - u \rangle + \int_\Gamma \xi \, (\gamma v - \gamma u) \, d\Gamma \geq 0, \quad \forall \, v \in K, \qquad (3.205)$$

where $\xi \in \partial j_2(\cdot, \cdot, \gamma u)$. Let us further assume $h \in L^q(Q)$.

The Sub-Supersolution Method

Let us recall the notion of sub- and supersolution introduced in Sect. 3.3.1 specified for the parabolic variational inequality (3.205).

Definition 3.66. *A function $\underline{u} \in W$ is called a **subsolution** of (3.205), if there is a $\underline{\xi} \in L^q(\Gamma)$ such that the following holds:*

(i) $\underline{u} \vee K \subseteq K, \quad \underline{u}(\cdot, 0) \leq 0 \;\; in \;\; \Omega,$

(ii) $\underline{\xi} \in \partial j_2(\cdot, \cdot, \gamma \underline{u}),$

(iii) $\langle \underline{u}_t + A\underline{u} - h, v - \underline{u} \rangle + \int_\Gamma \underline{\xi} \, (\gamma v - \gamma \underline{u}) \, d\Gamma \geq 0, \; \forall \, v \in \underline{u} \wedge K.$

Definition 3.67. *A function $\overline{u} \in W$ is called a **supersolution** of (3.205), if there is a $\overline{\xi} \in L^q(\Gamma)$ such that the following holds:*

(i) $\overline{u} \wedge K \subseteq K, \quad \overline{u}(\cdot, 0) \geq 0 \;\; in \;\; \Omega,$

(ii) $\overline{\xi} \in \partial j_2(\cdot, \cdot, \gamma \overline{u}),$

(iii) $\langle \overline{u}_t + A\overline{u} - h, v - \overline{u} \rangle + \int_\Gamma \overline{\xi} \, (\gamma v - \gamma \overline{u}) \, d\Gamma \geq 0, \; \forall \, v \in \overline{u} \vee K.$

In our approach to extend the sub-supersolution method to the parabolic variational inequality (3.205), the notion of the so-called penalty operator associated with K will play an important role.

Definition 3.68. *Let $K \neq \emptyset$ be a closed and convex subset of a reflexive Banach space Y. A bounded, hemicontinuous, and monotone operator $P : Y \to Y^*$ is called a **penalty operator** associated with K if*

$$P(u) = 0 \Longleftrightarrow u \in K. \qquad (3.206)$$

We impose the following hypothesis on the given closed and convex subset $K \subseteq X$.

(P) Assume the existence of a penalty operator $P : X \to X^*$ associated with K satisfying the following condition:
For each $u \in D(L)$, there exists $w = w(u) \in X$, $w \neq 0$ if $P(u) \neq 0$, such that

(i) $\langle u' + Au, w \rangle \geq 0$, and

(ii) $\langle P(u), w \rangle \geq D\|P(u)\|_{X^*}(\|w\|_{L^p(Q)} + \|\gamma w\|_{L^p(\Gamma)})$,

with some constant $D > 0$ independent of u and w.

With respect to a given ordered pair of sub-supersolutions $\underline{u} \leq \overline{u}$, we introduce the modification $\tilde{j}_2 : \Gamma \times \mathbb{R} \to \mathbb{R}$ by

$$
\tilde{j}_2(x,t,s) = \begin{cases} j_2(x,t,\gamma\underline{u}(x,t)) + \underline{\xi}(x,t)(s - \gamma\underline{u}(x,t)) & \text{if } s < \gamma\underline{u}(x,t), \\ j_2(x,t,s) & \text{if } \gamma\underline{u}(x,t) \leq s \leq \gamma\overline{u}(x,t), \\ j_2(x,t,\gamma\overline{u}(x,t)) + \overline{\xi}(x,t)(s - \gamma\overline{u}(x,t)) & \text{if } s > \gamma\overline{u}(x,t). \end{cases}
$$
(3.207)

Lemma 3.69. *Let hypothesis (P-j2) be satisfied. Then the function* $\tilde{j}_2 : \Gamma \times \mathbb{R} \to \mathbb{R}$ *has the following properties:*

(i) $(x,t) \mapsto \tilde{j}_2(x,t,s)$ *is measurable in* Γ *for all* $s \in \mathbb{R}$*, and* $s \mapsto \tilde{j}_2(x,t,s)$ *is Lipschitz continuous in* $\mathbb{R}$ *for a.e.* $(x,t) \in \Gamma$*.*

(ii) *Clarke's generalized gradient* $\partial\tilde{j}_2(x,t,s)$ *of* $s \mapsto \tilde{j}_2(x,t,s)$ *satisfies the estimate*

$$|\xi| \leq k_\Gamma(x,t), \quad \forall\, \xi \in \partial\tilde{j}_2(x,t,s)$$

for a.e. $(x,t) \in \Gamma$ *and for all* $s \in \mathbb{R}$*.*

(iii) *Clarke's generalized gradient of* $s \mapsto \tilde{j}_2(x,t,s)$ *is given by*

$$
\partial\tilde{j}_2(x,t,s) = \begin{cases} \underline{\xi}(x,t) & \text{if } s < \gamma\underline{u}(x,t), \\ \partial\tilde{j}_2(x,t,\gamma\underline{u}(x,t)) & \text{if } s = \gamma\underline{u}(x,t), \\ \partial j_2(x,t,s) & \text{if } \gamma\underline{u}(x,t) < s < \gamma\overline{u}(x,t), \\ \partial\tilde{j}_2(x,t,\gamma\overline{u}(x,t)) & \text{if } s = \gamma\overline{u}(x,t), \\ \overline{\xi}(x,t) & \text{if } s > \gamma\overline{u}(x,t), \end{cases}
$$
(3.208)

and the inclusions $\partial\tilde{j}_2(x,t,\gamma\underline{u}(x,t)) \subseteq \partial j_2(x,t,\gamma\underline{u}(x,t))$ *and* $\partial\tilde{j}_2(x,t,\gamma\overline{u}(x,t)) \subseteq \partial j_2(x,t,\gamma\overline{u}(x,t))$ *hold true.*

Proof: The proof follows immediately from the definition (3.207) of $\tilde{j}_2$, and using the assumptions (P-j2) on j_2, as well as from the fact that Clarke's generalized gradient $\partial j_2(x,t,s)$ is a convex set.   □

By means of $\tilde{j}_2$ we introduce the integral functional $\tilde{J}_2$ defined on $L^p(\Gamma)$ and and given by

$$\tilde{J}_2(v) = \int_\Gamma \tilde{j}_2(x,t,v(x,t))\, d\Gamma, \quad v \in L^p(\Gamma).$$

Due to hypotheses (P-j2), and (ii) of Lemma 3.69, by applying Lebourg's mean value theorem (see [62, Theorem 2.177]), the functional $\tilde{J}_2 : L^p(\Gamma) \to \mathbb{R}$ is well defined and Lipschitz continuous, so that Clarke's generalized gradient $\partial\tilde{J}_2 : L^p(\Gamma) \to 2^{L^q(\Gamma)}$ is well defined, too. Moreover, Aubin–Clarke theorem (cf. [80, p. 83]) provides the following characterization of the generalized gradient. For $v \in L^p(\Gamma)$ we have

$$\tilde{\xi} \in \partial \tilde{J}_2(v) \Longrightarrow \tilde{\xi} \in L^q(\Gamma) \text{ with } \tilde{\xi}(x,t) \in \partial \tilde{j}_2(x,t,v(x,t)) \text{ for a.a. } (x,t) \in \Gamma.$$
(3.209)

By means of Clarke's generalized gradient $\partial \tilde{J}_2$ we further introduce the following multi-valued operator:

$$\Phi(u) := (\gamma^* \circ \partial \tilde{J}_2 \circ \gamma)(u), \quad u \in X,$$

where $\gamma^* : L^q(\Gamma) \to X^*$ denotes the adjoint operator of the trace operator $\gamma : X \to L^p(\Gamma)$. If $\xi \in L^q(\Gamma)$ then

$$\langle \gamma^* \xi, \varphi \rangle = \int_\Gamma \xi \gamma \varphi \, d\Gamma, \quad \forall \, \varphi \in X.$$

In view of (ii) of Lemma 3.69, the multi-valued operator $\Phi : X \to 2^{X^*} \setminus \{\emptyset\}$ is uniformly bounded.

Lemma 3.70. *Let hypothesis (P-j2) be satisfied. Then the operator $\Phi : X \to 2^{X^*} \setminus \{\emptyset\}$ is uniformly bounded, and pseudomonotone w.r.t. $D(L)$.*

Proof: We only need to show that Φ is pseudomonotone w.r.t. $D(L)$. For any $u \in X$ the set $\partial \tilde{J}_2(\gamma u) \subset L^q(\Gamma)$ is nonempty, convex, weak-compact, and bounded, i.e.,

$$\|\xi\|_{L^q(\Gamma)} \le C, \quad \forall \, \xi \in \partial \tilde{J}_2(\gamma u),$$

cf. [80, Prop. 2.1.2]. Therefore, $\gamma^*(\partial \tilde{J}_2(\gamma u))$ is nonempty, convex, and bounded in X^*. To see that $\gamma^*(\partial \tilde{J}_2(\gamma u))$ is closed, let $(\gamma^* \xi_n)$ be a sequence such that

$$\gamma^* \xi_n \to w \text{ in } X^*, \quad \xi_n \in \partial \tilde{J}_2(\gamma u).$$

Since $\partial \tilde{J}_2(\gamma u) \subset L^q(\Gamma)$ is weak-compact there is a subsequence (ξ_{n_i}) such that $\xi_{n_i} \rightharpoonup \xi$ in $L^q(\Gamma)$ with $\xi \in \partial \tilde{J}_2(\gamma u)$, which implies $\gamma^* \xi_{n_i} \rightharpoonup \gamma^* \xi = w$, and thus the closedness of $\gamma^*(\partial \tilde{J}_2(\gamma u))$. By hypothesis (ii) of Lemma 3.69 the multifunction $\partial \tilde{J}_2 : L^p(\Gamma) \to 2^{L^q(\Gamma)} \setminus \{\emptyset\}$ is bounded, hence the mapping $\gamma^* \circ \partial \tilde{J}_2 \circ \gamma : X \to 2^{X^*} \setminus \{\emptyset\}$ is bounded as well. To verify condition (ii) of Definition 3.54 assume that $\gamma^* \circ \partial \tilde{J}_2 \circ \gamma$ fails to be upper semicontinuous in $u \in X$. Thus there is a sequence $(u_k) \subset X$ with $u_k \to u$ in X, and a sequence $(\xi_k) \subseteq \partial \tilde{J}_2(\gamma u_k)$ such that $\gamma^* \xi_k \rightharpoonup u^*$ in X^* and $u^* \notin \gamma^*(\partial \tilde{J}_2(\gamma u))$. This, however, yields a contradiction in the following way: $u_k \to u$ in X implies $\gamma u_k \to \gamma u$ in $L^p(\Gamma)$. Since $(\xi_k) \subseteq \partial \tilde{J}_2(\gamma u_k)$ is bounded there is a subsequence (ξ_{k_j}) such that $\xi_{k_j} \rightharpoonup \xi$ in $L^q(\Gamma)$, which implies

$$\gamma^* \xi_{k_j} \rightharpoonup \gamma^* \xi = u^*.$$

Applying the upper semicontinuity of $\tilde{J}_2^{\,o} : L^p(\Gamma) \times L^p(\Gamma) \to \mathbb{R}$ we obtain

$$\tilde{J}_2^{\,o}(\gamma u; \gamma v) \ge \limsup_{j \to \infty} \tilde{J}_2^{\,o}(\gamma u_{k_j}; \gamma v) \ge \limsup_{j \to \infty} \langle \xi_{k_j}, \gamma v \rangle = \langle \xi, \gamma v \rangle,$$

which shows that $\xi \in \partial \tilde{J}_2(\gamma u)$, and thus $\gamma^* \xi = u^* \in \gamma^*(\partial \tilde{J}_2(\gamma u))$, i.e., we have reached a contradiction and (ii) of Definition 3.54 is verified. To prove condition (iii) of Definition 3.54, let $(u_n) \subset D(L)$ with $u_n \rightharpoonup u$ in X, $Lu_n \rightharpoonup Lu$ in X^*, and $u_n^* := \gamma^* \xi_n$ with $\xi_n \in \partial \tilde{J}_2(\gamma u_n)$ such that $u_n^* \rightharpoonup u^*$ in X^*. We are going to show that these assumptions already imply the desired assertions, i.e., $\langle u_n^*, u_n \rangle \to \langle u^*, u \rangle$ and $u^* \in \gamma^*(\partial \tilde{J}_2(\gamma u))$. By the assumptions on (u_n) we have $u_n \rightharpoonup u$ in W, which due to the compactness of the trace operator $\gamma : W \to L^p(\Gamma)$ implies $\gamma u_n \to \gamma u$ in $L^p(\Gamma)$. Since $\tilde{J}_2 : L^p(\Gamma) \to \mathbb{R}$ is locally Lipschitz there is a constant C (depending on γu) such that for n sufficiently large we have

$$|\tilde{J}_2^{\,o}(\gamma u_n; v)| \leq C \|v\|_{L^p(\Gamma)}, \quad \forall\, v \in L^p(\Gamma),$$

and thus, in particular,

$$|\langle \xi_n, \gamma u_n - \gamma u \rangle| \leq C \|\gamma u_n - \gamma u\|_{L^p(\Gamma)}. \tag{3.210}$$

In view of $\gamma u_n \to \gamma u$ in $L^p(\Gamma)$ we get from (3.210)

$$\langle \gamma^* \xi_n, u_n - u \rangle \to 0 \quad \text{as } n \to \infty,$$

which implies

$$\langle u_n^*, u_n \rangle - \langle u_n^*, u \rangle \to 0. \tag{3.211}$$

Since $u_n^* \rightharpoonup u^*$, it follows

$$\langle u_n^*, u \rangle \to \langle u^*, u \rangle,$$

and thus by (3.211) we obtain $\langle u_n^*, u_n \rangle \to \langle u^*, u \rangle$. Finally, to prove that $u^* \in \gamma^*(\partial \tilde{J}_2(\gamma u))$ holds, we note first that according to $\xi_n \in \partial \tilde{J}_2(\gamma u_n)$ we have

$$\tilde{J}_2^{\,o}(\gamma u_n; v) \geq \langle \xi_n, v \rangle, \quad \forall\, v \in L^p(\Gamma),$$

which implies by the upper semicontinuity of $\tilde{J}_2^{\,o}$, and the boundedness of (ξ_n) in $L^q(\Gamma)$, and using $\gamma u_n \to \gamma u$ in $L^p(\Gamma)$, the inequality

$$\tilde{J}_2^{\,o}(\gamma u; v) \geq \langle \xi, v \rangle, \quad \forall\, v \in L^p(\Gamma), \tag{3.212}$$

where ξ is the weak limit of some weakly convergent subsequence (ξ_{n_j}) of (ξ_n). As $u_{n_j}^* = \gamma^* \xi_{n_j} \rightharpoonup u^*$ in X^* and also $\gamma^* \xi_{n_j} \rightharpoonup \gamma^* \xi$ we get $u^* = \gamma^* \xi$, which in view of (3.212) proves $u^* \in \gamma^*(\partial \tilde{J}_2(\gamma u))$. □

As in Sect. 3.3.2 we introduce the cut-off function $b : Q \times \mathbb{R} \to \mathbb{R}$ related to the ordered pair of sub- and supersolutions, and given by

$$b(x, t, s) = \begin{cases} (s - \overline{u}(x,t))^{p-1} & \text{if } s > \overline{u}(x,t), \\ 0 & \text{if } \underline{u}(x,t) \leq s \leq \overline{u}(x,t), \\ -(\underline{u}(x,t) - s)^{p-1} & \text{if } s < \underline{u}(x,t). \end{cases}$$

Then $b : Q \times \mathbb{R} \to \mathbb{R}$ is a Carathéodory function satisfying the estimate (3.156) and (3.157), which implies that the associated Nemytskij operator $B : L^p(Q) \to L^q(Q)$ is bounded and continuous, and $\hat{B} = i^* \circ B \circ i : X \to X^*$ is bounded and pseudomonotone w.r.t. $D(L)$ due to the compact embedding $W \hookrightarrow L^p(Q)$. By hypotheses (AP1)–(AP3), the operator $A : X \to X^*$ is continuous, bounded, and pseudomonotone w.r.t. $D(L)$ (cf. Theorem 9.109), which together with the properties of $\hat{B}$ and Φ yields the following result.

Lemma 3.71. *Let (AP1)–(AP3) and (P-j2) be fulfilled. Then the multi-valued operator $\mathcal{A} : X \to 2^{X^*}$ defined by $\mathcal{A} = A + \hat{B} + \Phi$ is bounded, coercive, and pseudomonotone w.r.t. $D(L)$.*

Proof: Apparently $\mathcal{A}$ is bounded, because each of the operators A, $\hat{B}$, and Φ is bounded. The coercivity of $\mathcal{A}$ readily follows from (AP3), (3.157), and the uniform boundedness of Φ. Thus it only remains to show that $\mathcal{A}$ is pseudomonotone w.r.t. $D(L)$, and for this only condition (iii) of Definition 3.54 needs to be checked, because (i) and (ii) of Definition 3.54 are obvious. To this end assume $(u_n) \subset D(L)$ with $u_n \rightharpoonup u$ in X, $Lu_n \rightharpoonup Lu$ in X^*, $u_n^* \in (A + \hat{B} + \Phi)(u_n)$ with $u_n^* \rightharpoonup u^*$ in X^*, and

$$\limsup_n \langle u_n^*, u_n - u \rangle \leq 0. \tag{3.213}$$

Due to $u_n^* \in (A + \hat{B} + \Phi)(u_n)$ we have $u_n^* = Au_n + \hat{B}(u_n) + \gamma^* \tilde{\xi}_n$ with $\tilde{\xi}_n \in \partial \tilde{J}_2(u_n)$. The compact embedding $W \hookrightarrow L^p(Q)$ implies that the operator $\hat{B} : W \to L^q(Q) \subset X^*$ is completely continuous, and hence it follows that $\hat{B}(u_n) \to \hat{B}(u)$ in X^*. As $(\tilde{\xi}_n)$ is uniformly bounded in $L^q(\Gamma)$, and $\gamma u_n \to \gamma u$ in $L^p(\Gamma)$, we get

$$\lim_{n \to \infty} \langle \gamma^* \tilde{\xi}_n, u_n - u \rangle = \lim_{n \to \infty} \int_\Gamma \tilde{\xi}_n (\gamma u_n - \gamma u) \, d\Gamma = 0.$$

From (3.213) we thus deduce

$$\limsup_n \langle Au_n, u_n - u \rangle \leq 0.$$

The sequence $(Au_n) \subset X^*$ is bounded, so that there is a subsequence with $Au_k \rightharpoonup v$. Since A is pseudomonotone w.r.t. $D(L)$, it follows that $v = Au$ and $\langle Au_k, u_k \rangle \to \langle Au, u \rangle$. It is clear that we have shown that

$$Au_n \rightharpoonup Au \quad \text{and} \quad \langle Au_n, u_n \rangle \to \langle Au, u \rangle. \tag{3.214}$$

From (3.214) and $u_n^* \rightharpoonup u^*$ we obtain $\gamma^* \tilde{\xi}_n \rightharpoonup u^* - Au - \hat{B}(u)$, which in view of the pseudomonotonicity of Φ, implies $u^* \in (A + \hat{B} + \Phi)(u)$ and $\langle u_n^*, u_n \rangle \to \langle u^*, u \rangle$. $\qquad \square$

Lemma 3.72. *Let (AP1)–(AP3) and (P-j2) be fulfilled. Then the multi-valued operator $\mathcal{A}_\varepsilon : X \to 2^{X^*}$ defined by $\mathcal{A}_\varepsilon = A + \hat{B} + \frac{1}{\varepsilon} P + \Phi$, is bounded, coercive, and pseudomonotone w.r.t. $D(L)$ for any $\varepsilon > 0$, where P is the penalty operator belonging to K.*

Proof: By Lemma 3.71, the multi-valued operator $A + \hat{B} + \Phi : X \to 2^{X^*}$ is pseudomonotone w.r.t. $D(L)$, bounded, and coercive. On the other hand, by definition, the penalty operator $P : X \to X^*$ is bounded, hemicontinuous, and monotone. This implies that $P : X \to X^*$ is pseudomonotone and bounded. Thus, by applying Theorem 3.57 with $A_1 = A + \hat{B} + \Phi$ and $A_2 = \frac{1}{\varepsilon}P$, $\mathcal{A}_\varepsilon : X \to 2^{X^*}$ is bounded and pseudomonotone w.r.t. $D(L)$. Finally, the coercivity of $\mathcal{A}_\varepsilon$ is obtained by a straightforward calculation employing the fact that P is a monotone operator. $\qquad\square$

A crucial point in the proof of the sub-supersolution method is the solvability of the following auxiliary multi-valued parabolic variational inequality.

Auxiliary Variational Inequality

Find $u \in K \cap D(L)$, and $\tilde{\xi} \in L^q(\Gamma)$ such that

$$\langle Lu + Au + \hat{B} - h, v - u \rangle + \int_\Gamma \tilde{\xi}\,(\gamma v - \gamma u)\,d\Gamma \geq 0, \quad \forall\, v \in K, \qquad (3.215)$$

where $\tilde{\xi} \in \partial \tilde{j}_2(\cdot, \cdot, \gamma u)$.

Lemma 3.73. *Let (AP1)–(AP3), (P-j2), and (P) be satisfied, and suppose $D(L) \cap K \neq \emptyset$. Then the auxiliary parabolic variational inequality (3.215) has a solution.*

Proof: We introduce the following Penalty problem related to (3.215): Find $u \in D(L)$, and $\tilde{\xi} \in L^q(\Gamma)$ such that

$$\langle Lu + Au + \hat{B}(u) + \frac{1}{\varepsilon}P(u) - h, v \rangle + \int_\Gamma \tilde{\xi}\gamma v\,d\Gamma = 0, \quad \forall\, v \in X, \qquad (3.216)$$

where $\varepsilon > 0$, P is the associated penalty operator, and $\tilde{\xi} \in \partial \tilde{j}_2(\cdot, \cdot, \gamma u)$.

Step 1: Existence of solutions of (3.216)

In view of Lemma 3.72, the operator $\mathcal{A}_\varepsilon = A + \hat{B} + \frac{1}{\varepsilon}P + \Phi$ is bounded, coercive, and pseudomonotone w.r.t. $D(L)$ for any $\varepsilon > 0$, and therefore, by applying Theorem 3.56, there exists a $u \in D(L)$ such that

$$h \in Lu + \mathcal{A}_\varepsilon u,$$

which means that there is a $\tilde{\xi} \in \partial \tilde{j}_2(\cdot, \cdot, \gamma u)$ such that

$$Lu + Au + \hat{B} + \frac{1}{\varepsilon}P(u) + \gamma^*\tilde{\xi} = h \quad \text{in } X^*. \qquad (3.217)$$

The latter shows that u solves (3.216).

Step 2: Boundedness of the penalty solutions of (3.216) in W

According to Step 1, for any $\varepsilon > 0$ problem (3.216) admits a solution u_ε. We show that the family $\{u_\varepsilon : \varepsilon > 0 \text{ small}\}$ is bounded with respect to the graph norm of $D(L)$. To this end let u_0 be a (fixed) element of $D(L) \cap K$. Using (3.217) with u replaced by u_ε, we get

$$\langle Lu_\varepsilon + Au_\varepsilon + \hat{B}(u_\varepsilon) + \frac{1}{\varepsilon}P(u_\varepsilon) + \gamma^*\tilde{\xi}_\varepsilon, u_\varepsilon - u_0 \rangle = \langle h, u_\varepsilon - u_0 \rangle,$$

where $\tilde{\xi}_\varepsilon \in \partial\tilde{j}_2(\cdot, \cdot, \gamma u_\varepsilon)$. On the basis of the monotonicity of L and because $Pu_0 = 0$, we derive

$$\langle h - u_0', u_\varepsilon - u_0 \rangle$$
$$= \langle u_\varepsilon' - u_0', u_\varepsilon - u_0 \rangle + \langle (A + \hat{B})(u_\varepsilon), u_\varepsilon - u_0 \rangle + \frac{1}{\varepsilon}\langle Pu_\varepsilon - Pu_0, u_\varepsilon - u_0 \rangle$$
$$+ \langle \gamma^*\tilde{\xi}_\varepsilon, u_\varepsilon - u_0 \rangle$$
$$\geq \langle (A + \hat{B})(u_\varepsilon) + \gamma^*\tilde{\xi}_\varepsilon, u_\varepsilon - u_0 \rangle.$$

Thus,

$$\frac{\langle (A + \hat{B})(u_\varepsilon) + \gamma^*\tilde{\xi}_\varepsilon, u_\varepsilon - u_0 \rangle}{\|u_\varepsilon - u_0\|_X} \leq \|h - u_0'\|_{X^*},$$

for all $\varepsilon > 0$. Since the operator $A + \hat{B} + \Phi : X \to 2^{X^*}$ is coercive, we have that $\|u_\varepsilon\|_X$ is bounded. As a consequence, the sequences (Au_ε), $(\hat{B}(u_\varepsilon))$, and $(\gamma^*\tilde{\xi}_\varepsilon)$ are bounded in X^*. Moreover, from the growth conditions of b, we readily see that $(B(u_\varepsilon))$ is a bounded sequences in $L^q(Q)$. All these facts make clear from (3.217) that (u_ε') is bounded if and only if $(\frac{1}{\varepsilon}P(u_\varepsilon))$ is bounded.

Next, we check that the sequence $(\frac{1}{\varepsilon}P(u_\varepsilon))$ is bounded in X^*, assuming $P(u_\varepsilon)) \neq 0$, because otherwise it is trivial. To see this, for each ε, we choose $w = w_\varepsilon$ to be an element satisfying (P) with $u = u_\varepsilon$. From (3.217), we have

$$\langle u_\varepsilon', w_\varepsilon \rangle + \langle (A + \hat{B})(u_\varepsilon) + \gamma^*\tilde{\xi}_\varepsilon, w_\varepsilon \rangle + \frac{1}{\varepsilon}\langle P(u_\varepsilon), w_\varepsilon \rangle = \langle h, w_\varepsilon \rangle.$$

Also, from (P) we know that $\langle u_\varepsilon', w_\varepsilon \rangle + \langle Au_\varepsilon, w_\varepsilon \rangle \geq 0$, therefore

$$\frac{1}{\varepsilon}\langle P(u_\varepsilon), w_\varepsilon \rangle \leq \langle h - \hat{B}(u_\varepsilon) - \gamma^*\tilde{\xi}_\varepsilon, w_\varepsilon \rangle. \tag{3.218}$$

Let $c > 0$ be some generic constant. By the definition of Φ and the boundedness properties of (u_ε), $(\hat{B}(u_\varepsilon))$, we see that there is $c > 0$ such that

$$|\langle h - \hat{B}(u_\varepsilon) - \gamma^*\tilde{\xi}_\varepsilon, w_\varepsilon \rangle| \leq c(\|w_\varepsilon\|_{L^p(Q)} + \|\gamma w_\varepsilon\|_{L^p(\Gamma)}), \ \forall \varepsilon > 0 \text{ small.}$$

This, (3.218), and (P) imply that

$$\frac{1}{\varepsilon}\|Pu_\varepsilon\|_{X^*} \leq \frac{c}{D}, \ \forall \varepsilon > 0 \text{ small.} \tag{3.219}$$

Consequently, (u_ε) is bounded in W, and thus for a relabeled subsequence (u_n) with $u_n := u_{\varepsilon_n}$ and $\varepsilon_n \to 0$ as $n \to \infty$ we obtain

$$u_n \rightharpoonup u \quad \text{in } X, \quad u_n' \rightharpoonup u' \quad \text{in } X^*.$$

Since $D(L)$ is closed in W and convex, it is true that $u \in D(L)$.

Step 3: The limit u solves (3.215)

First, note that by (3.219) we have $Pu_n \to 0$ in X^*. It follows from the monotonicity of P that

$$\langle Pv, v - u \rangle \geq 0, \ \forall v \in X.$$

As in the proof of Minty's lemma (cf. [151]), this leads to $\langle Pu, v \rangle \geq 0$ for all $v \in X$. Hence, $Pu = 0$ in X^*, so $u \in K$. If we test (3.216) satisfied by the penalty solutions u_n with $v = u_n - u$ we obtain

$$\langle u_n' + Au_n + \hat{B}(u_n) + \frac{1}{\varepsilon_n} P(u_n) - h, u_n - u \rangle + \int_\Gamma \tilde{\xi}_n \left(\gamma u_n - \gamma u \right) d\Gamma = 0. \quad (3.220)$$

Taking into account that

$$\langle u' - u_n', u - u_n \rangle \geq 0, \quad \frac{1}{\varepsilon_n} \langle P(u_n), u_n - u \rangle \geq 0,$$

we have from (3.220) the inequality

$$\langle Au_n, u_n - u \rangle \leq \langle u' + \hat{B}(u_n) - h, u - u_n \rangle + \int_\Gamma \tilde{\xi}_n \left(\gamma u - \gamma u_n \right) d\Gamma.$$

Due to the compact embedding $W \hookrightarrow L^p(Q)$, the compactness of the trace operator $\gamma : W \to L^p(\Gamma)$, and the uniform boundedness of $(\tilde{\xi}_n)$ in $L^q(\Gamma)$, we get

$$\limsup_{n \to \infty} \langle Au_n, u_n - u \rangle \leq 0.$$

Because A is an operator of class (S_+) with respect to $D(L)$ (cf., e.g., [20, 21, 44], or Theorem 9.109), we infer the strong convergence $u_n \to u$ in X. On the other hand, testing (3.216) satisfied by u_n with $v - u_n$ where $v \in K$, we obtain the following inequality for the penalty solutions u_n:

$$\langle u_n' + Au_n + \hat{B}(u_n) - h, v - u_n \rangle + \int_\Gamma \tilde{\xi}_n \left(\gamma v - \gamma u_n \right) d\Gamma = \langle -\frac{1}{\varepsilon_n} P(u_n), v - u_n \rangle \geq 0 \quad (3.221)$$

As $(\tilde{\xi}_n) \subset L^q(\Gamma)$ is uniformly bounded, there is a weakly convergent subsequence $(\tilde{\xi}_k)$, i.e.,

$$\tilde{\xi}_k \rightharpoonup \tilde{\xi} \quad \text{as } k \to \infty.$$

From $\tilde{\xi}_k \in \partial \tilde{j}_2(\cdot, \cdot, \gamma u_k)$ and $\gamma u_k \to \gamma u$ in $L^p(\Gamma)$, it follows that the weak limit $\tilde{\xi}$ satisfies $\tilde{\xi} \in \partial \tilde{j}_2(\cdot, \cdot, \gamma u)$. Replacing u_n in (3.221) by u_k and $\tilde{\xi}_n$ by $\tilde{\xi}_k$, and passing to the limit as $k \to \infty$ proves that u is a solution of (3.215). $\square$

Our main existence and comparison result is formulated in the following theorem.

Theorem 3.74. *Let the hypotheses of Lemma 3.73 be satisfied including that* $D(L) \cap K \neq \emptyset$, *and let* $(\underline{u}, \overline{u})$ *be a pair of sub-supersolutions of problem (3.205) with* $\underline{u} \leq \overline{u}$. *Then the multi-valued variational inequality (3.205) has at least one solution within the ordered interval* $[\underline{u}, \overline{u}]$.

Proof: In view of Lemma 3.73 the auxiliary problem (3.215) possesses solutions. To prove the assertion of Theorem 3.74 we only need to show that there are solutions u of (3.215) lying within the interval $[\underline{u}, \overline{u}]$. This is because in such a case $\hat{B}(u) = 0$, and $\tilde{\xi} \in \partial \tilde{j}_2(\cdot, \cdot, \gamma u)$, which due to Lemma 3.69 implies $\partial \tilde{j}_2(\cdot, \cdot, \gamma u) \subseteq \partial j_2(\cdot, \cdot, \gamma u)$, and thus u must be a solution of (3.205). In fact we are going to show that any solution u of (3.215) belongs to $[\underline{u}, \overline{u}]$. Let us show first that $u \leq \overline{u}$, where u is a solution of (3.215) and $\overline{u}$ is the given supersolution of (3.205)

By using $v = \overline{u} \wedge u = u - (u - \overline{u})^+ \in K$ in (3.215) and $v = \overline{u} \vee u = \overline{u} + (u - \overline{u})^+$ in Definition 3.67, we obtain

$$\langle Lu + Au + \hat{B}(u) - h, -(u - \overline{u})^+ \rangle + \int_\Gamma \tilde{\xi} \left(-(\gamma u - \gamma \overline{u})^+ \right) d\Gamma \geq 0, \quad (3.222)$$

where $\tilde{\xi} \in \partial \tilde{j}_2(\cdot, \cdot, \gamma u)$, and

$$\langle \overline{u}_t + A\overline{u} - h, (u - \overline{u})^+ \rangle + \int_\Gamma \overline{\xi} (\gamma u - \gamma \overline{u})^+ d\Gamma \geq 0, \quad (3.223)$$

where $\overline{\xi} \in \partial j_2(\cdot, \cdot, \gamma \overline{u})$. Adding (3.222) and (3.223) we get

$$\langle \overline{u}_t - u_t + A\overline{u} - Au - \hat{B}(u), (u - \overline{u})^+ \rangle + \int_\Gamma (\overline{\xi} - \tilde{\xi}) (\gamma u - \gamma \overline{u})^+ d\Gamma \geq 0. \quad (3.224)$$

Because of

$$\langle u_t - \overline{u}_t, (u - \overline{u})^+ \rangle \geq 0, \quad \langle Au - A\overline{u}, (u - \overline{u})^+ \rangle \geq 0,$$

and, by applying Lemma 3.69 (iii),

$$\int_\Gamma (\overline{\xi} - \tilde{\xi}) (\gamma u - \gamma \overline{u})^+ d\Gamma = 0$$

we obtain the following inequality from (3.224)

$$\langle \hat{B}(u), (u - \overline{u})^+ \rangle \leq 0,$$

which means

$$\int_Q b(\cdot, \cdot, u)(u - \overline{u})^+ dx dt = \int_Q \left((u - \overline{u})^+ \right)^p dx dt \leq 0.$$

From the last inequality we readily infer $(u - \overline{u})^+ = 0$, i.e., $u \leq \overline{u}$. Since the proof of $\underline{u} \leq u$ can be done in a similar way, this completes the proof of the theorem. □

Application: Obstacle Problem

We consider an obstacle problem, where the convex, closed set K is given by

$$K = \{u \in X : u \leq \psi \text{ a.e. on } Q\},$$

with any obstacle function ψ specified as follows:

(i) $\psi \in W$ and $\psi(\cdot, 0) \geq 0$ in Ω,
(ii) $\psi' + A\psi \geq 0$ in X^*, i.e., $\langle \psi' + A\psi, v \rangle \geq 0$, $\forall v \in X \cap L_+^p(Q)$.

The penalty operator $P : X \to X^*$ can be chosen as

$$\langle P(u), v \rangle = \int_Q [(u - \psi)^+]^{p-1} v \, dx dt + \int_\Gamma [(\gamma u - \gamma \psi)^+]^{p-1} \gamma v \, d\Gamma, \quad (3.225)$$

for all $u, v \in X$. Indeed, P is bounded, continuous, and monotone. Let us check that it satisfies (3.206), and thus is a penalty operator for K. If $P(u) = 0$, then $(u - \psi)^+ = 0$ a.e. in Q, i.e.,

$$u \leq \psi \text{ a.e. in } Q, \quad (3.226)$$

that is $u \in K$. Conversely, assume that u satisfies (3.226). Then, for a.a. $t \in (0, \tau)$, we have $u(\cdot, t) \leq \psi(\cdot, t)$ a.e. in Ω, which implies that

$$\gamma_{\partial \Omega} u(\cdot, t) \leq \gamma_{\partial \Omega} \psi(\cdot, t) \text{ a.e. on } \partial \Omega$$

($\gamma_{\partial \Omega}$ is the trace operator on $\partial \Omega$). This means that $\gamma u \leq \gamma \psi$ a.e. on Γ showing that $P(u) = 0$. To check (P), for each $u \in D(L)$ we choose $w = (u - \psi)^+$. Then, $w \in X$, and $w \neq 0$ whenever $P(u) \neq 0$. We justify that (P)(i) is satisfied. Since, according to assumption (i) for ψ, $(u - \psi)^+(\cdot, 0) = 0$, we have

$$\langle u' - \psi', (u - \psi)^+ \rangle = \frac{1}{2} \|(u - \psi)^+(\cdot, \tau)\|_{L^2(\Omega)}^2 \geq 0.$$

Combining with the inequality $\langle Au - A\psi, (u - \psi)^+ \rangle \geq 0$, we arrive at

$$\langle u' + Au, (u - \psi)^+ \rangle \geq \langle \psi' + A\psi, (u - \psi)^+ \rangle \geq 0$$

because $(u - \psi)^+ \in X \cap L_+^p(Q)$. So we have checked (i) of (P). To verify (P)(ii), we note that

$$\langle P(u), w \rangle = \|(u - \psi)^+\|_{L^p(Q)}^p + \|(\gamma u - \gamma \psi)^+\|_{L^p(\Gamma)}^p. \quad (3.227)$$

From (3.225) and Hölder's inequality, we derive, for a constant $c > 0$, that

$$|\langle P(u), v \rangle| \leq \|(u - \psi)^+\|_{L^p(Q)}^{p-1} \|v\|_{L^p(Q)} + \|(\gamma u - \gamma \psi)^+\|_{L^p(\Gamma)}^{p-1} \|v\|_{L^p(\Gamma)}$$
$$\leq c(\|(u - \psi)^+\|_{L^p(Q)}^{p-1} + \|(\gamma u - \gamma \psi)^+\|_{L^p(\Gamma)}^{p-1}) \|v\|_X,$$

for all $v \in X$. Hence,

$$\|P(u)\|_{X^*} \le c(\|(u - \psi)^+\|_{L^p(Q)}^{p-1} + \|(\gamma u - \gamma \psi)^+\|_{L^p(\Gamma)}^{p-1}), \ \forall u \in X.$$

This, together with (3.227), implies (P)(ii).

For our example of K, $\bar{u} \wedge K \subseteq K$ for every $\bar{u} \in W$, and $\underline{u} \vee K \subseteq K$ if $\underline{u} \le \psi$ on Q.

We conclude this chapter by several remarks.

3.4 Notes and Comments

The main goal of this chapter was to establish a general method of sub-supersolutions for multi-valued variational inequalities of elliptic and parabolic type that include as special cases, in particular, a wide range of specific elliptic and parabolic boundary value problems by specifying the closed convex set K to some subspace of V and X, respectively. Our presentation is based on and includes results recently obtained in [38, 39, 66, 69, 70, 72], which partly extend results of [62] that are based on some one-sided growth condition of Clarke's generalized gradient. In our treatment we only require some local L^q-growth condition on the Clarke's generalized gradient.

The treatment of the multi-valued parabolic variational inequality (3.126)–(3.128) is by no means a straightforward extension of the elliptic case considered in Sect. 3.2, and raises additional technical challenges. First, while V, V_0, X, and X_0 possess lattice structure, the solution spaces W and W_0 of (3.126)–(3.128) do not have lattice structure. Second, the solvability of (3.126)–(3.128) requires certain additional properties on the penalty operator associated with K, unless K is a closed subspace of X. The sub-supersolution method that has been established in that section also improves earlier results obtained in this direction in the monograph [62], where, as mentioned above, the generalized gradients $s \mapsto \partial j_k(x, t,, s)$ were required to satisfy a one-sided growth condition of the form

$$\eta_i \in \partial j_k(x, t, s_i): \quad \eta_1 \le \eta_2 + c(s_2 - s_1)^{p-1}, \tag{3.228}$$

for a.a. (x, t), for all $\eta_i \in \partial j_k(x, t, s_i)$, $i = 1, 2$, and for all s_1, s_2 with $s_1 < s_2$. As in Sect. 3.2, here we establish a comparison principle for (3.126)–(3.128) in terms of sub-supersolution without assuming condition (3.228), whose proof at the same time simplifies the proofs of comparison results for this kind of problem in recent works (see, e.g., [62, 64]), which were based on rather involved regularization techniques as well as passing to the limit procedures.

Regarding parabolic variational inequalities we note that under the hypotheses of Theorem 3.74, and employing the technique as in Sect. 3.3.2, one can prove that the solution set $\mathcal{S}$ of all solutions of (3.205) lying within the interval $[\underline{u}, \bar{u}]$ is directed provided K possesses the lattice structure in the form that $K \wedge K \subseteq K$ and $K \vee K \subseteq K$. The existence of extremal solutions can be proved if, in addition, S is bounded in W. The boundedness of S in X is readily obtained and can be proved in the same way as in the elliptic case. However, to get the boundedness in W, one needs, in addition, that

$$\left\| \frac{\partial u}{\partial t} \right\|_{X^*} \leq c, \quad \forall \, u \in S,$$

which requires additional conditions. As seen from the proof of the comparison result for parabolic variational inequalities we need to impose an additional assumption on the closed convex set K given in terms of conditions imposed on the penalty operator, which is the main difference between elliptic and parabolic variational inequalities. This was basically needed to get control of $\|\partial u / \partial t\|_{X^*}$.

Finally, it should be noted that the sub-supersolution method along with the characterization of the solution set within the sector of sub- and super-solution as developed in this chapter plays an important role in the qualitative study of (nonsmooth) elliptic and parabolic problems. In particular, this method when combined with variational methods provides an effective tool to prove existence and multiplicity results for such kind of problems, see, e.g., [40, 67, 68, 71].

4

Discontinuous Multi-Valued Elliptic Problems

In this chapter we study various multi-valued elliptic boundary value problems involving discontinuous and nonlocal nonlinearities. The basic tools in dealing with these kinds of problems are on the one hand the existence and comparison results of Sect. 3.2 and on the other hand the abstract fixed point results provided in Chap. 2.

4.1 Nonlocal and Discontinuous Elliptic Inclusions

We consider multi-valued elliptic problems with nonlocal nonlinearities. Assuming only certain growth conditions on the data, we are able to prove existence results for the problem under consideration. In particular, no continuity assumptions are imposed on the nonlocal term.

Let us use the notation of Sect. 3.2, i.e., let $\Omega \subset \mathbb{R}^N$ be a bounded domain with Lipschitz boundary $\partial\Omega$, and let $V = W^{1,p}(\Omega)$ and $V_0 = W_0^{1,p}(\Omega)$, $1 < p < \infty$. In this subsection we consider the following quasilinear elliptic inclusion problem: Find $u \in V_0$ and an $\eta \in L^q(\Omega)$ satisfying

$$\langle -\Delta_p u, v \rangle + \int_\Omega \eta\, v\, dx = \langle \mathcal{F}u, v \rangle, \quad \forall\, v \in V_0, \tag{4.1}$$

where $\eta(x) \in \partial j(x, u(x))$ for a.e. $x \in \Omega$, with $s \mapsto \partial j(x, s)$ denoting Clarke's generalized gradient of the locally Lipschitz function $j(x, \cdot) : \mathbb{R} \to \mathbb{R}$. Only for simplifying our presentation, the elliptic operator is $\Delta_p u = \mathrm{div}\,(|\nabla u|^{p-2}\nabla u)$, which is the p-Laplacian with $1 < p < \infty$. As usual, $\langle \cdot, \cdot \rangle$ stands for the duality pairing between V_0 and V_0^*. The mapping $\mathcal{F} : V_0 \to V_0^*$ on the right-hand side of (4.1) comprises the nonlocal term and is generated by a function $F : \Omega \times L^p(\Omega) \to \mathbb{R}$ through

$$\mathcal{F}u := F(\cdot, u). \tag{4.2}$$

Thus (4.1) stands for the formulation of the weak solution of the quasilinear elliptic inclusion

S. Carl and S. Heikkilä, *Fixed Point Theory in Ordered Sets and Applications:*
From Differential and Integral Equations to Game Theory,
DOI 10.1007/978-1-4419-7585-0_4, © Springer Science+Business Media, LLC 2011

$$-\Delta_p u(x) + \partial j(x, u(x)) \ni F(x, u) \quad \text{in } \Omega, \quad u = 0 \quad \text{on } \partial\Omega.$$

While multi-valued elliptic problems in the form (4.1) with $\mathcal{F}u$ replaced by a given element $h \in V_0^*$ have been treated in Sect. 3.2 under the assumption that appropriately defined sub- and supersolutions are available, the novelty of the problem under consideration is that the term on the right-hand side of (4.1) is nonlocal and not necessarily continuous in u. Moreover, we do not assume the existence of sub- and supersolutions.

Our main goal is to prove existence results for the problem (4.1) only under the assumption that certain growth conditions on the data are satisfied. Let us next consider a few important special cases that are included in (4.1).

(i)   For $s \mapsto j(x, s)$ smooth, (4.1) is the weak formulation of the nonlocal single-valued Dirichlet problem

$$u \in V_0: \quad -\Delta_p u + j'(x, u) = \mathcal{F}u \quad \text{in } V_0^*.$$

(ii) If $s \mapsto j(x, s)$ is locally Lipschitz, and $g : \Omega \times \mathbb{R} \to \mathbb{R}$ is a Carathéodory function with its Nemytskij operator G, then the multi-valued elliptic problem: Find $u \in V_0$ and $\eta \in L^q(\Omega)$ such that

$$\eta \in \partial j(\cdot, u), \quad \langle -\Delta_p u, v \rangle + \int_\Omega \eta\, v\, dx = \langle G(u), v \rangle, \quad \forall\, v \in V_0, \qquad (4.3)$$

is a special case of (4.1) by defining $F(x, u) := g(x, u(x))$, which has been treated in Sect. 3.2.

(iii) As for an example of a (discontinuous) nonlocal $\mathcal{F}$ that will be treated later we consider F defined by

$$F(x, u) = [|x|] + \gamma \int_\Omega [u(y)]\, dy,$$

where γ is some positive constant, and $[\cdot] : \mathbb{R} \to \mathbb{Z}$ is the integer-function that assigns to each $s \in \mathbb{R}$ the greatest integer $[s] \in \mathbb{Z}$ satisfying $[s] \le s$. Apparently, $u \mapsto F(\cdot, u)$ is nonlocal and discontinuous.

4.1.1 Hypotheses, Main Result, and Preliminaries

We denote the norms in $L^p(\Omega)$, V_0 and V_0^* by $\|\cdot\|_p$, $\|\cdot\|_{V_0}$, and $\|\cdot\|_{V_0^*}$, respectively, and by λ_1 the first Dirichlet eigenvalue of $-\Delta_p$, which is positive (see [171]) and variationally characterized by

$$\lambda_1 = \inf_{0 \ne u \in V_0} \frac{\int_\Omega |\nabla u|^p\, dx}{\int_\Omega |u|^p\, dx}.$$

As usual, let $L^p(\Omega)$ be equipped with the natural partial ordering of functions defined by $u \le w$ if and only if $w - u$ belongs to the positive cone $L_+^p(\Omega)$ of all nonnegative elements of $L^p(\Omega)$. We assume the following hypothesis for j and F:

(H1) The function $j : \Omega \times \mathbb{R} \to \mathbb{R}$ is a Carathéodory function, with $s \mapsto j(x, s)$ being locally Lipschitz for a.e. $x \in \Omega$, and Clarke's generalized gradient ∂j satisfies the following growth condition: There is a $\varepsilon \in (0, \lambda_1)$ and a $k_1 \in L_+^q(\Omega)$ such that

$$|\eta| \le k_1(x) + (\lambda_1 - \varepsilon)|s|^{p-1}, \quad \forall \, \eta \in \partial j(x, s), \quad \forall \, s \in \mathbb{R},$$

and for a.e. $x \in \Omega$.

(H2) The function $F : \Omega \times L^p(\Omega) \to \mathbb{R}$ is assumed to satisfy:
   (i) $(x, u) \mapsto F(x, u)$ is measurable in $x \in \Omega$ for all $u \in L^p(\Omega)$; and for a.e. $x \in \Omega$ the function $u \mapsto F(x, u)$ is increasing, i.e., $F(x, u) \le F(x, v)$ whenever $u \le v$.
   (ii) There exist constants $c_2 \ge 0$, $\mu \ge 0$ and $\alpha \in [0, p-1]$ such that

$$\|\mathcal{F}u\|_q \le c_2 + \mu \|u\|_p^\alpha, \quad \forall \, u \in L^p(\Omega),$$

   where $\mu \ge 0$ may be arbitrarily if $\alpha \in [0, p-1)$, and $\mu \in [0, \varepsilon)$ if $\alpha = p - 1$, where ε is the constant in (H1).

The main result of this subsection is given by the following theorem.

Theorem 4.1. *Let hypotheses (H1) and (H2) be satisfied. Then problem (4.1) possesses solutions, and the solution set of all solutions of (4.1) is bounded in V_0 and has minimal and maximal elements.*

Note that the existence result formulated in Theorem 4.1 only assumes certain growth conditions on Clarke's generalized gradient and on the nonlocal mapping $\mathcal{F}$. Further, the notion of *maximal* and *minimal* is understood in the set theoretical sense. The proof of Theorem 4.1 requires several preliminary results, which will be provided next.

Preliminaries

Let $h \in V_0^*$ be given. We consider first the following auxiliary multi-valued problem: Find $u \in V_0$ and $\eta \in L^q(\Omega)$ such that

$$\eta \in \partial j(\cdot, u), \quad \langle -\Delta_p u, v \rangle + \int_\Omega \eta v \, dx = \langle h, v \rangle, \quad \forall \, v \in V_0. \tag{4.4}$$

We are going to prove the existence of solutions of (4.4), the existence of *extremal solutions* of (4.4), and the monotone dependence on h of these extremal solutions. For the existence of solutions of (4.4) we make use of the following surjectivity result for multivalued pseudomonotone and coercive operators, see, e.g., [184, Theorem 2.6] or [229, Chapter 32], which we recall here for convenience (see also Chap. 9, Theorem 9.92).

Corollary 4.2. *Let X be a real reflexive Banach space with dual space X^*, and let the multivalued operator $\mathcal{A} : X \to 2^{X^*}$ be pseudomonotone, bounded, and coercive. Then $\mathcal{A}$ is surjective, i.e., range $(\mathcal{A}) = X^*$.*

By means of Corollary 4.2 the proof of the following result is basically a simple application of arguments used in the proof of the sub-supersolution method of Sect. 3.2.

Lemma 4.3. *The multi-valued elliptic problem (4.4) possesses solutions for each $h \in V_0^*$.*

Proof: We introduce the functional $J : L^p(\Omega) \to \mathbb{R}$ defined by

$$J(v) = \int_\Omega j(x, v(x))\, dx, \quad \forall\, v \in L^p(\Omega).$$

Using the growth condition (H1) and Lebourg's mean value theorem, we note that the function J is well-defined and Lipschitz continuous on bounded sets in $L^p(\Omega)$, thus locally Lipschitz. Moreover, the Aubin–Clarke theorem (see [80, p. 83]) ensures that, for each $u \in L^p(\Omega)$ we have

$$\eta \in \partial J(u) \Longrightarrow \eta \in L^q(\Omega) \text{ with } \eta(x) \in \partial j(x, u(x)) \text{ for a.e. } x \in \Omega.$$

Consider now the multi-valued operator $\mathcal{A} : V_0 \to 2^{V_0^*}$ defined by

$$\mathcal{A}(v) = -\Delta_p v + \Phi(v), \quad \forall\, v \in V_0,$$

where $\Phi(v) = (i^* \circ \partial J \circ i)(v)$. It is well known that $-\Delta_p : V_0 \to V_0^*$ is continuous, bounded, strictly monotone, and thus, in particular, pseudomonotone. From Sect. 3.2 we know that the multi-valued operator $\Phi : V_0 \to 2^{V_0^*}$ is pseudomonotone in the multi-valued sense, and bounded due to (H1). Thus $\mathcal{A} : V_0 \to 2^{V_0^*}$ is bounded and pseudomonotone, and due to Corollary 4.2, the operator $\mathcal{A}$ is surjective provided $\mathcal{A}$ is coercive. By making use of the equivalent norm in V_0, which is $\|u\|_{V_0}^p = \int_\Omega |\nabla u|^p\, dx$, and the variational characterization of the first eigenvalue of $-\Delta_p$, the coercivity can readily be seen as follows: For any $v \in V_0$ and any $\eta \in \Phi(v)$ we obtain by applying (H1) the estimate

$$\frac{1}{\|v\|_{V_0}}\langle -\Delta_p v + \eta, v\rangle \geq \frac{1}{\|v\|_{V_0}} \left[\int_\Omega |\nabla v|^p\, dx - \int_\Omega (k_1 + (\lambda_1 - \varepsilon)|v|^{p-1})|v|\, dx \right]$$

$$\geq \frac{1}{\|v\|_{V_0}} \left[\|v\|_{V_0}^p - \frac{\lambda_1 - \varepsilon}{\lambda_1}\|v\|_{V_0}^p - \|k_1\|_q \|v\|_p \right],$$

which proves the coercivity of $\mathcal{A}$. Applying Corollary 4.2 we obtain the existence of $u \in V_0$ such that $h \in \mathcal{A}(u)$, i.e., there is an $\eta \in \Phi(u)$ with $\eta \in L^q(\Omega)$ and $\eta(x) \in \partial j(x, u(x))$ for a.e. $x \in \Omega$ such that

$$-\Delta_p u + \eta = h \quad \text{in } V_0^*, \tag{4.5}$$

where

$$\langle \eta, \varphi \rangle = \int_\Omega \eta(x)\, \varphi(x)\, dx, \quad \forall\, \varphi \in V_0, \tag{4.6}$$

which proves that $u \in V_0$ is a solution of (4.4). $\qquad \square$

Lemma 4.4. *The multi-valued elliptic problem (4.4) possesses extremal solutions, i.e., (4.4) has the greatest solution u^* and the smallest solution u_*.*

Proof: Let us introduce the set S of all solutions of (4.4). The proof will be given in steps (a), (b), and (c).

(a) Claim: S is compact in V_0

First, let us show that S is bounded in V_0. To this end let u be any solution of (4.4). Testing equation (4.4) with $v = u$ we get

$$\langle -\Delta_p u, u \rangle = \langle h, u \rangle - \int_\Omega \eta\, u\, dx, \tag{4.7}$$

which yields by applying (H1) and taking into account $\eta \in \partial j(\cdot, u)$,

$$\|u\|_{V_0}^p \leq \|h\|_{V_0^*} \|u\|_{V_0} + \|k_1\|_q \|u\|_p + (\lambda_1 - \varepsilon)\|u\|_p^p.$$

By means of Young's inequality we get for any $\delta > 0$

$$\|u\|_{V_0}^p \leq \|h\|_{V_0^*} \|u\|_{V_0} + c(\delta) + \delta \|u\|_p^p + (\lambda_1 - \varepsilon)\|u\|_p^p,$$

which yields for $\delta < \varepsilon$ and setting $\tilde\varepsilon = \varepsilon - \delta$ the estimate

$$\|u\|_{V_0}^p \leq \|h\|_{V_0^*} \|u\|_{V_0} + c(\delta) + \frac{\lambda_1 - \tilde\varepsilon}{\lambda_1}\|u\|_{V_0}^p,$$

and hence the boundedness of S in V_0.

Let $(u_n) \subseteq S$. Then there is a subsequence (u_k) of (u_n) with

$$u_k \rightharpoonup u \text{ in } V_0, \quad u_k \to u \text{ in } L^p(\Omega), \quad \text{and } u_k(x) \to u(x) \text{ a.e. in } \Omega. \tag{4.8}$$

Since the u_k solve (4.4), we get with $v = u_k - u$ in (4.4)

$$\langle -\Delta_p u_k, u_k - u \rangle = \langle h, u_k - u \rangle - \int_\Omega \eta_k (u_k - u)\, dx, \tag{4.9}$$

where $\eta_k \in \partial j(\cdot, u_k)$, and $(\eta_k) \subset L^q(\Omega)$ is bounded due to the boundedness of (u_k) in V_0 and by applying (H1). We thus obtain from (4.8) and (4.9)

$$\limsup_k \langle -\Delta_p u_k, u_k - u \rangle = 0. \tag{4.10}$$

Since the operator $-\Delta_p$ enjoys the (S_+)-property, the weak convergence of (u_k) in V_0 along with (4.10) imply the strong convergence $u_k \to u$ in V_0, see, e.g., [44, Theorem D.2.1]. As $(\eta_k) \subset L^q(\Omega)$ is bounded, there is a subsequence (again denoted by (η_k)) that is weakly convergent in $L^q(\Omega)$, i.e.,

$$\eta_k \rightharpoonup \eta \text{ in } L^q(\Omega).$$

Similarly as in the proof of Lemma 3.29 of Sect. 3.2.3, one shows that $\eta \in \partial j(\cdot, u)$. Replacing u in (4.4) by u_k and η by η_k we may pass to the limit as $k \to \infty$, which proves that the limit u of $(u_k) \subseteq \mathcal{S}$ belongs to $\mathcal{S}$, and hence it follows that $\mathcal{S}$ is compact.

(b) Claim: $\mathcal{S}$ is a directed set

Let us show that $\mathcal{S}$ is upward directed, i.e., given $u_1, u_2 \in \mathcal{S}$ such that (u_1, η_1) and (u_2, η_2) satisfy (4.4), where $\eta_k \in \partial j(\cdot, u_k)$. We need to show that there is a $u \in \mathcal{S}$ satisfying $u_1 \leq u$ and $u_2 \leq u$. The proof adopts ideas used in Sect. 3.2.2 and relies on an appropriate construction of an auxiliary problem. To this end we consider the following auxiliary multi-valued problem: Find $u \in V_0$ and $\eta \in L^q(\Omega)$ such that

$$\tilde{\eta} \in \partial \tilde{j}(\cdot, u): \quad \langle -\Delta_p u - h + \hat{B}(u) - \hat{G}(u), v \rangle + \int_\Omega \tilde{\eta} v \, dx = 0, \quad \forall \, v \in V_0,$$
(4.11)

where $\hat{B} = i^* \circ B \circ i$ with B being the Nemytskij operator given by the following cut-off function $b : \Omega \times \mathbb{R} \to \mathbb{R}$:

$$b(x, s) = \begin{cases} 0 & \text{if } u_0(x) \leq s, \\ -(u_0(x) - s)^{p-1} & \text{if } s < u_0(x), \end{cases}$$
(4.12)

with $u_0 = \max\{u_1, u_2\}$. The function b is easily seen to be a Carathéodory function satisfying a growth condition of order $p - 1$, which implies that $B : L^p(\Omega) \to L^q(\Omega)$ is bounded and continuous, and thus $\hat{B} : V_0 \to V_0^*$ defines a completely continuous and bounded operator. The function $\tilde{j} : \Omega \times \mathbb{R} \to \mathbb{R}$ arises from the given j by the following construction:

$$\tilde{j}(x, s) = \begin{cases} j(x, u_0(x)) + \eta_0(x)(s - u_0(x)) & \text{if } s < u_0(x), \\ j(x, s) & \text{if } u_0(x) \leq s, \end{cases}$$
(4.13)

where η_0 is defined by

$$\eta_0(x) = \begin{cases} \eta_1(x) & \text{if } x \in \{u_1 \geq u_2\}, \\ \eta_2(x) & \text{if } x \in \{u_2 > u_1\}. \end{cases}$$

Thus Clarke's generalized gradient of $s \mapsto \tilde{j}(x, s)$ is given by

$$\partial \tilde{j}(x, s) = \begin{cases} \eta_0(x) & \text{if } s < u_0(x), \\ \partial \tilde{j}(x, u_0(x)) & \text{if } s = u_0(x), \\ \partial j(x, s) & \text{if } u_0(x) < s, \end{cases}$$
(4.14)

and the inclusions $\partial \tilde{j}(x, u_0(x)) \subseteq \partial j(x, u_0(x))$ holds true. If we introduce the functional $\tilde{J}$ by

$$\tilde{J}(v) = \int_\Omega \tilde{j}(x, v(x)) \, dx, \quad \forall \, v \in L^p(\Omega),$$

then
$$\tilde{\Phi} = i^* \circ \partial \tilde{J} \circ i$$

possesses qualitatively the same properties as Φ before. The operator $\hat{G}$ is given by
$$\hat{G} = i^* \circ G \circ i,$$

where G is the Nemytskij operator associated to $g : \Omega \times \mathbb{R} \to \mathbb{R}$ defined as follows
$$g(x, s) = \sum_{l=1}^{2} |g_l(x, s) - \eta_0(x)|, \tag{4.15}$$

with
$$g_l(x, s) = \begin{cases} \eta_l(x) & \text{if } s \le u_l(x), \\ \eta_l(x) + \frac{\eta_0(x) - \eta_l(x)}{u_0(x) - u_l(x)}(s - u_l(x)) & \text{if } u_l(x) < s < u_0(x), \\ \eta_0(x) & \text{if } s \ge u_0(x). \end{cases} \tag{4.16}$$

One easily verifies that $g : \Omega \times \mathbb{R} \to \mathbb{R}$ is a Carathéodory function, and its Nemytskij operator $G : L^p(\Omega) \to L^q(\Omega)$ is continuous and uniformly bounded. This allows us to apply the same arguments as in the proof of Lemma 4.3 to show the existence of solutions of problem (4.11) provided we are able to verify that the corresponding multi-valued operator related to (4.11) is coercive, i.e., we only need to verify the coercivity of $\mathcal{A} = -\Delta_p + \hat{B} - \hat{G} + \tilde{\Phi}$. This, however, readily follows from the proof of the coercivity of the operator $-\Delta_p + \Phi$, the uniform boundedness of $\hat{G}$, and the following estimate of $\langle \hat{B}(v), v \rangle$. In view of the definition (4.12) the function $s \mapsto b(x, s)$ is increasing and $b(\cdot, u_0) = 0$. Therefore we get by applying Young's inequality for any $\delta > 0$ the estimate

$$\langle \hat{B}(v), v \rangle = \int_\Omega b(\cdot, v)(v - u_0 + u_0)\, dx \ge \int_\Omega b(\cdot, v) u_0\, dx \ge -\delta \|v\|_p^p - c(\delta), \tag{4.17}$$

which implies the coercivity of $-\Delta_p + \hat{B} - \hat{G} + \tilde{\Phi}$ when δ is chosen sufficiently small, and hence the existence of solutions of the auxiliary problem (4.11). Now the set S is shown to be upward directed provided that any solution u of (4.11) with $\eta \in \partial \tilde{j}(\cdot, u)$ satisfies $u_k \le u$, $k = 1, 2$, because then $u_0 \le u$, and therefore $\hat{B}u = 0$, $\hat{G}u = 0$ and $\eta \in \partial \tilde{j}(\cdot, u) \subseteq \partial j(\cdot, u)$, and hence thus $u \in S$ exceeding u_k. Recall that $u_k \in S$ means that u_k satisfies

$$\eta_k \in \partial j(\cdot, u_k) : \quad \langle -\Delta_p u_k - h, v \rangle + \int_\Omega \eta_k v\, dx = 0, \quad \forall\, v \in V_0. \tag{4.18}$$

Taking the special test functions $v = (u_k - u)^+$ in (4.11) and in (4.18), and subtracting the resulting equations we obtain

$$\langle -\Delta_p u_k - (-\Delta_p u), (u_k - u)^+ \rangle - \langle \hat{B}(u), (u_k - u)^+ \rangle$$
$$+ \langle \hat{G}(u), (u_k - u)^+ \rangle + \int_\Omega (\eta_k - \tilde{\eta})(u_k - u)^+\, dx = 0, \tag{4.19}$$

where $\tilde{\eta} \in \partial \tilde{j}(\cdot, u)$. For the first term on the left-hand side of (4.19) we have the estimate

$$\langle -\Delta_p u_k - (-\Delta_p u), (u_k - u)^+ \rangle \geq 0. \tag{4.20}$$

For $x \in \{u_k > u\}$, from (4.16) it follows $g_k(x, u(x)) = \eta_k(x)$, hence we obtain by using the definition of g the equation (with $l \neq k$, and $l, k \in \{1, 2\}$)

$$g(x, u(x)) = |\eta_k(x) - \eta_0(x)| + |g_l(x, u(x)) - \eta_0(x)|, \quad x \in \{u_k > u\}. \tag{4.21}$$

Further, for $x \in \{u_k > u\} \subseteq \{u_0 > u\}$ from (4.14) it follows $\tilde{\eta}(x) \in \partial \tilde{j}(x, u(x)) = \eta_0(x)$, and thus by taking (4.21) into account we get

$$\langle \hat{G}(u), (u_k - u)^+ \rangle + \int_\Omega (\eta_k - \tilde{\eta})(u_k - u)^+ \, dx \geq 0. \tag{4.22}$$

In view of (4.20) and (4.22), from (4.19) we obtain

$$-\langle \hat{B}(u), (u_k - u)^+ \rangle \leq 0,$$

which by definition of $\hat{B}$ yields

$$0 \geq -\langle \hat{B}(u), (u_k - u)^+ \rangle = \int_\Omega (u_0 - u)^{p-1}(u_k - u)^+ \, dx \geq \int_\Omega ((u_k - u)^+)^p \, dx \geq 0.$$

The last inequality implies

$$\int_\Omega ((u_k - u)^+)^p \, dx = 0,$$

and hence it follows $(u_k - u)^+ = 0$, i.e., $u_k \leq u$, $k = 1, 2,$, which completes the proof of S being upward directed. The proof that S is downward directed is done similarly by changing the construction of the auxiliary problem in an obvious way.

(c) Claim: S possesses extremal solutions

The proof of this assertion is based on steps (a) and (b). We shall show the existence of the greatest element of S. Since V_0 is separable we have that $S \subset V_0$ is separable, too, so there exists a countable, dense subset $Z = \{z_n : n \in \mathbb{N}\}$ of S. By step (b), S is upward directed, so we can construct an increasing sequence $(u_n) \subseteq S$ as follows. Let $u_1 = z_1$. Select $u_{n+1} \in S$ such that

$$\max\{z_n, u_n\} \leq u_{n+1}.$$

The existence of u_{n+1} is due to step (b). By the compactness of S we find a subsequence of (u_n), denoted again by (u_n), and an element $u \in S$ such that $u_n \to u$ in V_0, and $u_n(x) \to u(x)$ a.e. in Ω. This last property of (u_n) combined with its increasing monotonicity implies that the entire sequence is convergent in V_0 and, moreover, $u = \sup_n u_n$. By construction, we see that

$$\max\{z_1, z_2, \ldots, z_n\} \leq u_{n+1} \leq u, \quad \forall n,$$

thus $Z \subseteq V_0^{\leq u} := \{w \in V_0 : w \leq u\}$. Since $V_0^{\leq u}$ is closed in V_0, we infer

$$\mathcal{S} \subseteq \overline{Z} \subseteq V_0^{\leq u},$$

which in conjunction with $u \in \mathcal{S}$ ensures that u is the greatest solution of (4.4).

The existence of the smallest solution of (4.4) can be proved in a similar way. This completes the proof of Lemma 4.4. $\qquad\square$

Monotone dependence of the extremal solutions of (4.4)

From Lemma 4.4 we know that for given $h \in V_0^*$ the multi-valued problem (4.4) has a smallest solution u_* and a greatest solution u^*. The purpose of this paragraph is to show that these extremal solutions depend monotonously on $h \in V_0^*$, where the dual order in V_0^* is defined by:

$$h_1, h_2 \in V_0^*: \quad h_1 \leq h_2 \iff \langle h_1, \varphi \rangle \leq \langle h_2, \varphi \rangle, \quad \forall \varphi \in V_0 \cap L_+^p(\Omega).$$

Lemma 4.5. *Let u_k^* be the greatest and $u_{k,*}$ the smallest solutions of (4.4) with right-hand sides $h_k \in V_0^*$, $k = 1, 2$, respectively. If $h_1 \leq h_2$ then it follows $u_1^* \leq u_2^*$ and $u_{1,*} \leq u_{2,*}$.*

Proof: We are going to prove $u_1^* \leq u_2^*$. To this end we consider the following auxiliary multi-valued problem: Find $u \in V_0$ and $\eta \in L^q(\Omega)$ such that

$$\eta \in \partial \hat{j}(\cdot, u): \quad \langle -\Delta_p u + \hat{B}(u), v \rangle + \int_\Omega \eta v \, dx = \langle h_2, v \rangle, \quad \forall v \in V_0, \quad (4.23)$$

where $\hat{B} = i^* \circ B \circ i$, and B is the Nemytskij operator related to the following cut-off function $b : \Omega \times \mathbb{R} \to \mathbb{R}$:

$$b(x, s) = \begin{cases} 0 & \text{if } u_1^*(x) \leq s, \\ -(u_1^*(x) - s)^{p-1} & \text{if } s < u_1^*(x), \end{cases} \quad (4.24)$$

which can be written as $b(x, s) = -[(u_1^*(x) - s)^+]^{p-1}$. The function $\hat{j} : \Omega \times \mathbb{R} \to \mathbb{R}$ is defined as the following modification of the given j: Let $\eta_1^* \in L^q(\Omega)$ belong to u_1^*, such that (u_1^*, η_1^*) is the greatest solution of (4.4) with right-hand side h_1, then we set

$$\hat{j}(x, s) = \begin{cases} j(x, u_1^*(x)) + \eta_1^*(x)(s - u_1^*(x)) & \text{if } s < u_1^*(x), \\ j(x, s) & \text{if } u_1^*(x) \leq s. \end{cases} \quad (4.25)$$

Clarke's generalized gradient of $\hat{j}$ is now given by:

$$\partial \hat{j}(x, s) = \begin{cases} \eta_1^*(x) & \text{if } s < u_1^*(x), \\ \partial \hat{j}(x, u_1^*(x)) & \text{if } s = u_1^*(x), \\ \partial j(x, s) & \text{if } u_1^*(x) < s, \end{cases} \quad (4.26)$$

and the inclusion $\partial \hat{j}(x, u_1^*(x)) \subseteq \partial j(x, u_1^*(x))$ holds true. With $\hat{J} : L^p(\Omega) \to \mathbb{R}$ given by

$$\hat{J}(v) = \int_\Omega \hat{j}(x, v(x)) \, dx$$

we define

$$\hat{\Phi}(u) = (i^* \circ \partial \hat{J} \circ i)(u).$$

The multi-valued operator

$$\mathcal{A} = -\Delta_p + \hat{B} + \hat{\Phi} : V_0 \to 2^{V_0^*}$$

is seen to satisfy Corollary 4.2, and thus $\mathcal{A} : V_0 \to 2^{V_0^*}$ is surjective. The proof follows basically the same arguments as in Lemma 4.3. Therefore, there exists a $u \in V_0$ such that $h_2 \in \mathcal{A}(u)$, i.e., there is an $\eta \in \partial \hat{J}(u)$ with $\eta(x) \in \hat{j}(x, u(x))$ such that

$$\langle -\Delta_p u + \hat{B}(u) + i^* \eta, v \rangle = \langle h_2, v \rangle, \quad \forall \, v \in V_0,$$

which shows that the auxiliary problem (4.23) has a solution. Let us show next that any solution u of (4.23) satisfies $u_1^* \le u$, where u_1^* is the greatest solution of (4.4) with right-hand side h_1, i.e.,

$$\eta_1^* \in \partial j(\cdot, u_1^*), \quad \langle -\Delta_p u_1^*, v \rangle + \int_\Omega \eta_1^* v \, dx = \langle h_1, v \rangle, \quad \forall \, v \in V_0. \tag{4.27}$$

Subtracting (4.23) from (4.27) and testing the resulting equation by $v = (u_1^* - u)^+ \in V_0 \cap L_+^p(\Omega)$ yields

$$\langle -\Delta_p u_1^* - (-\Delta_p u), (u_1^* - u)^+ \rangle - \langle \hat{B}(u), (u_1^* - u)^+ \rangle$$
$$+ \int_\Omega (\eta_1^* - \eta)(u_1^* - u)^+ \, dx = \langle h_1 - h_2, (u_1^* - u)^+ \rangle. \tag{4.28}$$

For $x \in \{u_1^* > u\}$ we get in view of (4.26) that $\eta(x) \in \partial \hat{j}(x, u(x)) = \eta_1^*(x)$, and therefore

$$\int_\Omega (\eta_1^* - \eta)(u_1^* - u)^+ \, dx = 0.$$

Since $h_1 \le h_2$, the right-hand side of (4.28) is nonpositive, and thus from (4.28) we deduce the inequality

$$-\langle \hat{B}(u), (u_1^* - u)^+ \rangle \le 0,$$

which by definition of $\hat{B}$ results in

$$- \int_\Omega -[(u_1^* - u)^+]^{p-1}(u_1^* - u)^+ \, dx = \int_\Omega [(u_1^* - u)^+]^p \, dx \le 0.$$

The last inequality implies $(u_1^* - u)^+ = 0$, i.e., $u_1^* \le u$, and hence $\hat{B}(u) = 0$. Furthermore, as $u_1^* \le u$, from (4.26) we see that $\eta \in \partial \hat{j}(x, u(x)) \subseteq \partial j(x, u(x))$, and hence any solution u of (4.23) is a solution of (4.4) with right-hand side h_2 as well, which satisfies $u_1^* \le u$. Because u_2^* is the greatest solution of (4.4) with right-hand side h_2, it follows $u_1^* \le u_2^*$. The proof for the monotone dependence of the smallest solutions follows by similar arguments and can be omitted. $\quad \square$

4.1.2 Proof of Theorem 4.1

The proof of Theorem 4.1 is based on the results of the preceding subsection, and, in particular, on the following consequence of the abstract fixed point result, Theorem 2.47, which has been proved in Chap. 2.

Theorem 4.6. *Let $E = (E, \| \cdot \|, \leq)$ be an ordered Banach space having the following properties:*

(E0) Bounded and monotone sequences of E have weak or strong limits.
(E1) $x^+ = \sup\{0, x\}$ exists, and $\|x^+\| \leq \|x\|$ for every $x \in E$.

If $\mathcal{G} : B_R(c) \to 2^{B_R(c)} \setminus \emptyset$ is an increasing multi-valued mapping whose values are weakly compact in E, then $\mathcal{G}$ has a maximal and minimal fixed point in $B_R(c)$, where $B_R(c)$ denotes the ball with radius R centered at $c \in E$.

Remark 4.7. (i) Note that according to the hypotheses of Theorem 4.6, no continuity assumption is imposed on the operator $\mathcal{G}$.

(ii) The notions *minimal* and *maximal* have to be understood in the usual set-theoretical sense. As for the definition of an increasing multi-valued mapping, we refer to Definition 2.4.

(iii) Theorem 4.6 remains true if E is merely an ordered normed space. As for examples of ordered Banach spaces E satisfying (E0) and (E1), we refer to Remark 2.50 of Chap. 2. In particular, the following two spaces, which are of importance here, are easily seen to have these properties:

(iv) $L^p(\Omega)$, $1 < p < \infty$, ordered a.e. pointwise, where Ω is a σ-finite measure space.
(v) The Sobolev spaces $W^{1,p}(\Omega)$, $W_0^{1,p}(\Omega)$, $1 < p < \infty$, ordered a.e. pointwise with Ω being a bounded Lipschitz domain in $\mathbb{R}^N$.

Proof of Theorem 4.1

We are going to relate our original nonlocal multi-valued elliptic problem (4.1) to the abstract setting of Theorem 4.6. For this purpose we need to transform problem (4.1) into a fixed point equation of the form

$$u \in \mathcal{G}u$$

for an appropriately defined multi-valued and increasing fixed point operator $\mathcal{G}$ acting on some ball B_R of $V_0 = W_0^{1,p}(\Omega)$ into itself. In view of hypothesis (H2), the operator $\mathcal{F} : V_0 \to L^q(\Omega)$ is well-defined and, moreover, increasing. Lemma 4.4 shows that for each $h \in L^q(\Omega)$, the elliptic inclusion (4.4) has the greatest and the smallest solution, which due to Lemma 4.5 depend monotonously on the right-hand side h. Now, let us define the operator $\mathcal{G} : v \mapsto \mathcal{G}v$ as follows:

For given $v \in V_0$, let $\mathcal{G}v$ denote the set of all solutions of the following multi-valued elliptic problem

$$\langle -\Delta_p u, \varphi \rangle + \int_\Omega \eta \, \varphi \, dx = \langle \mathcal{F}v, \varphi \rangle, \quad \forall \, \varphi \in V_0, \tag{4.29}$$

where $\eta \in L^q(\Omega)$ with $\eta(x) \in \partial j(x, u(x))$. Setting $h = \mathcal{F}v$, which is in $L^q(\Omega) \subseteq V_0^*$, from Lemma 4.3 we know that $\mathcal{G}v \neq \emptyset$, and, in view of Lemma 4.4, $\mathcal{G}v \subseteq V_0$ has the smallest and greatest element. Taking into account that $\mathcal{F} : V_0 \to L^q(\Omega)$ is an increasing operator, both the existence of extremal elements of $\mathcal{G}v$ and the monotone dependence of these extremal solutions on $h = \mathcal{F}v$ allow us to verify that $\mathcal{G} : V_0 \to 2^{V_0} \setminus \emptyset$ is an increasing multi-valued mapping according to Definition 2.4. Moreover, one readily observes that any fixed point of $\mathcal{G}$ is a solution of the original multi-valued elliptic problem (4.1) and vice versa. As V_0 satisfies (E0) and (E1) due to Remark 4.7, according to Theorem 4.6 it remains to show the existence of some ball $B_R \subset V_0$ with radius R (centered at $c = 0$) such that the multi-valued mapping $\mathcal{G}$ defined above is a mapping from B_R into itself whose values are weakly compact in V_0. Let us first prove the existence of a ball $B_R \subset V_0$ such that $\mathcal{G} : B_R \to 2^{B_R} \setminus \emptyset$. Let $u \in \mathcal{G}v$ be any solution of (4.29). Testing equation (4.29) with $\varphi = u$ and applying the growth condition of (H1) and (H2) on ∂j and $\mathcal{F}$, respectively, we obtain the estimate

$$
\begin{aligned}
\|u\|_{V_0}^p &\leq \|k_1\|_q \|u\|_p + (\lambda_1 - \varepsilon)\|u\|_p^p + \|\mathcal{F}v\|_q \|u\|_p \\
&\leq \|k_1\|_q \|u\|_p + (\lambda_1 - \varepsilon)\|u\|_p^p + \left(c_2 + \mu \|v\|_p^\alpha \right) \|u\|_p,
\end{aligned} \tag{4.30}
$$

where $\|u\|_{V_0}^p := \int_\Omega |\nabla u|^p \, dx$ for $u \in V_0$. By using the variational characterization of the first eigenvalue λ_1 we see that

$$\lambda_1 \|u\|_p^p \leq \|u\|_{V_0}^p, \quad \forall \, u \in V_0. \tag{4.31}$$

From (4.30) and (4.31) we deduce for some positive constant c_3 the following estimate:

$$
\begin{aligned}
\|u\|_{V_0}^p &\leq \left(c_3 + \mu \|v\|_p^\alpha \right) \|u\|_p + (\lambda_1 - \varepsilon)\|u\|_p^p \\
&\leq \left(c_3 + \mu \left(\frac{1}{\lambda_1} \right)^{\frac{\alpha}{p}} \|v\|_{V_0}^\alpha \right) \left(\frac{1}{\lambda_1} \right)^{\frac{1}{p}} \|u\|_{V_0} + \frac{\lambda_1 - \varepsilon}{\lambda_1} \|u\|_{V_0}^p,
\end{aligned}
$$

which results in

$$\frac{\varepsilon}{\lambda_1} \|u\|_{V_0}^{p-1} \leq c_4 + \mu \left(\frac{1}{\lambda_1} \right)^{\frac{\alpha+1}{p}} \|v\|_{V_0}^\alpha,$$

where $c_4 = c_3 (1/\lambda_1)^{1/p}$, and thus we get

$$\|u\|_{V_0}^{p-1} \leq c_5 + \frac{\mu}{\varepsilon} \lambda_1^{\frac{p-1-\alpha}{p}} \|v\|_{V_0}^\alpha, \tag{4.32}$$

where $c_5 = c_4 (\lambda_1/\varepsilon)$. According to hypothesis (H2) we have to consider two cases.

Case 1: $\alpha \in [0, p-1)$ and $\mu \geq 0$ arbitrary

First, by means of Young's inequality we get for any $\delta > 0$ not depending on v the inequality

$$\|v\|_{V_0}^{\alpha} \leq \delta \|v\|_{V_0}^{p-1} + C(\delta), \quad \forall\, v \in V_0. \tag{4.33}$$

Therefore, from (4.32) in conjunction with (4.33) we obtain

$$\|u\|_{V_0}^{p-1} \leq \tilde{\delta} \|v\|_{V_0}^{p-1} + C(\tilde{\delta}), \quad \forall\, v \in V_0, \tag{4.34}$$

for any $\tilde{\delta} > 0$. Thus, by choosing $\tilde{\delta}$ small enough such that $0 < \tilde{\delta} < 1$, from (4.34) we infer the existence of $R > 0$ satisfying

$$C(\tilde{\delta}) + \tilde{\delta} R^{p-1} = R^{p-1},$$

which proves that $\mathcal{G}$ is a mapping of B_R into itself.

Case 2: $\alpha = p - 1$ and $\mu \in [0, \varepsilon)$

In this case, from (4.32) we get

$$\|u\|_{V_0}^{p-1} \leq c_5 + \frac{\mu}{\varepsilon} \|v\|_{V_0}^{p-1},$$

and as $0 < \mu/\varepsilon < 1$, we readily find an $R > 0$ such that

$$c_5 + \frac{\mu}{\varepsilon} R^{p-1} = R^{p-1},$$

which shows again that $\mathcal{G}$ is a mapping from B_R into itself.

To apply the abstract fixed point theorem (Theorem 4.6) it remains to verify that the values of $\mathcal{G}v$ are weakly compact in V_0, where $v \in B_R \subset V_0$. To this end let us be given any sequence $(u_n) \subseteq \mathcal{G}v$, i.e., the u_n satisfy (4.29), which is

$$\langle -\Delta_p u_n, \varphi \rangle + \int_{\Omega} \eta_n \, \varphi \, dx = \langle \mathcal{F}v, \varphi \rangle, \quad \forall\, \varphi \in V_0, \tag{4.35}$$

where $\eta_n \in L^q(\Omega)$ with $\eta_n(x) \in \partial j(x, u_n(x))$ for a.e. $x \in \Omega$. We already know that $(u_n) \subset B_R \subset V_0$ for some $R > 0$, and therefore, due to (H1) the sequence (η_n) is bounded in $L^q(\Omega)$. Thus there exists a subsequence (u_k) of (u_n) and (η_k) of (η_n) with

$$u_k \rightharpoonup u \text{ in } V_0, \quad u_k \to u \text{ in } L^p(\Omega), \quad \eta_k \rightharpoonup \eta \text{ in } L^q(\Omega). \tag{4.36}$$

Replacing in (4.35) n by k, φ by $u_k - u$, and applying (4.36), one gets

$$\lim_{k \to \infty} \langle -\Delta_p u_k, u_k - u \rangle = 0,$$

which yields the strong convergence of $u_k \to u$ in V_0, since the operator $-\Delta_p$ enjoys the (S_+)-property. Replacing again n by k in (4.35), we may pass to the limit as $k \to \infty$, which results in

$$\langle -\Delta_p u, \varphi \rangle + \int_\Omega \eta \, \varphi \, dx = \langle \mathcal{F}v, \varphi \rangle, \quad \forall \, \varphi \in V_0, \tag{4.37}$$

where $\eta_k \rightharpoonup \eta$ in $L^q(\Omega)$ with $\eta_k(x) \in \partial j(x, u_k(x))$ for a.e. $x \in \Omega$. Similarly as in the proof of Lemma 3.29 of Sect. 3.2.3, we see that for the limit η we have $\eta(x) \in \partial j(x, u(x))$ for a.e. $x \in \Omega$. This proves the weak compactness of $\mathcal{G}v$. Thus we may apply the abstract fixed point result Theorem 4.6, which asserts the existence of minimal and maximal solutions of the original problem (4.1) in some ball B_R. To finish the proof of Theorem 4.1 we only need to verify that the set of all solutions of (4.1) belong to some ball B_R. The latter, however, readily follows from estimate (4.32) when replacing v on the right-hand side by u, which results in the following estimate for any solution of u of (4.1):

$$\|u\|_{V_0}^{p-1} \le c_5 + \frac{\mu}{\varepsilon} \lambda_1^{\frac{p-1-\alpha}{p}} \|u\|_{V_0}^\alpha. \tag{4.38}$$

In both cases of the hypothesis (H2)(ii) we readily see from (4.38) that all solutions are contained in some ball B_R. □

Example 4.8. Consider problem (4.1) with the nonlocal term $\mathcal{F}$ generated by the following F.

$$F(x, u) = [|x|] + \gamma \int_\Omega [u(y)] \, dy, \tag{4.39}$$

where $[\cdot] : \mathbb{R} \to \mathbb{Z}$ is the integer-function and γ is some positive constant. Let $|\Omega|$ denote the Lebesgue measure of the bounded domain $\Omega \subset \mathbb{R}^N$, and $c > 0$ some generic constant not depending on u. Then for $u \in L^p(\Omega)$ we get

$$|(\mathcal{F}u)(x)| \le c + \gamma \int_\Omega (|u(y)| + 1) \, dy \le c + \gamma |\Omega|^{1/q} \|u\|_p,$$

which yields the estimate

$$\|\mathcal{F}u\|_q \le c|\Omega|^{1/q} + \gamma |\Omega|^{2/q} \|u\|_p. \tag{4.40}$$

According to hypothesis (H2)(ii) we have the following correspondences: $c_2 = c|\Omega|^{1/q}$, $\mu = \gamma |\Omega|^{2/q}$, and $\alpha = 1$. Hence, under the assumption (H1) on j, by Theorem 4.1 the existence of solutions of (4.1) follows provided either $p > 2$ or $p = 2$ and $\gamma |\Omega|^{2/q} < \varepsilon$.

Remark 4.9. We note that the results obtained in this section can be extended to more general problems of the form (4.1) replacing the p-Laplacian by a general quasilinear elliptic operator of Leray–Lions type. By applying the definition of Clarke's generalized gradient, it is obvious that any solution of (4.1) is also a solution of the following so-called nonlocal hemivariational inequality:

$$u \in V_0: \quad \langle -\Delta_p u, v - u \rangle + \int_\Omega j^o(\cdot, u; v - u) \, dx \ge \langle \mathcal{F}u, v - u \rangle, \quad \forall \, v \in V_0, \tag{4.41}$$

where $j^o(x, s; r)$ denotes the generalized directional derivative of the locally Lipschitz function $s \mapsto j(x, s)$ at s in the direction r given by

$$j^o(x, s; r) = \limsup_{y \to s, \, t \downarrow 0} \frac{j(x, y + t\,r) - j(x, y)}{t}.$$

Problem (4.41) has been treated in [54], however, under an additional restriction on Clarke's generalized gradient $s \to \partial j(x, s)$ of the following form: There exists a constant $c \geq 0$ such that

$$\eta_1 \leq \eta_2 + c(s_2 - s_1)^{p-1}$$

for all $\eta_i \in \partial j(x, s_i)$, $i = 1, 2$, and for all s_1, s_2 with $s_1 < s_2$. Our treatment here allows us to completely drop this one-sided growth condition.

4.1.3 Extremal Solutions

Theorem 4.1 ensures the existence of minimal and maximal solutions of our original discontinuous, multi-valued, and nonlocal elliptic problem (4.1). The goal of this subsection is to show that (4.1) possesses in fact extremal solutions, i.e, (4.1) has the greatest and smallest solution. For this we only need to know that the solution set S of all solutions of (4.1) is a directed set, as will be seen by the following lemma.

Lemma 4.10. *If the solution set S of all solutions of (4.1) is directed, then S has the greatest and smallest element, which is the greatest and smallest solution of (4.1).*

Proof: Due to Theorem 4.1 there are minimal and maximal elements of S, which are minimal and maximal solutions of (4.1). Let us prove the existence of the greatest element of S. To this end we show that the maximal element is uniquely defined. Therefore, let u_1 and u_2 be two maximal elements of S with $u_1 \neq u_2$. Because S is supposed to be directed, it is, in particular, upward directed, and thus there is a $u \in S$ satisfying $u \geq u_1$ and $u \geq u_2$. However, this is a contradiction to the assumption of u_1 and u_2 being maximal elements. Let u^* be the uniquely defined maximal element. We easily can then see that u^* must be the greatest element of S. In fact, suppose there is a $u \in S$ such that $u \not\leq u^*$. Then, due to S being upward directed, there is a $v \in S$ such that $u \leq v$ and $u^* \leq v$. Apparently $v \neq u^*$, which is a contradiction to the fact that u^* is the uniquely defined maximal solution. Hence, it follows that u^* must be greatest element of S, i.e., u^* is the greatest solution of (4.1). Since the proof of the existence of the smallest solution follows obvious similar arguments, this completes the proof of the lemma. $\square$

Theorem 4.11. *Let hypotheses (H1) and (H2) be satisfied. Then the solution set S of all solutions of (4.1) is directed, and thus possesses extremal solutions.*

Proof: Let us prove only that $\mathcal{S}$ is upward directed, since the proof for $\underline{\mathcal{S}}$ being downward directed follows similar arguments. Assume u_1, $u_2 \in \mathcal{S}$, i.e., there are η_1, $\eta_2 \in L^q(\Omega)$ such that (u_k, η_k), $k = 1, 2$, satisfy (4.1), that is,

$$\eta_k \in \partial j(\cdot, u_k): \quad \langle -\Delta_p u_k, v \rangle + \int_\Omega \eta_k \, v \, dx = \langle \mathcal{F} u_k, v \rangle, \quad \forall \, v \in V_0. \quad (4.42)$$

The following appropriately constructed auxiliary problem is crucial for the proof: Find $u \in V_0$, and $\tilde{\eta} \in L^q(\Omega)$ satisfying

$$\tilde{\eta} \in \partial \tilde{j}(\cdot, u): \quad \langle -\Delta_p u + \hat{B}(u) - \hat{G}(u), v \rangle + \int_\Omega \tilde{\eta} \, v \, dx = \langle \mathcal{F}_{u_0} u, v \rangle, \quad \forall \, v \in V_0,$$
$$(4.43)$$

where $u_0 = \max\{u_1, u_2\}$, and $\tilde{j}$, $\hat{B}$, and $\hat{G}$ are constructed by means of u_k, u_0 and η_k, η_0 just like in step (b) of the proof of Lemma 4.4 in Sect. 4.1.1. The nonlocal term $\mathcal{F}_{u_0}$ on the right-hand side of (4.43) is defined as $\mathcal{F}_{u_0} u = F_{u_0}(\cdot, u)$, where $F_{u_0}(\cdot, u)$ arises from $F(\cdot, u)$ by truncation as follows:

$$F_{u_0}(x, u) = \begin{cases} F(x, u_0) & \text{if } x \in \{u < u_0\}, \\ F(x, u) & \text{if } x \in \{u \geq u_0\}. \end{cases}$$

If we set $A = -\Delta_p + \hat{B} - \hat{G}$, then the auxiliary problem (4.43) is equivalent to

$$\tilde{\eta} \in \partial \tilde{j}(\cdot, u): \quad \langle Au, v \rangle + \int_\Omega \tilde{\eta} \, v \, dx = \langle \mathcal{F}_{u_0} u, v \rangle, \quad \forall \, v \in V_0. \quad (4.44)$$

As $\hat{G} = i^* \circ G \circ i$, and $G : L^p(\Omega) \to L^q(\Omega)$ is continuous and uniformly bounded, we get

$$|\langle \hat{G}(u), v \rangle| \leq \int_\Omega |G(u)| \, |v| \, dx \leq c \, \|v\|_p.$$

Due to (4.17) we have for any $\delta > 0$ the estimate

$$\langle \hat{B}(v), v \rangle \geq -\delta \, \|v\|_p^p - c(\delta).$$

By the last two estimates we obtain for any $\hat{\delta} > 0$

$$\langle Av, v \rangle \geq \|v\|_{V_0}^p - \hat{\delta} \, \|v\|_p^p - c(\hat{\delta}). \quad (4.45)$$

The operator A may be considered as a compact perturbation of $-\Delta_p$, and thus behaves qualitatively basically like $-\Delta_p$. In view of (4.45), and because $\tilde{j}$ and $\mathcal{F}_{u_0}$ fulfil hypotheses (H1) and (H2), respectively, we may apply Theorem 4.1 to (4.44), which guarantees the existence of solutions of (4.44). The proof of the theorem is complete, provided one can show that any solution u of (4.44) satisfies $u_1 \leq u$ and $u_2 \leq u$, since then $u_0 \leq u$, and thus $\hat{B}(u) = 0$, $\hat{G}(u) = 0$, $\mathcal{F}_{u_0} u = \mathcal{F} u$, and $\tilde{\eta} \in \partial \tilde{j}(\cdot, u) \subseteq \partial j(\cdot, u)$, which shows that u is a solution of the original problem (4.1), i.e., $u \in \mathcal{S}$ and u exceeds u_1, $u_2 \in \mathcal{S}$. Taking the special functions $v = (u_k - u)^+$ in (4.42) and in (4.43), and subtracting (4.43) from (4.42) we obtain

$$\langle -\Delta_p u_k - (-\Delta_p u), (u_k - u)^+ \rangle - \langle \hat{B}(u), (u_k - u)^+ \rangle$$
$$+\langle \hat{G}(u), (u_k - u)^+ \rangle + \int_\Omega (\eta_k - \tilde{\eta})(u_k - u)^+) \, dx$$
$$= \langle \mathcal{F} u_k - \mathcal{F}_{u_0} u, (u_k - u)^+ \rangle. \tag{4.46}$$

Taking into account the definition of $\mathcal{F}_{u_0}$ and hypothesis (H2) (i), the right-hand side of (4.46) can be estimated as follows:

$$\langle \mathcal{F} u_k - \mathcal{F}_{u_0} u, (u_k - u)^+ \rangle = \int_\Omega (F(\cdot, u_k) - F_{u_0}(\cdot, u))(u_k - u)^+ \, dx$$
$$= \int_{\{u < u_k\}} (F(\cdot, u_k) - F(\cdot, u_0))(u_k - u) \, dx \le 0, \tag{4.47}$$

because $F(\cdot, u_k) - F(\cdot, u_0) \le 0$ due to (H2) (i). The left-hand side of (4.46) can be estimated below exactly the same way as in step (b) of the proof of Lemma 4.4, which yields

$$\langle -\Delta_p u_k - (-\Delta_p u), (u_k - u)^+ \rangle + \langle -\hat{B}(u) + \hat{G}(u), (u_k - u)^+ \rangle$$
$$\ge -\langle \hat{B}(u), (u_k - u)^+ \rangle. \tag{4.48}$$

From (4.46)–(4.48) we finally get

$$-\langle \hat{B}(u), (u_k - u)^+ \rangle \le 0,$$

which by definition of $\hat{B}$ yields

$$0 \ge -\langle \hat{B}(u), (u_k - u)^+ \rangle = \int_\Omega (u_0 - u)^{p-1}(u_k - u)^+ \, dx \ge \int_\Omega ((u_k - u)^+)^p \, dx \ge 0.$$

The last inequality implies

$$\int_\Omega ((u_k - u)^+)^p \, dx = 0,$$

and hence it follows $(u_k - u)^+ = 0$, i.e., $u_k \le u$, $k = 1, 2$, which completes the proof of $\mathcal{S}$ being upward directed. $\qquad \square$

In the next subsection, Theorem 4.1 and Theorem 4.11 are applied to elliptic inclusions whose multi-valued nonlinearity is given as the difference of Clarke's generalized gradient of some locally Lipschitz function and the subdifferential of a convex function.

4.1.4 Application: Difference of Clarke's Gradient and Subdifferential

We consider the Dirichlet boundary value problem for an elliptic inclusion governed by the p-Laplacian and a multivalued term that is given by the difference of Clarke's generalized gradient of some locally Lipschitz function and the subdifferential of some convex function. Our main goal is to characterize the solution set of the problem under consideration. In particular we are going to prove that the solution set possesses extremal elements with respect to the underlying natural partial ordering of functions, and that the solution set is compact.

Let $\Omega \subset \mathbb{R}^N$ be a bounded domain with Lipschitz boundary $\partial\Omega$. We consider the Dirichlet problem for the following elliptic inclusion

$$-\Delta_p u + \partial j(\cdot, u) - \partial\beta(\cdot, u) \ni 0 \quad \text{in} \quad \Omega, \quad u = 0, \quad \text{on} \quad \partial\Omega, \tag{4.49}$$

where $s \mapsto \partial j(x, s)$ is Clarke's generalized gradient of some locally Lipschitz function $j(x, \cdot) : \mathbb{R} \to \mathbb{R}$, which is measurable in $x \in \Omega$. The function $\beta : \Omega \times \mathbb{R} \to \mathbb{R}$ is assumed to be the primitive of some Borel measurable function $h : \Omega \times \mathbb{R} \to \mathbb{R}$, which is increasing in its second variable, i.e.,

$$\beta(x, s) = \int_0^s h(x, \tau)\, d\tau. \tag{4.50}$$

Thus $\beta(x, \cdot) : \mathbb{R} \to \mathbb{R}$ is convex, and $\partial\beta(x, \cdot) : \mathbb{R} \to 2^{\mathbb{R}} \setminus \emptyset$ is the usual subdifferential of convex functions. Then one has

$$\partial\beta(x, s) = [\underline{h}(x, s), \overline{h}(x, s)], \tag{4.51}$$

where $\underline{h}(x, s)$ and $\overline{h}(x, s)$ denote the left-sided and right-sided limits of h at s, respectively.

Definition 4.12. *A function $u \in V_0 = W_0^{1,p}(\Omega)$ is called a **solution** of the Dirichlet problem (4.49) if there are functions $\eta \in L^q(\Omega)$ and $\theta \in L^q(\Omega)$ such that the following holds:*

(i) $\eta(x) \in \partial j(x, u(x))$, $\quad \theta(x) \in \partial\beta(x, u(x))$ for a.e. $x \in \Omega$,

(ii) $\langle -\Delta_p u, \varphi \rangle + \displaystyle\int_\Omega (\eta(x) - \theta(x))\, \varphi(x)\, dx = 0, \quad \forall\, \varphi \in V_0.$

We assume the following hypotheses on j and h, where, as in Sect. 4.1.1, $\lambda_1 > 0$ is the first Dirichlet eigenvalue of $-\Delta_p$.

(D1) The function $j : \Omega \times \mathbb{R} \to \mathbb{R}$ satisfies hypothesis (H1) of Sect. 4.1.1, i.e., j is a Carathéodory function that is locally Lipschitz with respect to its second variable, and there is a $\varepsilon \in (0, \lambda_1)$ and a $k_1 \in L_+^q(\Omega)$ such that

$$|\eta| \le k_1(x) + (\lambda_1 - \varepsilon)|s|^{p-1}, \quad \forall\, \eta \in \partial j(x, s), \quad \forall\, s \in \mathbb{R},$$

for a.e. $x \in \Omega$.

(D2) The function $h : \Omega \times \mathbb{R} \to \mathbb{R}$ is Borel measurable, increasing in its second argument, and satisfies with some function $k_2 \in L^q_+(\Omega)$ the growth condition

$$|h(x, s)| \le k_2(x) + \mu |s|^{p-1}$$

for a.e. $x \in \Omega$ and for all $s \in \mathbb{R}$, and some constant $\mu \in [0, \varepsilon)$.

The main result of this subsection is given by the following theorem.

Theorem 4.13. *Let hypotheses (D1)–(D2) be satisfied. Then the multi-valued elliptic BVP (4.49) possesses extremal solutions, and the solution set of all solutions of (4.49) is a compact subset in V_0.*

Proof: The proof is carried out in two steps.

Step 1: Existence of extremal solutions

Consider the following auxiliary multi-valued problem related to (4.49):

$$-\Delta_p u + \partial j(\cdot, u) \ni \overline{h}(\cdot, u) \quad \text{in } \Omega, \quad u = 0, \quad \text{on } \partial\Omega. \tag{4.52}$$

A function $u \in V_0$ is called a solution of (4.52), if there is an $\eta \in L^q(\Omega)$ such that

$$\eta \in \partial j(\cdot, u) : \langle -\Delta_p u, v \rangle \int_\Omega \eta v \, dx = \int_\Omega \overline{h}(\cdot, u) v \, dx, \quad \forall \, v \in V_0. \tag{4.53}$$

If we define $F(x, u) = \overline{h}(x, u(x))$ then by hypothesis (D2) we get the estimate

$$\|\mathcal{F}u\|_q \le \|k_2\|_q + \mu \|u\|_p^{p-1}, \quad \forall \, u \in L^p(\Omega),$$

and thus (4.52) satisfies all assumptions of Theorem 4.11, which guarantees the existence of extremal solutions of (4.52). Apparently, any solution of (4.52) is a solution of (4.49) as well. Denote the greatest solution of (4.52) by u^*. We are going to show that any solution w of (4.49) satisfies $w \le u^*$, which proves the existence of the greatest solution of (4.49). Recall that $w \in V_0$ is a solution of (4.49) if there are functions $\eta_w \in L^q(\Omega)$ and $\theta_w \in L^q(\Omega)$ such that the following holds:

(i) $\eta_w(x) \in \partial j(x, w(x))$, $\theta_w(x) \in \partial\beta(x, w(x))$ for a.e. $x \in \Omega$,

(ii) $\langle -\Delta_p w, v \rangle + \displaystyle\int_\Omega \eta_w(x) v(x) = \int_\Omega \theta_w(x) v(x) \, dx, \quad \forall \, v \in V_0. \tag{4.54}$

To show $w \le u^*$, let us define an auxiliary problem that arises from (4.52) by truncation procedures as follows: Find $u \in V_0$, $\eta \in L^q(\Omega)$ such that

$$\eta \in \partial\hat{j}(\cdot, u) : \quad \langle -\Delta_p u + \hat{B}(u), v \rangle + \int_\Omega \eta v \, dx = \int_\Omega \overline{h}_w(\cdot, u) v \, dx, \quad \forall \, v \in V_0, \tag{4.55}$$

where $\hat{B} = i^* \circ B \circ i$ and B is the Nemytskij operator given by the following cut-off function $b : \Omega \times \mathbb{R} \to \mathbb{R}$:

$$b(x, s) = \begin{cases} 0 & \text{if } w(x) \leq s, \\ -(w(x) - s)^{p-1} & \text{if } s < w(x), \end{cases} \qquad (4.56)$$

which can be written as $b(x, s) = -[(w(x) - s)^+]^{p-1}$. The function $\hat{j} : \Omega \times \mathbb{R} \to \mathbb{R}$ is defined as the following modification of the given j:

$$\hat{j}(x, s) = \begin{cases} j(x, w(x)) + \eta_w(x)(s - w(x)) & \text{if } s < w(x), \\ j(x, s) & \text{if } w(x) \leq s. \end{cases} \qquad (4.57)$$

Clarke's generalized gradient of $\hat{j}$ is now given by:

$$\partial \hat{j}(x, s) = \begin{cases} \eta_w(x) & \text{if } s < w(x), \\ \partial \hat{j}(x, w(x)) & \text{if } s = w(x), \\ \partial j(x, s) & \text{if } w < s, \end{cases} \qquad (4.58)$$

and the inclusion $\partial \hat{j}(x, w(x)) \subseteq \partial j(x, w(x))$ holds true. The nonlinearity $\overline{h}_w : \Omega \times \mathbb{R} \to \mathbb{R}$ on the right-hand side of (4.55) is defined by

$$\overline{h}_w(x, s) = \begin{cases} \overline{h}(x, w(x)) & \text{if } s < w(x), \\ \overline{h}(x, s) & \text{if } w(x) \leq s. \end{cases} \qquad (4.59)$$

One verifies that (4.55) satisfies all assumptions of Theorem 4.11, which guarantees the existence of solutions, even extremal solutions of (4.55). Let u be any solution of (4.55). We next show that $w \leq u$. To this end we test (4.54) and (4.55) with $v = (w - u)^+$ and subtract (4.55) from (4.54) to get

$$\langle -\Delta_p w - (-\Delta_p u), (w - u)^+ \rangle - \langle \hat{B}(u), (w - u)^+ \rangle + \int_\Omega (\eta_w - \eta)(w - u)^+ \, dx$$

$$= \int_\Omega (\theta_w - \overline{h}_w(\cdot, u)) (w - u)^+ \, dx. \qquad (4.60)$$

For $x \in \{w > u\}$ we get in view of (4.58) that $\eta(x) \in \partial \hat{j}(x, u(x)) = \eta_w(x)$, and therefore

$$\int_\Omega (\eta_w - \eta)(w - u)^+ \, dx = 0.$$

The right-hand side of (4.60) is nonpositive, since $\theta_w(x) - \overline{h}_w(x, u(x))) \leq 0$ for $x \in \{w > u\}$. Thus from (4.60) we deduce the inequality

$$-\langle \hat{B}(u), (w - u)^+ \rangle \leq 0,$$

which by definition of $\hat{B}$ results in

$$-\int_\Omega -[(w - u)^+]^{p-1}(w - u)^+ \, dx = \int_\Omega [(w - u)^+]^p \, dx \leq 0.$$

The last inequality implies $(w - u)^+ = 0$, i.e., $w \leq u$, and hence $\hat{B}(u) = 0$, and $\overline{h}_w = \overline{h}$. Furthermore, as $w \leq u$, from (4.58) we see that $\eta \in \partial \hat{j}(x, u(x)) \subseteq \partial j(x, u(x))$, and hence any solution u of the auxiliary problem (4.55) must be a solution of (4.53). As u^* is the greatest solution of (4.53), it follows $w \leq u^*$, which completes the proof of the existence of the greatest solution. The existence of the smallest solution u_* of (4.49) can be shown in a similar way by using the auxiliary problem

$$-\Delta_p u + \partial j(\cdot, u) \ni \underline{h}(\cdot, u) \quad \text{in} \quad \Omega, \quad u = 0, \quad \text{on} \quad \partial \Omega.$$

Step 2: Compactness of the solution set

Let $\mathcal{S}$ denote the set of all solutions of (4.49), which has extremal solutions according to Step 1, i.e., there is the greatest and smallest solution u^* and u_*, respectively, and

$$u_* \leq u \leq u^*, \quad \forall \, u \in \mathcal{S}.$$

This implies that $\mathcal{S}$ is $L^p(\Omega)$-bounded. Let $(u_n) \subseteq \mathcal{S}$ be any sequence. Replacing u by u_n in (ii) of Definition 4.12, testing the relation with $\varphi = u_n$, and applying (D1)–(D2) we get the estimate

$$\|\nabla u_n\|_p^p \leq (\|k_1\|_q + \|k_2\|_q)\|u_n\|_p + (\lambda_1 - \varepsilon + \mu)\|u_n\|_p^p,$$

which shows that (u_n) is bounded in V_0, because the right-hand side of the last inequality is bounded. The corresponding sequences (η_n) and (θ_n) satisfying $\eta_n \in \partial j(\cdot, u_n)$ and $\theta_n \in \beta(\cdot, u_n)$, are bounded in $L^q(\Omega)$. Therefore, there are subsequences (u_k), (η_k), and (θ_k) with the following convergence properties:

$$u_k \rightharpoonup u \text{ in } V_0, \quad u_k \to u \text{ in } L^p(\Omega), \tag{4.61}$$

$$\eta_k \rightharpoonup \eta \text{ in } L^q(\Omega), \tag{4.62}$$

$$\theta_k \rightharpoonup \theta \text{ in } L^q(\Omega), \tag{4.63}$$

which satisfy the relation

$$\langle -\Delta_p u_k, \varphi \rangle + \int_\Omega (\eta_k(x) - \theta_k(x)) \, \varphi(x) \, dx = 0, \quad \forall \, \varphi \in V_0. \tag{4.64}$$

Testing (4.64) with $\varphi = u_k - u$, we readily observe that

$$\limsup_{k \to \infty} \langle -\Delta_p u_k, u_k - u \rangle = 0,$$

which in view of the $(S)_+$-property of $-\Delta_p$ yields the strong convergence of (u_k) in V_0. This along with the convergence properties above allows to pass to the limit as $k \to \infty$ in (4.64), which results in

$$\langle -\Delta_p u, \varphi \rangle + \int_\Omega (\eta(x) - \theta(x)) \, \varphi(x) \, dx = 0, \quad \forall \, \varphi \in V_0. \tag{4.65}$$

For the limits η and θ we have $\eta \in \partial j(\cdot, u)$ and $\theta \in \beta(\cdot, u)$, which can be shown in a standard way, e.g., like in Sect. 3.2.3. This completes the proof. □

Remark 4.14. (i) Theorem 4.13 can be extended to more general Leray–Lions operators A such as, e.g.,

$$Au(x) = -\sum_{i=1}^{N} \frac{\partial}{\partial x_i} a_i(x, u(x), \nabla u(x)).$$

Only for the sake of simplifying our presentation and in order to emphasize the main idea we have taken $A = -\Delta_p$.

(ii) Hypotheses (D1) and (D2) can be relaxed in case that one assumes the existence of an ordered pair $\underline{u} \le \overline{u}$ with $\underline{u} \le 0$ on $\partial\Omega$ and $\overline{u} \ge 0$ on $\partial\Omega$ that satisfies the following inequalities:

$$\underline{u} \in V: \quad -\Delta_p \underline{u} + \underline{\eta} \le \underline{h}(\cdot, \underline{u}), \quad \text{where } \underline{\eta} \in \partial j(\cdot, \underline{u}),$$

$$\overline{u} \in V: \quad -\Delta_p \overline{u} + \overline{\eta} \ge \overline{h}(\cdot, \overline{u}), \quad \text{where } \overline{\eta} \in \partial j(\cdot, \overline{u}).$$

In this case ∂j and h are only required to satisfy a local growth condition where $\lambda_1 - \varepsilon$ in (D1) and μ in (D2) may be replaced by any constant.

4.2 State-Dependent Clarke's Gradient Inclusion

We develop a flexible tool in terms of sub- and supersolutions that allows for obtaining existence, bounds, and multiplicity of solutions for quasilinear elliptic inclusions with a discontinuous multi-function that is given as the product of a discontinuous nonlinearity and Clarke's generalized gradient, i.e., a discontinuously state-dependent Clarke's gradient. The special feature of this kind of multi-function is that they need neither be upper nor lower semicontinuous, which is usually a least requirement for its theoretical treatment. Our approach is based on a combined use of abstract fixed point results for monotone mappings on partially ordered sets provided in Chap. 2, and on existence and comparison results for multi-valued quasilinear elliptic problems with Clarke's generalized gradient developed in Chap. 3.

4.2.1 Statement of the Problem

Let $\Omega \subset \mathbb{R}^N$ be a bounded domain with Lipschitz boundary $\partial\Omega$. As before, let $V = W^{1,p}(\Omega)$ and $V_0 = W_0^{1,p}(\Omega)$, $1 < p < \infty$. We consider the following quasilinear elliptic inclusion under homogeneous Dirichlet boundary conditions:

$$u \in V_0: \quad Au + f(\cdot, u) \in h(u)\partial j(\cdot, u) \quad \text{in } V_0^*, \tag{4.66}$$

where A is a second-order quasilinear elliptic differential operator of divergence form

$$Au(x) = -\sum_{i=1}^{N} \frac{\partial}{\partial x_i} a_i(x, \nabla u(x)), \quad \text{with } \nabla u = \left(\frac{\partial u}{\partial x_1}, \ldots, \frac{\partial u}{\partial x_N} \right).$$

The function $(x, s) \mapsto j(x, s)$ with $j : \Omega \times \mathbb{R} \to \mathbb{R}$ is assumed to be measurable in $x \in \Omega$ for all $s \in \mathbb{R}$, and locally Lipschitz in $s \in \mathbb{R}$ for a.a. $x \in \Omega$, and $s \mapsto \partial j(x, s)$ denotes Clarke's generalized gradient. Further, we assume that $\partial j(x, s) \subset \mathbb{R}_+$ (cf., e.g., Fig. 4.1), $h : \mathbb{R} \to \mathbb{R}$ is increasing and bounded (not necessarily continuous), and $f : \Omega \times \mathbb{R} \to \mathbb{R}$ is a Carathéodory function. As h is allowed to be discontinuous the multi-valued function $s \mapsto h(s)\partial j(\cdot, s)$ on the right-hand side of (4.66) is, in general, neither monotone nor continuous. Even more, $s \mapsto h(s)\partial j(\cdot, s)$ need not necessarily be an upper semicontinuous or a lower semicontinuous multi-valued function (cf., e.g., Fig. 4.2), where $h : \mathbb{R} \to \mathbb{R}$ is given by $h(s) = -1$ if $s \leq 0$, and $h(s) = 1$ if $s > 0$. Therefore, in general, $s \mapsto h(s)\partial j(\cdot, s)$ cannot be represented in the form of a Clarke's generalized gradient, which causes serious difficulties in the analytical treatment of problem (4.66). This is because for the analysis of our discontinuous multi-valued problem neither variational methods nor fixed point theorems for multi-valued operators or the theory of multi-valued pseudomonotone operators can be applied. Differential inclusions (4.66) with $h = 1$ have attracted increasing interest over the past decades, because, among others, they arise, e.g., in mechanical problems governed by nonconvex, possibly nonsmooth energy functionals that appear if nonmonotone, multi-valued constitutive laws are taken into account, see [184, 190]. As the function h in (4.66) may be discontinuous, we are in a position to model certain free boundary problems or threshold phenomena, i.e., to take into consideration the fact that a certain constitutive law is active only if u passes a specific threshold value.

Our main goal and the novelty here is to establish existence and comparison results for the discontinuous multi-valued problem (4.66) in terms of sub- and supersolutions as introduced in Chap. 3. More precisely, we are going to prove the existence of extremal (greatest and smallest) solutions within an ordered interval of sub- and supersolution. In this way we are able to provide existence and comparison results for (4.66) with a rather irregular and complicated multi-valued right-hand side.

If the function $j : \Omega \times \mathbb{R} \to \mathbb{R}$ is given by the primitive of some nonnegative, measurable, and locally bounded function $g : \Omega \times \mathbb{R} \to \mathbb{R}$, i.e.,

$$j(x, s) := \int_0^s g(x, t)\, dt, \tag{4.67}$$

then the function $s \mapsto j(x, s)$ is locally Lipschitz for a.a. $x \in \Omega$, and the multi-valued function $s \mapsto h(s)\partial j(\cdot, s)$ on the right-hand side of (4.66) can be characterized as follows:

$$h(s)\partial j(x, s) = h(s)[g_1(x, s), g_2(x, s)], \tag{4.68}$$

where

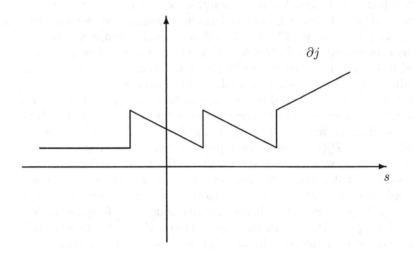

Fig. 4.1. Clarke's Gradient

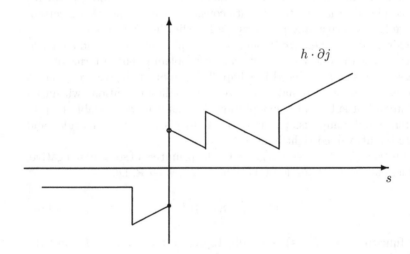

Fig. 4.2. State-Dependent Clarke's Gradient

$$g_1(x,t) := \lim_{\delta \to 0^+} \operatorname*{ess\,inf}_{|\tau-t|<\delta} g(x,\tau), \quad g_2(x,t) := \lim_{\delta \to 0^+} \operatorname*{ess\,sup}_{|\tau-t|<\delta} g(x,\tau),$$

see [181, Proposition 1.7]. In this case problem (4.66) includes the following special cases:

(i) If $h : \mathbb{R} \to \mathbb{R}$ is continuous then one can show that (4.66) reduces to the multi-valued elliptic boundary value problem

$$u \in V_0 : \quad Au \in \partial \tilde{j}(\cdot, u) \quad \text{in } V_0^*, \tag{4.69}$$

where $s \mapsto \partial \tilde{j}(x, s)$ is Clarke's generalized gradient of some appropriately modified locally Lipschitz function $s \mapsto \tilde{j}(x, s)$. Existence and comparison results for inclusions of the form (4.69) governed by Clarke's gradient have been studied in Chap. 3.

(ii) If $h : \mathbb{R} \to \mathbb{R}$ is continuous and $g : \Omega \times \mathbb{R} \to \mathbb{R}$ is a Carathéodory function then the function $s \mapsto j(x, s)$ is continuously differentiable, and $\partial j(x, s) = \{\partial j(x, s)/\partial s\} = \{g(x, s)\}$, i.e., $s \mapsto \partial j(x, s)$ is single-valued, and (4.66) reduces to the (single-valued) elliptic boundary value problem

$$u \in V_0 : \quad Au + \tilde{f}(\cdot, u) = 0 \quad \text{in } V_0^*, \tag{4.70}$$

where $(x, s) \mapsto \tilde{f}(x, s) := f(x, s) - h(s)g(x, s)$ is a Carathéodory function. This is a special case of the multi-valued elliptic problems studied in Chap. 3.

(iii) If $h : \mathbb{R} \to \mathbb{R}$ is only assumed to be increasing (not necessarily continuous), and $g : \Omega \times \mathbb{R} \to \mathbb{R}$ is a Carathéodory function, then (4.66) reduces to the single-valued discontinuous elliptic boundary value problem

$$u \in V_0 : \quad Au = \tilde{f}(\cdot, u, u) \quad \text{in } V_0^*, \tag{4.71}$$

where $(x, r, s) \mapsto \tilde{f}(x, r, s) := h(r)g(x, s) - f(x, s)$, i.e., $(x, s) \mapsto \tilde{f}(x, r, s)$ is a Carathéodory function for $r \in \mathbb{R}$ fixed, and $r \mapsto \tilde{f}(x, r, s)$ is increasing for fixed (x, s) (note: $g(x, s) \geq 0$). This kind of single-valued problem will be treated in Sect. 4.3, see also Sect. 4.1 and [44, Chap. 5].

It should be mentioned that problem (4.66) can be extended to include more general elliptic operators A of Leray–Lions type such as

$$Au(x) = -\sum_{i=1}^{N} \frac{\partial}{\partial x_i} a_i(x, u, \nabla u(x)) + a_0(x, u, \nabla u(x)).$$

Nonlinear mixed boundary conditions can be treated as well.

4.2.2 Notions, Hypotheses, and Preliminaries

Let q denote the Hölder conjugate real to p with $1 < p < \infty$. We assume the following hypotheses of Leray–Lions type on the coefficient functions a_i, $i = 1, ..., N$, of the operator A:

(A1) Each $a_i : \Omega \times \mathbb{R}^N \to \mathbb{R}$ satisfies the Carathéodory conditions, i.e., $a_i(x, \zeta)$ is measurable in $x \in \Omega$ for all $\zeta \in \mathbb{R}^N$, and continuous in ζ for a.a. $x \in \Omega$. There exist a constant $c_0 > 0$ and a function $k_0 \in L^q(\Omega)$ such that

$$|a_i(x, \zeta)| \le k_0(x) + c_0 \, |\zeta|^{p-1},$$

for a.a. $x \in \Omega$ and for all $\zeta \in \mathbb{R}^N$.

(A2) For a.a. $x \in \Omega$, and for all $\zeta, \zeta' \in \mathbb{R}^N$ with $\zeta \neq \zeta'$ the following monotonicity holds:

$$\sum_{i=1}^{N} (a_i(x, \zeta) - a_i(x, \zeta'))(\zeta_i - \zeta_i') > 0.$$

(A3) There is some constant $\nu > 0$ such that for a.a. $x \in \Omega$ and for all $\zeta \in \mathbb{R}^N$ the inequality

$$\sum_{i=1}^{N} a_i(x, \zeta)\zeta_i \ge \nu|\zeta|^p - k_1(x)$$

is satisfied for some function $k_1 \in L^1(\Omega)$.

Let $\langle \cdot, \cdot \rangle$ denote the dual pairing between V_0 and V_0^*. In view of (A1), (A2), the operator A defined by

$$\langle Au, \varphi \rangle := \int_\Omega \sum_{i=1}^{N} a_i(x, \nabla u)\frac{\partial \varphi}{\partial x_i} \, dx, \quad \forall \, \varphi \in V_0$$

is known to provide a continuous, bounded, and monotone (resp. strictly monotone) mapping from V (resp. V_0) into V_0^*.

As for the functions $j : \Omega \times \mathbb{R} \to \mathbb{R}$, $h : \mathbb{R} \to \mathbb{R}$, and $f : \Omega \times \mathbb{R} \to \mathbb{R}$ we assume the following hypotheses with $c \ge 0$ being some universal constant that may have different values at different places:

(H1) The function $x \mapsto j(x, s)$ is measurable in Ω for all $s \in \mathbb{R}$, and $s \mapsto j(x, s)$ is locally Lipschitz continuous in $\mathbb{R}$ for a.a. $x \in \Omega$. There exists a $k_2 \in L_+^q(\Omega)$ such that

$$0 \le \eta \le k_2(x) + c|s|^{p-1}, \quad \forall \, \eta \in \partial j(x, s),$$

for a.a. $x \in \Omega$ and for all $s \in \mathbb{R}$.

(H2) $h : \mathbb{R} \to \mathbb{R}$ is increasing and bounded, i.e.,

$$|h(s)| \le c, \quad \forall \, s \in \mathbb{R}.$$

(H3) $f : \Omega \times \mathbb{R} \to \mathbb{R}$ is a Carathéodory function satisfying for some $k_3 \in L_+^q(\Omega)$ the growth condition

$$|f(x, s)| \le k_3(x) + c|s|^{p-1}$$

for a.a. $x \in \Omega$ and for all $s \in \mathbb{R}$.

Notions

For convenience let us recall the notion of sub- and supersolution specified to the multi-valued problem (4.66)

Definition 4.15. *A function $u \in V_0$ is called a **solution** of (4.66) if there is an $\eta \in L^q(\Omega)$ satisfying $\eta(x) \in \partial j(x, u(x))$ for a.a. $x \in \Omega$, and*

$$\int_\Omega \sum_{i=1}^N a_i(x, \nabla u) \frac{\partial \varphi}{\partial x_i} \, dx + \int_\Omega f(x, u) \, \varphi \, dx = \int_\Omega h(u) \, \eta \, \varphi \, dx, \quad \forall \, \varphi \in V_0.$$

$$(4.72)$$

If F and H denote the Nemytskij operators related to f and h by $F(u)(x) := f(x, u(x))$ and $H(u)(x) := h(u(x))$, then equality (4.72) is equivalent to the following operator equation:

$$Au + i^* F(u) = i^* (H(u)\eta) \quad \text{in } V_0^*,$$

where $i^* : L^q(\Omega) \hookrightarrow V_0^*$ denotes the adjoint operator to the embedding $i : V \hookrightarrow L^p(\Omega)$, i.e., for $\alpha \in L^q(\Omega)$ we have

$$\langle i^* \alpha, \varphi \rangle = \int_\Omega \alpha \varphi \, dx, \quad \forall \, \varphi \in V_0.$$

(Note we write $iu = u \in L^p(\Omega)$ for $u \in V$.) Let us recall our basic notion of sub- and supersolution for (4.66), which can easily be deduced from the general notion introduced in Chap. 3

Definition 4.16. *A function $\underline{u} \in V$ is called a **subsolution** of (4.66) if $\underline{u}|_{\partial \Omega} \leq 0$, and if there is an $\underline{\eta} \in L^q(\Omega)$ satisfying $\underline{\eta}(x) \in \partial j(x, \underline{u}(x))$ for a.a. $x \in \Omega$, and*

$$\int_\Omega \sum_{i=1}^N a_i(x, \nabla \underline{u}) \frac{\partial \varphi}{\partial x_i} \, dx + \int_\Omega f(x, \underline{u}) \, \varphi \, dx \leq \int_\Omega h(\underline{u})\underline{\eta} \, \varphi \, dx, \quad \forall \, \varphi \in V_{0,+}.$$

$$(4.73)$$

Here we set $V_{0,+} := V_0 \cap L^p_+(\Omega)$ with $L^p_+(\Omega)$ denoting the (positive) cone of all nonnegative functions in $L^p(\Omega)$. In a similar way we define the supersolution.

Definition 4.17. *A function $\overline{u} \in V$ is called a **supersolution** of (4.66) if $\overline{u}|_{\partial \Omega} \geq 0$, and if there is an $\overline{\eta} \in L^q(\Omega)$ satisfying $\overline{\eta}(x) \in \partial j(x, \overline{u}(x))$ for a.a. $x \in \Omega$, and*

$$\int_\Omega \sum_{i=1}^N a_i(x, \nabla \overline{u}) \frac{\partial \varphi}{\partial x_i} \, dx + \int_\Omega f(x, \overline{u}) \, \varphi \, dx \geq \int_\Omega h(\overline{u})\overline{\eta} \, \varphi \, dx, \quad \forall \, \varphi \in V_{0,+}.$$

$$(4.74)$$

Remark 4.18. As already mentioned in Sect. 4.2.1, one can reduce problem (4.66) to the special case (4.69) in case that $h : \mathbb{R} \to \mathbb{R}$ is continuous and bounded. However, if h is allowed to be discontinuous, then, in general, $s \mapsto h(s)\partial j(x,s) - f(x,s)$ is neither upper nor lower semicontinous (see figures), and therefore (4.66) can no longer be reduced to (4.69).

Our main result of this section states that there exist extremal solutions of (4.66) within the ordered interval $[\underline{u}, \overline{u}]$ formed by a pair of sub- and supersolution. The proof of our main result is based on fixed point results for not necessarily continuous fixed point operators in partially ordered sets (see Sect. 2), and on comparison principles for multi-valued elliptic problems with Clarke's gradient proved in Chap. 3.

Preliminaries

First, we recall the comparison principle for elliptic inclusions with Clarke's generalized gradient of the form

$$u \in V_0 : \quad Au \in \partial \hat{j}(\cdot, u) \text{ in } V_0^*, \tag{4.75}$$

where $\hat{j} : \Omega \times \mathbb{R} \to \mathbb{R}$ is a function satisfying the following hypotheses:

(j1) The function $x \mapsto \hat{j}(x,s)$ is measurable in Ω for all $s \in \mathbb{R}$, and $s \mapsto \hat{j}(x,s)$ is locally Lipschitz continuous in $\mathbb{R}$ for a.a. $x \in \Omega$.
(j2) There exist $c > 0$, and $k \in L^q(\Omega)$ such that

$$|\eta| \le k(x) + c|s|^{p-1}, \quad \forall \eta \in \partial \hat{j}(x,s),$$

for a.a. $x \in \Omega$ and for all $s \in \mathbb{R}$.

In Sect. 3 the following result has been proved.

Theorem 4.19. *Let hypotheses (A1)–(A3) and (j1)–(j2) be satisfied, and assume the existence of sub- and supersolutions of (4.75), $\underline{u}$ and $\overline{u}$, respectively, satisfying $\underline{u} \le \overline{u}$. Then there exist the greatest and smallest solution of (4.75) within the ordered interval $[\underline{u}, \overline{u}]$.*

Furthermore, in the proof of our main result we make use of an abstract fixed point theorem for increasing (not necessarily continuous) operators in subsets of ordered normed spaces. Let $(E, \le)$ be an ordered normed space with partial order $\le$. The fixed point result we are going to apply is the following one.

Theorem 4.20. *Let $\mathcal{P}$ be a subset of an ordered normed space E, and let $G : \mathcal{P} \to \mathcal{P}$ be an increasing mapping, that is, if $x, y \in \mathcal{P}$ and $x \le y$, then $Gx \le Gy$. Then the following holds:*

(a) *If the image $G(\mathcal{P})$ has a lower bound in $\mathcal{P}$ and increasing sequences of $G(\mathcal{P})$ converge weakly in $\mathcal{P}$, then G has the smallest fixed point $x_* = \min\{x : Gx \le x\}$.*

(b) *If the image $G(\mathcal{P})$ has an upper bound in $\mathcal{P}$ and decreasing sequences of $G(\mathcal{P})$ converge weakly in $\mathcal{P}$, then G has the greatest fixed point $x^* = \max\{x : x \leq Gx\}$.*

Proof: As for the proof we note that Theorem 4.20 is a consequence of Proposition 2.39 of Chap. 2 (see also [44, Theorem 1.1.1]).   □

4.2.3 Existence and Comparison Result

Let $(\underline{u}, \overline{u})$ be an ordered pair of sub- and supersolutions of (4.66), i.e., $\underline{u} \leq \overline{u}$. Throughout this section we assume hypotheses (A1)–(A3) and (H1)–(H3). Consider the following auxiliary inclusion problem: Let $v \in [\underline{u}, \overline{u}]$ be fixed. Find $u \in V_0$ such that

$$Au + f(\cdot, u) \in h(v)\partial j(\cdot, u) \quad \text{in } V_0^*. \tag{4.76}$$

By means of (4.76) we are going to define a fixed point operator on some subset of the partially ordered normed space V, which will allow us to apply Theorem 4.20. For this purpose several auxiliary lemmas will be proved next.

Lemma 4.21. *Let $v \in [\underline{u}, \overline{u}]$ be any fixed supersolution of (4.66). Then problem (4.76) has the greatest solution v^* and the smallest solution v_* within $[\underline{u}, v]$. Analogously, if $w \in [\underline{u}, \overline{u}]$ is any fixed subsolution of (4.66), then problem (4.76) with v replaced by w has the greatest solution w^* and the smallest solution w_* within $[w, \overline{u}]$.*

Proof: We are going to prove the first part of the lemma. Let $v \in [\underline{u}, \overline{u}]$ be a fixed supersolution of (4.66). Defining

$$\hat{j}(x, s) := h(v(x))\, j(x, s) - \int_0^s f(x, t)\, dt, \tag{4.77}$$

we obtain $\partial \hat{j}(x, s) = h(v(x))\, \partial j(x, s) - f(x, s)$, and thus problem (4.76) can be rewritten in the form

$$u \in V_0 : \quad Au \in \partial \hat{j}(x, u) \quad \text{in } V_0^*. \tag{4.78}$$

By means of (H1)–(H3) one verifies that $\hat{j}$ satisfies hypotheses (j1)–(j2) of Theorem 4.19. Furthermore, since v is a supersolution of (4.66), it is also a supersolution of (4.76), and thus of (4.78). The given subsolution $\underline{u}$ of (4.66) is readily seen to be a subsolution of (4.78) due to (H1) and (H2). Applying Theorem 4.19 we obtain the existence of extremal solutions of problem (4.76) within $[\underline{u}, v]$. By similar arguments the second part of the lemma can be proved.   □

Lemma 4.22. *Let $v \in [\underline{u}, \overline{u}]$ be a supersolution of (4.66), and let v^* be the greatest solution of (4.76) within $[\underline{u}, v]$. Let $z \in [\underline{u}, \overline{u}]$ be a supersolution of (4.66), and let z^* be the greatest solution within $[\underline{u}, z]$ of (4.76) with v replaced by z. If $z \leq v$, then $z^* \leq v^*$.*

Proof: By definition z^* is the greatest solution of

$$z^* \in V_0 : \quad Az^* + f(\cdot, z^*) \in h(z)\partial j(\cdot, z^*) \quad \text{in } V_0^* \tag{4.79}$$

within $[\underline{u}, z]$. From $\partial j(x, s) \geq 0$, $s \mapsto h(s)$ increasing, and $z \leq v$ it follows that z^* is a subsolution for problem (4.76). As v is also a supersolution of (4.76) we infer by Theorem 4.19 the existence of extremal solutions of (4.76) within $[z^*, v]$, which implies $v^* \in [z^*, v]$, i.e., $z^* \leq v^*$. □

By obvious dual reasoning one can show the following result.

Lemma 4.23. *Let $w \in [\underline{u}, \overline{u}]$ be a subsolution of (4.66), and let w_* be the smallest solution within $[w, \overline{u}]$ of (4.76) with v replaced by w. Let $y \in [\underline{u}, \overline{u}]$ be a subsolution of (4.66), and let y_* be the smallest solution within $[y, \overline{u}]$ of (4.76) with v replaced by y. If $y \leq w$, then $y_* \leq w_*$.*

We define subsets $\mathcal{U}$ and $\mathcal{W}$ of V as follows:

$$\mathcal{U} := \{v \in V : v \in [\underline{u}, \overline{u}] \text{ and } v \text{ is a supersolution of (4.66)}\}, \tag{4.80}$$

$$\mathcal{W} := \{w \in V : w \in [\underline{u}, \overline{u}] \text{ and } w \text{ is a subsolution of (4.66)}\}, \tag{4.81}$$

and introduce operators G and L on $\mathcal{U}$ and $\mathcal{W}$, respectively. For $v \in \mathcal{U}$ let $Gv := v^*$ denote the greatest solution of (4.76) within $[\underline{u}, v]$. For $w \in \mathcal{W}$ let $Lw := w_*$ denote the smallest solution within $[w, \overline{u}]$ of (4.76) with v replaced by w. In view of Lemma 4.22 and Lemma 4.23, the operators $G : \mathcal{U} \to V_0$ and $L : \mathcal{W} \to V_0$ are well defined.

Lemma 4.24. *The operator $G : \mathcal{U} \to \mathcal{U}$ is increasing, and any fixed point of G is a solution of (4.66) within $[\underline{u}, \overline{u}]$, and vice versa.*

Proof: By Lemma 4.22 the operator $G : \mathcal{U} \to V_0$ is increasing, and for any $v \in \mathcal{U}$ we have $\underline{u} \leq Gv \leq v$ where $v^* := Gv$ is the greatest solution of (4.76) in $[\underline{u}, v]$, i.e.,

$$Av^* + f(\cdot, v^*) = h(v)\eta^* \quad \text{in } V_0^*,$$

where $\eta^* \in \partial j(\cdot, v^*)$. Due to (H1) and (H2), the last equation results in

$$Av^* + f(\cdot, v^*) \geq h(v^*)\eta^* \quad \text{in } V_0^*,$$

which shows that $Gv \in \mathcal{U}$, and thus $G : \mathcal{U} \to \mathcal{U}$ is increasing. The definition of G immediately implies that any fixed point u of G is a solution of (4.66) which belongs to $[\underline{u}, \overline{u}]$. Conversely, if $u \in [\underline{u}, \overline{u}]$ is a solution of (4.66) then u is, in particular, a supersolution, i.e., $u \in \mathcal{U}$, and it is trivially a solution of (4.76) with v replaced by u, and thus u is the greatest solution of (4.76) within $[\underline{u}, u]$, i.e., $u = Gu$. □

Lemma 4.25. *The range $G(\mathcal{U})$ of G has an upper bound in $\mathcal{U}$, and decreasing sequences of $G(\mathcal{U})$ converge weakly in $\mathcal{U}$.*

Proof: The given supersolution $\overline{u} \in \mathcal{U} \subseteq [\underline{u}, \overline{u}]$ is apparently an upper bound of $G(\mathcal{U})$. Let $(u_n) \subseteq G(\mathcal{U})$ be a decreasing sequence. Since $\mathcal{U}$ is L^p-bounded, we infer by means of (A3) and (H1)–(H3) that (u_n) is bounded in V_0, and thus by the monotonicity of the sequence and the compact embedding $V_0 \hookrightarrow L^p(\Omega)$ we get:

$$u_n \rightharpoonup u \text{ in } V_0, \quad u_n \to u \text{ in } L^p(\Omega). \tag{4.82}$$

To complete the proof we need to show that $u \in \mathcal{U}$. To this end we note that $u_n = Gv_n$ for some $v_n \in \mathcal{U}$, i.e.,

$$Au_n + f(\cdot, u_n) = h(v_n)\eta_n \text{ in } V_0^*, \tag{4.83}$$

where $\eta_n \in \partial j(\cdot, u_n)$. Testing (4.83) with $\varphi = u_n - u \in V_0$ we obtain

$$\langle Au_n, u_n - u \rangle = \int_\Omega \Big(h(v_n)\eta_n - f(\cdot, u_n) \Big)(u_n - u)\, dx. \tag{4.84}$$

Due to the boundedness of $\Big(h(v_n)\eta_n - f(\cdot, u_n) \Big)$ in $L^q(\Omega)$, the right-hand side of (4.84) tends to zero in view of (4.82), which by the (S_+)-property of A and the weak convergence of (u_n) results in the strong convergence of (u_n) in V_0, i.e.,

$$u_n \to u \text{ in } V_0.$$

As the limit u satisfies

$$\underline{u} \leq u \leq u_n \leq v_n \leq \overline{u},$$

we get by (H1) and (H2) $h(v_n)\eta_n \geq h(u)\eta_n$ for all $n \in \mathbb{N}$, which yields

$$\langle Au_n, \varphi \rangle + \int_\Omega f(\cdot, u_n)\,\varphi\, dx \geq \int_\Omega h(u)\eta_n\, \varphi\, dx, \quad \forall\, \varphi \in V_{0,+}. \tag{4.85}$$

From (H1) and (H2) it follows that (η_n) is bounded in $L^q(\Omega)$, and thus there is a weakly convergent subsequence of (η_n) (which is again denoted by (η_n)), i.e.,

$$\eta_n \rightharpoonup \eta \text{ in } L^q(\Omega). \tag{4.86}$$

The weak convergence of (η_n) and the strong convergence of (u_n) in V_0 allows us to pass to the limit in (4.85) (for some subsequence if necessary), which results in

$$\langle Au, \varphi \rangle + \int_\Omega f(\cdot, u)\,\varphi\, dx \geq \int_\Omega h(u)\eta\, \varphi\, dx, \quad \forall\, \varphi \in V_{0,+}. \tag{4.87}$$

As $u \in [\underline{u}, \overline{u}]$, the last inequality shows that the limit u is a supersolution of (4.66), i.e., $u \in \mathcal{U}$, provided the inclusion

$$\eta(x) \in \partial j(x, u(x)) \text{ for a.a. } x \in \Omega \tag{4.88}$$

holds. To complete the proof we are going to verify (4.88) by appropriately adopting the idea from Sect. 3. By definition of Clarke's gradient, (4.88) is proved provided that

$$\eta(x)\, r \le j^o(x, u(x); r) \qquad (4.89)$$

holds for a.a. $x \in \Omega$ and for all $r \in \mathbb{R}$. To this end let $r \in \mathbb{R}$ be fixed and E be any measurable subset of Ω. First we note that from the definition of j^o we infer that $x \mapsto j^o(x, v(x); r)$ is measurable in Ω whenever $x \mapsto v(x)$ is a measurable function in Ω, and thus

$$x \mapsto j^o(x, u(x); r), \quad x \mapsto j^o(x, u_n(x); r) \text{ are measurable.}$$

In view of (H1) and the inequality $\underline{u} \le u \le u_n \le v_n \le \overline{u}$ we get

$$|j^o(x, u(x); r)| \le |r|(k_2(x) + c|u(x)|^{p-1}) \le \tilde{k}(x)$$
$$|j^o(x, u_n(x); r)| \le |r|(k_2(x) + c|u_n(x)|^{p-1}) \le \tilde{k}(x),$$

for some $\tilde{k} \in L_+^q(\Omega)$, and thus $j^o(\cdot, u; r) \in L^q(\Omega)$ and $j^o(\cdot, u_n; r) \in L^q(\Omega)$ for all $n \in \mathbb{N}$. As $u_n(x) \to u(x)$ for a.a. $x \in \Omega$ and $s \mapsto j^o(x, s; r)$ is upper semicontinuous, we get by applying Fatou's lemma

$$\limsup_n \int_E j^o(x, u_n(x); r)\, dx \le \int_E \limsup_n j^o(x, u_n(x); r)\, dx$$

$$\le \int_E j^o(x, u(x); r)\, dx. \qquad (4.90)$$

The weak convergence (4.86) implies

$$\int_E \eta_n(x) r\, dx \to \int_E \eta(x) r\, dx \text{ as } n \to \infty. \qquad (4.91)$$

For each $n \in \mathbb{N}$ we have $\eta_n(x) \in \partial j(x, u_n(x))$, which by definition of Clarke's gradient results in

$$\eta_n(x) r \le j^o(x, u_n(x); r) \text{ for a.a. } x \in \Omega,$$

and thus

$$\int_E \eta_n(x) r\, dx \le \int_E j^o(x, u_n(x); r)\, dx, \quad \forall n \in \mathbb{N}. \qquad (4.92)$$

Passing to the lim sup in (4.92) and applying (4.90), (4.91), we finally get

$$\int_E \eta(x) r\, dx \le \int_E j^o(x, u(x); r)\, dx. \qquad (4.93)$$

Since inequality (4.93) holds for all measurable subsets E of Ω, it follows that (4.89) holds for a.a. $x \in \Omega$, which completes our proof. $\qquad \square$

As for the operator $L : W \to V_0$ defined above, we obtain analogous results that are summarized in the following lemma whose proof can be done by obvious modifications of the proofs of Lemmas 4.24 and 4.25.

Lemma 4.26. *The operator $L : \mathcal{W} \to \mathcal{W}$ is increasing, and any fixed point of L is a solution of (4.66) within the interval $[\underline{u}, \overline{u}]$ and vice versa. The range $L(\mathcal{W})$ of L has a lower bound in $\mathcal{W}$, and increasing sequences of $L(\mathcal{W})$ converge weakly in $\mathcal{W}$.*

Our main result here is the following existence and comparison theorem for problem (4.66).

Theorem 4.27. *Let hypotheses (A1)–(A3) and (H1)–(H3) be satisfied, and assume the existence of sub- and supersolutions of (4.66), $\underline{u}$ and $\overline{u}$, respectively, satisfying $\underline{u} \le \overline{u}$. Then problem (4.66) has the greatest and smallest solution within the ordered interval $[\underline{u}, \overline{u}]$.*

Proof: The proof is an immediate consequence of Lemma 4.24, Lemma 4.25, and Lemma 4.26 in conjunction with Theorem 4.20. First, let us focus on the existence of the greatest solution of (4.66) within $[\underline{u}, \overline{u}]$. By Lemma 4.24 any fixed point of $G : \mathcal{U} \to \mathcal{U}$ is a solution of (4.66) in $[\underline{u}, \overline{u}]$ and vice versa. Lemma 4.24 and Lemma 4.25 imply that Theorem 4.20 (b) can be applied, which yields the existence of the greatest fixed point of G in $\mathcal{U}$, and thus the greatest solution of (4.66) within $[\underline{u}, \overline{u}]$. Finally, by Lemma 4.26 and Theorem 4.20 (a) one infers the existence of the smallest solution of (4.66) within $[\underline{u}, \overline{u}]$. $\qquad\square$

4.2.4 Application: Multiplicity Results

Let us consider the following specific inclusion

$$-\Delta_p u \in h(u)\partial j(u) \text{ in } \Omega, \quad u = 0 \text{ on } \partial\Omega, \qquad (4.94)$$

where $\Delta_p u = \operatorname{div}(|\nabla u|^{p-2}\nabla u)$ is the p-Laplacian, and $2 \le p < \infty$. We assume that the multi-function $s \mapsto h(s)\partial j(s)$ on the right-hand side of (4.94) is like in the second figure of Sect. 4.2.1. By inspection we see that the multi-function admits an estimate in the form

$$-a_1 \le h(s)\partial j(s) \le a_1 + a_2|s|, \ \forall \, s \in \mathbb{R}, \qquad (4.95)$$

where a_1, a_2 are some positive constants. Let λ_1 be the first eigenvalue of $(-\Delta_p, V_0)$, which is positive and simple, see [6]. If $p > 2$ then by Young's inequality from (4.95) we get for any $\delta > 0$ the estimate

$$-a_1 \le h(s)\partial j(s) \le a_3(\delta) + \delta|s|^{p-1}, \ \forall \, s \in \mathbb{R}, \qquad (4.96)$$

where $a_3(\delta)$ is some positive constant that only depends on δ. In particular, δ may always be chosen in such a way that $\delta < \lambda_1$.

Corollary 4.28. *If $2 < p < \infty$, then problem (4.94) has a greatest and a smallest solution. If $p = 2$, then (4.94) has a greatest and a smallest solution provided that the constant a_2 in (4.95) satisfies $0 \le a_2 < \lambda_1$.*

Proof: In the case that $2 < p < \infty$ we consider the auxiliary (single-valued) problem

$$u \in V_0 : -\Delta_p u = a_3(\delta) + \delta|u|^{p-1} \quad \text{in } V_0^* \tag{4.97}$$

with $0 < \delta < \lambda_1$. One readily verifies that the operator S defined as $Su := -\Delta_p u - \delta|u|^{p-1}$ is a bounded, continuous, pseudomonotone, and coercive operator from V_0 into V_0^*. Hence it follows that $S : V_0 \to V_0^*$ is surjective, which ensures the existence of solutions of (4.97). Moreover, any solution u of (4.97) must be nonnegative. This can readily be seen by testing the equation (4.97) with $u^- := \max\{-u, 0\} \in V_0$, which results in

$$\int_\Omega |\nabla u|^{p-2} \nabla u \nabla(u^-) \, dx = \int_\Omega (a_3(\delta) + \delta|u|^{p-1})u^- \, dx \geq 0,$$

and thus

$$\int_\Omega |\nabla u^-|^p \, dx = \int_{\{x \in \Omega : u(x) \leq 0\}} |\nabla u|^p \, dx \leq 0,$$

which implies $u^- = 0$. Problem (4.97) may be considered as a special case of problem (4.49) treated in Sect. 4.1.4, and thus we can ensure the existence of greatest and smallest solutions of (4.97). Let $\overline{u}$ denote the greatest solution of (4.97). In view of (4.96) we infer that $\overline{u}$ is a supersolution for problem (4.94), where for $x \mapsto \overline{\eta}(x)$ one can take any measurable selection of $x \mapsto \partial j(\overline{u}(x))$. As $-\Delta_p : V_0 \to V_0^*$ is bounded, continuous, and even strongly monotone (note: $2 \leq p < \infty$) we readily conclude that the Dirichlet problem

$$u \in V_0 : -\Delta_p u = -a_1 \quad \text{in } V_0^* \tag{4.98}$$

has a unique solution denoted by $\underline{u}$, which apparently is nonpositive and a subsolution of (4.94). Therefore, by applying Theorem 4.27, problem (4.94) has the greatest solution u^* and the smallest solution u_* within $[\underline{u}, \overline{u}]$. Moreover, because any solution u of (4.94) is a subsolution of the auxiliary problem (4.97) and supersolution of the auxiliary problem (4.98), it must belong to the interval $[\underline{u}, \overline{u}]$, since $\overline{u}$ is the greatest solution of (4.97) and $\underline{u}$ is the unique solution of (4.98). This shows that the set $\mathcal{S}$ of all solutions of (4.94) satisfies $\mathcal{S} \subseteq [u_*, u^*] \subseteq [\underline{u}, \overline{u}]$, and u_*, u^* are the extremal elements of $\mathcal{S}$. The case $p = 2$ can be treated in just the same way.   □

A more detailed analysis can be carried out due to the behavior of $s \mapsto h(s)\partial j(s)$ for $s \leq 0$, which allows for an estimate in the form

$$-a_1 \leq h(s)\partial j(s) \leq -a_4, \ \forall \ s \leq 0, \tag{4.99}$$

with some positive constant a_4. By means of estimate (4.99) we can construct a nonpositive supersolution of (4.94) using the eigenfunction φ_1 of $(-\Delta_p, V_0)$ corresponding to λ_1, which satisfies $\varphi_1 \in \text{int}\,(C_0^1(\overline{\Omega})_+)$, where $\text{int}\,(C_0^1(\overline{\Omega})_+)$ is the interior of the positive cone $C_0^1(\overline{\Omega})_+$, which can be characterized by

$$\text{int}\,(C_0^1(\overline{\Omega})_+) = \{u \in C_0^1(\overline{\Omega}) \; : \; u(x) > 0, \; \forall x \in \Omega, \; \text{and} \; \frac{\partial u}{\partial n}(x) < 0, \; \forall x \in \partial\Omega\}.$$

Let us find a nonpositive supersolution $\overline{v}$ of (4.94) in the form $\overline{v} = -\varepsilon\,\varphi_1$ with $\varepsilon > 0$. Applying (4.99) we get for ε sufficiently small

$$-\Delta_p\overline{v} - h(\overline{v})\partial j(\overline{u}) \geq -\Delta_p\overline{v} + a_4 = -\lambda_1\varepsilon^{p-1}\varphi_1^{p-1} + a_4 \geq 0,$$

which shows that $\overline{v} = -\varepsilon\,\varphi_1$ is in fact a supersolution of (4.94). Let $\underline{u}$ be the unique solution of (4.98), which is a nonpositive subsolution of (4.94). The nonlinear regularity theory for the p-Laplacian and the nonlinear strong maximum principle due to [221] imply $\underline{u} \in -\text{int}\,(C_0^1(\overline{\Omega})_+)$. Therefore, for ε small enough one can always get $\underline{u} \leq \overline{v} = -\varepsilon\,\varphi_1$, which by applying Theorem 4.27 yields the following result.

Corollary 4.29. *The smallest solution u_* of problem (4.94) is nonpositive and satisfies $\underline{u} \leq u_* \leq -\varepsilon\,\varphi_1$ for ε sufficiently small, where $\underline{u}$ is the unique solution of (4.98).*

Next we are going to verify that the greatest solution u^* of (4.94) is nonnegative. As $h(0)\partial j(0) = \{-a_5\}$ for some positive constant a_5 we readily see that $u = 0$ is not a solution of problem (4.94). Thus (4.94) has at least two distinct solutions, namely u_* and u^* satisfying

$$\underline{u} \leq u_* \leq 0 \leq u^* \leq \overline{u}, \tag{4.100}$$

where $\underline{u}$ and $\overline{u}$ are given above. To prove (4.100) we are going to construct a nonnegative subsolution $\underline{v}$ of (4.94) in the form $\underline{v} = \varepsilon\varphi_1$ with $\varepsilon > 0$ and φ_1 as above. By inspection of the graph of $s \mapsto h(s)\partial j(s)$ (see second figure in Sect. 4.2.1) and in view of (4.96) we have the following estimate:

$$0 < a_6 \leq h(s)\partial j(s) \leq a_3(\delta) + \delta\, s^{p-1}, \quad \text{for } s > 0, \tag{4.101}$$

with $0 < \delta < \lambda_1$, where a_6 is some positive constant. Note, for $s = 0$ we have $h(0)\partial j(0) = \{-a_5\}$. If $\varphi \in V_{0,+}$ is any nonnegative test function, then we obtain with $\underline{v} = \varepsilon\varphi_1 \geq 0$

$$\langle -\Delta_p\underline{v} - h(\underline{v})\partial j(\underline{v}), \varphi \rangle = \int_\Omega \lambda_1(\varepsilon\varphi_1)^{p-1}\varphi\,dx - \int_{\{x \in \Omega : \underline{v}(x)=0\}} h(\underline{v})\partial j(\underline{v})\varphi\,dx$$

$$- \int_{\{x \in \Omega : \underline{v}(x)>0\}} h(\underline{v})\partial j(\underline{v})\varphi\,dx. \tag{4.102}$$

As $\varphi_1 \in \text{int}\,(C_0^1(\overline{\Omega})_+)$ we conclude that the N-dimensional Lebesgue measure of $\{x \in \Omega : \underline{v}(x) = 0\}$ is zero, and hence by means of (4.101) and (4.102) we get the estimate

$$\langle -\Delta_p\underline{v} - h(\underline{v})\partial j(\underline{v}), \varphi \rangle \leq \int_\Omega \left(\lambda_1(\varepsilon\varphi_1)^{p-1} - a_6\right)\varphi\,dx \leq 0 \tag{4.103}$$

for $\varepsilon > 0$ sufficiently small, which shows that $\underline{v} = \varepsilon\varphi_1 \geq 0$ is in fact a subsolution of (4.94). To complete the proof of (4.100) we need to make sure that the inequality $\underline{v} = \varepsilon\varphi_1 \leq \overline{u}$ holds true for some ε small. To this end we only note that $\underline{v}$ is readily seen to be also a subsolution of the auxiliary problem (4.97), and one can prove that (4.97) must have solutions above $\underline{v}$. Therefore, because $\overline{u}$ is the greatest solution of (4.97), it follows that $\underline{v} \leq \overline{u}$, and thus $(\underline{v}, \overline{u})$ is a pair of sub- and supersolution of (4.94). We summarize the obtained results in the following corollary.

Corollary 4.30. *The multi-valued boundary value problem (4.94) has a negative smallest solution u_* satisfying $\underline{u} \leq u_* \leq -\varepsilon\varphi_1$, and a greatest positive solution u^* satisfying $\varepsilon\varphi_1 \leq u^* \leq \overline{u}$, where $\underline{u}$ is the unique solution of (4.98) and $\overline{u}$ is the greatest solution of (4.97).*

4.3 Discontinuous Elliptic Problems via Fixed Points for Multifunctions

In this section we shall first recall fixed point results for multi-valued mappings $G : P \to 2^P \setminus \emptyset$ defined on a nonempty subset P of a lattice-ordered reflexive Banach space E with the property:

(N+) $\|u^+\| \leq \|u\|$ for each $x \in E$, where $u^+ := \sup\{u, 0\}$.

As is readily seen, the function spaces $L^p(\Omega)$, $W^{1,p}(\Omega)$ and $W_0^{1,p}(\Omega)$, $1 < p < \infty$, ordered by the positive cone $L_+^p(\Omega)$, possess property (N+). By means of these abstract fixed point theorems for multi-valued self-mappings we are able to treat discontinuous elliptic problems without assuming the existence of an ordered pair of sub-supersolutions. The problems considered here are single-valued generalizations of those treated in Sect. 4.1. Moreover, it should be mentioned that also elliptic problems with lack of compactness such as, e.g., elliptic equations involving nonlinearities with critical growth (critical Sobolev exponent) or elliptic problems in unbounded domains, can be treated using these abstract results, cf. [48]. The abstract fixed point theorems we are going to apply here are extensions of those already used in Sect. 4.1.

4.3.1 Abstract Fixed Point Theorems for Multi-Functions

Let E be a lattice-ordered reflexive Banach space. Before formulating the abstract fixed point theorems let us recall a few important notions in this respect.

Definition 4.31. *Given a nonempty subset P of E, we say that $v \in P$ is a* **maximal fixed point** *of a multi-valued mapping $F : P \to 2^P \setminus \emptyset$, if $v \in F(v)$, and if $u \in F(u)$ and $v \leq u$ implies $u = v$. If $u \in F(u)$ and $u \leq v$ implies $u = v$, we call v a* **minimal fixed point** *of F.*

The following notions of monotonicity for multi-valued mappings is used in what follows, see also Definition 2.4.

Definition 4.32. *We say that* $F : P \to 2^P \setminus \emptyset$ *is* **increasing upward** *if* $u, v \in P$, $u \leq v$, *and* $x \in F(u)$ *imply the existence of* $y \in F(v)$ *such that* $x \leq y$. *F is* **increasing downward** *if* $u, v \in P$, $u \leq v$, *and* $y \in F(v)$ *imply that* $x \leq y$ *for some* $x \in F(u)$. *If F is increasing upward and downward we say that F is* **increasing**.

Definition 4.33. *We say that a subset P of E has a* **sup-center** *(resp. an* **inf-center***)* c *if* $\sup\{c, u\}$ *(resp.* $\inf\{c, u\}$*) exists and belongs to P for each* $u \in P$. *If c is both a sup-center and an inf-center of P we say that c is an* **order center** *of P.*

Example 4.34. For instance, if E is a lattice-ordered Banach space, and if $\|u^+\| \leq \|u\|$ for each $u \in E$, that is (N+) holds, then the center of each closed ball B of E is its order center, which is justified by the the following lemma.

Lemma 4.35. *Let E be a lattice-ordered Banach space, and let* $\|u^+\| \leq \|u\|$ *for each* $u \in E$ *be satisfied. Then the center a of any closed ball* $B(a, r) = \{x \in E : \|x - a\| \leq r\}$ *with radius r is both a sup-center and inf-center.*

Proof: According to the definition of sup-center and inf-center we need to show that for any $y \in B(a, r)$ both $\sup\{a, y\}$ and $\inf\{a, y\}$ belong to $B(a, r)$. This, however, readily follows from the following representations:

$$\sup\{a, y\} = (y - a)^+ + a \quad \text{and} \quad \inf\{a, y\} = a - (a - y)^+,$$

so that by (N+), i.e., by applying $\|x^\pm\| \leq \|x\|$ we get

$$\|\sup\{a, y\} - a\| = \|(y - a)^+\| \leq \|(y - a)\| \leq r,$$

which shows that $\sup\{a, y\} \in B(a, r)$, and similarly $\inf\{a, y\} \in B(a, r)$. $\square$

The following theorem can be deduced from Chap. 2, Theorem 2.12, and Theorem 2.25.

Theorem 4.36. *Let E be a lattice-ordered reflexive Banach space. Assume that P is a bounded and weakly sequentially closed subset of E, and that* $G : P \to 2^P \setminus \emptyset$ *is an increasing mapping whose values are weakly sequentially closed. Then the following holds:*

(a) If P has a sup-center, then G has a minimal fixed point.
(b) If P has an inf-center, then G has a maximal fixed point.
(c) If P has an order center, then G has minimal and maximal fixed points.

An immediate consequence is the following theorem, which will be used later in our study of discontinuous elliptic problems.

Theorem 4.37. *Let E be a lattice-ordered reflexive Banach space with the property (N+) given above. If B is a closed and bounded ball in E, then each increasing mapping $G : B \to 2^B \setminus \emptyset$ with weakly sequentially closed values has minimal and maximal fixed points.*

Proof: The ball B, as a closed and convex set, is weakly sequentially closed. Moreover, property (N+) of E ensures that the center of B is its order center, see Example 4.34. Thus all the hypotheses of Theorem 4.36 hold, which implies the assertion. $\qquad\square$

Examples of spaces possessing property (N+) that are useful in applications are given in the following lemma whose proof is quite obvious and can therefore be omitted.

Lemma 4.38. *Each of the following spaces are lattice-ordered and reflexive Banach spaces having property (N+), when $1 < p < \infty$.*

(a) $L^p(\Omega)$, ordered a.e. pointwise, where $(\Omega, \mathcal{A}, \mu)$ is a σ-finite measure space.
(b) $W^{1,p}(\Omega)$, and $W_0^{1,p}(\Omega)$, ordered a.e. pointwise, where Ω is a bounded Lipschitz domain in $\mathbb{R}^N$.
(c) l^p, ordered coordinatewise and normed by the usual p-norm.
(d) $\mathbb{R}^N$, ordered coordinatewise and normed by the p-norm.

4.3.2 Discontinuous Elliptic Functional Equations

Let $\Omega \subset \mathbb{R}^N$ be bounded, and denote $V_0 = W_0^{1,p}(\Omega)$ with $1 < p < \infty$. By Lemma 4.38 it follows that V_0 is a lattice-ordered reflexive Banach space possessing property (N+) given at the beginning of Sect. 4.3. As an example of a quasilinear discontinuous boundary value problem we consider the following functional Dirichlet boundary value problem (BVP for short):

$$u \in V_0: \quad -\Delta_p u = f(\cdot, u(\cdot), u) \quad \text{in} \quad V_0^*, \qquad (4.104)$$

where Δ_p is as in the preceding sections the p-Laplacian. According to (4.104) we are seeking functions $u \in V_0$ that satisfy the following variational equation:

$$\int_\Omega \sum_{i=1}^N |\nabla u|^{p-2} \frac{\partial u}{\partial x_i} \frac{\partial \varphi}{\partial x_i} \, dx = \int_\Omega f(x, u(x), u) \, \varphi \, dx, \quad \forall \; \varphi \in V_0.$$

For the nonlinearity $f : \Omega \times \mathbb{R} \times V_0 \to \mathbb{R}$ we make the following assumptions:

(f1) The function f is superpositionally measurable (sup-measurable), i.e., $x \mapsto f(x, u(x), v)$ is measurable in Ω whenever $x \mapsto u(x)$ is measurable, $s \mapsto f(x, s, v)$ is continuous in $\mathbb{R}$, and $v \mapsto f(x, s, v)$ is increasing for a.e. $x \in \Omega$ and for all $s \in \mathbb{R}$ and $v \in V_0$.

(f2) f satisfies the growth condition

$$|f(x, s, v)| \leq k(x) + \mu |s|^{p-1} \quad \text{for a.e. } x \in \Omega, \ \forall s \in \mathbb{R}, \ \forall v \in V_0 \quad (4.105)$$

with $k \in L_+^q(\Omega)$, $q = p/p - 1$, and $0 \leq \mu < \lambda_1$, where λ_1 is the first Dirichlet eigenvalue of $-\Delta_p$.

Note that the right-hand side $(x, v) \mapsto f(x, s, v)$ of (4.104) is, in general, non-linear and discontinuous, and depends functionally on v. Therefore, problem (4.104) may be considered as a single-valued extension of the multi-valued problem treated in Sect. 4.1.

Let us denote the norms in V_0 and $L^p(\Omega)$ by $\|\cdot\|_{V_0}$ and $\|\cdot\|_p$, respectively. By applying the abstract fixed point theorems of the preceding subsection, we are going to prove first the following existence result.

Theorem 4.39. *Let hypotheses (f1) and (f2) be satisfied. Then the BVP (4.104) has minimal and maximal solutions in V_0.*

Proof: The proof will be given in three steps.

Step 1: A Priori Bound

Let u be any solution of (4.104). Then, by using the special test function $\varphi = u$, the growth condition (4.105), the variational characterization of the first eigenvalue λ_1, and Young's inequality, we obtain

$$\|\nabla u\|_p^p \leq \int_\Omega |k(x) u(x)| \, dx + \mu \int_\Omega |u(x)|^p \, dx \leq C(\varepsilon) + \frac{\varepsilon + \mu}{\lambda_1} \|\nabla u\|_p^p,$$

for any $\varepsilon > 0$. In view of (f2) and selecting ε sufficiently small, the last inequality implies the existence of a positive constant R such that

$$\|u\|_{V_0} \leq R.$$

Step 2: Multi-Valued Fixed Point Operator

Let $B(0, R)$ be the closed ball in V_0, and $v \in B(0, R)$ be fixed. Consider the following BVP:

$$u \in V_0: \quad -\Delta_p u = f(\cdot, u(\cdot), v) \quad \text{in } V_0^*. \quad (4.106)$$

For $v \in V_0$ fixed let $F_v(u)(x) = f(x, u(x), v)$. Due to (f2) the mapping $u \mapsto F_v(u)$ is bounded and continuous from $L^p(\Omega)$ into $L^q(\Omega)$, and thus $i^* \circ F_v \circ i : V_0 \to V_0^*$ is a bounded and completely continuous operator due to the compact embedding $V_0 \hookrightarrow L^p(\Omega)$. As $-\Delta_p : V_0 \to V_0^*$ is bounded and strictly monotone, it follows that

$$-\Delta_p - i^* \circ F_v \circ i : V_0 \to V_0^* \quad \text{is bounded and pseudomonotone.}$$

The BVP (4.106) is equivalent to the operator equation

$$u \in V_0 : \quad -\Delta_p u - (i^* \circ F_v \circ i)(u) = 0 \quad \text{in } V_0^*. \tag{4.107}$$

Therefore, the BVP (4.106) possesses solutions provided $-\Delta - i^* \circ F_v \circ i :$ $V_0 \to V_0^*$ is coercive. By applying (f2) we obtain

$$\langle -\Delta_p u - (i^* \circ F_v \circ i)(u), u \rangle \geq \|\nabla u\|_p^p - \|k\|_q \|u\|_p - \mu \|u\|_p^p$$

$$\geq \left(1 - \frac{\mu}{\lambda_1}\right) \|u\|_{V_0}^p - c\|u\|_{V_0},$$

which proves the coercivity. In view of Step 1, for $v \in B(0, R)$ fixed, any solution of (4.106) belongs to $B(0, R) \subset V_0$. Now we define the multi-valued mapping $G : B(0, R) \to 2^{B(0,R)}$ as $v \mapsto G(v)$, where $G(v)$ is the set of all solutions of (4.106). Apparently, any fixed point of G is a solution of the original problem (4.104) and vice versa.

Step 3: Existence of Maximal and Minimal Solutions

The existence of maximal and minimal solutions is proved provided G has maximal and minimal fixed points. By Theorem 4.37 the assertion is proved provided the multi-valued operator $G : B(0, R) \to 2^{B(0,R)}$ introduced in Step 2 is increasing, and $G(v)$ is a weakly sequentially closed subset in $B(0, R) \subset V_0$. Note the closed ball $B(0, R)$ of V_0 has property (N+) by Lemma 4.38. To show that $G(v)$ is weakly sequentially closed, assume that (u_n) is a sequence in $G(v)$, and that $u_n \rightharpoonup u$. Since $B(0, R)$ is weakly sequentially closed, then $u \in B(0, R)$. We only need to verify that $u \in G(v)$, which means that u satisfies (4.107). By definition the functions u_n satisfy

$$u \in V_0 : \quad -\Delta_p u_n - (i^* \circ F_v \circ i)(u_n) = 0 \quad \text{in } V_0^*. \tag{4.108}$$

From (4.108) we see that

$$\langle -\Delta_p u_n, u_n - u \rangle = \int_\Omega F_v(u_n)(u_n - u)\, dx \to 0, \quad \text{as } n \to \infty,$$

which implies in view of the (S)$_+$-property of $-\Delta_p : V_0 \to V_0^*$ that $u_n \to u$ is strongly convergent in V_0. Passing to the limit in (4.108) as $n \to \infty$ verifies that $u \in G(v)$, and thus $G(v)$ is weakly sequentially closed.

Next we are going to show that the multi-valued mapping $G : B(0, R) \to 2^{B(0,R)} \setminus \emptyset$ defined above is increasing in the sense of Definition 4.32. Let us first show that $G : B(0, R) \to 2^{B(0,R)} \setminus \emptyset$ is increasing upward. To this end let $v_1, v_2 \in B(0, R)$ be given with $v_1 \leq v_2$, and let $u_1 \in G(v_1)$, which means

$$u_1 \in V_0 : \quad -\Delta_p u_1 = f(\cdot, u_1(\cdot), v_1) \quad \text{in } V_0^*. \tag{4.109}$$

We need to show the existence of an $u_2 \in G(v_2)$ such that $u_1 \leq u_2$ holds. Consider the auxiliary BVP

$$u \in V_0 : \quad -\Delta_p u = \hat{f}(\cdot, u(\cdot), v_2) \quad \text{in } V_0^*, \tag{4.110}$$

where $\hat{f}$ is defined as follows:

$$\hat{f}(x,s,v) = \begin{cases} f(x,s,v) & \text{if} \quad s > u_1(x), \\ f(x,u_1(x),v) & \text{if} \quad s \le u_1(x). \end{cases}$$

Obviously, the truncated function $\hat{f}$ possesses the same regularity and growth conditions as f, and thus the existence of solutions of (4.110) can be shown in the same way as for (4.106). Let u_2 be any solution of (4.110). Then subtracting (4.110) from (4.109), and taking into account the monotonicity of $v \mapsto f(x,s,v)$, we obtain the inequality

$$-(\Delta_p u_1 - \Delta_p u_2) \le f(\cdot, u_1(\cdot), v_2) - \hat{f}(\cdot, u_2(\cdot), v_2), \quad \text{in } V_0^*. \tag{4.111}$$

Testing the last inequality with the nonnegative test function $\varphi = (u_1 - u_2)^+$, we obtain the inequality

$$\int_\Omega \sum_{i=1}^N \left(|\nabla u_1|^{p-2} \frac{\partial u_1}{\partial x_i} - |\nabla u_2|^{p-2} \frac{\partial u_2}{\partial x_i} \right) \frac{\partial(u_1 - u_2)^+}{\partial x_i} \, dx$$
$$\le \int_\Omega \left(f(x, u_1(x), v_2) - \hat{f}(x, u_2(x), v_2) \right) (u_1 - u_2)^+(x) \, dx. \tag{4.112}$$

Applying the definition of $\hat{f}$ one readily sees that the right-hand side of (4.112) is zero, which yields

$$0 \le \int_{\{u_1 \ge u_2\}} \sum_{i=1}^N \left(|\nabla u_1|^{p-2} \frac{\partial u_1}{\partial x_i} - |\nabla u_2|^{p-2} \frac{\partial u_2}{\partial x_i} \right) \frac{\partial(u_1 - u_2)}{\partial x_i} \, dx \le 0,$$

and thus $\nabla(u_1 - u_2)^+ = 0$, i.e., $\|(u_1 - u_2)^+\|_{V_0} = 0$, which is equivalent to $(u_1 - u_2)^+ = 0$, i.e., $u_1 \le u_2$. But then we have $\hat{f}(\cdot, u_2(\cdot), v_2) = f(\cdot, u_2(\cdot), v_2)$, and therefore the solution u_2 is actually a solution of the BVP

$$u_2 \in V_0 : \quad -\Delta_p u_2 = f(\cdot, u_2(\cdot), v_2) \text{ in } V_0^*,$$

which means $u_2 \in G(v_2)$. This proves that G is increasing upward. In a similar way one can also prove that G is increasing downward, i.e., for any $u_2 \in G(v_2)$ there exists a $u_1 \in G(v_1)$ such that $u_1 \le u_2$. This completes the proof for the multifunction G to be increasing. Now we are able to apply Theorem 4.37, which completes the proof of the existence of minimal and maximal solutions of the BVP (4.104). □

Remark 4.40. (i) Denoting the set of all solution of the BVP (4.104) by $\mathcal{S}$, one can show that $\mathcal{S}$ is a directed set, which implies that $\mathcal{S}$ has extremal solutions. The proof follows basically the approach used in Sect. 4.1.

(ii) Replacing $f(x,s,v)$ by $g(x,s,v(x))$ and μ by $\mu_1 + \mu_2$ we get the following existence theorem for the BVP

$$u \in V_0 : \quad -\Delta_p u = g(\cdot, u(\cdot), u(\cdot)) \text{ in } V_0^*. \tag{4.113}$$

Corollary 4.41. *Assume that the function* $g : \Omega \times \mathbb{R} \times \mathbb{R} \to \mathbb{R}$ *satisfies the following hypotheses.*

(g1) $g(\cdot, s, \cdot) : \Omega \times \mathbb{R} \to \mathbb{R}$ *is superpositionally measurable (sup-measurable),* $s \mapsto g(x, s, r)$ *is continuous, and* $r \mapsto g(x, s, r)$ *is increasing for a.e.* $x \in \Omega$ *and for all* $s, r \in \mathbb{R}$.

(g2) $|g(x, s, r)| \le k(x) + \mu_1 |s|^{p-1} + \mu_2 |r|^{p-1}$ *for a.e.* $x \in \Omega$, *for all* $s, r \in \mathbb{R}$, *with* $k \in L^q_+(\Omega)$, $q = p/p - 1$, $\mu_1 \ge 0$ *and* $\mu_2 \ge 0$.

(g3) $\mu_1 + \mu_2 < \lambda_1$, *where* λ_1 *is the first Dirichlet eigenvalue of* $-\Delta_p$.

Then the BVP (4.113) has minimal and maximal solutions, and even extremal ones in V_0.

Remark 4.42. In case that there exists an ordered pair of sub- and supersolutions for the discontinuous BVP (4.113), the assertion of Corollary 4.41 holds true without assuming hypotheses (g2) and (g3). Instead of (g2) and (g3) only a local $L^q(\Omega)$-bound with respect to the ordered interval of sub-supersolutions for the nonlinearity g is needed.

4.3.3 Implicit Discontinuous Elliptic Functional Equations

The abstract fixed point theorem Theorem 4.37 of Sect. 4.3.1 is used to prove first existence results for discontinuous functional equations in general L^p-spaces, which will then be employed to treat implicit and explicit elliptic boundary-value problems involving discontinuous nonlinearities.

Functional Equations in $\mathbf{L^p(\Omega)}$

Let $\Omega = (\Omega, \mathcal{A}, \mu)$ be a measure space, and assume that the space $L^p(\Omega)$ with norm $\|\cdot\|_p$, $1 < p < \infty$, is ordered by the positive cone $L^p_+(\Omega)$, i.e., $L^p(\Omega)$ is ordered a.e. pointwise. In the proof of our existence theorem for the functional equation

$$h(x) = f(x, \phi(h(x)), h(x)) \quad \text{a.e. in } \Omega, \tag{4.114}$$

we make use of the following fixed point result.

Lemma 4.43. *Assume that a mapping* $G : L^p(\Omega) \to L^p(\Omega)$ *is increasing, and that* $\|Gh\|_p \le M + \psi(\|h\|_p)$, *where* $\psi : \mathbb{R}_+ \to \mathbb{R}_+$ *is increasing, and* $M + \psi(R) \le R$ *for some* $R > 0$. *Then* G *has a fixed point.*

Proof: Choose an $R > 0$ such that $M + \psi(R) \le R$. Because ψ is increasing, then G maps the set $B(0, R) = \{h \in L^p(\Omega) : \|h\|_p \le R\}$ into itself. Due to Lemma 4.38 the space $L^p(\Omega)$ is a lattice-ordered and reflexive Banach space satisfying the property (N+). Thus G has a fixed point in $B(0, R)$ by Theorem 4.37. $\qquad\square$

For the functions $\phi : L^p(\Omega) \to L^p(\Omega)$ and $f : \Omega \times \mathbb{R} \times \mathbb{R} \to \mathbb{R}$ we impose the following hypotheses:

(ϕ) ϕ is increasing, and $\|\phi \circ h\|_p \leq m + \kappa\|h\|_p$ for some $m \geq 0$ and $\kappa > 0$.

(f1) f is sup-measurable, i.e., $x \mapsto f(x, u(x), v(x))$ is measurable in Ω whenever $u, v : \Omega \to \mathbb{R}$ are measurable.

(f2) $|f(x, y, z)| \leq k(x) + c_1(x)|y|^\alpha + c_2(x)|z|^\beta$ for a.e. $x \in \Omega$ and for all $y, z \in \mathbb{R}$, where $k \in L^p(\Omega)$, and either

   (i) $0 < \alpha, \beta < 1$, $c_1 \in L_+^{\frac{p}{1-\alpha}}(\Omega)$, $c_2 \in L_+^{\frac{p}{1-\beta}}(\Omega)$, and $f(x, \cdot, \cdot)$ is increasing for a.e. $x \in \Omega$, or

   (ii) $\alpha = \beta = 1$, $\kappa\|c_1\|_\infty + \|c_2\|_\infty < 1$, where κ is the constant in (ϕ), and the function $(y, z) \mapsto f(x, y, z) + \lambda z$ is increasing for a.e. $x \in \Omega$ and for some $\lambda \geq 0$.

Our existence result for the functional equation (4.114 reads as follows.

Theorem 4.44. *Under the assumptions (ϕ), (f1), and (f2), equation (4.114) has a solution h in $L^p(\Omega)$.*

Proof: The hypotheses (ϕ) and (f1) imply that for each $h \in L^p(\Omega)$ the relation

$$Gh := f(\cdot, \phi(h(\cdot)), h(\cdot)) \tag{4.115}$$

defines a measurable function $Gh : \Omega \to \mathbb{R}$. To show that (4.115) defines an increasing mapping $G : L^p(\Omega) \to L^p(\Omega)$ that satisfies the hypotheses of Lemma 4.43, we need to consider three cases depending on the assumptions of (f2):

(a) Case: (f2) (i)

$$\begin{aligned}
\|Gh\|_p &= \|f(\cdot, \phi(h), h)\|_p \leq \|k\|_p + \||c_1|\phi(h)|^\alpha\|_p + \|c_2|h|^\beta\|_p \\
&\leq \|k\|_p + (\|c_1^p\|_{\frac{1}{1-\alpha}} \| |\phi(h)|^{p\alpha}\|_{\frac{1}{\alpha}})^{\frac{1}{p}} + (\|c_2^p\|_{\frac{1}{1-\beta}} \| |h|^{p\beta}\|_{\frac{1}{\beta}})^{\frac{1}{p}} \\
&= \|k\|_p + \|c_1\|_{\frac{p}{1-\alpha}} \|\phi \circ h\|_p^\alpha + \|c_2\|_{\frac{p}{1-\beta}} \|h\|_p^\beta \\
&\leq \|k\|_p + \|c_1\|_{\frac{p}{1-\alpha}} (m + \kappa\|h\|_p)^\alpha + \|c_2\|_{\frac{p}{1-\beta}} \|h\|_p^\beta .
\end{aligned}$$

Thus $Gh \in L^p(\Omega)$, and

$$\|Gh\|_p \leq M + \psi(\|h\|_p), \tag{4.116}$$

where $M = \|k\|_p$ and $\psi(r) := \|c_1\|_{\frac{p}{1-\alpha}} (m + \kappa r)^\alpha + \|c_2\|_{\frac{p}{1-\beta}} r^\beta$. If $h_1, h_2 \in L^p(\Omega)$, $h_1 \leq h_2$, then $\phi(h_1) \leq \phi(h_2)$ by (ϕ). Since $f(x, \cdot, \cdot)$ is increasing, then for a.e. $x \in \Omega$,

$$Gh_1(x) = f(x, \phi(h_1(x)), h_1(x)) \leq f(x, \phi(h_2(x)), h_2(x)) = Gh_2(x).$$

This proves that G is increasing. Since $0 < \alpha, \beta < 1$, then the mapping $\psi : \mathbb{R}_+ \to \mathbb{R}_+$ defined in (4.116) is increasing, and $r - \psi(r) \to \infty$ as $r \to \infty$. Thus $M + \psi(R) \leq R$ when R is large enough, whence G has a fixed point by Lemma 4.43.

(b) Case: (f2) (ii) and $\lambda = 0$

In this case $f(x, \cdot, \cdot)$ is increasing, whence G is increasing by the above proof. Since $\alpha = \beta = 1$ we get:

$$\|Gh\|_p \le \|k\|_p + \|c_1\|_\infty \|\phi(h)\|_p + \|c_2\|_\infty \|h\|_p$$
$$\le M + \psi(\|h\|_p)$$

with ψ given by $\psi(r) = \|c_1\|_\infty (m + \kappa r) + \|c_2\|_\infty r$. If $\kappa \|c_1\|_\infty + \|c_2\|_\infty < 1$, then $M + \psi(R) \le R$ when R is sufficiently large. Thus G has a fixed point by Lemma 4.43.

The above proof shows that in the cases (a) and (b), G has a fixed point $h \in L^p(\Omega)$. This implies by (4.115) that $h(x) = Gh(x) = f(x, \phi(h(x)), h(x))$ a.e. in Ω.

(c) Case: (f2) (ii) and $\lambda > 0$

Assume finally that the hypothesis (f2) (ii) holds with $\lambda > 0$. Then a function $\tilde{f} : \Omega \times \mathbb{R} \times \mathbb{R}$, defined by

$$\tilde{f}(x, y, z) = \frac{f(x, y, z) + \lambda z}{1 + \lambda}, \quad x \in \Omega, \ y, z \in \mathbb{R}, \tag{4.117}$$

is sup-measurable, $\tilde{f}(x, \cdot, \cdot)$ is increasing, and

$$|\tilde{f}(x, y, z)| \le \|\tilde{k}\|_2 + \tilde{c}_1(x)|y| + \tilde{c}_2(x)|z|,$$

where $\tilde{k}_2 = \frac{k_2}{1+\lambda}$, $\tilde{c}_1 = \frac{c_1}{1+\lambda}$, $\tilde{c}_2 = \frac{c_2+\lambda}{1+\lambda}$. Since $\kappa \|c_1\|_\infty + \|c_2\|_\infty < 1$, then

$$\kappa \|\tilde{c}_1\|_\infty + \|\tilde{c}_2\|_\infty = \frac{\kappa \|c_1\|_\infty + \|c_2\|_\infty + \lambda}{1 + \lambda} < \frac{1 + \lambda}{1 + \lambda} = 1.$$

Thus $\tilde{f}$ satisfies the hypotheses (f1) and (f2) (ii) with $\lambda = 0$. The proof of the case (b) above implies the existence of a $h \in L^p(\Omega)$ such that $h(x) = \tilde{f}(x, \phi(h(x)), h(x))$, or equivalently, by (4.117), $h(x) = f(x, \phi(h(x)), h(x))$ a.e. in Ω. This concludes the proof.   □

As a consequence of Theorem 4.44 we obtain an existence result for the equation

$$h(x) = g(x, \phi(h(x))), \quad \text{a.e. in } \Omega, \tag{4.118}$$

under the following hypotheses on g:

(g1) g is sup-measurable, and $g(x, \cdot)$ is increasing for a.e. $x \in \Omega$.
(g2) $|g(x, y)| \le k(x) + c_1(x)|y|^\alpha$ for a.e. $x \in \Omega$ and for all $y \in \mathbb{R}$, where $k \in L^p(\Omega)$, and either $0 < \alpha < 1$ and $c_1 \in L_+^{\frac{p}{1-\alpha}}(\Omega)$, or $\alpha = 1$ and $\kappa \|c_1\|_\infty < 1$.

Corollary 4.45. *Assume that $\phi : L^p(\Omega) \to L^p(\Omega)$ satisfies the hypothesis (ϕ), and that $g : \Omega \times \mathbb{R} \to \mathbb{R}$ satisfies (g1) and (g2). Then (4.118) has a solution h in $L^p(\Omega)$.*

Remark 4.46. The hypotheses of Theorem 4.44 and Corollary 4.45 allow the functions f and g to be discontinuous in all their arguments. Even the mapping ϕ may be discontinuous.

Implicit Discontinuous Elliptic Problems

Let $\Omega \subset \mathbb{R}^N$, $N \geq 3$, be a bounded domain with Lipschitz boundary $\partial\Omega$. In this paragraph we study the existence of weak solutions of the implicit elliptic BVP

$$Au(x) = f(x, u(x), Au(x)) \quad \text{in } \Omega, \quad u = 0 \quad \text{on } \partial\Omega, \tag{4.119}$$

where A is the semilinear elliptic operator

$$Au(x) := -\sum_{i,j=1}^{N} \frac{\partial}{\partial x_i}\left(a_{ij}(x)\frac{\partial u(x)}{\partial x_j}\right) + q(x, u(x)).$$

Theorem 4.44 will be the main tool in our investigations. We assume that the coefficients $a_{ij} \in L^{\infty}(\Omega)$ satisfy the ellipticity condition

$$\sum_{i,j=1}^{N} a_{ij}(x)\zeta_i\zeta_j \geq \gamma \sum_{i=1}^{N} \zeta_i^2 \tag{4.120}$$

for a.e. $x \in \Omega$, all $\zeta = (\zeta_1, \ldots, \zeta_N) \in \mathbb{R}^N$, and some $\gamma > 0$.

Let $V_0 = W_0^{1,2}(\Omega)$ and V_0^* its dual space. We are going to introduce conditions that ensure that (4.119) has a weak solution in the following sense.

Definition 4.47. *A function $u \in V_0$ is called a* **solution** *of the BVP* (4.119) *if there exists a function $h \in L^2(\Omega)$ such that*

$$h(x) = f(x, u(x), h(x)) \quad \text{for a.e. } x \in \Omega, \tag{4.121}$$

and u is a (weak) solution of the semilinear BVP

$$Au(x) = h(x) \quad \text{in } \Omega, \quad u = 0 \quad \text{on } \partial\Omega. \tag{4.122}$$

We first prove an existence, uniqueness, and comparison result for the BVP (4.122), where we assume the following hypotheses for q:

(q1) q is a Carathéodory function, and $s \mapsto q(x, s)$ is increasing for a.e. $x \in \Omega$.
(q2) $|q(x, s)| \leq k_0(x) + c_0(x)|s|^{p_0-1}$ for a.e. $x \in \Omega$ and for all $s \in \mathbb{R}$, where $k_0 \in L^{\frac{p_0}{p_0-1}}(\Omega)$, $c_0 \in L_+^{\infty}(\Omega)$ and $1 < p_0 \leq 2^* := \frac{2N}{N-2}$ (critical Sobolev exponent).

Lemma 4.48. *Assume that $q : \Omega \times \mathbb{R} \to \mathbb{R}$ satisfies (q1) and (q2). Then* (4.122) *has a unique weak solution u for each $h \in L^2(\Omega)$. Moreover, u is increasing with respect to h and there exist constants $m \geq 0$ and $\kappa > 0$ such that*

$$\|u\|_{V_0} \leq m + \kappa\|h\|_2. \tag{4.123}$$

Proof: Due to (4.120) and hypotheses (q1), (q2), the semilinear operator $A : V_0 \to V_0^*$ is continuous, bounded, and strongly monotone, which implies that $A : V_0 \to V_0^*$ is bijective. Thus the BVP (4.122) has a uniquely defined solution, in particular, for each $h \in L^2(\Omega) \subset V_0^*$. Further, by standard comparison arguments one readily observes that the solution depends monotonically on h, i.e., the inverse $A^{-1} : V_0^* \to V_0$ is monotone increasing. To prove estimate (4.123), let $h \in L^2(\Omega)$ be given, and let $u \in V_0$ be the solution of (4.122). The monotonicity of $s \mapsto q(x, s)$ along with the continuous embedding $V_0 \hookrightarrow L^{p_0}(\Omega)$ and (4.120) yield the following estimate:

$$c \|u\|_{V_0}^2 \le \langle Au, u \rangle - \int_\Omega q(x, 0) u(x)\, dx$$

$$= \int_\Omega h(x) u(x)\, dx - \int_\Omega q(x, 0) u(x)\, dx$$

$$\le \|h\|_2 \|u\|_2 + \|k_0\|_{\frac{p_0}{p_0 - 1}} \|u\|_{p_0} \le (b \|k_0\|_{\frac{p_0}{p_0 - 1}} + \|h\|_2) \|u\|_{V_0},$$

for some positive constant b. Thus (4.123) holds with $m = \frac{b}{c} \|k_0\|_{\frac{p_0}{p_0 - 1}}$ and $\kappa = 1/c$. □

As an application of Theorem 4.44 and Lemma 4.48, we are going to prove the following existence result for (4.119).

Theorem 4.49. *Assume that $q : \Omega \times \mathbb{R} \to \mathbb{R}$ satisfies the hypotheses (q1) and (q2), and that $f : \Omega \times \mathbb{R} \times \mathbb{R} \to \mathbb{R}$ satisfies the hypotheses (f1) and (f2) with $p = 2$. Then the implicit BVP (4.119) possesses solutions.*

Proof: From Lemma 4.48 follows that the mapping $\phi : L^2(\Omega) \to L^2(\Omega)$, which assigns to each $h \in L^2(\Omega)$ the unique solution $u := \phi(h) \in V_0 \subset L^2(\Omega)$ of the BVP (4.122), is increasing. Moreover, the inequality (4.123) holds, whence

$$\|\phi \circ h\|_2 = \|u\|_2 \le \|u\|_{V_0} \le m + \kappa \|h\|_2, \quad h \in L^2(\Omega).$$

This proves that ϕ satisfies the hypothesis (ϕ). Thus the hypotheses of Theorem 4.44 hold when $p = 2$, whence there exists a function $h \in L^2(\Omega)$ such that

$$h(x) = f(x, \phi(h(x)), h(x)) = f(x, u(x), h(x)) \quad \text{a.e. in } \Omega,$$

and u is a solution of (4.122). This implies by Definition 4.47 that u is a solution of (4.119). □

As a consequence of Theorem 4.49, we obtain an existence result for the (explicit) BVP

$$Au(x) = g(x, u(x)) \quad \text{a.e. in } \Omega, \quad u = 0 \quad \text{on } \partial\Omega, \tag{4.124}$$

where, as before,

$$Au(x) := -\sum_{i,j=1}^N \frac{\partial}{\partial x_i} \left(a_{ij}(x) \frac{\partial u(x)}{\partial x_j} \right) + q(x, u(x)).$$

Corollary 4.50. *Assume that* $q : \Omega \times \mathbb{R} \to \mathbb{R}$ *satisfies the hypotheses (q1) and (q2), and that* $g : \Omega \times \mathbb{R} \to \mathbb{R}$ *satisfies the hypotheses (g1) and (g2) with* $p = 2$. *Then the BVP* (4.124) *has a weak solution.*

Remark 4.51. (i) The hypotheses of Theorem 4.49 and Corollary 4.50 allow both functions f and g to be discontinuous in all their arguments.

(ii) Theorem 4.49 and Corollary 4.50 also apply to problems in domains Ω of dimensions $N = 1$ and $N = 2$, since in these cases the critical exponent $2^* = \infty$ and Lemma 4.48 is valid with an exponent p_0 satisfying $1 < p_0 < \infty$.

(iii) If the coefficients a_{ij} are uniformly Lipschitz continuous, it follows by the regularity result [99, Theorem 8.8] that the weak solutions of problems (4.119) and (4.124) satisfy their differential equation a.e. pointwise. This holds, in particular, when $a_{ij} = \delta_{ij}$, which is the case in the following examples, where $[z]$ denotes the greatest integer $\leq z \in \mathbb{R}$.

Example 4.52. Assume that $\mathbb{R}^4$ is equipped with the Euclidean norm $|\cdot|$. Choose $\Omega = \{x \in \mathbb{R}^4 : \frac{1}{2} < |x| < 1\}$, and consider the BVP

$$Au(x) = 5 + [6|x|] + 7[10^9 u(x)]^{\frac{1}{3}} + 8[10^{10} Au(x)]^{\frac{1}{5}}, \quad \text{a.e. in } \Omega, \quad u = 0 \quad \text{on } \partial\Omega,$$
$$\tag{4.125}$$

where $Au(x) := -\Delta u(x) + u(x)^3$, for $x \in \Omega$. The BVP (4.125) is of the form (4.119), where

$$a_{ij}(x) \equiv \delta_{ij}, \ q(x,y) = y^3 \ \text{ and } \ f(x,y,z) = 5 + [6|x|] + 7[10^9 y]^{1/3} + 8[10^{10} z]^{1/5}.$$

The critical exponent here is $2^* = 4$, and it is easy to see that the hypotheses (q), (f1), and (f2) with $p = 2$ hold, whence the BVP (4.125) has a solution by Theorem 4.49.

Example 4.53. For $\Omega = (0,1)$, consider the boundary-value problem

$$-u''(x) = 2 + 2[2 - 2x] + 2\left[(2u(x) - 2x)^{\frac{1}{3}}\right] + \left[(-u''(x) - 1)^{\frac{1}{3}}\right] \quad \text{a.e. in } (0,1)$$
$$u(0) = u(1) = 0.$$
$$\tag{4.126}$$

Problem (4.126) is of the form (4.119), with

$$q(x,y) \equiv -1, \ \text{ and } \ f(x,y,z) = 1 + 2[2 - 2x] + 2\left[(2y - 2x)^{\frac{1}{3}}\right] + \left[(z - 1)^{\frac{1}{3}}\right].$$
$$\tag{4.127}$$

By elementary calculations one can show that for each $h \in L^2(\Omega)$ the function

$$u(x) = \phi(h(x)) = (1 - x)\int_0^x t(1 + h(t))dt + x\int_x^1 (1 - t)(1 + h(t))dt$$
$$= \frac{x - x^2}{2} + (1 - x)\int_0^x th(t)dt + x\int_x^1 (1 - t)h(t)dt, \quad x \in [0,1]$$
$$\tag{4.128}$$

is the unique solution of the BVP

$$Au(x) := -u''(x) - 1 = h(x) \ \text{in} \ (0,1), \ u(0) = u(1) = 0$$

in $W_0^{1,2}(0,1)$, and that for certain nonnegative constants m, κ, one has

$$\|u\|_2 = \|\phi \circ h\|_2 \leq m + \kappa \|h\|_2.$$

Thus the hypothesis (ϕ) holds. Obviously, f is sup-measurable, i.e., (f1) is fulfilled. Since

$$|f(x,y,z)| \leq 15 + 4|y|^{1/3} + |z|^{1/3},$$

then the hypothesis (f2) (i) is satisfied. It then follows from Theorem 4.49 that the BVP (4.126) has a solution.

4.4 Notes and Comments

The results and the presentation of this chapter are mainly based on the authors' joint work of recent years, see [48, 49, 51, 52, 57, 58]. The development of more efficient comparison results for multi-valued elliptic problems as provided in Chap. 3 allows for improvements of most of the results in the above cited works.

We emphasize that classical fixed point theorems for discontinuous operators on partially ordered sets such as, e.g., the Bourbaki-Kneser Fixed Point Theorem, the Amann or Tarski Fixed Point Theorem (see, e.g., [228, Chap. 11]) cannot be applied in the proof of Theorem 4.27, because these would require, e.g., that every chain of $\mathcal{U}$ has an infimum (or at least a lower bound), which need not be true. Moreover, the abstract fixed point results applied in this chapter can also be used as alternative tools in the study of elliptic boundary value problems that lack continuity and/or compactness of the operators involved, which has been demonstrated in [48]. The lack of continuity and/or compactness may be caused in various different ways such as, e.g., in the following cases: (i) the lower order terms have critical growth, or (ii) the lower order nonlinearity is discontinuous with respect to the unknown function, or (iii) the domain Ω in which the problem is defined is unbounded.

We note that one can extend the investigation of Sect. 4.2 on state-dependent Clarke's gradient inclusions to problems in the following form

$$Au + \partial j(\cdot, u, u) \ni h, \ \text{in} \ \Omega, \quad u = 0 \ \text{on} \ \partial\Omega,$$

where Clarke's generalized gradient is allowed to depend discontinuously on the state u in a much more general way. This, however, would require a more subtle treatment. Only for the sake of simplifying our presentation and in order to avoid too much technicalities have we restricted our presentation to problem (4.66).

5

Discontinuous Multi-Valued Evolutionary Problems

In this chapter we consider multi-valued evolutionary problems involving discontinuous data. Abstract fixed point results developed in Chap. 2 and the theory on multi-valued parabolic variational inequalities provided in Chap. 3 are the main tools used in the treatment of such kind of problems.

5.1 Discontinuous Parabolic Inclusions with Clarke's Gradient

Throughout this section we adopt the notions of Sect. 3.3. We are going to study here the following initial-Dirichlet problem:

$$u_t + Au + \partial j(\cdot, \cdot, u) \ni F(u) + h, \quad \text{in } Q \tag{5.1}$$

$$u(\cdot, 0) = 0 \quad \text{in } \Omega, \quad u = 0 \quad \text{on } \Gamma, \tag{5.2}$$

where $\Omega \subset \mathbb{R}^N$ is a bounded domain with Lipschitz boundary $\partial\Omega$, $Q = \Omega \times (0, \tau)$, and $\Gamma = \partial\Omega \times (0, \tau)$. Let $V = W^{1,p}(\Omega)$ and $V_0 = W_0^{1,p}(\Omega)$, and assume $2 \le p < \infty$ with its Hölder conjugate q. As in Sect. 3.3 the underlying solution space for the problem (5.1)–(5.2) is $W_0 = \{u \in X_0 : u_t \in X_0^*\}$, where $X_0 = L^p(0, \tau; V_0)$ and X_0^* denotes its dual space. We also use the spaces $X = L^p(0, \tau; V)$ and $W = \{u \in X : u_t \in X^*\}$. By $L : X_0 \to X_0^*$ we denote the time derivative $\partial/\partial t$ considered as a mapping form X_0 into its dual, with its domain of definition $D_0(L)$ given by

$$D_0(L) = \{u \in W_0 : u(\cdot, 0) = 0\}.$$

Let $\underline{u}, \overline{u} \in W$ be an ordered pair of sub-supersolutions for (5.1)–(5.2), and let $h \in X_0^*$. We assume the following hypotheses on A, j, and F:

(A) The operator A is a second-order quasilinear elliptic differential operator of the form

S. Carl and S. Heikkilä, *Fixed Point Theory in Ordered Sets and Applications:*
From Differential and Integral Equations to Game Theory,
DOI 10.1007/978-1-4419-7585-0_5, © Springer Science+Business Media, LLC 2011

$$Au(x,t) = -\sum_{i=1}^{N} \frac{\partial}{\partial x_i} a_i(x,t,u(x,t),\nabla u(x,t)),$$

whose coefficients a_i satisfy conditions (AP1)–(AP4) given in Sect. 3.3.

(j) The function $j : Q \times \mathbb{R} \to \mathbb{R}$ fulfills condition (P-j1) given in Sect. 3.3.1,
    i.e., let $[\underline{u}, \overline{u}]$ be the ordered interval formed by given sub- and superso-
    lutions of (5.1)–(5.2), then j satisfies:
    (i) $(x,t) \mapsto j(x,t,s)$ is measurable in Q for all $s \in \mathbb{R}$, and $s \mapsto j(x,t,s)$
        is locally Lipschitz continuous in $\mathbb{R}$ for a.e. $(x,t) \in Q$.
    (ii) There exists a function $k_Q \in L_+^q(Q)$ such that for a.e. $(x,t) \in Q$ and
        for all $s \in [\underline{u}(x,t), \overline{u}(x,t)]$ the growth condition for Clarke's general-
        ized gradient ∂j

$$|\eta| \le k_Q(x,t), \quad \forall\, \eta \in \partial j(x,t,s)$$

    is fulfilled.

(F) F denotes the Nemytskij operator related to some nonlinearity $f : Q \times \mathbb{R} \times \mathbb{R} \to \mathbb{R}$ by $F(u)(x,t) = f(x,t,u(x,t),u(x,t))$, where f satisfies the
    hypotheses:
    (i) For each $r \in \mathbb{R}$, $(x,t,s) \mapsto f(x,t,s,r)$ is a Carathéodory function in
        $Q \times \mathbb{R}$.
    (ii) For a.e. $(x,t) \in Q$, for all $s \in \mathbb{R}$, $r \mapsto f(x,t,s,r)$ is increasing, and
        $(x,t) \mapsto f(x,t,u(x,t),v(x,t))$ is measurable whenever the functions
        $(x,t) \mapsto u(x,t)$ and $(x,t) \mapsto v(x,t)$ are measurable, i.e., f is sup-
        measurable.
    (iii) There exists a function $k_f \in L_+^q(Q)$ such that for a.e. $(x,t) \in Q$ and
        for all $s, r \in [\underline{u}(x,t), \overline{u}(x,t)]$ the growth condition

$$|f(x,t,s,r)| \le k_f(x,t)$$

    is fulfilled.

Remark 5.1. (i) Without loss of generality we have assumed homogeneous initial and boundary conditions in (5.2). Nonhomogeneous initial and boundary
conditions can be transformed to homogeneous ones by translation provided
the nonhomogeneous initial and boundary values are the traces, respectively,
of some function from W. Moreover, mixed nonlinear boundary conditions of
Robin type can be treated as well by the method to be developed later.

   (ii) The Nemytskij operator F may be discontinuous as $r \mapsto f(x,t,s,r)$ is
only assumed to be increasing, i.e., not necessarily continuous.

Method of Sub-Supersolution

Let us first recall the notions for the (weak) solution, and the sub-supersolutions for the parabolic problem (5.1)–(5.2) introduced in Sect. 3.3.2.

Definition 5.2. *A function* $u \in D_0(L) \subset W_0$ *is called a* **solution** *of (5.1)–(5.2) if there is an* $\eta \in L^q(Q)$ *with* $\eta(x,t) \in \partial j(x,t,u(x,t))$ *for a.a.* $(x,t) \in Q$ *such that*

$$\langle Lu + Au, \varphi \rangle + \int_Q \eta \, \varphi \, dx \, dt = \int_Q F(u) \, \varphi \, dx \, dt + \langle h, \varphi \rangle, \quad \forall \, \varphi \in X_0. \quad (5.3)$$

Definition 5.3. *A function* $\underline{u} \in W$ *is called a* **subsolution** *of (5.1)–(5.2) if there is an* $\underline{\eta} \in L^q(Q)$ *satisfying* $\underline{\eta}(x,t) \in \partial j(x,t,\underline{u}(x,t))$ *for a.a.* $(x,t) \in Q$ *such that* $\underline{u}(x,0) \le 0$ *for a.a.* $x \in \Omega$, $\underline{u}|_\Gamma \le 0$, *and*

$$\langle \underline{u}_t + A\underline{u}, \varphi \rangle + \int_Q \underline{\eta} \, \varphi \, dx \, dt \le \int_Q F(\underline{u}) \, \varphi \, dx \, dt + \langle h, \varphi \rangle, \quad \forall \, \varphi \in X_0 \cap L^p_+(Q).$$
$$(5.4)$$

Definition 5.4. *A function* $\overline{u} \in W$ *is called a* **supersolution** *of (5.1)–(5.2) if there is an* $\overline{\eta} \in L^q(Q)$ *satisfying* $\overline{\eta}(x,t) \in \partial j(x,t,\overline{u}(x,t))$ *for a.a.* $(x,t) \in Q$ *such that* $\overline{u}(x,0) \ge 0$ *for a.a.* $x \in \Omega$, $\overline{u}|_\Gamma \ge 0$, *and*

$$\langle \overline{u}_t + A\overline{u}, \varphi \rangle + \int_Q \overline{\eta} \, \varphi \, dx \, dt \ge \int_Q F(\overline{u}) \, \varphi \, dx \, dt + \langle h, \varphi \rangle, \quad \forall \, \varphi \in X_0 \cap L^p_+(Q).$$
$$(5.5)$$

Our main goal in this subsection is to show that the discontinuous parabolic problem (5.1)–(5.2) has extremal solutions within the ordered interval of a given pair of sub-supersolutions. The main tools used to achieve this goal are Theorem 3.64 of Sect. 3.3 on the existence of extremal solutions for (5.1)–(5.2) when f is independent on r, and the abstract fixed point theorem, Theorem 4.20, of Sect. 4.2.2.

Theorem 5.5. *Let* $\underline{u}, \overline{u} \in W$ *be sub- and supersolution of (5.1)–(5.2) such that* $\underline{u} \le \overline{u}$. *Assume hypotheses (A), (j), and (F). Then problem (5.1)–(5.2) admits extremal solutions, i.e., a greatest solution* u^* *and a smallest solution* u_*, *within the interval* $[\underline{u}, \overline{u}]$.

Proof: In the proof we focus on the existence of the greatest solution u^* of (5.1)–(5.2) within $[\underline{u}, \overline{u}]$, because the existence of the smallest solution is then shown by obvious analog reasoning. To this end we first derive an equivalent fixed point equation that allows us to apply statement (b) of the abstract Theorem 4.20.

Step 1: Equivalent Fixed Point Problem

For any $v \in [\underline{u}, \overline{u}]$ fixed, let us introduce the operator F_v defined as follows

$$F_v(u)(x,t) = f(x,t,u(x,t),v(x,t))$$

which is the Nemytskij operator generated by the nonlinearity $(x,t,s) \mapsto f(x,t,s,v(x,t))$, which for v fixed is a Carathéodory function, and thus by hypothesis (f) it follows that $F_v : [\underline{u}, \overline{u}] \to L^q(Q)$ is bounded and continuous. For v fixed, consider next the auxiliary problem

$$u_t + Au + \partial j(\cdot, \cdot, u) \ni F_v(u) + h, \quad \text{in } Q$$
$$u(\cdot, 0) = 0 \quad \text{in } \Omega, \quad u = 0 \quad \text{on } \Gamma,$$

which is equivalent to: Find a function $u \in D_0(L) \subset W_0$ and an $\eta \in L^q(Q)$ with $\eta(x, t) \in \partial j(x, t, u(x, t))$ for a.a. $(x, t) \in Q$ such that

$$\langle Lu + Au, \varphi \rangle + \int_Q \eta \varphi \, dx \, dt = \int_Q F_v(u) \varphi \, dx \, dt + \langle h, \varphi \rangle, \quad \forall \varphi \in X_0. \quad (5.6)$$

We next prove the following assertion.

Lemma 5.6. *Let $v \in [\underline{u}, \overline{u}]$ be any fixed supersolution of (5.1)–(5.2). Then problem (5.6) has the greatest solution v^* and the smallest solution v_* within $[\underline{u}, v]$.*

If $v \in [\underline{u}, \overline{u}]$ is a supersolution of the original problem, then v is, in particular, a supersolution of (5.6). Due to the monotonicity of $r \mapsto f(x, t, s, r)$ one readily sees that the given subsolution $\underline{u}$ is also a subsolution of (5.6). Introducing $\hat{j} : Q \times \mathbb{R} \to \mathbb{R}$ defined by

$$\hat{j}(x, t, s) = j(x, t, s) - \int_0^s f(x, t, \varsigma, v(x, t)) \, d\varsigma,$$

problem (5.6) can equivalently be written in the form: Find a function $u \in D_0(L) \subset W_0$ and an $\hat{\eta} \in L^q(Q)$ with $\hat{\eta}(x, t) \in \partial \hat{j}(x, t, u(x, t))$ for a.a. $(x, t) \in Q$ such that

$$\langle Lu + Au, \varphi \rangle + \int_Q \hat{\eta} \varphi \, dx \, dt = \langle h, \varphi \rangle, \quad \forall \varphi \in X_0. \quad (5.7)$$

Since $\hat{j}$ has the same qualities as j, we may apply Theorem 3.64 of Sect. 3.3 to ensure the existence of the greatest solution v^* and the smallest solution v_* of the auxiliary problem (5.6) within $[\underline{u}, v]$, which proves Lemma 5.6.

Now we introduce the set $\mathcal{U}$ as

$$\mathcal{U} := \{v \in [\underline{u}, \overline{u}] : v \text{ is a supersolution of (5.1)–(5.2)}\}$$

and define an operator G on $\mathcal{U}$ as follows. For $v \in \mathcal{U}$ let $Gv := v^*$ denote the greatest solution of (5.6) within $[\underline{u}, v]$. In view of the monotonicity of $r \mapsto f(x, t, s, r)$ we readily verify that Gv is again a supersolution of the original problem (5.1)–(5.2). Thus by Lemma 5.6 the operator $G : \mathcal{U} \to \mathcal{U}$ is well defined, and, moreover, the following holds.

Lemma 5.7. *Any fixed point of $G : \mathcal{U} \to \mathcal{U}$ is a solution of (5.1)–(5.2) within the interval $[\underline{u}, \overline{u}]$, and vice versa.*

The definition of G implies that any fixed point u of G is a solution of (5.1)–(5.2), which belongs to $[\underline{u}, \overline{u}]$. Conversely, if $u \in [\underline{u}, \overline{u}]$ is a solution of (5.1)–(5.2), then u is, in particular, a supersolution, i.e., $u \in \mathcal{U}$, and it is trivially a solution of (5.6) with v replaced by u, and thus u is the greatest solution of (5.6) within $[\underline{u}, u]$, which proves the assertion of Lemma 5.7.

Step 2: G Has the Greatest Fixed Point

By Step 1, Lemma 5.7, the greatest fixed point of G corresponds to the greatest solution of (5.1)–(5.2) within $[\underline{u}, \overline{u}]$. For the proof of the existence of the greatest fixed point of G we are going to apply Theorem 4.20 (b). To this end we first verify that $G : \mathcal{U} \to \mathcal{U}$ is increasing. Let $v, w \in \mathcal{U}$ be given such that $v \leq w$ is fulfilled. By definition, $Gv = v^*$ is the greatest solution of (5.6) within $[\underline{u}, v]$, and $Gw = w^*$ is the greatest solution of (5.6) within $[\underline{u}, w]$, where v on the right-hand side of (5.6) is replaced by w. Since $v \leq w$ and $r \mapsto f(x, t, s, r)$ is increasing, we see that $v^* \in [\underline{u}, v]$ is a subsolution of (5.6) with v replaced by w, i.e., of the problem: Find $u \in D_0(L) \subset W_0$ and an $\eta \in L^q(Q)$ with $\eta(x, t) \in \partial j(x, t, u(x, t))$ for a.a. $(x, t) \in Q$ such that

$$\langle Lu + Au, \varphi \rangle + \int_Q \eta \varphi \, dx \, dt = \int_Q F_w(u) \, \varphi \, dx \, dt + \langle h, \varphi \rangle, \quad \forall \, \varphi \in X_0. \quad (5.8)$$

Because w is a supersolution of (5.8), and $w \geq v \geq v^*$, we infer again by applying Theorem 3.64 the existence of the greatest solution of (5.8) within $[v^*, w]$, which implies $v^* \leq w^*$, i.e., $G : \mathcal{U} \to \mathcal{U}$ is increasing. To complete the proof of the existence of the greatest fixed point we need to verify the following:

Lemma 5.8. *The range $G(\mathcal{U})$ of G has an upper bound in $\mathcal{U}$, and decreasing sequences of $G(\mathcal{U})$ converge weakly in $\mathcal{U}$.*

The given supersolution $\overline{u} \in \mathcal{U} \subseteq [\underline{u}, \overline{u}]$ is apparently an upper bound of $G(\mathcal{U})$. Let $(u_n) \subseteq G(\mathcal{U})$ be a decreasing sequence, i.e. $u_n = Gv_n$ for certain $v_n \in \mathcal{U}$, which means u_n is the greatest solution in $[\underline{u}, v_n]$ of the following problem: $u_n \in D_0(L) \subset W_0$, $\eta_n \in L^q(Q)$ with $\eta_n(x, t) \in \partial j(x, t, u_n(x, t))$ for a.a. $(x, t) \in Q$ such that

$$\langle Lu_n + Au_n, \varphi \rangle + \int_Q \eta_n \, \varphi \, dx \, dt = \int_Q F_{v_n}(u_n) \, \varphi \, dx \, dt + \langle h, \varphi \rangle, \quad \forall \, \varphi \in X_0.$$

$$(5.9)$$

Testing (5.9) by $\varphi = u_n$ and taking into account that $\mathcal{U}$ is $L^p(Q)$-bounded, and

$$\langle Lu_n, u_n \rangle \geq 0, \quad \forall \, n \in \mathbb{N},$$

we readily infer by means of (AP3), (j), and (F) that (u_n) is bounded in X_0, which in turn implies, by using the equation (5.9), that (Lu_n) is bounded in X_0^*, and thus

$$\|u_n\|_{W_0} \le c, \quad \forall\, n \in \mathbb{N}.$$

By the monotonicity of the sequence (u_n), and the compact embedding $W_0 \hookrightarrow L^p(Q)$, we get

$$u_n \rightharpoonup u \ \text{in}\ W_0, \quad u_n \to u \ \text{in}\ L^p(Q). \tag{5.10}$$

To complete the proof we need to show that $u \in \mathcal{U}$. Testing (5.9) with $\varphi = u_n - u \in X_0$, and using

$$\langle Lu_n - Lu, u_n - u \rangle \ge 0, \quad \forall\, n \in \mathbb{N},$$

as well as the boundedness of (η_n), $(F_{v_n}(u_n))$ in $L^q(Q)$, and the convergence properties (5.10), we obtain

$$\limsup_{n\to\infty}\langle Au_n, u_n - u \rangle \le 0.$$

Hence, by the (S_+)-property w.r.t. $D_0(L)$ of A (see, e.g., [62, Theorem 2.153]) we infer the strong convergence of (u_n) in X_0, i.e., $u_n \to u$ in X_0. The limit u satisfies

$$\underline{u} \le u \le u_n \le v_n \le \overline{u},$$

which due to the monotonicity of $r \mapsto f(x,t,s,r)$ implies

$$\langle Lu_n + Au_n, \varphi \rangle + \int_Q \eta_n\, \varphi\, dx\, dt \ge \int_Q F_u(u_n)\, \varphi\, dx\, dt + \langle h, \varphi \rangle, \ \forall\, \varphi \in X_0 \cap L^p_+(Q). \tag{5.11}$$

As (η_n) is bounded in $L^q(Q)$, there is a subsequence (η_{n_k}) of (η_n) converging weakly to η, i.e.,

$$\eta_{n_k} \rightharpoonup \eta \ \text{in}\ L^q(Q),$$

where the weak limit satisfies $\eta(x,t) \in \partial j(x,t,u(x,t))$ for a.a. $(x,t) \in Q$. The convergence properties of (u_n) and (η_{n_k}) allow to pass to the limit in (5.11) for some subsequence (u_{n_k}), which results in

$$\langle Lu + Au, \varphi \rangle + \int_Q \eta\, \varphi\, dx\, dt \ge \int_Q F_u(u)\, \varphi\, dx\, dt + \langle h, \varphi \rangle, \ \forall\, \varphi \in X_0 \cap L^p_+(Q), \tag{5.12}$$

and hence $u \in \mathcal{U}$, which completes the proof of Theorem 5.5. $\square$

5.2 Implicit Functional Evolution Equations

In this section we prove existence results for initial value problems of implicit functional evolution equations involving discontinuous nonlinearities. One of the main tools is the abstract fixed point result (see Theorem 4.37) for increasing self-mappings of closed balls in ordered Banach spaces, which is used already in Sect. 4.3. The obtained existence results are applied to implicit functional parabolic initial-boundary value problems.

5.2.1 Preliminaries

Let $V \subseteq H \subseteq V^*$ be an *ordered evolution triple*, i.e., the following properties hold:

(H) $H = (H, (\cdot|\cdot)_H, \leq)$ is a lattice-ordered separable Hilbert space, the mapping $x \mapsto x^+ := \sup\{0, x\}$ is continuous, and $\|x^+\| \leq \|x\|$ for all $x \in H$.

(V) V is a separable and reflexive Banach space that is continuously and densely embedded in H, and whose dual is denoted by V^*.

For instance, if $H = L^2(\Omega)$ is equipped with the a.e. pointwise ordering and the natural inner product, and if $V = W_0^{1,p}(\Omega)$, $2 \leq p < \infty$, where Ω is a bounded domain in $\mathbb{R}^N$, then V, H, and V^* form an ordered evolution triple. We consider the following implicit functional initial value problem (IVP)

$$u'(t) + A(t)u(t) = F(t, u, u' + \hat{A}u) \quad \text{a.e. in} \quad J = (0, \tau), \quad u(0) = 0, \quad (5.13)$$

where $A(t) : V \to V^*$, $F : J \times L^2(J, H) \times L^2(J, H) \to H$, and $\hat{A}$ is defined by

$$(\hat{A}u)(t) := A(t)u(t), \quad u \in V, \ t \in J.$$

We introduce the space $\mathcal{W}$ defined by

$$\mathcal{W} = \{u \in L^2(J, V) : u' \in L^2(J, V^*)\}$$

where the derivative u' is understood in the sense of vector-valued distributions. The solution space for (5.13) is given by

$$W := \{u \in \mathcal{W} : u(0) = 0 \ \text{and} \ u' + \hat{A}u \in L^2(J, H)\}. \quad (5.14)$$

The ordering of H and the a.e. pointwise ordering of $L^2(J, H)$ induce partial orderings to their subsets V and W, respectively.

The mappings $A(t)$, $t \in J$, and F are assumed to satisfy the following hypotheses:

(A1) Denoting by $\langle \cdot, \cdot \rangle_V$ the duality pairing between V and V^*, we define

$$\langle A(t)y, z \rangle_V := a(y, z; t), \quad t \in J, \ y, z \in V, \quad (5.15)$$

where $a(y, z; \cdot) : J \to \mathbb{R}$ is measurable for all y, $z \in V$, $a(\cdot, \cdot; t) : V \times V \to \mathbb{R}$ is bilinear for all $t \in J$, and for all y, $z \in V$ and for a.e. $t \in J$

$$a(y, y; t) \geq \kappa \|y\|_V^2 - \rho \|y\|_H^2 \quad \text{and} \quad |a(y, z; t)| \leq C \|y\|_V \|z\|_V \quad (5.16)$$

with constants C, $\kappa > 0$, and $\rho \geq 0$ being independent of t.

(A2) If $w \in W$ and $w'(t) + A(t)w(t) \leq 0$ for a.e. $t \in J$, then $w(t) \leq 0$ for a.e. $t \in J$.

(F1) The function $t \mapsto F(t, u, v)$ is measurable in J for all u, $v \in L^2(J, H)$, and the function $(u, v) \mapsto F(\cdot, u, v)$ is increasing.

(F2) There exist nonnegative constants M, μ, λ, α and β such that
$$\|F(\cdot,u,v)\|_{L^2(J,H)} \leq M + \mu\|u\|^{\alpha}_{L^2(J,H)} + \lambda\|v\|^{\beta}_{L^2(J,H)} \text{ for all } u, v \in L^2(J,H).$$

According to these hypotheses we are faced with the following difficulties in the treatment of (5.13). In view of (F1) the mapping F may depend discontinuously on all its arguments and, in addition, functionally on its last two arguments. The growth condition (F2) does not provide means to construct a priori subsolutions and/or supersolutions for the IVP (5.13) so that methods based on sub-supersolutions and related fixed point theorems are not directly applicable here. Moreover, the differential equation of (5.13) is implicit with respect to the evolution operator $u' + \hat{A}u$.

Our main goal is to show that the hypotheses (H), (V), (A1), (A2), (F1), and (F2), where either $0 \leq \alpha$, $\beta < 1$, or $\alpha = \beta = 1$ and the constants μ and λ in (F2) are small enough, or α, $\beta > 1$ and the constant M in (F2) is sufficiently small, ensure the existence of a solution of the IVP (5.13) in W. The proof is based on a fixed point result that states that *increasing self-mappings of closed balls of* $L^2(J,H)$ *have fixed points*.

Precise a priori estimate and the inverse monotonicity of the evolution operator are additional tools that allow us to define an increasing mapping $G : L^2(J,H) \to L^2(J,H)$ such that any fixed point h of G provides a solution of the IVP (5.13) by $u = L^{-1}h$, where L^{-1} is the inverse of the evolution operator.

Abstract Fixed Point Result

Assume that the spaces V, H, and V^* form an ordered evolution triple, i.e., the properties listed in (H) and (V) are satisfied. We recall here for convenience a single-valued version of Theorem 4.37 that will be used later.

Theorem 5.9. *Assume that E is a reflexive lattice-ordered Banach space, and that for all $x \in E$, $\|x^+\| \leq \|x\|$, where $x^+ = \sup\{0, x\}$. Then each increasing self-mapping $G : B \to B$ of a closed ball B of E has a fixed point.*

As an application of Theorem 5.9 we obtain the following lemma.

Lemma 5.10. *If $B = \{u \in L^2(J,H) : \|u\|_{L^2(J,H)} \leq R\}$, $R \geq 0$, and if a mapping $G : B \to B$ is increasing with respect to the a.e. pointwise ordering of B, then G has a fixed point.*

Proof: $L^2(J,H)$ is a lattice-ordered Hilbert space with respect to the a.e. pointwise ordering and the inner product
$$(u|v) = \int_0^\tau (u(t)|v(t))_H \, dt, \quad u, v \in L^2(J,H). \tag{5.17}$$

In particular, $L^2(J,H)$ is a reflexive and lattice-ordered Banach space with respect to the norm $\|u\|_{L^2(J,H)} = (u|u)^{\frac{1}{2}}$. Moreover, since $\|x^+\| \leq \|x\|$ for each $x \in H$, then

$$\|u^+\|^2_{L^2(J,H)} = \int_0^\tau \|u^+(t)\|^2_H \, dt = \int_0^\tau \|u(t)^+\|^2_H \, dt$$

$$\leq \int_0^\tau \|u(t)\|^2_H \, dt = \|u\|^2_{L^2(J,H)},$$

i.e., $\|u^+\|_{L^2(J,H)} \leq \|u\|_{L^2(J,H)}$. The assertion now follows from Theorem 5.9.

□

Lemma 5.11. *Let the hypotheses (A1) and (A2) be satisfied. Then for any $h \in L^2(J,H)$ the IVP*

$$u'(t) + A(t)u(t) = h(t) \quad a.e. \ in \ J, \quad u(0) = 0, \tag{5.18}$$

has a unique solution $u \in W$, which is increasing with respect to $h \in L^2(J,H)$.

Proof: The unique solvability follows from [229, Corollary 23.26]. Let h_1, $h_2 \in L^2(J,H)$, $h_1 \leq h_2$, be given, and let $u_j \in W$ denote the solution of (5.18) with $h = h_j$, $j = 1, 2$. Denoting $w = u_1 - u_2$ we then have

$$w'(t) + A(t)w(t) \leq 0 \quad \text{for a.e. } t \in J.$$

This implies by (A2) that $w \leq 0$, i.e., $u_1 \leq u_2$.

□

Lemma 5.12. *Let the hypothesis (A1) hold. Then for each $h \in L^2(J,H)$ the solution $u \in W$ of the IVP (5.18) satisfies the following inequality:*

$$\|u\|_{L^2(J,H)} \leq \sqrt{\frac{e^{\tau(2\rho+1)} - 1}{2\rho + 1}} \|h\|_{L^2(J,H)}. \tag{5.19}$$

Proof: Let $h \in L^2(J,H)$ be given, and let $u \in W$ be the solution of (5.18). Thus

$$\langle u'(t) + A(t)u(t), v \rangle_V = \langle h(t), v \rangle_V,$$

which means

$$\langle u'(t), v \rangle_V + a(u(t), v; t) = (h(t)|v)_H, \quad \text{for a.e. } t \in J, \ \forall \, v \in V.$$

Choosing $v = u(t)$ and applying the hypothesis (A1) we obtain

$$\langle u'(t), u(t) \rangle_V + \kappa \|u(t)\|^2_V - \rho \|u(t)\|^2_H \leq (h(t)|u(t))_H.$$

for a.e. $t \in J$. Deleting the second term from the left-hand side and applying Hölder and Young's inequality to the right-hand side of the above inequality, we get by applying the integration by parts formula (cf. [229, 23.6])

$$\frac{1}{2}\|u(t)\|^2_H \leq \int_0^t \left(\rho\|u(s)\|^2_H + \frac{1}{2}\|h(s)\|^2_H + \frac{1}{2}\|u(s)\|^2_H \right) ds$$

for all $t \in J$. Denoting

$$y(t) := \|u(t)\|_H^2, \quad c_1 := 2\rho + 1 \quad \text{and} \quad c_2 := \|h\|_{L^2(J,H)}^2, \qquad (5.20)$$

we obtain

$$y(t) \le c_1 \int_0^t y(s)\, ds + c_2, \qquad t \in J.$$

The function $z(t) = c_2 e^{c_1 t}$ is a unique solution of the integral equation

$$z(t) = c_1 \int_0^t z(s)\, ds + c_2, \quad t \in J.$$

Denoting $w(t) = \max\{0, y(t) - z(t)\}$, $t \in J$, we then have

$$w(t) \le c_1 \int_0^t w(s)\, ds, \quad t \in J,$$

whence $w(t) = 0$ a.e. on J (cf. [44, B7]). Thus $y(t) \le z(t)$ a.e. on J, i.e.,

$$y(t) \le c_2 e^{c_1 t} \quad \text{for a.e. } t \in J.$$

Integrating the last inequality over $J = (0, \tau)$ we get

$$\int_0^\tau y(t)\, dt \le \frac{c_2}{c_1}(e^{c_1 \tau} - 1).$$

In view of (5.20) this inequality can be rewritten as

$$\|u\|_{L^2(J,H)}^2 \le \frac{e^{\tau(2\rho+1)} - 1}{2\rho + 1} \|h\|_{L^2(J,H)}^2,$$

which implies (5.19).   □

Remark 5.13. Instead of Gronwall's lemma we have used integral inequality techniques in the proof of Lemma 5.12 in order to get the sharper estimate (5.19).

5.2.2 Main Result

By means of Lemmas 5.10–5.12 we are now able to prove the following existence result for the IVP (5.13).

Theorem 5.14. Assume that the spaces H and V have properties (H) and (V), that the operators $A(t) : V \to V^*$, $t \in J$, satisfy the hypotheses (A1) and (A2), and that $F : J \times L^2(J, H) \times L^2(J, H) \to H$ satisfies the hypotheses (F1) and (F2). Then the IVP (5.13) has a solution in W in the following cases:

(a) $\alpha = \beta = 1$ and $\mu c + \lambda < 1$, where α, β, μ, and λ are the constants in (F2), and $c = \sqrt{\frac{e^{\tau(2\rho+1)}-1}{2\rho+1}}$ with ρ as in (5.16).

(b) $0 \leq \alpha$, $\beta < 1$ *in the hypothesis (F2).*
(c) α, $\beta > 1$ *and the constant M in the hypothesis (F2) is small enough.*

Proof: Let us introduce the set U defined by

$$U := \{u \in W : u \text{ is the solution of (5.18) for some } h \in L^2(J,H)\},$$

and the operator L defined by

$$Lu(t) := u'(t) + A(t)u(t), \quad u \in U, \ t \in J. \tag{5.21}$$

From Lemma 5.11 it follows that $L : U \to L^2(J,H)$ is a bijection, and that its inverse L^{-1} is increasing. The growth condition (F2) and the inequality (5.19) ensure that the equation

$$Gh := F(\cdot, L^{-1}h, h), \quad h \in L^2(J,H), \tag{5.22}$$

defines a mapping $G : L^2(J,H) \to L^2(J,H)$. If $h_1, h_2 \in L^2(J,H)$, and $h_1 \leq h_2$, then $L^{-1}h_1 \leq L^{-1}h_2$ by Lemma 5.11. Thus the hypothesis (F1) and the definition (5.22) of G imply that

$$Gh_1 = F(\cdot, L^{-1}h_1, h_1) \leq F(\cdot, L^{-1}h_2, h_2) = Gh_2,$$

which shows that G is increasing. The growth condition (F2) and (5.19) imply that

$$\|Gh\|_{L^2(J,H)} = \|F(\cdot, L^{-1}h, h)\|_{L^2(J,H)} \leq M + \mu\|L^{-1}h\|^\alpha_{L^2(J,H)} + \lambda\|h\|^\beta_{L^2(J,H)}$$
$$\leq M + \mu c^\alpha\|h\|^\alpha_{L^2(J,H)} + \lambda\|h\|^\beta_{L^2(J,H)}$$
$$= M + \psi(\|h\|_{L^2(J,H)}), \tag{5.23}$$

where

$$\psi(r) = \mu c^\alpha r^\alpha + \lambda r^\beta, \quad r \geq 0. \tag{5.24}$$

Ad (a) Assume first that $\alpha = \beta = 1$. In this case the function ψ in (5.24) takes the form $\psi(r) = (\mu c + \lambda)r$. Hence, if $\mu c + \lambda < 1$ and $R \geq \frac{M}{1-(\mu c+\lambda)}$, then G maps the ball

$$B_R := \{h \in L^2(J,H) \mid \|h\|_{L^2(J,H)} \leq R\} \tag{5.25}$$

into itself.

Ad (b) Assume next that $0 \leq \alpha$, $\beta < 1$ in (F2). Since ψ given by (5.24) is increasing and $r - \psi(r) \to \infty$ as $r \to \infty$, then choosing $R > 0$ large enough such that $M + \psi(R) \leq R$, it follows from (5.23) that G maps the ball B_R defined in (5.25) into itself.

Ad (c) Assume finally that α, $\beta > 1$. In this case the function ψ given by (5.24) satisfies $\psi(R) < R$ for small $R > 0$. Thus $M + \psi(R) \leq R$ when both R

and M are sufficiently small, which implies $G[B_R] \subseteq B_R$, where B_R is given by (5.25).

The above proof shows that in all the cases (a)–(c), G is a self-mapping of a closed ball B_R of $L^2(J, H)$. Since G is also increasing, then G has by Lemma 5.10 a fixed point h. Denoting $u = L^{-1}h$, then $u \in U \subseteq W$ and

$$Lu = h = Gh = F(\cdot, L^{-1}h, h) = F(\cdot, u, Lu).$$

In view of this result and (5.21), u is a solution of the IVP (5.13) in W.   □

5.2.3 Generalization and Special Cases

Assume that the spaces H and V have properties (H) and (V), and that operators $A(t)$, $t \in J = (0, \tau)$, have properties (A1) and (A2). Consider the implicit functional evolution problem

$$u'(t) + A(t)u(t) = \mathcal{F}(t, u, u'(t) + A(t)u(t)) \quad \text{a.e. in } J, \quad u(0) = 0, \quad (5.26)$$

where $\mathcal{F} : J \times L^2(J, H) \times H \to H$ satisfies the following hypotheses:

($\mathcal{F}1$) $t \mapsto \mathcal{F}(t, u, v(t))$ is measurable for all u, $v \in L^2(J, H)$, and there is a constant $b \geq 0$ such that $(u, \zeta) \mapsto \mathcal{F}(t, u, \zeta) + b\zeta$ is increasing in u and in ζ for a.e. $t \in J$.
($\mathcal{F}2$) There exist constants $\hat{M}$, $\hat{\mu} \geq 0$, and $\hat{\lambda} \in [0, 1)$ such that for all u, $v \in L^2(J, H)$,

$$\|\mathcal{F}(\cdot, u, v(\cdot))\|_{L^2(J,H)} \leq \hat{M} + \hat{\mu}\|u\|_{L^2(J,H)} + \hat{\lambda}\|v\|_{L^2(J,H)}.$$

We are going to show that these hypotheses with $\hat{\mu}$ small enough are sufficient to ensure the existence of a solution to the IVP (5.26).

Theorem 5.15. *Let hypotheses (A1), (A2), ($\mathcal{F}1$), and ($\mathcal{F}2$) with*

$$\hat{\mu} < \frac{\sqrt{2\rho + 1}(1 - \hat{\lambda})}{\sqrt{e^{\tau(2\rho+1)} - 1}}$$

be satisfied. Then the IVP (5.26) has a solution.

Proof: Define F as follows:

$$F(t, u, v) := \frac{1}{1+b}(\mathcal{F}(t, u, v(t)) + bv(t)), \quad t \in J, \quad u, v \in L^2(J, H).$$

Then the IVP (5.26) is equivalent to (5.13). The hypotheses given for $\mathcal{F}$ imply that F satisfies the hypotheses (F1), and (F2) with $\alpha = \beta = 1$, $M = \frac{\hat{M}}{1+b}$, $\lambda = \frac{\hat{\lambda}+b}{1+b}$, and $\mu = \frac{\hat{\mu}}{1+b}$. The hypothesis given for $\hat{\mu}$ implies that $\hat{\mu}c + \hat{\lambda} < 1$, where $c = \sqrt{\frac{e^{\tau(2\rho+1)}-1}{2\rho+1}}$. Thus

$$\mu c + \lambda = \frac{\hat{\mu} c}{1+b} + \frac{\hat{\lambda}+b}{1+b} = \frac{\hat{\mu} c + \hat{\lambda} + b}{1+b} < \frac{1+b}{1+b} = 1.$$

The asserted existence result follows then from Theorem 5.14 (a).   □

Remark 5.16. The implicit functional evolution problem

$$\mathcal{H}(t, u, u'(t) + A(t)u(t)) = 0 \quad \text{a.e. in } J, \quad u(0) = 0, \tag{5.27}$$

where $\mathcal{H} : J \times L^2(J,H) \times H \to H$, can be converted to the form (5.26) by defining

$$\mathcal{F}(t, u, \zeta) = \zeta - (\nu \cdot \mathcal{H})(t, u, \zeta), \quad t \in J, \ u \in L^2(J,H), \ \zeta \in H,$$

where $\nu : J \times L^2(J,H) \times H \to (0, \infty)$. If ν can be chosen in such a way that the so defined function $\mathcal{F}$ has the properties assumed in Theorem 5.15, we obtain an existence result for (5.27).

The nonfunctional implicit evolution problem

$$u'(t) + A(t)u(t) = q(t, u(t), u'(t) + A(t)u(t)) \quad \text{a.e. in } J, \quad u(0) = 0, \tag{5.28}$$

where $q : J \times H \times H \to H$, can be reduced to problem (5.26) by defining

$$\mathcal{F}(t, u, \zeta) = q(t, u(t), \zeta), \quad t \in J, \quad u \in L^2(J,H), \ \zeta \in H.$$

The so defined function $\mathcal{F} : J \times L^2(J,H) \times H \to H$ satisfies the hypotheses $(\mathcal{F}1)$ and $(\mathcal{F}2)$ when we impose the following hypotheses on q:

(q1) The nonlinearity q is sup-measurable, and there is a constant $b \geq 0$ such that $(r, \zeta) \mapsto q(t, r, \zeta) + b\,\zeta$ is increasing for a.e. $t \in J$.

(q2) There exist constants $\hat{M}$, $\hat{\mu} \geq 0$, and $\hat{\lambda} \in [0,1)$ such that for all $u, v \in L^2(J,H)$ the following estimate holds:

$$\|q(\cdot, u(\cdot), v(\cdot))\|_{L^2(J,H)} \leq \hat{M} + \hat{\mu} \|u\|_{L^2(J,H)} + \hat{\lambda} \|v\|_{L^2(J,H)}.$$

Thus Theorem 5.15 implies the following existence results for the IVB (5.28).

Proposition 5.17. *Let hypotheses (A1), (A2), (q1), and (q2) with $\hat{\mu} < \frac{\sqrt{2\rho+1}(1-\hat{\lambda})}{\sqrt{e^{\tau(2\rho+1)}-1}}$ be satisfied. Then the IVP (5.28) has a solution.*

As a consequence of Theorem 5.14 we readily obtain an existence result for the explicit functional IVP in the form

$$u'(t) + A(t)u(t) = g(t, u) \quad \text{a.e. in } J, \quad u(0) = 0, \tag{5.29}$$

where $g : J \times L^2(J,H) \to H$ is assumed to satisfy the hypotheses:

(g1) g is sup-measurable, and $u \mapsto g(t, u)$ is increasing for a.e. $t \in J$.

(g2) $\|g(\cdot, u)\|_{L^2(J,H)} \le M + \mu\|u\|^{\alpha}_{L^2(J,H)}$ for all $u \in L^2(J, H)$, where M, $\mu \ge 0$,

and either $0 \le \alpha < 1$, or $\alpha = 1$ and $\mu \le \sqrt{\frac{2\rho+1}{e^{\tau(2\rho+1)}-1}}$, or $\alpha > 1$ and M is small enough.

Proposition 5.18. *If the hypotheses (g1), (g2), (A1), and (A2) are satisfied, then the IVP (5.29) has a solution.*

5.2.4 Application

We apply the result of Theorem 5.14 to the following implicit functional parabolic initial-boundary value problem (IBVP)

$$\Lambda u(x, t) = f(x, t, u, \Lambda u) \text{ in } Q, \quad u = 0 \text{ on } \Gamma, \quad u = 0 \text{ in } \Omega \times \{0\}, \quad (5.30)$$

where $\Omega \subset \mathbb{R}^N$ is a bounded domain, $Q = \Omega \times J$, $J = (0, \tau)$, $\Gamma = \partial\Omega \times (0, \tau)$, $f : \Omega \times J \times L^2(Q) \times L^2(Q) \to \mathbb{R}$, and the operator Λ is defined by

$$\Lambda u := \frac{\partial u}{\partial t} - \sum_{i,j=1}^{N} \frac{\partial}{\partial x_i}\left(a_{ij}(x, t)\frac{\partial u}{\partial x_j}\right) + \sum_{i=1}^{N} b_i(x, t)\frac{\partial u}{\partial x_i} + a(x, t)u. \quad (5.31)$$

As an immediate consequence of the weak maximum principle for linear parabolic equations, we obtain the following comparison result.

Lemma 5.19. *Assume that the coefficients in (5.31) have the properties:*

(C) a_{ij}, b_i, $a \in L^\infty(Q)$ and $\sum_{i,j=1}^{N} a_{ij}(x, t)\xi_i\xi_j \ge \gamma \sum_{i=1}^{N}(\xi_i)^2$ for a.e. $(x, t) \in Q$ for all $(\xi_1, ..., \xi_N) \in \mathbb{R}^N$, with some constant $\gamma > 0$.

Choose $H = L^2(\Omega)$ and $V = W_0^{1,2}(\Omega)$), and define the operators $A(t)$, $t \in J$, by (5.15), where the mapping $a : W_0^{1,2}(\Omega) \times W_0^{1,2}(\Omega) \times J \to \mathbb{R}$, is given by the bilinear form

$$a(y, z; t) := \int_\Omega \left(\sum_{i,j=1}^{N} a_{ij}(\cdot, t)D_iy D_jz + \sum_{i=1}^{N} b_i(\cdot, t)z D_iy + a(\cdot, t)\, yz\right)dx, \quad (5.32)$$

with $D_iy = \frac{\partial y}{\partial x_i}$ denoting the generalized partial derivative. If w belongs to the set

$$W = \{u \in L^2(J, V) : \frac{\partial u}{\partial t} \in L^2(J, V^*), \ u(\cdot, 0) = 0 \text{ and } \Lambda u \in L^2(Q)\}, \quad (5.33)$$

and if $w'(t) + A(t)w(t) \le 0$ for a.e. $t \in J$, then $w(t) \le 0$ for a.e. $t \in J$.

Applying Theorem 5.14 we obtain the following result.

Theorem 5.20. *Assume hypothesis (C) of Lemma 5.19, and let the mapping $f : Q \times L^2(Q) \times L^2(Q) \to \mathbb{R}$ satisfy the following hypotheses when $L^2(Q)$ is equipped with the a.e. pointwise ordering:*

(f1) The function $(x,t) \mapsto f(x,t,u,v)$ is measurable in Q for all u, $v \in L^2(Q)$, and the function $(u,v) \mapsto f(\cdot,\cdot,u,v)$ is increasing for a.e. $(x,t) \in Q$.
(f2) There exist nonnegative constants M, μ, λ, α, and β such that

$$\|f(\cdot,\cdot,u,v)\|_{L^2(Q)} \leq M + \mu \|u\|_{L^2(Q)}^\alpha + \lambda\|v\|_{L^2(Q)}^\beta$$

for all u, $v \in L^2(Q)$.

Then the IBVP (5.30) has a solution in W in the following cases.

(a) $\alpha = \beta = 1$ and $\mu c + \lambda < 1$, where $c = \sqrt{\frac{e^{\tau(2\rho+1)}-1}{2\rho+1}}$
*   with $\rho = \max\limits_{1 \leq i \leq N} \|b_i\|_\infty + \|a\|_\infty$.*
(b) $0 \leq \alpha$, $\beta < 1$.
(c) α, $\beta > 1$, and the constant M is small enough.

Proof: Choosing $H = L^2(\Omega)$ and $V = W_0^{1,2}(\Omega))$, one can transform IBVP (5.30) into an evolution problem of the form (5.13), where the operators $A(t)$, $t \in J$, are defined by (5.15) with $a : V \times V \times J \to \mathbb{R}$ given by (5.32), and where we set

$$u(t)(x) := u(x,t) \quad \text{and} \quad F(t,u,v)(x) := f(x,t,u,v), \quad x \in \Omega, \ t \in J. \quad (5.34)$$

Since $L^2(Q)$ can be identified with $L^2(J,H)$, the hypotheses (f1) and (f2) imply that the hypotheses (F1) and (F2) are satisfied for the mapping F defined in (5.34). By means of (C), one readily verifies that the bilinear form a defined by (5.32) has the properties listed in (A1) (see also [229, Proposition 23.30]), where ρ in inequality (5.16) is given by $\rho = \max\limits_{1 \leq i \leq N} \|b_i\|_\infty + \|a\|_\infty$. Moreover, the hypothesis (A2) is valid due to Lemma 5.19. Thus all the hypotheses of Theorem 5.14 are satisfied, which concludes the proof.   □

Example 5.21. Let $\Omega = \{x \in \mathbb{R}^4 : |x| < 1\}$ be the unit sphere in $\mathbb{R}^4$, equipped with the Euclidean norm $|\cdot|$, $Q = \Omega \times (0,1)$ and $\Gamma = \partial\Omega \times (0,1)$, and let $[z]$ denote the greatest integer $\leq z \in \mathbb{R}$. Consider the IBVP

$$\begin{cases} \Lambda u(x,t) = [t + |x|] + \frac{1}{11}[u(x,t)] + \frac{1}{11}[\int_Q \Lambda u(x,t)\,dx\,dt] \ \text{in } Q, \\ u = 0 \ \text{in } \Omega \times \{0\}, \quad u = 0 \ \text{on } \Gamma, \end{cases} \quad (5.35)$$

where

$$\Lambda u(x,t) := \frac{\partial u(x,t)}{\partial t} - \Delta u(x,t) - \sum_{i=1}^4 \frac{\partial u(x,t)}{\partial x_i} + u(x,t).$$

The IBVP (5.35) is of the form (5.30), where

$$f(x, t, u, v) = [t + |x|] + \frac{1}{11}[u(x, t)] + \frac{1}{11}\left[\int_Q v(x, t)\, dx\, dt\right],$$

and the operator Λ is of the form (5.31), where $a_{ij}(x, t) \equiv \begin{cases} 1, & i = j, \\ 0, & i \neq j, \end{cases}$
$b_i(x, t) \equiv -1$ and $a(x, t) \equiv 1$. In particular, the hypothesis (C) of Lemma 5.19 is satisfied. The function $f(\cdot, \cdot, u, v)$ is obviously measurable, and $f(x, t, \cdot, \cdot)$ is increasing, whence the hypothesis (f1) holds, and by elementary calculations one gets

$$\|f(\cdot, \cdot, u, v)\|_{L^2(Q)} \leq 5 + \frac{1}{11}\|u\|_{L^2(Q)} + \frac{\pi^2}{22}\|v\|_{L^2(Q)}$$

for all u, $v \in L^2(Q)$. Because $\mu = \frac{1}{11}$, $\lambda = \frac{\pi^2}{22}$ and $\rho = 2$, we have

$$\mu c + \lambda = \frac{1}{11}\sqrt{\frac{e^5 - 1}{5}} + \frac{\pi^2}{22} \approx .94 < 1.$$

Thus the hypotheses of Theorem 5.20 (a) are satisfied, whence the IBVP (5.35) has a solution.

5.3 Notes and Comments

The presentation of this chapter is mainly based on the authors' joint work, see, e.g., [43, 45, 47, 50, 55].

By means of Theorem 5.5 on extremal solutions for discontinuous parabolic inclusions, we are able to deal with more general parabolic inclusions, such as, e.g., the following:

$$u_t + Au + \partial j(\cdot, \cdot, u) - \partial\beta(\cdot, \cdot, u) \ni h, \quad \text{in } Q \tag{5.36}$$
$$u(\cdot, 0) = 0 \text{ in } \Omega, \quad u = 0 \text{ on } \Gamma, \tag{5.37}$$

where A, j, and h are the same as before, and $s \mapsto \partial\beta(x, t, s)$ is the usual subdifferential of some Carathéodory function $\beta : Q \times \mathbb{R} \to \mathbb{R}$, which, in addition, is convex with respect to s. Although $s \mapsto j(x, t, s) - \beta(x, t, s)$ is a locally Lipschitz function, in general, one only has a strict inclusion of the form (see Sect. 9.6)

$$\partial(j - \beta)(x, t, s) \subseteq \partial j(x, t, s) - \partial\beta(x, t, s), \tag{5.38}$$

and equality in (5.38) holds if at least one of the functions $s \mapsto j(x, t, s)$ or $s \mapsto \beta(x, t, s)$ is strictly differentiable in s. In this sense the solution set for (5.36)–(5.37) is larger than for the parabolic inclusion

$$u_t + Au + \partial(j - \beta)(\cdot, \cdot, u) \ni h, \quad \text{in } Q \tag{5.39}$$
$$u(\cdot, 0) = 0 \text{ in } \Omega, \quad u = 0 \text{ on } \Gamma. \tag{5.40}$$

But still we can prove existence of extremal solutions of (5.36)–(5.37) by making use of Theorem 5.5. A special case of problem (5.36)–(5.37) has been considered in [65] under an additional one-sided growth condition on Clarke's generalized gradient. Due to our general comparison results of Chap. 3, we are now able to completely drop this one-sided growth condition.

We note that in most of the problems treated in this chapter, we do not suppose the existence of sub-supersolutions. Moreover, it won't be even possible to construct ordered pairs of sub-supersolutions. But still the abstract fixed point theorems, Theorem 5.9 and Lemma 5.10, allow us to ensure the existence of solutions of implicitly defined problems with governing nonlinearities that may be discontinuous in all their arguments.

Another advantage of the existence result of Theorem 5.20 is that the assumptions can easily be verified in practice, as seen from Example 5.21. The results of Sect. 5.2.3 have analogous applications in the theory of parabolic IBVPs. As for other existence results for implicit parabolic differential equations, see, e.g., [27, 44, 47].

The hypothesis (A2) of Sect. 5.2.1 is needed only in the proof of Lemma 5.11. All the above results can also be proved when (A2) is replaced by the following, more general assumption:

(A3) For each $h \in L^2(J, H)$ the IVP (5.18) has in W the smallest solution, which is increasing with respect to h.

The only change is to replace the subset U of W defined in the proof of Theorem 5.14 by

$$U := \{u \in W : u \text{ is the smallest solution of (5.18) for some } h \in L^2(J, H)\}.$$

The assumption (A3) allows to treat cases where the operators $A(t)$ are nonlinear, and where the corresponding IVP (5.18) is not necessarily uniquely solvable. As for conditions that ensure (A3) for parabolic operators, see, e.g., [44].

Finally we remark that discontinuously coupled systems of evolution variational inequalities have been treated in [59] by employing the abstract fixed point theory of Chap. 2.

6

Banach-Valued Ordinary Differential Equations

The main purpose of this chapter is to derive well-posedness, extremality, and comparison results for solutions of discontinuous ordinary differential equations in Banach spaces. A novel feature is that functions in considered differential equations are allowed to be Henstock–Lebesgue (HL) integrable with respect to the independent variable. HL integrability can be replaced also by Bochner integrability, although it is assumed explicitly only in the last section.

In Sect. 6.1 we study existence and uniqueness of solutions and their continuous dependence on the initial values. By assuming that the underlying Banach space is ordered, we derive in Sect. 6.2 existence and comparison results for extremal, i.e., the smallest and greatest solutions of first order discontinuous nonlocal semilinear differential equations, equipped with discontinuous and nonlocal initial conditions.

In Sect. 6.3 we apply results of Sects. 6.1 and 6.2 to derive well-posedness, existence, and comparison results for solutions of higher order differential equations in Banach spaces. Section 6.4 deals with the existence and comparison of the smallest and greatest solutions of first and second order initial value problems as well as for second order boundary value problem in an ordered Banach space. In the problems under consideration the dependence of the nonlinearities involved upon the unknown function is allowed to be implicit (except in Sect. 6.4.1), discontinuous, and nonlocal.

Finally, in Sect. 6.5 we study the existence and comparison of solutions of first order implicit functional differential equations containing Bochner integrable functions. A novel feature is that the existence of subsolutions and/or supersolutions is not assumed explicitly.

The main tools used in the treatment of the above problems are the abstract existence and comparison results derived in Chap. 2 for solutions of equations in ordered spaces, and the results presented in Chap. 9 for Bochner integrable and HL integrable vector-valued functions. Moreover, concrete problems are solved by using symbolic programming.

S. Carl and S. Heikkilä, *Fixed Point Theory in Ordered Sets and Applications:*
From Differential and Integral Equations to Game Theory,
DOI 10.1007/978-1-4419-7585-0_6, © Springer Science+Business Media, LLC 2011

6.1 Cauchy Problems

In this section we study Cauchy problems in a Banach space. The functions in the considered differential equations are HL integrable with respect to the independent variable. Recall that $^K\!\!\int$ denotes the Henstock–Kurzweil integral.

6.1.1 Preliminaries

Let E be a Banach space, and let J be a compact interval in $\mathbb{R}$. Denote by $C(J, E)$ the space of all continuous functions from J to E. This space is a Banach space with respect to the usual addition and scalar multiplication of functions and the uniform norm: $\|y\|_0 = \sup\{\|y(t)\| : t \in J\}$.

Recall that a function $y : [a, b] \to E$ satisfies the *Strong Lusin Condition* if for each $\epsilon > 0$ and for each null set Z of $[a, b]$ there exists a function $\delta : [a, b] \to (0, \infty)$ such that $\sum_{i=1}^m \|y(t_{2i}) - y(t_{2i-1})\| < \epsilon$ if the sequence $(t_i)_{i=1}^{2m}$ of $[a, b]$ is increasing, $\{\xi_i\}_{i=1}^m \subseteq Z$ and $\xi_i - \delta(\xi_i) < t_{2i-1} \leq \xi_i \leq t_{2i} < \xi_i + \delta(\xi_i)$ for every $i = 1, \ldots, m$.

Consider the Cauchy problem

$$y'(t) = f(t, y(t)) \quad \text{for a.e.} \ \ t \in J := [a, b], \quad y(a) = x_0, \tag{6.1}$$

where $a < b$, $f : J \times E \to E$, and $y'(t)$ is the derivative of a function y at t. We are looking for solutions of (6.1) from the set

$$W^1_{SL}(J, E) = \left\{ \begin{array}{l} y : J \to E : y \ \text{is a.e. differentiable and} \\ \text{satisfies the Strong Lusin Condition.} \end{array} \right\} \tag{6.2}$$

From Theorem 9.18 we deduce the following auxiliary result.

Lemma 6.1. *A function $y \in W^1_{SL}(J, E)$ is a solution of the Cauchy problem (6.1) if and only if $f(\cdot, y(\cdot))$ is HL integrable on J and y is a solution of the integral equation*

$$y(t) = x_0 + {}^K\!\!\int_a^t f(s, y(s))ds, \quad t \in J. \tag{6.3}$$

6.1.2 A Uniqueness Theorem of Nagumo Type

As an application of Lemma 6.1 we prove the following uniqueness result for the Cauchy problem (6.1).

Theorem 6.2. *Problem (6.1) has at most one solution in $W^1_{SL}(J, E)$ provided $f : J \times E \to E$ satisfies the following properties.*

(fi) f is continuous at (a, x_0), and $\|f(t, y) - f(t, z)\| \leq \frac{\|y-z\|}{t-a}$ for all $y, z \in E$ and for a.e. $t \in J$.

(fii) $f(\cdot, x(\cdot))$ *is strongly measurable for all* $x \in C(J, E)$.

Proof: Assume that x and y are solutions of (6.1) in $W^1_{SL}(J, E)$. The function $s \mapsto f(s, x(s)) - f(s, y(s))$ is by (fii) strongly measurable. Let $\beta \in (a, b)$ be fixed. In view of the hypothesis (fi) we obtain

$$\|f(s, x(s)) - f(s, y(s))\| \le \frac{\|x(s) - y(s)\|}{s - a} \quad \text{for a.e.} \quad s \in [\beta, b]. \tag{6.4}$$

Thus $s \mapsto f(s, x(s)) - f(s, y(s))$ is Bochner integrable, and hence also HL integrable on $[\beta, b]$, which in conjunction with Lemma 6.1 implies that

$$x(t) - y(t) = x(\beta) - y(\beta) + {}^K\!\!\int_\beta^t (f(s, x(s)) - f(s, y(s)))\, ds, \quad \beta \le t \le b. \tag{6.5}$$

Applying (6.4) we get

$$\left\| {}^K\!\!\int_\beta^t (f(s, x(s)) - f(s, y(s)))\, ds \right\| \le \int_\beta^t \frac{\|x(s) - y(s)\|}{s - a}\, ds, \quad \beta \le t \le b. \tag{6.6}$$

From (6.5) and (6.6) it follows that

$$\|x(t) - y(t)\| \le \|x(\beta) - y(\beta)\| + \int_\beta^t \frac{\|x(s) - y(s)\|}{s - a}\, ds, \quad \beta \le t \le b. \tag{6.7}$$

The greatest solution of

$$u(t) \le \|x(\beta) - y(\beta)\| + \int_\beta^t \frac{u(s)}{s - a}\, ds, \quad \beta \le t \le b, \tag{6.8}$$

is

$$u_\beta(t) = \|x(\beta) - y(\beta)\| \frac{t - a}{\beta - a}, \quad \beta \le t \le b. \tag{6.9}$$

Thus, by (6.7), (6.8), and (6.9) we obtain

$$\|x(t) - y(t)\| \le \|x(\beta) - y(\beta)\| \frac{b - a}{\beta - a}, \quad \beta \le t \le b. \tag{6.10}$$

This last result is valid for all $\beta \in (a, b)$. Moreover, we have

$$\frac{\|x(\beta) - y(\beta)\|}{\beta - a} = \frac{1}{\beta - a} \left\| {}^K\!\!\int_a^\beta (f(s, x(s)) - f(s, y(s)))\, ds \right\|, \quad a < \beta \le b. \tag{6.11}$$

Since f is continuous at (a, x_0), then $f(s, x(s)) - f(s, y(s)) \to 0$ as $s \to a$, and hence

$$\frac{1}{\beta - a} \left\| {}^K\!\!\int_a^\beta (f(s, x(s)) - f(s, y(s)))\, ds \right\| \to 0 \quad \text{as} \quad \beta \to a. \tag{6.12}$$

It then follows from (6.11) and (6.12) that $\frac{\|x(\beta) - y(\beta)\|}{\beta - a} \to 0$ as $\beta \to a$. Thus $x(t) \equiv y(t)$ by (6.10), so that the Cauchy problem (6.1) can have only one solution in $W^1_{SL}(J, E)$. □

6.1.3 Existence Results

In this subsection we prove existence results for the Cauchy problem

$$y'(t) = g(t, y(t), y) \text{ for a.e. } t \in J := [a, b], \quad y(a) = x_0, \tag{6.13}$$

where $a < b$ and $g : J \times E \times C(J, E) \to E$.

A first existence result is based on the following fixed point theorem, which follows from Proposition 2.39, and also from [44, Proposition 1.1.1].

Lemma 6.3. *Let $[\underline{y}, \overline{y}] = \{x \in X : \underline{y} \leq x \leq \overline{y}\}$ be a nonempty order interval in an ordered normed space X, and let $G : [\underline{y}, \overline{y}] \to [\underline{y}, \overline{y}]$ be increasing. If G maps monotone sequences to convergent sequences, then G has the smallest and greatest fixed points in $[\underline{y}, \overline{y}]$, and they are increasing with respect to G.*

Lemma 6.3 is used to prove the following existence result for the Cauchy problem (6.13).

Theorem 6.4. *Let E be a Banach space ordered by a regular order cone. Assume that $C(J, E)$ is ordered pointwise, and that $g : J \times E \times C(J, E) \to E$ satisfies the following hypotheses:*

(gi) $g(\cdot, x(\cdot), x)$ is strongly measurable for all $x \in C(J, E)$.
(gii) $g(t, x, y)$ is increasing in x and in y for a.e. $t \in J$, and there exist HL integrable functions $f_{\pm} : J \to E$ such that $f_{-}(t) \leq g(t, x, y) \leq f_{+}(t)$ for a.e. $t \in J$ and for all $x \in E$ and $y \in C(J, E)$.

Then the Cauchy problem (6.13) has the smallest and greatest solutions in $W^1_{SL}(J, E)$.

Proof: The hypotheses (gi) and (gii) imply by Proposition 9.14 that $g(\cdot, y(\cdot), y)$ is HL integrable for every $y \in C(J, E)$. Applying the hypothesis (gii) and the result of Lemma 9.11, it is easy to show that if $x, y \in C(J, E)$, and $x(s) \leq y(s)$ for all $s \in J$, and if $a \leq \underline{t} \leq t \leq b$, then

$$^K\!\!\int_{\underline{t}}^t f_{-}(s)\,ds \leq {}^K\!\!\int_{\underline{t}}^t g(s, x(s), x)\,ds \leq {}^K\!\!\int_{\underline{t}}^t g(s, y(s), y)\,ds \leq {}^K\!\!\int_{\underline{t}}^t f_{+}(s)\,ds. \tag{6.14}$$

Define functions $\underline{y}, \overline{y}$ of $C(J, E)$ by

$$\underline{y}(t) = x_0 + {}^K\!\!\int_a^t f_{-}(s)\,ds, \quad \overline{y}(t) = x_0 + {}^K\!\!\int_a^t f_{+}(s)\,ds \quad t \in J. \tag{6.15}$$

From (6.14) and (6.15) it follows that the integral operator G, defined by

$$Gx(t) = x_0 + {}^K\!\!\int_a^t g(s, x(s), x)\,ds, \quad t \in J, \tag{6.16}$$

is increasing and maps $C(J, E)$ into its order interval $[\underline{y}, \overline{y}]$. Moreover, (6.14), (6.15), and (6.16) imply that for all $x \in [\underline{y}, \overline{y}]$,

$$0 \le Gx(t) - Gx(\underline{t}) - (\underline{y}(t) - \underline{y}(\underline{t})) \le \overline{y}(t) - \overline{y}(\underline{t}) - (\underline{y}(t) - \underline{y}(\underline{t})), \qquad (6.17)$$

whenever $a \le \underline{t} \le t \le b$. Because the order cone of E is regular, it is also normal by Lemma 9.3. So from (6.16) and (6.17) it then follows that for all $x \in [\underline{y}, \overline{y}]$,

$$\|Gx(t) - Gx(\underline{t})\| \le (\lambda + 1)\|\underline{y}(t) - \underline{y}(\underline{t})\| + \lambda \|\overline{y}(t) - \overline{y}(\underline{t})\|, \ a \le \underline{t} \le t \le b. \ (6.18)$$

Let (x_n) be a monotone sequence in $[\underline{y}, \overline{y}]$. Then for every $t \in J$, $(Gx_n(t))$ is a monotone sequence in the order interval $[\underline{y}(t), \overline{y}(t)]$ of E. Since the order cone of E is regular, then $(Gx_n(t))$ converges in E for every $t \in J$. Moreover, from (6.18) we see that the sequence (Gx_n) is equicontinuous, which in view of Lemma 9.44 implies that (Gx_n) converges uniformly on J, and hence with respect to the uniform norm of $C(J, E)$.

The above proof shows that the hypotheses of Lemma 6.3 are valid for the restriction to $[\underline{y}, \overline{y}]$ of mapping G defined by (6.16), where X is replaced by $C(J, E)$ equipped with the pointwise ordering and the uniform norm, and the functions $\underline{y}, \overline{y}$ of $C(J, E)$ are defined by (6.15). Thus by Lemma 6.3, the mapping G has the smallest fixed point y_* and the greatest fixed point y^* in $[\underline{y}, \overline{y}]$, and hence in $C(J, E)$, because G maps $C(J, E)$ into $[\underline{y}, \overline{y}]$. This result implies by Lemma 6.1 that y_* and y^* are the smallest and greatest solutions of the Cauchy problem (6.13) in $W^1_{SL}(J, E)$. □

Remark 6.5. According to Theorem 2.16 the smallest fixed point y_* of G : $[\underline{y}, \overline{y}] \rightarrow [\underline{y}, \overline{y}]$ in Lemma 6.3 is the maximum of the w-o chain C of $\underline{y}G$-iterations. The greatest fixed point y^* of G is by Theorem 2.16 the minimum of the i.w-o chain D of $\overline{y}G$ iterations. Since $\underline{y} \le G\underline{y}$, the first elements of C are iterations $G^n\underline{y}$. These iterations form an increasing sequence because G is increasing. If $G^n\underline{y} = G^{n+1}\underline{y}$ for some n, then $G^n\underline{y} = \max C = y_*$. Similarly, the first elements of D are iterations $G^n\overline{y}$. These iterations form a decreasing sequence. If $G^m\overline{y} = G^{m+1}\overline{y}$ for some m, then $G^m\overline{y} = \min D = y^*$ (see Sect. 2.2.4).

Example 6.6. Determine the smallest and greatest solutions of the system

$$\begin{cases} u'(t) = \frac{1}{t}\sin\frac{1}{t} + 10^{-4}[2 \cdot 10^4 \arctan(\int_1^4 v(s)ds)](1 + \cos t) \\ \qquad\qquad \text{for a.e. } t \in J, \qquad u(0) = 0, \\ v'(t) = -\frac{1}{t}\sin\frac{1}{t} + 10^{-4}[3 \cdot 10^4 \tanh(\int_0^1 u(s)ds)](1 + \cos t) \\ \qquad\qquad \text{for a.e. } t \in J, \qquad v(0) = 0, \end{cases} \quad (6.19)$$

where $J = [0, 4]$. Note, the brackets in the equations stand for the integer function $s \mapsto [s] := \max\{n \in \mathbb{Z} : n \le s\}$.

Solution: Problem (6.19) can be converted to the Cauchy problem (6.13), where the components of $g = (g_1, g_2) : J \times C(J, \mathbb{R}^2) \to \mathbb{R}^2$, $J = [0, 4]$ are defined by

$$
\begin{aligned}
g_1(t, (u, v)) &= \tfrac{1}{t} \sin \tfrac{1}{t} + 10^{-4} [2 \cdot 10^4 \arctan(\int_1^4 v(s) ds)](1 + \cos t), \\
g_2(t, (u, v)) &= -\tfrac{1}{t} \sin \tfrac{1}{t} + 10^{-4} [3 \cdot 10^4 \tanh(\int_0^1 u(s) ds)](1 + \cos t).
\end{aligned}
\tag{6.20}
$$

It is easy to see that G satisfies the hypotheses (gi) and (gii) of Theorem 6.4 when

$$
\begin{aligned}
f_-(t) &= (\tfrac{1}{t} \sin \tfrac{1}{t} - 3.1416(1 + \cos t), -\tfrac{1}{t} \sin \tfrac{1}{t} - 3(1 + \sin t)), \\
f_+(t) &= (\tfrac{1}{t} \sin \tfrac{1}{t} + 3.1415(1 + \cos t), -\tfrac{1}{t} \sin \tfrac{1}{t} + 3(1 + \sin t)).
\end{aligned}
$$

The functions $\underline{y} = (u_-, v_-), \overline{y} = (u_+, v_+)$, defined in (6.15), have the following components, which are continuous on J and differentiable on $(0, 4]$:

$$
\begin{aligned}
u_-(t) &= -Si(\tfrac{1}{t}) - 3.1416(t + \sin t) + \tfrac{\pi}{2}, \\
v_-(t) &= Si(\tfrac{1}{t}) - 3(1 - t - \cos(t)) - \tfrac{\pi}{2}, \\
u_+(t) &= -Si(\tfrac{1}{t}) + 3.1415(t + \sin t) + \tfrac{\pi}{2}, \\
v_+(t) &= Si(\tfrac{1}{t}) + 3(1 - t - \cos(t)) - \tfrac{\pi}{2},
\end{aligned}
$$

where

$$
Si(x) = \int_0^x \frac{\sin s}{s} ds
$$

is the **sine integral**. Thus the hypotheses of Lemma 1.12 hold for the functions $u_\pm$ and $v_\pm$.

Calculating the iterations $G^n \underline{y}$, $n \in \mathbb{N}$, where G is defined by (6.16), we see that $G^5 \underline{y} = G^6 \underline{y}$, so that $y_* := G^5 \underline{y}$ is the smallest fixed point of G by Remark 6.5. The fixed point $y_* = (u_*, v_*)$ is the smallest solution of the Cauchy problem (6.19) by the proof of Theorem 6.4. Similarly, we see that $y^* := G^5 \overline{y}$ is the greatest fixed point of G, and hence the greatest solution of the Cauchy problem (6.19). The exact expressions of the components of $y_* = (u_*, v_*)$ and $y^* = (u^*, v^*)$ are:

$$
\begin{aligned}
u_*(t) &= -Si(\tfrac{1}{t}) - 3.0909(t + \sin t) + \tfrac{\pi}{2}, \\
v_*(t) &= Si(\tfrac{1}{t}) - 2.9799(1 - t - \cos(t)) - \tfrac{\pi}{2}, \\
u^*(t) &= -Si(\tfrac{1}{t}) + 3.0806(t + \sin t) + \tfrac{\pi}{2}, \\
v^*(t) &= Si(\tfrac{1}{t}) + 2.9872(1 - t - \cos(t)) - \tfrac{\pi}{2}.
\end{aligned}
\tag{6.21}
$$

Calculations are carried out by constructing appropriate Maple programs.

As a special case of Theorem 6.4 we get the following corollary.

Corollary 6.7. *Let E be a Banach space ordered by a regular order cone. Assume that $f : J \times E \to E$ satisfies the hypothesis (fii) of Theorem 6.2 and the following hypothesis:*

(fiii) *The function $x \mapsto f(t,x)$ is increasing for a.e. $t \in J$, and there exist HL integrable functions $f_\pm : J \to E$ such that $f_-(t) \le f(t,x) \le f_+(t)$ for a.e. $t \in J$ and for all $x \in E$.*

Then problem (6.1) has the smallest and greatest solutions in $W^1_{SL}(J,E)$.

Theorem 6.9 given below provides another type of existence results for (6.13). Its novel feature is that neither order boundedness of the right-hand side nor sub-supersolutions are assumed. The proof of Theorem 6.9 is based on a fixed point result that is given by the next lemma. Before we formulate and prove this lemma, recall that an ordered Banach space E is *lattice ordered* if $\sup\{x,y\}$ and $\inf\{x,y\}$ exist for all $x, y \in E$. When $x \in E$, denote

$$|x| = \sup\{x,-x\}, \quad x^+ = \sup\{0,x\} \quad \text{and} \quad x^- = \sup\{-x,0\} = (-x)^+. \quad (6.22)$$

We call E a *Banach lattice* if

$$|x| \le |y| \text{ in } E \text{ implies } \|x\| \le \|y\|. \quad (6.23)$$

The following fixed point result is a special case of Proposition 2.40.

Lemma 6.8. *Given a Banach lattice E and a nonempty compact real interval J, we assume that $C(J,E)$ is equipped with the pointwise ordering $\le$ and the uniform norm $\|\cdot\|_0$. If a mapping $G : C(J,E) \to C(J,E)$ is increasing, and if a sequence $(Gy_n)_{n=0}^\infty$ converges in $(C(J,E),\|\cdot\|_0)$ whenever $(y_n)_{n=0}^\infty$ is a monotone sequence in $C(J,E)$, then G has*

(a) *minimal and maximal fixed points;*
(b) *smallest and greatest fixed points y_* and y^* in $[\underline{y},\overline{y}]$, where $\underline{y}$ is the greatest solution of $y = -(-Gy)^+$ and $\overline{y}$ is the smallest solution of $y = (Gy)^+$.*
(c) *All the solutions $\underline{y}$, $\overline{y}$, y_*, and y^* are increasing with respect to G.*

Proof: Since E is a Banach lattice, it can be shown (see, e.g., [204]) that $|x^\pm - y^\pm| \le |x - y|$ for all $x, y \in E$. Thus, by (6.23),

$$\|x^\pm - y^\pm\| \le \|x - y\|, \quad x, y \in E.$$

In particular, the mappings $E \ni x \mapsto x^\pm$ are continuous. Hence, if $y \in C(J,E)$, then the mappings $y^\pm := t \mapsto y(t)^\pm$ belong to $C(J,E)$. Consequently, the zero function is an order center of $C(J,E)$. Moreover, $C(J,E)$ is an ordered normed space with respect to the pointwise ordering and the sup-norm. Noticing also that $\sup\{0,y\} = y^+$ and $\inf\{0,y\} = -(-y)^+$, the conclusions follow from Proposition 2.40. $\qquad\square$

Applying Lemma 6.8, we shall prove existence and comparison results for the Cauchy problem (6.13) when the function $g : J \times E \times C(J,E) \to E$ satisfies the following hypotheses.

(g) $g(\cdot,y(\cdot),y)$ is HL integrable for all $y \in C(J,E)$.

(gg) If $x \le y$ in $C(J, E)$, then $g(s, x(s), x) \le g(s, y(s), y)$ for a.e. $s \in J$.
(ggg) The sequence $\left(K \int_a^b g(s, y_n(s), y_n)\, ds \right)_{n=0}^\infty$ is bounded whenever $(y_n)_{n=0}^\infty$ is a monotone sequence in $C(J, E)$.

Theorem 6.9. *Let E be a weakly sequentially complete Banach lattice, and assume hypotheses (g)–(ggg). Then the Cauchy problem (6.13) has*

(a) minimal and maximal solutions;
(b) least and greatest solutions y_* and y^* in $[\underline{y}, \overline{y}]$, where $\underline{y}$ is the greatest solution of the equation

$$y(t) = -\left(-x_0 - {}^K\!\!\int_a^t g(s, y(s), y)ds \right)^+ , \quad t \in J,$$

and $\overline{y}$ is the smallest solution of the equation

$$y(t) = \left(x_0 + {}^K\!\!\int_a^t g(s, y(s), y)ds \right)^+ , \quad t \in J.$$

(c) All the solutions $\underline{y}$, $\overline{y}$, y_*, and y^* are increasing with respect to g and x_0.

Proof: According to hypotheses (g) and (gg), relation (6.16) defines the mapping $G : C(J, E) \to C(J, E)$, which is increasing by Lemma 9.11. In order to show that G maps monotone sequences of $C(J, E)$ to convergent sequences, let $(y_n)_{n=0}^\infty$ be an increasing sequence in $C(J, E)$. Since G is increasing, the sequence $(Gy_n)_{n=0}^\infty$ is increasing, too. Hypotheses (gg), (ggg), and (6.16) imply that the sequence $(Gy_n(b))_{n=0}^\infty$ is bounded in E and increasing. Moreover, the order cone of E is normal by (6.23), and hence fully regular by Lemma 9.3. Thus the sequence $(Gy_n(b))_{n=0}^\infty$ converges. By using

$$0 \le Gy_m(t) - Gy_n(t) \le Gy_m(b) - Gy_n(b), \quad t \in J, \ n \le m,$$

and (6.23), we obtain

$$\|Gy_m(t) - Gy_n(t)\| \le \|Gy_m(b) - Gy_n(b)\|, \quad t \in J, \ n \le m.$$

This result implies

$$\|Gy_m - Gy_n\|_0 = \sup_{t \in J} \|Gy_m(t) - Gy_n(t)\| \le \|Gy_m(b) - Gy_n(b)\|, \quad n \le m. \tag{6.24}$$

Because the sequence $(Gy_n(b))_{n=0}^\infty$ converges, it is a Cauchy sequence in E, which along with (6.24) implies that $(Gy_n)_{n=0}^\infty$ is a Cauchy sequence in $(C(J, X), \| \cdot \|_0)$. Therefore, $(Gy_n)_{n=0}^\infty$ converges whenever $(y_n)_{n=0}^\infty$ is an increasing sequence in $C(J, E)$. The proof that the sequence $(Gy_n)_{n=0}^\infty$ converges in $(C(J, E), \| \cdot \|_0)$ whenever $(y_n)_{n=0}^\infty$ is a decreasing sequence in $C(J, E)$ is similar. The above proof shows that G satisfies the hypotheses of Lemma 6.8.

Moreover, G is increasing with respect to g and x_0. Thus the conclusions follow from Lemma 6.8 since solutions of the Cauchy problem (6.13) and the fixed points of G are same. $\qquad\square$

The following spaces are examples of weakly sequentially complete Banach lattices:

- $\mathbb{R}^m$, ordered coordinatewise and normed by a p-norm, $1 \le p < \infty$.
- Reflexive Banach lattices.
- Separable Hilbert spaces whose order cones are generated by orthonormal bases.
- Sequence spaces l^p, $1 \le p < \infty$, normed by p-norm and ordered componentwise.
- Function spaces $L^p(\Omega)$, $1 \le p < \infty$, normed by p-norm and ordered a.e. pointwise, where Ω is a measure space.
- Function spaces $L^p([a,b], X)$, $1 \le p < \infty$, ordered a.e. pointwise, where X is any of the spaces listed above.

6.1.4 Existence and Uniqueness Results

Combining the results of Theorem 6.2 and Corollary 6.7 we obtain the following existence and uniqueness result.

Proposition 6.10. *Let E be an ordered Banach space, ordered by a regular order cone, and $f : J \times E \to E$. If the hypotheses (fi), (fii), and (fiii) of Theorem 6.2 and Corollary 6.7 are valid, then the Cauchy problem (6.1) has exactly one solution in $W_{SL}^1(J, E)$.*

In what follows, E is a Banach space. We are going to present conditions for a function $f : J \times E \to E$ under which the Cauchy problem (6.1) has a uniquely determined solution that can be obtained by the method of successive approximations. In the proof we make use of the following auxiliary result.

Lemma 6.11. *Assume that the function $q : J \times [0, r] \to \mathbb{R}_+$, $r > 0$, satisfies the following condition.*

(q) *$q(\cdot, x)$ is measurable for all $x \in [0, r]$, $q(\cdot, r) \in L^1(J, \mathbb{R}_+)$, $q(t, \cdot)$ is increasing and right-continuous for a.e. $t \in J$, and the zero-function is the only absolutely continuous (AC) solution with $u_0 = 0$ of the Cauchy problem*

$$u'(t) = q(t, u(t)) \quad a.e. \ on \ J, \quad u(a) = u_0. \tag{6.25}$$

Then there exists an $r_0 > 0$ such that the Cauchy problem (6.25) has for every $u_0 \in [0, r_0]$ the smallest AC solution $u = u(\cdot, u_0)$, which is increasing with respect to u_0. Moreover, $u(t, u_0) \to 0$ uniformly over $t \in J$ when $u_0 \to 0+$.

Proof: Defining

$$\hat{q}(t,x) = q(t,\max\{0,\min\{x,r\}\}), \quad t \in J, \, x \in \mathbb{R}, \qquad (6.26)$$

we obtain a function $\hat{q} : J \times \mathbb{R} \to \mathbb{R}_+$ that is L^1-bounded, $\hat{q}(\cdot,x)$ is measurable for every $x \in \mathbb{R}$, and $\hat{q}(t,\cdot)$ is increasing and right-continuous for a.e. $t \in J$. Moreover, for every $u_0 \in [0,r]$ the zero-function $\underline{u}(t) \equiv 0$ is a lower AC solution of

$$u'(t) = \hat{q}(t,u(t)) \quad \text{a.e. on } \, J, \quad u(a) = u_0, \qquad (6.27)$$

and the function $\overline{u}(t) = r + \int_a^t q(s,r)\,ds, \, t \in J$ is an upper AC solution of (6.27). It then follows from [44, Theorem 2.1.4] that (6.27) has for every $u_0 \in [0,r]$ the smallest AC solution $u = u(\cdot,u_0)$, and

$$u(t,u_0) = \min\{u_+(t) : u_+ \text{ is an upper AC solution of (6.27)}\}. \qquad (6.28)$$

If $0 \le \hat{u}_0 \le u_0 \le r$, then $u(\cdot,u_0)$ is an upper AC solution of (6.27) with u_0 replaced by $\hat{u}_0$. This implies by (6.28) that $u(\cdot,u_0)$ is increasing with respect to u_0.

To prove the assertions, choose $k \ge \frac{1}{r}$ and denote $u_n = u(\cdot,\frac{1}{n})$, $n \ge k$. The so obtained sequence $(u_n)_{n=k}^\infty$ is decreasing by the above proof, and bounded from below by the zero function, whence the limit

$$v(t) = \lim_{n \to \infty} u_n(t) \qquad (6.29)$$

exists for each $t \in J$. Since $\hat{q}(\cdot,x)$ is measurable for every $x \in \mathbb{R}$ and $\hat{q}(t,\cdot)$ is right-continuous for a.e. $t \in J$, then $\hat{q}(\cdot,u_n(\cdot))$ is measurable for every $n \ge k$ (cf. the proof of [161, Proposition 1.1.4]). Because $\hat{q}(\cdot,u_n(\cdot))$ is nonnegative-valued and bounded from above by a Lebesgue integrable function $q(\cdot,r)$, then $\hat{q}(\cdot,u_n(\cdot))$ is also Lebesgue integrable. Thus

$$u_n(t) = \frac{1}{n} + \int_a^t \hat{q}(s,u_n(s))\,ds, \quad t \in J, \, n \ge k. \qquad (6.30)$$

Since $\hat{q}(s,\cdot)$ is increasing for a.e. $s \in J$, and since (u_n) is decreasing, it follows from (6.30) that

$$0 \le u_n(t) - u_m(t) \le \frac{1}{n} - \frac{1}{m} + \int_a^b (\hat{q}(s,u_n(s)) - \hat{q}(s,u_m(s)))\,ds = u_n(b) - u_m(b)$$

whenever $a \le t \le b$ and $k \le n \le m$, so that the convergence in (6.29) is uniform. In particular, $v \in C(J,\mathbb{R}_+)$. Because (u_n) is decreasing and $\hat{q}(t,\cdot)$ is right-continuous, then

$$\lim_{n \to \infty} \hat{q}(t,u_n(t)) = \hat{q}(t,v(t)) \quad \text{for a.e. } \, t \in J.$$

It then follows from (6.30) by the dominated convergence theorem, as $n \to \infty$, that

$$v(t) = \int_a^t \hat{q}(s, v(s)) \, ds, \quad t \in J.$$

But this implies that v is an AC solution of the Cauchy problem (6.27) with $u_0 = 0$. Applying the definition of $\hat{q}$ and the hypothesis that $u(t) \equiv 0$ is the only AC solution of (6.25) when $u_0 = 0$, it is elementary to verify that $v(t) \equiv 0$. In particular $u_n(t) = u(t, \frac{1}{n}) \to 0$ uniformly over $t \in J$ as $n \to \infty$. This result and the fact that $u(\cdot, u_0)$ is increasing with respect to u_0 imply that $u(t, u_0) \to 0$ uniformly over $t \in J$ when $u_0 \to 0+$. In particular, there exists an $r_0 > 0$ such that $u(t, u_0) \le r$ for all $t \in J$, and $u_0 \in [0, r_0]$. Thus $u(\cdot, u_0)$ is by (6.26) the smallest AC solution of (6.25) when $u_0 \in [0, r_0]$.   □

Denoting $|y| = \|y(\cdot)\|$, $y \in C(J, E)$, we have the following fixed point result.

Proposition 6.12. *([133, Theorem 1.4.9]) Let $F : C(J, E) \to C(J, E)$ satisfy the hypothesis:*

(F) There exists a $v \in C(J, \mathbb{R}_+)$ and an increasing mapping $Q : [0, v] \to [0, v]$ satisfying $Qv(t) < v(t)$ and $Q^n v(t) \to 0$ for each $t \in J$, such that

$$|Fy - F\bar{y}| \le Q|y - \bar{y}| \quad \text{for } \ y, \bar{y} \in C(J, E), \ |y - \bar{y}| \le v. \tag{6.31}$$

Then for each $y_0 \in C(J, E)$ the sequence $(F^n y_0)_{n=0}^\infty$ converges uniformly on J to a unique fixed point of F.

Now we are able to prove an existence and uniqueness theorem for the solution of the Cauchy problem (6.1).

Theorem 6.13. *Assume that $f : J \times E \to E$ has the following properties:*

(f1) $f(\cdot, x)$ is strongly measurable for all $x \in E$ and HL integrable for some $x \in E$.

(f2) There exists an $r > 0$ such that

$$\|f(t, y) - f(t, z)\| \le q(t, \|y - z\|)$$

for all $y, z \in E$ with $\|y - z\| \le r$ and for a.e. $t \in J$, where $q : J \times [0, r] \to \mathbb{R}_+$ satisfies the hypothesis (q) of Lemma 6.11.

Then for each $x_0 \in E$ the Cauchy problem (6.1) has a unique solution y in $W_{SL}^1(J, E)$. Moreover, y is the uniform limit of the sequence $(y_n)_{n=0}^\infty$ of the successive approximations

$$y_{n+1}(t) = x_0 + {}^K\!\!\int_a^t f(s, y_n(s)) \, ds, \quad t \in J, \ n \in \mathbb{N}_0, \tag{6.32}$$

for each choice of $y_0 \in C(J, E)$.

Proof: According to Lemma 6.11 the Cauchy problem (6.25) has for some $u_0 = r_0 > 0$ the smallest AC solution $v = u(\cdot, r_0)$, and $r_0 \le v(t) \le r$ for each $t \in J$. Since $q(s, \cdot)$ is increasing and right-continuous in $[0, r]$ for a.e. $s \in J$, and because $q(\cdot, x)$ is measurable for all $x \in [0, r]$ and $q(\cdot, r)$ is Lebesgue integrable, it follows that $q(\cdot, u(\cdot))$ is Lebesgue integrable whenever u belongs to the order interval $[0, v] = \{u \in C(J, \mathbb{R}) : 0 \le u(t) \le v(t), \ t \in J\}$. Thus the equation

$$Qw(t) = \int_a^t q(s, w(s)) \, ds, \quad t \in J \tag{6.33}$$

defines a mapping $Q : [0, v] \to C(J, \mathbb{R}_+)$. Condition (q) ensures that Q is increasing, and the choice of r_0 and v that

$$r_0 + Qv = v. \tag{6.34}$$

Thus $Qv(t) < v(t)$ for every $t \in J$. The sequence $(Q^n v)_{n=0}^\infty$ is decreasing because $q(t, \cdot)$ is increasing for a.e. $t \in J$. The reasoning similar to that applied to the sequence (u_n) in the proof of Lemma 6.11 shows that $(Q^n v)_{n=0}^\infty$ converges uniformly on J to the zero function. Since this function satisfies by (q) the equation $u'(t) = q(t, u(t))$ for a.e. $t \in J$, then $q(t, 0) = 0$ for a.e. $t \in J$. This result and hypotheses (f1) and (f2) imply that f is a Carathéodory function. Thus by [133, Theorem 1.4.3] $f(\cdot, y(\cdot))$ is strongly measurable on J for all $y \in C(J, E)$. Let $y \in C(J, E)$ be fixed, and choose by (f1) a $z \in E$ so that $f(\cdot, z)$ is HL integrable. Choosing

$$y_i(t) = z + \frac{i}{m}(y(t) - z), \quad i = 0, \ldots, m \ge \frac{\|y - z\|}{r_0},$$

we have $\|y_i(t) - y_{i-1}(t)\| \le r_0 \le v(t)$ on J for each $i = 1, \ldots, m$, whence

$$\|f(t, y(t)) - f(t, z)\| \le \sum_{i=1}^m \|f(t, y_i(t)) - f(t, y_{i-1}(t))\|$$

$$\le \sum_{i=1}^m q(t, \|y_i(t) - y_{i-1}(t)\|) \le \sum_{i=1}^m q(t, v(t)) = m \, v'(t)$$

for a.e. $t \in J$. This result and the strong measurability of $f(\cdot, y(\cdot))$ and $f(\cdot, z)$ imply that $f(\cdot, y(\cdot)) - f(\cdot, z)$ is Bochner integrable, and hence also HL integrable on J. Thus $f(\cdot, y(\cdot)) = f(\cdot, z) + f(\cdot, y(\cdot)) - f(\cdot, z)$ is HL integrable on J. In particular, for each fixed $x_0 \in E$ the equation

$$Fy(t) = x_0 + {}^K\!\!\int_a^t f(s, y(s)) ds, \quad t \in J, \tag{6.35}$$

defines a mapping $F : C(J, E) \to C(J, E)$.

Next we show that the mappings Q and F defined by (6.33) and (6.35) satisfy the hypotheses of Proposition 6.12. To verify that (6.31) holds, let $y, \bar{y} \in$

$C(J, E)$ be given. The functions $f(\cdot, y(\cdot)) - f(\cdot, z)$ and $f(\cdot, \bar{y}(\cdot)) - f(\cdot, z)$ are Bochner integrable by the above proof. Thus the function $f(\cdot, y(\cdot)) - f(\cdot, \bar{y}(\cdot))$ is Bochner integrable. This result implies that the function $\|f(\cdot, \bar{y}(\cdot)) - f(\cdot, z)\|$ is Lebesgue integrable. Moreover, for all $t \in J$ we have the estimate

$$\left\| {}^K\!\!\int_a^t f(s, y(s))\, ds - {}^K\!\!\int_a^t f(s, \bar{y}(s))\, ds \right\| = \left\| \int_a^t (f(s, y(s))\, ds - f(s, \bar{y}(s)))\, ds \right\|$$

$$\leq \int_a^t \|f(s, y(s))\, ds - f(s, \bar{y}(s))\|\, ds.$$

Applying this result, the hypotheses (f1) and (f2), and definitions (6.33) and (6.35) we see that

$$|Fy - F\bar{y}| \leq Q|y - \bar{y}| \quad \text{for} \quad y, \bar{y} \in C(J, E), \ |y - \bar{y}| \leq v.$$

The above proof shows that the operators F and Q satisfy the hypotheses of Proposition 6.12. Thus the iteration sequence $(F^n y_0)_{n=0}^\infty$, which equals to the sequence $(y_n)_{n=0}^\infty$ of successive approximations (6.32), converges for every choice of $y_0 \in C(J, E)$ uniformly in J to a unique fixed point y of F. This result and the definition of F imply by Lemma 6.1 that y is the uniquely determined solution of the Cauchy problem (6.1) in $W_{SL}^1(J, E)$.   □

6.1.5 Dependence on the Initial Value

We shall first prove that under the hypotheses (f1) and (f2), the difference of solutions y of (6.1) belonging to initial values x_0 and $\hat{x}_0$, respectively, can be estimated by the *smallest solution* of the comparison problem (6.25) with initial value $u_0 = \|x_0 - \hat{x}_0\|$. This estimate implies by Lemma 6.11 the continuous dependence of y on x_0.

Proposition 6.14. *Let $f : J \times E \to E$ satisfy the hypotheses (f1) and (f2). If $y = y(\cdot, x_0)$ denotes the solution of the Cauchy problem (6.1) and $u = u(\cdot, u_0)$ the smallest solution of the Cauchy problem (6.25), then for all $x_0, \hat{x}_0 \in E$, with $\|x_0 - \hat{x}_0\|$ small enough,*

$$\|y(t, x_0) - y(t, \hat{x}_0))\| \leq u(t, \|x_0 - \hat{x}_0\|), \qquad t \in J. \tag{6.36}$$

In particular, $y(\cdot, x_0)$ depends continuously on x_0 in the sense that $y(t, \hat{x}_0) \to y(t, x_0)$ uniformly over $t \in J$ as $\hat{x}_0 \to x_0$.

Proof: Assume that $x_0, \hat{x}_0 \in E$, and that $\|x_0 - \hat{x}_0\| \leq r_0$, where r_0 is chosen as in Lemma 6.11. The solutions $y = y(\cdot, x_0)$ and $\hat{y} = y(\cdot, \hat{x}_0)$ exist by Theorem 6.13, and they satisfy

$$y(t) = Fy(t) = x_0 + {}^K\!\!\int_a^t f(s, y(s))\, ds, \quad t \in J,$$

and

$$\hat{y}(t) = \hat{F}\hat{y}(t) = \hat{x}_0 + {}^K\!\!\int_a^t f(s, \hat{y}(s))\, ds, \quad t \in J.$$

Moreover, F satisfies the hypotheses of Proposition 6.12 with Q defined by

$$Qw(t) = \int_a^t q(s, w(s))\, ds, \quad t \in J,$$

and $u = u(\cdot, \|x_0 - \hat{x}_0\|)$ is the smallest AC solution of

$$u = \|x_0 - \hat{x}_0\| + Qu.$$

Denote

$$W = \{y \in C(J, E) : |y - \hat{y}| \le u\}.$$

Since Q is increasing, and since

$$F\hat{y}(t) - \hat{y}(t) = F\hat{y}(t) - \hat{F}\hat{y}(t) = x_0 - \hat{x}_0$$

for all $t \in J$, we have for every $y \in W$,

$$|Fy - \hat{y}| \le |F\hat{y} - \hat{y}| + |Fy - F\hat{y}| \le |F\hat{y} - \hat{y}| + Q|y - \hat{y}|$$
$$\le \|x_0 - \hat{x}_0\| + Qu = u.$$

Thus $F[W] \subseteq W$. Since $\hat{y} \in W$, then $(F^n \hat{y}) \in W$ for every $n \in \mathbb{N}_0$. The uniform limit $y = \lim_n F^n \hat{y}$ exists by Theorem 6.13 and is the solution of (6.1). Because W is closed, then $y \in W$, so that $|y - \hat{y}| \le u$. This proves (6.36). According to Lemma 6.11, $u(t, \|x_0 - \hat{x}_0\|) \to 0$ uniformly over $t \in J$ as $\|x_0 - \hat{x}_0\| \to 0$. This result and (6.36) imply that the last assertion of the proposition is true.   □

Remark 6.15. If $r = \infty$ in condition (f2), then (6.36) holds for all $x_0, \hat{x}_0 \in E$.

The hypotheses imposed on $q : J \times [0, r] \to \mathbb{R}_+$ in (q) hold if $q(t, \cdot)$ is increasing for a.e. $t \in J$, and if q is an L^1-bounded Carathéodory function such that the following local Kamke's condition holds.

$u \in C(J, [0, r])$ and $u(t) \le \int_a^t q(s, u(s))\, ds$ for all $t \in J$ imply $u(t) \equiv 0$.

6.1.6 Well-Posedness of a Semilinear Cauchy Problem

The main goal of this subsection is to study the well-posedness of the following semilinear Cauchy problem

$$y'(t) = A(t)\, y(t) + g(t, y(t)) \quad \text{for a.e. } t \in J := [a, b], \qquad y(a) = x_0. \quad (6.37)$$

Given a Banach space E, denote by $L(E)$ the space of all bounded linear mappings $T : E \to E$. We impose the following hypotheses on $A : J \to L(E)$ and $g : J \times E \to E$.

(A) $A(\cdot)x$ is strongly measurable for each $x \in E$, and $\|A(\cdot)\| \leq p_1 \in L^1(J, \mathbb{R}_+)$.

(g1) $g(\cdot, x)$ is strongly measurable for every $x \in E$ and HL integrable for some $x \in E$.

(g2) There exists an $r > 0$ such that for all $x, y \in E$ with $\|x - y\| \leq r$ and for a.e. $t \in J$ the estimate

$$\|g(t, x) - g(t, y)\| \leq p(t)\,\phi(\|x - y\|)$$

holds, where $p \in L^1(J, \mathbb{R}_+)$, $\phi : [0, r] \to \mathbb{R}_+$ is increasing and right-continuous, and $\int_0^r \frac{dv}{\phi(v)} = \infty$.

Theorem 6.16. *If the hypotheses (A) (g1) and (g2) are satisfied, then the Cauchy problem (6.37) has for each $x_0 \in E$ a unique solution in $W^1_{SL}(J, E)$ that continuously depends on x_0.*

Proof: It suffices to show that the hypotheses (f1), (f2), and (q) hold for the functions $f : J \times E \to E$ and $q : J \times \mathbb{R}_+ \to \mathbb{R}_+$ defined by

$$f(t, x) = A(t)x + g(t, x), \quad t \in J, \, x \in E \tag{6.38}$$

and

$$q(t, u) = p_1(t)u + p(t)\phi(u), \quad t \in J, \, u \in \mathbb{R}_+. \tag{6.39}$$

If $y \in C(J, E)$, it follows from hypothesis (A) by means of [133, Corollary 1.4.4] that $A(\cdot)y(\cdot)$ is strongly measurable. This result and the hypothesis (g1) imply that $f(\cdot, y(\cdot)) = A(\cdot)y(\cdot) + g(\cdot, y(\cdot))$ is strongly measurable. By hypothesis (g1) there exists also such a $z \in E$ that $g(\cdot, z)$ is HL integrable on J. Since $A(\cdot)z$ is strongly measurable, and $\|A(\cdot)z\| \leq \|z\|\, p_1 \in L^1(J, \mathbb{R}_+)$, then $A(\cdot)z$ is Bochner integrable, and hence also HL integrable. Thus $f(\cdot, z) = A(\cdot)z + g(\cdot, z)$ is HL integrable. Consequently, the hypothesis (f1) is valid.

The hypotheses (A) and (g2) imply that the functions $f : J \times E \to E$ and $q : J \times \mathbb{R}_+ \to \mathbb{R}_+$ defined by (6.38) and (6.39) satisfy

$$\|f(t, x) - f(t, y)\| \leq q(t, \|x - y\|)$$

for all $x, y \in E$ with $\|x - y\| \leq r$ and for a.e. $t \in J$. Moreover, the properties of p_1 and ϕ given in the hypothesis (g2) imply by the proof of [133, Proposition 5.1.1] that the function $q : J \times \mathbb{R}_+ \to \mathbb{R}_+$ defined by (6.39) satisfies the hypothesis (q) of Lemma 6.11. Consequently, the function $f : J \times E \to E$ defined by (6.38) satisfies the hypothesis (f2). Thus by applying Theorem 6.13 and Proposition 6.14, the Cauchy problem (6.1) with f defined by (6.38), or equivalently, the Cauchy problem (6.37) has for every choice of $x_0 \in E$ a unique solution in $W^1_{SL}(J, E)$ that continuously depends on x_0.   □

Remark 6.17. Let $\ln_n$ and $\exp_n$ denote n-fold iterated logarithm and exponential functions, respectively, and $\exp_0(1) := 1$. The functions $\phi_n : \mathbb{R}_+ \to \mathbb{R}_+$, $n \in \mathbb{N}$, defined by $\phi_n(0) = 0$, and

$$\phi_n(x) = x \prod_{j=1}^{n} \ln_j \frac{1}{x}, \quad 0 < x \le exp_n(1)^{-1},$$

have properties assumed in (g2) for the function ϕ.

Example 6.18. Let E be the Banach space c_0 of the sequences $(x_n)_{n=1}^{\infty}$ of real numbers that converge to 0, ordered componentwise and normed by $\|x\|_0 = \sup_n |x_n|$. The mapping $h = (h_n)_{n=1}^{\infty} : [0,1] \to c_0$, whose components h_n are defined by $h_n(0) = 0$,

$$h_n(t) = \frac{2t}{\sqrt{n}} \cos\left(\frac{1}{t^2}\right) + \frac{2}{\sqrt{nt}} \sin\left(\frac{1}{t^2}\right), \quad t \in (0,1], \; n \in \mathbb{N} \tag{6.40}$$

belongs to $HL([0,1], c_0)$ by Lemma 1.12. The solutions of the initial value problems

$$u'(t)) + \frac{1}{(e+t)\ln(e+t)} u(t) = \frac{h(t)}{\ln(e+t)} \text{ for a.e. } t \in [0,1], \; u(0) = \left(\frac{1}{\sqrt{n}}\right)_{n=1}^{\infty}, \tag{6.41}$$

are

$$u_{\pm}(t) = \left(\frac{1}{\sqrt{n}\log(e+t)} \left(t^2 \cos\left(\frac{1}{t^2}\right) + 1\right)\right)_{n=1}^{\infty}. \tag{6.42}$$

More generally, the infinite system of Cauchy problems defined in $J = [0,1]$

$$\begin{cases} y_n'(t) + \dfrac{1}{(e+t)\ln(e+t)} y_n(t) = \dfrac{1}{\ln(e+t)}(h_n(t) + g_n(y(t))) \text{ for a.e. } t \in J, \\ y_n(0) = x_n, \; n \in \mathbb{N}, \end{cases} \tag{6.43}$$

where $g = (g_n)_{n=1}^{\infty} : c_0 \to c_0$, has by Theorem 6.16 for each $x_0 = (x_n)_{n=1}^{\infty} \in c_0$ a unique solution $(y_n)_{n=1}^{\infty}$, which depends continuously on x_0, if there exists a function $p \in L^1([0,1], \mathbb{R}_+)$ such that

$$\sup_n |g_n(x) - g_n(y)| \le p(t)\|x-y\|_0 \ln \frac{1}{\|x-y\|_0} \text{ whenever } 0 < \|x-y\|_0 \le \frac{1}{e}.$$

Remark 6.19. No component of the mapping h defined in (6.40) is Lebesgue integrable on $[0,t]$ for any $t \in (0,1]$. Consequently, the mapping h is not Bochner integrable on $[0,t]$ for any $t \in (0,1]$.

6.2 Nonlocal Semilinear Differential Equations

In this section we derive existence and comparison results for extremal solutions of first order discontinuous nonlocal semilinear differential equations containing non-absolutely integrable functions, and equipped with discontinuous and nonlocal initial conditions. In the proofs we apply results proved in Chap. 9 for HL integrable functions from a real interval to an ordered Banach space and fixed point results proved in Chap. 2.

6.2.1 Existence and Comparison Results

We study first the semilinear initial function problem

$$\begin{cases} y'(t) = A(t)y(t) + f(t, y, y(t)) & \text{for a.e. } t \in J := [0, b], \\ y(t) = B(t, y) & \text{in } J_0 := [-r, 0], \end{cases} \tag{6.44}$$

where $0 < b < \infty$, $0 \le r < \infty$, $A : J \to L(E) = \{T : E \to E : T \text{ is linear and bounded}\}$, $f : J \times C([-r, b], E) \times E \to E$ and $B : J_0 \times C([-r, b], E) \to E$. E is a Banach space ordered by a regular order cone.

We present conditions under which problem (6.44) has solutions in the set

$$S = \{y \in C([-r, b], E) : y|_J \in W^1_{SL}(J, E)\}. \tag{6.45}$$

We also study the dependence of the solutions of (6.44) on the functions f and B.

Definition 6.20. *We say that a function $y \in S$ is a* **subsolution** *of problem (6.44) if*

$$y'(t) \le A(t)y(t) + f(t, y, y(t)) \quad \text{for a.e. } t \in J, \quad y(t) \le B(t, y) \text{ in } J_0, \tag{6.46}$$

If reversed inequalities hold in (6.46), we say that y is a **supersolution** *of (6.44). If equalities hold in (6.46), then y is called a* **solution** *of (6.44).*

Assuming that the spaces $C([-r, b], E)$ and $C(J_0, E)$ are equipped with pointwise ordering, we impose the following hypotheses on the functions A, f, and B.

(A_0) $A(\cdot)x$ is strongly measurable for every $x \in E$.
(A_1) There is a $\hat{p} \in L^1(J, \mathbb{R}_+)$ such that $x \mapsto A(t)x + \hat{p}(t)x$ is increasing for a.e. $t \in J$.
(f_0) $f(\cdot, y, y(\cdot))$ is strongly measurable on J for every $y \in C([-r, b], E)$.
(f_1) There is an $\bar{p} \in L^1(J, \mathbb{R}_+)$ such that $(y, x) \mapsto f(t, y, x) + \bar{p}(t)x$ is increasing for a.e. $t \in J$.
(B_0) If (y_n) is a monotone and order bounded sequence in $C([-r, b], E)$, then the sequence $(B(\cdot, y_n))$ is equicontinuous on J_0.
(B_1) $B(t, y)$ is increasing with respect to y for all $t \in J_0$.
(lu) Problem (6.44) has a subsolution $\underline{y} \in S$ and a supersolution $\overline{y} \in S$, and $\underline{y} \le \overline{y}$.

Denote by $HL(J, E)$ the set of all HL integrable functions $y : J \to E$, and equip $HL(J, E)$ with the a.e. pointwise ordering.

Lemma 6.21. *Let the hypotheses (A_0), (A_1), (f_0), (f_1), (B_0), (B_1), and (lu) be satisfied. Denote $[\underline{y}, \overline{y}] = \{y \in C([-r, b], E) : \underline{y} \le y \le \overline{y}\}$, where $\underline{y}, \overline{y} \in S$ are sub- and supersolutions of (6.44), and $\underline{y} \le \overline{y}$.*

(a) The equation

$$Fy(t) := A(t)y(t) + f(t, y, y(t)) + p(t)y(t), \ y \in [\underline{y}, \overline{y}], \ t \in J, \ p = \hat{p} + \overline{p}, \tag{6.47}$$

defines an increasing mapping $F : [\underline{y}, \overline{y}] \to HL(J, E)$.

(b) Denoting $t_- = \min\{0, t\}$, $t_+ = \max\{0, t\}$, $t \in \mathbb{R}$, *and* $P(t) = \int_0^t p(s) \, ds$, $t \in J$, *the equation*

$$Gy(t) := e^{-P(t_+)} \left(B(t_-, y) + {}^K\!\!\int_0^{t_+} e^{P(s)} Fy(s) \, ds \right) \tag{6.48}$$

defines an increasing mapping $G : [\underline{y}, \overline{y}] \to [\underline{y}, \overline{y}]$.

Proof: Ad (a) The given hypotheses ensure that (6.47) defines for every $y \in [\underline{y}, \overline{y}]$ a strongly measurable mapping $Fy : J \to E$, and that

$$\underline{y}'(t) + p(t)\underline{y}(t) \leq Fy_1(t) \leq Fy_2(t) \leq \overline{y}'(t) + p(t)\overline{y}(t) \tag{6.49}$$

for a.e. $t \in J$, and for $\underline{y} \leq y_1 \leq y_2 \leq \overline{y}$. Because $\underline{y}' + p\underline{y}$ and $\overline{y}' + p\overline{y}$ are HL integrable on J, the above result implies by Proposition 9.14 and Lemma 9.11 that the functions Fy_1 and Fy_2 belong to $HL(J, E)$, and that $Fy_1 \leq Fy_2$.

Ad (b) Let $y \in [\underline{y}, \overline{y}]$ be fixed. Since Fy is HL integrable by (a), then (6.48) defines a mapping $Gy : [-r, b] \to E$ for every $y \in [\underline{y}, \overline{y}]$. Because $Gy(t) = B(t, y)$ for $t \in J_0$, then Gy is continuous on J_0 by (B$_0$). If $t \in J$, then $Gy(t) = e^{-P(t)} B(0, y) + {}^K\!\!\int_0^t e^{P(s)} Fy(s) \, ds$, whence Gy is continuous on J. Moreover, $Gy(t) \to B(0, y)$ as $t \to 0+$, whence Gy is continuous also at 0. Consequently, $Gy \in C([-r, b], E)$.

Since $\overline{y}$ is a supersolution of (6.44), it follows from (6.47) and (6.48) by Lemma 9.22 that for every $t \in J$ the following holds:

$$G\overline{y}(t) = e^{-P(t)} \left(B(0, \overline{y}) + {}^K\!\!\int_0^t e^{P(s)} (A(s)\overline{y}(s) + f(s, \overline{y}, \overline{y}(s)) + p(s)\overline{y}(s)) \, ds \right)$$

$$\leq e^{-P(t)} \left(B(0, \overline{y}) + {}^K\!\!\int_0^t e^{P(s)} (\overline{y}'(s) + p(s)\overline{y}(s)) \, ds \right)$$

$$\leq e^{-P(t)} \left(B(0, \overline{y}) + e^{P(t)}\overline{y}(t) - \overline{y}(0) \right) \leq \overline{y}(t).$$

When $t \in J_0$, then

$$G\overline{y}(t) = B(t, \overline{y}) \leq \overline{y}(t).$$

The above proof shows that $G\overline{y} \leq \overline{y}$. Similarly it can be shown that $\underline{y} \leq G\underline{y}$. If $\underline{y} \leq y_1 \leq y_2 \leq \overline{y}$, then $B(\cdot, y_1) \leq B(\cdot, y_2)$ by (B$_1$), and $Fy_1 \leq Fy_2$ by (a). It then follows from (6.48) by Lemma 9.11 and by the above proof that

$$\underline{y} \leq G\underline{y} \leq Gy_1 \leq Gy_2 \leq G\overline{y} \leq \overline{y}.$$

This result proves (b). $\qquad\qquad\qquad\qquad\qquad\qquad\qquad\qquad\qquad$ $\square$

Lemma 6.22. *Assume hypotheses (A_0), (A_1), (f_0), (f_1), (B_0), (B_1), and (lu). Then $y \in [\underline{y}, \overline{y}]$ is a solution of problem (6.44) if and only if y is a fixed point of the mapping $G : [\underline{y}, \overline{y}] \to [\underline{y}, \overline{y}]$ defined by (6.48).*

Proof: Assume first that $y \in [\underline{y}, \overline{y}]$ is a solution of problem (6.44). Then $y \in S$, so that y' is HL integrable on J, and

$$y'(t) = A(t)y(t) + f(t, y, y(t)) \quad \text{for a.e. } t \in J, \quad y(0) = B(0, y).$$

Applying this result, and (6.47), (6.48) as well as Lemma 9.22, we have for every $t \in J$ (Note, $P(t) = \int_0^t p(s)\, ds$),

$$Gy(t) = e^{-P(t)} \left(B(0, y) + {}^{K}\!\!\int_0^t e^{P(s)} (A(s)y(s) + f(s, y, y(s)) + p(s)y(s))\, ds \right)$$

$$= e^{-P(t)} \left(B(0, y) + {}^{K}\!\!\int_0^t e^{P(s)} (y'(s) + p(s)y(s))\, ds \right)$$

$$= e^{-P(t)} \left(B(0, y) + e^{P(t)}y(t) - y(0) \right) = y(t).$$

Moreover, if $t \in J_0$, then

$$y(t) = B(t, y) = Gy(t).$$

Thus $y = Gy$.

Conversely, assume that y is a fixed point of the mapping $G : [\underline{y}, \overline{y}] \to [\underline{y}, \overline{y}]$ defined by (6.48). Since $t \mapsto e^{-P(t)}$ is absolutely continuous on J and $t \mapsto B(0, y) + {}^{K}\!\int_0^t e^{P(s)} Fy(s)\, ds$ is a primitive of a HL integrable function on J, from

$$y(t) = Gy(t) = e^{-P(t)} \left(B(0, y) + {}^{K}\!\!\int_0^t e^{P(s)} Fy(s)\, ds \right), \quad t \in J$$

and by Lemma 9.22 it follows that $y|_J \in W^1_{SL}(J, E)$, and that

$$y'(t) = -p(t)y(t) + Fy(t) = A(t)y(t) + f(t, y, y(t))$$

for a.e. $t \in J$. Moreover, (6.48) implies that $y(t) = Gy(t) = B(t, y)$ for $t \in J_0$. Thus $y \in S$ and y is a solution of problem (6.44). □

Now we are in the position to prove an existence and comparison theorem for solutions of problem (6.44).

Theorem 6.23. *Assume hypotheses (A_0), (A_1), (f_0), (f_1), (B_0), and (B_1), and let $\underline{y}, \overline{y} \in S$ be sub- and supersolutions of (6.44) such that $\underline{y} \leq \overline{y}$. Then problem (6.44) has the smallest and greatest solutions within the order interval $[\underline{y}, \overline{y}]$. Moreover, these extremal solutions are increasing with respect to f and B.*

Proof: We shall first show that the mapping $G : [\underline{y}, \overline{y}] \to [\underline{y}, \overline{y}]$, defined by (6.48), satisfies the hypotheses of Lemma 6.3. G is increasing by Lemma 6.21. To show that $(Gy_n)_{n=0}^{\infty}$ is equicontinuous on $[-r, b]$ whenever $(y_n)_{n=0}^{\infty}$ is an increasing sequence in $[\underline{y}, \overline{y}]$, let $(y_n)_{n=0}^{\infty}$ be such a sequence. The definition (6.48) of G implies that $Gy_n(t) = B(t, y_n)$ for all $t \in J_0$ and $n \in \mathbb{N}_0$. Thus the sequence $(Gy_n)_{n=0}^{\infty}$ is equicontinuous on J_0 by the hypothesis (B_0).

Next we prove $(Gy_n)_{n=0}^{\infty}$ is equicontinuous also on J when $(y_n)_{n=0}^{\infty}$ is an increasing sequence in $[\underline{y}, \overline{y}]$. Denote

$$\begin{cases} v_n(s) = e^{P(s)} Fy_n(s)), & s \in J, \ n \in \mathbb{N}_0, \\ \underline{v}(s) = e^{P(s)}(\underline{y}'(s) + p(s)\underline{y}(s)), & s \in J \\ \overline{v}(s) = e^{P(s)}(\overline{y}'(s) + p(s)\overline{y}(s)), & s \in J. \end{cases} \quad (6.50)$$

It follows from (6.49) that $(v_n(s))_{n=0}^{\infty}$ is for a.e. $s \in J$ an increasing sequence within the order interval $[\underline{v}(s), \overline{v}(s))]$. Both the functions $\underline{v}$ and $\overline{v}$ are HL integrable, and every v_n is strongly measurable on J. Thus every v_n is HL integrable on J by Proposition 9.14. Denote

$$w_n(t) = {}^K\!\!\int_0^t v_n(s)\,ds = {}^K\!\!\int_0^t e^{P(s)} Fy_n(s))\,ds, \quad t \in J, \ n \in \mathbb{N}_0. \quad (6.51)$$

The sequence $(w_n)_{n=0}^{\infty}$ belongs to $C(J, E)$ by [207, Theorem 7.4.1], and is increasing by Lemma 9.11. According to Lemma 9.11 we have

$$0 \le {}^K\!\!\int_a^t (v_n(s) - \underline{v}(s))\,ds \le {}^K\!\!\int_a^t (\overline{v}(s) - \underline{v}(s))\,ds, \quad 0 \le a \le t \le b.$$

This result and the normality of the order cone E_+ imply that

$$\left\| {}^K\!\!\int_a^t (v_n(s) - \underline{v}(s))\,ds \right\| \le \lambda \left\| {}^K\!\!\int_a^t (\overline{v}(s) - \underline{v}(s))\,ds \right\|, \quad 0 \le a \le t \le b.$$

The last inequality and (6.51) result in

$$\|w_n(t) - w_n(a)\| \le \left\| {}^K\!\!\int_a^t \underline{v}(s)\,ds \right\| + \lambda \left\| {}^K\!\!\int_a^t (\overline{v}(s) - \underline{v}(s))\,ds \right\|, \quad 0 \le a \le t \le b$$

for all $n \in \mathbb{N}_0$. This result, the notations (6.51), and the definition (6.48) imply that the sequence of functions

$$Gy_n(t) := e^{-P(t+)}\left(B(0, y_n) + {}^K\!\!\int_0^{t_+} e^{P(s)} Fy_n(s)\,ds \right)$$

is equicontinuous on J.

The above proof shows that $(Gy_n)_{n=0}^{\infty}$ is equicontinuous on $J_0 \cup J = [-r, b]$ when $(y_n)_{n=0}^{\infty}$ is an increasing sequence in $[\underline{y}, \overline{y}]$. The proof that $(Gy_n)_{n=0}^{\infty}$ is equicontinuous on $[-r, b]$ when $(y_n)_{n=0}^\infty$ is a decreasing sequence in $[\underline{y}, \overline{y}]$ is similar. This result and Lemma 9.44 imply that (Gy_n) converges uniformly on $[-r, b]$, and hence with respect to the uniform norm of $C([-r, b], E)$.

The above proof shows that the hypotheses of Lemma 6.3 are valid for the mapping G defined by (6.48), when X is $C([-r, b], E)$ equipped with the pointwise ordering and the uniform norm, and the functions $\underline{y}, \overline{y}$ of $C([-r, b], E)$ are defined by (6.15), because G is also increasing due to Lemma 6.21. Thus by Lemma 6.3, G has the smallest and greatest fixed points y_* and y^*. In view of Lemma 6.22, y_* and y^* are the smallest and greatest solutions of problem (6.44) in $[\underline{y}, \overline{y}]$.

It follows from (6.48) and by Lemma 9.11 that G is increasing with respect to f and B. This result implies the last conclusion of the theorem because y_* and y^* are increasing with respect to G by Lemma 6.3. $\square$

Example 6.24. Let E be the Banach space c_0 of the sequences $(x_n)_{n=1}^\infty$ of real numbers that converge to 0, ordered componentwise and normed by $\|x\|_0 = \sup_n |x_n|$. The mapping $h : [0, 1] \to c_0$, defined by $h(0) = (0, 0, \dots)$,

$$h(t) = \left(\frac{2t}{\sqrt{n}} \cos\left(\frac{1}{t^2}\right) + \frac{2}{\sqrt{nt}} \sin\left(\frac{1}{t^2}\right) \right)_{n=1}^\infty, \quad t \in (0, 1], \tag{6.52}$$

belongs to $HL([0, 1], c_0)$ by Lemma 1.12. The solutions of the initial value problems

$$\begin{cases} y'(t) + \dfrac{y(t)}{(e+t)\ln(e+t)} = \dfrac{h(t) \pm (\frac{1}{\sqrt{n}})_{n=1}^\infty}{\ln(e+t)} \text{ for a.e. } t \in [0, 1], \\ \\ y(0) = \left(\pm\dfrac{1}{\sqrt{n}} \right)_{n=1}^\infty, \end{cases} \tag{6.53}$$

are

$$y_\pm(t) = \left(\frac{1}{\sqrt{n}\ln(e+t)} \left(t^2 \cos\left(\frac{1}{t^2}\right) \pm (t+1) \right) \right)_{n=1}^\infty. \tag{6.54}$$

More generally, consider the problem

$$\begin{cases} y'(t) + \dfrac{1}{(e+t)\ln(e+t)} y(t) = h(t) + \left(\dfrac{1}{\sqrt{n}} f_n(y) \right)_{n=1}^\infty \text{ a.e. on } [0, 1], \\ \\ y(0) = \left(\dfrac{B_n(y)}{\sqrt{n}} \right)_{n=1}^\infty, \end{cases} \tag{6.55}$$

where each $f_n, B_n : C([0, 1], c_0) \to \mathbb{R}$, is increasing and $-1 \leq B_n(y), f_n(y) \leq 1$ for all $y \in C([0, 1], c_0)$ and $n \in \mathbb{N}$. The functions $y_\pm$ are sub- and supersolutions of (6.55). It is easy to verify that all the hypotheses of Theorem 6.23 are valid when

$$\begin{cases} r = 0, \ b = 1, \ A(t) = \frac{1}{(e+t)\ln(e+t)} I, \ B(0, y) = \left(\frac{B_n(y)}{\sqrt{n}}\right)_{n=1}^\infty, \\ f(t, y, y(t)) = h(t) + \left(\frac{1}{\sqrt{n}} f_n(y)\right)_{n=1}^\infty. \end{cases} \tag{6.56}$$

Thus (6.55) has the smallest and greatest solutions $y_* = (y_{*n})_{n=1}^\infty$ and $y^* = (y_n^*)_{n=1}^\infty$ within the order interval $[y_-, y_+]$, where $y_\pm$ are given by (6.54). Because interval $[y_-, y_+]$ contains all the solutions of (6.55), then y_* and y^* are the smallest and greatest solutions of (6.55).

Remark 6.25. No component of the mapping h defined in (6.52) is Lebesgue integrable on $[0, t]$ for any $t \in (0, 1]$. Consequently, the mapping h is not Bochner integrable on $[0, t]$ for any $t \in (0, 1]$.

6.2.2 Applications to Multipoint Initial Value Problems

The main result of Sect. 6.2.1 can be applied to the following multipoint initial value problem:

$$
\begin{cases}
y'(t) = A(t)y(t) + f(t, y, y(t)) & \text{for a.e. } t \in J := [0, b], \\
y(0) = \sum_{k=1}^m b_k y(t_k),
\end{cases}
\tag{6.57}
$$

where $0 < t_k < b$, $b_k \in \mathbb{R}_+$, $k = 1, \ldots, m$, $\sum_{k=1}^m b_k = 1$, $A : J \to L(E)$ and $f : J \times C(J, E) \times E \to E$. E is a Banach space ordered by a regular order cone.

Definition 6.26. *We say that a function $y \in W_{SL}^1(J, E)$ is a **subsolution** of problem (6.57), if*

$$
y'(t) \le A(t)y(t) + f(t, y, y(t)) \quad \text{for a.e. } t \in J, \quad y(0) \le \sum_{k=1}^m b_k y(t_k). \tag{6.58}
$$

*If reversed inequalities hold in (6.58), we say that y is a **supersolution** of (6.57). If equalities hold in (6.58), then y is called a **solution** of (6.57).*

As a consequence of Theorem 6.23 we obtain the following result.

Proposition 6.27. *Let the hypotheses (A_0), (A_1), (f_0), and (f_1) hold, and let $\underline{y}, \overline{y} \in S$ be sub- and supersolutions of (6.57) such that $\underline{y} \le \overline{y}$. Then problem (6.57) has the smallest and greatest solutions within the order interval $[\underline{y}, \overline{y}]$, and these extremal solutions are increasing with respect to f.*

Proof: Choosing $r = 0$ and $B(0, y) = \sum_{k=1}^m b_k y(t_k)$, $y \in C(J, E)$, we see that the hypotheses (B_0) and (B_1) are valid. The conclusions follow then from Theorem 6.23. $\qquad\square$

Remark 6.28. (i) The functional dependence on the unknown function y of f and B may occur, e.g., as bounded, linear, and positive operators, such as integral operators of Volterra and/or Fredholm type with nonnegative kernels. Thus the results derived in this section can be applied also to integro-differential equations.

(ii) Functional differential equations containing non-absolutely integrable functions are studied also in [92, 93, 132].

6.3 Higher Order Differential Equations

In this section we shall study higher order differential equations in Banach spaces. We shall first apply results of Sect. 6.1 to derive existence, uniqueness, and estimation results for an mth order Cauchy problem, as well as sufficient conditions for the continuous dependence of its solution and its lower order derivatives on the initial values. By assuming that the underlying Banach space is ordered, we then prove existence and comparison results for the smallest and greatest solutions of mth order semilinear initial function problems by using results of Sect. 6.2. The data of the considered differential equations are allowed to be non-absolutely integrable with respect to the independent variable and may depend discontinuously on the unknown (dependent) variables.

6.3.1 Well-Posedness Results

Given a Banach space E and $J = [0, b]$, $b > 0$, consider the mth order Cauchy problem

$$\begin{cases} y^{(m)}(t) = g(t, y(t), y'(t), \ldots, y^{(m-1)}(t)) & \text{for a.e. } t \in J, \\ y(0) = x_{01}, \ y'(0) = x_{02}, \ \ldots, \ y^{(m-1)}(0) = x_{0m}, \end{cases} \quad (6.59)$$

where $g : J \times E^m \to E$ and $x_{0i} \in E$, $i = 1, \ldots, m$. In this section we assume that the product space E^m is equipped with the norm

$$\|x\| = \|x_1\| + \cdots + \|x_m\|, \quad x = (x_1, \ldots, x_m) \in E^m. \quad (6.60)$$

Definition 6.29. *We say that* $y : J \to E$ *is a* **solution** *of (6.59) if* $y^{(m-1)}$ *belongs to* $W^1_{SL}(J, E)$, *and if (6.59) holds.*

By using these notations and definitions the Cauchy problem (6.59) can be converted into the first order Cauchy problem as follows:

Lemma 6.30. $y : J \to E$ *is a solution of the Cauchy problem (6.59) if and only if*
$x = (x_1, \ldots, x_m) = (y, y', \ldots, y^{(m-1)})$ *is a solution of the Cauchy problem*

$$x'(t) = f(t, x(t)) \quad \text{for a.e. } t \in J, \quad x(0) = x_0, \quad (6.61)$$

where $f : J \times E^m \to E^m$ *is defined by*

$$f(t, x) = (x_2, x_3, \ldots, x_m, g(t, x)), \quad (6.62)$$

with $t \in J$ *and* $x = (x_1, \ldots, x_m) \in E^m$.

We shall first prove that the Cauchy problem (6.59) has a uniquely determined solution if $g : J \times E^m \to E$ has the following properties.

(g1) $g(\cdot, x)$ is strongly measurable for each $x \in E^m$, and $g(\cdot, 0)$ is HL integrable.

(g2) There is $r > 0$ such that $\|g(t, x) - g(t, y)\| \le q(t, \|x - y\|)$ for all $x, y \in E^m$ with $\|x - y\| < r$ and for a.e. $t \in J$, where $q : J \times [0, r] \to \mathbb{R}_+$, $q(\cdot, x)$ is measurable for all $x \in [0, r]$, $q(\cdot, r) \in L^1(J, \mathbb{R}_+)$, $q(t, \cdot)$ is increasing and right-continuous for a.e. $t \in J$, and the zero-function is for $u_0 = 0$ the only absolutely continuous solution of the Cauchy problem

$$u'(t) = u(t) + q(t, u(t)), \quad \text{for a.e. } t \in J, \qquad u(0) = u_0, \qquad (6.63)$$

Theorem 6.31. *If the hypotheses (g1) and (g2) hold, then the Cauchy problem (6.59) has a unique solution y on J for each $x_0 = (x_{01}, \dots, x_{0m}) \in E^m$. Moreover, y can be obtained by the method of successive approximations.*

Proof: The hypotheses (g1) and (g2) imply that the function f, defined in (6.62), satisfies the hypotheses of Theorem 6.13 when E is replaced by E^m and $q(t, u)$ by $u + q(t, u)$. Thus by Theorem 6.13 the Cauchy problem (6.61) has a unique solution $x = (x_1, \dots, x_m)$, which can be obtained as the uniform limit of the successive approximations. This result and Lemma 6.30 prove the assertions.   □

Next we shall consider the dependence of the solution of the Cauchy problem (6.59) on the initial values $x_{01}, \dots, x_{0m}$.

Proposition 6.32. *Assume that $g : J \times E^m \to E$ satisfies the hypotheses (g1) and (g2). Let y and $\hat{y}$ denote the solutions of the Cauchy problem (6.59) corresponding to initial values $x_0 = (x_{01}, \dots, x_{0m})$ and $\hat{x}_0 = (\hat{x}_{01}, \dots, \hat{x}_{0m})$, and let $u = u(\cdot, u_0)$ denote the smallest solution of the Cauchy problem (6.63). Then for $\|x_0 - \hat{x}_0\|$ sufficiently small and $t \in J$ we have*

$$\|y(t) - \hat{y}(t)\| + \|y'(t) - \hat{y}'(t)\| + \dots + \|y^{(m-1)}(t) - \hat{y}^{(m-1)}(t)\| \le u(t, \|x_0 - \hat{x}_0\|).$$
$$(6.64)$$

Moreover, $y, y', \dots, y^{(m-1)}$ depend continuously on $x_{01}, \dots, x_{0m}$.

Proof: The assertions are immediate consequences of Lemma 6.30, Proposition 6.14 and (6.60).   □

6.3.2 Semilinear Problem

Consider the mth order semilinear Cauchy problem

$$\begin{cases} y^{(m)}(t) = A_m(t)y^{(m-1)}(t) + \dots + A_1(t)y(t) \\ \qquad\qquad + f(t, y(t), y'(t), \dots, y^{(m-1)}(t)) \text{ a.e. on } J, & (6.65) \\ y(0) = x_{01}, \ y'(0) = x_{02}, \ \dots, \ y^{(m-1)}(0) = x_{0m}, \end{cases}$$

where $J := [0, b]$, $A_i : J \to L(E)$, $i = 1, \dots, m$, and $f : J \times E^m \to E$. Problem (6.65) can be converted into the first order semilinear Cauchy problem as follows:

Lemma 6.33. *$y : J \to E$ is a solution of the Cauchy problem (6.65) if and only if*
$$x = (x_1, \ldots, x_m) = (y, y', \ldots, y^{(m-1)})$$ *is a solution of the semilinear Cauchy problem*

$$x'(t) = A(t)x(t) + g(t, x(t)) \text{ a.e. on } J \qquad x(0) = x_0, \qquad (6.66)$$

where $g : J \times E^m \to E^m$ and $A : J \to L(E^m)$ are defined by

$$g(t, x) = (x_2, x_3, \ldots, x_m, f(t, x)), \quad A(t)x = (0, \ldots, 0, \sum_{i=1}^{m} A_i(t)x_i), \quad (6.67)$$

when $t \in J$ and $x = (x_1, \ldots, x_m) \in E^m$.

We shall assume that the functions $A_i : J \to L(E)$ and $f : J \times E^m \to E$ have the following properties.

(A0) For each $i = 1, \ldots, m$ and $x \in E$, $A_i(\cdot)x$ is strongly measurable, and there is $p_i \in L^1(J, \mathbb{R}_+)$ such that $\|A_i(t)\| \leq p_i(t)$ for a.e. $t \in J$.
(f1) $f(\cdot, x)$ is strongly measurable for each $x \in E^m$, and $f(\cdot, 0)$ is HL integrable.
(f2) There is $r > 0$ and $p_0 \in L^1(J, \mathbb{R}_+)$ such that

$$\|f(t, x) - f(t, y)\| \leq p_0(t) \, \phi(\|x - y\|) \qquad (6.68)$$

for all $x, y \in E^m$ with $\|x - y\| < r$ and for a.e. $t \in J$, where $\phi : [0, r] \to \mathbb{R}_+$ is increasing and right-continuous, and $\int_0^r \frac{dv}{\phi(v)} = \infty$.

Theorem 6.34. *If the hypotheses (A0), (f1), and (f2) are satisfied, then the Cauchy problem (6.65) has a unique solution y on J for each $x_0 = (x_{01}, \ldots, x_{0m}) \in E^m$. Moreover, $y, y', \ldots, y^{(m-1)}$ depend continuously on $x_{01}, \ldots, x_{0m}$.*

Proof: The hypotheses (A0), (f1), and (f2) imply that the functions A and g in (6.66) satisfy the hypotheses of Theorem 6.16 when E is replaced by E^m, p by $1 + p_0$, $\phi(u)$ by $u + \phi(u)$, and p_1 by $\max\{p_i\}_{i=1}^{m}$. Thus by Theorem 6.16, the Cauchy problem (6.66) has a unique solution $x = (x_1, \ldots, x_m)$, which depends continuously on x_0. This result and Lemma 6.33 imply the assertions. $\square$

Corollary 6.35. *Let the hypotheses (A0) and (f1) hold. Assume there exist $r > 0$ and $p_0 \in L^1(J, \mathbb{R}_+)$ such that*

$$\|f(t, x) - f(t, y)\| \leq p_0(t) \|x - y\| \qquad (6.69)$$

for all $x, y \in E^m$ with $\|x - y\| < r$ and for a.e. $t \in J$. Then the Cauchy problem (6.65) has for each $x_0 = (x_{01}, \ldots, x_{0m}) \in E^m$ exactly one solution y. Moreover, $y, y', \ldots, y^{(m-1)}$ depend continuously on $x_{01}, \ldots, x_{0m}$.

In particular, we have.

Corollary 6.36. *If* $A_i : J \to L(E)$, $i = 1, \ldots, m$, *satisfy condition (A0), and if* $h \in HL(J, E)$, *then the Cauchy problem*

$$\begin{cases} y^{(m)} = A_m(t)y^{(m-1)} + \cdots + A_1(t)y + h(t) \text{ a.e. on } J, \\ y(0) = x_{01}, \ y'(0) = x_{02}, \ldots, y^{(m-1)}(0) = x_{0m} \end{cases}$$

has for each $x_0 = (x_{01}, \ldots, x_{0m}) \in E^m$ *a unique solution, which together with its first* $m - 1$ *derivatives depend continuously on* $x_{01}, \ldots, x_{0m}$.

6.3.3 Extremal Solutions

In this subsection we shall consider the existence of extremal solutions of mth order semilinear initial function problems when E is an ordered Banach space. We shall assume in this subsection that the order cone E_+ of E is regular. Obviously, E_+^m is a regular order cone in E^m, whose norm is defined by (6.60). Given $b > 0$, denote $J := [0, b]$. We shall first study the mth order semilinear initial function problem

$$\begin{cases} y^{(m)}(t) = A_m(t)y^{(m-1)}(t) + \cdots + A_1(t)y(t) \\ \qquad + g(t, y(t), y'(t), \ldots, y^{(m-1)}(t)) \text{ for a.e. } t \in J, \qquad (6.70) \\ y(0) = B_1(y), \ y'(0) = B_2(y'), \ \ldots, \ y^{(m-1)}(0) = B_m(y^{(m-1)}), \end{cases}$$

where $A_i : J \to L(E)$, $B_i : C(J, E) \to E$, $i = 1, \ldots, m$, and $g : J \times E^m \to E$.

We shall present conditions under which problem (6.70) has solutions in the set

$$S = \{y \in C(J, E) : y^{(m-1)} \in W_{SL}^1(J, E)\}. \qquad (6.71)$$

We study also dependence of solutions of (6.70) on the functions g and B_i.

Definition 6.37. *We say that a function* $y \in S$ *is a* **subsolution** *of problem (6.70) if*

$$\begin{cases} y^{(m)}(t) \le A_m(t)y^{(m-1)}(t) + \cdots + A_1(t)y(t) \\ \qquad + g(t, y(t), y'(t), \ldots, y^{(m-1)}(t)) \quad \text{for a.e. } t \in J, \qquad (6.72) \\ y(0) \le B_1(y), \ y'(0) \le B_2(y'), \ldots, y^{(m-1)}(0) \le B_m(y^{(m-1)}). \end{cases}$$

If reversed inequalities hold in (6.72), we say that y *is a* **supersolution** *of (6.70). If equalities hold in (6.72), then* y *is called a* **solution** *of (6.70).*

Lemma 6.38. *The function* $y : J \to E$ *is a subsolution, a supersolution, or a solution of problem (6.70) if and only if* $x = (x_1, \ldots, x_m) = (y, y', \ldots, y^{(m-1)})$ *is a subsolution, a supersolution, or a solution of the semilinear problem*

$$x'(t) = A(t)x(t) + f(t, x(t)), \quad for \ a.e. \ t \in J, \quad x(0) = B(x), \tag{6.73}$$

where $f : J \times E^m \to E^m$, $A : J \to L(E^m)$ and $B : C(J, E^m) \to E^m$ are defined by

$$f(t, x) = (x_2, x_3, \ldots, x_m, g(t, x)), \quad A(t)x = (0, \ldots, 0, \sum_{i=1}^{m} A_i(t)x_i), \ t \in J,$$

$$B(x) = (B_1(x_1), \ldots, B_m(x_m)), \quad x = (x_1, \ldots, x_m).$$
$$\tag{6.74}$$

Assuming that the space $C(J, E)$ is equipped with pointwise ordering, we impose the following hypotheses on the functions A_i, g, and B_i.

(A$_0$) $A_i(\cdot)x$ is strongly measurable for all $x \in E$ and $i = 1, \ldots, m$.
(A$_1$) For every $i = 1, \ldots, m$ $A_i(t)E_+ \subseteq E_+$ for a.e. $t \in J$.
(g$_0$) $g(\cdot, y_1(\cdot), \ldots, y_m(\cdot))$ is strongly measurable on J for all $y_i \in C(J, E)$, $i = 1, \ldots, m$.
(g$_1$) $g(t, \cdot)$ is increasing for a.e. $t \in J$.
(B$_0$) B_i is increasing for every $i = 1, \ldots, m$.
(lu) Problem (6.70) has a subsolution $\underline{y} \in S$ and a supersolution $\overline{y} \in S$, and
$$\underline{y} \le \overline{y}, \underline{y}' \le \overline{y}', \ldots, \underline{y}^{(m-1)} \le \overline{y}^{(m-1)}.$$

As a consequence of Theorem 6.23 we obtain an existence comparison theorem for solutions of problem (6.70).

Theorem 6.39. *Let the hypotheses (A$_0$), (A$_1$), (g$_0$), (g$_1$), (B$_0$), and (lu) hold, and let $\underline{y}, \overline{y} \in S$ be sub- and supersolutions of (6.70) assumed in (lu). Then problem (6.70) has the smallest and greatest solutions in the order interval $[\underline{y}, \overline{y}]$ of $C(J, E)$, and they are increasing with respect to g and B_i, $i = 1, \ldots, m$.*

Proof: Let $A : J \to L(E^m)$ and $f : J \times E^m \to E^m$ be defined by (6.74). Denoting $\underline{x} = (\underline{y}, \underline{y}', \ldots, \underline{y}^{(m-1)})$ and $\bar{x} = (\overline{y}, \overline{y}', \ldots, \overline{y}^{(m-1)})$, it is easy to see that the hypotheses (A$_0$), (A$_1$), (f$_0$), (f$_1$), (B$_0$), and (B$_1$) given in Sect. 6.2.1 hold when $\overline{p} = \hat{p} \equiv 0$. Thus by Theorem 6.23, problem (6.70) has the smallest and greatest solutions in $[\underline{x}, \bar{x}]$, and they are increasing with respect to g and B. This result and Lemma 6.38 imply the assertions.   □

We apply the results of Theorem 6.39 to the following multipoint initial value problem, where $t_i \in J$, $i = 1, \ldots, m$.

$$\begin{cases} y^{(m)}(t) = A_m(t)y^{(m-1)}(t) + \cdots + A_1(t)y(t) \\ \qquad\qquad + g(t, y(t), y'(t), \ldots, y^{(m-1)}(t)) \quad for \ a.e. \ t \in J, \\ y(0) = y(t_1), \ y'(0) = y'(t_2), \ \ldots, \ y^{(m-1)}(0) = y^{(m-1)}(t_m). \end{cases} \tag{6.75}$$

As a consequence of Theorem 6.39 we obtain the following result.

Proposition 6.40. *Assume that the hypotheses (A_0), (A_1), (g_0), and (g_1) hold. Suppose that problem (6.75) has a subsolution $\underline{y} \in S$ and a supersolution $\overline{y} \in S$, and that $\underline{y} \le \overline{y}$, $\underline{y}' \le \overline{y}', \ldots, \underline{y}^{(m-1)} \le \overline{y}^{(m-1)}$. Then problem (6.75) has the smallest and greatest solutions within the order interval $[\underline{y}, \overline{y}]$ of $C(J, E)$, and these extremal solutions are increasing with respect to g.*

Proof: The hypothesis (B_0) holds when $B_i(x_i) = x_i(t_i)$, $x_i \in C(J, E)$, $t_i \in J$, $i = 1, \ldots, m$. □

Next we consider the existence of the smallest and greatest solutions of the Cauchy problem

$$\begin{cases} y^{(m)} = A_m(t)y^{(m-1)} + \cdots + A_1(t)y + g(t, y, y', \ldots, y^{(m-1)}), \\ y(0) = x_{01}, \ y'(0) = x_{02}, \ \ldots, \ y^{(m-1)}(0) = x_{0m}. \end{cases} \tag{6.76}$$

Proposition 6.41. *Assume that $A_i : J \to L(E)$, $i = 1, \ldots, m$, and $g, \underline{g}, \overline{g} : J \times E^m \to E$ satisfy the hypotheses (A_0), (A_1), (g_0), and (g_1). Further assume that the hypotheses (A0), (f1), and (f2) of Theorem 6.34 hold for functions A_i, $i = 1, \ldots, m$, $f = \underline{g}$ and $f = \overline{g}$, and that*

$$\underline{g}(t, x) \le g(t, x) \le \overline{g}(t, x) \text{ for all } x \in E^m \text{ and for a.e. } t \in J. \tag{6.77}$$

Then for every choice of $x_{0i} \in E$, $i = 1, \ldots, m$, the Cauchy problem (6.76) has the smallest and greatest solutions, and these extremal solutions are increasing with respect to g and $x_{0i} \in E$, $i = 1, \ldots, m$.

Proof: Let $x_{0i} \in E$, $i = 1, \ldots, m$, be given. The hypotheses (A0), (f1), and (f2) imposed on the functions A_i, $i = 1, \ldots, m$, $f = \underline{g}$ and $f = \overline{g}$ ensure by Theorem 6.31 that the Cauchy problem (6.76) has uniquely determined solutions $\underline{y}$ and $\overline{y}$ when g is replaced by $\underline{g}$ and $\overline{g}$, respectively. Because the functions A_i, $i = 1, \ldots, m$, and $g, \underline{g}, \overline{g}$ satisfy also the hypotheses (A_0), (A_1), (g_0), and (g_1), the conclusions follow from Theorem 6.39. □

As special case of the above result we obtain the following corollary.

Corollary 6.42. *Assume that $A_i : J \to L(E)$, $i = 1, \ldots, m$, and $g : J \times E^m \to E$ satisfy the hypotheses of Proposition 6.41. Assume also that*

$$\underline{h}(t) \le g(t, x) \le \overline{h}(t) \text{ for all } x \in E^m \text{ and for a.e. } t \in J, \tag{6.78}$$

where $\underline{h}, \overline{h} \in HL(J, E)$. Then for every choice of $x_{0i} \in E$, $i = 1, \ldots, m$, the Cauchy problem (6.76) has the smallest and greatest solutions, which are increasing with respect to g and $x_{0i} \in E$, $i = 1, \ldots, m$.

Remark 6.43. The following spaces are examples of Banach spaces that have regular order cones (cf [133]):

1. A reflexive (e.g., a uniformly convex) Banach space ordered by a normal order cone.
2. A finite-dimensional normed space ordered by any closed cone.
3. A separable Hilbert space whose order cone is generated by an orthonormal basis.
4. A Hilbert space H with such an order cone H_+ that $(x|y) \geq 0$ for all $x \in H_+$.
5. A Hilbert space H whose order cone is $H_+ = \{x \in H : (x|\bar{e}) \geq c\|x\|_2\}$, where $\bar{e}$ is an unit vector of H and $c \in (0,1)$.
6. A function space $L^p(\Omega)$, $1 \leq p < \infty$, normed by p-norm and ordered a.e. pointwise, where Ω is a measure space.
7. A function space $L^p([a,b], X)$, $1 \leq p < \infty$, normed by p-norm and ordered a.e. pointwise, where X is any of the spaces listed above.
8. A function space $HL([a,b], E)$, normed by Alexiewicz norm and ordered a.e. pointwise, where E is any of the spaces listed above.
9. A sequence space l^p, $1 \leq p < \infty$, normed by p-norm and ordered componentwise.
10. The sequence space c_0 of the sequences of real numbers converging to zero, normed by sup-norm, and ordered componentwise.

6.4 Singular Differential Equations

In this section we derive existence and comparison results for the smallest and greatest solutions of first and second order initial value problems as well as for a second order boundary value problem in an ordered Banach space E whose order cone is regular. The right-hand sides of differential equations comprise locally HL integrable vector-valued functions. The following special types are included in the considered problems:

– differential equations may be singular;
– both the differential equations and the initial or boundary conditions may depend functionally on the unknown function and/or on its derivatives;
– both the differential equations and the initial or boundary conditions may be implicit and contain discontinuous nonlinearities;
– problems of random type.

In case that E is the sequence space c_0, we obtain results for infinite systems of initial and boundary value problems, as shown in examples. Moreover, concrete finite systems are solved to illustrate the effects of non-absolutely integrable data to the solutions of such problems.

6.4.1 First Order Explicit Initial Value Problems

In this section we study the explicit initial value problem (IVP)

$$\begin{cases} \dfrac{d}{dt}(p(t)u(t)) = g(t, u(t), u) \text{ for a.e. } t \in J := (a, b), \\[2mm] \lim_{t \to a+} (p(t)u(t)) = x_0, \end{cases} \qquad (6.79)$$

where $-\infty < a < b \leq \infty$, $x_0 \in E$, $p : J \to \mathbb{R}_+$, and $g : J \times E \times HL_{loc}(J, E) \to E$. We assume that E is a Banach space ordered by a regular order cone. We are looking for extremal solutions of (6.79) from the subset S of the space $HL_{loc}(J, E)$ of locally HL integrable functions from J to E, defined by

$$S := \left\{ \begin{array}{l} u \in HL_{loc}(J, E) : p \cdot u \in W^1_{SL}(I, E) \\ \text{for every closed subinterval } I \text{ of } [a, b) \end{array} \right\}. \qquad (6.80)$$

We shall first convert the IVP (6.79) to an integral equation.

Lemma 6.44. *Assume that $x_0 \in E$, that $\frac{1}{p} \in L^1_{loc}(J, \mathbb{R}_+)$, and that $g(\cdot, u(\cdot), u)$ belongs to $HL_{loc}([a, b), E)$ for all $u \in HL_{loc}(J, E)$. Then u is a solution of the IVP (6.79) in S if and only if u is a solution of the following integral equation in $HL_{loc}(J, E)$:*

$$u(t) = \frac{1}{p(t)} \left(x_0 + {}^K\!\!\int_a^t g(s, u(s), u)\, ds \right), \quad t \in J. \qquad (6.81)$$

Proof: Assume that u is a solution of (6.79) in S. The definition (6.80) of S and (6.79) ensure by Corollary 9.19 that

$${}^K\!\!\int_r^t g(s, u(s), u)\, ds = {}^K\!\!\int_r^t \frac{d}{ds}(p(s)u(s))\, ds = p(t)u(t) - p(r)u(r), \ a < r \leq t < b.$$

This result and the initial condition of (6.79) imply that (6.81) is valid. Conversely, let u be a solution of (6.81) in $HL_{loc}(J, E)$. According to (6.81) we have

$$p(t)u(t) = x_0 + {}^K\!\!\int_a^t g(s, u(s), u)\, ds, \quad t \in J. \qquad (6.82)$$

This equation implies by Theorem 9.18 that $u \in S$, that the initial condition of (6.79) is valid, and that

$$\frac{d}{dt}(p(t)u(t)) = g(t, u(t), u) \ \text{ for a.e. } t \in J.$$

Thus u is a solution of the IVP (6.79). $\qquad \square$

To prove our main existence and comparison result for the IVP (6.79), assume that S is ordered a.e. pointwise, and that the functions p and g satisfy the following hypotheses:

(p) $\frac{1}{p} \in L^1_{loc}(J, \mathbb{R}_+)$.

(g0) $g(\cdot, u(\cdot), u) \in HL_{loc}(J, E)$ for every $u \in HL_{loc}(J, E)$.

(g1) There exist $w_\pm \in S$ with $w_- \leq w_+$ such that for all $u, v \in S$ satisfying $w_- \leq u \leq v \leq w_+$, the inequality

$$p(t)w_-(t) - x_0 \leq K \int_a^t g(s, u(s), u)\, ds$$

$$\leq K \int_a^t g(s, v(s), v)\, ds \leq p(t)w_+(t) - x_0$$

holds true for all $t \in J$.

Proposition 6.45. *Let the hypotheses (p), (g0), and (g1) hold. Then the IVP (6.79) has the smallest and greatest solutions within the order interval $[w_-, w_+]$ of S. Moreover, these solutions are increasing with respect to g.*

Proof: Let $x_0 \in E$ be given. The given hypotheses ensure that the relation

$$G(u)(t) := \frac{1}{p(t)} \left(x_0 + K \int_a^t g(s, u(s), u)\, ds \right), \tag{6.83}$$

defines an increasing mapping $G : [w_-, w_+] \to [w_-, w_+]$. Let W be a well-ordered or an inversely well-ordered chain in the range of G. It follows from Proposition 9.39 that $\sup W$ and $\inf W$ exist in $HL_{loc}(J, E)$. The above proof shows that the operator G defined by (6.83) satisfies the hypotheses of Proposition 2.18. Thus G has the smallest fixed point u_* and the greatest fixed point u^*. According to Lemma 6.44, u_* and u^* belong to S, and they are solutions of the IVP (6.79). To prove that u_* and u^* are the smallest and greatest of all solutions of (6.79) within the order interval $[w_-, w_+]$ of S, let $u \in [w_-, w_+]$ be any solution of (6.79). In view of Lemma 6.44 and the definition (6.83) of G, u is a fixed point of mapping G. Because u_* and u^* are the smallest and greatest fixed points of G, then $u_* \leq u \leq u^*$. In particular, u_* and u^* are the smallest and greatest of all solutions of the IVP (6.79) in $[w_-, w_+]$. The last assertion is a consequence of (6.83) and the last conclusion of Proposition 2.18. □

Example 6.46. Consider the following system of initial value problems:

$$\begin{cases} t\, u_i'(t) + u_i(t) = p_i(t)\, \mathrm{sgn}(u_i(t)) + q_i(t)\, h_i(u_{1-i}) & \text{for a.e. } t \in J := (0, \infty), \\ \lim_{t \to 0+} t\, u_i(t) = 0, & i = 0, 1, \end{cases}$$

$$\tag{6.84}$$

where

$$p_0(t) = \left| \cos\left(\frac{1}{t} \right) \right| + \frac{1}{t} \mathrm{sgn}\left(\cos\left(\frac{1}{t} \right) \right) \sin\left(\frac{1}{t} \right),$$

$$p_1(t) = \left| \sin\left(\frac{1}{t} \right) \right| - \frac{1}{t} \mathrm{sgn}\left(\sin\left(\frac{1}{t} \right) \right) \cos\left(\frac{1}{t} \right),$$

$$h_0(y) = \frac{1}{10^6} \left[2 \cdot 10^6 \arctan\left(K \int_1^4 y(s)\, ds \right) \right],$$

$$h_1(y) = \frac{1}{10^6}\left[3 \cdot 10^6 \tanh\left(K\int_1^2 y(s)\,ds\right)\right],$$

$$q_0(t) = 1 + \cos(t), \quad q_1(t) = 1 + \sin(t),$$

$$[x] = \max\{n \in \mathbb{Z} : n \le x\} \text{ and } \operatorname{sgn}(x) = \begin{cases} 1, & x > 0, \\ 0, & x = 0, \\ -1, & x < 0. \end{cases}$$

Note, that the greatest integer function $[\cdot]$ occurs in the functions h_i, $i = 0, 1$.

Problem (6.84) is of the form (6.79), where $x_0 = (0,0)$, $p(t) = t$, $u = (u_0, u_1)$ and

$$g(t, u(t), u) = (p_0(t)\operatorname{sgn}(u_0(t)) + q_0(t) h_0(u_1), p_1(t)\operatorname{sgn}(u_1(t)) + q_1(t) h_1(u_0)). \tag{6.85}$$

We shall first determine $\mathbb{R}_+^2$-valued solutions of (6.84). Denote

$$u_+(t) = \left(t\left|\cos\left(\frac{1}{t}\right)\right| + 8, t\left|\sin\left(\frac{1}{t}\right)\right| + 6\right), \quad w_+(t) = \frac{1}{t} u_+(t), \quad t \in J.$$

It can be shown by applying Lemmas 1.12 and 9.11 (see also the reasoning used in Example 1.14) that

$$0 \le K\int_0^t g(s, u(s), u)\,ds \le K\int_0^t g(s, v(s), v)\,ds \le u_+(t)$$

for all $t \in J$ whenever $u, v \in HL_{loc}(J, \mathbb{R}^2)$ as well as $(0,0) \le u(t) \le v(t)$ for all $t \in J$. Thus the hypotheses of Proposition 6.45 are valid when $w_- = (0,0)$. Moreover, relation

$$Gu(t) := \frac{1}{t} K\int_0^t g(s, u(s), u)\,ds, \quad t \in J \tag{6.86}$$

defines an increasing mapping G from $HL_{loc}((0, \infty), \mathbb{R}_+^2)$ to its order interval $[(0,0), w_+]$. By the proof of Theorem 2.16, the greatest fixed point of G is the minimum of the inversely well-ordered chain D of G-iterations of w_+. The greatest elements of D are iterations $G^n w_+$, $n \in \mathbb{N}$. Calculating these iterations we see that $u^* := G^5 w_+ = Gu^*$. This means (cf. 6.5) that $u^* = \min D$, whence u^* is the greatest fixed point of G in $[(0,0), w_+]$, and hence also in $HL_{loc}((0, \infty), \mathbb{R}_+^2)$. According to the proof of Proposition 6.45, u^* is also the greatest nonnegative-valued solution, and hence the greatest solution, of the initial value problem (6.84). The exact expression of the components of $u^* = (u_0^*, u_1^*)$ are:

$$\begin{cases} u_0^*(t) = \left|\cos\left(\frac{1}{t}\right)\right| + \dfrac{3016003}{1000000}\left(1 + \dfrac{\sin(t)}{t}\right), \\[2ex] u_1^*(t) = \left|\sin\left(\frac{1}{t}\right)\right| + \dfrac{2999941}{1000000}\left(1 + \dfrac{1 - \cos(t)}{t}\right). \end{cases}$$

Other solutions of problem (6.84) whose components are nonnegative-valued are of the form $u = (u_0, u_1)$, where:

$$u_0(t) = \begin{cases} 0, & \text{if } 0 < t \leq t_0, \\ \left|\cos\left(\dfrac{1}{t}\right)\right| - \dfrac{t_0}{t}\left|\cos\left(\dfrac{1}{t_0}\right)\right| + \dfrac{b_0(t_0)}{t}(t - t_0 + \sin(t) - \sin(t_0)), \\ & \text{if } t_0 < t < \infty, \end{cases}$$

and

$$u_1(t) = \begin{cases} 0, & \text{if } 0 < t \leq t_0, \\ \left|\sin\left(\dfrac{1}{t}\right)\right| - \dfrac{t_0}{t}\left|\sin\left(\dfrac{1}{t_0}\right)\right| + \dfrac{b_1(t_0)}{t}(t - t_0 - \cos(t) + \cos(t_0)), \\ & \text{if } t_0 < t < \infty, \end{cases}$$

where t_0 goes through all the points of the interval $(0, \infty)$, and $b_0(t_0) \in \{0, 1, 2, 3\}$ and $b_1(t_0) \in \{0, 1, 2\}$ are constants that depend on t_0.

To determine solutions of (6.84) whose components are negative-valued, denote

$$u_-(t) = \left(-t\left|\cos\left(\frac{1}{t}\right)\right| - 8, -t\left|\sin\left(\frac{1}{t}\right)\right| - 6\right), \quad w_-(t) = \frac{1}{t}u_-(t), \quad t \in J.$$

It can be shown that $w_-(t) \leq Gu(t) \leq Gv(t) \leq 0$ for all $t \in J$ whenever $u, v \in HL_{loc}(J, \mathbb{R}^2)$ and $u(t) \leq v(t) \leq (0, 0)$ for all $t \in J$. Calculating the iterations $G^n w_-$, $n \in \mathbb{N}$, we obtain $u_* := G^5 w_- = Gu_*$, so that u_* is a fixed point of G. In fact, it is the smallest fixed point of G. It is also the smallest solution of the Cauchy problem (6.84). The exact expressions of the components of $u_* = (u_{0*}, u_{1*})$ are:

$$\begin{cases} u_{0*}(t) = -\left|\cos\left(\dfrac{1}{t}\right)\right| - \dfrac{754001}{250000}\left(1 + \dfrac{\sin(t)}{t}\right), \\ u_{1*}(t) = -\left|\sin\left(\dfrac{1}{t}\right)\right| - \dfrac{1499971}{500000}\left(1 + \dfrac{1 - \cos(t)}{t}\right). \end{cases}$$

Components of other negative-valued solutions $u = (u_0, u_1)$ of (6.84) are of the form:

$$u_0(t) = \begin{cases} 0, & \text{if } 0 < t \leq t_0, \\ -\left|\cos\left(\dfrac{1}{t}\right)\right| - \dfrac{t_0}{t}\left|\cos\left(\dfrac{1}{t_0}\right)\right| - \dfrac{b_0(t_0)}{t}(t - t_0 + \sin(t) - \sin(t_0)), \\ & \text{if } t_0 < t < \infty, \end{cases}$$

and

$$u_1(t) = \begin{cases} 0, & \text{if } 0 < t \le t_0, \\ -\left|\sin\left(\dfrac{1}{t}\right)\right| + \dfrac{t_0}{t}\left|\sin\left(\dfrac{1}{t_0}\right)\right| - \dfrac{b_1(t_0)}{t}(t - t_0 - \cos(t) + \cos(t_0)), \\ & \text{if } t_0 < t < \infty, \end{cases}$$

where t_0 goes through all the points of the interval $(0, \infty)$, and $b_i(t_0) \in \{-4, -3, -2, -1\}$ are constants that depend on t_0. Notice also that the zero function is a solution of (6.84).

The above results imply that every point $(t_0, (0, 0))$, $t_0 \in [0, \infty)$, is a bifurcation point for at least three solutions of (6.84).

The function $(t, x, y) \mapsto g(t, x, y)$, defined in (6.85), has the following features:

- It is locally HL integrable by Lemma 1.12, but it is neither Lebesgue integrable nor continuous with respect to the independent variable t on any interval $[0, a]$, $a > 0$, if $x \neq 0$, because the functions p_i are not Lebesgue integrable.
- Its dependence on all the variables t, x, and y is discontinuous, since the signum function sgn, the greatest integer function $[\cdot]$, and the functions p_i are discontinuous.
- Its dependence on x is not monotone, since the functions p_i change their signs infinitely many times. For instance, $u_i^*(t) > u_{i*}(t)$ for all $t \in (0, \infty)$, but the difference functions $t \mapsto g_i(t, u^*(t), u^*) - g_i(t, u_*(t), u_*)$, $i = 0, 1$, are neither nonnegative-valued nor Lebesgue integrable on any interval $[0, a]$, $a > 0$.

6.4.2 First Order Implicit Initial Value Problems

In this subsection we study the implicit initial value problem

$$\begin{cases} Lu(t) := \dfrac{d}{dt}(p(t)u(t)) = f(t, u, Lu) \text{ for a.e. } t \in J := (a, b), \\ \lim_{t \to a+}(p(t)u(t)) = c(u, Lu) \end{cases} \quad (6.87)$$

in the space $HL_{loc}(J, E)$ of locally HL integrable functions from J to a Banach space E ordered by a regular order cone.

Given $-\infty < a < b \le \infty$, $p : J \to \mathbb{R}_+$, $f : J \times HL_{loc}(J, E) \times HL_{loc}(J, E) \to E$, and $c : HL_{loc}(J, E) \times HL_{loc}(J, E) \to E$, we are looking for extremal solutions of (6.87) from the set

$$S := \left\{ \begin{array}{l} u \in HL_{loc}(J, E) : p \cdot u \in W_{SL}^1(I, E) \\ \text{for every closed subinterval } I \text{ of } [a, b) \end{array} \right\}. \quad (6.88)$$

A first step in the treatment is to equivalently convert the IVP (6.87) to a system of two equations.

Lemma 6.47. *Assume that $\frac{1}{p} \in L^1_{loc}(J, \mathbb{R}_+)$, and that $f(\cdot, u, v)$ belongs to $HL_{loc}([a, b), E)$ for all $u, v \in HL_{loc}(J, E)$. Then u is a solution of the IVP (6.87) in S if and only if $(u, Lu) = (u, v)$, where (u, v) is a solution of the following system in $HL_{loc}(J, E) \times HL_{loc}(J, E)$:*

$$\begin{cases} u(t) = \dfrac{1}{p(t)}\left(c(u, v) + {}^K\!\!\int_a^t v(s)\, ds\right), & t \in J, \\ v(t) = f(t, u, v), & \text{for a.e. } t \in J. \end{cases} \tag{6.89}$$

Proof: Assume that u is a solution of (6.87) in S. Denote

$$v(t) = Lu(t) = \frac{d}{dt}(p(t)u(t)), \quad t \in J. \tag{6.90}$$

The differential equation of (6.87) implies that the second equation of (6.89) holds. The definition (6.88) of S and (6.90) ensure by Corollary 9.21 that

$${}^K\!\!\int_r^s v(t)\, dt = {}^K\!\!\int_r^s \frac{d}{dt}(p(t)u(t))dt = p(s)u(s) - p(r)u(r), \quad a < r \le s < b.$$

This result and the initial condition of (6.87) imply that the first equation of (6.89) is valid.

Conversely, let (u, v) be a solution of the system (6.89) in $HL_{loc}(J, E) \times HL_{loc}(J, E)$. According to (6.89) we have

$$p(t)u(t) = c(u, v) + {}^K\!\!\int_a^t v(s)\, ds, \quad t \in J. \tag{6.91}$$

This equation implies by Theorem 9.18 that $u \in S$, and that

$$v(t) = \frac{d}{dt}(p(t)u(t)) = Lu(t) \quad \text{for a.e. } t \in J.$$

This result, the equation (6.91), and the second equation of (6.89) imply that u is a solution of the IVP (6.87). $\qquad\square$

To prove our main existence and comparison result for the IVP (6.87), assume that $HL_{loc}(J, E)$, $HL_{loc}([a, b), E)$ and S are ordered a.e. pointwise, and that the functions p, f, and c satisfy the following hypotheses:

(p) $\frac{1}{p} \in L^1_{loc}(J, \mathbb{R}_+)$.

(fa) $f(\cdot, u, v)$ is strongly measurable for all u, $v \in HL_{loc}(J, E)$, and there exist functions $h_-, h_+ \in HL_{loc}([a, b), E)$ such that $h_- \le f(\cdot, u, v) \le h_+$ for all $u, v \in HL_{loc}(J, E)$.

(fb) There is a $\lambda \ge 0$ such that $f(\cdot, u_1, v_1) + \lambda v_1 \le f(\cdot, u_2, v_2) + \lambda v_2$ whenever u_i, $v_i \in HL_{loc}(J, E)$, $i = 1, 2$, $u_1 \le u_2$ and $v_1 \le v_2$.

(c) There are functions $c_-, c_+ \in E$ such that $c_- \le c(u_1, v_1) \le c(u_2, v_2) \le c_+$ whenever u_i, $v_i \in HL_{loc}(J, E)$, $i = 1, 2$, $u_1 \le u_2$ and $v_1 \le v_2$.

Theorem 6.48. *Assume that the hypotheses (p), (fa), (fb), and (c) hold. Then the IVP (6.87) has the smallest and greatest solutions, and they are increasing with respect to f and c.*

Proof: Assume that $P = HL_{loc}(J, E)^2$ is ordered componentwise. The relations

$$x_\pm(t) := \left(\frac{1}{p(t)} \left(c_\pm + {}^K\!\!\int_a^t h_\pm(s)\, ds \right), h_\pm(t) \right) \tag{6.92}$$

define functions x_-, $x_+ \in P$ (cf. the proof of Theorem 6.54). By Proposition 9.14 $v : [a, b) \to E$ is in $HL_{loc}([a, b), E)$ whenever v is strongly measurable and $h_- \le v \le h_+$. Hence, if $(u, v) \in [x_-, x_+]$, then $v \in HL_{loc}([a, b), E)$. By applying this result, and taking into account the given hypotheses, Lemma 9.11, and Theorem 9.18, one can verify that the relations

$$G_1(u, v)(t) := \frac{1}{p(t)} \left(c(u, v) + {}^K\!\!\int_a^t v(s)\, ds \right), \quad G_2(u, v)(t) := \frac{f(t, u, v) + \lambda v(t)}{1 + \lambda}, \tag{6.93}$$

define an increasing mapping $G = (G_1, G_2) : [x_-, x_+] \to [x_-, x_+]$.

Let W be a well-ordered chain in the range of G. The sets $W_1 = \{u : (u, v) \in W\}$ and $W_2 = \{v : (u, v) \in W\}$ are well-ordered and order-bounded chains in $HL_{loc}(J, E)$. It then follows from Proposition 9.39 that $\sup W_1$ and $\sup W_2$ exist in $HL_{loc}(J, E)$. Obviously, $(\sup W_1, \sup W_2)$ is a supremum of W in P. Similarly one can show that each inversely well-ordered chain of the range of G has an infimum in P.

The above proof shows that the operator $G = (G_1, G_2)$ defined by (6.93) satisfies the hypotheses of Proposition 2.18, whence we conclude that G has the smallest fixed point $x_* = (u_*, v_*)$ and the greatest fixed point $x^* = (u^*, v^*)$. From (6.93) it follows that (u_*, v_*) and (u^*, v^*) are solutions of the system (6.89). According to Lemma 6.47, u_* and u^* belong to S and both are solutions of the IVP (6.87).

To prove that u_* and u^* are the smallest and greatest of all solutions of (6.87) in S, let $u \in S$ be any solution of (6.87). In view of Lemma 6.47, $(u, v) = (u, Lu)$ is a solution of the system (6.89). Applying the hypotheses (fa) and (c) it is easy to show that $x = (u, v) \in [x_-, x_+]$, where $x_\pm$ are defined by (6.92). Thus $x = (u, v)$ is a fixed point of mapping G. Because $x_* = (u_*, v_*)$ and $x^* = (u^*, v^*)$ are the smallest and greatest fixed points of G, then $(u_*, v_*) \le (u, v) \le (u^*, v^*)$. In particular, $u_* \le u \le u^*$, whence u_* and u^* are the smallest and greatest of all solutions of the IVP (6.87).

The last assertion is an easy consequence of the last conclusion of Proposition 2.18 and the definition of G.   □

As a special case we obtain an existence result for the IVP

$$\begin{cases} \dfrac{d}{dt}(p(t)u(t)) = g(t, u(t), \dfrac{d}{dt}(p(t)u(t))) & \text{for a.e. } t \in J := (a, b), \\ \lim_{t \to a+}(p(t)u(t)) = c. \end{cases} \tag{6.94}$$

Proposition 6.49. *Let the hypothesis (p) hold, and let $g : J \times E \times E \to E$ satisfy the following hypotheses:*

(ga) $g(\cdot, u(\cdot), v(\cdot))$ is strongly measurable, and $h_- \leq g(\cdot, u(\cdot), v(\cdot)) \leq h_+$ for all u, $v \in HL_{loc}(J, E)$ and for some $h_\pm \in HL_{loc}([a, b), E)$.
(gb) There is a $\lambda \geq 0$ such that $g(t, x, z) + \lambda z \leq g(t, y, w) + \lambda w$ for a.e. $t \in J$ whenever $x \leq y$ and $z \leq w$ in E.

Then the IVP (6.94) has for each choice of $c \in E$ the smallest and greatest solutions in S. Moreover, these solutions are increasing with respect to g and c.

Proof: If $c \in E$, the IVP (6.94) is reduced to (6.87) when we define

$$\begin{cases} f(t, u, v) = g(t, u(t), v(t)), & t \in J, \ u, \ v \in HL_{loc}(J, E), \\ c(u, v) \equiv c, \ u, \ v \in HL_{loc}(J, E). \end{cases} \tag{6.95}$$

The hypotheses (ga) and (gb) imply that f satisfies the hypotheses (fa) and (fb). The hypothesis (c) is also valid. Therefore, (6.87) with f and c defined by (6.95), and hence also (6.94), has the smallest and greatest solutions due to Theorem 6.48. The last assertion follows from the last assertion of Theorem 6.48. □

Example 6.50. Determine the smallest and greatest solutions of the following system of IVPs:

$$\begin{cases} L_1 u_1(t) := \dfrac{d}{dt}(\sqrt{t}u_1(t)) = -\dfrac{1}{t}\sin\dfrac{1}{t} + \dfrac{[\int_1^{t^2}(u_2(s) + L_2 u_2(s))\,ds]}{1 + |[\int_1^{t^2}(u_2(s) + L_2 u_2(s))\,ds]|}, \\ L_2 u_2(t) := \dfrac{d}{dt}(\sqrt{t}u_2(t)) = \dfrac{1}{t}\sin\dfrac{1}{t} + \dfrac{[\int_1^{t^2}(u_1(s) + L_1 u_1(s))\,ds]}{1 + |[\int_1^{t^2}(u_1(s) + L_1 u_1(s))\,ds]|}, \\ \lim_{t \to 0+}\sqrt{t}u_1(t) = \dfrac{2[u_2(1)]}{1 + |[u_2(1)]|}, \quad \lim_{t \to 0+}\sqrt{t}u_2(t) = \dfrac{3[u_1(1)]}{1 + |[u_1(1)]|}, \end{cases} \tag{6.96}$$

where the differential equations in (6.96) hold for a.e. $t \in (0, \infty)$, and where $s \mapsto [s]$ denotes the integer function, i.e., $[s]$ is the greatest integer $\leq s$.

Solution: System (6.96) is a special case of (6.87) when $E = \mathbb{R}^2$, ordered coordinatewise, $a = 0$, $b = \infty$, $p(t) = \sqrt{t}$, and the components of $f(t, (u_1, u_2), (v_1, v_2)) = (f_1(t, (u_1, u_2), (v_1, v_2)), f_2(t, (u_1, u_2), (v_1, v_2)))$ as well as c are given by

$$\begin{aligned} f_1(t, (u_1, u_2), (v_1, v_2)) &= -\dfrac{1}{t}\sin\dfrac{1}{t} + \dfrac{[\int_1^{t^2}(u_2(s) + v_2(s))\,ds]}{1 + |[\int_1^{t^2}(u_2(s) + v_2(s))\,ds]|}, \\ f_2(t, (u_1, u_2), (v_1, v_2)) &= \dfrac{1}{t}\sin\dfrac{1}{t} + \dfrac{[\int_1^{t^2}(u_1(s) + v_1(s))\,ds]}{1 + |[\int_1^{t^2}(u_1(s) + v_1(s))\,ds]|}, \end{aligned} \tag{6.97}$$

$$c((u_1, u_2), (v_1, v_2)) = \left(\frac{2\,[u_2(1)]}{1 + |[u_2(1)]|}, \frac{3\,[u_1(1)]}{1 + |[u_1(1)]|} \right).$$

The hypotheses (fa), (fb), and (c) are satisfied when setting $h_\pm(t) = (-\frac{1}{t}\sin\frac{1}{t} \pm 1, \frac{1}{t}\sin\frac{1}{t} \pm 1)$, $\lambda = 0$, and $c_\pm = (\pm 2, \pm 3)$. Thus (6.96) has the smallest and greatest solutions. The functions x_- and x_+ defined by (6.92) can be calculated, and one obtains

$$x_-(t) = \left(\left(-\frac{2}{\sqrt{t}} + \frac{Si(\frac{1}{t})}{\sqrt{t}} - \frac{\pi}{2\sqrt{t}} - \sqrt{t}, -\frac{3}{\sqrt{t}} - \frac{Si(\frac{1}{t})}{\sqrt{t}} + \frac{\pi}{2\sqrt{t}} - \sqrt{t} \right), h_-(t) \right)$$

$$x_+(t) = \left(\left(\frac{2}{\sqrt{t}} + \frac{Si(\frac{1}{t})}{\sqrt{t}} - \frac{\pi}{\sqrt{t}} + \sqrt{t}, \frac{3}{\sqrt{t}} - \frac{Si(\frac{1}{t})}{\sqrt{t}} + \frac{\pi}{2\sqrt{t}} + \sqrt{t} \right), h_+(t) \right),$$

where

$$Si(x) = \int_0^x \frac{\sin t}{t}\, dt$$

is the sine integral function. According to Lemma 6.47 the smallest solution of (6.96) is equal to the first component of the smallest fixed point of $G = (G_1, G_2)$, defined by (6.93), with f and c given by (6.97), and $p(t) = \sqrt{t}$. By the proof of Proposition 2.14 the smallest fixed point of G is the maximum of a well-ordered chain of x_-G-iterations, whose smallest elements are iterations $G^n x_-$. Calculating these iterations it turns out that $G^3 x_- = G^4 x_-$. Thus $G_1^3 x_-$ is the smallest solution of (6.96). Similarly, one can show that $G^3 x_+ = G^4 x_+$, which implies that $G_1^3 x_+$ is the greatest solution of (6.96). The exact expressions of the components of these solutions are

$$u_{1*}(t) = -\frac{3}{2\sqrt{t}} + \frac{Si\left(\frac{1}{t}\right)}{\sqrt{t}} - \frac{\pi}{2\sqrt{t}} - \frac{3}{4}\sqrt{t},$$

$$u_{2*}(t) = -\frac{9}{4\sqrt{t}} - \frac{Si\left(\frac{1}{t}\right)}{\sqrt{t}} + \frac{\pi}{2\sqrt{t}} - \frac{5}{6}\sqrt{t},$$

$$u_1^*(t) = \frac{4}{3\sqrt{t}} + \frac{Si\left(\frac{1}{t}\right)}{\sqrt{t}} - \frac{\pi}{2\sqrt{t}} + \frac{3}{4}\sqrt{t},$$

$$u_2^*(t) = \frac{3}{2\sqrt{t}} - \frac{Si\left(\frac{1}{t}\right)}{\sqrt{t}} + \frac{\pi}{2\sqrt{t}} - \frac{1}{2}\sqrt{t}.$$

Example 6.51. Let E be the space c_0 of the sequences of real numbers converging to zero, ordered componentwise and normed by the sup-norm. The mappings $h_\pm : (0, \infty) \to c_0$, defined by

$$h_\pm(t) = \left(\frac{1}{n} \left| \cos\left(\frac{\pi}{t}\right) \right| + \frac{\pi}{nt} \operatorname{sgn}\left(\cos\left(\frac{\pi}{t}\right)\right) \sin\left(\frac{\pi}{t}\right) \pm \frac{1}{n} \right)_{n=1}^{\infty} \qquad (6.98)$$

belong to $HL_{loc}([0, \infty), c_0)$ by Lemma 1.12. Thus these mappings are possible upper and lower boundaries for f in the hypothesis (fa) of Theorem 6.48

and for g in the hypothesis (ga) of Proposition 6.49 when $E = c_0$. Choosing $c_\pm = (\pm n^{-1})_{n=1}^\infty$ and $p(t) := t$, the solutions of the initial value problems

$$\frac{d}{dt}(p(t)u(t)) = h_\pm(t) \text{ for a.e. } t \in (0,\infty), \ \lim_{t\to 0+}p(t)u(t) = c_\pm, \quad (6.99)$$

are

$$u_\pm(t) = \left(\frac{1}{nt}\left(t\left|\cos\left(\frac{\pi}{t}\right)\right| \pm (t+1)\right)\right)_{n=1}^\infty. \quad (6.100)$$

Consider, in particular, the infinite system of initial value problems

$$\begin{cases} \dfrac{d}{dt}(tu_n(t)) = \dfrac{1}{n}\left(\left|\cos\left(\dfrac{\pi}{t}\right)\right| + \dfrac{\pi}{t}\operatorname{sgn}\left(\cos\left(\dfrac{\pi}{t}\right)\right)\sin\left(\dfrac{\pi}{t}\right) + f_n(t,u)\right) \\ \text{for a.e. } t \in (0,\infty), \\ \lim_{t\to 0+}(tu_n(t)) = \dfrac{c_n}{n}, \ n \in \mathbb{N}. \end{cases} \quad (6.101)$$

Setting $u = (u_n)_{n=1}^\infty$ and $f = (f_n)_{n=1}^\infty : (0,\infty) \times HL_{loc}((0,\infty),c_0) \to c_0$, and assuming that $f(\cdot,u)$ is strongly measurable for each $u \in c_0$, that $f[t,\cdot]$ is increasing, as well as that $-1 \le c_n \le 1$ and $-1 \le f(\cdot,u) \le 1$ for all $u \in HL_{loc}((0,\infty),c_0)$ and $n \in \mathbb{N}$, then (6.101) has by Theorem 6.48 the smallest and greatest solutions $u_* = (u_{*n})_{n=1}^\infty$ and $u^* = (u_n^*)_{n=1}^\infty$, respectively, and they belong to the order interval $[u_-,u_+]$, where $u_\pm$ are given by (6.100).

Remark 6.52. No component of the mappings $h_\pm$ defined in (6.98) belongs to $L^1((0,t),\mathbb{R})$ for any $t > 0$. Consequently, the mappings $h_\pm$ don't belong to $L^1((0,t),c_0)$ for any $t > 0$. Notice also that if f in Theorem 6.48 and g in Proposition 6.49 are norm-bounded by a function h_0 that belongs to $L^1((a,t),\mathbb{R}_+)$ for every $t \in (a,b)$, then the mappings $f(\cdot,u,v)$ and $g(\cdot,u(\cdot),v(\cdot))$ belong to $L^1((a,t),E)$ for all $t \in (a,b)$.

6.4.3 Second Order Initial Value Problems

Next we study the second order initial value problem

$$\begin{cases} Lu(t) := \dfrac{d}{dt}(p(t)u'(t)) = f(t,u,u',Lu) \text{ for a.e. } t \in J := (a,b), \\ \lim_{t\to a+}(p(t)u'(t)) = c(u,u',Lu), \quad \lim_{t\to a+}u(t) = d(u,u',Lu), \end{cases} \quad (6.102)$$

where $-\infty < a < b \le \infty$, $f : J \times HL_{loc}(J,E)^3 \to E$, $c, d : HL_{loc}(J,E)^3 \to E$, and $p : J \to \mathbb{R}_+$. Now we are looking for the smallest and greatest solutions of (6.102) from the set

$$Y := \left\{\begin{array}{l} u : J \to E : u \text{ and } pu' \text{ are in } W_{SL}^1(I,E) \\ \text{for compact intervals } I \text{ of } J \end{array}\right\}. \quad (6.103)$$

The IVP (6.102) can be converted to a system of equations that does not contain derivatives.

Lemma 6.53. *Assume that $\frac{1}{p} \in L_{loc}^1([a,b), \mathbb{R}_+)$, and that $f(\cdot, u, v, w) \in HL_{loc}([a,b), E)$ for all $u, v, w \in HL_{loc}(J, E)$. Then u is a solution of the IVP (6.102) in Y if and only if $(u, u', Lu) = (u, v, w)$, where $(u, v, w) \in HL_{loc}(J, E)^3$ is a solution of the system*

$$\begin{cases} u(t) = d(u, v, w) + {}^K\!\!\int_a^t v(s)\,ds, \quad t \in J, \\[2mm] v(t) = \dfrac{1}{p(t)}\left(c(u, v, w) + {}^K\!\!\int_a^t w(s)\,ds\right), \quad t \in J, \\[2mm] w(t) = f(t, u, v, w) \quad \text{for a.e. } t \in J. \end{cases} \tag{6.104}$$

Proof: Assume that u is a solution of (6.102) in Y, and denote

$$w(t) = Lu(t) = \frac{d}{dt}(p(t)v(t)), \quad v(t) = u'(t). \tag{6.105}$$

The differential equation, the initial conditions of (6.102), the definition (6.103) of Y, and the notations (6.105) ensure by Corollary 9.21 that the third equation of (6.104) is satisfied, and that

$$\begin{aligned} {}^K\!\!\int_a^t w(s)\,ds &= \lim_{r \to a+} {}^K\!\!\int_r^t w(s)\,ds = \lim_{r \to a+} {}^K\!\!\int_r^t \frac{d}{ds}(p(s)v(s))\,ds \\ &= \lim_{r \to a+} (p(t)v(t) - p(r)v(r)) = p(t)v(t) - c(u, v, w), \quad t \in J, \end{aligned}$$

and

$$\begin{aligned} u(t) - d(u, v, w) &= \lim_{r \to a+} (u(t) - u(r)) = \lim_{r \to a+} {}^K\!\!\int_r^t u'(s)\,ds \\ &= {}^K\!\!\int_a^t u'(s)\,ds = {}^K\!\!\int_a^t v(s)\,ds, \quad t \in J. \end{aligned}$$

Thus the first and second equations of (6.104) hold.

Conversely, let (u, v, w) be a solution of the system (6.104) in $HL_{loc}(J, E)^3$. The first equation of (6.104) implies by Theorem 9.18 that $v = u'$, that $u \in W_{SL}^1(I, E)$ for every closed interval I of J, and that the second initial condition of (6.102) is fulfilled. Since $v = u'$, it follows from the second equation of (6.104) that

$$p(t)u'(t) = c(u, u', w) + {}^K\!\!\int_a^t w(s)\,ds, \quad t \in J. \tag{6.106}$$

By Theorem 9.18 the equation (6.106) implies that $p \cdot u' \in W_{SL}^1(I, E)$ for every closed interval I of J, and thus $u \in Y$, as well as that

$$w(t) = \frac{d}{dt}(p(t)u'(t)) = Lu(t) \quad \text{for a.e. } t \in J. \tag{6.107}$$

The last relation and (6.106) imply that the first initial condition of (6.102) holds. The validity of the differential equation of (6.102) is a consequence of the third equation of (6.104), the equation (6.107), and the fact that $v = u'$.

$\square$

Assume that $HL_{loc}(J, E)$ and $HL_{loc}([a, b), E)$ are ordered a.e. pointwise, that Y is ordered pointwise, and that the functions p, f, c, and d satisfy the following hypotheses:

(p0) $\frac{1}{p} \in L^1_{loc}([a, b), \mathbb{R}_+)$.

(f0) $f(\cdot, u, v, w)$ is strongly measurable, and there exist such $h_-, h_+ \in HL_{loc}([a, b), E)$ that $h_- \leq f(\cdot, u, v, w) \leq h_+$ for all $u, v, w \in HL_{loc}(J, E)$.

(f1) There exists a $\lambda \geq 0$ such that $f(\cdot, u_1, v_1, w_1) + \lambda w_1 \leq f(\cdot, u_2, v_2, w_2) + \lambda w_2$ whenever $u_i, v_i, w_i \in HL_{loc}(J, E)$, $i = 1, 2$, $u_1 \leq u_2$, $v_1 \leq v_2$, and $w_1 \leq w_2$.

(c0) $c_\pm \in \mathbb{R}$, and $c_- \leq c(u_1, v_1, w_1) \leq c(u_2, v_2, w_2) \leq c_+$ whenever $u_i, v_i, w_i \in HL_{loc}(J, E)$, $i = 1, 2$, $u_1 \leq u_2$, $v_1 \leq v_2$, and $w_1 \leq w_2$.

(d0) $d_\pm \in \mathbb{R}$, and $d_- \leq d(u_1, v_1, w_1) \leq d(u_2, v_2, w_2) \leq d_+$ whenever $u_i, v_i, w_i \in HL_{loc}(J, E)$, $i = 1, 2$, $u_1 \leq u_2$, $v_1 \leq v_2$, and $w_1 \leq w_2$.

Our main existence and comparison result for the IVP (6.102) reads as follows.

Theorem 6.54. *Assume that the hypotheses (p0), (f0), (f1), (c0), and (d0) hold. Then the IVP (6.102) has the smallest and greatest solutions in Y, and they are increasing with respect to f, c, and d.*

Proof: Assume that $P = HL_{loc}(J, E)^3$ is ordered componentwise. We shall first show that the vector-functions x_+, x_- given by

$$x_\pm(t) := \begin{pmatrix} d_\pm + {}^K\!\!\int_a^t \frac{1}{p(s)} \left(c_\pm + {}^K\!\!\int_a^s h_\pm(\tau)\, d\tau \right) ds \\ \frac{1}{p(t)} \left(c_\pm + {}^K\!\!\int_a^t h_\pm(s)\, ds \right) \\ h_\pm(t) \end{pmatrix} \qquad (6.108)$$

define functions $x_\pm \in P$. The third components of $x_\pm$ belong to $HL_{loc}(J, E)$ by the hypothesis (f0). Since $1/p$ is locally Lebesgue integrable and the functions $t \mapsto c_\pm + {}^K\!\!\int_a^t h_\pm(s)\, ds$ are continuous on $[a, b)$, then the second components of $x_\pm$ are strongly measurable by [133, Theorem 1.4.3]. Moreover, if $t_1 \in J$ then for each $t \in [a, t_1]$, $\|\frac{1}{p(t)} (c_\pm + {}^K\!\!\int_a^t h_\pm(s)\, ds)\| \leq M_\pm \frac{1}{p(t)}$, where $M_\pm = \max\{\|c_\pm + {}^K\!\!\int_a^t h_\pm(s)\, ds\| : t \in [a, t_1]\}$. Thus the second components of $x_\pm$ are locally Bochner integrable, and belong to $HL_{loc}(J, E)$. This result implies that the first components of $x_\pm$ are defined and continuous, whence they belong to $HL_{loc}(J, E)$.

Similarly, by applying the given hypotheses in conjunction with Lemma 9.11, Proposition 9.14, and Theorem 9.18, one can verify that $G = (G_1, G_2, G_3)$:

$[x_-, x_+] \to [x_-, x_+]$ defines an increasing mapping, where the components G_i, $i = 1, 2, 3$ are given by the following relations:

$$\begin{cases} G_1(u, v, w)(t) := d(u, v, w) + {}^K\!\!\int_a^t v(s)\, ds,\ t \in J, \\ G_2(u, v, w)(t) := \dfrac{1}{p(t)}\Big(c(u, v, w) + {}^K\!\!\int_a^t w(s)\, ds\Big),\ t \in J, \\ G_3(u, v, w)(t) := \dfrac{f(t, u, v, w) + \lambda w(t)}{1 + \lambda},\ t \in J, \end{cases} \qquad (6.109)$$

Let W be a well-ordered chain in the range of G. The sets $W_1 = \{u : (u, v, w) \in W\}$, $W_2 = \{v : (u, v, w) \in W\}$, and $W_3 = \{w : (u, v, w) \in W\}$ are well-ordered and order-bounded chains in $HL_{loc}(J, E)$. It then follows from Proposition 9.39 that the supremums of W_1, W_2, and W_3 exist in $HL_{loc}(J, E)$. Obviously, $(\sup W_1, \sup W_2, \sup W_3)$ is the supremum of W in P. Similarly one can show that each inversely well-ordered chain of the range of G has the infimum in P.

The above proof shows that the operator $G = (G_1, G_2, G_3)$ defined by (6.109) satisfies the hypotheses of Proposition 2.18, and therefore G has the smallest fixed point $x_* = (u_*, v_*, w_*)$ and the greatest fixed point $x^* = (u^*, v^*, w^*)$. It follows from (6.109) that (u_*, v_*, w_*) and (u^*, v^*, w^*) are solutions of the system (6.104). According to Lemma 6.53, u_* and u^* belong to Y and are solutions of the IVP (6.102).

To prove that u_* and u^* are the smallest and greatest of all solutions of (6.102) in Y, let $u \in Y$ be any solution of (6.102). In view of Lemma 6.53, $(u, v, w) = (u, u', Lu)$ is a solution of the system (6.104). Applying the hypotheses (f0), (c0), and (d0), it is easy to show that $x = (u, v, w) \in [x_-, x_+]$, where $x_\pm$ are defined by (6.108). Thus $x = (u, v, w)$ is a fixed point of $G = (G_1, G_2, G_3) : [x_-, x_+] \to [x_-, x_+]$, defined by (6.109). Because $x_* = (u_*, v_*, w_*)$ and $x^* = (u^*, v^*, w^*)$ are the smallest and greatest fixed points of G, then $(u_*, v_*, w_*) \le (u, v, w) \le (u^*, v^*, w^*)$. In particular, $u_* \le u \le u^*$, whence u_* and u^* are the smallest and greatest of all solutions of the IVP (6.102).

The last assertion is an easy consequence of the last conclusion of Proposition 2.18, Lemma 9.11, and the definition (6.109) of $G = (G_1, G_2, G_3)$. □

As a special case we obtain an existence result for the IVP

$$\begin{cases} \dfrac{d}{dt}(p(t)u'(t)) = g(t, u(t), u'(t), \dfrac{d}{dt}(p(t)u'(t)))\ \text{for a.e. } t \in J, \\ \lim_{t \to a+}(p(t)u'(t)) = c,\quad \lim_{t \to a+} u(t) = d. \end{cases} \qquad (6.110)$$

Corollary 6.55. *Assume hypothesis (p0), and let $g : J \times E \times E \times E \to E$ satisfy the following hypotheses:*

(g0) $g(\cdot, u(\cdot), v(\cdot), w(\cdot))$ is strongly measurable and $h_- \le g(\cdot, u(\cdot), v(\cdot), w(\cdot)) \le h_+$ for all $u, v, w \in HL_{loc}(J, E)$ and for some $h_\pm \in HL_{loc}([a, b), E)$.

(g1) There exists a $\lambda \geq 0$ such that $g(t, x_1, x_2, x_3) + \lambda x_3 \leq g(t, y_1, y_2, y_3) + \lambda y_3$ for a.e. $t \in J$ and whenever $x_i \leq y_i$ in E, $i = 1, 2, 3$.

Then the IVP (6.110) has for each choice of c, $d \in E$ *the smallest and greatest solutions in* Y. *Moreover, these solutions are increasing with respect to* g, c, *and* d.

Proof: If c, $d \in E$, the IVP (6.110) is reduced to (6.102) when we define

$$f(t, u, v, w) = g(t, u(t), v(t), w(t)), \quad t \in J, \ u, v, w \in HL_{loc}(J, E),$$
$$c(u, v, w) \equiv c, \ d(u, v, w) \equiv d, \ u, v, w \in HL_{loc}(J, E).$$

The hypotheses (g0) and (g1) imply that f satisfies the hypotheses (f0) and (f1). The hypotheses (c0) and (d0) are also valid, whence we conclude that (6.102), with f, c, and d defined above, and hence also (6.110), has the smallest and greatest solutions due to Theorem 6.54. The last assertion follows from the last assertion of Theorem 6.54. □

Example 6.56. Determine the smallest and greatest solutions of the following system of implicit singular IVPs in $J = (0, \infty)$

$$\begin{cases} L_1 u_1(t) = \dfrac{d}{dt}\left(t \sin \dfrac{1}{t}\right) + \dfrac{[\int_1^2 (u_2(s) + u_2'(s) + L_2 u_2(s))\, ds]}{1 + |[\int_1^2 (u_2(s) + u_2'(s) + L_2 u_2(s))\, ds]|}, \\[3mm] L_2 u_2(t) = \dfrac{d}{dt}\left(t \cos \dfrac{1}{t}\right) + \dfrac{[\int_1^2 (u_1(s) + u_1'(s) + L_1 u_1(s))\, ds]}{1 + |[\int_1^2 (u_1(s) + u_1'(s) + L_1 u_1(s))\, ds]|}, \quad (6.111)\\[3mm] \lim\limits_{t \to 0+} \sqrt{t} u_1'(t) = \dfrac{[u_2'(1)])}{1 + |[u_2'(1)]|}, \quad \lim\limits_{t \to 0+} u_1(t) = \dfrac{[u_2(1)]}{1 + |[u_2(1)]|}, \\[3mm] \lim\limits_{t \to 0+} \sqrt{t} u_2'(t) = \dfrac{[u_1'(1)]}{1 + |[u_1'(1)]|}, \quad \lim\limits_{t \to 0+} u_2(t) = \dfrac{[u_1(1)]}{1 + |[u_1(1)]|}, \end{cases}$$

where

$$L_1 u_1(t) := \frac{d}{dt}(\sqrt{t} u_1'(t)) \quad \text{and} \quad L_2 u_2(t) := \frac{d}{dt}(\sqrt{t} u_2'(t)).$$

Solution: System (6.111) is a special case of (6.102) by setting $E = \mathbb{R}^2$, $a = 0$, $b = \infty$, $p(t) = \sqrt{t}$, and $f = (f_1, f_2)$, c, d given by

$$f_1(t, (u_1, u_2), (v_1, v_2), (w_1, w_2)) = \frac{d}{dt}\left(t \sin \frac{1}{t}\right)$$
$$+ \frac{[\int_1^2 (u_2(s) + v_2(s) + w_2(s))\, ds]}{1 + |[\int_1^2 (u_2(s) + v_2(s) + w_2(s))\, ds]|},$$

$$f_2(t, (u_1, u_2), (v_1, v_2), (w_1, w_2)) = \frac{d}{dt}\left(t \cos \frac{1}{t}\right)$$
$$+ \frac{[\int_1^2 (u_1(s) + v_1(s) + w_1(s))\, ds]}{1 + |[\int_1^2 (u_1(s) + v_1(s) + w_1(s))\, ds]|},$$

$$c((u_1, u_2), (v_1, v_2), (w_1, w_2)) = \left(\frac{[v_2(1)]}{1 + |[v_2(1)]|}, \frac{[v_1(1)])}{1 + |[v_1(1)]|} \right),$$

$$d((u_1, u_2), (v_1, v_2), (w_1, w_2)) = \left(\frac{[u_2(1)]}{1 + |[u_2(1)]|}, \frac{[u_1(1)])}{1 + |[u_1(1)]|} \right).$$

(6.112)

In view of Lemma 1.12 and Lemma 9.11, the hypotheses (f0), (f1), (c0), and (d0) hold when $h_\pm(t) = (\frac{d}{dt}(t \sin \frac{1}{t}) \pm 1, \frac{d}{dt}(t \cos \frac{1}{t}) \pm 1)$, $\lambda = 0$, and $c_\pm = d_\pm = (\pm 1, \pm 1)$. Thus (6.111) has the smallest and greatest solutions. The functions x_- and x_+ defined by (6.93) can be calculated, and their first components are

$$u_-(t) = -1 - \frac{2\sqrt{2\pi}}{3} - 2\sqrt{t} + \frac{2t\sqrt{t}}{3} \sin \frac{1}{t}$$
$$+ \frac{4\sqrt{t}}{3} \cos \frac{1}{t} + \frac{4\sqrt{2\pi}}{3} FrC(\sqrt{\frac{2}{\pi t}}) - \frac{2t\sqrt{t}}{3},$$

$$v_-(t) = -1 - \frac{2\sqrt{2\pi}}{3} - 2\sqrt{t} + \frac{2t\sqrt{t}}{3} \cos \frac{1}{t}$$
$$- \frac{4\sqrt{t}}{3} \sin \frac{1}{t} + \frac{4\sqrt{2\pi}}{3} FrC(\sqrt{\frac{2}{\pi t}}) - \frac{2t\sqrt{t}}{3},$$

$$u_+(t) = 1 - \frac{2\sqrt{2\pi}}{3} + 2\sqrt{t} + \frac{2t\sqrt{t}}{3} \sin \frac{1}{t}$$
$$+ \frac{4\sqrt{t}}{3} \cos \frac{1}{t} + \frac{4\sqrt{2\pi}}{3} FrC(\sqrt{\frac{2}{\pi t}}) + \frac{2t\sqrt{t}}{3},$$

$$v_+(t) = 1 - \frac{2\sqrt{2\pi}}{3} + 2\sqrt{t} + \frac{2t\sqrt{t}}{3} \cos \frac{1}{t}$$
$$- \frac{4\sqrt{t}}{3} \sin \frac{1}{t} + \frac{4\sqrt{2\pi}}{3} FrC(\sqrt{\frac{2}{\pi t}}) + \frac{2t\sqrt{t}}{3},$$

where

$$FrC(x) = \int_0^x \cos \left(\frac{\pi}{2} t^2 \right) dt$$

is the **Fresnel cosine integral**. According to Lemma 6.53 the smallest solution of (6.111) is equal to the first component of the smallest fixed point of $G = (G_1, G_2, G_3)$, defined by (6.109), with f, c, and d given by (6.112) and $p(t) = \sqrt{t}$. Calculating the iterations $G^n x_-$ it turns out that $G^2 x_- = G^3 x_-$, whence $G_1^2 x_-$ is the smallest solution of (6.111). Similarly, one can show that $G_1^4 x_+$ is the greatest solution of (6.111). The exact expressions of the components of these solutions are

$$u_{1*}(t) = -\frac{3}{4} - \frac{2\sqrt{2\pi}}{3} - \sqrt{t} + \frac{2t\sqrt{t}}{3} \sin \frac{1}{t}$$
$$+ \frac{4\sqrt{t}}{3} \cos \frac{1}{t} + \frac{4\sqrt{2\pi}}{3} FrC(\sqrt{\frac{2}{\pi t}}) - \frac{t\sqrt{t}}{2},$$

$$u_{2*}(t) = -\frac{2}{3} - \frac{2\sqrt{2\pi}}{3} - \sqrt{t} + \frac{2t\sqrt{t}}{3}\cos\frac{1}{t}$$
$$+ \frac{4\sqrt{t}}{3}\sin\frac{1}{t} + \frac{4\sqrt{2\pi}}{3}FrC(\sqrt{\frac{2}{\pi t}}) - \frac{8t\sqrt{t}}{15},$$

$$u_1^*(t) = \frac{2}{3} - \frac{2\sqrt{2\pi}}{3} + \frac{4\sqrt{t}}{3} + \frac{2t\sqrt{t}}{3}\sin\frac{1}{t}$$
$$+ \frac{4\sqrt{t}}{3}\cos\frac{1}{t} + \frac{4\sqrt{2\pi}}{3}FrC(\sqrt{\frac{2}{\pi t}}) + \frac{16t\sqrt{t}}{27},$$

$$u_2^*(t) = \frac{3}{4} - \frac{2\sqrt{2\pi}}{3} + \frac{4\sqrt{t}}{3} + \frac{2t\sqrt{t}}{3}\cos\frac{1}{t}$$
$$- \frac{4\sqrt{t}}{3}\sin\frac{1}{t} + \frac{4\sqrt{2\pi}}{3}FrC(\sqrt{\frac{2}{\pi t}}) + \frac{7t\sqrt{t}}{12}.$$

Example 6.57. Let E be the space (c_0), ordered coordinatewise and normed by the sup-norm. The mappings $h_\pm : [0, \infty) \to c_0$, defined by $h_\pm(0) = (0, 0, \dots)$ and

$$h_\pm(t) = \left(\frac{1}{nt}\sin\frac{1}{t} + \frac{1}{n}\cos\frac{1}{t} \pm \frac{1}{n}\right)_{n=1}^\infty, \quad t \in (0, \infty), \quad (6.113)$$

belong to $HL_{loc}([0, \infty), E)$ by Lemma 1.12. Thus these mappings are possible upper and lower boundaries for f in the hypothesis (f0) of Theorem 6.54 and for g in the hypothesis (g0) of Corollary 6.55 when $E = c_0$. Choosing $c_\pm = (\pm n^{-1})_{n=1}^\infty$, $d_\pm = (\pm n^{-1})_{n=1}^\infty$, and $p(t) := \sqrt{t}$, the solutions of the initial value problems

$$\begin{cases} \frac{d}{dt}(\sqrt{t}\,u'(t)) = h_\pm(t) \text{ for a.e. } t \in (0, \infty), \\ \lim_{t \to 0+}(\sqrt{t}\,u'(t)) = c_\pm, \ \lim_{t \to 0+}u(t) = d_\pm \end{cases} \quad (6.114)$$

are

$$\begin{cases} u_+(t) = \left(\frac{1}{n}\left(\frac{2}{3}t\sqrt{t}\cos\left(\frac{1}{t}\right) - \frac{4}{3}\sin\left(\frac{1}{t}\right) + \frac{4\sqrt{2\pi}}{3}FrC\left(\sqrt{\frac{2}{\pi t}}\right)\right. \right. \\ \qquad \left. \left. +2\sqrt{t} + \frac{2}{3}t\sqrt{t} + 1 - \frac{2}{3}\sqrt{2\pi}\right)\right)_{n=1}^\infty, \\ u_-(t) = \left(\frac{1}{n}\left(\frac{2}{3}t\sqrt{t}\cos\left(\frac{1}{t}\right) - \frac{4}{3}\sin\left(\frac{1}{t}\right) + \frac{4\sqrt{2\pi}}{3}FrC\left(\sqrt{\frac{2}{\pi t}}\right)\right. \right. \\ \qquad \left. \left. -2\sqrt{t} - \frac{2}{3}t\sqrt{t} - 1 - \frac{2}{3}\sqrt{2\pi}\right)\right)_{n=1}^\infty. \end{cases} \quad (6.115)$$

Consider, in particular, the infinite system of initial value problems on $(0, \infty)$

$$\begin{cases} L_n u_n(t) := \dfrac{d}{dt}(\sqrt{t}\, u'_n(t)) = \dfrac{1}{n}\left(\dfrac{1}{t}\sin\dfrac{1}{t} + \cos\dfrac{1}{t} + f_n(u, u', Lu)\right) & \text{a.e.,} \\[2mm] \lim_{t\to 0+}(\sqrt{t}\, u'_n(t)) = \dfrac{c_n}{n}, \quad \lim_{t\to 0+} u_n(t) = \dfrac{d_n}{n}, \quad n \in \mathbb{N}. \end{cases}$$

$$(6.116)$$

Setting $u = (u_n)_{n=1}^\infty$, $Lu = (L_n u_n)_{n=1}^\infty$, and assuming that each $f_n : HL_{loc}((0,\infty), c_0)^3 \to \mathbb{R}$ is increasing with respect to every argument, and $-1 \le c_n, d_n, f_n(u, v, w) \le 1$ for all $u, v, w \in HL_{loc}((0,\infty), c_0)$ and $n \in \mathbb{N}$, then (6.116) has the smallest and greatest solutions $u_* = (u_{*n})_{n=1}^\infty$ and $u^* = (u_n^*)_{n=1}^\infty$, and they belong to the order interval $[u_-, u_+]$, where $u_\pm$ are given by (6.115).

6.4.4 Second Order Boundary Value Problems

This section is devoted to the study of the boundary value problem

$$\begin{cases} Lu(t) := -\dfrac{d}{dt}(p(t)u'(t)) = f(t, u, u', Lu) \text{ for a.e. } t \in J := [a, b], \\[2mm] \lim_{t\to a+}(p(t)u'(t)) = c(u, u', Lu), \quad \lim_{t\to b-} u(t) = d(u, u', Lu), \end{cases} \quad (6.117)$$

where $-\infty < a < b < \infty$, $f : J \times HL(J, E)^3 \to E$, $c, d : HL(J, E)^3 \to E$ and $p : J \to \mathbb{R}_+$. Now we are looking for the smallest and greatest solutions of (6.117) from the set

$$Z := \{u : [a, b) \to E : u \text{ and } pu' \text{ are in } W^1_{SL}(J, E)\}. \qquad (6.118)$$

As in Sect. 6.4.3 we first convert the BVP (6.117) to an equivalent system of three equations.

Lemma 6.58. *Assume that $\frac{1}{p} \in L^1(J, \mathbb{R}_+)$, and that $f(\cdot, u, v, w) \in HL(J, E)$ for all $u, v, w \in HL(J, E)$. Then u is a solution of the BVP (6.117) in Z, defined by (6.118) if and only if $(u, u', Lu) = (u, v, w)$, where $(u, v, w) \in HL(J, E)^3$ is a solution of the system*

$$\begin{cases} u(t) = d(u, v, w) - K\displaystyle\int_t^b v(s)\, ds, \quad t \in J, \\[3mm] v(t) = \dfrac{1}{p(t)}\left(c(u, v, w) - K\displaystyle\int_a^t w(s)\, ds\right), \quad t \in J, \\[3mm] w(t) = f(t, u, v, w) \quad \text{for a.e. } t \in J. \end{cases} \qquad (6.119)$$

Proof: Assume that u is a solution of (6.117) in Z, and denote

$$w(t) = Lu(t) = -\dfrac{d}{dt}(p(t)v(t)), \quad v(t) = u'(t), \quad t \in J. \qquad (6.120)$$

The differential equation and the boundary conditions of (6.117), the definition (6.118) of Z, and notations (6.120) ensure by Corollary 9.21 that the third equation of (6.119) is satisfied, and that

$$- {}^K\!\int_a^t w(s)\,ds = -\lim_{r\to a+} {}^K\!\int_r^t w(s)\,ds = \lim_{r\to a+} {}^K\!\int_r^t \frac{d}{ds}(p(s)v(s))\,ds$$
$$= \lim_{r\to a+}(p(t)v(t) - p(r)v(r)) = p(t)v(t) - c(u,v,w), \quad t\in J,$$

and

$$- {}^K\!\int_t^b v(s)\,ds = -\lim_{r\to b-} {}^K\!\int_t^r v(s)\,ds = -\lim_{r\to b-} {}^K\!\int_t^r u'(s)\,ds$$
$$= -\lim_{r\to b-}(u(r) - u(t)) = u(t) - d(u,v,w), \quad t\in J.$$

Thus the first and second equations of (6.119) hold.

Conversely, let (u,v,w) be a solution of the system (6.119) in $HL(J,E)^3$. The first equation of (6.119) implies by Theorem 9.18 that u is in $W^1_{SL}(J,E)$, that $v = u'$, and that the second boundary condition of (6.117) holds. Since $v = u'$, it follows from the second equation of (6.119) that

$$p(t)u'(t) = c(u,u',w) - {}^K\!\int_a^t w(s)\,ds, \quad t\in J. \tag{6.121}$$

This equation implies by Theorem 9.18 that $p\cdot u'$ is in $W^1_{SL}(J,E)$, and thus $u\in Z$, and that

$$w(t) = -\frac{d}{dt}(p(t)u'(t)) = Lu(t) \quad \text{for a.e. } t\in J. \tag{6.122}$$

The last equation and (6.121) imply that the first boundary condition of (6.117) is fulfilled. The validity of the differential equation of (6.117) is a consequence of the third equation of (6.119), the equation (6.122), and the fact that $v = u'$. $\qquad\square$

Assuming that $HL(J,E)$ is ordered a.e. pointwise, we shall impose the following hypotheses for the functions p, f, c, and d.

(p$_1$) $\frac{1}{p}\in L^1(J,\mathbb{R}_+)$.

(f_0) $f(\cdot,u,v,w)$ is strongly measurable, and there are functions $h_-, h_+ \in HL(J,E)$ such that $h_- \le f(\cdot,u,v,w) \le h_+$ for all $u,v,w\in HL(J,E)$.

(f_1) There exists a $\lambda \ge 0$ such that $f(\cdot,u_1,v_1,w_1) + \lambda w_1 \le f(\cdot,u_2,v_2,w_2) + \lambda w_2$ whenever $u_i,v_i,w_i \in HL(J,E)$, $i=1,2$, $u_1 \le u_2$, $v_1 \ge v_2$, and $w_1 \le w_2$.

(c_1) There are $c_\pm \in E$ with $c_- \le c(u_2,v_2,w_2) \le c(u_1,v_1,w_1) \le c_+$ whenever $u_i,v_i,w_i \in HL(J,E)$, $i=1,2$, $u_1 \le u_2$, $v_1 \ge v_2$, and $w_1 \le w_2$.

(d_1) There exist $d_\pm \in E$ such that and $d_- \le d(u_1,v_1,w_1) \le d(u_2,v_2,w_2) \le d_+$ whenever $u_i,v_i,w_i \in HL(J,E)$, $i=1,2$, $u_1 \le u_2$, $v_1 \ge v_2$, and $w_1 \le w_2$.

The next theorem is our main existence and comparison result for the BVP (6.117).

Theorem 6.59. *Assume that the hypotheses* (p_1), (f_0), (f_1), (c_1), *and* (d_1) *hold. Then the BVP (6.117) has the smallest and greatest solutions in* Z, *and they are increasing with respect to* f *and* d *and decreasing with respect to* c.

Proof: Assume that $P = HL(J, E)^3$ is ordered by

$$(u_1, v_1, w_1) \preceq (u_2, v_2, w_2) \text{ if and only if } u_1 \le u_2, v_1 \ge v_2, \text{ and } w_1 \le w_2. \tag{6.123}$$

The following triples

$$\left(d_- - \int_t^b \frac{1}{p(s)} \left(c_+ - {}^K\!\!\int_a^s h_-(\tau)\,d\tau \right) ds, \frac{1}{p(t)} \left(c_+ - {}^K\!\!\int_a^t h_-(s)\,ds \right), h_-(t) \right)$$

and

$$\left(d_+ - \int_t^b \frac{1}{p(s)} \left(c_- - {}^K\!\!\int_a^s h_+(\tau)\,d\tau \right) ds, \frac{1}{p(t)} \left(c_- - {}^K\!\!\int_a^t h_+(s)\,ds \right), h_+(t) \right)$$

$$\tag{6.124}$$

define functions $x_\pm \in P$ satisfying $x_- \preceq x_+$. To show that $x_\pm \in P$, notice first that the third components of $x_\pm$ are in $HL(J, E)$ by the hypothesis (f_0). Since $1/p$ is Lebesgue integrable and the function $t \mapsto c_+ - {}^K\!\int_a^t h_-(s)\,ds$ is continuous on J, then the second component of x_+ is strongly measurable by [133, Theorem 1.4.3]. Moreover, for each $t \in J$, $\|\frac{1}{p(t)}(c_+ - {}^K\!\int_a^t h_-(s)\,ds)\| \le M\frac{1}{p(t)}$, where $M = \max\{\|c_+ - {}^K\!\int_a^t h_-(s)\,ds\| : t \in J\}$. Thus the second component of x_+ is Bochner integrable, and hence also HL integrable on J. Similarly one can show that the second component of x_- belongs to $HL(J, E)$. These results ensure that the first components of $x_\pm$ are defined and continuous in t, and hence are in $HL(J, E)$.

Similarly, by applying the given hypotheses in conjunction with Lemma 9.11, Proposition 9.14, and Theorem 9.18, one can verify that the relations

$$\begin{cases} G_1(u, v, w)(t) := d(u, v, w) - {}^K\!\!\int_t^b v(s)\,ds, \ t \in J, \\[2mm] G_2(u, v, w)(t) := \frac{1}{p(t)} \left(c(u, v, w) - {}^K\!\!\int_a^t w(s)\,ds \right), \ t \in J, \\[2mm] G_3(u, v, w)(t) := \frac{f(t, u, v, w) + \lambda w(t)}{1 + \lambda}, \ t \in J \end{cases} \tag{6.125}$$

define an increasing mapping $G = (G_1, G_2, G_3) : [x_-, x_+] \to [x_-, x_+]$.

Let W be a well-ordered chain in the range of G. The sets $W_1 = \{u : (u, v, w) \in W\}$ and $W_3 = \{w : (u, v, w) \in W\}$ are well-ordered, $W_2 = \{v : (u, v, w) \in W\}$ is inversely well-ordered, and all three are order-bounded in $HL(J, E)$. It then follows from Proposition 9.39 that the supremums of W_1 and W_3 and the infimum of W_2 exist in $HL(J, E)$. Obviously, $(\sup W_1, \inf W_2, \sup W_3)$ is the supremum of W in $(P, \preceq)$. Similarly one can show that each inversely well-ordered chain of the range of G has the infimum in $(P, \preceq)$.

The above proof shows that the operator $G = (G_1, G_2, G_3)$ defined by (6.125) satisfies the hypotheses of Proposition 2.18, whence G has the smallest fixed point $x_* = (u_*, v_*, w_*)$ and a greatest fixed point $x^* = (u^*, v^*, w^*)$. It follows from (6.125) that (u_*, v_*, w_*) and (u^*, v^*, w^*) are solutions of the system (6.119). According to Lemma 6.58, u_* and u^* belong to Z and are solutions of the BVP (6.117).

To prove that u_* and u^* are the smallest and greatest of all solutions of (6.117) in Z, let $u \in Z$ be any solution of (6.117). In view of Lemma 6.58, $(u, v, w) = (u, u', Lu)$ is a solution of the system (6.119). Applying the hypotheses (f_1), (c_1), and (d_1) it is easy to show that $x = (u, v, w) \in [x_-, x_+]$, where $x_\pm$ are defined by (6.124). Thus $x = (u, v, w)$ is a fixed point of $G = (G_1, G_2, G_2) : [x_-, x_+] \to [x_-, x_+]$, defined by (6.125). Because $x_* = (u_*, v_*, w_*)$ and $x^* = (u^*, v^*, w^*)$ are the smallest and greatest fixed points of G, respectively, then $(u_*, v_*, w_*) \preceq (u, v, w) \preceq (u^*, v^*, w^*)$. In particular, $u_* \leq u \leq u^*$, whence u_* and u^* are the smallest and greatest of all solutions of the IVP (6.117).

The last assertion is an easy consequence of the last conclusions of Proposition 2.18, Lemma 9.11, and the definition (6.125) of $G = (G_1, G_2, G_3)$. □

As a special case we obtain an existence result for the BVP

$$\begin{cases} -\dfrac{d}{dt}(p(t)u'(t)) = g(t, u(t), u'(t), -\dfrac{d}{dt}(p(t)u'(t))) \text{ for a.e. } t \in J, \\ \lim_{t \to a+}(p(t)u'(t)) = c, \quad \lim_{t \to b-} u(t) = d. \end{cases} \quad (6.126)$$

Corollary 6.60. Let the hypothesis (p_1) hold, and let $g : J \times E \times E \to E$ satisfy the following hypotheses:

(g_0) $g(\cdot, u(\cdot), v(\cdot), w(\cdot))$ is strongly measurable and $h_- \leq g(\cdot, u(\cdot), v(\cdot), w(\cdot)) \leq h_+$ for all $u, v, w \in HL(J, E)$ and for some $h_\pm \in HL(J, E)$.

(g_1) There exists a $\lambda \geq 0$ such that $g(t, x_1, y_1, z_1) + \lambda z_1 \leq g(t, x_2, y_2, x_2) + \lambda z_2$ for a.e. $t \in J$ and whenever $x_1 \leq x_2$, $y_1 \geq y_2$, and $z_1 \leq z_2$ in E.

Then the BVP (6.126) has for each choice of c, $d \in E$ the smallest and greatest solutions in Z. Moreover, these solutions are increasing with respect to g and d and decreasing with respect to c.

Proof: If c, $d \in E$, then the BVP (6.126) is reduced to (6.117) when we define

$$\begin{cases} f(t, u, v, w) = g(t, u(t), v(t), w(t)), \ t \in J, \ u, v, w \in HL(J, E), \\ c(u, v, w) \equiv c, \quad d(u, v, w) \equiv d, \ u, v, w \in HL(J, E). \end{cases} \quad (6.127)$$

The hypotheses (g_0) and (g_1) imply that f satisfies the hypotheses (f_0) and (f_1). The hypotheses (c_1) and (d_1) are satisfied as well, whence (6.117) with f, c, and d defined by (6.127), and hence also (6.126), has the smallest and

greatest solutions by Theorem 6.59. The last assertion follows from the last
assertion of Theorem 6.59.   □

Example 6.61. Determine the smallest and greatest solutions of the following
system of BVPs in $J = [0, 3]$:

$$
\begin{cases}
L_1 u_1(t) := -\dfrac{d}{dt}(\sqrt{t}\, u_1'(t)) \\[2mm]
\quad = \dfrac{d}{dt}(t \sin \tfrac{1}{t}) + \left[10 \tanh\left(\dfrac{1}{100}\displaystyle\int_1^2 (3u_2(s) - 2u_2'(s) + L_2 u_2(s))\, ds\right)\right], \\[4mm]
L_2 u_2(t) := -\dfrac{d}{dt}(\sqrt{t}\, u_2'(t)) \\[2mm]
\quad = \dfrac{d}{dt}(t \cos \tfrac{1}{t}) + \left[10 \arctan\left(\dfrac{1}{100}\displaystyle\int_1^2 (2u_1(s) - u_1'(s) + 3L_1 u_1(s))\, ds\right)\right], \\[4mm]
\displaystyle\lim_{t\to 0+} \sqrt{t}\, u_1'(t) = \dfrac{[u_2'(1)]}{1 + |[u_2'(1)]|}, \quad u_1(3) = \dfrac{[u_2(1)]}{1 + |[u_2(1)]|}, \\[4mm]
\displaystyle\lim_{t\to 0+} \sqrt{t}\, u_2'(t) = \dfrac{[u_1'(1)]}{1 + |[u_1'(1)]|}, \quad u_2(3) = \dfrac{[u_1(1)]}{1 + |[u_1(1)]|}.
\end{cases}
$$

$$(6.128)$$

Solution: System (6.128) is a special case of (6.117) when setting $E = \mathbb{R}^2$,
$a = 0$, $b = 3$, $p(t) = \sqrt{t}$, and $f = (f_1, f_2)$, c, and d defined by

$$
\begin{cases}
f_1(t, (u_1, u_2), (v_1, v_2), (w_1, w_2)) \\[2mm]
\quad = \dfrac{d}{dt}(t \sin \tfrac{1}{t}) + \left[10 \tanh\left(\displaystyle\int_1^2 (3u_2(s) - 2v_2(s) + w_2(s))\, ds/100\right)\right], \\[4mm]
f_2(t, (u_1, u_2), (v_1, v_2), (w_1, w_2)) \\[2mm]
\quad = \dfrac{d}{dt}(t \cos \tfrac{1}{t}) + \left[10 \arctan\left(\displaystyle\int_1^2 (2u_1(s) - v_1(s) + 3w_1(s))\, ds/100\right)\right], \\[4mm]
c((u_1, u_2), (v_1, v_2), (w_1, w_2)) = \left(\dfrac{[v_2(1)]}{1 + |[v_2(1)]|}, \dfrac{[v_1(1)]}{1 + |[v_1(1)]|}\right), \\[4mm]
d((u_1, u_2), (v_1, v_2), (w_1, w_2)) = \left(\dfrac{[u_2(1)]}{1 + |[u_2(1)]|}, \dfrac{[u_1(1)]}{1 + |[u_1(1)]|}\right),
\end{cases}
$$

$$(6.129)$$

where again $s \mapsto [s]$ denotes the integer function. The hypotheses (f$_0$), (f$_1$),
(c$_1$), and (d$_1$) hold when $h_\pm(t) = (\frac{d}{dt}(t \sin \frac{1}{t}) \pm 10, \frac{d}{dt}(t \cos \frac{1}{t}) \pm 16)$, $\lambda = 0$, and
$c_\pm = d_\pm = (\pm 1, \pm 1)$. Thus (6.128) has the smallest and greatest solutions.
The first components of the functions x_- and x_+ defined by (6.124) are

$$
u_-(t) = \begin{cases}
-1 + 2\sqrt{t} - \frac{2t\sqrt{t}}{3} \sin \frac{1}{t} - \frac{4\sqrt{t}}{3} \cos \frac{1}{t} - \frac{4\sqrt{2\pi}}{3} FrS\left(\frac{\sqrt{2}}{\sqrt{t\pi}}\right) + \frac{20t\sqrt{t}}{3} \\[2mm]
-22\sqrt{3} + 2\sqrt{3} \sin \frac{1}{3} + \frac{4\sqrt{3}}{3} \cos \frac{1}{3} + \frac{4\sqrt{2\pi}}{3} FrS\left(\frac{\sqrt{6}}{3\sqrt{\pi}}\right),
\end{cases}
$$

$$v_-(t) = \begin{cases} -1 + 2\sqrt{t} - \frac{2t\sqrt{t}}{3}\cos\frac{1}{t} + \frac{4\sqrt{t}}{3}\sin\frac{1}{t} - \frac{4\sqrt{2\pi}}{3}FrC\left(\frac{\sqrt{2}}{\sqrt{t\pi}}\right) + \frac{32t\sqrt{t}}{3} \\ -34\sqrt{3} + 2\sqrt{3}\cos\frac{1}{3} - \frac{4\sqrt{3}}{3}\sin\frac{1}{3} + \frac{4\sqrt{2\pi}}{3}FrC\left(\frac{\sqrt{6}}{3\sqrt{\pi}}\right), \end{cases}$$

$$u_+(t) = \begin{cases} 1 - 2\sqrt{t} - \frac{2t\sqrt{t}}{3}\sin\frac{1}{t} - \frac{4\sqrt{t}}{3}\cos\frac{1}{t} - \frac{4\sqrt{2\pi}}{3}FrS\left(\frac{\sqrt{2}}{\sqrt{t\pi}}\right) - \frac{20t\sqrt{t}}{3} \\ +22\sqrt{3} + 2\sqrt{3}\sin\frac{1}{3} + \frac{4\sqrt{3}}{3}\cos\frac{1}{3} + \frac{4\sqrt{2\pi}}{3}FrS\left(\frac{\sqrt{6}}{3\sqrt{\pi}}\right), \end{cases}$$

$$v_+(t) = \begin{cases} 1 - 2\sqrt{t} - \frac{2t\sqrt{t}}{3}\cos\frac{1}{t} + \frac{4\sqrt{t}}{3}\sin\frac{1}{t} - \frac{4\sqrt{2\pi}}{3}FrC\left(\frac{\sqrt{2}}{\sqrt{t\pi}}\right) - \frac{32t\sqrt{t}}{3} \\ +34\sqrt{3} + 2\sqrt{3}\cos\frac{1}{3} - \frac{4\sqrt{3}}{3}\sin\frac{1}{3} + \frac{4\sqrt{2\pi}}{3}FrC\left(\frac{\sqrt{6}}{3\sqrt{\pi}}\right), \end{cases}$$

where

$$FrS(x) = \int_0^x \sin\left(\frac{\pi}{2}t^2\right)dt \quad \text{and} \quad FrC(x) = \int_0^x \cos\left(\frac{\pi}{2}t^2\right)dt$$

are the *Fresnel sine and cosine integrals.*

According to Lemma 6.58 the smallest solution of (6.128) is equal to the first component of the smallest fixed point of $G = (G_1, G_2, G_3)$, defined by (6.125), with f, c, and d given by (6.129) and $p(t) = \sqrt{t}$. Calculating the first iterations $G^n x_-$ it turns out that $G^6 x_- = G^7 x_-$. Thus $G_1^6 x_-$ is the smallest solution of (6.128). Similarly, one can show that $G^4 x_+ = G^5 x_+$, whence $G_1^4 x_+$ is the greatest solution of (6.128). The exact expressions of the components of these solutions are

$$u_1^*(t) = \begin{cases} \frac{15}{16} - \frac{12\sqrt{t}}{7} - \frac{2t\sqrt{t}}{3}\sin\frac{1}{t} - \frac{4\sqrt{t}}{3}\cos\frac{1}{t} - \frac{4\sqrt{2\pi}}{3}FrS\left(\frac{\sqrt{2}}{\sqrt{t\pi}}\right) - \frac{10t\sqrt{t}}{3} \\ +\frac{82}{7}\sqrt{3} + 2\sqrt{3}\sin\frac{1}{3} + \frac{4\sqrt{3}}{3}\cos\frac{1}{3} + \frac{4\sqrt{2\pi}}{3}FrS\left(\frac{\sqrt{6}}{3\sqrt{\pi}}\right), \end{cases}$$

$$u_2^*(t) = \begin{cases} \frac{17}{18} - \frac{7\sqrt{t}}{4} - \frac{2t\sqrt{t}}{3}\cos\frac{1}{t} + \frac{4\sqrt{t}}{3}\sin\frac{1}{t} - \frac{4\sqrt{2\pi}}{3}FrC\left(\frac{\sqrt{2}}{\sqrt{t\pi}}\right) - \frac{8t\sqrt{t}}{3} \\ +\frac{39}{4}\sqrt{3} + 2\sqrt{3}\cos\frac{1}{3} - \frac{4\sqrt{3}}{3}\sin\frac{1}{3} + \frac{4\sqrt{2\pi}}{3}FrC\left(\frac{\sqrt{6}}{3\sqrt{\pi}}\right), \end{cases}$$

$$u_{1*}(t) = \begin{cases} -\frac{14}{15} + \frac{5\sqrt{t}}{3} - \frac{2t\sqrt{t}}{3}\sin\frac{1}{t} - \frac{4\sqrt{t}}{3}\cos\frac{1}{t} - \frac{4\sqrt{2\pi}}{3}FrS\left(\frac{\sqrt{2}}{\sqrt{t\pi}}\right) + \frac{10t\sqrt{t}}{3} \\ -\frac{35}{3}\sqrt{3} + 2\sqrt{3}\sin\frac{1}{3} + \frac{4\sqrt{3}}{3}\cos\frac{1}{3} + \frac{4\sqrt{2\pi}}{3}FrS\left(\frac{\sqrt{6}}{3\sqrt{\pi}}\right), \end{cases}$$

$$u_{2*}(t) = \begin{cases} -\frac{15}{16} + \frac{8\sqrt{t}}{5} - \frac{2t\sqrt{t}}{3}\cos\frac{1}{t} + \frac{4\sqrt{t}}{3}\sin\frac{1}{t} - \frac{4\sqrt{2\pi}}{3}FrC\left(\frac{\sqrt{2}}{\sqrt{t\pi}}\right) + \frac{10t\sqrt{t}}{3} \\ -\frac{58}{5}\sqrt{3} + 2\sqrt{3}\cos\frac{1}{3} - \frac{4\sqrt{3}}{3}\sin\frac{1}{3} + \frac{4\sqrt{2\pi}}{3}FrC\left(\frac{\sqrt{6}}{3\sqrt{\pi}}\right). \end{cases}$$

Example 6.62. Let E be the space (c_0), ordered coordinatewise and normed by the sup-norm. The mappings $h_\pm : [0,1] \to c_0$, defined by $h_{pm}(0) = h_\pm(1) = 0$,

$$h_\pm(t) = \left(\frac{1}{nt}\sin\frac{1}{t} + \frac{1}{n}\cos\frac{1}{t} + \frac{1}{n\sqrt{1-t}}\sin\left(\frac{1}{1-t}\right) \pm \frac{1}{n}\right)_{n=1}^\infty, \quad t \in (0,1),$$

$$(6.130)$$

belong to $HL([0,1], c_0)$ by Lemma 1.12. Thus these mappings are possible upper and lower boundaries for f in the hypothesis (f_0) of Theorem 6.59 and for g in the hypothesis (g_0) of Corollary 6.60 when $E = c_0$. Choosing $c = (n^{-1})_{n=1}^{\infty}$, $d = (n^{-1})_{n=1}^{\infty}$, and $p(t) \equiv 1$, the solutions of the boundary value problems

$$\begin{cases} -u''(t)) = h_{\pm}(t) \text{ for a.e. } t \in [0,3], \\ \lim_{t \to 0+} u'(t) = c, \ \lim_{t \to 1-} u(t) = d \end{cases} \tag{6.131}$$

are $u_+(t) = (\frac{1}{n} v(t))_{n=1}^{\infty}$ and $u_-(t) = (\frac{1}{n}(v(t) + t^2 - 1))_{n=1}^{\infty}$, where

$$\begin{cases} v(t) = -\dfrac{t^2}{2} \cos\left(\dfrac{1}{t}\right) + \dfrac{t}{2} \sin\left(\dfrac{1}{t}\right) - \dfrac{1}{2} Ci\left(\dfrac{1}{t}\right) - \dfrac{4}{3}\sqrt{1-t}\sin\left(\dfrac{1}{t-1}\right) \\ \quad + \dfrac{4t}{3}\sqrt{1-t}\sin\left(\dfrac{1}{t-1}\right) - \dfrac{4}{3}\sqrt{1-t}\cos\left(\dfrac{1}{t-1}\right) \\ \quad - \dfrac{4}{3}\sqrt{2\pi} FrS\left(\dfrac{\sqrt{2}}{\sqrt{\pi(1-t)}}\right) - 2\sqrt{2\pi} FrC\left(\dfrac{\sqrt{2}}{\sqrt{\pi(1-t)}}\right) \\ \quad + 2t\sqrt{2\pi} FrC\left(\dfrac{\sqrt{2}}{\sqrt{\pi(1-t)}}\right) - \dfrac{t^2}{2} + t + 2t\sin 1 - 2FrC\left(\dfrac{\sqrt{2}}{\sqrt{\pi}}\right)\sqrt{2\pi t} \\ \quad - \dfrac{1}{2} + \dfrac{1}{2}\cos 1 - \dfrac{5}{2}\sin 1 + \dfrac{1}{2} Ci(1) + \dfrac{2}{3}\sqrt{2\pi} + 2FrC\left(\dfrac{\sqrt{2}}{\sqrt{\pi}}\right)\sqrt{2\pi}, \end{cases}$$

and

$$Ci(x) = \int_0^x \frac{\cos t - 1}{t}\, dt + \gamma + \ln x, \quad \gamma = \lim_{n \to \infty}\left(\sum_{i=1}^n \frac{1}{i} - \ln n\right)$$

is the **cosine integral**.

Consider, in particular, the infinite system of boundary value problems

$$\begin{cases} -u_n''(t) = \dfrac{1}{n}\left(\dfrac{1}{t}\sin\dfrac{1}{t} + \cos\dfrac{1}{t} + \dfrac{1}{n\sqrt{1-t}}\sin\left(\dfrac{1}{1-t}\right) + f_n(u, u', u'')\right) \\ \text{a.e. on } [0,1], \ \lim_{t \to 0+} u_n'(t)) = \dfrac{1}{n}, \ \lim_{t \to 1-} u_n(t) = \dfrac{1}{n}, \ n \in \mathbb{N}. \end{cases}$$
$$\tag{6.132}$$

Set $u = (u_n)_{n=1}^{\infty}$, and suppose that each function $f_n : HL([0,1], c_0)^3 \to \mathbb{R}$ is increasing with respect the first and third argument, and decreasing with respect to the second argument. Further let $-1 \leq f_n(u, v, w) \leq 1$ be satisfied for all u, v, $w \in HL_{loc}((0,1), c_0)$, and $n \in \mathbb{N}$. Then (6.132) has the smallest and greatest solutions $u_* = (u_{*n})_{n=1}^{\infty}$ and $u^* = (u_n^*)_{n=1}^{\infty}$, respectively, and they belong to the order interval $[u_-, u_+]$, where $u_{\pm}$ are given above.

Remark 6.63. Examples of ordered Banach spaces whose order cones are regular are given in Remark 6.43. In particular, we can choose E to be one of these spaces in the above considerations.

Problems of the form (6.79), (6.102), and (6.117) include many kinds of special types, and may be, e.g.,

- singular, because a case $\lim_{t\to a+} p(t) = 0$ is allowed, and since the limits $\lim_{t\to a+} f(t, u, v)$ and/or $\lim_{t\to b-} f(t, u, v)$ need not exist;
- functional, because the functions c, d, and f may depend functionally on u, u', and/or Lu;
- discontinuous, because the dependencies of c, d, and f on u, u', and/or Lu can be discontinuous;
- a finite system when $E = \mathbb{R}^m$;
- an infinite system when E is l^p or c_0-space;
- of random type when $E = L^p(\Omega)$ and Ω is a probability space.

The solutions of the above examples have been calculated by using simple Maple programming.

6.5 Functional Differential Equations Containing Bochner Integrable Functions

In this subsection we apply Theorem 2.26 to derive existence and comparison results for solutions of first order implicit functional differential equations in an ordered Banach space $E = (E, \|\cdot\|, \leq)$ that has the following properties.

(E0) Bounded and monotone sequences of E have weak limits.
(E1) E is lattice-ordered and $\|x^+\| \leq \|x\|$ for all $x \in E$, where $x^+ = \sup\{0, x\}$.
(E2) The mapping $E \ni x \to x^+$ is continuous.

We shall first study solvability of the implicit functional problem

$$\begin{cases} \dfrac{d}{dt}(\varphi(t)u(t)) = p(t)u(t) + f\left(t, u, u(t), \dfrac{d}{dt}(\varphi(t)u(t)) - p(t)u(t)\right) \text{ a.e. on } J, \\ u(t) = B(t, u, u(t)) \text{ in } J_0, \end{cases}$$
(6.133)

where $J = [a, b]$ and $J_0 = [a - r, a]$, $a \leq b$, $r \geq 0$, $\varphi \in C(J, (0, \infty))$, $f : J \times X \times E \times E \to E$, and $B : J_0 \times X \times E \to E$, $X = C([a - r, b], E)$. We study also dependence of the solutions of (6.133) on the data f and B. As a special case we get existence and comparison results for implicit initial value problems when $r = 0$ and implicit functional equations when $b = a$.

Solutions are assumed to be in the set

$$W = \{u \in C([a - r, b], E) : \varphi \cdot u | J \in W^{1,1}(J, E)\},$$
(6.134)

where

$$W^{1,1}(J, E) = \{v : J \to E : v \text{ is absolutely continuous and a.e. differentiable.}\}$$

6.5.1 Hypotheses and Preliminaries

Assuming that the spaces $X = C([a-r, b], E)$ and $C(J_0, E)$ are equipped with pointwise ordering, and $L^1(J, E)$ with a.e. pointwise ordering, we impose the following hypotheses on the functions f and B.

(f_1) $f(\cdot, u, u(\cdot), v(\cdot))$ is strongly measurable whenever $u \in X$ and $v \in L^1(J, E)$.

(f_2) $f(t, u, x, y)$ is increasing in u, x, and y for a.e. $t \in J$.

(f_3) $\|f(t, u, x, y)\| \le h_1(t) + p_1(t)\|x\| + \lambda_1\|y\|$ for a.e. $t \in J$ and all $u \in X$ and $x, y \in E$, where $p_1, h_1 \in L^1(J, \mathbb{R}_+)$ and $\lambda_1 \in [0, 1)$.

(B_1) The chains of the set $\{t \mapsto B(t, u, u(t)) : u \in X\}$ are equicontinuous.

(B_2) $B(t, u, x)$ is increasing in u and x for a.e. $t \in J_0$.

(B_3) $\|B(t, u, x)\| \le \psi(t, \|x\|)$ for all $u \in X$, $x \in E$, and $t \in J_0$, where $\psi \in C(J_0 \times \mathbb{R}_+, \mathbb{R}_+)$, $\psi(t, \cdot)$ is increasing for all $t \in J_0$, and there exists a function $R \in C(J_0, \mathbb{R}_+)$ such that $R(\cdot) = \psi(\cdot, R(\cdot))$, and $r \le R$ whenever $r \in C(J_0, \mathbb{R}_+)$ and $r(\cdot) \le \psi(\cdot, r(\cdot))$.

In our considerations we need the following existence, uniqueness, and comparison result.

Lemma 6.64. *If $p \in L^1(J, \mathbb{R}_+)$, $h \in L^1(J, E)$, and $\alpha \in C(J_0, E)$, then the linear initial function problem*

$$\frac{d}{dt}(\varphi(t)u(t)) = p(t)u(t) + h(t) \quad \text{a.e. on} \quad J, \quad u(t) = \alpha(t) \quad \text{in} \quad J_0 \quad (6.135)$$

has a unique solution $u = \phi(h, \alpha)$ in W that is given by

$$u(t) = \begin{cases} \alpha(t), & t \in J_0, \\ \dfrac{e^{\int_a^t \frac{p(s)}{\varphi(s)} ds}}{\varphi(t)} \left(\varphi(a)\alpha(a) + \int_a^t e^{-\int_a^s \frac{p(\tau)}{\varphi(\tau)} d\tau} h(s) \right) ds, & t \in J. \end{cases}$$

Moreover, u is increasing with respect to h and α.

In case $E = \mathbb{R}$, Lemma 6.64 implies that if the functions h_1, p_1, and R and the constant λ_1 are as in the hypotheses (f_3) and (B_3), then problem

$$\begin{cases} \dfrac{d}{dt}(\varphi(t)w(t)) = \dfrac{h_1(t) + p_1(t)w(t)}{(1 - \lambda_1)} + p(t)w(t) \quad \text{a.e. on} \quad J, \\ w(t) = R(t) \quad \text{in} \quad J_0, \end{cases} \quad (6.136)$$

has a unique solution w. Denote

$$Y = L^1(J, E) \times C(J_0, E), \quad q(t) = \frac{h_1(t) + p_1(t)w(t)}{1 - \lambda_1}, \quad t \in J, \quad (6.137)$$

$$P = \{(h, \alpha) \in Y : \|h(t)\| \le q(t) \text{ a.e. on } J, \|\alpha(t)\| \le R(t) \text{ in } J_0\}, \quad (6.138)$$

and define a partial ordering on Y by

$$(h, \alpha) \le (\hat{h}, \hat{\alpha}) \quad \text{iff } h(t) \le \hat{h}(t) \text{ a.e. on } J \text{ and } \alpha(t) \le \hat{\alpha}(t) \text{ in } J_0. \quad (6.139)$$

Lemma 6.65. *The following equations*

$$\begin{cases} L_1u(t) := \frac{d}{dt}(\varphi(t)u(t)) - p(t)u(t), & t \in J, \\ L_2u(t) := u(t), & t \in J_0, \end{cases} \qquad (6.140)$$

define mappings $L_1 : W \to L^1(J, E)$ and $L_2 : W \to C(J_0, E)$. Setting

$$Lu := (L_1u, L_2u), \quad u \in V := \phi[P], \qquad (6.141)$$

where $\phi : Y \to W$ is as in Lemma 6.64, we obtain a bijection $L : V \to P$, whose inverse is increasing. Moreover, $(0,0)$ is an order-center of P.

Proof: By Lemma 6.64, ϕ assigns to each $(h, \alpha) \in P$ a unique solution $u = \phi(h, \alpha)$ of the IFP (6.135) in $V = \phi[P] \subseteq W$, i.e., $L_1u = h$, $L_2u = \alpha$. Thus the mapping L, defined by (6.141) is surjective. Moreover, since ϕ is increasing, then $u \leq \hat{u}$ in V whenever $Lu \leq L\hat{u}$ in P. This implies that L is also injective, and that its inverse is increasing.

The proofs of Propositions 9.49 and 9.50 imply that $(h^+, \alpha^+) \in P$ whenever $(h, \alpha) \in P$. Moreover, it is easy to see that $(h^+, \alpha^+) = \sup\{(0,0), (h, \alpha)\}$ with respect to the partial ordering of P defined in (6.138). Consequently, $c = (0, 0)$ is a sup-center of P. Similarly one can show that $(0, 0)$ is an inf-center of P. $\qquad\square$

Lemma 6.66. *Let w be the solution of (6.136). If $u \in V$, then $\|u(t)\| \leq w(t)$, for each $t \in J_0 \cup J$.*

Proof: Let $u \in V$ be given. Since $Lu = (h, \alpha) \in P$ by Lemma 6.65, we get by applying (6.138), (6.140), and (6.141) that for each $t \in J_0$,

$$\|u(t)\| = \|\alpha(t)\| \leq R(t) = w(t).$$

Moreover, it follows from (6.138), (6.139), and (6.141) that

$$\|L_1u(t)\| = \|h(t)\| \leq \frac{h_1(t) + p_1(t)w(t)}{1 - \lambda_1} \quad \text{for a.e. } t \in J.$$

This inequality and the definition (6.140) of $L_1u(t)$ imply that

$$\left\|\frac{d}{dt}(\varphi(t)u(t))\right\| \leq \|L_1u(t)\| + \|p(t)u(t)\| \leq \frac{h_1(t) + p_1(t)w(t)}{1 - \lambda_1} + p(t)\|u(t)\|.$$

Noticing also that $\|u(a)\| \leq w(a)$, we get for $t \in J$

$$\begin{aligned} \|\varphi(t)u(t)\| &\leq \|\varphi(a)u(a)\| + \int_a^t \left\|\frac{d}{ds}(\varphi(s)u(s))\right\| ds \\ &\leq \varphi(a)w(a) + \int_a^t \left(\frac{h_1(s) + p_1(s)w(s)}{(1 - \lambda_1)} + p(s)\|u(s)\|\right) ds. \end{aligned} \qquad (6.142)$$

In view of (6.136), we have for each $t \in J$,

$$\varphi(t)w(t) = \varphi(a)w(a) + \int_a^t \left(\frac{h_1(s) + p_1(s)w(s)}{(1-\lambda_1)} + p(s)w(s) \right) ds.$$

It follows from this equality and from inequality (6.142) that

$$v(t) := \int_a^t p(s)(w(s) - \|u(s)\|)ds \leq \varphi(t)(w(t) - \|u(t)\|), \quad t \in J, \quad (6.143)$$

so that

$$v'(t) = p(t)(w(t) - \|u(t)\|) \geq \frac{p(t)}{\varphi(t)} v(t) \quad \text{a.e. on } J, \quad v(a) = 0.$$

This implies that for each $t \in J$,

$$\int_a^t e^{-\int_a^s \frac{p(\tau)}{\varphi(\tau)} d\tau} (v'(s) - \frac{p(s)}{\varphi(s)} v(s)) ds = e^{-\int_a^t \frac{p(\tau)}{\varphi(\tau)} d\tau} v(t) \geq 0.$$

Applying the last inequality and (6.143) we obtain

$$0 \leq v(t) \leq \varphi(t)(w(t) - \|u(t)\|), \quad \text{i.e., } \|u(t)\| \leq w(t) \text{ on } J.$$

The above proof shows that $\|u(t)\| \leq w(t)$ for each $t \in J_0 \cup J$.   □

Lemma 6.67. *Let the hypotheses (f_1)–(f_3) and (B_1)–(B_3) be satisfied. Relations*

$$\begin{cases} Nu := (N_1 u, N_2 u), \quad u \in V = \phi[P], \quad where \\ N_1 u(t) := f(t, u, u(t), L_1 u(t)), \quad t \in J, \\ N_2 u(t) := B(t, u, L_2 u(t)), \quad t \in J_0, \end{cases} \quad (6.144)$$

define an operator $N : V \to P$. Moreover, N is increasing with respect to the product ordering (6.139) of P and the graph ordering $\preceq$ of V, defined by

$$u \preceq v \quad \text{iff } u \leq v \text{ and } Lu \leq Lv. \quad (6.145)$$

Proof: The given hypotheses imply that (6.144) defines a mapping $N : V \to Y$. To prove that $N[V] \subseteq P$, let $u \in V$ be given. Applying the inequality $\|u(t)\| \leq w(t)$ on $J_0 \cup J$, proved in Lemma 6.66, definitions (6.137) and (6.144), and conditions (f_3) and (B_3), we get

$$\|N_1 u(t)\| = \|f(t, u, u(t), L_1 u(t))\| \leq h_1(t) + p_1(t)\|u(t)\| + \lambda_1 \|L_1 u(t)\|$$
$$\leq h_1(t) + p_1(t)w(t) + \lambda_1 q(t) = q(t), \quad \text{a.e. on } J,$$

and

$$\|N_2 u(t)\| = \|B(t, u, L_2 u(t))\| \leq \psi(t, \|L_2 u(t)\|) \leq \psi(t, R(t)) = R(t), \quad t \in J_0.$$

These inequalities and the definition (6.138) of P imply that $Nu = (N_1 u, N_2 u)$ belongs to P. To prove that N is increasing, assume that $u \preceq \hat{u}$, i.e., $u \leq \hat{u}$

and $Lu \leq L\hat{u}$. These inequalities, (6.144) and the hypotheses (f_2) and (B_2) yield

$$Nu = (f(\cdot, u, u(\cdot)), L_1 u(\cdot)), B(\cdot, u, L_2 u(\cdot)))$$
$$\leq (f(\cdot, \hat{u}, \hat{u}(\cdot)), L_1 \hat{u}(\cdot)), B(\cdot, \hat{u}, L_2 \hat{u}(\cdot))) = N\hat{u},$$

which proves that N is increasing.   □

The next result implies that the implicit functional problem (IFP) (6.133) has the same solutions as the operator equation $Lu = Nu$.

Lemma 6.68. *Let the hypotheses (f_1)–(f_3) and (B_1)–(B_3) be fulfilled. Then $u \in W$ is a solution of the IFP (6.133) if and only if $u \in V$ and u is a solution of the operator equation $Lu = Nu$.*

Proof: Assume that $u \in W$ is a solution of the IFP (6.133). Then, by first equations of (6.133) and (6.140), we get for a.e. $t \in J$

$$\|L_1 u(t)\| = \|f(t, u, u(t), L_1 u(t))\| \leq h_1(t) + p_1(t)\|u(t)\| + \lambda_1 \|L_1 u(t)\|,$$

whence

$$\|L_1 u(t)\| = \|f(t, u, u(t), L_1 u(t))\| \leq \frac{h_1(t) + p_1(t)\|u(t)\|}{1 - \lambda_1}. \tag{6.146}$$

This inequality and the definition (6.140) of $L_1 u(t)$ result in

$$\|\frac{d}{dt}(\varphi(t)u(t))\| \leq \|L_1 u(t)\| + \|p(t)u(t)\| \leq \frac{h_1(t) + p_1(t)\|u(t)\|}{1 - \lambda_1} + p(t)\|u(t)\|.$$

Noticing also that $\|u(a)\| \leq w(a)$, we get for $t \in J$

$$\|\varphi(t)u(t)\| \leq \|\varphi(a)u(a)\| + \int_a^t \|\frac{d}{ds}(\varphi(s)u(s))\|ds$$
$$\leq \varphi(a)w(a) + \int_a^t \left(\frac{h_1(s) + p_1(s)\|u(s)\|}{(1 - \lambda_1)} + p(s)\|u(s)\| \right) ds. \tag{6.147}$$

In view of (6.136) we have for each $t \in J$,

$$\varphi(t)w(t) = \varphi(a)w(a) + \int_a^t \left(\frac{h_1(s) + p_1(s)w(s)}{(1 - \lambda_1)} + p(s)w(s) \right) ds.$$

From the last equality and from inequality (6.147), it follows that

$$v(t) := \int_a^t \left(\frac{p_1(s)}{1 - \lambda_1} + p(s) \right) (w(s) - \|u(s)\|)ds \leq \varphi(t)(w(t) - \|u(t)\|), \tag{6.148}$$

so that $v(a) = 0$, and for a.e. $t \in J$

$$v'(t) = \left(\frac{p_1(t)}{1-\lambda_1} + p(t) \right) (w(t) - \|u(t)\|) \geq \frac{\left(\frac{p_1(t)}{1-\lambda_1} + p(t) \right)}{\varphi(t)} v(t).$$

This implies that for each $t \in J$,

$$\int_a^t e^{-\int_a^s \left(\frac{p_1(\tau)}{1-\lambda_1} + p(\tau) \right) \frac{d\tau}{\varphi(\tau)}} \left(v'(s) - \left(\frac{p_1(s)}{1-\lambda_1} + p(s) \right) \frac{v(s)}{\varphi(s)} \right) ds$$

$$= e^{-\int_a^t \left(\frac{p_1(\tau)}{1-\lambda_1} + p(\tau) \right) \frac{d\tau}{\varphi(\tau)}} v(t) \geq 0.$$

By means of the last inequality and (6.148) we obtain

$$0 \leq v(t) \leq \varphi(t)(w(t) - \|u(t)\|), \quad \text{i.e., } \|u(t)\| \leq w(t) \text{ on } J.$$

The above proof shows that $\|u(t)\| \leq w(t)$ for each $t \in J_0 \cup J$. This result and (6.146) imply that

$$\|L_1 u(t)\| = \|f(t, u, u(t), L_1 u(t))\| \leq \frac{h_1(t) + p_1(t)w(t)}{1-\lambda_1} = q(t) \text{ for a.e. } t \in J.$$

Moreover, it follows from (6.133), (6.140), and (B$_3$) that for each $t \in J_0$,

$$\|L_2 u(t)\| = \|B(t, u, L_2 u(t))\| \leq \psi(t, \|L_2 u(t)\|),$$

which by means of hypothesis (B$_3$) yields

$$\|L_2 u(t)\| \leq R(t), \quad t \in J_0.$$

By the above proof and the definition (6.138) of P we infer that $Lu = (L_1 u, L_2 u) \in P$, whence $u \in V$ in view of the definition of V. It then follows that u is a solution of the operator equation $Lu = Nu$. Conversely, if $u \in V$ and $Lu = Nu$, it follows from (6.133), (6.140), (6.141), and (6.144) that $u \in W$, and u is a solution of the IFP (6.133).   □

6.5.2 Existence and Comparison Results

Applying the results of Sect. 6.5.1 and Theorem 2.26 we shall prove an existence and comparison result for the initial function problem (6.133).

Theorem 6.69. *Assume that mappings f and B satisfy the hypotheses (f_1)–(f_3) and (B_1)–(B_3), and let $\preceq$ be defined by (6.139). Then the IFP (6.133) has*

(a) *minimal and maximal solutions in $(W, \preceq)$;*
(b) *the smallest and greatest solutions u_* and u^* within the order interval $[\underline{u}, \overline{u}]$ of $(W, \preceq)$, where $\overline{u}$ is the smallest solution of the IFP*

$$\begin{cases} L_1 u(t) = f(t, u, u(t), L_1 u(t))^+ & \text{a.e. on } J = [a, b], \\ L_2 u(t) = B(t, u, L_2 u(t))^+ & \text{in } J_0 = [a - r, a], \end{cases} \quad (6.149)$$

and $\underline{u}$ is the greatest solution of the IFP

$$\begin{cases} L_1 u(t) = -(-f(t, u, u(t), L_1 u(t)))^+ & \text{a.e. on } J = [a, b], \\ L_2 u(t) = -(-B(t, u, L_2 u(t)))^+ & \text{in } J_0 = [a - r, a]. \end{cases} \quad (6.150)$$

Moreover, both u_* and u^* are increasing with respect to f and B.

Proof: According to Lemmas 6.65 and 6.67, the relations (6.138), (6.141), and (6.144) define function spaces $V \subseteq W$ and $P \subset Y = L^1(J, E) \times C(J_0, E)$ and mappings $L, N : V \to P$, which satisfy the hypotheses

(L1) L is bijective, and $L^{-1} : (P, \leq) \to (V, \leq)$ is increasing.
(N) $N : (V, \preceq) \to (P, \leq)$ is increasing.

In order to apply Theorem 2.26 we have to show that chains of $N[V]$ have supremums and infimums in P. Let C be a well-ordered subset of $N[V]$. Since $N[V] \subseteq P \subseteq L^1(J, E) \times C(J_0, E)$, then the projections

$$C_1 = \{f(\cdot, u, u(\cdot), L_1 u(\cdot)) : u \in L^{-1}[C]\},$$
$$C_2 = \{B(\cdot, u, L_2 u(\cdot)) : u \in L^{-1}[C]\}$$

of C into $L^1(J, E)$ and $C(J_0, E)$, respectively, are well-ordered. Since $C \subset P$, then C_1 is a well-ordered chain of $L^1(J, E)$, which is a.e. pointwise bounded by an L^1-function q, given by (6.138). Thus $y = \sup C_1$ exists by Lemma 9.32. C_2 is a bounded and well-ordered subset of $C(J_0, E)$, and also equicontinuous by hypothesis (B_1), whence $b = \sup C_2$ exists by Proposition 9.41. It is easy to see that

$$(y, b) = (\sup C_1, \sup C_2) = \sup C.$$

Moreover, the above cited results imply the existence of increasing sequences $(y_n^j)_{n=0}^\infty$ in C_j such that

$$y_n^1(t) \rightharpoonup y(t) \text{ for a.e. } t \in J, \qquad y_n^2(t) \rightharpoonup b(t), \quad t \in J_0,$$

which along with (6.138) yields

$$\|y(t)\| \leq \liminf_{n \to \infty} \|y_n^1(t)\| \leq q(t) \text{ a.e. on } J,$$

$$\|b(t)\| \leq \liminf_{n \to \infty} \|y_n^2(t)\| \leq R(t), \quad t \in J_0.$$

The last inequalities and (6.138) imply that $\sup C = (y, b) \in P$. Similarly, it can be shown that each inversely well-ordered subset of $N[V]$ has the infimum in P. Because $(0, 0)$ is by Lemma 6.65 an order center of P, and since N is increasing in $(V, \preceq)$, the operator equation $Lu = Nu$ has by Theorem 2.26 minimal and maximal solutions in $(V, \preceq)$, as well as the smallest and greatest solutions u_* and u^* within the order interval $[\underline{u}, \overline{u}]$ of $(V, \preceq)$, where $\overline{u}$ is the smallest solution of $Lu = (Nu)^+$ and $\underline{u}$ is the smallest solution of $Lu = -(-Nu)^+$. These results imply the assertions (a) and (b), because by Lemma

6.68 the IFP (6.133) has the same solutions as the operator equation $Lu = Nu$, and the same holds also for problems (6.149) and (6.150) and corresponding operator equations $Lu = (Nu)^+$ and $Lu = -(-Nu)^+$. Moreover, both u_* and u^* are increasing with respect to N, which implies by (6.144) that they are increasing with respect to f and B.   □

The hypothesis (f_2) of Theorem 6.69 can be generalized as follows.

Proposition 6.70. *The results of Theorem 6.69 remain true if condition (f_2) is replaced by the following hypothesis*

(f_2') *There exists a constant $\beta \geq 0$ such that $f(t, u, x, y) + \beta y$ is increasing in u, x, and y for a.e. $t \in J$.*

Proof: Consider the problem

$$\frac{d}{dt}(\varphi(t)u(t)) = p(t)u(t)) + \hat{f}(t, u, u(t), \frac{d}{dt}(\varphi(t)u(t)) - p(t)u(t))) \quad \text{a.e. on } J,$$

$$u(t) = B(t, u, u(t)) \qquad \text{in } J_0,$$
$$(6.151)$$

where $\hat{f} : J \times X \times E \times E \to E$ is defined by

$$\hat{f}(t, u, x, y) = \frac{f(t, u, x, y) + \beta y}{1 + \beta}. \qquad (6.152)$$

The function $\hat{f}$ satisfies the hypotheses (f_1)–(f_3) with λ_1 replaced by $\frac{\lambda_1 + \beta}{1 + \beta}$. Thus the results of Theorem 6.69 hold for the IFP (6.151) with $\hat{f}$ defined by (6.152). In view of (6.151) and (6.152), the solutions of (6.151) are the same as those of (6.133). This implies the assertions because f increases if and only if $\hat{f}$ increases.   □

As a special case of Theorem 6.69 we obtain the following result.

Proposition 6.71. *Assume that $g : J \times E \to E$ satisfies the following hypotheses.*

(g_1) *$g(\cdot, u(\cdot))$ is strongly measurable whenever $u \in C(J, E)$.*
(g_2) *$g(t, x)$ is increasing in x for a.e. $t \in J$.*
(g_3) *$\|g(t, x)\| \leq p(t)\|x\|$ for a.e. $t \in J$ and all $x \in E$, where $p \in L^1(J, \mathbb{R}_+)$.*

Assume that $p \in L^1(J, \mathbb{R}_+)$, $h \in L^1(J, E)$, and $\alpha \in C(J_0, E)$, and let $\preceq$ be defined by (6.139). Then the initial function problem

$$\frac{d}{dt}(\varphi(t)u(t)) = g(t, u(t)) + h(t) \quad \text{a.e. on } J, \quad u(t) = \alpha(t) \text{ in } J_0 \quad (6.153)$$

has

(a) minimal and maximal solutions in $(W, \preceq)$;

(b) the smallest and greatest solutions u_- and u_+ in the order interval $[\underline{u}, \overline{u}]$ of $(W, \preceq)$, where $\overline{u}$ is the smallest solution of the problem

$$\frac{d}{dt}(\varphi(t)u(t)) = (g(t, u(t)) + h(t))^+ \quad a.e. \ on \ J, \quad u(t) = \alpha(t)^+ \quad in \ J_0$$

$$(6.154)$$

and $\underline{u}$ is the greatest solution of the following problem

$$\frac{d}{dt}(\varphi(t)u(t)) = -(-g(t, u(t)) - h(t))^+ \quad a.e. \ on \ J,$$

$$u(t) = -(-\alpha(t))^+ \quad in \ J_0.$$

$$(6.155)$$

Moreover, both u_- and u_+ are increasing with respect to α and h.

The result of Proposition 6.71 is applied in the proof of the following theorem.

Theorem 6.72. *Assume that the functions* $f : J \times X \times E \times E \to E$, $g : J \times E \to E$ *and* $B : J_0 \times X \times E \to E$ *satisfy the hypotheses* (f_1)–(f_3), (g_1)–(g_3), *and* (B_1)–(B_3). *Then the IFP*

$$\begin{cases} \dfrac{d}{dt}(\varphi(t)u(t)) = g(t, u(t)) + f(t, u, u(t), \dfrac{d}{dt}(\varphi(t)u(t)) - g(t, u(t))), \quad a.e. \ on \ J, \\ u(t) = B(t, u, u(t)) \quad in \ J_0. \end{cases}$$

$$(6.156)$$

has solutions in the set W, defined by (6.134).

Proof: The equations

$$\begin{cases} L_1 u(t) := \frac{d}{dt}(\varphi(t)u(t)) - g(t, u(t)), & t \in J, \\ L_2 u(t) := u(t), & t \in J_0, \end{cases}$$

$$(6.157)$$

define mappings $L_1 : W \to L^1(J, E)$ and $L_2 : W \to C(J_0, E)$. Let $\phi_+ : Y \to W$ be the mapping that assigns to each $(h, \alpha) \in Y$ the solution u_+ of the IFP (6.153) determined in Proposition 6.71. Defining

$$Lu := (L_1 u, L_2 u), \quad u \in V_+ := \phi_+[P],$$

$$(6.158)$$

we obtain a bijection $L : V_+ \to P$, whose inverse is increasing due to Proposition 6.71. The hypotheses (g_1)–(g_3) ensure that we can apply the proof of Lemma 6.65 to show that if $u \in V_+$, then $\|u(t)\| \leq w(t)$, for each $t \in J_0 \cup J$, where w is the solution of (6.136). This result and the proof of Lemma 6.67 show that equations (6.144) define a mapping $N : V_+ \to P$, which is increasing when V_+ is ordered by the graph ordering $\preceq$ defined by (6.145) and P is ordered by the product ordering (6.139). As in the proof of Theorem 6.69 one can verify that the chains of $N[V_+]$ have supremums and infimums in P. Thus by Theorem 2.26 the operator equation $Lu = Nu$ has the smallest and greatest solutions u_* and u^* within the order interval $[\underline{u}, \overline{u}]$ of $(V_+, \preceq)$, where

$\overline{u}$ is the smallest solution of $Lu = (Nu)^+$ and $\underline{u}$ is the smallest solution of $Lu = -(-Nu)^+$. In view of (6.144), (6.157), and (6.158), both u_* and u^* are solutions of the IFP (6.156).

Similar results are obtained also when ϕ_+ is replaced above by the mapping $\phi_- : Y \to W$ that assigns to each $[h, \alpha] \in Y$ the solution u_- of the IFP (6.153) determined in Proposition 6.71. □

Example 6.73. Consider the IFP

$$
\begin{cases}
u'(t) = H(u(t) - 3t) + \dfrac{[3u(t-1)]}{2 + |[3u(t-1)]|} + \dfrac{[2u(t) - 4t]}{4} \\
\qquad + \dfrac{[u'(t) - H(u(t) - 3t)]}{2}, \quad \text{a.e. on } J = [0,1], \\
u(t) = \dfrac{1}{2} - \dfrac{[\int_{-1}^{1} u(t)dt]}{1 + |[\int_{-1}^{1} u(t)dt]|} t, \quad \text{in } J_0 = [-1,0],
\end{cases}
\tag{6.159}
$$

where H is the Heaviside function:

$$
H(x) = \begin{cases} 1 & \text{if } x \geq 0, \\ 0 & \text{if } x < 0, \end{cases}
$$

and $[x]$ denotes the greatest integer less than or equal to x, i.e., $x \mapsto [x]$ is the integer function. The IFP (6.147) is of the form (6.156) with $E = \mathbb{R}$, $\varphi(t) \equiv 1$, and

$$
\begin{cases}
f(t, u, x, y) = H(x - 3t) + \dfrac{[3u(t-1)]}{2 + |[3u(t-1)]|} + \dfrac{[2x - 4t]}{4} + \dfrac{[y]}{2}, \\
B(t, u, x) = \dfrac{1}{2} - \dfrac{[\int_{-1}^{1} u(t)dt]}{1 + |[\int_{-1}^{1} u(t)dt]|} t.
\end{cases}
\tag{6.160}
$$

Obviously the hypotheses (f$_1$)–(f$_3$) and (B$_1$)–(B$_3$) are satisfied. Thus by Theorem 6.72, the IFP (6.159) has solutions u^* and u_*. In this case the algorithm (iii) of Corollary 2.27 can be used to approximate the solution u^* of (6.159). Using simple numerical integration methods to calculate the functions u_n of the algorithm (iii) with $c = 0$, we get the following estimate for u^* ($\chi_{[a,b]}$ is the characteristic function of the interval $[a, b]$):

$$
\begin{aligned}
u^*(t) \approx{} & (1 - .5(t+1))\chi_{[-1,0]} + (.5 + 1.5t)\chi_{[0,.333]} \\
& + (.998 + .5(t - .333))\chi_{[.333,.555]} + (1.108 + .25(t - .555))\chi_{[.555,.667]} \\
& + (1.136 + .083(t - .667))\chi_{[.667,.825]} + (1.15 - .667(t - .825)\chi_{[.825,1]}.
\end{aligned}
$$

This approximation yields also $\int_{-1}^{1} u^*(t)\, dt \approx 1.73$. With the help of these estimates and noticing that u is continuous at the discontinuity points of u', one can infer that $H(u(t) - 3t)$ jumps from 1 to 0 at $t = \frac{1}{3}$, that $[3u(t-1)]$ jumps from 2 to 1 at $t = \frac{2}{3}$, that $[2u(t) - 4t]$ jumps from 0 to -1 at $t = \frac{5}{9}$ and from -1 to -2 at $t = \frac{19}{23}$, and that the exact formula for u^* is

$$
u^*(t) = \left(\frac{1}{2} - \frac{1}{2}t\right)\chi_{[-1,0]} + \left(\frac{1}{2} + \frac{3}{2}t\right)\chi_{[0,\frac{1}{3}]} + \left(\frac{5}{6} + \frac{1}{2}t\right)\chi_{[\frac{1}{3},\frac{5}{9}]}
$$
$$
+ \left(\frac{35}{36} + \frac{1}{4}t\right)\chi_{[\frac{5}{9},\frac{2}{3}]} + \left(\frac{13}{12} + \frac{1}{12}t\right)\chi_{[\frac{2}{3},\frac{19}{23}]} + \left(\frac{235}{138} - \frac{2}{3}t\right)\chi_{[\frac{19}{23},1]}.
$$

The algorithm (iv) of Corollary 2.27 can be used to calculate another solution u_* of (6.159). On the basis of the so obtained estimate, one can infer that

$$
u_*(t) = \left(\frac{1}{2} - \frac{1}{2}t\right)\chi_{[-1,0]} + \left(\frac{1}{2} + \frac{3}{2}t\right)\chi_{[0,\frac{1}{3}]} + \left(\frac{5}{6} + \frac{1}{2}t\right)\chi_{[\frac{1}{3},\frac{5}{9}]}
$$
$$
+ \left(\frac{5}{4} - \frac{1}{4}t\right)\chi_{[\frac{5}{9},\frac{2}{3}]} + \left(\frac{49}{36} - \frac{5}{12}t\right)\chi_{[\frac{2}{3},\frac{67}{87}]} + \left(\frac{811}{522} - \frac{2}{3}t\right)\chi_{[\frac{67}{87},\frac{87}{100}]}
$$
$$
+ \left(\frac{103807}{52200} - \frac{7}{6}t\right)\chi_{[\frac{87}{100},\frac{156007}{165300}]} + \left(\frac{4412687}{1983600} - \frac{17}{12}t\right)\chi_{[\frac{156007}{165300},1]}.
$$

Moreover, denoting

$$
A = \left\{(t,x) : t \in \left[\frac{5}{9}, 1\right],\ u_*(t) \le x \le u^*(t)\right\},
$$

it is easy to show that each point of A is a bifurcation point for solutions of (6.159). Thus between u_* and u^* there is a continuum of chaotically behaving solutions of problem (6.159).

6.6 Notes and Comments

In Chap. 6 we have studied discontinuous differential equations in Banach spaces. The material of Sect. 6.1 is adopted from [126]. The results of Sects. 6.2 and 6.3 are new. Section 6.4 is based on [128] as well as on recently submitted manuscripts by the second author jointly with M. Kumpulainen. Problems that include some of the types considered in Sect. 6.4 when $E = \mathbb{R}$ are studied, e.g., in [3, 16, 17, 44, 137, 138, 174, 187, 225]. Initial and boundary value problems in ordered Banach spaces are studied, e.g., in [53, 56, 106, 129, 132, 133]. In Sect. 6.4 no continuity hypotheses are imposed on functions f, c, b, and g. Existence and uniqueness results for first order discontinuous implicit differential equations with or without impulses are derived, e.g., in [43, 113, 114, 130, 173, 208]. The material of Sect. 6.5 is adopted from [117].

7

Banach-Valued Integral Equations

In this chapter we derive existence, uniqueness, extremality, and comparison results for solutions of discontinuous nonlinear integral equations in various spaces of functions with values in an ordered Banach space E.

In Sect. 7.1 we study integral equations of Fredholm and Volterra types. We assume that E is ordered by a regular order cone, which means that order bounded and monotone sequences of E have strong limits. The solutions are assumed to be locally HL integrable functions, except in Sect. 7.1.4, where solutions are Bochner integrable. Integrals that occur in the equations are of Henstock–Kurzweil type.

In Sect. 7.2 we derive existence and comparison results for functional integral equations of Urysohn, Fredholm, and Volterra types in L^p-spaces. The data functions of the considered integral equations may be discontinuous in all their arguments.

Sect. 7.3 deals with the existence of continuous solutions of discontinuous functional evolution type integral equations. Well-posedness results are derived for equations containing HL integrable functions. The obtained results are applied to a Cauchy problem. Furthermore, extremality results are derived for solutions of equations containing HL integrable or Bochner integrable functions.

We also study cases where ordinary iterative methods are applicable and use them in conjunction with symbolic programming to solve concrete problems involving discontinuous nonlinearities that may even allow for functional dependence on the solution. The obtained results are applied to impulsive functional differential equations and to partial differential equations of parabolic type.

S. Carl and S. Heikkilä, *Fixed Point Theory in Ordered Sets and Applications:*
From Differential and Integral Equations to Game Theory,
DOI 10.1007/978-1-4419-7585-0_7, © Springer Science+Business Media, LLC 2011

7.1 Integral Equations in HL-Spaces

7.1.1 Fredholm Integral Equations

In this subsection we apply the following fixed point result, which is a consequence of Propositions 2.16 and 9.39.

Theorem 7.1. *Let E be a Banach space ordered by a regular order cone E_+ and (a,b) an open real interval, and let $[w_-, w_+]$ be a nonempty order interval in the a.e. pointwise ordered space $HL_{loc}((a,b), E)$ of locally HL integrable functions from (a,b) to E. Then every increasing mapping $G : HL_{loc}((a,b), E) \to [w_-, w_+]$ has the smallest and greatest fixed points, and they are increasing with respect to G.*

Denote by $L(E)$ the set of all bounded linear operators on E. We shall first study the functional Fredholm integral equation

$$u(t) = h(t, u) + {}^K\!\!\int_a^b T(t, s) f(s, u)\, ds, \quad t \in [a, b], \tag{7.1}$$

where $-\infty < a < b < \infty$, $h, f : [a, b] \times HL_{loc}((a,b), E) \to E$, and $T : [a, b] \times [a, b] \to L(E)$. Assuming that $HL_{loc}((a,b), E)$ is equipped with a.e. pointwise ordering, we impose the following hypotheses on the functions h, f, and T.

(h1) $h(t, \cdot)$ is increasing for a.e. $t \in [a, b]$, $h(\cdot, u)$ is strongly measurable for all $u \in HL_{loc}((a,b), E)$, and there exist $h_\pm \in HL_{loc}((a,b), E)$ such that $h_- \le h(\cdot, u) \le h_+$ for all $u \in HL_{loc}((a,b), E)$.

(f1) $f(\cdot, u)$ is strongly measurable for each $u \in HL_{loc}((a,b), E)$.

(f2) $f(s, \cdot)$ is increasing for a.e. $s \in [a, b]$.

(f3) $f_- \le f(\cdot, u) \le f_+$ for all $u \in HL_{loc}((a,b), E)$, where $f_\pm : [a, b] \to E$ are strongly measurable.

(T1) $(t, s) \mapsto T(t, s)x$ is continuous for each $x \in E$.

(T2) $T(t, s)E_+ \subseteq E_+$ for all (t, s) in the domain of T.

(Tf$_\pm$)   The mappings $s \mapsto T(t, s)f_\pm(s)$, $t \in [a, b]$ are HL integrable on $[a, b]$

Our main existence and comparison result for the integral equation (7.1) reads as follows.

Theorem 7.2. *Let E be a Banach space ordered by a regular order cone. Assume that the hypotheses (h1), (f1), (f2), (f3), (T1), (T2), and (Tf$_\pm$) are satisfied. Then the equation (7.1) has the smallest and greatest solutions in $HL_{loc}((a,b), E)$. Moreover, these solutions are increasing with respect to h and f.*

Proof: The hypotheses (h1), (f3), and (Tf$_\pm$) imply that the equations

$$w_\pm(t) = h_\pm(t) + {}^K\!\!\int_a^b T(t,s) f_\pm(s)\, ds, \quad t \in [a,b], \tag{7.2}$$

define functions $w_\pm : [a,b] \to E$. Since the integrand on the right-hand side of (7.2) is by (T1) continuous in t for fixed s, one can show by applying the Dominated Convergence Theorem 9.26 that the integral term on the right-hand side of (7.2) is continuous in t. Thus the functions $w_\pm$ belong to $HL_{loc}((a,b),E)$. Similarly, it follows from the hypotheses (h1), (f1), (f3), (T1), and (Tf$_\pm$) and Theorem 9.26 that the relation

$$Gu(t) = h(t,u) + {}^K\!\!\int_a^b T(t,s) f(s,u)\, ds, \qquad t \in [a,b] \tag{7.3}$$

defines a mapping $G : HL_{loc}((a,b),E) \to HL_{loc}((a,b),E)$. Applying the hypotheses (h1), (f2), and (T2) we see that if $u, v \in HL_{loc}((a,b),E)$ and $u \le v$, then

$$w_-(t) \le Gu(t) \le Gv(t) \le w_+(t) \text{ for a.e. } t \in [a,b].$$

Thus (7.3) defines an increasing mapping $G : HL_{loc}((a,b),E) \to [w_-,w_+]$. The above proof shows that all the hypotheses of Theorem 7.1 are valid for the operator G defined by (7.3). Thus G has the smallest and greatest fixed point u_* and u^*. Noticing that fixed points of G defined by (7.3) are solutions of (7.1) and vice versa, then u_* and u^* are the smallest and greatest solutions of (7.1). It follows from (7.3) that G is increasing with respect to h and f, whence the last assertion of the theorem follows from the last assertion of Theorem 7.1. □

Next we consider a situation where the extremal solutions, i.e., the smallest and greatest solutions, of the integral equation (7.1) can be obtained by ordinary iterations.

Proposition 7.3. *Assume the hypotheses of Theorem 7.2, and let G be defined by (7.3). Then the following assertions hold.*

(a) *The sequence $(u_n)_{n=0}^\infty = (G^n w_-)_{n=0}^\infty$ is increasing and converges a.e. pointwise to a function $u_* \in HL_{loc}((a,b),E)$. Moreover, u_* is the smallest solution of (7.1) if $h(t,u_n) \to h(t,u_*)$ for a.e. $t \in [a,b]$ and $f(s,u_n) \to f(s,u_*)$ for a.e. $s \in [a,b]$;*

(b) *The sequence $(v_n)_{n=0}^\infty = (G^n w_+)_{n=0}^\infty$ is decreasing and converges a.e. pointwise to a function $u^* \in HL_{loc}((a,b),E)$. Moreover, u^* is the greatest solution of (7.1) if $h(t,v_n) \to h(t,u^*)$ for a.e. $t \in [a,b]$ and $f(s,v_n) \to f(s,u^*)$ for a.e. $s \in [a,b]$.*

Proof: Ad (a) It is easy to see that the sequence $(G^n w_-)$ is increasing and is contained in the order interval $[w_-,w_+]$. It then follows from Proposition 9.39

that the asserted a.e. pointwise limit $u_* \in HL_{loc}((a,b), E)$ exists. Defining
the sequence of successive approximations $u_n : [a,b] \to E$ by

$$u_{n+1}(t) = h(t, u_n) + {}^K\!\!\int_a^b T(t,s) f(s, u_n)\, ds, \ u_0(t) = w_-(t), \ t \in [a,b], \quad (7.4)$$

then $(u_n) = (G^n w_-)$, whence (u_n) converges a.e. pointwise to u_*. In view of
this result, the hypotheses of (a), and Theorem 9.26 it follows from (7.4) that
u_* is a solution of (7.1). If u is any solution of (7.1), then $u = Gu \in [w_-, w_+]$.
By induction one can show that $u_n = G^n w_- \in [w_-, u]$ for each n. Thus
$u_* = \sup_n u_n \leq u$, which proves that u_* is the smallest solution of (7.1).
    Ad (b) the proof of (b) is similar to that of (a).   □

The above results can be applied to the functional Fredholm integral equation

$$u(t) = Gu(t) := h(t, u) + {}^K\!\!\int_a^b k(t,s) f(s, u)\, ds, \quad t \in [a,b]. \quad (7.5)$$

Corollary 7.4. *Assume that the hypotheses (h1), (f1), (f2), and (f3) hold,
and that the hypotheses (T1), (T2), and (Tf$_\pm$) are replaced by the hypothesis*

*(k1) $k : [a,b] \times [a,b] \to \mathbb{R}_+$ is continuous and the mappings $s \mapsto k(t,s) f_\pm(s)$
    are HL integrable on $[a,b]$ for each $t \in [a,b]$.*

*Then the equation (7.5) has the smallest and greatest solutions in the space
$HL_{loc}((a,b), E)$. Moreover, the solutions u_* and u^* are increasing with respect
to h and f.*

Proof: The hypothesis (k1) implies that the equation

$$T(t,s)x = k(t,s)x, \quad t, s \in [a,b], \ x \in E,$$

defines a mapping $T : [a,b] \times [a,b] \to L(E)$, and that the hypotheses (T1),
(T2), and (Tf$_\pm$) are valid. Thus the conclusions follow from Theorem 7.2.   □

The following corollary is a consequence of Proposition 7.3.

Corollary 7.5. *Assume that the hypotheses of Corollary 7.4 hold, and let $w_\pm$
and G be defined by (7.2) and (7.3) with $T(t,s)$ replaced by $k(t,s)$. Then the
following holds.*

(a) *The sequence $(u_n)_{n=0}^\infty = (G^n w_-)_{n=0}^\infty$ is increasing and converges a.e.
    pointwise to a function $u_* \in HL_{loc}((a,b), E)$. Moreover, u_* is the smallest
    solution of (7.5) if $h(t, u_n) \to h(t, u_*)$ for a.e. $t \in [a,b]$ and $f(s, u_n) \to
    f(s, u_*)$ for a.e. $s \in [a,b]$;*
(b) *The sequence $(v_n)_{n=0}^\infty = (G^n w_+)_{n=0}^\infty$ is decreasing and converges a.e.
    pointwise to a function $u^* \in HL_{loc}((a,b), E)$. Moreover, u^* is the greatest
    solution of (7.5) if $h(t, v_n) \to h(t, u^*)$ for a.e. $t \in [a,b]$ and $f(s, v_n) \to
    f(s, u^*)$ for a.e. $s \in [a,b]$.*

Remark 7.6. The functional dependence of h and f on u in equations (7.1) and (7.5) can be given, e.g., in terms of bounded, linear, and positive operators, such as integral operators of Volterra and/or Fredholm type with nonnegative kernels. Thus the results derived in this subsection can be applied also to integro-differential equations. The continuity hypotheses imposed on T and k, and Theorem 9.26 ensure that the integrals in (7.1) and (7.5) are continuous in t. If also h is continuous in t in these equations, then their solutions are continuous. Fredholm integral equations in Banach spaces involving HL integrable functions has been studied, e.g., in [91].

The first example is an application of Corollaries 7.4 and 7.5.

Example 7.7. Determine the smallest and greatest solutions of the following system of impulsive boundary value problems:

$$\begin{cases} -u_1''(t) = \dfrac{1}{t\sqrt{t}}\sin\dfrac{1}{t} + \arctan([\int_{\frac{1}{2}}^1 u_2(x)\,dx]), & \text{a.e. on } (0,1), \\[2mm] -u_2''(t) = \dfrac{1}{(1-t)^2}\sin(\dfrac{1}{1-t}) + 2\tanh([\int_{\frac{1}{2}}^1 u_1(x)\,dx]), & \text{a.e. on } (0,1), \\[2mm] u_1(0+) = u_2(0+) = 0, \\[2mm] u_1(1-) = \Delta u_1(\dfrac{1}{2}) = 3\tanh([2u_2(1-)]), \\[2mm] u_2(1-) = \Delta u_2(\dfrac{1}{2}) = 2\arctan([2u_1(1-)]). \end{cases}$$

$$(7.6)$$

(Hint: $s \mapsto [s]$ denotes the integer function.)
Solution. System (7.6) can be converted to the Fredholm integral equation (7.5) with $E = \mathbb{R}^2$, ordered coordinatewise, $a = 0$, $b = 1$, and setting

$$f(t,(u_1,u_2)) = (f_1(t,u_2), f_2(t,u_1)), \quad h(t,(u_1,u_2)) = (h_1(t,u_2), h_2(t,u_1)),$$

$$f_1(t,u_2) = \frac{1}{t\sqrt{t}}\sin\frac{1}{t} + \arctan\left(\left[\int_{\frac{1}{2}}^1 u_2(x)\,dx\right]\right),$$

$$f_2(t,u_1) = \frac{1}{(1-t)^2}\sin\left(\frac{1}{1-t}\right) + 2\tanh\left(\left[\int_{\frac{1}{2}}^1 u_1(x)\,dx\right]\right),$$

$$h_1(t,u_2) = 3H(t - \frac{1}{2})\tanh([2u_2(1-)]),$$

$$h_2(t,u_1) = 2H(t - \frac{1}{2})\arctan([2u_1(1-)]),$$

$$k(t,s) = (1-t)s,\ 0 \le s \le t,\ t(1-s),\ t \le s \le 1.$$

$$(7.7)$$

The hypotheses (h1), (f1), (f2), and (f3) hold when

$$f_{\pm}(t) = \left(\frac{1}{t\sqrt{t}}\sin\left(\frac{1}{t}\right) \pm \frac{\pi}{2}, \frac{1}{(1-t)^2}\sin\left(\frac{1}{1-t}\right) \pm 2\right),$$

$$h_{\pm}(t) = H\left(t - \frac{1}{2}\right)(\pm 3, \pm\pi).$$

Also hypothesis (k1) is fulfilled due to Lemma 1.12. Thus (7.6) has by Corollary 7.4 the smallest and greatest solutions. To determine these solutions, notice that G, defined by (7.5), can be rewritten as

$$Gu(t) = (G_1 u_2(t), G_2 u_1(t)), \tag{7.8}$$

where

$$G_1 u_2(t) = h_1(t, u_2) + (1-t)^K\!\!\int_0^t s f_1(s, u_2)\, ds + t^K\!\!\int_t^1 (1-s) f_1(s, u_2)\, ds,$$

$$G_2 u_1(t) = h_2(t, u_1) + (1-t)^K\!\!\int_0^t s f_2(s, u_1)\, ds + t^K\!\!\int_t^1 (1-s) f_2(s, u_1)\, ds.$$

The functions $w_- = (w_1^-, w_2^-)$ and $w_+ = (w_1^+, w_2^+)$ defined by (7.2) with $T(t, s)$ replaced by $k(t, s)$ defined above can be calculated, and one obtains

$$w_1^-(t) = -3H(t - \frac{1}{2}) + \sqrt{2\pi}\left(t\, FrS\left(\sqrt{\frac{2}{\pi t}}\right) - 2\, FrC\left(\sqrt{\frac{2}{\pi t}}\right)\right)$$

$$+ \sqrt{2\pi}\left(2\, FrC\left(\sqrt{\frac{2}{\pi}}\right) - FrS\left(\sqrt{\frac{2}{\pi}}\right)\right)t + 2\sqrt{t}\sin(\frac{1}{t})$$

$$- 2\sin(1)t + \sqrt{2\pi}(1-t) + \frac{\pi}{4}(t^2 - t),$$

$$w_2^-(t) = -\pi H(t - \frac{1}{2}) + (t-1)\cos(\frac{1}{t-1})$$

$$+ Si(\frac{1}{t-1}) + (\cos(1) + Si(1))(1-t) + (\frac{\pi}{2} - 1)t + t^2,$$

$$w_1^+(t) = 3H(t - \frac{1}{2}) + \sqrt{2\pi}\left(t\, FrS\left(\sqrt{\frac{2}{\pi t}}\right) - 2\, FrC\left(\sqrt{\frac{2}{\pi t}}\right)\right)$$

$$+ \sqrt{2\pi}\left(FrS\left(\sqrt{\frac{2}{\pi}}\right) - FrC\left(\sqrt{\frac{2}{\pi}}\right)\right)t + 2\sqrt{t}\sin(\frac{1}{t})$$

$$- 2\sin(1)t + \sqrt{2\pi}(1-t) + \frac{\pi}{4}(t - t^2),$$

$$w_2^+(t) = \pi H(t - \frac{1}{2}) + (t-1)\cos(\frac{1}{t-1})$$

$$+ Si(\frac{1}{t-1}) + (\cos(1) + Si(1))(1-t) + (\frac{\pi}{2} + 1)t - t^2,$$

where Si is the sine integral function, FrS is the Fresnel sine integral, and FrC is the Fresnel cosine integral, i.e.,

$$Si(x) = \int_0^x \frac{\sin t}{t}\, dt, \ FrS(x) = \int_0^x \sin(\frac{\pi}{2}t^2)\, dt, \ FrC(x) = \int_0^x \cos(\frac{\pi}{2}t^2)\, dt.$$

The smallest solution of (7.6) is equal to the smallest fixed point of $G = (G_1, G_2)$, defined by (7.8). Calculating the iterations $G^n w_-$ it turns out that $G^2 w_- = G^3 w_-$. Thus $u_* = G^2 w_-$ is by Corollary 7.5 the smallest solution of (7.6). Similarly, one can show that $G^2 w_+ = G^3 w_+$, which implies that $u^* = G^2 w_+$ is the greatest solution of (7.6). The exact expression of the components of these solutions $u_* = (u_{*1}, u_{*2})$ and $u^* = (u_1^*, u_2^*)$ are

$$u_{*1}(t) = -3 \tanh(3) H(t - \frac{1}{2}) + \sqrt{2\pi} \left(t\, FrS \left(\sqrt{\frac{2}{\pi t}} \right) - 2\, FrC \left(\sqrt{\frac{2}{\pi t}} \right) \right)$$

$$+ \sqrt{2\pi} \left(2\, FrC \left(\sqrt{\frac{2}{\pi}} \right) - FrS \left(\sqrt{\frac{2}{\pi}} \right) \right) t + 2\sqrt{t} \sin(\frac{1}{t}) - 2 \sin(1)t$$

$$+ \sqrt{2\pi}(1 - t) + \frac{1}{2} \arctan(2)(t^2 - t),$$

$$u_{*2}(t) = -2 \arctan(6) H(t - \frac{1}{2}) + (t - 1) \cos(\frac{1}{t - 1})$$

$$+ Si(\frac{1}{t - 1}) + (\cos(1) + Si(1))(1 - t) + \frac{\pi}{2}t + \tanh(2)(t^2 - t),$$

$$u_1^*(t) = 3 \tanh(2) H(t - \frac{1}{2}) + \sqrt{2\pi} \left(t\, FrS \left(\sqrt{\frac{2}{\pi t}} \right) - 2\, FrC \left(\sqrt{\frac{2}{\pi t}} \right) \right)$$

$$+ \sqrt{2\pi} \left(2\, FrC \left(\sqrt{\frac{2}{\pi}} \right) - FrS \left(\sqrt{\frac{2}{\pi}} \right) \right) t$$

$$+ 2\sqrt{t} \sin(\frac{1}{t}) - 2 \sin(1)t + \sqrt{2\pi}(1 - t) + \frac{\pi}{8}(t - t^2),$$

$$u_2^*(t) = 2 \arctan(5) H(t - \frac{1}{2}) + (t - 1) \cos(\frac{1}{t - 1}) + Si(\frac{1}{t - 1})$$

$$+ (\cos(1) + Si(1))(1 - t) + \frac{\pi}{2}t + \tanh(1)(t - t^2).$$

Example 7.8. The infinite system of impulsive boundary value problems

$$
\begin{cases}
-w_{2n-1}''(t) = \dfrac{1}{2n - 1} \left(-\dfrac{\sin(\frac{1}{t})}{t^2} \pm 1 \right), \\[2ex]
-w_{2n}''(t) = \dfrac{1}{2n} \left(\dfrac{\sin(\frac{1}{1-t})}{(1 - t)^2} \pm 1 \right), \\[2ex]
w_{2n-1}(0+) = 0, \quad w_{2n-1}(1-) = \Delta w_{2n-1} \left(t - \dfrac{2n - 1}{2n + 1} \right) = \pm \dfrac{1}{2n - 1}, \\[2ex]
w_{2n}(0+) = 0, \quad w_{2n}(1-) = \Delta w_{2n} \left(t - \dfrac{2n}{2n + 1} \right) = \pm \dfrac{1}{2n - 1}, \quad n \in \mathbb{N},
\end{cases}
$$

$$(7.9)$$

have unique solutions $w_\pm(t) = \left(\frac{1}{n}w_n^\pm(t)\right)_{n=1}^\infty$, where

$$
\begin{aligned}
w_{2n-1}^+(t) &= H\left(t - \tfrac{2n-1}{2n+1}\right) + (1-t)\left(Si\left(t^{-1}\right) - \tfrac{1}{2}\pi + \tfrac{1}{2}t^2\right) \\
&\quad + t\left(-\cos(1) - Si(1) + \cos\left(t^{-1}\right) + Si\left(t^{-1}\right) - t + \tfrac{1}{2} + \tfrac{1}{2}t^2\right), \\
w_{2n-1}^-(t) &= -H\left(t - \tfrac{2n-1}{2n+1}\right) + (1-t)\left(Si\left(t^{-1}\right) - \tfrac{1}{2}\pi - \tfrac{1}{2}t^2\right) \\
&\quad + t\left(-\cos(1) - Si(1) + \cos\left(t^{-1}\right) + Si\left(t^{-1}\right) + t - \tfrac{1}{2} - \tfrac{1}{2}t^2\right), \\
w_{2n}^+(t) &= H\left(t - 2\tfrac{n}{2n+1}\right) + t\left(Si\left(\tfrac{1}{t-1}\right) + \tfrac{1}{2}\pi - t + \tfrac{1}{2} + \tfrac{1}{2}t^2\right) \\
&\quad + (1-t)\left(\cos(1) + Si(1) - \cos\left(\tfrac{1}{t-1}\right) + Si\left(\tfrac{1}{t-1}\right) + \tfrac{1}{2}t^2\right), \\
w_{2n}^-(t) &= -H\left(t - 2\tfrac{n}{2n+1}\right) + t\left(Si\left(\tfrac{1}{t-1}\right) + \tfrac{1}{2}\pi + t - \tfrac{1}{2} - \tfrac{1}{2}t^2\right) \\
&\quad + (1-t)\left(\cos(1) + Si(1) - \cos\left(\tfrac{1}{t-1}\right) + Si\left(\tfrac{1}{t-1}\right) - \tfrac{1}{2}t^2\right).
\end{aligned}
$$

Since $w_\pm \in HL_{loc}((0,1), c_0)$, Corollary 7.4 can be applied to show that the smallest and greatest solutions $u_* = (u_{*n})_{n=1}^\infty$ and $u^* = (u_n^*)_{n=1}^\infty$ of the infinite system of impulsive boundary value problems

$$
\begin{cases}
-u_{2n-1}''(t) = \dfrac{1}{2n-1}\left(-\dfrac{\sin(\frac{1}{t})}{t^2} + g_{2n-1}(u)\right), \\[2mm]
-u_{2n}''(t) = \dfrac{1}{2n}\left(\dfrac{\sin(\frac{1}{1-t})}{(1-t)^2} + g_{2n}(u)\right), \\[2mm]
u_{2n-1}(0+) = 0, \quad u_{2n-1}(1-) = \Delta u_{2n-1}\left(t - \dfrac{2n-1}{2n+1}\right) = \dfrac{1}{2n-1}q_{2n-1}(u), \\[2mm]
u_{2n}(0+) = 0, \quad u_{2n}(1-) = \Delta u_{2n}\left(t - \dfrac{2n}{2n+1}\right) = \dfrac{1}{2n}q_{2n}(u), \ n \in \mathbb{N}
\end{cases}
$$

$$(7.10)$$

exist in $HL_{loc}((0,1), c_0)$ and belong to its order interval $[w_-, w_+]$, if we assume that all the functions $g_n, q_n : HL_{loc}((0,1), c_0) \to \mathbb{R}$, are increasing, and if $-1 \le g_n(u), q_n(u) \le 1$ for all $u \in HL_{loc}((0,1), c_0)$ and $n \in \mathbb{N}$.

Remark 7.9. (i) In Example 7.7 the functions $f_\pm$ are not HL integrable. However, $k(t,s)$ is continuous, and the integrals ${}^K\!\int_0^1 k(t,s)f_\pm(s)\,ds$ exist for each $t \in (0,1)$ by Lemma 1.12, whence the hypothesis (k1) is valid.

(ii) The calculations needed in Examples 7.7 and 7.8 are carried out by using a simple Maple programming.

(iii) Differential equations in Banach spaces with HL integrable data is studied, e.g., in [92], [200].

7.1.2 Volterra Integral Equations

Throughout this subsection, $E = (E, \le, \|\cdot\|)$ is an ordered Banach space with a regular order cone and $-\infty < a < b \le \infty$. Here we first study the functional Volterra integral equation

$$u(t) = h(t, u) + {}^K\!\!\int_a^t T(t, s) f(s, u) \, ds, \quad t \in J := (a, b), \qquad (7.11)$$

where $h, f : J \times HL_{loc}(J, E) \to E$ and T is a mapping from $\Lambda = \{(t, s) : a < s \le t < b\}$ to $L(E)$. The following conditions are imposed on T.

(T0) $T(t, t) = I$, and $T(t, r) = T(t, s)T(s, r)$ whenever $a < r \le s \le t < b$.
(T1) $(t, s) \mapsto T(t, s)x$ is continuous for each $x \in E$.
(T2) $T(t, s)E_+ \subseteq E_+$ for all (t, s) in the domain of T.

Lemma 7.10. *Let the hypotheses (T0) and (T1) hold. Assume that $q : J \to E$, that $s \mapsto T(t, s)q(s)$ is HL integrable on $[a, t]$ for every $t \in J$, and that $t \mapsto {}^K\!\int_a^t T(t, s)q(s)ds$ is locally bounded. Then the equation*

$$u(t) = {}^K\!\!\int_a^t T(t, s)q(s)ds, \quad t \in J, \qquad (7.12)$$

defines a left-continuous function $u : J \to E$.

Proof: If $a < \bar{t} \le t < b$, it follows from (7.12) that

$$\|u(t) - u(\bar{t})\| \le \left\| {}^K\!\!\int_a^{\bar{t}} (T(t, s) - T(\bar{t}, s))q(s) \, ds \right\|$$
$$+ \left\| {}^K\!\!\int_{\bar{t}}^t T(t, s)q(s) \, ds \right\| = I_1 + I_2, \qquad (7.13)$$

where

$$I_1 = \left\| {}^K\!\!\int_a^{\bar{t}} (T(t, s) - T(\bar{t}, s))q(s) \, ds \right\|$$
$$\le \|T(t, \bar{t}) - I\| \left\| {}^K\!\!\int_a^{\bar{t}} T(\bar{t}, s)q(s) \, ds \right\|, \qquad (7.14)$$
$$I_2 = \left\| {}^K\!\!\int_{\bar{t}}^t T(t, s)q(s) \, ds \right\|.$$

It follows from (7.14) by (T0) and (T1) that $I_1 + I_2 \to 0$ as $\bar{t} \to t-$. This implies by (7.13) that $\|u(t) - u(\bar{t})\| \to 0$ as $t \to \bar{t}-$. Thus u is left-continuous on J. □

Assuming that $HL_{loc}(J, E)$ is equipped with a.e. pointwise ordering, we impose the following hypotheses on the functions h, f, and T.

(h0) $h(t, \cdot)$ is increasing for a.e. $t \in J$, $h(\cdot, u)$ is strongly measurable for all $u \in HL_{loc}(J, E)$, and there exist $h_\pm \in HL_{loc}(J, E)$ such that $h_- \le h(\cdot, u) \le h_+$ for all $u \in HL_{loc}(J, E)$.
(f0) There exist strongly measurable functions $f_\pm : J \to E$ such that $f_- \le f(\cdot, u) \le f_+$ for all $u \in HL_{loc}(J, E)$.

(f1) The mapping $f(\cdot, u)$ is strongly measurable for each $u \in HL_{loc}(J, E)$.
(f2) $f(s, \cdot)$ is increasing for a.e. $s \in J$.
(Tf) The functions $s \mapsto T(t, s)f_{\pm}(s)$ are HL integrable on $[a, t]$ for every
      $t \in J$, and the functions $t \mapsto {}^K\!\int_a^t T(t, s)f_{\pm}(s)ds$ are locally bounded.

Our main existence and comparison result for the integral equation (7.11)
reads as follows.

Theorem 7.11. *Assume that the hypotheses (h0), (f0), (f1), (f2) (T0), (T1),
(T2), and (Tf) are satisfied. Then the equation (7.11) has the smallest and
greatest solutions in $HL_{loc}((a, b), E)$. Moreover, these solutions u_* and u^* are
increasing with respect to h and f.*

Proof: The hypotheses (h0), (f0), and (Tf) ensure that the equations

$$w_{\pm}(t) = h_{\pm}(t) + {}^K\!\int_a^t T(t, s)f_{\pm}(s)\, ds, \quad t \in J, \tag{7.15}$$

define functions $w_{\pm} : J \to E$. The integral term on the right-hand side of
(7.15) is left-continuous in t by Lemma 7.10. Thus the functions $w_{\pm}$ belong
to $HL_{loc}(J, E)$. By using the hypotheses, Lemmas 7.10 and 9.11, Proposition
9.14, and Theorem 9.26, it can be shown that the equation

$$Gu(t) = h(t, u) + {}^K\!\int_a^t T(t, s)f(s, u)\, ds, \quad t \in J, \tag{7.16}$$

defines an increasing mapping $G : HL_{loc}(J, E) \to [w_-, w_+]$. From Theorem
7.1 it then follows that G has the smallest and greatest fixed points u_* and u^*.
Noticing that fixed points of G defined by (7.16) are solutions of (7.11) and
vice versa, then u_* and u^* are the smallest and greatest solutions of (7.11).
By Lemma 9.11, from (7.16) it follows that G is increasing with respect to h
and f, and thus the last assertion of the theorem is a consequence of the last
assertion of Theorem 7.1.   □

Next we study a special case where the extremal solutions of the integral
equation (7.11) can be obtained by ordinary iterations.

Proposition 7.12. *Assume that the hypotheses of Theorem 7.11 hold, and let
G be defined by (7.16). Then the following assertions are true.*

(a) *The sequence $(u_n)_{n=0}^{\infty} := (G^n w_-)_{n=0}^{\infty}$ is increasing and converges a.e.
    pointwise to a function $u_* \in HL_{loc}(J, E)$. Moreover, u_* is the smallest
    solution of (7.11) provided that $h(t, u_n) \to h(t, u_*)$ for a.e. $t \in J$ and
    $f(s, u_n) \to f(s, u_*)$ for all $t \in J$ and for a.e. $s \in [a, t]$;*
(b) *The sequence $(v_n)_{n=0}^{\infty} := (G^n w_+)_{n=0}^{\infty}$ is decreasing and converges a.e.
    pointwise to a function $u^* \in HL_{loc}((a, b), E)$. Moreover, u^* is the great-
    est solution of (7.11) provided that $h(t, v_n) \to h(t, u^*)$ for a.e. $t \in J$ and
    $f(s, v_n) \to f(s, u^*)$ for a.e. $s \in J$.*

Proof: Ad (a) The sequence $(u_n) := (G^n w_-)$ is increasing and contained in the order interval $[w_-, w_+]$. Hence the asserted a.e. pointwise limit $u_* \in HL_{loc}(J, E)$ exists by Proposition 9.39. Moreover, (u_n) equals to the sequence of successive approximations $u_n : J \to E$ defined by

$$u_{n+1}(t) = h(t, u_n) + \; {}^K\!\!\int_a^t T(t, s) f(s, u_n) \, ds, \quad u_0(t) = w_-(t), \quad t \in J, \; n \in \mathbb{N}_0.$$
(7.17)

In view of these results, the hypotheses of (a) and Theorem 9.26, it follows from (7.17) as $n \to \infty$ that u_* is a solution of (7.11).

If u is any solution of (7.11), then $u = Gu \in [w_-, w_+]$. By induction one can show that $u_n = G^n w_- \in [w_-, u]$ for each n. Thus $u_* = \sup_n u_n \le u$, which proves that u_* is the smallest solution of (7.11).

Ad (b) The proof of (b) is similar to that of (a).   □

The next result is a special case of Theorem 7.11.

Corollary 7.13. *The results of Theorem 7.11 remain true for solutions of the Volterra integral equation*

$$u(t) = h(t, u) + \; {}^K\!\!\int_a^t k(t, s) f(s, u) \, ds, \quad t \in J, \tag{7.18}$$

where $h, f : J \times HL_{loc}(J, E) \to E$ satisfy the hypotheses (h0), (f0), (f1), and (f2), and $k : \Lambda \to \mathbb{R}_+$ has the following properties.

(k0) k is continuous, $k(t, t) = 1$, and $k(t, r) = k(t, s) k(s, r)$ whenever $a < r \le s \le t < b$.

(kf$_\pm$) The functions $s \mapsto k(t, s) f_\pm(s)$ are HL integrable on $[a, t]$ for each $t \in J$, and the functions $t \mapsto {}^K\!\!\int_a^t k(t, s) f_\pm(s) ds$ are locally bounded.

Proof: The hypotheses (k0) and (kf$_\pm$) imply that the equation

$$T(t, s) x = k(t, s) x, \quad (t, s) \in \Lambda, \; x \in E,$$

defines a mapping $T : \Lambda \to L(E)$, and that the hypotheses (T0), (T1), (T2), and (Tf) are valid. Thus the conclusions follow from Theorem 7.2.   □

The next result is an application of Proposition 7.12.

Corollary 7.14. *Assume that the hypotheses of Corollary 7.13 hold, and let $w_\pm$ and G be defined by (7.15) and (7.16) with $T(t, s)$ replaced by $k(t, s)$.*

(a) The sequence $(u_n)_{n=0}^\infty = (G^n w_-)_{n=0}^\infty$ is increasing and converges a.e. pointwise to a function $u_* \in HL_{loc}(J, E)$. Moreover, u_* is the smallest solution of (7.18) provided that $h(t, u_n) \to h(t, u_*)$ for a.e. $t \in J$ and $f(s, u_n) \to f(s, u_*)$ for a.e. $s \in J$;

(b) The sequence $(v_n)_{n=0}^\infty = (G^n w_+)_{n=0}^\infty$ *is decreasing and converges a.e. pointwise to a function* $u^* \in HL_{loc}(J, E)$. *Moreover,* u^* *is the greatest solution of (7.18) provided that* $h(t, v_n) \to h(t, u^*)$ *for a.e.* $t \in J$ *and* $f(s, v_n) \to f(s, u^*)$ *for a.e.* $s \in J$.

Example 7.15. Let E be the space c_0 of all sequences $(c_n)_{n=1}^\infty$ of real numbers converging to zero, ordered componentwise and normed by the sup-norm. Define $f_n, h_n : [0, \infty) \to \mathbb{R}$ and $k : (0, \infty) \times [0, \infty) \to \mathbb{R}_+$ by the equations

$$\begin{cases} f_n(t) = \dfrac{2}{\sqrt{n}} \cos\left(\dfrac{1}{t^2}\right) + \dfrac{2}{\sqrt{n}t^2} \sin\left(\dfrac{1}{t^2}\right), \ t > 0, \ f_n(0) = 0, \\[2mm] h_n(t) = \dfrac{1}{\sqrt{nt}} H\left(t - \dfrac{2n-1}{2n}\right), \ t > 0, \ h_n(0) = 0, \ n \in \mathbb{N}, \qquad (7.19) \\[2mm] k(t, s) = \dfrac{s}{t}, t > 0, \ s \geq 0. \end{cases}$$

The solutions of the infinite system of integral equations

$$w_n(t) = \pm h_n(t) + {}^K\!\!\int_0^t k(t, s) \left(f_n(s) \pm \frac{1}{\sqrt{n}}\right) ds, \ n \in \mathbb{N}, \qquad (7.20)$$

in $HL_{loc}((0, \infty), c_0)$ are given by

$$w_\pm(t) = (w_{n\pm}(t))_{n=1}^\infty = \left(\pm \frac{1}{\sqrt{nt}} H\left(t - \frac{2n-1}{2n}\right) + \frac{t}{\sqrt{n}} \cos\left(\frac{1}{t^2}\right) \pm \frac{t}{2\sqrt{n}}\right)_{n=1}^\infty.$$
$$(7.21)$$

In particular, Corollary 7.13 can be applied to show that the infinite system of integral equations

$$u_n(t) = q_n(u)h_n(t) + {}^K\!\!\int_0^t k(t, s) \left(f_n(s) + \frac{1}{\sqrt{n}}g_n(u)\right) ds, \qquad (7.22)$$

has the smallest and greatest solutions $u_* = (u_{*n})_{n=1}^\infty$ and $u^* = (u_n^*)_{n=1}^\infty$ in $HL_{loc}((0, \infty), c_0)$, provided that all the functions $q_n, g_n : HL_{loc}([0, \infty), c_0) \to \mathbb{R}$ are increasing, and $-1 \leq g_n(u), q_n(u) \leq 1$ for all $u \in HL_{loc}((0, \infty), c_0)$ and $n \in \mathbb{N}$. Moreover, both u_* and u^* belong to the order interval $[w_-, w_+]$ of $HL_{loc}(0, \infty), c_0)$, where the functions $w_\pm$ are given by (7.21).

Remark 7.16. (i) The functions f_n in Example 7.15 are not HL integrable on $[0, t]$ for any $t > 0$. However, $k(t, s) = \frac{s}{t}$ satisfies the hypothesis (k0), and the functions $k(t, \cdot)f_n$ satisfy hypothesis (kf$_\pm$) in view of Lemma 1.12.

7.1.3 Application to Impulsive IVP

Let E be a Banach space ordered by a regular order cone. The result of Corollary 7.13 will now be applied to the following impulsive initial value problem

$$\begin{cases} u'(t) + p(t)u(t) = f(t, u) \text{ a.e. on } J = (a, b), \\ u(a) = x_0, \ \Delta u(\lambda) = D(\lambda, u), \ \lambda \in W, \end{cases} \quad (7.23)$$

where $p \in L^1([a, t], \mathbb{R})$ for every $t \in J$, $f : J \times HL_{loc}(J, E) \to E$, $x_0 \in E$, $\Delta u(\lambda) = u(\lambda + 0) - u(\lambda)$, $D : W \times HL_{loc}(J, E) \to E$, and W is a well-ordered (and hence countable) subset of (a, b).

Denoting $W^{<t} = \{\lambda \in W : \lambda < t\}$, $t \in J$, and by $W^1_{SL,loc}(J, E)$ the set of all functions from J to E that are a.e. differentiable and satisfy the Strong Lusin Condition on every compact interval of J, we say that $u : J \to E$ is a solution of problem (7.23) if it satisfies the equations of (7.23), and if it belongs to the set

$$V = \begin{cases} u : J \to E : \sum_{\lambda \in W} \|\Delta u(\lambda)\| < \infty, \text{ and} \\ t \mapsto u(t) - \sum_{\lambda \in W^{<t}} \Delta u(\lambda) \in W^1_{SL,loc}(J, E) \end{cases}.$$

It is easy to verify that V is a subset of $HL_{loc}(J, E)$.

The following result, which is a generalization of [46, Lemma 3.1], allows us to transform problem (7.23) into a Volterra integral equation.

Lemma 7.17. *Let $p : J \to \mathbb{R}$, $g : J \to E$, $x_0 \in E$, and $c : W \to E$ be given. Assume that $p \in L^1_{loc}([a, b), \mathbb{R})$, $g \in HL_{loc}([a, b), E)$, and that $\sum_{\lambda \in W} \|c(\lambda)\| < \infty$. Then the problem*

$$\begin{cases} u'(t) + p(t)u(t) = g(t) \text{ a.e. on } J, \\ u(a) = x_0, \ \Delta u(\lambda) = c(\lambda), \ \lambda \in W, \end{cases} \quad (7.24)$$

has a unique solution u. This solution can be represented as

$$u(t) = e^{-\int_a^t p(s)ds} x_0 + \sum_{\lambda \in W^{<t}} e^{-\int_\lambda^t p(s)ds} c(\lambda) + {}^K\!\!\int_a^t e^{-\int_s^t p(\tau)d\tau} g(s)ds, \ t \in J. \quad (7.25)$$

Moreover, u is increasing with respect to g, c, and x_0.

Proof: Let $u : J \to E$ be defined by (7.25). Given a compact subinterval $I = [a, t_1]$ of J, define a mapping $\Gamma : I \to I$ by

$$\Gamma(s) = \min\{t \in W \cup \{t_1\} : s < t\}, \ s \in [a, t_1), \quad \Gamma(t_1) = t_1.$$

Denote by C the well-ordered chain of Γ-iterations of a, i.e., (cf. [133], Theorem 1.1.1) C is the only well-ordered subset of I with the following properties:
$$a = \min C, \text{ and if } s > a, \text{ then } s \in C \text{ iff } s = \sup \Gamma\{t \in C | t < s\}.$$
It follows from [133], Corollary 1.1.1 that $W \subset C$, and I is a disjoint union of C and open intervals $(s, \Gamma(s))$, $s \in C$. Moreover, C is countable as a well-ordered set of real numbers. Hence, rewriting (7.25) as

$$u(t) = e^{-\int_a^t p(s)ds} \left[x_0 + \sum_{\lambda \in W^{<t}} e^{-\int_\lambda^a p(s)ds} c(\lambda) + {}^K\!\!\int_a^t e^{-\int_s^a p(\tau)d\tau} g(s)ds \right],$$

it is easy to verify that

$$u'(t) + p(t)u(t) = g(t) \quad \text{for a.e. } t \in I, \quad u(a) = x_0. \tag{7.26}$$

For each $\lambda \in W$ the open interval $(\lambda, \Gamma(\lambda))$ does not contain any point of W, so that

$$\Delta u(\lambda) = u(\lambda + 0) - u(\lambda) = \lim_{t \to \lambda + 0} e^{-\int_\lambda^t p(s)ds} c(\lambda) = c(\lambda), \quad \lambda \in W. \tag{7.27}$$

It follows from (7.25) and (7.27) that

$$u(t) - \sum_{\lambda \in W^{<t}} \Delta u(\lambda) = u(t) - \sum_{\lambda \in W^{<t}} c(\lambda) = v(t) + w(t), \tag{7.28}$$

where

$$v(t) = e^{-\int_a^t p(s)ds} x_0 + K\!\int_a^t e^{-\int_s^t p(\tau)d\tau} g(s)ds, \quad t \in I,$$
$$w(t) = \sum_{\lambda \in W^{<t}} (e^{-\int_\lambda^t p(s)ds} - 1)c(\lambda), \quad t \in I.$$

Thus, for $a \leq \bar{t} < t \leq t_1$ we obtain

$$w(t) - w(\bar{t}) = \sum_{\lambda \in W \cap (a,\bar{t})} (e^{-\int_\lambda^t p(s)ds} - e^{-\int_\lambda^{\bar{t}} p(s)ds})c(\lambda)$$
$$+ \sum_{\lambda \in W \cap [\bar{t},t)} (e^{-\int_\lambda^t p(s)ds} - 1)c(\lambda) = \sum_{\lambda \in W \cap (a,\bar{t})} \int_{\bar{t}}^t -p(s)e^{-\int_\lambda^s p(\tau)d\tau} ds\, c(\lambda)$$
$$+ \sum_{\lambda \in W \cap [\bar{t},t)} \int_\lambda^t -p(s)e^{-\int_\lambda^s p(\tau)d\tau} ds\, c(\lambda).$$

Applying this representation and denoting $M = e^{\int_a^{t_1} |p(s)|ds} \sum_{\lambda \in W} \|c(\lambda)\|$, it follows that

$$\|w(t) - w(\bar{t})\| \leq M \int_{\bar{t}}^t |p(s)|ds \quad \text{for } a \leq \bar{t} < t \leq t_1.$$

This implies that w is absolutely continuous. Obviously, w is a.e. differentiable and the function v belongs to $W^1_{SL,loc}(J, E)$.

The above result holds for every $t_1 \in (a, b)$, so that $u \in V$ by (7.28). This, (7.26), and (7.27) imply that u is a solution of problem (7.24).

If $v \in V$ is a solution of (7.24), then $w = u - v$ is a function of V and $\Delta w(\lambda) = 0$ for each $\lambda \in W$, whence $w \in W^1_{SL,loc}(J, E)$ and w is a solution of the initial value of problem

$$w'(t) + p(t)w(t) = 0 \quad \text{a.e. on } J, \quad w(a) = 0. \tag{7.29}$$

For every fixed $t \in J$ the function

$$h(s) = e^{\int_a^s p(\tau)d\tau}, \quad s \in I = [a, t],$$

is absolutely continuous on I and real-valued. It then follows from Lemma 9.22 that

$$h(t)w(t) - h(a)w(a) = K\!\int_a^t (h'(s)w(s) + h(s)w'(s))\, ds, \quad t \in J,$$

or equivalently,

$$h(t)w(t) - h(a)w(a) = {}^K\!\!\int_a^t \left(e^{\int_a^s p(\tau)d\tau}(p(s)w(s) + w'(s))\right)ds, \ t \in J.$$

This equation and (7.29) imply that $h(t)w(t) \equiv 0$, so that $w(t) \equiv 0$, whence $u = v$.

The last assertion of the lemma is a direct consequence from the representation (7.25) and Lemma 9.11. $\qquad\square$

Let us impose the following hypotheses on the function D of (7.23).

(D0) $D(\lambda, \cdot)$ is increasing for all $\lambda \in W$, and there exist $c_\pm : W \to E$ such that $c_-(\lambda) \le D(\lambda, u) \le c_+(\lambda)$ for all $\lambda \in W$ and $u \in HL_{loc}(J, E)$, and that $\sum_{\lambda \in W} \|c_\pm(\lambda)\| < \infty$.

As an application of Corollary 7.13 we get the following existence and comparison result for problem (7.23).

Theorem 7.18. *Let the functions f and D in (7.23) satisfy the hypotheses (f0)–(f2) and (D0). If $p \in L^1_{loc}([a, b), \mathbb{R})$, and if the functions $f_\pm$ are locally HL integrable on $[a, b)$, then problem (7.23) has for each $x_0 \in E$ the smallest and greatest solutions u_* and u^* in V. Moreover, these solutions are increasing with respect to x_0, D, and f.*

Proof: The hypotheses given for D and p ensure that for each $x_0 \in E$ the relations

$$\begin{cases} h(t, u) = e^{-\int_a^t p(s)ds}x_0 + \sum_{\lambda \in W^{<t}} e^{-\int_\lambda^t p(s)ds} D(\lambda, u), & t \in J, \\ k(t, s) = e^{-\int_s^t p(\tau)d\tau}, & (t, s) \in \Lambda, \end{cases} \quad (7.30)$$

define mappings $h : J \times HL_{loc}(J, E) \to E$ and $k : \Lambda \to \mathbb{R}_+$, which satisfy the hypotheses (h0) and (k0) of Corollary 7.13. Then the integral equation (7.18), which by (7.30) can be rewritten as a fixed point equation

$$u(t) = Gu(t) := e^{-\int_a^t p(s)ds}x_0 + \sum_{\lambda \in W^{<t}} e^{-\int_\lambda^t p(s)ds} D(\lambda, u)$$

$$+ {}^K\!\!\int_a^t e^{-\int_s^t p(\tau)d\tau} f(s, u)ds, \quad (7.31)$$

has by Corollary 7.13 the smallest and greatest solutions u_* and u^*, and they are increasing with respect to h and f. Because by Lemma 7.17 the solutions of problem (7.23) are the same as the solutions of the integral equation (7.31), then u_* and u^* are the smallest and greatest solutions of problem (7.23), and they are increasing with respect to x_0, D, and h. $\qquad\square$

The next result is a consequence of Corollary 7.14.

Proposition 7.19. *Assume the hypotheses of Theorem 7.18, and let G be defined by (7.31). Then the following assertions hold.*

(a) *The sequence $(u_n)_{n=0}^{\infty} = (G^n w_-)_{n=0}^{\infty}$ is increasing and converges a.e. pointwise to a function $u_* \in HL_{loc}(J, E)$. Moreover, u_* is the smallest solution of (7.23) provided that $D(\lambda, u_n) \to D(\lambda, u_*)$ for each $\lambda \in W$ and $f(s, u_n) \to f(s, u_*)$ for a.e. $s \in J$;*

(b) *The sequence $(v_n)_{n=0}^{\infty} = (G^n w_+)_{n=0}^{\infty}$ is decreasing and converges a.e. pointwise to a function $u^* \in HL_{loc}(J, E)$. Moreover, u^* is the greatest solution of (7.23) provided that $D(\lambda, v_n) \to D(\lambda, u^*)$ for each $\lambda \in W$ and $f(s, v_n) \to f(s, u^*)$ for a.e. $s \in J$.*

Example 7.20. Let E be, as in Example 7.15, the space c_0 of the sequences of real numbers converging to zero, ordered componentwise and normed by the sup-norm. The solutions of the infinite system of problems

$$\begin{cases} w_n'(t) + \dfrac{1}{1+t} w_n(t) = \dfrac{2}{\sqrt{n}(1+t)} \left(\cos\left(\dfrac{1}{t^2}\right) + \dfrac{2}{t} \sin\left(\dfrac{1}{t^2}\right) \right) \pm \dfrac{1}{\sqrt{n}}, \\ w_n(0+) = 0, \quad \Delta w_n\left(t - \dfrac{2n-1}{2n}\right) = \pm\dfrac{1}{\sqrt{n}}, \quad n \in \mathbb{N}, \end{cases} \tag{7.32}$$

in $HL_{loc}((0, 2), c_0)$ are

$$(w_{n\pm}(t))_{n=1}^{\infty} = \left(\dfrac{1}{2\sqrt{n}(1+t)} \left(\pm\dfrac{4n-1}{n} H\left(t - \dfrac{2n-1}{2n}\right) \right.\right.$$
$$\left.\left. +2t^2 \cos\left(\dfrac{1}{t^2}\right) \pm 2t \pm t^2 \right) \right)_{n=1}^{\infty}. \tag{7.33}$$

Thus Theorem 7.18 can be applied to show that the smallest and greatest solutions $u_* = (u_{*n})_{n=1}^{\infty}$ and $u^* = (u_n^*)_{n=1}^{\infty}$ of infinite system of problems

$$\begin{cases} u_n'(t) + \dfrac{1}{1+t} u_n(t) = \dfrac{1}{\sqrt{n}(1+t)} \left(\cos\left(\dfrac{1}{t^2}\right) + \dfrac{2}{t} \sin\left(\dfrac{1}{t^2}\right) \right) + \dfrac{1}{\sqrt{n}} g_n(u), \\ w_n(0+) = 0, \quad \Delta w_n\left(t - \dfrac{2n-1}{2n}\right) = \dfrac{1}{\sqrt{n}} D_n(u), \quad n \in \mathbb{N}, \end{cases}$$
$$\tag{7.34}$$

exist in $HL_{loc}((0, 2), c_0)$ and belong to its order interval $[w_-, w_+]$, if we assume that all the functions $D_n, g_n : HL_{loc}([0, 2), c_0) \to \mathbb{R}$, are increasing, and if $-1 \le D_n(u), g_n(u) \le 1$ for all $u \in HL_{loc}((0, 2), c_0)$ and $n = 1, 2, \ldots$.

Example 7.21. Choose $J = [0, 1]$, and define functions $h_i : J \to \mathbb{R}$ and $q_i : \mathbb{R} \to \mathbb{R}$, $i = 1, 2, \ldots$, by $h_i(0) = 0$,

$$h_i(t) = \dfrac{1}{2^i} \sum_{m=1}^{i} \sum_{k=1}^{\infty} \dfrac{2 + [k^{\frac{1}{m}} t] - k^{\frac{1}{m}} t}{(km)^2} \dfrac{1}{t} \sin\dfrac{1}{t}, \quad t \in (0, 1], \; i = 1, 2, \ldots,$$

where $[x]$ denotes the greatest integer $\le x$, and

$$q_i(s) = \frac{1}{2^i} \sum_{m=1}^{i} \sum_{k=1}^{\infty} \frac{\frac{\pi}{2} + \arctan(k^{\frac{1}{m}}t)}{(km)^2}, \quad s \in \mathbb{R}, \; i = 1, 2, \ldots.$$

Let $E = l^1$ be ordered by the cone l_+^1 of those elements of l^1 with nonnegative coordinates. For $x = (x_1, x_2, \ldots) \in l^1$, define

$$g = (g_1, g_2, \ldots), \quad \text{where} \quad g_i(t, x) = h_i(t) + q_i\left(\sum_{j=1}^{i} x_j\right)$$

for $i = 1, 2, \ldots, t \in J$. Then one can easily verify that $f(t, u) = g(t, u(t))$ satisfies hypotheses (f0)–(f2).

It can be shown (cf. [133, Example 1.1.1]) that the set

$$W = \left\{ \begin{array}{l} 1 - 2^{-m-1} - \sum_{k=0}^{m} 2^{-k-m-2} \prod_{j=0}^{k} 2^{-n_j} - 2^{-2m-2} \prod_{j=0}^{m} 2^{-n_j} : \\ m, n_0, \ldots, n_m \in \mathbb{N}_0 \end{array} \right\}$$

is a well-ordered subset of J. Define a mapping $\Gamma : \mathbb{R} \to J$ by

$$\Gamma(s) = \begin{cases} \min\{t \in W \cup \{1\} : s < t\}, & s < 1, \\ 1, & s \geq 1. \end{cases}$$

Denoting

$$c(\lambda) = (c_1(\lambda), c_2(\lambda), \ldots), \quad \text{where} \quad c_i(\lambda) = 2^{-i}(\Gamma(\lambda) - \lambda), \quad \lambda \in W, \; i = 1, 2, \ldots,$$

the series $\sum_{\lambda \in \Lambda} c(\lambda)$ is seen to be absolutely convergent. Thus the function $D(\cdot, u) \equiv c$ has the properties assumed in (D0). With c and g defined before, consider the problem

$$\begin{cases} u'(t) = g(t, u(t)) \quad \text{a.e. on} \; [0, 1], \\ u(t_0) = x_0, \quad \Delta u(\lambda) = c(\lambda), \; \lambda \in W. \end{cases} \tag{7.35}$$

The above proof shows that the hypotheses of Theorem 7.18 are valid, when $f(t, u) = g(t, u(t))$ and $D(\lambda, u) = c(\lambda)$. Thus problem (7.35) has smallest and greatest solutions for each $x_0 \in l^1$.

Remark 7.22. The functional dependence of h, f, and D on u may be given, e.g., in terms of bounded, linear, and positive operators, such as integral operators of Volterra and/or Fredholm type with nonnegative kernels. Thus the results derived in this section can be applied also to integro-differential equations.

7.1.4 A Volterra Equation Containing HL Integrable Functions

In this subsection we assume that E is a weakly sequentially complete Banach lattice. We shall prove existence and comparison results for the Volterra integral equation

$$u(t) = h(t, u) + {}^K\!\!\int_a^t g(s, u(s), u)ds, \quad t \in J := [a, b], \qquad (7.36)$$

where $h : J \times L^1(J, E) \to E$ and $g : J \times E \times L^1(J, E) \to E$. In the proof we apply the following special case of Proposition 2.40.

Lemma 7.23. *Let $G : L^1(J, E) \to L^1(J, E)$ be increasing with respect to the a.e. pointwise ordering, and assume that G maps monotone sequences to convergent sequences in $(L^1(J, E), \| \cdot \|_1)$. Then G has*

(a) minimal and maximal fixed points;
(b) the smallest and greatest fixed points u_* and u^*, respectively, within the order interval $[\underline{u}, \overline{u}]$ of $L^1(J, E)$, where $\underline{u}$ is the greatest solution of $u = -(-Gu)^+$ and $\overline{u}$ is the smallest solution of $u = (Gu)^+$.
(c) All the solutions $\underline{u}$, $\overline{u}$, u_* and u^* are increasing with respect to G.

Proof: As shown in Lemma 6.8, the mappings $E \ni x \mapsto x^\pm$ are continuous. Hence, if $u \in L^1(J, E)$, then the mappings $u^\pm := t \mapsto u(t)^\pm$ belong to $L^1(J, E)$. Consequently, the zero function is an order center of $L^1(J, E)$. Moreover, $L^1(J, E)$ is an ordered normed space. Noticing also that $\sup\{0, u\} = u^+$ and $\inf\{0, u\} = -(-u)^+$, the conclusions follow from Proposition 2.40.   □

On the functions h and g of (7.36) we impose the following hypotheses:

(h)   The function $(t, u) \mapsto h(t, u)$ is increasing in u for a.e. $t \in J$, and strongly measurable in t for all $u \in L^1(J, E)$. There exists an $\alpha \in L^1(J, \mathbb{R}_+)$ such that $\|h(t, u)\| \le \alpha(t)$ for a.e. $t \in J$ and for all $u \in L^1(J, E)$.
(ga) $g(\cdot, u(\cdot), u)$ is HL integrable for all $u \in L^1(J, E)$.
(gb) If $u \le v$ in $L^1(J, E)$, then $g(s, u(s), u) \le g(s, v(s), v)$ for a.e. $s \in J$.
(gc) The sequence $\left({}^K\!\!\int_a^b g(s, u_n(s), u_n)\, ds \right)_{n=0}^\infty$ is bounded whenever $(u_n)_{n=0}^\infty$ is a monotone sequence in $L^1(J, E)$.

Theorem 7.24. *Under hypotheses (h), (ga), (gb), and (gc), the equation (7.36) has*

(a) minimal and maximal solutions;
(b) the smallest and greatest solutions u_* and u^*, respectively, within the order interval $[\underline{u}, \overline{u}]$ of $L^1(J, E)$, where $\underline{u}$ is the greatest solution of the equation

$$u(t) = - \left(-h(t, u) - {}^K\!\!\int_a^t g(s, u(s), u)ds \right)^+, \quad t \in J,$$

and $\overline{u}$ is the smallest solution of the equation

$$u(t) = \left(h(t, u) + {}^K\!\!\int_a^t g(s, u(s), u)ds \right)^+, \quad t \in J.$$

(c) All the solutions $\underline{u}$, $\bar{u}$, u_* and u^* are increasing with respect to g and h.

Proof: According to hypotheses (h), (ga), and (gb), the relation

$$Gu(t) = h(t, u) + {}^K\!\!\int_a^t g(s, u(s), u)ds, \quad t \in J, \tag{7.37}$$

defines a mapping $G : L^1(J, E) \to L^1(J, E)$, which is increasing by Lemma 9.11. To prove that G maps monotone sequences to convergent sequences, let $(u_n)_{n=0}^\infty$ be an increasing sequence in $L^1(J, E)$. Denoting

$$v_n(t) = g(t, u_n(t), u_n), \quad w_n(t) = {}^K\!\!\int_a^t v_n(s)\, ds, \quad t \in J, \ n \in \mathbb{N}, \tag{7.38}$$

we get increasing sequences of HL integrable functions $v_n : J \to E$, and continuous functions $w_n : J \to E$. Thus

$$0 \le w_m(t) - w_n(t) = {}^K\!\!\int_a^t \big(v_m(s) - v_n(s)\big)\, ds \le {}^K\!\!\int_a^b \big(v_m(s) - v_n(s)\big)\, ds$$

whenever $t \in J$, and $n \le m$. This result and (6.23) imply the estimate

$$\|w_m(t) - w_n(t)\| \le \left\| {}^K\!\!\int_a^b \big(v_m(s) - v_n(s)\big)\, ds \right\|, \quad n \le m. \tag{7.39}$$

Hypothesis (gc) implies that the sequence $({}^K\!\!\int_a^b v_n(s)\, ds)_{n=0}^\infty$ is bounded in E. This sequence is also increasing in view of Lemma 9.11. Moreover, the order cone of E is normal by (6.23), and hence fully regular due to Lemma 9.3. Thus the sequence $({}^K\!\!\int_a^b v_n(s)\, ds)_{n=0}^\infty$ converges, whence it is a Cauchy sequence in E. This result and (7.39) imply that $(w_n)_{n=0}^\infty$ converges uniformly on J, and hence also in $(L^1(J, E), \|\cdot\|_1)$. In view of hypothesis (h), the sequence $(h(\cdot, u_n))_{n=0}^\infty$ is increasing and a.e. pointwise bounded by $\alpha \in L^1(J, \mathbb{R}_+)$. Thus $(h(\cdot, u_n))_{n=0}^\infty$ converges a.e. pointwise, since the order cone of E is fully regular. These results and the dominated convergence theorem ensure that $(h(\cdot, u_n))_{n=0}^\infty$ converges in $(L^1(J, E), \|\cdot\|_1)$. Since $Gu_n(t) = h(t, u_n) + w_n(t)$ for all $t \in J$ and $n \in \mathbb{N}$, the above proof shows that the sequence $(Gu_n)_{n=0}^\infty$ converges in $(L^1(J, E), \|\cdot\|_1)$ whenever $(u_n)_{n=0}^\infty$ is an increasing sequence in $L^1(J, E)$.

The proof that $(Gu_n)_{n=0}^\infty$ converges in $(L^1(J, E), \|\cdot\|_1)$ whenever $(u_n)_{n=0}^\infty$ is a decreasing sequence in $L^1(J, E)$ is similar.

The above proof shows that G satisfies the hypotheses of Lemma 7.23. Since solutions of the equation (7.36) and the fixed points of G are the same, and since G given by (7.37) is increasing with respect to h and g, the conclusions of the theorem follow from Lemma 7.23. $\square$

Remark 7.25. The result of Theorem 7.24 can be applied, for instance, to the following impulsive initial value problem

$$\begin{cases} u'(t) = g(t, u(t), u) \ \text{ a.e. on } \ J = [a, b], \\ u(a) = x_0, \ \Delta u(\lambda) = D(\lambda, u), \ \lambda \in W, \end{cases} \tag{7.40}$$

where $g : J \times E \times L^1(J, E) \to E$, $x_0 \in E$, $\Delta u(\lambda) = u(\lambda + 0) - u(\lambda)$, $D : W \times L^1(J, E) \to E$, and W is a well-ordered subset of (a, b).

7.2 Integral Equations in L^p-Spaces

In this section we derive existence and comparison results for functional integral equations of Urysohn, Fredholm, and Volterra type in L^p-spaces. The obtained results are applied to initial and boundary value problems of first and second order impulsive functional differential equations. We also study cases where ordinary iterative methods are applicable, and use them to solve concrete impulsive boundary value problems involving discontinuities and functional dependencies. The main features of this section are:

- The functions in the considered integral equations may be discontinuous in all their arguments.
- The usual hypotheses dealing with equations in ordered Banach spaces, such as normality, (full) regularity and/or solidity of their order cones, are not assumed in our main theorems.
- Neither subsolutions nor supersolutions are required in our study.

7.2.1 Preliminaries

Let $\Omega = (\Omega, \mathcal{A}, \mu)$ be a σ-finite measure space, and let E be a lattice-ordered Banach space that has the following properties.

(E0) Bounded and increasing sequences of E have weak limits.
(E1) The mapping $E \ni x \mapsto x^+ := \sup\{0, x\}$ is demicontinuous, and $\|x^+\| \le \|x\|$, $x \in E$.

Recall that since E is lattice-ordered, then $x^- := \sup\{0, -x\}$ can be represented in the form $x^- = x^+ - x$. Consequently, if E has property (E1), then also the mapping $x \mapsto x^-$ is demicontinuous. Applying Proposition 9.2, property (E1), and the definitions of the p-norms, we obtain the following result.

Lemma 7.26. *Let E be a lattice-ordered Banach space with property (E1). If $u \in L^p(\Omega, E)$, $1 \le p \le \infty$, then the mappings $u^\pm := t \mapsto u(t)^\pm$ are also in $L^p(\Omega, E)$. Moreover, $\|u^\pm\|_p \le \|u\|_p$.*

Assume that $L^p(\Omega, E)$ is ordered a.e. pointwise, i.e.,

$$u \le v \text{ if and only if } u(t) \le v(t) \text{ for a.e. } t \in \Omega. \tag{7.41}$$

Given $w \in L^p(\Omega, \mathbb{R}_+)$, denote

$$P = \{u \in L^p(\Omega, E) : \|u(t)\| \le w(t) \text{ for a.e. } t \in \Omega\}. \tag{7.42}$$

Our main existence and comparison results are based on the following fixed point result.

Proposition 7.27. *Let E be a lattice-ordered Banach space with properties (E0) and (E1), let P be given by (7.42) with $w \in L^p(\Omega, \mathbb{R}_+)$ and $p \in [1, \infty]$, and assume that $G : P \to P$ is an increasing mapping. Then*

(a) G has minimal and maximal fixed points;
(b) Equation $u = \inf\{0, Gu\}$ has the greatest solution $\underline{u}$, and equation $u = \sup\{0, Gu\}$ has the smallest solution $\overline{u}$.
(c) G has the smallest and greatest fixed points u_* and u^* in $[\underline{u}, \overline{u}] \cap P$.

Moreover, $\underline{u}$, $\overline{u}$, u_*, and u^* are increasing with respect to G.

Proof: If C is a well-ordered subset of $G[P]$, then C is norm-bounded and a.e. pointwise bounded subset of P. Thus C contains by Proposition 9.34 a sequence (u_n) that converges a.e. pointwise weakly to $u = \sup C$. Moreover,

$$\|u(t)\| \le \liminf_{n \to \infty} \|u_n(t)\| \le w(t) \text{ for a.e. } t \in \Omega,$$

whence $u = \sup C \in P$. Because $\inf C = \min C$, then $\inf C \in P$. If D is a nonempty and inversely well-ordered subset of P, then $C = -D$ is nonempty, well-ordered, norm-bounded and a.e. pointwise bounded subset of $-P = P$, whence $\sup C$ exists in P by the above proof. Thus $\inf D = -\sup C$ exists and belongs to P. Moreover, $\sup D = \max D \in P$. If $u \in P$, then $u^+ = t \mapsto \sup\{0, u(t)\}$ belongs to $L^p(\Omega, E)$ by Lemma 7.26. Property (E1) and (7.42) imply that

$$\|u^+(t)\| = \|u(t)^+\| \le \|u(t)\| \le w(t) \quad \text{for a.e. } t \in \Omega.$$

Thus $u^+ = \sup\{0, u\} \in P$. The proof that $\inf\{0, u\}$ exists in $L^p(\Omega, E)$ and belongs to P is similar, whence 0 is an order center of P. The assertions follow then from Theorem 2.17. □

7.2.2 Urysohn Equations

Let E be an ordered Banach space, and let $\Omega = (\Omega, \mathcal{A}, \mu)$ be a σ-finite measure space. In this section we derive existence and comparison results for the functional integral equation

$$u(t) = h(t, u) + \lambda \int_\Omega f(t, s, u(s), u) \, d\mu(s), \quad t \in \Omega, \qquad (7.43)$$

where $h : \Omega \times L^p(\Omega, E) \to E$, $f : \Omega \times \Omega \times E \times L^p(\Omega, E) \to E$, $1 \le p \le \infty$, and $\lambda \ge 0$. Assuming that $L^p(\Omega, E)$ is equipped with a.e. pointwise ordering (7.41), we impose the following hypotheses on the functions h and f.

(f0) $f(t, \cdot, u(\cdot), u)$ is integrable for each $t \in \Omega$ and $\int_\Omega f(\cdot, s, u(s), u) \, d\mu(s)$ is μ-measurable whenever $u \in L^p(\Omega, E)$.

(f1) $f(t, s, z, u)$ is increasing with respect to z and u for a.e. $(t, s) \in \Omega \times \Omega$.

(f2) There exists a $\beta \in L^p(\Omega, \mathbb{R}_+)$ such that $\| \int_\Omega f(t, s, u(s), u) \, d\mu(s) \| \le \beta(t)$ for all $u \in L^p(\Omega, E)$ and for a.e. $t \in \Omega$.

(h0) $h(t, \cdot)$ is increasing for a.e. $t \in \Omega$, and $h(\cdot, u)$ is μ-measurable for all $u \in L^p(\Omega, E)$.

(h1) There exists an $\alpha \in L^p(\Omega, \mathbb{R}_+)$ such that $\|h(t, u)\| \le \alpha(t)$ for all $u \in L^p(\Omega, E)$ and for a.e. $t \in \Omega$.

As a consequence of Proposition 7.27 we get the following results.

Theorem 7.28. *Let E be a lattice-ordered Banach space with properties (E0) and (E1), and assume that the hypotheses (f0)–(f2), (h0), and (h1) are satisfied. Then the equation (7.43) has minimal and maximal solutions for every $\lambda \ge 0$. It has also the smallest and greatest solutions u_- and u_+ in $[\underline{u}, \overline{u}]$, where $\underline{u}$ is the greatest solution of the integral equation*

$$u(t) = -(h(t, u) + \lambda \int_\Omega f(t, s, u(s), u) \, d\mu(s))^-, \quad t \in \Omega, \qquad (7.44)$$

and $\overline{u}$ is the smallest solution of the integral equation

$$u(t) = (h(t, u) + \lambda \int_\Omega f(t, s, u(s), u) \, d\mu(s))^+, \quad t \in \Omega. \qquad (7.45)$$

Moreover, the solutions $\underline{u}$, $\overline{u}$, u_-, and u_+ are increasing with respect to h and f.

Proof: Let $\lambda \ge 0$ be given, and let P be defined by (7.42), with $w = \alpha + \lambda \beta$, where α and β are as in (f2) and (h1). The given hypotheses imply that the relation

$$Gu(t) = h(t, u) + \lambda \int_\Omega f(t, s, u(s), u) \, d\mu(s), \quad t \in \Omega, \qquad (7.46)$$

defines a mapping $G : L^p(\Omega, E) \to P$. Because $\lambda \ge 0$, the hypotheses (f1) and (h0) and Lemma 9.4 imply that if $u, v \in P$ and $u \le v$, then

$$Gu(t) = h(t, u) + \lambda \int_\Omega f(t, s, u(s), u) \, d\mu(s)$$

$$\le h(t, v) + \lambda \int_\Omega f(t, s, v(s), v) \, d\mu(s) = Gv(t)$$

for a.e. $t \in \Omega$. This proves that G is increasing. Thus the hypotheses of Proposition 7.27 are valid. Since $G[L^p(\Omega, E)] \subseteq P$, then the following results hold:

(a) G has minimal and maximal fixed points.
(b) Equation $u = \inf\{0, Gu\}$ has the greatest solution $\underline{u}$, and equation $u = \sup\{0, Gu\}$ has the smallest solution $\overline{u}$ in P.
(c) G has the smallest and greatest fixed points u_- and u_+ in $[\underline{u}, \overline{u}]$.
(d) The solutions $\underline{u}$, $\overline{u}$, u_-, and u_+ are increasing with respect to G.

From (7.46), (7.44), and (7.45), it follows that minimal and maximal fixed points of G are minimal and maximal solutions of (7.43) in P, that $\underline{u}$ is the greatest solution of (7.44) in P, that $\overline{u}$ is the smallest solution of (7.45) in P, and that u_- and u_+ are the smallest and greatest solutions of (7.43) in $[\underline{u}, \overline{u}] \cap P$. Moreover, by the hypotheses (f1) and (h0) and Lemma 9.4, the operator G is increasing with respect to the functions h and f. This result along with (d) imply that the solutions $\underline{u}$, $\overline{u}$, u_-, and u_+ of (7.43) are increasing with respect to h and f. □

Next we consider the cases when the extremal solutions of the integral equation (7.43) can be obtained by successive approximations.

Proposition 7.29. *Let E be a lattice-ordered Banach space with properties (E0) and (E1), assume that the hypotheses (f0)–(f2), (h0), and (h1) are satisfied, and that the following hypothesis holds:*

(D) $h(t, u_n) \rightharpoonup h(t, u)$ and $f(t, s, u_n(s), u_n) \rightharpoonup f(t, s, u(s), u)$ for a.e. $t, s \in \Omega$ if (u_n) is a bounded and monotone sequence in P and $u_n(s) \rightharpoonup u(s)$ for a.e. $s \in \Omega$.

Then the successive approximations:

(a) $a_{n+1}(t) = -(h(t, a_n) + \lambda \int_\Omega f(t, s, a_n(s), a_n) \, d\mu(s))^-$, $t \in \Omega$, $a_0 = 0$, converge weakly a.e. pointwise to the greatest solution $\underline{u}$ of (7.44);
(b) $b_{n+1}(t) = (h(t, b_n) + \lambda \int_\Omega f(t, s, b_n(s), b_n) \, d\mu(s))^+$, $t \in \Omega$, $b_0 = 0$, converge weakly a.e. pointwise to the smallest solution $\overline{u}$ of (7.45);
(c) $u_{n+1}(t) = h(t, u_n) + \lambda \int_\Omega f(t, s, u_n(s), u_n) \, d\mu(s)$, $t \in \Omega$, $u_0 = \underline{u}$, converge weakly a.e. pointwise to the smallest solution u_- of (7.43) in $[\underline{u}, \overline{u}]$;
(d) $v_{n+1}(t) = h(t, v_n) + \lambda \int_\Omega f(t, s, v_n(s), v_n) \, d\mu(s)$, $t \in \Omega$, $v_0 = \overline{u}$, converge weakly a.e. pointwise to the greatest solution u_+ of (7.43) in $[\underline{u}, \overline{u}]$.

Proof: Let us prove assertion (b). It follows from (7.45) and (b) that $b_n(t) = \sup\{0, G^n 0(t)\}$ for each $n \in \mathbb{N}_0$, where $G : L^p(\Omega, E) \to P$ is defined by (7.46). Since G is increasing, then

$$0 \equiv b_0(t) \le b_1(t) \le \cdots \le b_n(t)$$

for a.e. $t \in \Omega$. In particular, (b_n) is increasing, bounded, and a.e. pointwise bounded. Thus the a.e. pointwise weak limit $\overline{u}$ of (b_n) exists. Moreover, $\overline{u} = \sup_n b_n$ belongs to P. The hypothesis (D) implies that

$$\varphi(f(t, s, b_n(s), b_n)) \to \varphi(f(t, s, \overline{u}(s), \overline{u})) \quad \text{for a.e. } t,\ s \in \Omega \text{ and for all } \varphi \in E'.$$

This result and the Dominated Convergence Theorem yield

$$\int_\Omega \varphi(f(t, s, b_n(s), b_n)\, d\mu(s)) \to \int_\Omega \varphi(f(t, s, \overline{u}(s), \overline{u}))\, d\mu(s),$$

or equivalently,

$$\varphi\left(\int_\Omega f(t, s, b_n(s), b_n)\, d\mu(s)\right) \to \varphi\left(\int_\Omega f(t, s, \overline{u}(s), \overline{u})\, d\mu(s)\right)$$

for a.e. $t \in \Omega$ and for all $\varphi \in E'$. In view of this result and the hypothesis (D) we have $h(t, b_n) \rightharpoonup h(t, \overline{u})$ and

$$\int_\Omega f(t, s, b_n(s), b_n)\, d\mu(s) \rightharpoonup \int_\Omega f(t, s, \overline{u}(s), \overline{u})\, d\mu(s)$$

for a.e. $t \in \Omega$. By property (E1), and passing to the limit as $n \to \infty$ in (b), it then follows that $\overline{u}$ is a solution of (7.45). By standard arguments one can show that $\overline{u}$ is the smallest solution of (7.45). Similar arguments apply to show the following results:

- The sequence (a_n) defined in (a) is decreasing, equals to $(\inf\{0, G^n 0\})$, where $G : L^p(\Omega, E) \to P$ is defined by (7.46), and converges weakly a.e. pointwise to the greatest solution $\underline{u}$ of (7.44);
- The sequence (u_n) defined in (c) is increasing, equals to $(G^n \underline{u})$, where $G : L^p(\Omega, E) \to P$ is defined by (7.46), and converges weakly a.e. pointwise to the smallest solution u_* of (7.43) in $[\underline{u}, \overline{u}]$;
- The sequence (v_n) defined in (d) is decreasing, equals to $(G^n \overline{u})$, where $G : L^p(\Omega, E) \to P$ is defined by (7.46), and converges weakly a.e. pointwise to the greatest solution u^* of (7.43) in $[\underline{u}, \overline{u}]$.

$\square$

7.2.3 Fredholm Integral Equations

In this section we derive existence and comparison result for the functional Fredholm integral equation

$$u(t) = h(t, u) + \lambda \int_\Omega k(t, s) g(s, u(s), u)\, d\mu(s), \quad t \in \Omega, \tag{7.47}$$

where $h : \Omega \times L^p(\Omega, E) \to E$, $g : \Omega \times E \times L^p(\Omega, E) \to E$, $1 \le p \le \infty$, $k : \Omega \times \Omega \to \mathbb{R}_+$, and $\lambda \ge 0$. Assuming that $L^p(\Omega, E)$ is equipped with a.e. pointwise ordering (7.41), we impose the following hypotheses on the functions g, h, and k.

(g0) $g(\cdot, u(\cdot), u)$ is μ-measurable for each $u \in L^p(\Omega, E)$.

(g1) $g(s, x, u)$ is increasing with respect to x and u for a.e. $s \in \Omega$.

(g2) There exists an $m \in L^p(\Omega, \mathbb{R})$ such that $\|g(s, x, u)\| \le \|x\| + m(s)$ for a.e. $s \in \Omega \times \Omega$ and all $x \in E$, $u \in L^p(\Omega, E)$,

(h0) $h(t, \cdot)$ is increasing for a.e. $t \in \Omega$, and $h(\cdot, u)$ is μ-measurable for all $u \in L^p(\Omega, E)$.

(h1) There exists an $\alpha \in L^p(\Omega, \mathbb{R}_+)$ such that $\|h(t, u)\| \le \alpha(t)$ for a.e. $t \in \Omega$ and all $u \in L^p(\Omega, E)$.

(k0) $k : \Omega \times \Omega \to \mathbb{R}_+$ is product measurable, and there exists a $K \ge 0$ such that $\int_\Omega k(t, s)\, d\mu(s) \le K$ for each $t \in \Omega$ and $\int_\Omega k(t, s)\, d\mu(t) \le K$ for each $s \in \Omega$.

Lemma 7.30. *Assume that $0 \le \lambda\, \rho(T) < 1$, where $\rho(T)$ is the spectral radius of the operator $T : L^p(\Omega, \mathbb{R}) \to L^p(\Omega, \mathbb{R})$ defined by*

$$Tw(t) = \int_\Omega k(t, s) w(s)\, d\mu(s), \quad t \in \Omega. \tag{7.48}$$

Then the integral equation

$$w(t) = \alpha(t) + \lambda \int_\Omega k(t, s)(w(s) + m(s))\, d\mu(s) \tag{7.49}$$

has a unique solution $w \in L^p(\Omega, \mathbb{R}_+)$.

Proof: The hypothesis (k0) implies by [164, VII, Theorem 5.6] that (7.48) defines an operator $T : L^p(\Omega, \mathbb{R}) \to L^p(\Omega, \mathbb{R})$, and that

$$\|Tw\|_p \le K \|w\|_p \quad \text{for each } w \in L^p(\Omega, \mathbb{R}).$$

Thus T is a bounded and linear operator. If $0 \le \lambda\, \rho(T) < 1$, where $\rho(T)$ is the spectral radius of T, then for each $v \in L^p(\Omega, \mathbb{R})$ the function $w = \sum_{n=0}^\infty (\lambda T)^n v$ is the unique solution of equation

$$w = v + \lambda Tw.$$

This result and (7.48) imply, by choosing

$$v(t) = \alpha(t) + \lambda \int_\Omega k(t, s) m(s)\, d\mu(s),$$

where α and m are as in (h1) and (g2), that w is the unique solution of the integral equation (7.49). $\qquad \square$

As an application of Proposition 7.27 and Lemma 7.30 we prove the following existence and comparison result for the integral equation (7.47).

Proposition 7.31. *Assume that E is a lattice-ordered Banach space with properties (E0) and (E1), and that g, h, and k satisfy the hypotheses (g0)–(g2), (h0), (h1), and (k0). Assume that $0 \le \lambda\, \rho(T) < 1$, where $\rho(T)$ is the spectral radius of the operator $T : L^p(\Omega, \mathbb{R}) \to L^p(\Omega, \mathbb{R})$, defined by (7.48). Then the equation (7.47) has*

(a) *minimal and maximal solutions,*

(b) *the smallest and greatest solutions* u_- *and* u_+ *in* $[\underline{u}, \overline{u}]$, *where* $\underline{u}$ *is the greatest solution of the integral equation*

$$u(t) = -(h(t, u) + \lambda \int_\Omega k(t, s)g(s, u(s), u)\, d\mu(s))^-, \quad t \in \Omega,$$

and $\overline{u}$ *is the smallest solution of the integral equation*

$$u(t) = (h(t, u) + \lambda \int_\Omega k(t, s)g(s, u(s), u)\, d\mu(s))^+, \quad t \in \Omega.$$

Moreover, the solutions u_- *and* u_+ *are increasing with respect to* h *and* f.

Proof: Let $w \in L^p(\Omega, \mathbb{R})$ be the unique solution of the integral equation (7.49), and let $P \subset L^p(\Omega, E)$ be defined by (7.49). If $u \in P$, the hypotheses (h1) and (g2) imply that

$$\|h(t, u)\| \le \alpha(t), \quad \text{for a.e. } t \in \Omega, \text{ and}$$
$$\|g(s, u(s), u)\| \le \|u(s)\| + m(s) \le w(s) + m(s) \text{ for a.e. } s \in \Omega.$$

The above inequalities and the hypotheses (g0), (h0), and (k0) imply that the relation

$$Gu(t) = h(t, u) + \lambda \int_\Omega k(t, s)g(s, u(s), u)\, d\mu(s) \tag{7.50}$$

defines a mapping $G : P \to L^p(\Omega, E)$, and the following estimate holds:

$$\|Gu(t)\| \le \|h(t, u)\| + \lambda \int_\Omega k(t, s)\|g(s, u(s), u)\|\, d\mu(s)$$
$$\le \alpha(t) + \lambda \int_\Omega k(t, s)(\|u(s)\| + m(s))\, d\mu(s)$$
$$\le \alpha(t) + \lambda \int_\Omega k(t, s)(w(s) + m(s))\, d\mu(s) = w(t)$$

for a.e. $t \in \Omega$. This proves that $G[P] \subset P$. Because $\lambda \ge 0$ and k is nonnegative-valued, the hypotheses (g1) and (h0) and Lemma 9.4 imply that if $u, v \in P$ and $u \le v$, then

$$Gu(t) = h(t, u) + \lambda \int_\Omega k(t, s)g(s, u(s), u)\, d\mu(s)$$
$$\le h(t, v) + \lambda \int_\Omega k(t, s)g(s, v(s), v)\, d\mu(s) = Gv(t)$$

for a.e. $t \in \Omega$, which shows that G is increasing. Thus the hypotheses of Proposition 7.27 are valid. Hence it is enough to show that

– all the solutions of (7.48) are contained in P.

So, let $u \in L^p(\Omega, E)$ be a solution of (7.47). Applying (h1) and (f2) we get

$$\|u(t)\| \leq \|h(t, u\| + \lambda \int_\Omega k(t, s)\|g(s, u(s), u)\| \, d\mu(s)$$

$$\leq \alpha(t) + \lambda \int_\Omega k(t, s)(\|u(s)\| + m(s)) \, d\mu(s)$$

for a.e. $t \in \Omega$. Thus the function $q = t \mapsto \|u(t)\|$ satisfies the inequality

$$q(t) \leq \alpha(t) + \lambda \int_\Omega k(t, s)(q(s) + m(s)) \, d\mu(s)$$

for a.e. $t \in \Omega$. Denoting $v(t) = \alpha(t) + \lambda \int_\Omega k(t, s)m(s) \, d\mu(s)$, then the above inequality can be rewritten as

$$q \leq v + \lambda Tq,$$

where T is defined by (7.48). Since k is nonnegative-valued, then the operator T is positive. Thus the above inequality and the Abstract Gronwall Lemma (cf. [228, Proposition 7.15]) imply that $q \leq w$. Thus $\|u(t)\| \leq w(t)$ for a.e. $t \in \Omega$, so that $u \in P$. The assertions follow now from Proposition 7.27 and from (7.50). □

Assume next that E is a *weakly sequentially complete Banach lattice*, i.e., E is lattice-ordered, its weak Cauchy sequences posses weak limits, and the norm $\| \cdot \|$ of E and its valuation $E \ni x \mapsto |x| = \sup\{x, -x\}$ satisfy

(E) $\|x\| \leq \|y\|$ whenever $x, y \in E$ and $|x| \leq |y|$.

As an application of Proposition 2.18 we shall prove that the integral equation (7.47) has the smallest and greatest solution when the hypotheses (f2) and (h1) in Proposition 7.31 are replaced by the following hypotheses.

(f3) There exists an $m \in L^p(\Omega, E)$ such that $|g(s, x, u)| \leq |x| + m(s)$ for a.e. $s \in \Omega$ and all $x \in E$, $u \in L^p(\Omega, E)$.
(h2) There exists an $\alpha \in L^p(\Omega, E)$ such that $|h(t, u)| \leq \alpha(t)$ for a.e. $t \in \Omega$ and all $u \in L^p(\Omega, E)$.

Proposition 7.32. *Assume that E is a weakly sequentially complete Banach lattice, and that g, h, and k satisfy the hypotheses (g0), (g1), (g3), (h0), (h2), and (k0). Then the integral equation (7.43) has the smallest and greatest solutions in $L^p(\Omega, E)$ whenever $0 \leq \lambda \rho(T) < 1$, where $\rho(T)$ is the spectral radius of the operator $T : L^p(\Omega, E) \to L^p(\Omega, E)$, defined by*

$$Tw(t) = \int_\Omega k(t, s)w(s) \, d\mu(s), \quad t \in \Omega. \tag{7.51}$$

Moreover, these extremal solutions of (7.43) are increasing with respect to h and f.

Proof: As in the proof of Lemma 7.30 one can show that the following results hold.

- The equation (7.51) defines a bounded and linear operator $T : L^p(\Omega, E) \to L^p(\Omega, E)$.

If $0 \leq \lambda\, \rho(T) < 1$, then for each $v \in L^p(\Omega, E)$ the function $b = \sum_{n=0}^{\infty}(\lambda\, T)^n v$ is the unique solution of the integral equation

$$u(t) = \alpha(t) + \lambda \int_\Omega k(t,s)(u(s) + m(s))\, d\mu(s) \tag{7.52}$$

in $L^p(\Omega, E_+)$. If $u \in [-b, b]$, or equivalently, if $|u(t)| \leq b(t)$ for a.e $t \in \Omega$, the hypothesis (g3) implies that

$$|g(s, u(s), u)| \leq |u(s)| + m(s) \leq b(s) + m(s) \text{ for a.e. } s \in \Omega.$$

By means of the last inequality, and hypothesis (h2) as well as (E), we get

$$\|h(t, u\| \leq \|\alpha(t)\|, \quad \text{for a.e. } t \in \Omega, \text{ and}$$
$$\|g(s, u(s), u)\| \leq \|u(s)\| + \|m(s)\| \leq \|b(s)\| + \|m(s)\| \text{ for a.e. } s \in \Omega.$$

By the last inequalities along with hypotheses (g0), (h0), and (k0), we infer that the relation (7.50) defines a mapping $G : [-b, b] \to L^p(\Omega, E)$. If $u \in [-b, b]$, then

$$|Gu(t)| \leq |h(t, u| + \lambda \int_\Omega k(t,s)|g(s, u(s), u)|\, d\mu(s)$$

$$\leq \alpha(t) + \lambda \int_\Omega k(t,s)(|u(s)| + m(s))\, d\mu(s)$$

$$\leq \alpha(t) + \lambda \int_\Omega k(t,s)(b(s) + m(s))\, d\mu(s) = b(t)$$

for a.e. $t \in \Omega$. This proves that $G[-b, b] \subset [-b, b]$. Moreover, if $u \in [-b, b]$, then $|Gu(t)| \leq b(t)$ by the above proof, and hence $\|Gu(t)\| \leq \|b(t)\|$ for a.e. $t \in \Omega$ by (E). This result implies that $\|Gu\|_p \leq \|b\|_p$ for each $u \in [-b, b]$, so that $G[-b, b]$ is a bounded subset of $[-b, b]$. The order cone of a weakly sequentially complete Banach lattice is regular by Lemma 9.3. Thus all chains of $[-b, b]$ have supremums and infimums in $[-b, b]$ by [133, Propositions 1.3.2, 5.8.7, and 5.8.8]. The above proof shows that all the hypotheses of Proposition 2.18 hold when $P = [-b, b]$. Thus G has the smallest and greatest fixed points u_* and u^*, which by (7.50) are the smallest and greatest solutions of the integral equation (7.47) in $[-b, b]$. To prove that u_* and u^* are the smallest and greatest of all the solutions of (7.47) in $L^p(\Omega, E)$, it is enough to show that all the solutions of (7.47) are contained in $[-b, b]$. So, let $u \in L^p(\Omega, E)$ be a solution of (7.47). Applying (h2) and (g3) we get

$$|u(t)| \leq |h(t, u| + \lambda \int_\Omega k(t, s)|g(s, u(s), u)| \, d\mu(s)$$

$$\leq \alpha(t) + \lambda \int_\Omega k(t, s)(|u(s)| + m(s)) \, d\mu(s)$$

for a.e. $t \in \Omega$. Thus the function $q = t \mapsto |u(t)|$ satisfies the inequality

$$q(t) \leq \alpha(t) + \lambda \int_\Omega k(t, s)(q(s) + m(s)) \, d\mu(s)$$

for a.e. $t \in \Omega$. Denoting $v(t) = \alpha(t) + \lambda \int_\Omega k(t, s)m(s) \, d\mu(s)$, then the above inequality can be rewritten as

$$q \leq v + \lambda T q,$$

where T is defined by (7.51). Since $\lambda \geq 0$ and k is nonnegative-valued, then the operator T is positive. Thus the above inequality and the Abstract Gronwall Lemma (cf. [228, Proposition 7.15]) imply that $q \leq b$. Thus $|u(t)| \leq b(t)$ for a.e. $t \in \Omega$, so that $u \in [-b, b]$.

The last assertion follows from the definition (7.50) of G since the fixed points u_* and u^* of G are increasing with respect to G by Proposition 2.18.

$\square$

Proposition 7.33. *If the hypotheses of Proposition 7.32 and the hypothesis*

(D') $h(t, u_n) \rightharpoonup h(t, u)$ and $g(s, u_n(s), u_n) \rightharpoonup g(s, u(s), u)$ for a.e. $t, s \in \Omega$ if (u_n) is a bounded and monotone sequence in P and $u_n(s) \rightharpoonup u(s)$ for a.e. $s \in \Omega$.

hold, and if $0 \leq \lambda \rho(T) < 1$, where $\rho(T)$ is the spectral radius of the operator T defined by (7.51), then the smallest solution u_* and the greatest solution u^* of the integral equation (7.47) are strong limits of the successive approximations (u_n) and (v_n) defined by

$$u_{n+1}(t) = h(t, u_n) + \lambda \int_\Omega k(t, s)g(s, u_n(s), u_n) \, d\mu(s), \quad t \in \Omega, \quad u_0 = -b,$$

$$v_{n+1}(t) = h(t, v_n) + \lambda \int_\Omega k(t, s)g(s, v_n(s), v_n) \, d\mu(s), \quad t \in \Omega, \quad v_0 = b,$$

where $b \in L^p(\Omega, E)$ is the solution of the integral equation (7.52).

Remark 7.34. Propositions 7.31 and 7.32 imply existence and comparison results for the ordinary Fredholm integral equation

$$u(t) = h(t) + \int_\Omega k(t, s)q(s, u(s)) \, d\mu(s), \quad t \in \Omega.$$

When $\lambda = 0$ we obtain existence and comparison results for the functional equation

$$u(t) = h(t, u).$$

The hypothesis $0 \leq \lambda \rho(T) < 1$, where $\rho(T)$ is the spectral radius of T, defined by (7.51), cannot be improved, in general. For instance, if $E = \mathbb{R}$, $\Omega = J = [0, 1]$, and

$$Tr(t) = \int_J k(t, s) r(s) \, ds, \quad t \in J, \quad \text{where } k(t, s) = \begin{cases} (1-t)s, & 0 \leq s \leq t, \\ t(1-s), & t \leq s \leq 1, \end{cases}$$

then the integral equation

$$r(t) = t + \lambda \int_J k(t, s) r(s) \, ds, \quad t \in J,$$

or equivalently, the BVP

$$-r''(t)) = \lambda r(t) \quad \text{a.e. on } J = [0, 1], \quad r(0) = 0, \ r(1) = 1,$$

has no solutions when $\lambda = \pi^2 = \frac{1}{\rho(T)}$.

The following example illustrates the applicability of the above results when $E = \mathbb{R}$, $\Omega = J = [0, 1]$ and μ is Lebesgue measure.

Example 7.35. Let H be the Heaviside function, and let $[z]$ denote the greatest integer $\leq z$. Consider the following impulsive boundary value problem in $J = [0, 1]$:

$$\begin{cases} -u''(t) = H(1 - 2t) + \dfrac{1}{2}[1 + t + 2u(t)] + \dfrac{[5 \int_0^1 u(s) ds]}{5 + 5 |[5 \int_0^1 u(s) ds]|} & \text{a.e. on } J, \\[2ex] u(0) = 0, \ u(1) = u(\frac{1}{2}+) - u(\frac{1}{2}), \\[2ex] u(\frac{1}{2}+) - u(\frac{1}{2}) = \dfrac{1}{100} + \dfrac{[5 \int_0^1 u(s) ds]}{5 + 5 |[5 \int_0^1 u(s) ds]|}, \end{cases}$$

$$(7.53)$$

where $s \mapsto [s]$ again denotes the integer function. The BVP (7.53) can be converted to the integral equation

$$u(t) = h(t, u) + \int_J k(t, s) g(s, u(s), u) \, ds, \quad t \in J = [0, 1], \qquad (7.54)$$

where

$$\begin{cases} h(t, u) = (1 - H(2t - 1)) \left(\dfrac{1}{100} + \dfrac{[5 \int_0^1 u(s) ds]}{5 + 5 |[5 \int_0^1 u(s) ds]|} \right), \\[2ex] g(t, x, u) = H(1 - 2t) + \dfrac{1}{2}[1 + t + 2x] + \dfrac{[5 \int_0^1 u(s) ds]}{5 + 5 |[5 \int_0^1 u(s) ds]|}, \\[2ex] k(t, s) = \begin{cases} (1-t)s, & 0 \leq s \leq t, \\ t(1-s), & t \leq s \leq 1, \end{cases} \end{cases}$$

$$(7.55)$$

in the sense that (7.53) and (7.54) have the same solutions in the set

$$Y = \{u \in L^\infty(J, \mathbb{R}) : t \mapsto u(t) - h(t, u) \in W\},$$

where

$$W = \{w : J \to \mathbb{R} : w' \text{ is absolutely continuous}\}.$$

It is easy to see that the functions $g : J \times \mathbb{R} \times L^\infty(J, \mathbb{R}) \to \mathbb{R}$, $h : J \times L^\infty(J, \mathbb{R}) \to \mathbb{R}$ and $k : J \times J \to \mathbb{R}_+$, defined by (7.55), satisfy the hypotheses (g0), (g1), and (g3) with $m(s) \equiv 4$, (h0), (h1), and (k0). If T is defined by (7.51), with k defined by (7.55), then $\rho(T) = \frac{1}{\pi^2} < 1$. Thus the integral equation (7.54), and hence also the BVP (7.53), has by Proposition 7.32 the smallest and greatest solutions, and they belong to the order interval $[-b, b]$, where b is the solution of the integral equation (7.52) with $\alpha(t) \equiv 1$ and k given by (7.55), or equivalently, the solution of the BVP

$$-u''(t) = u(t) + 4 \quad t \in J = [0, 1], \quad u(0) = u(1) = 1.$$

Thus

$$b(t) = 5 \frac{1 - \cos 1}{\sin 1} \sin t + 5 \cos t - 4, \quad t \in J. \tag{7.56}$$

Because the Heaviside function H and the greatest integer function $[\cdot]$ are right-continuous, then also the hypothesis (D') of Proposition 7.33, restricted to decreasing sequences (u_n) of $L^\infty(J, \mathbb{R})$, holds for the functions h and g defined by (7.55). Thus the greatest solution u^* of (7.53) can be obtained as the limit of the sequence of successive approximations v_n defined in Proposition 7.33. In the current case these approximations can be rewritten as

$$v_{n+1}(t) = h(t, v_n) + (1-t) \int_0^t s\, g(s, v_n(s), v_n)\, ds + t \int_t^1 (1-s) g(s, v_n(s), v_n)\, ds,$$

$t \in J$, $v_0 = b$, where g, h, and b are given by (7.55) and (7.56). Calculating these approximations numerically by the Simpson rule, one obtains the following estimates: $\int_0^1 u^*(s)\, ds \approx .2008$ and

$$(u^*)''(t) \approx \begin{cases} -2, & 0 < t < .5, \\ -1.5, & .5 < t < 1. \end{cases}$$

In view of these estimates and (7.53), one can infer that the greatest solution of (7.53) is of the form

$$u^*(t) = \begin{cases} u_1(t) = -t^2 + b_1 t, & 0 \le t \le \frac{1}{2}, \\ u_2(t) = -\frac{3}{4} t^2 + b_2 t + \frac{3}{4} - b_2 + \frac{1}{100} + \frac{1}{10}, & \frac{1}{2} < t \le 1. \end{cases} \tag{7.57}$$

It remains to determine the constants b_1 and b_2. Because the function $t \mapsto u^*(t) - h(t, u^*)$ and its derivative are continuous at $t = \frac{1}{2}$, we get two equations, from which we get

$$b_1 = \frac{15}{16}, \quad b_2 = \frac{11}{16}.$$

Inserting these values to (7.57) we get the following exact formula for the greatest solution of the BVP (7.53):

$$u^*(t) = \begin{cases} -t^2 + \frac{15}{16}t, & 0 \le t \le \frac{1}{2}, \\ -\frac{3}{4}t^2 + \frac{11}{16}t + \frac{69}{400}, & \frac{1}{2} < t \le 1. \end{cases} \tag{7.58}$$

In this case also the smallest solution u_* of (7.53) can be obtained as the limit of successive approximations

$$u_{n+1}(t) = h(t, u_n) + (1-t) \int_0^t s\, g(s, u_n(s), u_n)\, ds + t \int_t^1 (1-s) g(s, u_n(s), u_n)\, ds,$$

$t \in J$, $u_0 = -w$, where g, h, and w are given by (7.55) and (7.56). In this case we get the following estimates: $\int_0^1 u_*(s)\, ds \approx -.0033$ and

$$(u_*)''(t) \approx \begin{cases} -1, & 0 < t < .5, \\ 0 & .5 < t < 1. \end{cases}$$

From these estimates we can infer, as above, the following exact formula for u_* :

$$u_*(t) = \begin{cases} -\frac{1}{2}t^2 + \frac{3}{8}t, & 0 \le x \le \frac{1}{2}, \\ -\frac{1}{8}t + \frac{7}{200}, & \frac{1}{2} < t \le 1. \end{cases} \tag{7.59}$$

7.2.4 Volterra Integral Equations

Let E be a lattice-ordered Banach space that has properties (E0) and (E1), and let J be a real interval with t_0 its left endpoint. In this subsection we study the functional Volterra integral equation

$$u(t) = h(t, u) + \int_{t_0}^t f(t, s, u(s), u)\, ds, \quad t \in J, \tag{7.60}$$

where $h : J \times L^p(J, E) \to E$, $f : \Lambda \times E \times L^p(J, E) \to E$, $1 \le p \le \infty$, and $\Lambda = \{(t, s) \in J \times J : a \le s \le t\}$. For given $w \in L^p(J, \mathbb{R}_+)$, we introduce P by

$$P = \{u \in L^p(J, E) : \|u(t)\| \le w(t) \text{ for a.e. } t \in J\}. \tag{7.61}$$

Hypothesis (E1) implies that the mapping $v^+ = \sup\{0, v\} = t \mapsto \sup\{0, v(t)\}$ belongs to $L^p(J, E)$ for each $v \in L^p(J, E)$, and $\|v^+(t)\| \le \|v(t)\|$ for all $t \in J$. These properties ensure that $v^+ = \sup\{0, v\}$, and hence also $v^- = \sup\{0, -v\}$ as well as $\inf\{0, v\} = -v^-$ belong to P for each $v \in P$.

Assuming that $L^p(J, E)$ is equipped with the a.e. pointwise ordering, we impose the following hypotheses on the functions h and f.

(h0) $h(t, \cdot)$ is increasing for a.e. $t \in J$, $h(\cdot, u)$ is strongly measurable for all $u \in L^p(J, E)$, and there exists an $\alpha \in L^p(J, \mathbb{R})$ such that $\|h(t, u)\| \le \alpha(t)$ for a.e. $t \in J$ and all $u \in L^p(J, E)$.

(f0) The mappings $f(t, \cdot, u(\cdot), u)$, $t \in J$, and $t \mapsto \int_{t_0}^{t} f(t, s, u(s), u) \, ds$ are strongly measurable for each $u \in L^p(J, E)$.

(f1) $f(t, s, z, u)$ is increasing with respect to z and u for a.e. $(t, s) \in \Lambda$.

(f2) $\|f(t, s, x, u)\| \leq g(t, s, \|x\|)$ for a.e. $(t, s) \in \Lambda$ and all $x \in E$, $u \in L^p(J, E)$, where $g : \Lambda \times \mathbb{R}_+ \to \mathbb{R}_+$, $g(t, s, r)$ is increasing in r for a.e. $(t, s) \in \Lambda$, the functions $g(t, \cdot, w(\cdot))$ and $t \mapsto \int_{t_0}^{t} g(t, s, w(s)) \, ds$ are Lebesgue integrable for each $w \in L^p(J, \mathbb{R})$, and the integral equation

$$w(t) = \beta(t) + \int_{t_0}^{t} g(t, s, w(s)) \, ds, \quad t \in J \qquad (7.62)$$

has for each $\beta \in L^p(J, \mathbb{R}_+)$ the greatest solution in $L^p(J, \mathbb{R}_+)$.

As an application of Proposition 7.27 we shall first prove an existence and comparison result for the integral equation (7.60) under the hypotheses given above.

Theorem 7.36. *Let E be a lattice-ordered Banach space with properties (E0) and (E1), and assume that the hypotheses (f0), (f1), (f2), and (h0) are satisfied. Then the equation (7.60) has*

(a) minimal and maximal solutions in $L^p(J, E)$;

(b) the smallest and greatest solutions u_* and u^* in $[\underline{u}, \overline{u}]$, where $\underline{u}$ is the greatest solution of the integral equation

$$u(t) = -\left(h(t, u) + \int_{t_0}^{t} f(t, s, u(s), u) \, ds \right)^{-}, \quad t \in J, \qquad (7.63)$$

and $\overline{u}$ is the smallest solution of the integral equation

$$u(t) = \left(h(t, u) + \int_{t_0}^{t} f(t, s, u(s), u) \, ds \right)^{+}, \quad t \in J \qquad (7.64)$$

in $L^p(J, E)$.

Moreover, the solutions u_* and u^* are increasing with respect to h and f.

Proof: Let P be given by (7.61), where $w \in L^p(J, \mathbb{R}_+)$ is the greatest solution of (7.62) with $\beta = \alpha$. We shall first show that the relation

$$Gu(t) = h(t, u) + \int_{t_0}^{t} f(t, s, u(s), u) \, ds, \quad t \in J, \qquad (7.65)$$

defines a mapping $G : P \to P$. If $u \in P$, then $\|u(t)\| \leq w(t)$ for a.e. $t \in J$. Applying the hypotheses (h0) and (f2) we obtain

$$\|Gu(t)\| \leq \|h(t, u)\| + \int_{t_0}^{t} \|f(t, s, u(s), u)\| \, ds$$

$$\leq \alpha(t) + \int_{t_0}^{t} g(t, s, w(s)) \, ds = w(t)$$

for a.e. $t \in J$. This result implies that G maps P into P. If $u, v \in L^p(J, E)$ and $u \leq v$, then by means of hypotheses (f1) and (h0) we get

$$Gu(t) = h(t, u) + \int_{t_0}^{t} f(t, s, u(s), u) \, ds$$

$$\leq h(t, v) + \int_{t_0}^{t} f(t, s, v(s), v) \, ds = Gv(t)$$

for a.e. $t \in J$. This proves that G is increasing. Thus the hypotheses of Proposition 7.27 hold for $G : P \to P$ defined by (7.65). Assume next that u is a solution of (7.60) in $L^p(J, E)$, and let $\overline{w}$ denote the greatest solution of (7.62) with $\beta(t) = \max\{\|u(t)\|, \alpha(t)\}$. Then for $t \in J$ we obtain

$$\|u(t)\| \leq \|h(t, u)\| + \int_{t_0}^{t} \|f(t, s, u(s), u)\| \, ds \leq \alpha(t) + \int_{t_0}^{t} g(t, s, \|u(s)\|) \, ds$$

$$\leq \alpha(t) + \int_{t_0}^{t} g(t, s, \overline{w}(s)) \, ds \leq \beta(t) + \int_{t_0}^{t} g(t, s, \overline{w}(s)) \, ds = \overline{w}(t).$$

Thus, denoting $\underline{w} = t \mapsto \|u(t)\|$, the relation

$$Qv(t) = \alpha(t) + \int_{t_0}^{t} g(t, s, v(s)) \, ds, \quad t \in J \tag{7.66}$$

defines an increasing mapping Q from the order interval $[\underline{w}, \overline{w}]$ of $L^p(J, \mathbb{R})$ into itself. From [133, Theorem 1.2.3 and Proposition 5.8.9] it follows that Q has a fixed point in $[\underline{w}, \overline{w}]$. But w, as the greatest solution of (7.62), is the greatest fixed point of Q, whence $\|u(t)\| = \underline{w}(t) \leq w(t)$ for a.e. $t \in J$. This proves that all the solutions of (7.60) are contained in P. Because of the property (E1) of E, by a similar reasoning one shows that all the solutions of (7.63) and (7.64) belong to P. Noticing also that fixed points of G defined by (7.65) are solutions of (7.60) and vice versa, the assertions follow from Proposition 7.27. □

Next we consider situations in which the extremal solutions of the integral equation (7.60) can be obtained by successive approximations.

Proposition 7.37. *Let E be a lattice-ordered Banach space with properties (E0) and (E1). Assume that the hypotheses (f0), (f1), (f2), and (h0) hold, and, moreover, that the following hypothesis is fulfilled:*

(B) $h(t, u_n) \rightharpoonup h(t, u)$ *for a.e. $t \in J$ and $f(t, s, u_n(s), u_n) \rightharpoonup f(t, s, u(s), u)$ for a.e. $(t, s) \in \Lambda$ provided that (u_n) is a monotone sequence in $L^p(J, E)$ and $u_n(s) \rightharpoonup u(s)$ for a.e. $s \in J$.*

Then the successive approximations:

(a) $a_{n+1}(t) = -(h(t, a_n) + \int_{t_0}^{t} f(t, s, a_n(s), a_n) \, ds)^-, \quad t \in J, \quad a_0 = 0,$
converge weakly a.e. pointwise to the greatest solution $\underline{u}$ of (7.63);

(b) $b_{n+1}(t) = (h(t, b_n) + \int_{t_0}^t f(t, s, b_n(s), b_n)\, ds)^+, \quad t \in J, \quad b_0 = 0,$
   converge weakly a.e. pointwise to the smallest solution $\overline{u}$ of (7.64);

(c) $u_{n+1}(t) = h(t, u_n) + \int_{t_0}^t f(t, s, u_n(s), u_n)\, ds, \quad t \in J, \quad u_0 = \underline{u},$
   converge weakly a.e. pointwise to the smallest solution u_* of (7.60) in $[\underline{u}, \overline{u}]$;

(d) $v_{n+1}(t) = h(t, v_n) + \int_{t_0}^t f(t, s, v_n(s), v_n)\, ds, \quad t \in J, \quad v_0 = \overline{u},$
   converge weakly a.e. pointwise to the greatest solution u^* of (7.60) in $[\underline{u}, \overline{u}]$.

Proof: The hypotheses (h0) and (f1) imply that the sequences (b_n) and (u_n) are increasing, and that the sequences (a_n) and (v_n) are decreasing. Moreover, all these sequences are contained in P, defined by (7.61), whence they are a.e. pointwise bounded. Thus it follows from the hypothesis (E0) that all these sequences possess the asserted a.e. pointwise weak limits a, u_*, b, and u^*. It is easy to see that these limits belong to P. The hypothesis (B) implies that for a.e. $(t, s) \in \Lambda$

$$\varphi(f(t, s, u_n(s), u_n)) \to \varphi(f(t, s, u_*(s), u_*)) \quad \text{for all } \varphi \in E'.$$

This result and the Dominated Convergence Theorem imply that

$$\varphi\left(\int_{t_0}^t f(t, s, u_n(s), u_n)\, ds\right) = \int_{t_0}^t \varphi(f(t, s, u_n(s), u_n))\, ds$$
$$\to \int_{t_0}^t \varphi(f(t, s, u_*(s), u_*))\, ds = \varphi\left(\int_{t_0}^t f(t, s, u_*(s), u_*)\, ds\right)$$

for all $t \in J$ and $\varphi \in E'$. From this result and the hypothesis (B) we obtain

$$h(t, u_n) \rightharpoonup h(t, u_*) \quad \text{and} \quad \int_{t_0}^t f(t, s, u_n(s), u_n)\, ds \rightharpoonup \int_{t_0}^t f(t, s, u_*(s), u_*)\, ds$$

for a.e. $t \in J$. It then follows from (c) as $n \to \infty$ that u_* is a solution of (7.60). Similar reasoning shows that u^* is also a solution of (7.60), that $\underline{u}$ is a solution of (7.63), and that $\overline{u}$ is a solution of (7.64). By standard arguments one can show that $\underline{u}$ is the greatest solution of (7.63), that $\overline{u}$ is the smallest solution of (7.64), and that u_* and u^* are the smallest and greatest solutions of (7.60) in $[\underline{u}, \overline{u}]$. $\qquad\square$

Remark 7.38. (i) The hypothesis (B) is required to hold only for those iteration sequences that are defined in Proposition 7.37.

(ii) If the values of h and f are contained in the order cone E_+ of E, then in Theorem 7.36 and in Proposition 7.37, $u^* = \overline{u}$ is the smallest solution of (7.60). Similarly, if the values of h and f are in $-E_+$, then $u_* = \underline{u}$ is the greatest solution of (7.60). Thus the lower and upper bounds $\underline{u}$ and $\overline{u}$ of the solutions u_* and u^* cannot be improved, in general.

7.2.5 Application to Impulsive IVP

The results of Theorem 7.36 will now be applied to the following impulsive initial value problem

$$
\begin{cases}
u'(t) + p(t)u(t) = F(t, u(t), u) & \text{a.e. on } J = [t_0, t_1], \\
u(t_0) = x_0, \ \Delta u(\lambda) = H(\lambda, u), \ \lambda \in W,
\end{cases}
\tag{7.67}
$$

where $p \in L^1(J, \mathbb{R})$, $F : J \times E \times L^1(J, E) \to E$, $x_0 \in E$, $\Delta u(\lambda) = u(\lambda + 0) - u(\lambda)$, $H : W \times L^1(J, E) \to E$, and W is a well-ordered (and hence countable) subset of (t_0, t_1). Denoting by $W^{<t} = \{\lambda \in W : \lambda < t\}$, $t \in J$, and by $W^{1,1}(J, E)$ the set of all absolutely continuous and a.e. differentiable functions $v : J \to E$, we say that $u : J \to E$ is a solution of problem (7.67) if it satisfies the equations of (7.67), and if it is contained in the set

$$
V = \left\{
\begin{matrix}
u : J \to E : \sum_{\lambda \in W} \|\Delta u(\lambda)\| < \infty, \text{ and} \\
t \mapsto u(t) - \sum_{\lambda \in W^{<t}} \Delta u(\lambda) \in W^{1,1}(J, E)
\end{matrix}
\right\}.
$$

It is easy to verify that V is a subset of $L^1(J, E)$.

We impose the following hypotheses on the functions H and F.

(H0) $H(\lambda, \cdot)$ is increasing for all $\lambda \in W$, and there exists an $M > 0$ such that $\sum_{\lambda \in W} \|H(\lambda, u)\| \le M$ for all $u \in L^1(J, E)$.
(F0) The mapping $F(\cdot, u(\cdot), u)$ is Bochner integrable for each $u \in L^1(J, E)$.
(F1) $F(s, z, u)$ is increasing with respect to z and u for a.e. $s \in J$.
(F2) $\|F(s, x, u)\| \le q(s)\psi(\|x\|)$ for a.e. $s \in J$ and all $x \in E$, $u \in L^1(J, E)$, where $q \in L^1(J, \mathbb{R}_+)$, $\psi : \mathbb{R}_+ \to (0, \infty)$ is increasing, and $\int_0^\infty \frac{dx}{\psi(x)} = \infty$.

Theorem 7.39. *Let E be a lattice-ordered Banach space with properties (E0) and (E1), and assume that the hypotheses (F0), (F1), (F2), and (H0) are satisfied. Then problem (7.67) has for each $x_0 \in E$ and $p \in L^1(J, \mathbb{R})$,*

(a) minimal and maximal solutions;
(b) the smallest and greatest solutions u_* and u^* in $[\underline{u}, \overline{u}]$, where $\underline{u}$ is the greatest solution of the integral equation

$$
u(t) = -\left(e^{-\int_{t_0}^t p(s)ds} x_0 + \sum_{\lambda \in W^{<t}} e^{-\int_\lambda^t p(s)ds} H(\lambda, u) \right.
$$
$$
\left. + \int_{t_0}^t e^{-\int_s^t p(\tau)d\tau} F(s, u(s), u)ds \right)^-,
\tag{7.68}
$$

and $\overline{u}$ is the smallest solution of the integral equation

$$
u(t) = \left(e^{-\int_{t_0}^t p(s)ds} x_0 + \sum_{\lambda \in W^{<t}} e^{-\int_\lambda^t p(s)ds} H(\lambda, u) \right.
$$
$$
\left. + \int_{t_0}^t e^{-\int_s^t p(\tau)d\tau} F(s, u(s), u)ds \right)^+.
\tag{7.69}
$$

Moreover, the solutions u_* and u^* are increasing with respect to x_0, H, and F.

Proof: The given hypotheses ensure that for each $x_0 \in E$ the relations

$$h(t, u) = e^{-\int_{t_0}^t p(s)ds} x_0 + \sum_{\lambda \in W^{<t}} e^{-\int_{\lambda}^t p(s)ds} H(\lambda, u), \quad t \in J, \ u \in L^1(J, E),$$

$$f(t, s, x, u) = e^{-\int_s^t p(\tau)d\tau} F(s, x, u), \quad (t, s) \in \Lambda, \ u \in L^1(J, E)$$

$$(7.70)$$

define mappings $h : J \times L^1(J, E) \to E$ and $f : \Lambda \times E \times L^1(J, E)$. Denoting

$$K = e^{\int_{t_0}^{t_1} |p(s)|ds}, \quad \alpha(t) = (\|x_0\| + M)K, \quad t \in J,$$
$$g(t, s, r) = Kq(s)\psi(r), \quad (t, s) \in \Lambda, \ r \geq 0,$$

$$(7.71)$$

it follows that the hypotheses (h0), (f0), (f1), and also (f2) hold with the exception that $\beta \in L^1(J, \mathbb{R}_+)$ is now replaced in (7.61) by a constant $w_0 \geq 0$. By this replacement along with (7.71) we can rewrite (7.66) as

$$w(t) = w_0 + \int_{t_0}^t Kq(s)\psi(w(s)) \, ds, \quad t \in J. \tag{7.72}$$

Hypothesis (F2) in conjunction with [44, Lemma B.7.1] ensure that (7.72) has a unique absolutely continuous solution. In the proof of Theorem 7.36 we used the functions $\beta = \alpha$ which is now constant (see (7.71)), and $\beta(t) = \max\{\alpha(t), \|u(t)\|\}$, $t \in J$, where u is a fixed point of G, i.e., a solution of (7.60), which in view of (7.70) can be rewritten as

$$u(t) = e^{-\int_{t_0}^t p(s)ds} x_0 + \sum_{\lambda \in W^{<t}} e^{-\int_{\lambda}^t p(s)ds} H(\lambda, u)$$
$$+ \int_{t_0}^t e^{-\int_s^t p(\tau)d\tau} F(s, u(s), u)ds.$$

$$(7.73)$$

Thus we get

$$\|u(t)\| \leq w_0 := (\|x_0\| + M)K + \int_{t_0}^{t_1} K\|F(s, u(s), u)\|ds, \quad t \in J,$$

and so we can replace the function $\beta(t) = \max\{\alpha(t), \|u(t)\|\}$, $t \in J$ in the proof of Theorem 7.36 by w_0. Consequently, the results of Theorem 7.36 hold for (7.60), or equivalently, for (7.73), which implies the assertions, because the solutions of problem (7.67) and the solutions of the integral equation (7.73) are the same due to Lemma 7.17.   □

Remark 7.40. (i) The result of Proposition 7.37 implies that some solutions of problem (7.67) can be obtained via successive approximations provided H and F satisfy also the following hypothesis.

(A) $H(\lambda, u_n) \rightharpoonup H(\lambda, u)$ for all $\lambda \in W$ and $F(s, u_n(s), u_n) \rightharpoonup F(s, u(s), u)$ for a.e. $s \in J$ if (u_n) is a monotone sequence in $L^1(J, E)$ and $u_n(s) \rightharpoonup u(s)$ for a.e. $s \in J$.

(ii) The functional dependence of h, f, H, and F on u may occur, e.g., in the form of bounded, linear, and positive operators, such as integral operators of Volterra and/or Fredholm type with nonnegative kernels. Thus the results derived in this section can be applied also to integro-differential equations.

7.3 Evolution Equations

Let $J = [a, b]$, $a < b$, be a real interval and E a Banach space. Denote by $L(E)$ the set of all bounded linear operators on E. Given mappings $f : J \times E \to E$ and $T : J \times J \to L(E)$, we derive in Sects. 7.3.1–7.3.3 well-posedness results for continuous solutions of the evolution type integral equation

$$u(t) = T(t, a)x_0 + {}^K\!\!\int_a^t T(t, s)f(s, u(s))ds, \quad t \in J. \tag{7.74}$$

The obtained results are applied in Sect. 7.3.5 to a Cauchy problem. In Sect. 7.3.6 we study the nonlocal integral equation

$$u(t) = T(t, a)x_0 + {}^K\!\!\int_a^t T(t, s)g(s, u(s), u)ds, \quad t \in J, \tag{7.75}$$

where $g : J \times E \times Y \to E$ with $Y = C(J, E)$ and E an ordered Banach space. Our main goal here is to show the existence of continuous extremal solutions of (7.75) and their dependence on the data of the problem. Section 7.3.7 deals with equation (7.75) in case that $Y = L^1(J, E)$ where the integral is understood as Bochner integral, and T is defined in the set $\Lambda = \{(t, s) \in J \times J : s \leq t\}$. As an application we obtain in Sect. 7.3.8 an existence result for an initial value problem of a second order partial differential equation involving discontinuous nonlinearities.

7.3.1 Well-Posedness Results

We are going to study first the integral equation (7.74).

Preliminaries

Assume that $T : J \times J \to L(E)$ satisfies the following conditions.

(T0) $T(t, t) = I$, and $T(t, r) = T(t, s)T(s, r)$ whenever $a \leq r \leq s \leq t \leq b$.
(T1) $(t, s) \mapsto T(t, s)x$ is continuous for each $x \in E$.

From (T1) and the uniform boundedness principle it follows that

$$M = \sup\{\|T(t,s)\| : a \le s \le t \le b\} < \infty. \tag{7.76}$$

The following result ensures continuity of the solutions of (7.74) and (7.75).

Lemma 7.41. *Let the hypotheses (T0) and (T1) be satisfied. If* $h : J \to E$, *if* $s \mapsto T(t,s)h(s)$ *is HL integrable on* J *for every* $t \in J$, *and if the function* $t \mapsto {}^{K}\!\int_a^t T(t,s)h(s)ds$ *is bounded, then for each* $x_0 \in E$ *the equation*

$$u(t) = T(t,a)x_0 + {}^{K}\!\int_a^t T(t,s)h(s)ds, \quad t \in J, \tag{7.77}$$

defines a continuous function $u : J \to E$.

Proof: If $a \le \bar{t} \le t \le b$, from (7.77) it follows that

$$\|u(t) - u(\bar{t})\| \le \|T(t,a)x_0 - T(\bar{t},a)x_0\| + \left\| {}^{K}\!\int_a^{\bar{t}} (T(t,s) - T(\bar{t},s))h(s)\,ds \right\|$$
$$+ \left\| {}^{K}\!\int_{\bar{t}}^t T(t,s)h(s)\,ds \right\|$$
$$= I_1 + I_2 + I_3, \tag{7.78}$$

where

$$I_1 = \|T(t,a)x_0 - T(\bar{t},a)x_0\|,$$
$$I_2 = \left\| {}^{K}\!\int_a^{\bar{t}} (T(t,s) - T(\bar{t},s))h(s)\,ds \right\| \le \|T(t,\bar{t}) - I\| \left\| {}^{K}\!\int_a^{\bar{t}} T(\bar{t},s)h(s)\,ds \right\|$$
$$\le c\|T(t,\bar{t}) - I\|, \tag{7.79}$$
$$I_3 = \left\| {}^{K}\!\int_{\bar{t}}^t T(t,s)h(s)\,ds \right\|.$$

From (7.79) and applying (T0) and (T1) and the assumption that $t \mapsto {}^{K}\!\int_a^t T(t,s)h(s)ds$ is bounded, it follows that $I_1 + I_2 + I_3 \to 0$ as $\bar{t} \to t-$. This implies by (7.78) that $\|u(t) - u(\bar{t})\| \to 0$ as $t \to \bar{t}-$. Thus u is left-continuous. To prove right-continuity of u, notice that I_3 can be estimated by

$$I_3 = \left\| {}^{K}\!\int_{\bar{t}}^t T(t,s)h(s)\,ds \right\| = \left\| T(t,\bar{t}) {}^{K}\!\int_{\bar{t}}^t T(\bar{t},s)h(s)\,ds \right\|$$
$$\le M \left\| {}^{K}\!\int_{\bar{t}}^t T(\bar{t},s)h(s)\,ds \right\|. \tag{7.80}$$

Thus $I_1 + I_2 + I_3 \to 0$ as $t \to \bar{t}+$, so that u is also right-continuous.   □

The following hypotheses are imposed on the function $f : J \times E \to E$.

(fa) $f(\cdot, x)$ is strongly measurable for all $x \in E$.
(fb) There exists an $r > 0$ such that

$$\|f(t, y) - f(t, z)\| \le q(t, \|y - z\|)$$

for all y, $z \in E$ with $\|y - z\| \le r$ and for a.e. $t \in J$, where
(q) $q : J \times [0, r] \to \mathbb{R}_+$, $q(\cdot, x)$ is measurable for all $x \in [0, r]$, $q(\cdot, r) \in L^1(J, \mathbb{R}_+)$, $q(t, \cdot)$ is increasing and right-continuous for a.e. $t \in J$, and the zero-function is for $w_0 = 0$ the only absolutely continuous (a.c.) solution of the Cauchy problem

$$w'(t) = Mq(t, w(t)) \quad \text{a.e. on} \quad J, \quad w(a) = w_0. \tag{7.81}$$

(Tf) There exists a $z \in J$ such that $s \mapsto T(t, s)f(s, z)$ is HL integrable on J for every $t \in J$, and $t \mapsto {}^K\!\int_a^t T(t, s)f(s, z)ds$ is bounded.

The following result is a consequence of Lemma 6.11 with q replaced by Mq.

Lemma 7.42. *Let the hypothesis (q) hold. Then there exists an $r_0 > 0$ such that for every $u_0 \in [0, r_0]$ the Cauchy problem (7.81) has the smallest a.c. solution $w = w(\cdot, w_0)$, which is increasing with respect to w_0. Moreover, $w(t, w_0) \to 0$ uniformly in $t \in J$ as $w_0 \to 0+$.*

7.3.2 Existence and Uniqueness Result

Our main existence and uniqueness result for the integral equation (7.74) reads as follows.

Theorem 7.43. *If the hypotheses (T0), (T1), (Tf), (fa), and (fb) hold, then for each $x_0 \in E$ the integral equation (7.74) has a unique solution u in $C(J, E)$. Moreover, u is the uniform limit of the sequence $(u_n)_{n=0}^\infty$ of the successive approximations*

$$u_{n+1}(t) = T(t, a)x_0 + {}^K\!\int_a^t T(t, s)f(s, u_n(s))\, ds, \quad t \in J, \ n \in \mathbb{N}_0, \tag{7.82}$$

for any choice of $u_0 \in C(J, E)$.

Proof: The hypothesis (q) imposed on q in (fb) implies by Lemma 7.42 that the Cauchy problem (7.81) has for some $w_0 = r_0 > 0$ the smallest a.c. solution $v = w(\cdot, r_0)$, and $r_0 \le v(t) \le r$ for each $t \in J$. Since $q(s, \cdot)$ is increasing and right-continuous in $[0, r]$ for a.e. $s \in J$, and $q(\cdot, x)$ is measurable for all $x \in [0, r]$ as well as $q(\cdot, r)$ is Lebesgue integrable, then $q(\cdot, w(\cdot))$ is Lebesgue integrable whenever w belongs to the order interval $[0, v] = \{w \in C(J, \mathbb{R}) : 0 \le w(t) \le v(t), \ t \in J\}$. Thus the equation

$$Qw(t) = \int_a^t Mq(s, w(s))\,ds, \quad t \in J \tag{7.83}$$

defines a mapping $Q : [0, v] \to C(J, \mathbb{R}_+)$. Condition (q) ensures that Q is increasing, and the choice of r_0 and v shows that the equation

$$r_0 + Qv = v \tag{7.84}$$

is valid. Thus $Qv(t) < v(t)$ for every $t \in J$. The sequence $(Q^n v)_{n=0}^\infty$ is decreasing because $q(t, \cdot)$ is increasing for a.e. $t \in J$. The reasoning similar to that used in the proof of Lemma 6.11 shows that $(Q^n v)_{n=0}^\infty$ converges uniformly in J to the zero function. Since this function satisfies by (q) the equation $w'(t) = Mq(t, w(t))$ for a.e. $t \in J$, then $q(t, a) = 0$ for a.e. $t \in J$. This result and hypotheses (fa) and (fb) imply that f is a Carathéodory function. Thus $f(\cdot, u(\cdot))$ is strongly measurable in J for all $u \in C(J, E)$ due to [133, Theorem 1.4.3]. Let $u \in C(J, E)$ and $t \in J$ be fixed, and choose by (fa) a $z \in E$ so that $s \mapsto T(t, s)f(s, z)$ is HL integrable on J. Defining

$$y_i(s) = z + \frac{i}{m}(u(s) - z), \quad i = 0, \ldots, m \geq \frac{\max\{\|u(t) - z\| : t \in J\}}{r_0},$$

we have $\|y_i(s) - y_{i-1}(s)\| \leq r_0 \leq v(s)$ in J for each $i = 1, \ldots, m$, whence

$$\|f(s, u(s)) - f(s, z)\| \leq \sum_{i=1}^m \|f(s, y_i(s)) - f(s, y_{i-1}(s))\|$$

$$\leq \sum_{i=1}^m q(s, \|y_i(s) - y_{i-1}(s)\|) \leq \sum_{i=1}^m q(t, v(s)) = \frac{m}{M} v'(s)$$

for a.e. $s \in J$. This result and the strong measurability of $f(\cdot, u(\cdot))$ and $f(\cdot, z)$ imply that $f(\cdot, u(\cdot)) - f(\cdot, z)$ is Bochner integrable. Thus $T(t, \cdot)f(\cdot, u(\cdot)) = T(t, \cdot)f(\cdot, z) + T(t, \cdot)f(\cdot, u(\cdot)) - T(t, \cdot)f(\cdot, z)$ is HL integrable on J. Moreover, for each $u \in C(J, E)$ and $t \in J$,

$$\left\| {}^K\!\!\int_a^t T(t, s)f(s, u(s))ds \right\| \leq \left\| {}^K\!\!\int_a^t T(t, s)f(s, z)ds \right\|$$

$$+ M {}^K\!\!\int_a^b \|f(s, u(s)) - f(s, z)\|ds.$$

This result and the hypothesis (Tf) imply that $t \mapsto {}^K\!\!\int_a^t T(t, s)f(s, u(s))\,ds$ is bounded, and hence continuous by Lemma 7.41. Consequently, for each fixed $x_0 \in E$, the equation

$$Fu(t) = T(t, a)x_0 + {}^K\!\!\int_a^t T(t, s)f(s, u(s))ds, \quad t \in J, \tag{7.85}$$

defines a mapping $F : C(J, E) \to C(J, E)$.

Next we shall show that denoting $|u| = \|u(\cdot)\|$, $u \in C(J,E)$, the mappings Q and F defined by (7.83) and (7.85), respectively, satisfy the hypotheses of Proposition 6.12. To show that (6.31) holds, let $u, \bar{y} \in C(J,E)$ be given. The functions $T(t,\cdot)f(\cdot,u(\cdot)) - T(t,\cdot)f(\cdot,z)$ and $T(t,\cdot)f(\cdot,\bar{u}(\cdot)) - f(\cdot,z)$ are Bochner integrable over J for every $t \in J$, see the above proof. Thus the function $T(t,\cdot)f(\cdot,u(\cdot)) - T(t,\cdot)f(\cdot,\bar{u}(\cdot))$ is Bochner integrable on J, for each $t \in J$. This result implies that the function $\|T(t,\cdot)f(\cdot,\bar{u}(\cdot)) - T(t,\cdot)f(\cdot,z)\|$ is Lebesgue integrable on J for every $t \in J$. Moreover, for each $t \in J$ we have the following estimate

$$\left\| {}^K\!\!\int_a^t T(t,s)f(s,u(s))\,ds - {}^K\!\!\int_a^t T(t,s)f(s,\bar{u}(s))\,ds \right\|$$
$$= \left\| \int_a^t T(t,s)(f(s,u(s))\,ds - f(s,\bar{u}(s)))\,ds \right\|$$
$$\leq M \int_a^t \|f(s,u(s))\,ds - f(s,\bar{u}(s))\|\,ds.$$

Applying this result and the hypotheses (fa) and (fb) as well as taking into account the definitions (7.83) and (7.85), we see that

$$|Fu - F\bar{u}| \leq Q|u - \bar{u}| \quad \text{for} \quad u, \bar{u} \in C(J,E),\ |y - \bar{y}| \leq v.$$

The above proof shows that the operators F and Q satisfy all hypotheses of Proposition 6.12. Therefore, the iteration sequence $(F^n u_0)_{n=0}^\infty$, which equals to the sequence $(u_n)_{n=0}^\infty$ of successive approximations (7.82), converges for every choice of $u_0 \in C(J,E)$ uniformly in J to a unique fixed point u of F. This result and the definition of F imply that u is the uniquely determined solution of the integral equation (7.74) in $C(J,E)$. □

7.3.3 Continuous Dependence on x_0

We are going to prove that under the hypotheses (T0), (T1), (Tf), (fa), and (fb), the dependence of the solution u of (7.74) upon the initial value x_0 can be estimated in terms of the *smallest solution* of the Cauchy problem (7.81). In view of Lemma 7.42, this estimate implies the continuous dependence of u on x_0.

Proposition 7.44. *Let the hypotheses (T0), (T1), (Tf), (fa), and (fb) be satisfied. If $u = u(\cdot,x_0)$ denotes the solution of the integral equation (7.74) and $w = w(\cdot,w_0)$ the smallest solution of the Cauchy problem (7.81), then for all $x_0, \hat{x}_0 \in E$ with $\|x_0 - \hat{x}_0\|$ small enough, the following estimate holds:*

$$\|u(t,x_0) - u(t,\hat{x}_0))\| \leq w(t, M\|x_0 - \hat{x}_0\|), \quad t \in J. \tag{7.86}$$

In particular, $u(\cdot,x_0)$ depends continuously on x_0 in the sense that $u(t,\hat{x}_0) \to u(t,x_0)$ uniformly in $t \in J$ as $\hat{x}_0 \to x_0$.

Proof: Assume that x_0, $\hat{x}_0 \in E$, and that $\|x_0 - \hat{x}_0\| \leq r_0$, where r_0 is chosen as in Lemma 7.42. The solutions $u = u(\cdot, x_0)$ and $\hat{u} = u(\cdot, \hat{x}_0)$ exist by Theorem 7.43, and they satisfy

$$u(t) = Fu(t) = T(t, a)x_0 + {}^K\!\!\int_a^t T(t, s)f(s, u(s))\, ds, \quad t \in J,$$

$$\hat{u}(t) = \hat{F}\hat{u}(t) = T(t, a)\hat{x}_0 + {}^K\!\!\int_a^t T(t, s)f(s, \hat{u}(s))\, ds, \quad t \in J.$$

Moreover, F satisfies the hypotheses of Proposition 6.12 with Q defined by

$$Qw(t) = \int_a^t Mq(s, w(s))\, ds, \quad t \in J,$$

and $w = w(\cdot, M\|x_0 - \hat{x}_0\|)$ is the smallest solution of

$$w = M\|x_0 - \hat{x}_0\| + Qw.$$

Denote

$$W = \{u \in C(J, E) : |u - \hat{u}| \leq w\}.$$

Since Q is increasing, and since

$$F\hat{u}(t) - \hat{u}(t) = F\hat{u}(t) - \hat{F}\hat{u}(t) = T(t, a)(x_0 - \hat{x}_0)$$

for all $t \in J$, we have for every $u \in W$,

$$|Fu - \hat{u}| \leq |F\hat{u} - \hat{u}| + |Fu - F\hat{u}| \leq |F\hat{u} - \hat{u}| + Q|u - \hat{u}|$$
$$\leq M\|x_0 - \hat{x}_0\| + Qw = w.$$

Thus $F[W] \subseteq W$. Since $\hat{u} \in W$, then $(F^n \hat{u}) \in W$ for every $n \in \mathbb{N}_0$. The uniform limit $u = \lim_n F^n \hat{u}$ exists by Theorem 7.43 and is the solution of (7.74). Because W is closed, then $u \in W$, so that $|u - \hat{u}| \leq w$. This proves (7.86). According to Lemma 6.11, $w(t, M\|x_0 - \hat{x}_0\|) \to 0$ uniformly over $t \in J$ as $\|x_0 - \hat{x}_0\| \to 0$. This result and (7.86) imply that the last assertion of the proposition is true. □

Remark 7.45. If the Cauchy problem (7.81) has for some positive value of M the zero function as the only solution when $w_0 = 0$, the same does not necessarily hold for all positive M, as we see from the following example (cf. [37], p. 676):

Define $q \in C(J \times \mathbb{R}_+, \mathbb{R}_+)$, $J = [0, 1]$, by

$$q(t, r) = \begin{cases} 2t, & \text{for } r \geq t^2,\ t \in J, \\ \frac{2r}{t}, & \text{for } 0 \leq r < t^2,\ 0 < t \leq 1. \end{cases}$$

It is easy to show that $w(t) \equiv 0$ is the only solution of (7.81) when $M = \frac{1}{2}$ and $w_0 = 0$, whereas $w(t) = \gamma t^2$, $t \in J$ is for each $\gamma \in [0, 1]$ a solution of (7.81) when $M = 1$ and $w_0 = 0$.

7.3.4 Special Cases

The following result shows that the above kind of counter-example (see Remark 7.45) does not exist if q is of Osgood type.

Proposition 7.46. *Let the hypotheses (T0), (T1), (Tf), and (fa) hold, and assume that there is $r > 0$ and $p \in L^1(J, \mathbb{R}_+)$ such that for all $x, y \in E$ with $\|x - y\| < r$ and for a.e. $t \in J$,*

$$\|g(t, x) - g(t, y)\| \leq p(t)\,\phi(\|x - y\|),$$

where $\phi : [0, r] \to \mathbb{R}_+$ is increasing and right-continuous, and $\int_0^r \frac{dv}{\phi(v)} = \infty$. Then the integral equation (7.74) has for each $x_0 \in E$ a unique solution $u = u(\cdot, x_0)$, and it depends continuously on x_0.

Proof: It is easy to verify that condition (q) of hypothesis (fb) holds when the function $q : J \times \mathbb{R}_+ \to \mathbb{R}_+$ is defined by

$$q(t, v) = p(t)\phi(v), \quad t \in J, v \in \mathbb{R}_+.$$

Thus the hypotheses of Theorem 7.43 are satisfied with q given above.   □

Each of the functions ϕ_m, $m \in \mathbb{N}$, defined in Remark 6.17 satisfy the hypotheses given for ϕ in Proposition 7.46. In particular, when $\phi(u) = u$, we obtain the following corollary.

Corollary 7.47. *Let the hypotheses (T0), (T1), (Tf), and (fa) be fulfilled. Assume there is a $p_1 \in L^1(J, \mathbb{R}_+)$ such that*

$$\|f(t, x) - f(t, y)\| \leq p_1(t)\|x - y\|$$

for all $x, y \in E$ and for a.e. $t \in J$. Then the integral equation (7.74) has for each $x_0 \in E$ exactly one solution u. Moreover, u depends continuously on x_0.

Corollary 7.48. *The results of Theorem 7.43 and Proposition 7.44 remain true for solutions of the following Volterra integral equation*

$$u(t) = k(t, a)x_0 + {}^K\!\!\int_a^t k(t, s)f(s, u(s))\, ds, \quad t \in J = [a, b], \tag{7.87}$$

where $f : J \times E \to E$ satisfies the hypotheses (fa) and (fb), and $k : J \times J \to \mathbb{R}_+$ has the following properties:

(ka) *k is continuous, $k(t, t) = 1$, and $k(t, r) = k(t, s)k(s, r)$ whenever $a \leq r \leq s \leq t \leq b$.*

(kf) *There exists a $z \in J$ such that $s \mapsto k(t, s)f(s, z)$ is HL integrable on J for every $t \in J$, and the function $t \mapsto {}^K\!\!\int_a^t k(t, s)f(s, z)ds$ is bounded.*

Proof: The hypotheses (ka) and (kf) imply that the equation

$$T(t, s)x = k(t, s)x, \quad t, s \in J, x \in E,$$

defines a mapping $T : J \times J \to L(E)$, and that the hypotheses (T0), (T1), (T2), and (Tf) are valid.   □

7.3.5 Application to a Cauchy Problem

The result of Corollary 7.48 are now be applied to the Cauchy problem

$$u'(t) = p(t)u(t) + f(t, u(t)) \quad \text{a.e. on} \quad J = [a, b], \quad u(a) = x_0, \tag{7.88}$$

where $p \in L^1(J, \mathbb{R})$, $f : J \times E \to E$, and $x_0 \in E$.

Proposition 7.49. *Assume that $f : J \times E \to E$ satisfies (fb) and the following hypothesis.*

(fc) $f(\cdot, z)$ is strongly measurable for all $z \in E$ and HL integrable for some $z \in E$.

Then for each $x_0 \in E$ the Cauchy problem (7.88) has exactly one solution $u \in W^1_{SL}(J, E)$. Moreover, u depends continuously on x_0.

Proof: Applying Proposition 9.10, Lemma 9.22, and Lemma 9.24, one can show that a function $u \in W^1_{SL}(J, E)$ is a solution of the Cauchy problem (7.88) if and only if u is a continuous solution of the integral equation (7.87), where

$$k(t, s) = \exp(\textstyle\int_a^t p(x)\,dx) \exp(-\textstyle\int_a^s p(x)\,dx) \quad t, s \in J. \tag{7.89}$$

The so defined function $k : J \times J \to \mathbb{R}_+$ satisfies hypothesis (ka) of Corollary 7.48. Hypothesis (fc) ensures that the hypothesis (fa) holds. In view of Proposition 9.10 it ensures also that the hypothesis (kf) of Corollary 7.48 holds when k is defined by (7.89). Thus the integral equation (7.86), with k defined by (7.89), has a unique continuous solution $u : J \to E$ by Corollary 7.48, which depends continuously on x_0. This concludes the proof. $\qquad\square$

7.3.6 Extremal Solutions of Evolution Equations

Let $J = [a, b]$, $a < b$ be a real interval, and E a Banach space ordered by a regular order cone E_+. Given mappings $g : J \times E \times C(J, E) \to E$ and $T : J \times J \to L(E)$, we consider the integral equation

$$u(t) = T(t, a)x_0 + {}^K\!\!\int_a^t T(t, s)g(s, u(s), u)ds, \quad t \in J. \tag{7.90}$$

We assume that T has the following properties.

(T0) $T(t, t) = I$, and $T(t, r) = T(t, s)T(s, r)$ whenever $a \leq r \leq s \leq t \leq b$.
(T1) $(t, s) \mapsto T(t, s)x$ is continuous for each $x \in E$.
(T2) $T(t, s)E_+ \subseteq E_+$ for all (t, s) in the domain of T.

Condition (T2) ensures that the mapping $x \mapsto T(t, s)x$ is increasing, that is, $T(t, s)x \leq T(t, s)y$ whenever $a \leq s \leq t \leq b$, $x, y \in E$ and $x \leq y$.

Definition 7.50. *We say that a function* $u \in C(J, E)$ *is a* **subsolution** *of the integral equation (7.90) if* $s \mapsto T(t, s)g(s, u(s), u)$ *is HL integrable on* $[a, t]$ *for every* $t \in J$, *and if*

$$u(t) \leq T(t, a)x_0 + {}^{K}\!\!\int_{a}^{t} T(t, s)g(s, u(s), u)ds, \quad t \in J. \tag{7.91}$$

A **supersolution** *of (7.90) is defined similarly by reversing the inequality sign in (7.91). If equality holds in (7.91), we say that* u *is a* **solution** *of (7.90).*

We are going to show that the integral equation (7.90) has the smallest and greatest solutions within an order interval of $C(J, E)$ if $T : J \times J \to L(E)$ has properties (T0)–(T2), and if $g : J \times E \times C(J, E) \to E$ satisfies the following hypotheses.

(g0) For every $x_0 \in E$ the integral equation (7.90) has a subsolution $\underline{u}$ and a supersolution $\overline{u}$, and $\underline{u}(t) \leq \overline{u}(t)$ for all $t \in J$.

(g1) $g(\cdot, u(\cdot), u)$ is strongly measurable for each $u \in C(J, E)$.

(g2) $g(t, x, u)$ is increasing in x and u for a.e. $t \in J$.

Theorem 7.51. *Under the hypotheses (T0)–(T2) and (g0)–(g2), the integral equation (7.90) has the smallest and greatest solutions in* $[\underline{u}, \overline{u}]$. *Moreover, they are increasing with respect to* x_0 *and* g.

Proof: Let $x_0 \in E$ be given. By definition, $\underline{u}$ and $\overline{u}$ are continuous. If u belongs to the order interval $[\underline{u}, \overline{u}]$ of $C(J, E)$, then (g1) and (T1) imply by [133, Corollary 1.4.4] that $s \mapsto T(t, s)g(s, u(s), u)$ is strongly measurable on $[a, t]$ for each $t \in J$. By conditions (T2) and (g2) we have

$$T(t, s)g(s, \underline{u}(s), \underline{u}) \leq T(t, s)g(s, u(s), u) \leq T(t, s)g(s, \overline{u}(s), \overline{u}) \tag{7.92}$$

for all $t \in J$ and for a.e. $s \in [a, t]$. Because the functions $s \mapsto T(t, s)g(s, \underline{u}(s), \underline{u})$ and $s \mapsto T(t, s)g(s, \overline{u}(s), \overline{u})$ are HL integrable on $[a, t]$ for every $t \in J$, it then follows from (7.92) that $s \mapsto T(t, s)g(s, u(s), u)$ is HL integrable on $[a, t]$ for every $t \in J$ due to Proposition 9.14. Thus the equation

$$Gu(t) = T(t, a)x_0 + {}^{K}\!\!\int_{a}^{t} T(t, s)g(s, u(s), u)\, ds, \quad t \in J, \tag{7.93}$$

defines a function $Gu : J \to E$. In view of conditions (T2) and (g2), from (7.92) and by Lemma 9.11 it follows that $Gu(t) \leq Gv(t)$ for all $t \in J$ whenever $u, v \in [\underline{u}, \overline{u}]$ and $u \leq v$. Moreover, (7.93) and the definition of sub- and supersolutions of (7.90) imply that $\underline{u}(t) \leq G\underline{u}(t)$ and $G\overline{u}(t) \leq \overline{u}(t)$ for all $t \in J$. In particular, for all $u \in [\underline{u}, \overline{u}]$

$$\underline{u}(t) \leq T(t, a)x_0 + {}^{K}\!\!\int_{a}^{t} T(t, s)g(s, u(s), u)\, ds \leq \overline{u}(t), \quad t \in J. \tag{7.94}$$

Since the order cone of E is also normal, from (7.94) we get the estimate

$$\left\| ^K\!\!\int_a^t T(t,s)g(s,u(s),u)\,ds \right\| \le M\|x_0\| + (\lambda+1)\|\underline{u}(t)\| + \lambda\|\overline{u}(t)\|, \quad (7.95)$$

where $\lambda \ge 1$. In particular, $K = \sup_{t \in J} \| ^K\!\!\int_a^t T(t,s)g(s,u(s),u)\,ds\| < \infty$. Thus the hypotheses of Lemma 7.41 hold true for $s \mapsto T(t,s)g(s,u(s),u)$ for all $u \in [\underline{u},\overline{u}]$, whence $Gu \in C(J,E)$ for every $u \in [\underline{u},\overline{u}]$. The above proof shows that (7.93) defines an increasing mapping G from $[\underline{u},\overline{u}]$ to $[\underline{u},\overline{u}]$. In order to apply Proposition 2.39, we show that the functions Gu, $u \in [\underline{u},\overline{u}]$, form an equicontinuous set. Let $u \in [\underline{u},\overline{u}]$ be fixed. Denote

$$h(s) = g(s,u(s),u), \;\; h_-(s) = g(s,\underline{u}(s),\underline{u}), \;\; h_+(s) = g(s,\overline{u}(s),\overline{u}), \;\; s \in J. \quad (7.96)$$

By Lemma 9.11 from (7.92) it follows that if $t,\overline{t} \in J$ and $\overline{t} \le t$, then

$$^K\!\!\int_{\overline{t}}^t T(t,s)h_-(s)\,ds \le {}^K\!\!\int_{\overline{t}}^t T(t,s)h(s)\,ds \le {}^K\!\!\int_{\overline{t}}^t T(t,s)h_+(s)\,ds,$$
$$^K\!\!\int_{\overline{t}}^t T(\overline{t},s)h_-(s)\,ds \le {}^K\!\!\int_{\overline{t}}^t T(\overline{t},s)h(s)\,ds \le {}^K\!\!\int_{\overline{t}}^t T(\overline{t},s)h_+(s)\,ds. \quad (7.97)$$

Since the order cone of E is normal, from (7.97) we obtain the estimates

$$\left\| ^K\!\!\int_{\overline{t}}^t T(t,s)h(s)\,ds \right\| \le (\lambda+1)\left\| ^K\!\!\int_{\overline{t}}^t T(t,s)h_-(s)\,ds \right\|$$
$$+ \lambda\left\| ^K\!\!\int_{\overline{t}}^t T(t,s)h_+(s)\,ds \right\|,$$
$$\left\| ^K\!\!\int_{\overline{t}}^t T(\overline{t},s)h(s)\,ds \right\| \le (\lambda+1)\left\| ^K\!\!\int_{\overline{t}}^t T(\overline{t},s)h_-(s)\,ds \right\|$$
$$+ \lambda\left\| ^K\!\!\int_{\overline{t}}^t T(\overline{t},s)h_+(s)\,ds \right\|. \quad (7.98)$$

In view of (7.78), (7.79), and (7.80), we have

$$\|Gu(t) - Gu(\overline{t})\| \le \|T(t,a)x_0 - T(\overline{t},a)x_0\| + K\|T(t,\overline{t}) - I\|$$
$$+ \max\{\| ^K\!\!\int_{\overline{t}}^t T(t,s)h(s)\,ds\|, M\| ^K\!\!\int_{\overline{t}}^t T(\overline{t},s)h(s)\,ds\|\}. \quad (7.99)$$

It then follows from (7.96), (7.98), and (7.99) that the set $\{Gu : u \in [\underline{u},\overline{u}]\}$ is equicontinuous. Let (u_n) be a monotone sequence in $[\underline{u},\overline{u}]$. Then for every $t \in J$, $(Gu_n(t))$ is a monotone sequence within the order interval $[\underline{u}(t),\overline{u}(t)]$ of E. Since the order cone of E is regular, then $(Gu_n(t))$ converges in E for every $t \in J$. Moreover, it follows from the above proof that the sequence (Gu_n) is equicontinuous. This result implies by Lemma 9.44 that (Gu_n) converges uniformly on J, and hence with respect to the uniform norm of $C(J,E)$.

The above proof shows that the hypotheses of Proposition 2.39 are valid for the mapping G defined by (7.93), when X is replaced by $C(J, E)$ equipped with the pointwise ordering and the uniform norm, and the functions $\underline{u}$, $\overline{u}$ of $C(J, E)$ are sub- and supersolutions of (7.90) in the sense of Definition 7.50. Thus by Proposition 2.39, G has the smallest fixed point u_* and the greatest fixed point u^*. Because they belong to $C(J, E)$ and satisfy (7.90), they are solutions of (7.90) in the sense of Definition 7.50.

If $u \in [\underline{u}, \overline{u}]$ is a solution of (7.90), it is also a fixed point of G, whence $u_* \leq u \leq u^*$. Thus u_* and u^* are the smallest and greatest solutions of (7.90). Moreover, they are increasing with respect to G due to Proposition 2.39. This result, the hypotheses (T2), (g2), and (7.93) imply that u_* and u^* are increasing with respect to x_0 and g in view of Lemma 9.11.   □

As for the existence of the smallest and greatest solutions of (7.90) in the entire space $C(J, E)$, we have the following result.

Proposition 7.52. *Assume that $g : J \times E \times C(J, E) \to E$ satisfies the hypotheses (g1), (g2), and that $T : J \times J \to L(E)$ has properties (T0)–(T2). Further suppose that*

$$f_1(\cdot, x) \leq g(\cdot, x, u) \leq f_2(\cdot, x) \quad \text{for all } x \in E, \ u \in C(J, E) \text{ and } t \in J, \quad (7.100)$$

where $f_i : J \times E \to E$, $i = 1, 2$, satisfy conditions (fa), (fb), and (Tf) given in Sect. 7.3.1, and the following condition.

(fc) $f_i(s, \cdot)$ is increasing for a.e. $s \in J$.

Then for each choice of $x_0 \in E$, the integral equation (7.90) has the smallest and greatest solutions, and they are increasing with respect to x_0.

Proof: Let $x_0 \in E$ be given. Theorem 7.43 implies that the integral equation

$$u(t) = T(t, a)x_0 + {}^{K}\!\!\int_a^t T(t, s) f_i(t, u(s)) \, ds \qquad (7.101)$$

has a unique solution $\underline{u}$ when $i = 1$ and $\overline{u}$ when $i = 2$. From (7.100) and (7.101) it follows that $\underline{u}$ is a subsolution of (7.101), and $\overline{u}$ is its supersolution for both values of i. Moreover, the functions $g_i(t, x, u) = f_i(t, x)$, $i = 1, 2$ satisfy the hypotheses (g1) and (g2). Thus the monotone dependence result of Theorem 7.51 can be applied to ensure that $\underline{u} \leq \overline{u}$. Both functions $T(t, \cdot) f_i(\cdot, \underline{u}(\cdot))$ and $T(t, \cdot) f_2(\cdot, \overline{u}(\cdot))$ are HL integrable on J for every $t \in J$, and

$$T(t, \cdot) f_1(\cdot, x) \leq T(t, \cdot) g(\cdot, x, u) \leq T(t, \cdot) f_2(\cdot, x), \quad x \in E, \ u \in C(J, E).$$

These results and (g1) imply that the functions $T(t, \cdot) g(\cdot, \underline{u}(\cdot), \underline{u})$ and $T(t, \cdot) g(\cdot, \overline{u}(\cdot), \overline{u})$ are HL integrable.
    Since

$$\underline{u}(t) \le T(t,a)x_0 + {}^K\!\!\int_a^t T(t,s)g(s,\underline{u}(s),u)\,ds \quad \text{for a.e. } t \in J,$$

then $\underline{u}$ is a subsolution of (7.90). Similarly, it can be shown that $\overline{u}$ is a supersolution of (7.90). Thus g satisfies also condition (g0). By Theorem 7.51 the integral equation (7.90) has for each $x_0 \in E$ the smallest solution u_* and greatest solution u^* in $[\underline{u},\overline{u}]$.

If u is a solution of (7.90), from (7.100) and (7.101) it follows that u is a supersolution of (7.101) for $i = 1$ and a subsolution of (7.101) for $i = 2$, whence $\underline{u} \le u \le \overline{u}$. Thus all the solutions of (7.90) lie between $\underline{u}$ and $\overline{u}$, whence u_* is the smallest and u^* the greatest of all the solutions of (7.90). □

Corollary 7.53. *Assume that* $g : J \times E \times C(J,E) \to E$ *satisfies (g1) and (g2), that* $T : J \times J \to L(E)$ *satisfy (T0)–(T2), and that for all* $x \in E$ *and* $u \in HL(J,E)$ *and for a.e.* $t \in J$,

$$h_1(t) \le g(t,x,u) \le h_2(t), \tag{7.102}$$

where $h_i : J \to E$ *and* $T(t,\cdot)h_i(\cdot) \in HL([a,t],E)$ *for every* $t \in J$, *and*

$$\sup\left\{ \left\| {}^K\!\!\int_a^t T(t,s)h_i(s)\,ds \right\| : t \in J \right\} < \infty, \quad i = 1,2.$$

Then the integral equation (7.90) has for each choice of $x_0 \in E$ *the smallest and greatest solutions, both of which are increasing with respect to* x_0 *and* g.

Proof: It is easy to see that the hypotheses of Proposition 7.52 hold when $f_i(t,x) = h_i(t)$, $i = 1,2$. □

7.3.7 Evolution Equations Containing Bochner Integrable Functions

Let $J = [a,b]$, $a < b$ be a real interval, and E a lattice-ordered Banach space. In this subsection we consider nonlocal integral equations of the form

$$u(t) = T(t,a)x_0 + \int_a^t T(t,s)g(s,u(s),u)\,ds, \quad t \in J, \tag{7.103}$$

where the integral on the right-hand side stands for the Bochner integral. We assume that $g : J \times J \times L^1(J,E) \to E$, that $T : \Lambda \to L(E)$, where $\Lambda = \{(t,s) \in J \times J : s \le t\}$.

Denoting by E_+ the order cone of E, we assume that $T : \Lambda \to L(E)$ has properties (T0)–(T2). By the uniform boundedness principle it follows from (T1) that (7.76) holds. Condition (T2) ensures that the mapping $x \mapsto T(t,s)x$ is increasing, that is, $T(t,s)x \le T(t,s)y$ whenever $(t,s) \in \Lambda$, $x, y \in E$ and $x \le y$.

Let $\{A(t) : t \in J\}$, $J = [a, b]$, be a family of closed linear operators from a dense subspace $D(A)$ of E into E satisfying

$$-A(t)x = \lim_{\Delta t \to 0+} \frac{T(t + \Delta t, t)x - x}{\Delta t} \quad \text{for a.e. } t \in J \text{ and for all } x \in D(A).$$
(7.104)

It can be shown that if u is a solution of the integral equation (7.90), and if $x_0 \in D(A)$, then u is a solution of the initial value problem

$$u'(t) + A(t)u(t) = g(t, u(t), u) \quad \text{for a.e. } t \in J, \quad u(0) = x_0,$$
(7.105)

provided that $\int_a^t T(t, s)g(s, u(s), u)ds$ and $T(t, a)x_0$ belong to $D(A)$ for each $t \in J$, and that $u'(t)$ in (7.105) is considered as the right derivative of u at t. In general, when $D(A) \neq E$, the validity of these extra conditions requires certain smoothness properties for T and g (cf., e.g., [163]). On the other hand, as we shall see, the integral equation (7.90) may have continuous solutions also when g is not continuous. Therefore we say that $u \in C(J, E)$ is a *mild solution* of the IVP (7.105) if u is a solution of the integral equation (7.90).

Assuming that $L^1(J, E)$ is ordered a.e. pointwise, and that the positive constant M is defined by (7.76), we shall impose the following hypotheses on the function $g : J \times E \times L^1(J, E) \to E$.

(g0) $g(\cdot, u(\cdot), u)$ is strongly measurable for each $u \in L^1(J, E)$.
(g1) $g(t, x, u)$ is increasing in x and u for a.e. $t \in J$.
(g2) $\|g(t, x, u)\| \leq h(t, \|x\|)$ for a.e. $t \in J$ and all $x \in E$, $u \in L^1(J, E)$, where $h : J \times \mathbb{R}_+ \to \mathbb{R}_+$ is sup-measurable, $h(t, \cdot)$ is increasing for a.e. $t \in J$, and the IVP

$$w'(t) = M h(t, w(t)) \quad \text{for a.e. } t \in J, \quad w(t_0) = w_0$$
(7.106)

has for each $w_0 \geq 0$ the greatest absolutely continuous solution.

In the proof of our main existence result for the integral equation (7.90), we make use of the following auxiliary result.

Lemma 7.54. *Let the hypotheses given for h in (g2) be satisfied, and let $w_0 \geq 0$ be given. If $v \in C(J, \mathbb{R})$ satisfies the inequality*

$$v(t) \leq w_0 + \int_a^t M h(s, v(s)) ds, \quad t \in J,$$
(7.107)

then $v(t) \leq w(t)$ for each $t \in J$, where w is the greatest solution of (7.106).

Proof: The space $C(J, \mathbb{R})$ is an ordered metric space with respect to the sup-norm $\|\cdot\|_0$ and the pointwise ordering. Let $\overline{w}$ be the greatest solution of (7.106) with w_0 replaced by $\overline{w}_0 = \max\{w_0, \|v\|_0\}$. Denoting $[v, \overline{w}] = \{w \in C(J, \mathbb{R}) : v \leq w \leq \overline{w}\}$, the hypotheses given for h in (g2) imply that the relation

$$Qw(t) = w_0 + \int_a^t M\,h(s, w(s))\,ds, \quad t \in J \qquad (7.108)$$

defines an increasing mapping $Q : [v, \overline{w}] \to [v, \overline{w}]$. Moreover, the range of Q is equicontinuous because

$$Qw(t) - Qw(\bar{t}) \le \int_{\bar{t}}^t M\,h(s, \overline{w}(s))\,ds = \int_{\bar{t}}^t \overline{w}'(s)\,ds = \overline{w}(t) - \overline{w}(\bar{t})$$

for all $t, \bar{t} \in J$, $\bar{t} \le t$ and for all $w \in [v, \overline{w}]$. Thus (Qw_n) converges in $C(J, \mathbb{R})$ whenever (w_n) is a monotone sequence in $[v, \overline{w}]$. This implies by Proposition 2.39 that Q has fixed points, or equivalently, (7.106) has solutions in $[v, \overline{w}]$. But w was the greatest solution of (7.106), whence $v \le w$. □

The space E is assumed to posses the following properties.

(E0) Bounded and monotone sequences of E converge weakly.
(E1) The mapping $E \ni x \mapsto x^+ := \sup\{0, x\}$ is demicontinuous, and $\|x^+\| \le \|x\|$ for all $x \in E$.

As an application of Proposition 7.27 and Lemma 7.54, we are now able to prove existence results for continuous solutions of the integral equation (7.90).

Theorem 7.55. *Let E be a lattice-ordered Banach space with properties (E0) and (E1). Assume that $T : \Lambda \to L(E)$ and $g : J \times E \times L^1(J, E) \to E$ satisfy the hypotheses (T0)–(T2) and (g0)–(g2), respectively. Then for each $x_0 \in E$ all the solutions of (7.90) are continuous. Moreover,*

(a) (7.90) has maximal and minimal solutions;
(b) the equation

$$u(t) = -\left(T(t, a)x_0 + \int_a^t T(t, s)g(s, u(s), u)ds\right)^-, \quad t \in J \qquad (7.109)$$

has the greatest solution $\underline{u}$, and the equation

$$u(t) = \left(T(t, a)x_0 + \int_a^t T(t, s)g(s, u(s), u)ds\right)^+, \quad t \in J \qquad (7.110)$$

has the smallest solution $\overline{u}$;
(c) (7.90) has the smallest and greatest solutions in $[\underline{u}, \overline{u}]$.

Proof: Let $x_0 \in E$ be given, and let P be defined by

$$P = \{u \in L^p(J, E) : \|u(t)\| \le w(t) \text{ for a.e. } t \in J\}, \qquad (7.111)$$

where w is the greatest solution of (7.106) with $w_0 = M\|x_0\|$. If $u \in P$, then from (7.76), (7.106), (7.111), and (g2) it follows that for all $t \in J$,

$$\|T(t,s)g(s,u(s),u)\| \le M \|g(s,u(s),u)\|$$
$$\le M\,h(s,\|u(s)\|) \le M\,h(s,w(s)) = w'(s) \quad (7.112)$$

for a.e. $s \in [a,t]$. In particular, when $t = s$, we obtain

$$\|g(s,u(s),u)\| \le \frac{w'(s)}{M} \quad \text{for a.e.} \quad s \in J. \tag{7.113}$$

This result and the hypothesis (g0) imply that the function

$$q(s) = g(s,u(s),u), \quad s \in J \tag{7.114}$$

is Bochner integrable. By hypothesis (T1) we infer that for each $t \in J$ the function $f(s,x) = T(t,s)x$, $a \le s \le t$, $x \in E$, is a Carathéodory function. Thus the function $s \mapsto T(t,s)q(s)$ is strongly measurable on $[a,t]$ for each $t \in J$. In view of (7.76), the function $s \mapsto T(t,s)q(s)$ is also norm-bounded on $[a,t]$ by the Lebesgue integrable function $s \mapsto M\|q(s)\|$, whence it is Bochner-integrable on $[a,t]$. Thus the equation

$$Gu(t) = T(t,a)x_0 + \int_a^t T(t,s)g(s,u(s),u)\,ds, \quad t \in J, \tag{7.115}$$

defines a function $Gu: J \to E$. If $a \le \bar{t} \le t \le b$, then from (7.115) it follows by applying (T0), (7.111), and (7.114) that

$$\|Gu(t) - Gu(\bar{t})\| \le \|T(t,a)x_0 - T(\bar{t},a)x_0\| + \left\|\int_a^{\bar{t}} (T(t,s) - T(\bar{t},s))q(s)\,ds\right\|$$

$$+ \int_{\bar{t}}^t \|T(t,s)q(s)\|ds$$

$$\le \|(T(t,\bar{t}) - I)T(\bar{t},a)x_0\| + \left\|(T(t,\bar{t}) - I)\int_a^{\bar{t}} T(\bar{t},s)q(s)\,ds\right\|$$

$$+ \int_{\bar{t}}^t M\|q(s)\|\,ds$$

$$\le M\|T(t,\bar{t}) - I\|\|x_0\| + M\|T(t,\bar{t}) - I\| \int_a^b \|q(s)\|\,ds$$

$$+ M \int_{\bar{t}}^t \|q(s)\|\,ds.$$

This implies by (T0) and (T1) that $Gu \in C(J,E)$. Moreover, from (7.111), (7.112), and (7.115) it follows that

$$\|Gu(t)\| \le \|T(t,a)x_0\| + \int_a^t \|T(t,s)g(s,u(s),u)\|\,ds$$

$$\le M\,\|x_0\| + \int_a^t w'(s)\,ds$$

$$= w_0 + w(t) - w(a) = w(t), \quad t \in J,$$

whence $Gu \in P$. By hypotheses (T2) and (g2) we have $Gu \leq Gv$ whenever $u, v \in P$ and $u \leq v$.

The above proof shows that (7.115) defines an increasing mapping G from P to P. To show that all the solutions of (7.90) belong to P, let $u \in C(J, E)$ satisfy (7.90). Because $w_0 = M \|x_0\|$, from (7.76), (7.103), and (g2) it follows that

$$\|u(t)\| \leq \|T(t,a)x_0\| + \int_a^t \|T(t,s)g(s,u(s),u)\| \, ds$$

$$\leq M \|x_0\| + \int_a^t M \, h(s, \|u(s)\|) \, ds$$

$$= w_0 + \int_a^t M \, h(s, \|u(s)\|) \, ds, \quad t \in J.$$

Denoting $v(t) = \|u(t)\|$, $t \in J$, we then have

$$v(t) \leq w_0 + \int_{t_0}^t M \, h(s, v(s)) \, ds, \quad t \in J.$$

By Lemma 7.54, the last inequality implies that $v(t) \leq w(t)$, i.e., $\|u(t)\| \leq w(t)$ for all $t \in J$. Thus $u \in P$. The above result shows that u is a solution of (7.90) if and only if u is a fixed point of an increasing mapping $G : P \to P$, defined by (7.115). Thus the assertions follow from Proposition 7.27. $\qquad \Box$

Corollary 7.56. *The results of Theorem 7.55 remain true if $T : \Lambda \to L(E)$ and $g : J \times L^1(J, E) \times E \to E$ satisfy conditions (T0)–(T2) and (g0), (g1), respectively, and g, in addition, fulfills*

(g3) $\|g(t, u, v)\| \leq p(t)\psi(\|v\|)$ for a.e. $t \in J$ and all $v \in E$, $u \in L^1(J, E)$, where $p \in L^1(J, \mathbb{R}_+)$, $\psi : \mathbb{R}_+ \to (0, \infty)$ is increasing, and $\int_0^\infty \frac{dx}{\psi(x)} = \infty$.

Proof: The properties given for q and ψ in the hypothesis (g3) ensure by [44, Lemma B.7.1] that the IVP

$$w'(t) = M \, p(t)\psi(w(t)) \quad \text{for a.e. } t \in J, \quad w(t_0) = w_0 \qquad (7.116)$$

has for each $w_0 \geq 0$ a unique absolutely continuous solution. Thus the hypothesis (g2) holds with $h(t, x) = p(t)\psi(x))$, and the conclusion follows. $\qquad \Box$

7.3.8 Application

Consider the n-dimensional problem

$$y_t(x, t) - \Delta y(x, t) = f(x, t, y(x, t)) \quad \text{in } \mathbb{R}^n, \quad y(x, 0) = y_0(x). \qquad (7.117)$$

The next result follows from [89, Theorems 2.3.1 and 2.3.2].

Lemma 7.57. *Assume that $x_0 \in C(\mathbb{R}^n) \cap L^\infty(\mathbb{R}^n)$, and that $q \in C_1^2(\mathbb{R}^n \times \mathbb{R}_+)$ has a compact support. Then the IVP*

$$y_t(x,t) - \Delta y(x,t) = q(x,t) \quad in \ \ \mathbb{R}^n, \quad y(x,0) = y_0(x), \tag{7.118}$$

has the solution

$$y(x,t) = \int_{\mathbb{R}^n} K(x-z,t)y_0(z)\,dz + \int_0^t \int_{\mathbb{R}^n} K(x-z,t-s)q(z,s)\,dz\,ds, \tag{7.119}$$

where

$$K(z,t) = \frac{1}{(4\pi t)^{\frac{n}{2}}} e^{\frac{-\|z\|^2}{4t}}, \quad z \in \mathbb{R}^n, \ t > 0. \tag{7.120}$$

Moreover, defining

$$S(t)y_0(x) = \int_{\mathbb{R}^n} K(x-z,t)y_0(z)\,dz, \quad x \in \mathbb{R}^n, \ t > 0, \ S(0)y_0 = y_0,$$

we obtain a contraction semigroup $\{S(t)\}_{t \geq 0}$ on $L^2(\mathbb{R}^n)$ (cf. [89, p.427]). In particular, when $\Lambda = \{(t,s) : 0 \leq s \leq t \leq 1\}$, the relation

$$T(t,s)u = S(t-s)u, \quad u \in L^2(\mathbb{R}^n), \ (t,s) \in \Lambda$$

defines a family of operators $T(t,s)$ that has properties (T0)–(T2). Thus we get the following result as a consequence of Corollary 7.56.

Proposition 7.58. *Let K be defined by (7.120), and assume that the function $f : Q \times \mathbb{R} \to \mathbb{R}$, $Q = \mathbb{R}^n \times [0,1]$, has the following properties:*

(f0) $f(\cdot,\cdot,y(\cdot,\cdot))$ is measurable on Q whenever $y : Q \to \mathbb{R}$ is measurable.
(f1) $f(x,t,\cdot)$ is increasing for a.e. $(x,t) \in Q$.
(f2) $\|f(\cdot,t,v(\cdot))\|_2 \leq p(t)\psi(\|v\|_2)$ for all $t \in J = [0,1]$ and $v \in L^2(\mathbb{R}^n)$, where $p \in L^1(J,\mathbb{R}_+)$, $\psi : \mathbb{R}_+ \to (0,\infty)$ is increasing, and $\int_0^\infty \frac{dx}{\psi(x)} = \infty$.

Then for each $u_0 \in L^2(\mathbb{R}^n)$ the integral equation

$$y(x,t) = \int_{\mathbb{R}^n} K(x-z,t)y_0(z)\,dz + \int_0^t \int_{\mathbb{R}^n} K(x-z,t-s)f(z,s,y(z,s))\,dz\,ds, \tag{7.121}$$

has solutions, which are also mild solution of the IVP (7.117) in the set of those measurable functions $y : \mathbb{R}^n \times J \to \mathbb{R}$ for which $y(\cdot,t) \in L^2(\mathbb{R}^n)$ for each $t \in J$ and $\lim_{t \to t_0} \int_{\mathbb{R}^n} |y(x,t) - y(x,t_0)|^2 dx = 0$ for each $t_0 \in J$.

Remark 7.59. (i) The functional dependence on the second argument of g can be formed, e.g., by bounded, linear, and positive operators, such as integral operators of Volterra and/or Fredholm type with nonnegative kernels. Thus the results of this section can be applied also to integro-differential equations.

(ii) Weakly complete Banach lattices have properties (E0) and (E1) (cf. [170]). Examples of such spaces are, for instance, Banach-lattices with uniformly monotone norm (UMB-lattices) defined in [22, XV,14], the spaces $\mathbb{R}^m$, $m = 1, 2, \ldots$, and l^p, $p \in [1, \infty)$, ordered coordinatewise and normed by p-norm, and spaces $L^p(\Omega)$, where $p \in [1, \infty)$ and $\Omega = (\Omega, \mathcal{A}, \mu)$ is a measure space, equipped with p-norm and a.e. pointwise ordering. Moreover, the Sobolev spaces $W^{1,p}(\Omega)$ and $W^{1,p}(\Omega)$, $p \in (1, \infty)$ ordered a.e. pointwise, where Ω is a bounded domain in $\mathbb{R}^m$, posses properties (E0) and (E1) (cf. [44]). In particular, we can choose E to be one of these spaces in the above considerations.

7.4 Notes and Comments

In this chapter we have studied integral equations in ordered Banach spaces. The results presented in Sect. 7.1.4 are new, whereas all the other material of Sect. 7.1 dealing with Fredholm and Volterra integral equations in HL spaces is a slight generalization to that presented in [132] and [141]. Related results for improper integral equations are derived in [129] and [131]. As for other papers dealing with functional Volterra integral equations and differential equations via non-absolute integrals, see, e.g., [93, 92, 201]. Non-absolute integral equations in Banach spaces are considered also in [91, 211]. The results presented in Sect. 7.2 for integral equations in L^p-spaces and impulsive differential equations in abstract spaces are from [120] and [134]. The material of Sects. 7.3.1–7.3.6, where well-posedness and extremality results are derived for discontinuous integral equations containing non-absolutely integrable functions, is new. Sections 7.3.7 and 7.3.8, adopted from [50], deal with discontinuous evolution integral equations containing Bochner integrable functions, and an application to a second order partial differential equation of parabolic type. As for other existence results for integral equations in abstract spaces, see, e.g., [105, 107, 156, 170, 188, 189].

8

Game Theory

The main goal of this chapter is to study the existence of extremal Nash equilibria for normal-form games. Our interest is focused on games with strategic complementarities, which means roughly speaking that the best responses of players are increasing in actions of the other players. Properties, known as 'increasing differences,' '(quasi)supermodular,' and 'single crossing property,' are used to formalize, or even to define strategic complementarities, cf. [9, 12, 74, 75, 146, 179, 218]. In the last section of this chapter we consider the existence of winning strategies in a pursuit and evasion game. In addition we obtain new fixed point results for set-valued and single-valued mappings.

In Sect. 8.1 we prove existence results for the smallest and greatest pure Nash equilibria of a normal-form game $\Gamma = \{S_i, u_i\}_{i=1}^N$ of N players whose strategy spaces S_i are finite sets of real numbers, and whose utility functions u_i are real-valued. We also present conditions that ensure that the utilities of the obtained greatest or smallest pure Nash equilibria majorize the utilities of all pure Nash equilibria. An application to a pricing game is given.

In Sect. 8.2 we derive existence and comparison results for Nash equilibria of normal-form games when strategy spaces are finite posets and the utility functions are vector-valued. In the case when the game is supermodular and its strategy spaces are one-dimensional, in [86] extremal pure Nash equilibria are shown to exist and to be lower and upper bounds to all mixed Nash equilibria when they are ordered by the first-order stochastic dominance. The difficulties that the author of [86] encountered when the strategy spaces are multidimensional gave rise to the following conclusion by this author (see [87, p.16]):

> "It may be desirable to order both pure and mixed strategies in a way that is consistent (for example by using the first-order stochastic dominance order on mixed strategies) but unless all strategy spaces are chains, this is incompatible with strategic complementarities."

In spite of this, we are able to show by assuming very weak forms of strategic complementarities that all mixed Nash equilibria of a finite normal-form game, ordered by the first-order stochastic dominance order, lie between the extremal pure Nash equilibria. In particular, this result is shown to hold for a supermodular normal-form game $\Gamma = \{S_i, u_i\}_{i=1}^N$ when strategy spaces S_i are finite lattices and the ranges of the functions $u_i(\cdot, s_{-i})$ are upward directed sets in ordered vector spaces for every $s_{-i} = (s_1, \ldots, s_{i-1}, s_{i+1}, \ldots, s_N)$. Moreover, we show that if the utility functions $u_i(s_i, s_{-i})$ are monotone with respect to s_{-i}, then one of the extremal Nash equilibria gives the best utilities. The proofs are constructive and provide finite algorithms to determine those equilibria and their corresponding utilities. These algorithms and Maple programming are applied to concrete games. The obtained results are applied also to a multiproduct pricing game.

In Sect. 8.3 we derive existence results for smallest, greatest, minimal, and maximal pure Nash equilibria of normal-form games with strategic complements. A novel feature is that the ranges of utility functions are posets. This allows us to evaluate the utilities of different players in different ordinal scales. In the real-valued case we obtain generalizations to some results derived in [180, 218]. In Sect. 8.3.4 the obtained results are applied to a multiproduct version of Bertrand oligopoly pricing game. Algorithmic methods and concrete examples are also presented.

To ensure the existence of required maximums of utility functions, we prove new extreme value theorems in Sect. 8.3.1. It is well-known that an upper semicontinuous real-valued function attains its maximum if its domain is a compact topological space (cf., e.g., [29]). On the other hand, monotone functions defined on compact lattices and piecewise strictly monotone functions defined on compact chains have maximums, although they are not necessarily upper semicontinuous. One purpose is to find such a hypothesis that ensures the existence of maximums for all types of functions presented above. Another purpose is to formulate this hypothesis in such a way that, in addition to topological properties of the domains, it depends only on the orderings of the ranges of functions in question. Such a hypothesis allows us to find maximums for functions having ranges in posets. Noticing also that a lattice is complete if and only if it is compact in the order interval topology (cf. [23]), we obtain extreme value results under hypotheses that depend only on orderings of the domains and ranges of the considered functions.

The results of Sect. 8.2 are generalized in Sect. 8.4 to normal-form games whose strategy spaces are complete, separable, and ordered metric spaces. For instance, those results of Sect. 8.2 dealing with the existence of most profitable pure Nash equilibria of supermodular games have extensions when the utility functions are real-valued. These results are obtained as special cases of the corresponding results proved for normal-form games with vector-valued utilities having quasisupermodular mixed extensions. The results of Theorems 2.20 and 2.21 and their duals provide tools to overcome the difficulties caused by the existence of unordered pure strategies.

In Sect. 8.5 we derive necessary and sufficient conditions for the existence of undominated strategies, weakly dominating strategies, and weakly dominating pure Nash equilibria for normal-form games whose strategy spaces as well as ranges of utility functions are posets.

Finally, in Sect. 8.6 the Chain Generating Recursion Principle and generalized iteration methods presented in Chap. 2 are applied to prove the existence of winning strategies for a pursuit and evasion game. The obtained results are used to study the solvability of equations and inclusions in ordered spaces. The monotonicity hypotheses are weaker than those assumed in Chap. 2.

8.1 Pure Nash Equilibria for Finite Simple Normal-Form Games

As an introduction to the subject we study in this section a normal-form game $\Gamma = \{S_i, u_i\}_{i=1}^{N}$ of N players whose strategy spaces S_i are finite sets of real numbers, and the utility functions u_i are real-valued. We present conditions that ensure the existence of the smallest and greatest pure Nash equilibria for Γ. If the utilities $u_i(s_1, \ldots, s_N)$ are also increasing (respectively decreasing) in s_j, $j \neq i$, the utilities of the obtained greatest (respectively the smallest) pure Nash equilibrium are shown to majorize the utilities of all pure Nash equilibria. The obtained results are then applied to a **Bertrand oligopoly** model for firms that compete in prices.

8.1.1 Preliminaries

Definition 8.1. *We say that* $\Gamma = \{S_i, u_i\}_{i=1}^{N}$ *is a* **finite simple normal-form game** *of players* $i = 1, \ldots, N$, *if for each* i *the* **strategy space** S_i *for player* i *is a finite nonempty set of real numbers, and* u_i *is a real-valued* **utility function** *of player* i, *defined on* $S = S_1 \times \cdots \times S_N$.

We use the following notations: $s_{-i} = (s_1, \ldots, s_{i-1}, s_{i+1}, \ldots, s_N)$, $u_i(s_1, \ldots, s_N) = u_i(s_i, s_{-i})$, and $S_{-i} = S_1 \times \cdots \times S_{i-1} \times S_{i+1} \times \cdots \times S_N$.

Definition 8.2. *Let* $\Gamma = \{S_i, u_i\}_{i=1}^{N}$ *be a finite simple normal-form game. We say that strategies* $s_1^*, \ldots, s_N^*$ *form a* **pure Nash equilibrium** *for* Γ *if*

$$u_i(s_i^*, s_{-i}^*) = \max_{s_i \in S_i} u_i(s_i, s_{-i}^*) \quad \text{for all} \quad i = 1, \ldots, N.$$

This definition implies that *the strategies of players form a pure Nash equilibrium if and only if no player can improve his/her utility by changing the strategy when all the other players keep their strategies fixed.*

In what follows we assume that $\Gamma = \{S_i, u_i\}_{i=1}^{N}$ is a finite simple normal-form game. All products of sets of real numbers are assumed to be ordered coordinatewise. Since the strategy spaces S_i are finite, the functions $s_i \mapsto u_i(s_i, s_{-i})$, $s_{-i} \in S_{-i}$, have only a finite number of values. Thus the sets of their maximum points

$$F_i(s_{-i}) = \{x_i \in S_i : u_i(x_i, s_{-i}) = \max_{s_i \in S_i} u_i(s_i, s_{-i})\}, \quad i = 1, \dots, N, \quad (8.1)$$

are nonempty. These sets are in turn subsets of S_i, and thus finite sets of real numbers, whence they have maxima and minima.

8.1.2 Existence and Comparison Results

We shall first prove that a pure Nash equilibrium exists for a finite simple normal-form game $\Gamma = \{S_i, u_i\}_{i=1}^N$ if the following hypothesis is valid:

(h0) $\max F_i(s_{-i}) \le \max F_i(\hat{s}_{-i})$ whenever $s_{-i} \le \hat{s}_{-i}$ in S_{-i}, $i = 1, \dots, N$.

Theorem 8.3. *Assume that (h0) is satisfied. Then there exist strategies* $\overline{s}_1^*, \dots, \overline{s}_N^*$ *such that for every* $i = 1, \dots, N$ *the strategy* $\overline{s}_i^*$ *is the greatest among all the strategies of player i that maximizes its utility when the strategies of the other players are* $\overline{s}_1^*, \dots, \overline{s}_{i-1}^*, \overline{s}_{i+1}^*, \dots \overline{s}_N^*$. *In particular, the strategies* $\overline{s}_1^*, \dots, \overline{s}_N^*$ *form a pure Nash equilibrium for* Γ.

Proof: In the proof we use a method of successive approximations. The maxima $b_i = \max S_i$, $i = 1, \dots, N$, exist because the sets S_i are finite. Denote

$$s_i^0 := b_i \quad \text{and} \quad s_{-i}^0 := (b_1, \dots, b_{i-1}, b_{i+1}, \dots, b_N), \quad i = 1, \dots, N.$$

The above notations imply that

$$s_i^1 := \max F_i(s_{-i}^0) \le b_i = s_i^0 \quad \text{for all } i = 1, \dots, N.$$

By this result we also have $s_{-i}^1 := (s_1^1, \dots, s_{i-1}^1, s_{i+1}^1, \dots, s_N^1) \le s_{-i}^0$ for all $i = 1, \dots N$, which by (h0) implies that

$$s_i^2 := \max F_i(s_{-i}^1) \le \max F_i(s_{-i}^0) = s_i^1 \quad \text{for all } i = 1, \dots, N, \text{ etc.}$$

Because the set $\{\max F_i(s_{-i}) : s_{-i} \in S_{-i}\}$ is finite for each $i = 1, \dots, N$, then continuing the above reasoning a finite number of times, say k times, we get

$$s_i^{k+1} := \max F_i(s_{-i}^k) = \max F_i(s_{-i}^{k-1}) = s_i^k \quad \text{for all } i = 1, \dots, N.$$

Denoting $\overline{s}_i^* := s_i^k$ and $\overline{s}_{-i}^* := s_{-i}^k := (s_1^k, \dots, s_{i-1}^k, s_{i+1}^k, \dots, s_N^k)$, $i = 1, \dots, N$, we then have

$$\overline{s}_i^* = \max F_i(\overline{s}_{-i}^*) \quad \text{for all } i = 1, \dots, N. \quad (8.2)$$

According to this result the strategies $\overline{s}_i^*$ are the greatest strategies that maximize the utility $u_i(s_i, s_{-i})$ when $s_{-i} = \overline{s}_{-i}^*$. In particular,

$$u_i(\overline{s}_i^*, \overline{s}_{-i}^*) = \max_{s_i \in S_i} u_i(s_i, \overline{s}_{-i}^*), \quad i = 1, \dots, N,$$

so that the strategies $\overline{s}_1^*, \dots, \overline{s}_N^*$ form a pure Nash equilibrium for Γ. □

Next we show that the pure Nash equilibrium constructed in Theorem 8.3 majorizes all pure Nash equilibria for Γ, and that it gives best utilities if each $u_i(s_1, \dots, s_N)$ is increasing in s_j, $j \neq i$.

Theorem 8.4. *Assume that (h0) is valid, and let $\bar{s}_1^*, \ldots, \bar{s}_N^*$ be the pure Nash equilibrium constructed in Theorem 8.3. If $s_1^*, \ldots, s_N^*$ is a pure Nash equilibrium for Γ, then $s_i^* \leq \bar{s}_i^*$ for each $i = 1, \ldots, N$. Assume moreover that each $u_i(s_1, \ldots, s_N)$ is increasing in s_j, $j \neq i$, i.e.,*

(h1) $u_i(s_i, s_{-i}) \leq u_i(s_i, \hat{s}_{-i})$ whenever $s_{-i} \leq \hat{s}_{-i}$, $s_i \in S_i$ and $i = 1, \ldots, N$.

Then $u_i(s_i^*, s_{-i}^*) \leq u_i(\bar{s}_i^*, \bar{s}_{-i}^*)$ for all $i = 1, \ldots, N$.

Proof: Let $s_1^*, \ldots, s_N^*$ be a pure Nash equilibrium for Γ. Then $s_i^* \in F_i(s_{-i}^*)$, whence $s_i^* \leq \max F_i(s_{-i}^*)$ for each $i = 1, \ldots N$. According to the notations used in the proof of Theorem 8.3 we have

$$s_i^* \leq \max F_i(s_{-i}^*) \leq b_i = s_i^0 \quad \text{for all } i = 1, \ldots, N.$$

By this result we have also $s_{-i}^* \leq s_{-i}^0$ for all $i = 1, \ldots N$, which by the above result and (h0) implies that

$$s_i^* \leq \max F_i(s_{-i}^*) \leq \max F_i(s_{-i}^0) = s_i^1 \quad \text{for all } i = 1, \ldots, N.$$

The preceding inequalities imply that $s_{-i}^* \leq s_{-i}^1$ for all $i = 1, \ldots N$, which in view of (h0) yields

$$s_i^* \leq \max F_i(s_{-i}^*) \leq \max F_i(s_{-i}^1) = s_i^2 \text{ for all } = 1, \ldots, N, \text{ etc.}$$

Repeating the above process k times, where k is as in the proof of Theorem 8.3, we get

$$s_i^* \leq \max F_i(s_{-i}^*) \leq \max F_i(s_{-i}^k) = s_i^k = \bar{s}_i^* \quad \text{for all } i = 1, \ldots, N,$$

which proves the first assertion.

Assume next that (h1) holds. Since $s_i^* \leq \bar{s}_i^*$ for all $i = 1, \ldots, N$, then also $s_{-i}^* \leq \bar{s}_{-i}^*$ in s_{-i} for all $i = 1, \ldots, N$. From (h1) and from Definition 8.2 it then follows that for all $i = 1, \ldots, N$,

$$u_i(s_i^*, s_{-i}^*) \leq u_i(s_i^*, \bar{s}_{-i}^*) \leq u_i(\bar{s}_i^*, \bar{s}_{-i}^*),$$

which proves the last assertion.   □

Replacing in the above proof maxima by minima we get the following dual results to Theorems 8.3 and 8.4.

Theorem 8.5. *Assume that condition*

(h2) $\min F_i(s_{-i}) \leq \min F_i(\hat{s}_{-i})$ whenever $s_{-i} \leq \hat{s}_{-i}$ in S_{-i}, $i = 1, \ldots, N$

is satisfied. Then there exist strategies $\underline{s}_1^*, \ldots, \underline{s}_N^*$ such that for every $i = 1, \ldots, N$ the strategy $\underline{s}_i^*$ is the smallest among all the strategies of player i that maximizes its utility when the strategies of the other players are $\underline{s}_1^*, \ldots, \underline{s}_{i-1}^*, \underline{s}_{i+1}^*, \ldots \underline{s}_N^*$. In particular, the strategies $\underline{s}_1^*, \ldots, \underline{s}_N^*$ form a pure Nash equilibrium for Γ.

Proof: The minima $a_i = \min S_i$, $i = 1, \ldots, N$, exist because the sets S_i are finite. The notations

$$s_i^0 := a_i \quad \text{and} \quad s_{-i}^0 := (a_1, \ldots, a_{i-1}, a_{i+1}, \ldots, a_N), \quad i = 1, \ldots, N$$

imply that

$$a_i = s_i^0 \leq s_i^1 := \min F_i(s_{-i}^0) \quad \text{for all } i = 1, \ldots, N.$$

By this result we also have $s_{-i}^0 \leq s_{-i}^1 := (s_1^1, \ldots, s_{i-1}^1, s_{i+1}^1, \ldots, s_N^1)$ for all $i = 1, \ldots N$, which due to (h2) implies that

$$s_i^1 = \min F_i(s_{-i}^0) \leq s_i^2 := \min F_i(s_{-i}^1) \quad \text{for all } i = 1, \ldots, N, \text{ etc.}$$

After a finite number of steps, say j steps, we get

$$s_i^j := \min F_i(s_{-i}^{j-1}) = \min F_i(s_{-i}^j) = s_i^{j+1} \quad \text{for all } i = 1, \ldots, N.$$

Denoting $\underline{s}_i^* := s_i^j$ and $\underline{s}_{-i}^* := s_{-i}^j := (s_1^j, \ldots, s_{i-1}^j, s_{i+1}^j, \ldots, s_N^j)$, $i = 1, \ldots, N$, we then have

$$\underline{s}_i^* = \min F_i(\underline{s}_{-i}^*) \quad \text{for all } i = 1, \ldots, N. \tag{8.3}$$

Thus the strategies $\underline{s}_i^*$ are the smallest strategies that maximize the utility $u_i(s_i, s_{-i})$ when $s_{-i} = \underline{s}_{-i}^*$. In particular,

$$u_i(\underline{s}_i^*, \underline{s}_{-i}^*) = \max_{s_i \in S_i} u_i(s_i, \underline{s}_{-i}^*), \quad i = 1, \ldots, N.$$

Consequently, the strategies $\underline{s}_1^*, \ldots, \underline{s}_N^*$ form a pure Nash equilibrium for Γ.

□

The proof of the following theorem is dual to that of Theorem 8.4.

Theorem 8.6. *Assume that (h2) holds, and let $\underline{s}_1^*, \ldots, \underline{s}_N^*$ be the pure Nash equilibrium constructed in Theorem 8.5. If $s_1^*, \ldots, s_N^*$ is a pure Nash equilibrium for Γ, then $\underline{s}_i^* \leq s_i^*$ for each $i = 1, \ldots, N$. Assume, moreover, that each $u_i(s_1, \ldots, s_N)$ is decreasing in s_j, $j \neq i$, i.e.,*

(h3) $u_i(s_i, \hat{s}_{-i}) \leq u_i(s_i, s_{-i})$ whenever $s_{-i} \leq \hat{s}_{-i}$, $s_i \in S_i$ and $i = 1, \ldots, N$.

Then $u_i(s_i^*, s_{-i}^*) \leq u_i(\underline{s}_i^*, \underline{s}_{-i}^*)$ for all $i = 1, \ldots, N$.

To study the validity of hypotheses (h0) and (h2) we introduce the following definition.

Definition 8.7. *We say that a function $f : S_i \times S_{-i} \to \mathbb{R}$ has **increasing differences** in (s_i, s_{-i}) if $f(y_i, s_{-i}) - f(x_i, s_{-i}) \leq f(y_i, \hat{s}_{-i}) - f(x_i, \hat{s}_{-i})$ whenever $x_i < y_i$ and $s_{-i} < \hat{s}_{-i}$.*

Proposition 8.8. *Let $\Gamma = \{S_i, u_i\}_{i=1}^N$ be a finite simple normal-form game. Assume that each $u_i(s_i, s_{-i})$ has increasing differences in (s_i, s_{-i}). Then Γ has the smallest and greatest pure Nash equilibria. Moreover, the greatest (respectively the smallest) one gives the best utilities for the players provided (h1) (respectively (h3)) is satisfied.*

Proof: Denoting $x_i = \max F(s_{-i})$ and $y_i = \max F(\hat{s}_{-i})$ and applying (8.1) and Definition 8.7 we obtain

$$0 \le u_i(x_i, s_{-i}) - u_i(\min\{x_i, y_i\}, s_{-i}) = u_i(\max\{x_i, y_i\}, s_{-i}) - u_i(y_i, s_{-i})$$
$$\le u_i(\max\{x_i, y_i\}, \hat{s}_{-i}) - u_i(y_i, \hat{s}_{-i}) \le 0 \quad \text{whenever } s_{-i} \le \hat{s}_{-i} \text{ in } S_{-i}.$$

Thus all the inequalities above are equalities. In particular,

$$u_i(\max\{x_i, y_i\}, \hat{s}_{-i}) = u_i(y_i, \hat{s}_{-i}),$$

so that $\max\{x_i, y_i\} \in F(\hat{s}_{-i})$. Since $y_i = \max F_i(\hat{s}_{-i})$, then $\max\{x_i, y_i\} = y_i$, whence $x_i \le y_i$. This proves that (h0) is valid. Similarly one can show that condition (h2) holds. The assertions follow then from Theorems 8.4, 8.5, and 8.6. $\qquad\qquad\square$

8.1.3 An Application to a Pricing Game

To present an application, assume that $c_i \ge 0$, $i = 1, \ldots, N$, and that

$$u_i(s_i, s_{-i}) = d_i(s_i, s_{-i})(s_i - c_i), \quad \text{where } d_i : S_i \times S_{-i} \to \mathbb{R}_+, \ i = 1, \ldots, N. \tag{8.4}$$

Players i are considered to stand for firms, each of which sell the substitute products e_i. The feasible selling prices s_i per unit of e_i are assumed to form a finite subset S_i of $\mathbb{R}$, which is bounded from below by c_i, the 'unit cost' of e_i. The functions d_i stand for demands of products e_i, which tell how many units of e_i the firm i sells during a fixed time period. The real-valued utilities u_i defined by (8.4) are considered as profits of firms i.

As a consequence of Proposition 8.8 we obtain the following result.

Corollary 8.9. *Let Γ be a pricing game in which profits satisfy (8.4), and the demand functions $d_i(s_i, s_{-i})$ have increasing differences in (s_i, s_{-i}) and are increasing in s_{-i}. Then Γ has the smallest and greatest price equilibria, and the greatest one gives best profits among all price equilibria for Γ.*

Proof: Assume that $s_i < \hat{s}_i$ and $s_{-i} < \hat{s}_{-i}$. Since each $d_i(s_i, s_{-i})$ has increasing differences in (s_i, s_{-i}), we obtain

$$u_i(\hat{s}_i, \hat{s}_{-i}) - u_i(s_i, \hat{s}_{-i}) - (u_i(\hat{s}_i, s_{-i}) - u_i(s_i, s_{-i}))$$
$$= d_i(\hat{s}_i, \hat{s}_{-i})(\hat{s}_i - c_i) - d_i(s_i, \hat{s}_{-i})(s_i - c_i) - d_i(\hat{s}_i, s_{-i})(\hat{s}_i - c_i)$$
$$+ d_i(s_i, s_{-i})(s_i - c_i) = (d_i(\hat{s}_i, \hat{s}_{-i}) - d_i(\hat{s}_i, s_{-i}))(\hat{s}_i - s_i)$$
$$+ (d_i(\hat{s}_i, \hat{s}_{-i}) - d_i(s_i, \hat{s}_{-i}) - (d_i(\hat{s}_i, s_{-i}) - d_i(s_i, s_{-i})))(s_i - c_i) \ge 0.$$

This proves that each $u_i(s_i, s_{-i})$ has increasing differences in (s_i, s_{-i}). Since each $d_i(s_i, s_{-i})$ is increasing in s_{-i}, it follows from (8.4) that each $u_i(s_i, s_{-i})$ is increasing in s_{-i}. Thus the assertions follow from Proposition 8.8.   □

Remark 8.10. The model considered above is a Bertrand oligopoly model, where firms compete in prices. It is adopted from [218], which is an excellent source to the subject. The condition that the demand functions $d_i(s_i, s_{-i})$ are increasing in s_{-i} means that the product of firm i is equally or more competitive when one or several other firms raise their prices. In Example 8.11 the demands $d_i(s_i, s_{-i})$ are decreasing in s_i, meaning that the demand of any product is not increasing if its price is raised. If this condition holds, the assumption that demands $d_i(s_i, s_{-i})$ have increasing differences in (s_i, s_{-i}) means that (cf. [218]) "the demand of any product is more sensitive to its price when any other product is more competitive by virtue of its lower price."

If the functions d_i are positive-valued, and if the functions $\log d_i(s_i, s_{-i})$ have increasing differences in (s_i, s_{-i}), it follows from Proposition 8.60 that the hypotheses (h0) and (h2) are valid for profit functions u_i. In fact, these hypotheses present one formulation to strategic complementarities: either maximums or minimums of best responses of any player are increasing in actions of the other players.

Example 8.11. Assume that $N = 3$, and that the demands d_i, $i = 1, 2, 3$ are of the form

$$
\begin{aligned}
d_1(s_1, (s_2, s_3)) &= 370 + 213(s_2 + s_3) + 60s_1 - 230s_1^2, \\
d_2(s_2, (s_1, s_3)) &= 360 + 233(s_1 + s_3) + 55s_2 - 220s_2^2, \\
d_3(s_3, (s_1, s_2)) &= 375 + 226(s_1 + s_2) + 50s_3 - 200s_3^2.
\end{aligned}
\tag{8.5}
$$

Assume moreover that $c_1 := 1.10$, $c_2 := 1.2$, $c_3 := 1.25$, $a_i = 1.30$ and $b_i := 2.10$, $i = 1, 2, 3$, that the smallest price shift is five cents, and that profits are counted by whole euros. Show that the smallest and greatest Nash equilibria for prices exist. Calculate these equilibria and the corresponding profits.

Solution. Convert the problem into a pricing game $\{S_1, S_2, S_3, u_1, u_2, u_3\}$, where the utilities are

$$
u_i = d_i(s_i, s_{-i})(s_i - c_i), \quad i = 1, 2, 3,
$$

and strategy sets are

$$
S_1 = S_2 = S_3 = \{\frac{j}{20}, \ 26 \le j \le 42\}.
$$

From (8.5) it follows that the d_i's are of the form

$$
d_i(s_i, s_{-i}) = f_i(s_i) + g_i(s_{-i}), \quad i = 1, 2, 3,
$$

where every $f_i : S_i \to \mathbb{R}_+$ is decreasing and every $g_i : S_{-i} \to \mathbb{R}_+$ is increasing. Thus the demand functions $d_i(s_i, s_{-i})$ are increasing in s_{-i}. Since for each $i = 1, 2, 3$,

$$d_i(y_i, s_{-i}) - d_i(x_i, s_{-i}) = f_i(y_i) - f_i(x_i),$$

then the demand functions $d_i(s_i, s_{-i})$ have increasing differences in (s_i, s_{-i}). In view of Corollary 8.9 the pricing game $\Gamma = \{S_1, S_2, S_3, u_1, u_2, u_3\}$ has the smallest pure Nash equilibrium $\underline{s}_1, \underline{s}_2, \underline{s}_3$ and the greatest pure Nash equilibrium $\overline{s}_1, \overline{s}_2, \overline{s}_3$.

Applying successive approximations used in the proofs of Theorems 8.3 and 8.5, one can calculate the smallest and greatest pure Nash equilibria for product prices:

$$\underline{s}_1^* = 1.80, \ \underline{s}_2^* = 1.90 \ \text{and} \ \underline{s}_3^* = 1.95,$$

$$\overline{s}_1^* = 1.80, \ \overline{s}_2^* = 1.90, \ \text{and} \ \overline{s}_3^* = 2.00.$$

The corresponding profits are:

$$\underline{u}_1^* = 386, \ \underline{u}_2^* = 380, \ \underline{u}_3^* = 383,$$

$$\overline{u}_1^* = 394, \ \overline{u}_2^* = 388, \ \overline{u}_3^* = 383.$$

By Corollary 8.9 the profits $\overline{u}_1^*$ and $\overline{u}_2^*$ majorize the profits of all other pure Nash equilibria of Γ. Calculations are carried out by making use of simple Maple-programming.

8.2 Pure and Mixed Nash Equilibria for Finite Normal-Form Games

In this section we derive existence and comparison results for Nash equilibria of normal-form games $\Gamma = \{S_i, u_i\}_{i=1}^N$ of N players whose strategy spaces S_i are finite posets and the utility functions u_i are vector-valued. In particular, any supermodular game in which strategy spaces are finite lattices and the ranges of the functions $u_i(\cdot, s_{-i})$ are upward directed sets in ordered vector spaces is shown to posses the smallest and greatest Nash equilibria formed by pure strategies. Moreover, if the utilities $u_i(s_1, \ldots, s_N)$ are also increasing (respectively decreasing) in s_j, $j \neq i$, the utilities of the greatest (respectively the smallest) pure Nash equilibrium are shown to be best in the sense that they majorize both the utilities of all pure Nash equilibria and the expected utilities of all mixed Nash equilibria. The proofs are constructive and provide finite algorithms to determine the extremal pure Nash equilibria for Γ. These algorithms and Maple programming are applied to calculate such Nash equilibria and the corresponding utilities for concrete games.

8.2.1 Preliminaries

A real vector space E equipped with a partial ordering $\leq$ satisfying: $x \leq y$ implies $x + z \leq y + z$ for all $z \in E$, and $cx \leq cy$ for all $c \geq 0$, is called an **ordered vector space**, and denoted by $E = (E, \leq)$.

Definition 8.12. *We say that* $\Gamma = \{S_i, u_i\}_{i=1}^N$ *is a **finite normal-form game** of players* i, $i = 1, \ldots, N$, *if each **strategy set** S_i is a finite nonempty subset of a poset* $X_i = (X_i, \leq_i)$, *and u_i is a **utility function** of player i defined on $S = S_1 \times \cdots \times S_N$ and having values in an ordered vector space* $E_i = (E_i, \leq_i)$.

Mixed strategies are obtained when every player i is allowed to choose independently any randomization of strategies of S_i. The **mixed strategy** σ_i of player i is thus a probability measure over S_i. Denote by Σ_i the space of all mixed strategies of player i. The **expected utilities** $\mathcal{U}_i(\sigma_1, \ldots, \sigma_N)$, $i = 1, \ldots, N$, are defined by:

$$\mathcal{U}_i(\sigma_1, \ldots, \sigma_N) = \sum_{(s_1, \ldots, s_N) \in S} \sigma_1(\{s_1\}) \cdots \sigma_N(\{s_N\}) u_i(s_1, \ldots, s_N). \qquad (8.6)$$

Strategies of the form $\delta_{s_i}(x_i) = \begin{cases} 1, \ x_i = s_i \\ 0, \ x_i \neq s_i \end{cases}$, $s_i \in S_i$, are called **pure strategies**. The set of these pure strategies is denoted by P_i. In the following we use notations: $s_{-i} = (s_1, \ldots, s_{i-1}, s_{i+1}, \ldots, s_N)$, $u_i(s_1, \ldots, s_N) = u_i(s_i, s_{-i})$, $\sigma_{-i} = (\sigma_1, \ldots, \sigma_{i-1}, \sigma_{i+1}, \ldots, \sigma_N)$, $\mathcal{U}_i(\sigma_1, \ldots, \sigma_N) = \mathcal{U}_i(\sigma_i, \sigma_{-i})$, $S_{-i} = S_1 \times \cdots \times S_{i-1} \times S_{i+1} \times \cdots \times S_N$, $\Sigma_{-i} = \Sigma_1 \times \cdots \times \Sigma_{i-1} \times \Sigma_{i+1} \times \cdots \times \Sigma_N$, and $P_{-i} = P_1 \times \cdots \times P_{i-1} \times P_{i+1} \times \cdots \times P_N$.

Definition 8.13. *Let* $\Gamma = \{S_i, u_i\}_{i=1}^N$ *be a finite normal-form game. We say that mixed strategies* $\sigma_1^*, \ldots, \sigma_N^*$ *form a **Nash equilibrium** for Γ if*

$$\mathcal{U}_i(\sigma_i^*, \sigma_{-i}^*) = \max_{\sigma_i \in \Sigma_i} \mathcal{U}_i(\sigma_i, \sigma_{-i}^*) \quad \text{for all} \ \ i = 1, \ldots, N.$$

8.2.2 Existence Result for the Greatest Nash Equilibrium

In what follows we assume that $\Gamma = \{S_i, u_i\}_{i=1}^N$ is a finite normal-form game. All products of posets are assumed to be ordered by componentwise ordering. Assume also that for each fixed $i = 1, \ldots, N$, the set Σ_i of probability measures on S_i is ordered by **first order stochastic dominance** $\preceq_i$, defined as follows:

(SD) $\sigma_i \preceq_i \tau_i$ if $\sigma_i(A) \leq \tau_i(A)$ whenever $A \subseteq S_i$ is increasing, i.e., $x_i \in A$ and $x_i \leq_i y_i$ imply $y_i \in A$.

It follows from (8.6) that

$$\mathcal{U}_i(\sigma_i, \sigma_{-i}) = \sum_{s_i \in S_i} \sigma_i(\{s_i\})U_i(s_i, \sigma_{-i}), \quad i = 1, \dots, N, \text{ where}$$

$$U_i(s_i, \sigma_{-i}) = \sum_{s_{-i} \in S_{-i}} \prod_{j \neq i} \sigma_j(\{s_j\})u_i(s_i, s_{-i}). \tag{8.7}$$

Since each S_i is finite, and hence each function $s_i \mapsto U_i(s_i, \sigma_{-i})$ has only a finite number of values, then the hypothesis

(H0) For every fixed $i = 1, \dots, N$ and $\sigma_{-i} \in \Sigma_{-i}$, the set $\{U_i(s_i, \sigma_{-i}) : s_i \in S_i\}$ is directed upward;

ensures that the sets

$$F_i(\sigma_{-i}) = \{x_i \in S_i : U_i(x_i, \sigma_{-i}) = \max_{s_i \in S_i} U_i(s_i, \sigma_{-i})\}, \quad i = 1, \dots, N, \tag{8.8}$$

are nonempty. The following hypothesis

(H1) For every fixed $i = 1, \dots, N$ and $\sigma_{-i} \in \Sigma_{-i}$, the set $F_i(\sigma_{-i})$ is directed upward;

implies that the maxima of $F_i(\sigma_{-i})$ exist. Assume that these maxima have the following properties.

(H2) The set $\{\max F_i(\sigma_{-i}) : \sigma_{-i} \in \Sigma_{-i}\}$ has an upper bound in S_i, $i = 1, \dots, N$.

(H3) If $\sigma_{-i} \leq \tau_{-i}$ in Σ_{-i} and $\tau_{-i} \in P_{-i}$, then $\max F_i(\sigma_{-i}) \leq_i \max F_i(\tau_{-i})$, $i = 1, \dots, N$.

Lemma 8.14. *Let the hypothesis (H0) hold. Then for all fixed $i \in \{1, \dots, N\}$ and $\sigma_{-i} \in \Sigma_{-i}$ the set*

$$\mathcal{F}_i(\sigma_{-i}) = \{\tau_i \in \Sigma_i : \mathcal{U}_i(\tau_i, \sigma_{-i}) = \max_{\sigma_i \in \Sigma_i} \mathcal{U}_i(\sigma_i, \sigma_{-i})\} \tag{8.9}$$

is nonempty. If the hypotheses (H0) and (H1) are fulfilled, then

$$\max \mathcal{F}_i(\sigma_{-i}) = \delta_{\overline{s}_i(\sigma_{-i})}, \tag{8.10}$$

where $\overline{s}_i(\sigma_{-i}) = \max F_i(\sigma_{-i})$, $i = 1, \dots, N$, $\sigma_{-i} \in \Sigma_{-i}$.

Proof: Let the hypothesis (H0) be satisfied, and let $i \in \{1, \dots, N\}$ and $\sigma_{-i} \in \Sigma_{-i}$ be fixed. Denoting $c_i = \max_{s_i \in S_i} U_i(s_i, \sigma_{-i})$, it follows from (8.7) and (8.8) that for each $\tau_i \in \Sigma_i$,

$$\mathcal{U}_i(\tau_i, \sigma_{-i}) = c_i \cdot \tau_i(F_i(\sigma_{-i})) + \sum_{s_i \in S_i \setminus F_i(\sigma_{-i})} \tau_i(\{s_i\})U_i(s_i, \sigma_{-i}).$$

This result implies that $\mathcal{U}_i(\tau_i, \sigma_{-i}) \leq_i c_i$, and equality holds if and only if $\tau_i(F_i(\sigma_{-i})) = 1$. Thus

$$\mathcal{F}_i(\sigma_{-i}) = \{\tau_i \in \Sigma_i : \tau_i(F_i(\sigma_{-i})) = 1\}, \quad i = 1, \ldots, N, \ \sigma_{-i} \in \Sigma_{-i}. \quad (8.11)$$

The latter implies that each $\mathcal{F}_i(\sigma_{-i})$ is nonempty. Assuming also hypothesis (H1), then from (8.11), (H1), and from the definition (SD) of $\preceq_i$ it follows that (8.10) holds. $\square$

Denote $\Sigma = \Sigma_1 \times \cdots \times \Sigma_N$, and equip Σ with componentwise ordering $\preceq$.

Definition 8.15. *Let $\overline{\sigma}_1^*, \ldots, \overline{\sigma}_N^*$ be a Nash equilibrium for a normal-form game Γ. It is called the* **greatest Nash equilibrium** *for Γ if $\sigma_i^* \preceq_i \overline{\sigma}_i^*$ for each $i = 1, \ldots, N$ whenever $\sigma_1^*, \ldots, \sigma_N^*$ is a Nash equilibrium for Γ. The* **smallest Nash equilibrium** *for Γ is defined similarly by reversing the inequalities.*

Now we are ready to prove our main existence result.

Theorem 8.16. *Under the hypotheses (H0)–(H3) a finite normal-form game has the greatest Nash equilibrium, and it is pure one.*

Proof: The hypothesis (H3), the definition (SD) of $\preceq_i$, and (8.10) imply that for each $i = 1, \ldots, N$,

$$\max \mathcal{F}_i(\sigma_{-i}) \preceq_i \max \mathcal{F}_i(\hat{\sigma}_{-i}) \text{ whenever } \Sigma_{-i} \ni \sigma_{-i} \leq \hat{\sigma}_{-i} \in P_{-i}. \quad (8.12)$$

By the hypothesis (H2) we can choose upper bounds $b_i \in S_i$ for the sets $\{\max F_i(\sigma_{-i}) : \sigma_{-i} \in \Sigma_{-i}\}$, $i = 1, \ldots, N$. Then the elements $\delta_{b_i} \in P_i$ are upper bounds of the sets $\{\max \mathcal{F}_i(\sigma_{-i}) : \sigma_{-i} \in \Sigma_{-i}\}$, $i = 1, \ldots, N$. Defining G_+ by

$$G_+(\sigma) := (\max \mathcal{F}_1(\sigma_{-1}), \ldots, \max \mathcal{F}_N(\sigma_{-N})), \quad \sigma = (\sigma_1, \ldots, \sigma_N) \in \Sigma, \quad (8.13)$$

we obtain a mapping $G_+ : \Sigma \to P = P_1 \times \cdots \times P_N$ that has the following properties: $G_+(\sigma) \preceq G_+(\hat{\sigma})$ if $\Sigma \ni \sigma \preceq \hat{\sigma} \in P$, and $\overline{\sigma} = (\delta_{b_1}, \ldots, \delta_{b_N})$ is an upper bound of $G_+[\Sigma]$. Thus the iteration sequence $(G_+^n(\overline{\sigma}))$ is decreasing.

Because the set $\{\max F_i(\sigma_{-i}) : \sigma_{-i} \in P_{-i}\}$ is a subset of a finite set S_i for each $i = 1, \ldots, N$, then the set $\{\delta_{\overline{s}_i(\sigma_{-i})} = \max \mathcal{F}(\sigma_{-i}) : \sigma_{-i} \in P_{-i}\}$ is finite. It then follows from (8.13) that $G_+[P]$ is finite. In particular, the iteration sequence $(G_+^n(\overline{\sigma}))$ is finite. Because it is also decreasing, then $\overline{\sigma}^* := G_+^k(\overline{\sigma}) = G_+^{k+1}(\overline{\sigma}) = G_+(\overline{\sigma}^*)$ for some $k \in \mathbb{N}_0$. In particular, $\overline{\sigma}^* = G_+(\overline{\sigma}^*)$. This result, definition (8.13) of G_+ and notations $\overline{\sigma}^* = (\overline{\sigma}_1^*, \ldots, \overline{\sigma}_N^*)$ and $\overline{\sigma}_{-i}^* := (\sigma_1^k, \ldots, \sigma_{i-1}^k, \sigma_{i+1}^k, \ldots, \sigma_N^k)$, $i = 1, \ldots, N$, imply that

$$\overline{\sigma}_i^* = \max \mathcal{F}_i(\overline{\sigma}_{-i}^*) \text{ for all } i = 1, \ldots, N. \quad (8.14)$$

According to this result the strategies $\overline{\sigma}_i^*$ are the greatest strategies that maximize the utility $\mathcal{U}_i(\sigma_i, \sigma_{-i})$ when $\sigma_{-i} = \overline{\sigma}_{-i}^*$. In particular,

$$\mathcal{U}_i(\overline{\sigma}_i^*, \overline{\sigma}_{-i}^*) = \max_{\sigma_i \in \Sigma_i} \mathcal{U}_i(\sigma_i, \overline{\sigma}_{-i}^*), \quad i = 1, \ldots, N,$$

so that the strategies $\overline{\sigma}_1^*, \ldots, \overline{\sigma}_N^*$ satisfy the Nash equilibrium condition. Moreover, they are pure strategies by (8.10) and (8.14).

To prove that $\overline{\sigma}_1^*, \ldots, \overline{\sigma}_N^*$ is the greatest Nash equilibrium for Γ, let $\sigma = (\sigma_1^*, \ldots, \sigma_N^*)$ be an element of Σ whose components form a Nash equilibrium for Γ. Then $\sigma_i^* \in \mathcal{F}_i(\sigma_{-i}^*)$, whence $\sigma_i^* \preceq_i \max \mathcal{F}_i(\sigma_{-i}^*)$ for each $i = 1, \ldots N$. This result implies by (8.13) and by the choice of $\overline{\sigma}$ that

$$\sigma \preceq G_+(\sigma) \preceq \overline{\sigma}.$$

When we map all the elements of the preceding inequality k times and notice that $G_+(\sigma) \preceq G_+(\hat{\sigma})$ if $\Sigma \ni \sigma \preceq \hat{\sigma} \in P$, we obtain

$$\sigma \preceq G_+(\sigma) \preceq \cdots \preceq G_+^{k+1}(\sigma) \preceq G_+^k(\overline{\sigma}) = \overline{\sigma}^*.$$

Thus $\sigma = (\sigma_1, \ldots, \sigma_N) \preceq \overline{\sigma} = (\overline{\sigma}_1^*, \ldots, \overline{\sigma}_N^*)$, i.e., $\sigma_i^* \preceq_i \overline{\sigma}_i^*$ for each $i = 1, \ldots, N$. This proves that $\overline{\sigma}_1^*, \ldots, \overline{\sigma}_N^*$ is the greatest Nash equilibrium for Γ. □

8.2.3 Comparison Result for Utilities

In the proof of our main comparison results of this section we make use of the following lemma.

Lemma 8.17. *Let the hypotheses (H0)–(H3) be satisfied. Assume also that*

(H4) $u_i(s_i, s_{-i}) \leq_i u_i(s_i, \hat{s}_{-i})$ whenever $s_{-i} \leq \hat{s}_{-i}$ in S_{-i}, $s_i \in S_i$ and $i = 1, \ldots, N$.

Let $\overline{\sigma}_1^*, \ldots, \overline{\sigma}_N^*$ be the pure Nash equilibrium constructed in Theorem 8.16. If $\sigma_1, \ldots, \sigma_N$ is any Nash equilibrium for Γ, then

$$U_i(s_i, \sigma_{-i}) \leq_i U_i(s_i, \overline{\sigma}_{-i}^*) \quad \text{for all } i = 1, \ldots, N \text{ and } s_i \in S_i. \tag{8.15}$$

Proof: Let $\sigma_1, \ldots, \sigma_N$ be a Nash equilibrium for Γ. By Lemma 8.14 and (8.14) we have

$$\overline{\sigma}_i^* = \max \mathcal{F}_i(\overline{\sigma}_{-i}^*) = \delta_{\overline{s}_i}, \text{ where } \overline{s}_i = \max F_i(\overline{\sigma}_{-i}^*), \ i = 1, \ldots, N. \tag{8.16}$$

Since $\sigma_i \preceq_j \overline{\sigma}_i^*$, $i = 1, \ldots, N$ by Theorem 8.16, then $\sigma_j \preceq_j \delta_{\overline{s}_j}$ for $j \neq i$. If $s_j \in S_j$ and $s_j \not\preceq_j \overline{s}_j$, then $\overline{s}_j \notin A_j = \{x_j \in S_j : s_j \leq_j x_j\}$. Because A_j is increasing, from the definition (SD) of $\preceq_j$ it follows that $\sigma_j(\{s_j\}) \leq \sigma_j(A_j) \leq \delta_{\overline{s}_j}(A_j) = 0$. Thus $s_j \leq_j \overline{s}_j$ for all $s_j \in S_j$ for which $\sigma_j(\{s_j\}) > 0$. This result and the hypothesis (H4) imply that

$$U_i(s_i, \sigma_{-i}) = \sum_{s_{-i} \in S_{-i}} \prod_{j \neq i} \sigma_j(\{s_j\}) u_i(s_i, s_{-i})$$

$$\leq_i \sum_{s_{-i} \in S_{-i}} \prod_{j \neq i} \sigma_j(\{s_j\}) u_i(s_i, \overline{s}_{-i}) = u_i(s_i, \overline{s}_{-i})$$

$$= \sum_{s_{-i} \in S_{-i}} \prod_{j \neq i} \overline{\sigma}_j^*(\{s_j\}) u_i(s_i, s_{-i}) = U_i(s_i, \overline{\sigma}_{-i}^*).$$

This proves the validity of (8.15).   □

As a consequence of Lemma 8.17 we get our main comparison result.

Theorem 8.18. *Assume hypotheses (H0)–(H4), and let $\overline{\sigma}_1^*, \ldots, \overline{\sigma}_N^*$ be the Nash equilibrium constructed in Theorem 8.16. If $\sigma_1^*, \ldots, \sigma_N^*$ is a Nash equilibrium for Γ, then $\mathcal{U}_i(\sigma_i^*, \sigma_{-i}^*) \leq_i \mathcal{U}_i(\overline{\sigma}_i^*, \overline{\sigma}_{-i}^*)$ for all $i = 1, \ldots, N$.*

Proof: Let $\sigma_1^*, \ldots, \sigma_N^*$ be a Nash equilibrium for Γ. Since $\sigma_i^* \preceq_i \overline{\sigma}_i^*$ for all $i = 1, \ldots, N$, by Theorem 8.16, then also $\sigma_{-i}^* \leq \overline{\sigma}_{-i}^*$ in Σ_{-i} for all $i = 1, \ldots, N$. Then from (8.7), (8.15), and Definition 8.13 it follows that

$$\mathcal{U}_i(\sigma_i^*, \sigma_{-i}^*) = \sum_{s_i \in S_i} \sigma_i^*(\{s_i\}) U_i(s_i, \sigma_{-i}^*) \leq_i \sum_{s_i \in S_i} \sigma_i^*(\{s_i\}) U_i(s_i, \overline{\sigma}_{-i}^*)$$

$$= \mathcal{U}_i(\sigma_i^*, \overline{\sigma}_{-i}^*) \leq_i \mathcal{U}_i(\overline{\sigma}_i^*, \overline{\sigma}_{-i}^*) \quad \text{for all} \quad i = 1, \ldots, N.$$

This proves the assertion.   □

8.2.4 Dual Results

Here we present an existence and comparison result for the smallest Nash equilibrium of a finite normal-form game $\Gamma = \{S_i, u_i\}_{i=1}^N$. The hypotheses (H1)–(H3) will be replaced by the following hypotheses.

(Ha) For every fixed $i = 1, \ldots, N$ the sets $F_i(\sigma_{-i})$, $\sigma_{-i} \in \Sigma_{-i}$, are directed downward.

(Hb) The set $\{\min F_i(\sigma_{-i}) : \sigma_{-i} \in \Sigma_{-i}\}$ has a lower bound in S_i, $i = 1, \ldots, N$.

(Hc) If $\sigma_{-i} \leq \tau_{-i}$ in Σ_{-i} and $\sigma_{-i} \in P_{-i}$, then $\min F_i(\sigma_{-i}) \leq_i \min F_i(\tau_{-i})$, $i = 1, \ldots, N$.

The following existence and comparison results are dual to those of Theorems 8.16 and 8.18. Their proofs are similar and can be omitted.

Theorem 8.19. *Under the hypotheses (H0), (Ha), (Hb), and (Hc), Γ has the smallest Nash equilibrium $\{\underline{\sigma}_1^*, \ldots, \underline{\sigma}_N^*\}$, and it is pure. Assume, moreover, that*

(Hd) $u_i(s_i, \hat{s}_{-i}) \leq_i u_i(s_i, s_{-i})$ whenever $s_{-i} \leq \hat{s}_{-i}$ in S_{-i}, $s_i \in S_i$ and $i = 1, \ldots, N$.

If $\sigma_1^*, \ldots, \sigma_N^*$ is any Nash equilibrium for Γ, then $\mathcal{U}_i(\sigma_i^*, \sigma_{-i}^*) \leq_i \mathcal{U}_i(\underline{\sigma}_i^*, \underline{\sigma}_{-i}^*)$ for all $i = 1, \ldots, N$.

Remark 8.20. (i) The comparison results of Theorems 8.18 and 8.19 cannot be obtained by the methods used, e.g., in [86, 180, 217, 218, 222, 223], because the mixed strategies need not form complete lattices, not even lattices, and because the utility mappings u_i need not be chain-valued. The strategic complementarity properties contained in conditions (H3) and (Hc) are also weaker than those assumed in the above cited papers. Moreover, the strategy spaces S_i are not assumed to be lattice-ordered.

(ii) If $\sigma_i = \delta_{s_i}$ and $\hat{\sigma}_i = \delta_{\hat{s}_i}$, $i = 1, \ldots, N$, then $\sigma_i \preceq_i \hat{\sigma}_i$ in P_i if and only if $s_i \leq_i \hat{s}_i$ in S_i, so that the spaces P_i of pure strategies, ordered by first order stochastic dominance, are order isomorphic with the strategy spaces S_i. Moreover, if $\sigma_i = \delta_{s_i}$, $i = 1, \ldots, N$, from (8.6) and (8.7) it follows that

$$\mathcal{U}_i(\sigma_1, \ldots, \sigma_N) = u_i(s_1, \ldots, s_N) \text{ and } U_i(s_i, \sigma_{-i}) = u_i(s_i, s_{-i})$$

for $i = 1, \ldots, N$. Thus we can equalize δ_{s_i}'s with s_i's, P_i's with S_i's, and U_i's and $\mathcal{U}_i$'s with u_i's. The iteration sequence $(G_+^n(\overline{\sigma}))$, $\overline{\sigma} = (\delta_{b_1}, \ldots, \delta_{b_N})$, used in the proof of Theorem 8.16, can be rewritten as finite strictly decreasing sequences of successive approximations:

$$\overline{s}_i^{n+1} = \max F_i(\overline{s}_{-i}^n), \ n = 0, 1, \ldots, \quad \overline{s}_i^0 = b_i, \ i = 1, \ldots, N, \qquad (8.17)$$

where

$$F_i(s_{-i}) = \{x \in S_i : u_i(x, s_{-i}) = \max_{t \in S_i} u_i(t, s_{-i})\}, \quad i = 1 = 1, \ldots, N. \quad (8.18)$$

Hence, if the hypotheses (H0)–(H3) are valid, then the greatest pure Nash equilibrium of Γ is $(\overline{s}_1^k, \ldots, \overline{s}_N^k)$, where k is the smallest integer for which $s_i^k = s_i^{k+1}$ for every $i = 1, \ldots, N$. Similarly, under the hypotheses (H0), (Ha), (Hb), and (Hc), the components of the smallest pure Nash equilibrium of Γ are the last elements of finite strictly increasing sequences of the successive approximations:

$$\underline{s}_i^{n+1} = \min F_i(\underline{s}_{-i}^n), \ n = 0, 1, \ldots, \quad \underline{s}_i^0 = a_i, \ i = 1, \ldots, N. \qquad (8.19)$$

8.2.5 Applications to Finite Supermodular Games

In this subsection we apply the results of the previous subsections to finite supermodular normal-form games that have vector-valued utility functions. A subset S of a poset X is called a **sublattice** if $x \vee y := \sup\{x, y\}$ and $x \wedge y := \inf\{x, y\}$ exist in X and belong to S for all $x, y \in S$. Every chain is a sublattice and every sublattice is directed.

Definition 8.21. *A normal-form game* $\Gamma = \{S_i, u_i\}_{i=1}^N$ *is a* **finite super-modular normal-form game** *if for all* $i = 1, \ldots, N$, *(h1) S_i is a finite sublattice of a poset $X_i = (X_i, \leq_i)$, and the values of u_i are in an ordered vector space $E_i = (E_i, \leq_i)$;*

*(h2) $u_i(s_i, s_{-i})$ is **supermodular** in s_i, i.e., if $x_i, y_i \in S_i$ and $s_{-i} \in S_{-i}$, then $u_i(x_i, s_{-i}) + u_i(y_i, s_{-i}) \leq_i u_i(x_i \wedge y_i, s_{-i}) + u_i(x_i \vee y_i, s_{-i})$;*

*(h3) $u_i(s_i, s_{-i})$ has **increasing differences** in (s_i, s_{-i}), i.e., if $x_i <_i y_i \in S_i$ and $s_{-i} < \hat{s}_{-i}$ in S_{-i}, then $u_i(y_i, s_{-i}) - u_i(x_i, s_{-i}) \leq_i u_i(y_i, \hat{s}_{-i}) - u_i(x_i, \hat{s}_{-i})$.*

Lemma 8.22. *The hypotheses (H1)–(H3), (Ha), (Hb), and (Hc) are valid for any finite supermodular normal-form game that satisfies the hypothesis (H0).*

Proof: Let $\Gamma = \{S_i, u_i\}_{i=1}^N$ be a finite supermodular normal-form game, and assume that the hypothesis (H0) holds. To prove the validity of (H1) and (Ha), let $i \in \{1, \ldots, N\}$, $\sigma_{-i} \in \Sigma_{-i}$ and $x_i, y_i \in F_i(\sigma_{-i})$ be given. Applying (8.7), (8.8), and condition (h2) we obtain

$$
\begin{aligned}
0 &\leq U_i(x_i, \sigma_{-i}) - U_i(x_i \wedge y_i, \sigma_{-i}) \\
&= \sum_{s_{-i} \in S_{-i}} \prod_{j \neq i} \sigma_j(\{s_j\})(u_i(x_i, s_{-i}) - u_i(x_i \wedge y_i, s_{-i})) \\
&\leq_i \sum_{s_{-i} \in S_{-i}} \prod_{j \neq i} \sigma_j(\{s_j\})(u_i(x_i \vee y_i, s_{-i}) - u_i(y_i, s_{-i})) \\
&= U_i(x_i \vee y_i, \sigma_{-i}) - U_i(y_i, \sigma_{-i}) \leq 0.
\end{aligned}
$$

The above result, (8.8), and the choice of x_i and y_i from $F_i(\sigma_{-i})$ imply that $x_i \wedge y_i$ and $x_i \vee y_i$ belong to $F_i(\sigma_{-i})$. Thus $F_i(\sigma_{-i})$ is a sublattice. In particular, the hypotheses (H1) and (Ha) are verified. Because each S_i is a finite sublattice, then $a_i = \min S_i$ and $b_i = \max S_i$ exist; b_i is an upper bound of $\{\max F_i(\sigma_{-i}) : \sigma_{-i} \in \Sigma_{-i}\}$ in S_i, and a_i is a lower bound of $\{\min F_i(\sigma_{-i}) : \sigma_{-i} \in \Sigma_{-i}\}$ in S_i. This proves that the hypotheses (H2) and (Hb) are valid.

To prove that (H3) holds, let $i \in \{1, \ldots, N\}$ be given, and assume that $\sigma_{-i} \leq \tau_{-i}$ in Σ_{-i}, and that $\tau_{-i} = (\delta_{\hat{s}_1}, \ldots, \delta_{\hat{s}_{i-1}}, \delta_{\hat{s}_{i+1}}, \ldots, \delta_{\hat{s}_N}) \in P_{-i}$. Since $\sigma_{-i} \leq \tau_{-i}$ in Σ_{-i}, then $\sigma_j \preceq_j \delta_{\hat{s}_j}$ for $j \neq i$. As in the proof of Lemma 8.17 one can show that $s_j \leq_j \hat{s}_j$ for all $s_j \in S_j$ for which $\sigma_j(\{s_j\}) > 0$. Denoting $x_i = \max F(\sigma_{-i})$ and $y_i = \max F(\tau_{-i})$, applying conditions (h2) and (h3) of Definition 8.21, and noticing that $\sigma_{-i}(\{s_{-i}\}) > 0$ only if $s_{-i} \leq \hat{s}_{-i}$, we get

$$
\begin{aligned}
0 &\leq_i U_i(x_i, \sigma_{-i}) - U_i(x_i \wedge y_i, \sigma_{-i}) \\
&= \sum_{s_{-i} \in S_{-i}} \prod_{j \neq i} \sigma_j(\{s_j\})(u_i(x_i, s_{-i}) - u_i(x_i \wedge y_i, s_{-i})) \\
&\leq_i \sum_{s_{-i} \in S_{-i}} \prod_{j \neq i} \sigma_j(\{s_j\})(u_i(x_i, \hat{s}_{-i}) - u_i(x_i \wedge y_i, \hat{s}_{-i})) \\
&= u_i(x_i, \hat{s}_{-i}) - u_i(x_i \wedge y_i, \hat{s}_{-i}) \leq_i u_i(x_i \vee y_i, \hat{s}_{-i}) - u_i(y_i, \hat{s}_{-i}) \\
&= \sum_{s_{-i} \in S_{-i}} \prod_{j \neq i} \tau_j(\{s_j\})(u_i(x_i \vee y_i, s_{-i}) - u_i(y_i, s_{-i})) \\
&= U_i(x_i \vee y_i, \tau_{-i}) - U_i(y_i, \tau_{-i}) \leq 0.
\end{aligned}
$$

Thus all the inequalities above are equalities. In particular, $U_i(x_i \vee y_i, \tau_{-i}) = U_i(y_i, \tau_{-i})$, so that $x_i \vee y_i \in F(\tau_{-i})$. Since $y_i = \max F_i(\tau_{-i})$, then $x_i \vee y_i = y_i$, whence $x_i \leq y_i$. This proves that the hypothesis (H3) is valid. The proof for (Hc) to hold can be done in a similar way. □

The following proposition is a consequence of Lemma 8.22 and Theorems 8.16–8.19.

Proposition 8.23. *Let $\Gamma = \{S_i, u_i\}_{i=1}^N$ be a finite supermodular normal-form game. If the hypothesis (H0) holds, then Γ has the smallest and greatest Nash equilibria formed by pure strategies. If, in addition, hypothesis (H4) (respectively (Hd)) is satisfied, then the utilities of the greatest (respectively the smallest) Nash equilibrium majorize the utilities of all other Nash equilibria for Γ.*

The hypothesis (H0) holds if u_i's are real-valued, whence we obtain the results stated in Chap. 1.

Corollary 8.24. *A supermodular normal-form game, where strategy spaces are finite lattices, and utilities $u_i(s_i, s_{-i})$ are real-valued and increasing (respectively decreasing) in s_{-i}, has the smallest and greatest Nash equilibria formed by pure strategies, and the greatest (respectively the smallest) one gives the best utilities among all mixed Nash equilibria.*

Remark 8.25. (i) The proof of Lemma 8.22 shows that each $F_i(\sigma_{-i})$ is a sublattice. However, *this does not imply that $\mathcal{F}_i(\sigma_{-i})$ is a sublattice.* As for the definition of $\mathcal{F}_i(\sigma_{-i})$ see (8.9).
(ii) Log-supermodularity (cf., e.g., [218]) is not available because the utility functions u_i are vector-valued.

Consider next the case where the utility functions are of the form

$$\begin{cases} u_i(s_i, s_{-i}) = (u_{i1}(s_{i1}, s_{-i}), \ldots, u_{im_i}(s_{im_i}, s_{-i})), \\ s_i = (s_{i1}, \ldots, s_{im_i}) \in S_i = \times_{j=1}^{m_i} S_{ij}, \ s_{-i} \in S_{-i}, \text{ where} \\ u_{ij}(s_{ij}, s_{-i}) = d_{ij}(s_{ij}, s_{-i}) q_{ij}(s_{ij}), \ i = 1, \ldots, N, \ j = 1, \ldots, m_i, \\ d_{ij} : S_{ij} \times S_{-i} \to \mathbb{R}, \text{ and } q_{ij} : S_{ij} \to \mathbb{R}_+. \end{cases} \quad (8.20)$$

The following hypotheses are imposed.

(ha) Each S_{ij} is a finite subset of $\mathbb{R}$.
(hb) Each $d_{ij}(s_{ij}, s_{-i})$ has increasing differences in (s_{ij}, s_{-i}).

Proposition 8.26. *Let $\Gamma = \{S_i, u_i\}_{i=1}^N)$ be a normal-form game whose utilities are given by (8.20). Assume that the hypotheses (ha) and (hb) hold, and that all the functions q_{ij} and $d_{ij}(s_{ij}, \cdot)$ are increasing (respectively decreasing). Then Γ has extremal Nash equilibria formed by pure strategies, and the greatest (respectively the smallest) Nash equilibrium gives the best utilities among all Nash equilibria for Γ.*

Proof: We are going to show that Γ is a supermodular game. Let $i \in \{1,\ldots,N\}$ be fixed. The hypothesis (ha) and the definitions of u_i imply that condition (h1) of Definition 8.21 is satisfied for $X_i = E_i = \mathbb{R}^{m_i}$, ordered coordinatewise. Because each function is supermodular in its real variable, the components $u_{ij}(s_{ij}, s_{-i})$ of $u_i(s_i, s_{-i}) = (u_{i1}(s_{i1}, s_{-i}),\ldots,u_{im_i}(s_{im_i}, s_{-i}))$ are supermodular in s_{ij}. Thus $u_i(s_i, s_{-i})$ is supermodular in $s_i = (s_{i1},\ldots,s_{im_i})$.

Assume next that the functions q_{ij} and $d_{ij}(s_{ij}, \cdot)$ are all increasing or all decreasing, and that $s_{ij} < \hat{s}_{ij}$ and $s_{-i} < \hat{s}_{-i}$. Since each $d_{ij}(s_{ij}, s_{-i})$ has increasing differences in (s_{ij}, s_{-i}), we have the following estimate

$$u_{ij}(\hat{s}_{ij}, \hat{s}_{-i}) - u_{ij}(s_{ij}, \hat{s}_{-i}) - (u_{ij}(\hat{s}_{ij}, s_{-i}) - u_{ij}(s_{ij}, s_{-i}))$$
$$= d_{ij}(\hat{s}_{ij}, \hat{s}_{-i})q_{ij}(\hat{s}_{ij}) - d_{ij}(s_{ij}, \hat{s}_{-i})q_{ij}(s_{ij}) - d_{ij}(\hat{s}_{ij}, s_{-i})q_{ij}(\hat{s}_{ij})$$
$$\quad + d_{ij}(s_{ij}, s_{-i})q_{ij}(s_{ij})$$
$$= (d_{ij}(\hat{s}_{ij}, \hat{s}_{-i}) - d_{ij}(s_{ij}, \hat{s}_{-i}))(q_{ij}(\hat{s}_{ij}) - q_{ij}(s_{ij}))$$
$$\quad + (d_{ij}(\hat{s}_{ij}, \hat{s}_{-i}) - d_{ij}(s_{ij}, \hat{s}_{-i}) - (d_{ij}(\hat{s}_{ij}, s_{-i}) - d_{ij}(s_{ij}, s_{-i})))q_{ij}(s_{ij}) \geq 0.$$

This proves that each $u_{ij}(s_{ij}, s_{-i})$ has increasing differences in (s_{ij}, s_{-i}). Consequently, each $u_i(s_i, s_{-i})$ has increasing differences in (s_i, s_{-i}). The above proof shows that conditions (h2) and (h3) of Definition 8.21 hold true. The given hypotheses and the definition of u_i ensure that also the hypothesis (H0) is valid. Thus the assertions follow from Proposition 8.23. $\square$

Assume that $c_{ij} \geq 0$, $i = 1,\ldots,N$, and that

$$\begin{cases} u_i(s_i, s_{-i}) = (u_{i1}(s_{i1}, s_{-i}),\ldots,u_{im_i}(s_{im_i}, s_{-i})), \text{ where} \\ u_{ij}(s_{ij}, s_{-i}) = d_{ij}(s_{ij}, s_{-i})k_{ij}(s_{ij})(s_{ij} - c_{ij}), \\ d_{ij} : S_{ij} \times S_{-i} \to \mathbb{R}_+, \ k_{ij} : S_{ij} \to \mathbb{R}_+, \ i = 1,\ldots,N, \ j = 1,\ldots,m_i. \end{cases} \tag{8.21}$$

The following corollary is a consequence of Proposition 8.26.

Corollary 8.27. *Let Γ be a finite normal-form game with utilities given by (8.21), and let the hypotheses (ha) and (hb) be satisfied. Assume also that all the functions q_{ij} given by (8.25) and all the functions $d_{ij}(s_{ij}, \cdot)$ are increasing. Then Γ has the smallest and greatest Nash equilibria formed by pure strategies. Moreover, the utilities of the greatest Nash equilibrium majorize the utilities of all mixed Nash equilibria for Γ.*

Let us consider the following special case: For $i = 1,\ldots,N$, $j = 1,\ldots,m_i$, define

$$d_{ij}(s_{ij}, s_{-i}) = f_{ij}(s_{ij}) + g_{ij}(s_{-i}), \ s_{ij} \in S_{ij}, \ s_{-i} \in S_{-i}, \tag{8.22}$$

where $f_{ij} : S_{ij} \to \mathbb{R}$, and $g_{ij} : S_{-i} \to \mathbb{R}$. In this case we have

$$d_{ij}(\hat{s}_{ij}, \hat{s}_{-i}) - d_{ij}(s_{ij}, \hat{s}_{-i}) = f_{ij}(\hat{s}_{ij}) - f_{ij}(s_{ij}) = d_{ij}(\hat{s}_{ij}, s_{-i}) - d_{ij}(s_{ij}, s_{-i})$$

for all fixed $i = 1,\ldots,N$ and $j = 1,\ldots,m_i$. Thus the hypotheses given for d_{ij} in Corollary 8.27 are satisfied if each g_{ij} is increasing.

Example 8.28. Consider a two-person normal form game $\Gamma = \{S_1, S_2, u_1, u_2\}$. Assume that the strategy spaces are $S_i = (S_{i1}, S_{i2})$, where $S_{ij} = \{\frac{3}{2} + \frac{k}{10^{31}} : 0 \leq k \leq 10^{31}\}$, $i, j = 1, 2$, and that the utility functions are $u_i = (u_{i1}, u_{i2})$, where the components u_{ij}, $i, j = 1, 2$, are defined by (*sgn* denotes the sign function)

$$u_{11}(s_{11}, s_{-1}) = (52 - 22s_{11} + s_{21} + 4s_{22} + sgn(s_{21} \cdot s_{22} - 4))(s_{11} - 1),$$

$$u_{12}(s_{12}, s_{-1}) = (51 - 21s_{12} - sgn(s_{12} - \frac{11}{10}) + 2s_{21})(s_{12} - \frac{11}{10})$$

$$+ (3s_{22} + sgn(s_{21} + s_{22} - 4)))(s_{12} - \frac{11}{10}),$$

$$u_{21}(s_{21}, s_{-2}) = (50 - 20s_{21} - sgn(s_{21} - \frac{11}{10}) + 3s_{11})(s_{12} - \frac{11}{10})$$

$$+ (2s_{12} + sgn(s_{11} + s_{12} - 4))(s_{21} - 1),$$

$$u_{22}(s_{22}, s_{-2}) = (49 - 19s_{22} + 4s_{11} + s_{12} + sgn(s_{11} \cdot s_{12} - 4))(s_{22} - \frac{11}{10}),$$

$$\text{for } s_{ij} \in S_{ij}, \ i, j = 1, 2.$$

$$(8.23)$$

Show that the smallest and greatest Nash equilibria for Γ exist. Calculate them along with the corresponding utilities.

Solution. The utilities u_{ij} are of the form (8.21), where $k_{ij}(s_{ij}) \equiv 1$ and d_{ij}'s are of the form (8.22), where the functions g_{ij} are increasing. Thus the hypotheses given for d_{ij} in Corollary 8.27 hold, which ensures the existence of the smallest and greatest Nash equilibria for Γ. Moreover, they are pure. It remains to calculate them. The sets $S_{ij}^m = \{\frac{3}{2} + \frac{k}{10^m} : 0 \leq k \leq 10^m\}$, $m = 1, \ldots, 30$ form partitions of the sets S_{ij}, $i, j = 1, 2$. The games $\Gamma^m = \{S_1^m, S_2^m, u_1^m, u_2^m\}$, where $S_i^m = (S_{i1}^m, S_{i2}^m)$ and components of u_i^m equal to u_{ij} at points of S_{ij}^m. Thus every game Γ^m has the smallest and greatest Nash equilibria. Denote

$$F_{ij}^m(s_{-i}) = \{x \in S_{ij}^m : u_{ij}(x, s_{-i}) = \max\{u_{ij}^m(t, s_{-i}) : t \in S_{ij}^m\}\}, \quad s_{-i} \in S_{-i}.$$

Since the functions $u_{ij}(\cdot, s_{-i})$ are strictly convex, they are strictly increasing (resp. strictly decreasing) on all intervals $[\frac{k-1}{10^m}, \frac{k}{10^m}]$ that are on the left (resp. right) hand side of the interval $[\max F_{ij}^m(s_{-i}) - \frac{1}{10^m}, \max F_{ij}^m(s_{-i}) + \frac{1}{10^m}]$. Thus $\max F_{ij}^{m+1}(s_{-i}) \in [\max F_{ij}^m(s_{-i}) - \frac{1}{10^m}, \max F_{ij}^m(s_{-i}) + \frac{1}{10^m}]$. Consequently, if $((\overline{s}_{11}^m, \overline{s}_{12}^m), (\overline{s}_{21}^m, \overline{s}_{22}^m))$ denotes the greatest Nash equilibrium of Γ^m, then the components $\overline{s}_{ij}^{m+1}$ of the greatest Nash equilibrium of Γ^{m+1} are in $[\overline{s}_{ij}^m - \frac{1}{10^m}, \overline{s}_{ij}^m + \frac{1}{10^m}]$. Thus to find it we can shorten calculations restricting partitions S_{ij}^{m+1} to intervals $[\overline{s}_{ij}^m - \frac{1}{10^m}, \overline{s}_{ij}^m + \frac{1}{10^m}]$. Successive approximations (8.17), rewritten as

$$s_{ij}(n+1) = \max F_{ij}^{m+1}(s_{-i}(n)), \ n = 0, 1, \ldots, \ s_{ij}(0) = \overline{s}_{ij}^m + \frac{1}{10^m}, \ i, j = 1, 2,$$

$$(8.24)$$

can be used to calculate the components $\overline{s}_{ij}^{m+1}$ of the greatest Nash equilibrium of Γ^{m+1}.

Carrying out calculations when m takes values from 1 to 30, we obtain the greatest Nash equilibrium $(\overline{s}_1^{31}, \overline{s}_2^{31}) = ((\overline{s}_{11}^{31}, \overline{s}_{12}^{31}), (\overline{s}_{21}^{31}, \overline{s}_{22}^{31}))$ of $\Gamma^{31} = \Gamma$. Using a simple Maple programming we obtain:

$$\overline{s}_{11}^{31} = 1.9589554880776350471495641123733,$$
$$\overline{s}_{12}^{31} = 2.0758076467020979992037171143840,$$
$$\overline{s}_{21}^{31} = 2.1507120439409275284964031664672,$$
$$\overline{s}_{22}^{31} = 2.2608323578687536365211045669441.$$

The approximations of components corresponding utilities are:

$$\overline{u}_{11}^{31} = 19.9962118306080166726519158345 7,$$
$$\overline{u}_{12}^{31} = 20.231103818512735567241982435 51,$$
$$\overline{u}_{21}^{31} = 22.079915985650432428962751845 46,$$
$$\overline{u}_{22}^{31} = 30.204266458328735962454495810 63.$$

By similar calculations one obtains the smallest Nash equilibrium $(\underline{s}_1^{31}, \underline{s}_2^{31})$ $= ((\underline{s}_{11}^{31}, \underline{s}_{12}^{31}), (\underline{s}_{21}^{31}, \underline{s}_{22}^{31}))$ of Γ:

$$\underline{s}_{11}^{31} = 1.8267403137904112645456323 95766,$$
$$\underline{s}_{12}^{31} = 1.8691845405408141974123338 50086,$$
$$\underline{s}_{21}^{31} = 1.9371585286341796279601598 20695,$$
$$\underline{s}_{22}^{31} = 1.9393937618723712630846956 38778.$$

The components of the corresponding utilities have the following components:

$$\underline{u}_{11}^{31} = 12.98649138247908596033916270233,$$
$$\underline{u}_{12}^{31} = 14.01668804129889104127227753592,$$
$$\underline{u}_{21}^{31} = 14.79621963687527336835895565230,$$
$$\underline{u}_{22}^{31} = 16.62059884133321814018013777847.$$

These utilities majorize the utilities of all pure and mixed Nash equilibria for Γ.

8.2.6 Application to a Multiproduct Pricing Game

The values of utilities u_i defined by (8.21) can be considered as profit vectors of firms i that sell products e_{ij}, $j = 1, \ldots, m_i$. The selling prices s_{ij} per unit of e_{ij} are assumed to form a subset S_{ij} of $\mathbb{R}$, which is bounded from below by a unit purchase price c_{ij} of e_{ij}. The values of utilities u_i defined by (8.21) are considered as profit vectors of firms i. The functions

$$D_{ij}(s_{ij}, s_{-i}) = d_{ij}(s_{ij}, s_{-i}) k_{ij}(s_{ij})$$

stand for demands of products e_{ij}. The utilities u_i defined in (8.21) are of the form (8.20), where

$$q_{ij}(s_{ij}) = k_{ij}(s_{ij})(s_{ij} - c_{ij}), \quad i = 1, \ldots, N, \quad j = 1, \ldots, m_i. \qquad (8.25)$$

The result of Corollary 8.27 can be applied to find the greatest and smallest equilibria for the above described pricing game.

Example 8.29. Consider a pricing game, where firms 1 and 2 sell two different products. Costs per item in euros are $(c_{11}, c_{12}) = (1, 1.10)$ for firm 1, and $(c_{21}, c_{22}) = (1.10, 1)$ for firm 2. Prices s_{ij} may vary from 1.20, with five cents differences, to 2.40. Profits per day are calculated in whole euros. Assume that demands are $d_i = (d_{i1}, d_{i2})$, where the components d_{ij}, $i, j = 1, 2$, are defined by

$$\begin{aligned}
d_{11}(s_{11}, s_{-1}) &= 30(53 - 22s_{11} + 4s_{21} + s_{22}), \\
d_{12}(s_{12}, s_{-1}) &= 31(52 - 21s_{12} + 3s_{21} + 2s_{22}), \\
d_{21}(s_{21}, s_{-2}) &= 32(48 - 19s_{21} + 2s_{11} + 3s_{12}), \\
d_{22}(s_{22}, s_{-2}) &= 30(49 - 20s_{22} + s_{11} + 4s_{12}).
\end{aligned} \qquad (8.26)$$

Show that the smallest and greatest Nash equilibria for prices exist. Calculate these equilibria and the corresponding profits.

Solution. Convert the problem to a normal-form game $\{S_1, S_2, u_1, u_2\}$, where utilities are

$$u_i = (d_{i1}(s_{i1}, s_{-i})(s_{i1} - c_{i1}), d_{i2}(s_{i2}, s_{-2})(s_{i2} - c_{i2})), \quad i = 1, 2,$$

with $c_{11} = c_{22} = 1$, $c_{12} = c_{21} = \frac{11}{10}$, and strategy sets are $S_i = S_{i1} \times S_{i2}$, where

$$S_{11} = S_{22} = S_{12} = S_{21} = \{\frac{j}{20}, \, 24 \leq j \leq 48\}.$$

In this case $X_i = E_i = \mathbb{R}^2$, $i = 1, 2$. From (8.26) it follows that the demands are of the form (8.22), where every g_{ij} is increasing on S_{-i}. Thus Γ has by Corollary 8.27 the smallest and greatest Nash equilibria formed by pure strategies. In view of Remark 8.20 the components of the greatest price equilibrium $(\bar{s}_1^*, \bar{s}_2^*)$ are the last elements of finite strictly decreasing sequences of successive approximations:

$$\bar{s}_i^{n+1} = \max F_i(\bar{s}_{-i}^n), \quad n = 1, \ldots, k-1, \quad \bar{s}_i^0 = 2.40, \quad i = 1, 2,$$

where

$$F_i(s_{-i}) = \{x_i \in S_i : u_i(x_i, s_{-i}) = \max_{s_i \in S_i} u_i(s_i, s_{-i})\}, \quad i = 1, 2.$$

The smallest price equilibrium can be obtained as the last element $(\underline{s}_1^*, \underline{s}_2^*) = (\underline{s}_1^j, \underline{s}_2^j)$ of the following finite and strictly increasing sequences of successive approximations:

$$\underline{s}_i^{n+1} = \min F_i(\underline{s}_{-i}^n), \quad n = 1, \ldots, j-1, \quad \underline{s}_i^0 = 1.20, \quad i = 1, 2.$$

Using a simple Maple programming we obtain:

$$\bar{s}_1^* = (1.95, 2.05), \ \bar{s}_2^* = (2.10, 2.00), \ \underline{s}_1^* = (1.95, 2.00), \ \underline{s}_2^* = (2.05, 1.95).$$

Corresponding profits are

$$\bar{u}_1^* = (584, 566), \ \bar{u}_2^* = (580, 574), \ \underline{u}_1^* = (577, 559), \ \underline{u}_2^* = (576, 568).$$

Since the demand functions d_{ij} are increasing with respect to s_{-i}, then the profits $(\bar{u}_i^*, \bar{u}_2^*)$ majorize by Corollary 8.27 the profits obtained by all pure and mixed Nash equilibria of prices.

Example 8.30. Assume that $N = 3$, and that the demands d_i, $i = 1, 2, 3$ are of the form

$$\begin{aligned}
d_1(s_1, (s_2, s_3)) &= 370 + 213(s_2 + s_3) + 60s_1 - 230s_1^2, \\
d_2(s_2, (s_1, s_3)) &= 360 + 233(s_1 + s_3) + 55s_2 - 220s_2^2, \qquad (8.27) \\
d_3(s_3, (s_1, s_2)) &= 375 + 226(s_1 + s_2) + 50s_3 - 200s_3^2.
\end{aligned}$$

Assume moreover that $c_1 := 1.10$, $c_2 := 1.2$, $c_3 := 1.25$, $a_i = 1.30$ and $b_i := 2.10$, $i = 1, 2, 3$, that the smallest price shift is five cents, and that profits are counted by whole euros. Show that the smallest and greatest Nash equilibria for prices exist. Calculate these equilibria and the corresponding profits.

Solution. Transform the problem to a pricing game $\{S_1, S_2, S_3, u_1, u_2, u_3\}$, where the utilities are given by

$$u_i = d_i(s_i, s_{-i})(s_i - c_i), \ i = 1, 2, 3,$$

and strategy sets are given by

$$S_1 = S_2 = S_3 = \{\frac{j}{20}, \ 26 \leq j \leq 42\}.$$

From (8.27) we see that d_i's are of the form $d_i(x, y) = f_i(x) + g_i(y)$, where every $f_i : S_i \to (0, \infty)$ is decreasing and every $g_i : S_{-i} \to (0, \infty)$ is increasing. Then from Corollary 8.27 it follows that the pricing game $\{S_1, S_2, S_3, u_1, u_2, u_3\}$ has the smallest and greatest Nash equilibria, and that they are formed by pure strategies. These extremal Nash equilibria for prices and the corresponding profits are presented in Example 8.11.

Remark 8.31. In the special case where $k_{ij}(s_{ij}) \equiv 1$ and $m_i = 1$ in (8.21) we obtain the Bertrand oligopoly model considered in [218, Subsection 4.4.1].

8.3 Pure Nash Equilibria for Normal-Form Games

In this section we study the existence of pure Nash equilibria for a normal-form game $\Gamma = \{S_i, u_i\}_{i=1}^N$ in a more general setting as in preceding sections.

Both the strategy spaces S_i and the ranges of the utility functions u_i may be nonempty posets.

The existence of the smallest and greatest pure Nash equilibria is studied in Sect. 8.3.2. For instance, our main results imply that Γ has the smallest and greatest pure Nash equilibria $\underline{s}_1, \ldots, \underline{s}_N$ and $\overline{s}_1, \ldots, \overline{s}_N$, respectively, if Γ has (ordinal) strategic complementarities defined in [180]. Moreover, utilities at the smallest (greatest) pure Nash equilibrium majorize the utilities at each other Nash equilibrium of Γ when $u_i(s_1, \ldots, s_N)$ is decreasing (increasing) in s_j, $j \neq i$. Monotone comparative statics, i.e., monotone dependence on the parameter, is also studied. The poset-valued functions $s_j \mapsto u_i(s_1, \ldots, s_N)$, $j \neq i$, are not necessarily continuous, as assumed in the real-valued case in [180]. New extreme value results are proved in Sect. 8.3.1 to ensure the existence of maximum points of the functions $s_i \mapsto u_i(s_1, \ldots, s_N)$.

In Sect. 8.3.5 we present sufficient conditions for the existence of minimal and maximal pure Nash equilibria for normal-form games whose strategy spaces and the ranges of utility functions are posets. Applications and concrete examples are presented.

8.3.1 Extreme Value Results

For a normal-form game $\Gamma = \{S_i, u_i\}_{i=1}^N$ the existence of a Nash equilibrium requires that the functions $u_i(\cdot, , s_{-i})$ have maximum points. In this subsection we prove extreme value results that are applied to ensure the existence of these maximum points when the values of the utility functions are in posets. If y and z are elements of a poset $Y = (Y, \preceq)$, denote

$$[y) = \{x \in Y : y \preceq x\}, \ (z] = \{x \in Y : x \preceq z\} \ \text{ and } \ [y, z] = [y) \cap (z].$$

Definition 8.32. *Let S be a topological space, and let Y be a poset. A mapping $f : S \to Y$ is called* **upper closed** *if $f^{-1}[[y)]$ is closed for all $y \in f[S]$, and* **directedly upper closed** *if every element pair w, z of the range $f[S]$ of f has such an upper bound y in $f[S]$ that the set $f^{-1}[[y)]$ is closed.*

The next lemma presents conditions that ensure that an upper closed function is directedly upper closed.

Lemma 8.33. *An upper closed function f from a topological space S to a poset Y is directedly upper closed if the range $f[S]$ of f is directed upward. In particular, if Y is a chain, then every upper closed function is directedly upper closed.*

A function f from a topological space S to $\mathbb{R}$ is **upper semicontinuous** if the set $f^{-1}[[y)]$ is closed for every $y \in \mathbb{R}$. Since $\mathbb{R}$ is also a chain, then Lemma 8.33 implies following result.

Corollary 8.34. *If S is a topological space, then every upper semicontinuous function from S to $\mathbb{R}$ is upper closed and directedly upper closed.*

In [111] the above defined concept of upper closeness was called upper semicontinuity. However, there exist upper closed real functions that are not upper semicontinuous.

Example 8.35. Define $f : [-1,1] \to \mathbb{R}$ by $f(x) = 1 - x^2$, for $0 < |x| < 1$, $f(0) = 2$, and $f(\pm 1) = -1$. Both f and $-f$ are upper closed but neither of them is upper semicontinuous. Notice also that $f : [-1,1] \to \mathbb{R}$ is continuous if and only if both f and $-f$ are upper semicontinuous.

Given posets $S = (S, \leq)$, and $Y = (Y, \preceq)$, recall that a function $f : S \to Y$ is *increasing, strictly increasing, decreasing, or strictly decreasing* in a subset B of S if $x < y$ in B implies that $f(x) \preceq f(y)$, $f(x) \prec f(y)$, $f(y) \preceq f(x)$, or $f(y) \prec f(x)$, respectively. If f is (strictly) increasing or (strictly) decreasing in B, then f is called *(strictly) monotone* in B. If S is a chain having maximum and minimum, we say that f is **piecewise (strictly) monotone** if there is such a partition $\min S = t_0 < t_1 < \cdots < t_m = \max S$ of S that the restriction of f to any order interval $[t_{k-1}, t_k]$ of S is (strictly) monotone.

Remark 8.36. (i) Piecewise strictly monotone functions are not necessarily upper closed (define $f(1) = 0$ in Example 8.35).

(ii) Piecewise monotone functions are not necessarily directedly upper closed (define $f(0) = -1$ and $f(x) = 0$ for $0 < |x| \leq 1$).

(iii) Every piecewise strictly monotone function from a chain S having smallest and greatest elements to a chain is directedly upper closed in every T1-topology of S.

Our main extreme value theorem reads as follows.

Theorem 8.37. *Assume that S is a compact topological space, that Y is a poset, and that $f : S \to Y$ is directedly upper closed. Then the set of maximum points of f is nonempty and compact.*

Proof. Let B be the set of those $y \in f[S]$ for which the set $f^{-1}[[y]]$ is closed. The sets $f^{-1}[[y]]$, $y \in B$ form a nonempty family of closed subsets of S. The intersection of every finite subfamily is nonempty because f is directedly upper closed. Since S is compact, then the intersection of the whole family contains at least one point x. In particular, $x \in f^{-1}[[y]]$, i.e., $y \preceq f(x)$ for every $y \in B$. Denoting $w = f(x)$, then w is an upper bound of B. Since $w \in f[S]$, there exists by Definition 8.32 such a z in $f[S]$ that the set $f^{-1}[[z]]$ is closed and $w \preceq z$. On the other hand, since $z \in B$, and w is an upper bound of B, then $z \preceq w$. Thus $z = w$, so that $w = \max B$. Since every element of $f[S]$ is majorized by an element of B due to Definition 8.32, we infer $w = \max f[S]$, so that $f^{-1}[[w]]$ is the set of all maximum points of f. Because $w \in B$, it follows that $f^{-1}[[w]]$ is closed, and hence also compact because S is compact. □

The converse holds true in case that S is a Hausdorff space.

Proposition 8.38. *Assume that S is a compact Hausdorff space, and that Y is a poset. Then a function $f : S \to Y$ has a nonempty and compact set of maximum points if and only if f is directedly upper closed.*

Proof: The sufficiency part follows from Theorem 8.37. Conversely, assume that the set A of maximum points of f is nonempty and compact. Because S is a Hausdorff space, then A is also nonempty and closed. Moreover, $A = f^{-1}[[w)]$, where $w = \max f[S]$, and w is an upper bound for every element pair y, z of $f[S]$. Thus f is directedly upper closed.   □

In case that S is a subset of a Euclidean space we have the following result.

Corollary 8.39. *Let S be a nonempty subset of a Euclidean space. A function f from S to a poset has a nonempty and compact set of maximum points if and only if f is directedly upper closed and the set $f^{-1}[[y)]$ is bounded for some $y \in f[S]$.*

Proof: The hypotheses imposed on f imply that the set $f^{-1}[[y)]$ is closed and bounded, and hence compact for some $y \in f[S]$. Thus the restriction of f to $f^{-1}[[y)]$ satisfies the hypotheses of Theorem 8.37. This proves the sufficiency part. Necessity part follows from Proposition 8.38.   □

By an **order interval topology** of a poset S we mean a topology whose subbasis is formed by order intervals $[x)$ and $(x]$, $x \in S$. In Euclidean spaces the order interval topology coincides with the topology induced by the Euclidean metric. A poset S, equipped with a topology that is finer than the order interval topology, is called an **ordered topological space**.

The next results are also consequences of Theorem 8.37.

Corollary 8.40. *Let S be a compact ordered topological space. Then the following holds.*

(a) $\max S$ *exists if and only if S is directed upward.*
(b) $\min S$ *exists if and only if S is directed downward.*

Proof: Ad (a) If S is directed upward, then the hypotheses of Theorem 8.37 hold when $Y = S$ and $f(x) \equiv x$. Thus $\max f[S] = \max S$ exists. Conversely, if $\max S$ exists, then S is directed upward.

Ad (b) The assertion of (b) follows from (a) when the partial ordering of S is replaced by its dual ordering.   □

Definition 8.41. *A function f from a lattice S to a poset Y is called **quasi-supermodular** if for all $x, \hat{x} \in S$, $f(x \wedge \hat{x}) \preceq f(x)$ implies $f(\hat{x}) \preceq f(x \vee \hat{x})$, and $f(x \wedge \hat{x}) \prec f(x)$ implies $f(\hat{x}) \prec f(x \vee \hat{x})$.*

The following lemma is a useful tool to prove that the set of maximum points of f is directed.

Lemma 8.42. *Assume that S is a compact lattice in an ordered topological space, that Y is a poset, and that $f : S \to Y$ is directedly upper closed and quasisupermodular. Then the set A of maximum points of f is a nonempty compact sublattice of S.*

Proof: According to Theorem 8.37, A is nonempty and compact. To prove that A is a sublattice of S, let x and $\hat{x}$ be in A. Then $f(x \wedge \hat{x}) \preceq f(x)$. If $f(x \wedge \hat{x}) \prec f(x)$, then $f(\hat{x}) \prec f(x \vee \hat{x})$ because f is quasisupermodular. However, this is impossible, because $\hat{x}$ is a maximum point of f. Thus $f(x \wedge \hat{x}) = f(x)$, so that $x \wedge \hat{x}$ is in A. Because $f(x \wedge \hat{x}) = f(x)$ and f is quasisupermodular, then $f(\hat{x}) \preceq f(x \vee \hat{x})$, whence $x \vee \hat{x}$ is in A. Thus A is a sublattice of S. $\square$

According to [218] a sublattice S of a lattice X is subcomplete if and only if S is compact with respect to the order interval topology of X. Thus the following result is a consequence of Lemma 8.42.

Lemma 8.43. *Given a poset Y, a complete lattice S that is equipped with the order interval topology, and a function $f : S \to Y$ that is quasisupermodular and directedly upper closed, then the set of maximum points of f is a nonempty and subcomplete sublattice of S.*

As an application of Lemma 8.43 we obtain the following result.

Proposition 8.44. *Given posets Y_j, complete chains S_j, and functions $f_j : S_j \to Y_j$, $j = 1, \ldots, n$, denote $S = S_1 \times \cdots S_n$ and $Y = Y_1 \times \cdots \times Y_n$. Define a mapping $f : S \to Y$ by*

$$f(s_1, \ldots, s_n) = (f_1(s_1), \ldots, f_n(s_n)), \quad (s_1, \ldots, s_n) \in S. \qquad (8.28)$$

Assume that both S and Y are ordered componentwise, and that for each $j = 1, \ldots, n$ the function f_j is directedly upper closed in the order interval topology. Then the set of maximum points of f is a nonempty and subcomplete sublattice of S.

Proof: Every f_j is also quasisupermodular because S_j is a chain. Thus for every $j = 1, \ldots, n$ the set $M(f_j)$ of the maximum points of f_j is by Lemma 8.43 nonempty and compact, and hence also a subcomplete subchain of S_j. The product set $M(f) = M(f_1) \times \cdots \times M(f_n)$ is a nonempty and subcomplete sublattice in S and is the set of maximum points of f. $\square$

In view of [180], Theorem A3 (see also [158]), an order upper semicontinuous function from a lattice S to $\mathbb{R}$ is upper semicontinuous with respect to the order interval topology of S. Thus the following result is a consequence of Corollary 8.34 and Lemma 8.43.

Corollary 8.45. *Let S be a complete lattice. If a function $f : S \to \mathbb{R}$ is quasisupermodular and (order) upper semicontinuous, then the set of maximum points of f is a nonempty and subcomplete sublattice of S.*

8.3.2 Smallest and Greatest Pure Nash Equilibria

Consider a normal-form game $\Gamma = \{S_i, u_i\}_{i=1}^{N}$, where the utility functions u_i map the product space $S_1 \times \cdots \times S_N$ to posets $Y_i = (Y_i, \preceq_i)$. We impose the following hypotheses on the strategy spaces S_i and on utility functions u_i, $i = 1, \ldots, N$.

(HI) S_i is a nonempty compact and directed subset of an ordered topological space X_i.

(HII) For all fixed s_{-i} the function $u_i(\cdot, s_{-i})$ is directedly upper closed.

These hypotheses imply by Theorem 8.37 that the sets

$$F_i(s_{-i}) = \{t_i \in S_i : u_i(t_i, s_{-i}) = \max_{s_i \in S_i} u_i(s_i, s_{-i})\}, \quad i = 1, \ldots, N,$$

are nonempty and compact. The following hypothesis:

(HIII) The sets $F_i(s_{-i})$, $s_{-i} \in S_{-i}$ are directed;

ensures by Corollary 8.40 the existence of $\max F_i(s_{-i})$ and $\min F(s_{-i})$ for each $i = 1, \ldots, N$. Denote $S = S_1 \times \cdots \times S_N$, and define

$$\begin{aligned}
G_+(s) &= (\max F_1(s_{-1}), \ldots, \max F_N(s_{-N})), \quad s \in S, \\
G_-(s) &= (\min F_1(s_{-1}), \ldots, \min F_N(s_{-N})), \quad s \in S.
\end{aligned} \tag{8.29}$$

The strategic complementarity hypothesis:

(HIV) The functions $s_{-i} \mapsto \max F_i(s_{-i})$ and $s_{-i} \mapsto \min F_i(s_{-i})$ are increasing for each $i = 1, \ldots, N$;

and (8.29) imply that the mappings $G_\pm : S \to S$ are increasing with respect to the componentwise ordering of S.

Given a poset S and a function $G : S \to S$, we say that $x \in S$ is a *fixed point* of G if $x = G(x)$. If the set of all fixed points of G has a smallest and a greatest element, then they are called smallest and greatest fixed points of G, respectively. Definition 8.2, extended to all normal-form games implies that $(s_1^*, \ldots, s_N^*)$ is a pure Nash equilibrium for Γ if and only if $s_i^* \in F_i(s_{-i}^*)$ for all $i = 1, \ldots, N$. From (8.29) it follows that $s^* = (s_1^*, \ldots, s_N^*)$ is a pure Nash equilibrium for Γ if s^* is a fixed point of one of the mappings $G_\pm$. The following fixed point theorem is an immediate consequence of Theorem 2.16.

Theorem 8.46. *Let S be a poset and let $G : S \to S$ be an increasing mapping.*

(a) If $a \leq G(a)$, and if $\sup G[C]$ exists whenever C is a nonempty chain in $[a)$, then G has the smallest fixed point x_* in $[a)$, and

$$x_* = \min\{y \in [a) : G(y) \leq y\}. \tag{8.30}$$

(b) If $G(b) \leq b$, and if $\inf G[D]$ exists whenever D is a nonempty chain in $(b]$, then G has the greatest fixed point x^* in $(b]$, and

$$x^* = \max\{y \in (b] : y \leq G(y)\}. \tag{8.31}$$

Now we are ready to prove the existence of extreme pure Nash equilibria of a normal-form game.

Theorem 8.47. *If hypotheses (HI)–(HIV) are valid, then the smallest and greatest pure Nash equilibria exist for the normal-form game $\Gamma = \{S_i, u_i\}_{i=1}^N$.*

Proof: The hypothesis (HIV) implies that the mappings $G_\pm : S \to S$, defined by (8.29) are increasing. The hypothesis (HI) ensures by Corollary 8.40 that $a_i = \min S_i$ and $b_i = \max S_i$ exist. Denoting $a = (a_1, \ldots, a_N)$ and $b = (b_1, \ldots, b_N)$, then $a \leq G_\pm(s) \leq b$ for each $s \in S$. In particular, $a \leq G_\pm(a)$ and $G_\pm(b) \leq b$. Assume next that C is a nonempty chain in S. Since G_- is increasing, then $G_-[C]$ is a chain. Because S is ordered componentwise, then the set $C_i = \{s_i \in F_i[S_{-i}] : (s_1, \ldots, s_N) \in G_-[C]\}$ is a nonempty chain in $F_i[S_{-i}]$. It then follows from the hypothesis (HI) and [133, Proposition 1.1.4] that $z_i = \sup C_i$ exists for each $i = 1, \ldots, N$. The definitions of C_i and (8.29) imply that $(z_i, \ldots, z_N) = \sup G_-[C]$.

The above proof shows that G_- satisfies the hypotheses of Theorem 8.46 (a), whence we conclude that G_- has the smallest fixed point $\underline{s}^* = (\underline{s}_1^*, \ldots, \underline{s}_N^*)$. Similar reasoning proves that G_+ satisfies the hypotheses of Theorem 8.46 (b), whence G_+ has the greatest fixed point $\overline{s}^* = (\overline{s}_1^*, \ldots, \overline{s}_N^*)$. According to the definition (8.29) of $G_\pm$, $\underline{s}_i^* \in F_i(\underline{s}_{-i}^*)$ and $\overline{s}_i^* \in F_i(\overline{s}_{-i}^*)$ for each $i = 1, \ldots, N$, whence both $(\underline{s}_1^*, \ldots, \underline{s}_N^*)$ and $(\overline{s}_1^*, \ldots, \overline{s}_N^*)$ are pure Nash equilibria for Γ. If $s^* = (s_1^*, \ldots, s_N^*)$ is a pure Nash equilibrium for Γ, then $s_i^* \in F_i(s_{-i}^*)$ for each $i = 1, \ldots, N$. Thus $\min F_i(s_{-i}^*) \leq s_i^* \leq \max F_{-i}(s_{-i}^*)$ for each $i = i, \ldots, N$, so that $G_-(s^*) \leq s^* \leq G_+(s^*)$ by (8.29). This result implies by (8.30) and (8.31) that $\underline{s}^* \leq s^* \leq \overline{s}^*$.   □

Definition 8.48. *Given posets X, Y, and T, and a function $f : X \times T \to Y$, we say that $f(x, t)$ satisfies the **single crossing property** in (x, t) if $f(x, t) \leq f(\hat{x}, t)$ implies $f(x, \hat{t}) \leq f(\hat{x}, \hat{t})$, and $f(x, t) < f(\hat{x}, t)$ implies $f(x, \hat{t}) < f(\hat{x}, \hat{t})$ whenever $x < \hat{x}$ in X and $t < \hat{t}$ in T.*

Applying Lemma 8.42 and Corollary 8.34, one obtains the following special case of Theorem 8.47.

Proposition 8.49. *Let $\Gamma = \{S_i, u_i\}_{i=1}^N$ be a normal-form game. Assume that every S_i is a compact lattice in an ordered topological space X_i and that the values of utility functions u_i are in a poset $Y_i = (Y_i, \leq_i)$. Assume moreover that the following hypotheses are valid for every $i = 1, \ldots, N$.*

(i) $u_i(\cdot, s_{-i})$ is for each $s_{-i} \in S_{-i}$ directedly upper closed and quasisupermodular.

(ii) Each $u_i(s_i, s_{-i})$ has the single crossing property in (s_i, s_{-i}).

Then Γ has the smallest and greatest pure Nash equilibria.

Proof: Let $i \in \{1, \dots, N\}$ and $s_{-i} \in S_{-i}$ be fixed. The hypothesis (i) implies by Lemma 8.42 that the set $F_i(s_{-i})$ of all the maximum points of $u_i(\cdot, s_{-i})$ is a nonempty and compact sublattice of S_i. In particular, $F_i(s_{-i})$ is directed, whence $\max F_i(s_{-i})$ and $\min F_i(s_{-i})$ exist by Corollary 8.34. It remains to show that mappings $s_{-i} \mapsto \max F_i(s_{-i})$ and $s_{-i} \mapsto \min F_i(s_{-i})$ are increasing.

Assume that $s_{-i} < \hat{s}_{-i}$ in S_{-i}, and let $x \in F_i(s_{-i})$ and $y \in F_i(\hat{s}_{-i})$ be given. The definition of $F_i(s_{-i})$ implies that $u_i(x \wedge y, s_{-i}) \preceq_i u_i(x, s_{-i})$. If strict inequality holds, it follows from the quasisupermodularity hypothesis that $u_i(y, s_{-i}) \prec_i u_i(x \vee y, s_{-i})$. This result and the hypothesis (ii) imply that $u_i(y, \hat{s}_{-i}) \prec_i u_i(x \vee y, \hat{s}_{-i})$. But this is impossible, since $y \in F_i(\hat{s}_{-i})$. Thus $u_i(x \wedge y, s_{-i}) = u_i(x, s_{-i})$, so that $x \wedge y \in F_i(s_{-i})$ and $u_i(y, s_{-i}) \preceq_i u_i(x \vee y, s_{-i})$ by quasisupermodularity. Thus $u_i(y, \hat{s}_{-i}) \preceq_i u_i(x \vee y, \hat{s}_{-i})$ by (ii). Equality must hold since $y \in F_i(\hat{s}_{-i})$, whence also $x \vee y \in F_i(\hat{s}_{-i})$.

The above proof shows that if $s_{-i} < \hat{s}_{-i}$ in S_{-i}, and if $x \in F_i(s_{-i})$ and $y \in F_i(\hat{s}_{-i})$, then $x \wedge y \in F_i(s_{-i})$ and $x \vee y \in F_i(\hat{s}_{-i})$. In particular, when $x = \min F_i(s_{-i})$ and $y = \min F_i(\hat{s}_{-i})$, then $x \wedge y = x$. Thus $x = \min F_i(s_{-i}) \leq_i y = \min F_i(\hat{s}_{-i})$, whence $s_{-i} \mapsto \min F_i(s_{-i})$ is increasing. Choosing $x = \max F_i(s_{-i})$ and $y = \max F_i(\hat{s}_{-i})$ then $x \vee y = y$, so that $x = \max F_i(s_{-i}) \leq_i y = \max F_i(\hat{s}_{-i})$. This proves that $s_{-i} \mapsto \max F_i(s_{-i})$ is increasing.

The above proof implies that all the hypotheses (HI)–(HIV) are valid, whence the conclusion follows from Theorem 8.47. $\square$

Remark 8.50. In view of Lemma 8.42 and Corollary 8.45, condition (i) is a generalization to condition

(i)' $u_i(\cdot, s_{-i})$ is for each $s_{-i} \in S_{-i}$ (order) upper semicontinuous, real-valued, and quasisupermodular.

Conditions (i') and (ii) are the same as the corresponding properties of quasi-supermodularity and of single crossing property in the definition of a normal-form game with (ordinal) strategic complementarities (cf. [180]). Another generalization is obtained in that real-valued utility functions are replaced by poset-valued ones.

Assuming monotonicity for the functions $u_i(s_i, \cdot)$ we obtain comparison results for utilities of pure Nash equilibria.

Proposition 8.51. *Let $\Gamma = \{S_i, u_i\}_{i=1}^N$ be a normal-form game that satisfies the hypotheses of Theorem 8.47 or Proposition 8.49.*

(a) If each $u_i(s_i, s_{-i})$ is increasing in s_{-i}, then the utilities of the greatest pure Nash equilibrium for Γ majorize the utilities of all its pure Nash equilibria.

*(b) If each $u_i(s_i, s_{-i})$ is decreasing in s_{-i}, then the utilities of the smallest
pure Nash equilibrium for Γ majorize the utilities of all its pure Nash
equilibria.*

Proof: The greatest pure Nash equilibrium $\bar{s}^* = (\bar{s}_1^*, \ldots, \bar{s}_N^*)$ exists by Theorem 8.47 or Proposition 8.49. Let $s^* = (s_1^*, \ldots, s_N^*)$ be any pure Nash equilibrium for Γ. Then $s^* \leq \bar{s}^*$ in S. Because S is ordered componentwise, then $s_{-i}^* \leq \bar{s}_{-i}^*$ in S_{-i} for every $i = 1, \ldots, N$. Hence, if each $u_i(s_i, s_{-i})$ is increasing in s_{-i}, then the first inequality of

$$u_i(s_i^*, s_{-i}^*) \preceq_i u_i(s_i^*, \bar{s}_{-i}^*) \preceq_i u_i(\bar{s}_i^*, \bar{s}_{-i}^*)$$

holds for every $i = 1, \ldots, N$. This is true also for the second inequality because $\bar{s}^*$ is a pure Nash equilibrium for Γ. This proves (a). The proof of (b) is similar. $\quad\square$

Next we present a result dealing with monotone comparative statics of normal-form games.

Proposition 8.52. *Let $\{\Gamma^t = \{S_i, u_i^t\}_{i=1}^N : t \in T\}$, where T is a poset, be a family of normal-form games satisfying the hypotheses of Theorem 8.47 or Proposition 8.49 and the following hypothesis.*

(iii) Each $u_i^t(s_i, s_{-i})$ has the single crossing property in (s_i, t).

Then each Γ^t has the smallest and greatest pure Nash equilibria, which are increasing in t.

Proof: The existence of the smallest and greatest pure Nash equilibria for every Γ^t follows from Theorem 8.47 or Proposition 8.49. Define

$$G_+^t(s) = (\max F_1^t(s_{-1}), \ldots, \max F_N^t(s_{-N})), \ s \in S,$$
$$G_-^t(s) = (\min F_1^t(s_{-1}), \ldots, \min F_N^t(s_{-N})), \ s \in S, \text{ where} \qquad (8.32)$$
$$F_i^t(s_{-i}) = \{t_i \in S_i : u_i^t(t_i, s_{-i}) = \max_{s_i \in S_i} u_i^t(s_i, s_{-i})\}, \ i = 1, \ldots, N.$$

The monotonicity of mappings $t \mapsto \max F_i^t(s_{-i})$ and $t \mapsto \min F_i^t(s_{-i})$ can be proved similarly as the monotonicity of mappings $s_{-i} \mapsto \max F_i(s_{-i})$ and $s_{-i} \mapsto \min F_i(s_{-i})$ in the proof of Proposition 8.49, by using the hypothesis (iii) instead of (ii). Thus the mappings $G_\pm^t$ are increasing in t. It then follows from (8.30) that the smallest fixed point of G_-^t, which is the smallest pure Nash equilibrium for Γ^t, is increasing in t. Similarly the greatest fixed point of G_+^t, which is the greatest pure Nash equilibrium of Γ^t, is by (8.31) increasing in t. This proves the last conclusion. $\quad\square$

Example 8.53. Consider a two-person normal form game $\Gamma = \{S_1, S_2, u_1, u_2\}$, where the strategy spaces are $S_i = [\frac{3}{2}, \frac{5}{2}] \times [\frac{3}{2}, \frac{5}{2}]$. The utility functions are of the form $u_i((s_{i1}, s_{i2}), s_{-i}) = (u_{i1}(s_{i1}, s_{-i}), u_{i2}(s_{i2}, s_{-i}))$, where the components $u_{ij} : [\frac{3}{2}, \frac{5}{2}] \times [\frac{3}{2}, \frac{5}{2}]^2 \to \mathbb{R}$, $i, j = 1, 2$, are defined as follows ('sgn' denotes the sign function):

$$u_{11}(s, s_{-1}) = (52 - 21s + s_{21} + 4s_{22} + 8sgn(s_{21}s_{22} - 4))(s - 1),$$

$$u_{12}(s, s_{-1}) = \begin{cases} (51 - 21s - sgn(s - \frac{11}{5}) + 2s_{21} + 3s_{22})(s - \frac{11}{10}) \\ + 4sgn(s_{21} + s_{22} - 4))(s - \frac{11}{10}), \end{cases}$$

$$u_{21}(s, s_{-2}) = \begin{cases} (50 - 20s - sgn(s - \frac{11}{5}) + 3s_{11} + 2s_{12})(s - \frac{11}{10}) \\ + 2sgn(s_{11} + s_{21} - 4))(s - \frac{11}{10}), \end{cases}$$

$$u_{22}(s, s_{-2}) = (49 - 20s + 4s_{11} + s_{12} + sgn(s_{11}s_{12} - 4))(s - 1).$$

Show that the smallest and greatest Nash equilibria for Γ exist and calculate them and the corresponding utilities.

Solution. The spaces S_i are compact and directed subsets of $\mathbb{R}^2$, whence the hypothesis (HI) holds. It is also easy to show that the hypothesis (HII) is valid. Moreover, the functions $u_{ij}(\cdot, s_{-i})$ have unique maximum points $f_{ij}(s_{-i})$, which are solutions of equations $\frac{d}{ds}u_{ij}(s, s_{-i}) = 0$:

$$\begin{aligned} f_{11}(s_{-1}) &= \frac{73}{42} + \frac{1}{42}s_{21} + \frac{2}{21}s_{22} + \frac{4}{21}sgn(s_{21}s_{22} - 4), \\ f_{12}(s_{-1}) &= \frac{247}{140} + \frac{1}{21}s_{21} + \frac{1}{14}s_{22} + \frac{2}{21}sgn(s_{21} + s_{22} - 4), \\ f_{21}(s_{-2}) &= \frac{9}{5} + \frac{3}{40}s_{11} + \frac{1}{20}s_{12} + \frac{1}{20}sgn(s_{11} + s_{21} - 4), \\ f_{22}(s_{-2}) &= \frac{69}{40} + \frac{1}{10}s_{11} + \frac{1}{40}s_{12} + \frac{1}{40}sgn(s_{11}s_{12} - 4). \end{aligned} \qquad (8.33)$$

Thus $F_i(s_{-i}) = \{(f_{i1}(s_{-i}), f_{i2}(s_{-i}))\}$, $i = 1, 2$, whence the hypothesis (HIII) is valid. It follows from (8.33) that the functions $s_{-i} \mapsto f_{ij}(s_{-i})$ are increasing, which implies that the hypothesis (HIV) holds. Thus all the hypotheses of Theorem 8.47 are valid, so that Γ has the smallest and greatest pure Nash equilibria. Since $\min F_i(s_{-i}) = \max F_i(s_{-i}) = F_i(s_{-i})$, the mappings $G_{\pm}$, defined by (8.29), are equal to the function $G : [\frac{3}{2}, \frac{5}{2}]^2 \times [\frac{3}{2}, \frac{5}{2}]^2 \to \mathbb{R}^2 \times \mathbb{R}^2$, defined as follows:

$$G((s_{11}, s_{12}), (s_{21}, s_{22})) = ((f_{11}(s_{-1}), f_{12}(s_{-1})), (f_{21}(s_{-2}), f_{22}(s_{-2}))). \quad (8.34)$$

According to the proof of Theorem 8.47, the components of the smallest (respectively the greatest) fixed point of G form the smallest (respectively the greatest) Nash equilibrium for Γ. It follows from (8.33) and (8.34) that $((s_{11}, s_{12}), (s_{21}, s_{22}))$ is a fixed point of G if and only if $(s_{11}, s_{12}, s_{21}, s_{22})$ is a solution of the following system of four equations:

$$s_{11} = \frac{73}{42} + \frac{1}{42}s_{21} + \frac{2}{21}s_{22} + \frac{4}{21}sgn(s_{21}s_{22} - 4),$$

$$s_{12} = \frac{247}{140} + \frac{1}{21}s_{21} + \frac{1}{14}s_{22} + \frac{2}{21}sgn(s_{21} + s_{22} - 4),$$

$$s_{21} = \frac{9}{5} + \frac{3}{40}s_{11} + \frac{1}{20}s_{12} + \frac{1}{20}sgn(s_{11} + s_{21} - 4),$$

$$s_{22} = \frac{69}{40} + \frac{1}{10}s_{11} + \frac{1}{40}s_{12} + \frac{1}{40}sgn(s_{11}s_{12} - 4).$$

(8.35)

To determine the smallest solution of (8.35), notice that if $s_{ij} < 2$, $i, j = 1, 2$, then all the sign functions in (8.35) attain the value -1. Inserting these values in (8.35), the obtained system has the unique solution:

$$\underline{s}_{11} = \frac{4940854}{2778745} \approx 1.77808830965057966815954684579,$$

$$\underline{s}_{12} = \frac{5281784}{2778745} \approx 1.90078038826880480216788514239,$$

$$\underline{s}_{21} = \frac{5497457}{2778745} \approx 1.97839564263723371522036027055,$$

$$\underline{s}_{22} = \frac{10699993}{5557490} \approx 1.92532834067177808687015181314.$$

Since $\underline{s}_{ij} < 2$, $i, j = 1, 2$, the smallest solution of (8.35) is $(\underline{s}_{11}, \underline{s}_{12}, \underline{s}_{21}, \underline{s}_{22})$. Thus the smallest Nash equilibrium for Γ is $((\underline{s}_{11}, \underline{s}_{12}), (\underline{s}_{21}, \underline{s}_{22}))$. The corresponding utilities are $\underline{u}_i = (\underline{u}_{i1}, \underline{u}_{i2})$, where

$$\underline{u}_{11} = \frac{98169021885501}{7721423775025} \approx 12.71384976991282333234461357898,$$

$$\underline{u}_{12} = \frac{415913992373061}{30885695100100} \approx 13.46623383495469320411203932846,$$

$$\underline{u}_{21} = \frac{4766150161125}{308856951001} \approx 15.43157810008157602352226774092,$$

$$\underline{u}_{22} = \frac{26445337105009}{1544284755005} \approx 17.12465076100772408790305087200.$$

If $s_{ij} > 2$, $i, j = 1, 2$, then all the sign functions in (8.35) attain the value 1. Inserting these values in (8.35), the obtained system has the unique solution:

$$\overline{s}_{11} = \frac{6033654}{2778745} \approx 2.17135937266643754644632738880,$$

$$\overline{s}_{12} = \frac{5848294}{2778745} \approx 2.10465299982546077455829880036,$$

$$\overline{s}_{21} = \frac{5885617}{2778745} \approx 2.11808460294125585471138949418,$$

$$\overline{s}_{22} = \frac{11224753}{5557490} \approx 2.01975226226228027400859020889.$$

Because every $\overline{s}_{ij}$ is > 2, then the greatest Nash equilibrium for Γ is $((\overline{s}_{11}, \overline{s}_{12}), (\overline{s}_{21}, \overline{s}_{22}))$. The corresponding utilities are $\overline{u}_i = (\overline{u}_{i1}, \overline{u}_{i2})$, where

$$\bar{u}_{11} = \frac{222483084563901}{7721423775025} \approx 28.81373837860371244039573 61075,$$

$$\bar{u}_{12} = \frac{654649507171821}{30885695100100} \approx 21.1958806512242430294171224820,$$

$$\bar{u}_{21} = \frac{6402581484005}{308856951001} \approx 20.729925174921091786679843 6025,$$

$$\bar{u}_{22} = \frac{32117869911169}{1544284755005} \approx 20.79789352778076898930540 57605.$$

The functions $u_{ij}(s, s_{-i})$ are increasing with respect to s_{-i}, whence the utility functions u_i satisfy also the hypotheses of Proposition 8.51. Thus the utilities $\bar{u}_1 = (\bar{u}_{11}, \bar{u}_{12})$ and $\bar{u}_2 = (\bar{u}_{21}, \bar{u}_{22})$ majorize the utilities of all other pure Nash equilibria of Γ.

8.3.3 Special Cases

Consider next the special case when the strategy spaces are the product of chains.

Proposition 8.54. *Let* $\Gamma = \{S_i, u_i\}_{i=1}^N$ *be a normal-form game, which satisfies the following hypotheses.*

(S0) Each strategy set S_i is the product of nonempty compact chains S_{ij} of ordered topological spaces X_{ij}, $j = 1, \ldots, m_i$.

(u0) The utility functions are of the form $u_i = (u_{i1}, \ldots, u_{im_i})$, where u_{ij} is a mapping from $S_{ij} \times S_{-i}$ to a poset Y_{ij}, and every $u_{ij}(\cdot, s_{-i})$ is directedly upper closed.

(u1) Every $u_{ij}(x, y)$ satisfies the single crossing property in (x, y).

Then the smallest and greatest pure Nash equilibria exist for Γ.

Proof: The products $S_i = S_{i1} \times \cdots \times S_{im_i}$ of compact chains are compact lattices in the product spaces $X_i = X_{i1} \times \cdots \times X_{im_i}$. Thus the hypothesis (HI) of Theorem 8.47 is valid. From the hypothesis (u0) it follows that for all fixed s_{-i} the function $u_i(\cdot, s_{-i})$ is directedly upper closed. Thus the hypothesis (HII) of Theorem 8.47 is also valid. To prove the validity of the hypotheses (HIII) and (HIV), let $i \in \{1, \ldots, N\}$, $j \in \{1, \ldots, m_i\}$ and $s_{-i} \in S_{-i}$ be fixed. Noticing that S_{ij} is a compact chain, then it is also a compact lattice, and $u_{ij}(\cdot, s_{-i})$ is quasisupermodular. It then follows from the proof of Proposition 8.49 that the set $F_{ij}(s_{-i})$ of all the maximum points of $u_{ij}(\cdot, s_{-i})$ is non-empty and compact subset of a chain S_{ij}, that $\max F_{ij}(s_{-i})$ and $\min F_{ij}(s_{-i})$ exist, and that the mappings $s_{-i} \mapsto \max F_{ij}(s_{-i})$ and $s_{-i} \mapsto \min F_{ij}(s_{-i})$ are increasing. The above proof is valid for all $i = 1, \ldots, N$ and $j = 1, \ldots m_i$. Thus the sets $F_i(s_{-i}) = F_{i1}(s_{-i}) \times \cdots \times F_{im_i}(s_{-i})$ of maximum points of the functions $u_i(\cdot, s_{-i})$ are compact and directed. This proves the validity of hypothesis (HIII). Moreover, the above proof implies that the mappings

$$s_{-i} \mapsto \max F_i(s_{-i}) = (\max F_{i1}(s_{-i}), \ldots, \max F_{im_i}(s_{-i}))$$

and

$$s_{-i} \mapsto \min F_i(s_{-i}) = (\min F_{i1}(s_{-i}), \ldots, \min F_{im_i}(s_{-i}))$$

are increasing in S_{-i}. Thus the hypothesis (HIV) is valid. The conclusion follows then from Theorem 8.47. $\qquad\square$

The next result is a consequence of Proposition 8.54 and Remark 8.36.

Corollary 8.55. *Let* $\Gamma = \{S_i, u_i\}_{i=1}^{N}$ *be a normal-form game, which satisfies the hypotheses (S0) and (u1) of Proposition 8.54 and the following hypothesis.*

(u2) *The utility functions are of the form* $u_i = (u_{i1}, \ldots, u_{im_i})$, *where* u_{ij} *is a mapping from* $S_{ij} \times S_{-i}$ *to a chain* Y_{ij}, *and every* $u_{ij}(\cdot, s_{-i})$ *is piecewise strictly monotone.*

Then Γ *has the smallest and greatest pure Nash equilibria.*

Consider next the following special case of Proposition 8.54.

Proposition 8.56. *Assume that condition (S0) of Proposition 8.54 is satisfied, and let the components of the utility functions* $u_i = (u_{i1}, \ldots, u_{im_i})$ *have the following form:*

$$u_{ij}(s_{ij}, s_{-i}) = v_{ij}(s_{ij}, s_{-i}) h_{ij}(s_{-i}) q_{ij}(s_{ij}), \quad i = 1, \ldots, N, \ j = 1, \ldots, m_i, \tag{8.36}$$

where $v_{ij} : S_{ij} \times S_{-i} \to \mathbb{R}_+$, $h_{ij} : S_{-i} \to (0, \infty)$, *and* $q_{ij} : S_{ij} \to \mathbb{R}_+$. *The hypotheses*

(v0) *For every* $y \in S_{-i}$ *the function* $v_{ij}(\cdot, y) q_{ij}(\cdot)$ *is directedly upper closed;*
(v1) $v_{ij}(x, y) q_{ij}(x)$ *satisfies the single crossing property in* (x, y);

imply that $\Gamma = \{S_i, u_i\}_{i=1}^{N}$ *has the smallest and greatest pure Nash equilibria.*

Proof: It suffices to verify hypotheses (u0) and (u1) of Proposition 8.54. The hypothesis (v0) implies that the hypothesis (u0) is valid. To prove that hypothesis (u1) holds, suppose that $y < \hat{y}$ in S_{-i} and $x <_{ij} \hat{x}$ in S_{ij}, and that

$$u_{ij}(x, y) \le u_{ij}(\hat{x}, y), \tag{8.37}$$

or, by (8.36),

$$v_{ij}(x, y) h_{ij}(y) q_{ij}(x) \le v_{ij}(\hat{x}, y) h_{ij}(y) q_{ij}(\hat{x}).$$

Multiplying both sides of the last inequality by $1/h_{ij}(y)$, which is positive, we obtain

$$v_{ij}(x, y) q_{ij}(x) \le v_{ij}(\hat{x}, y) q_{ij}(\hat{x}).$$

This inequality, the hypothesis (v1), and the choices $y < \hat{y}$ and $x <_{ij} \hat{x}$ imply that

$$v_{ij}(x, \hat{y}) q_{ij}(x) \le v_{ij}(\hat{x}, \hat{y}) q_{ij}(\hat{x}).$$

Multiplying both sides of the last inequality by $h_{ij}(\hat{y})$, we get in view of (8.36)

$$u_{ij}(x, \hat{y}) \leq u_{ij}(\hat{x}, \hat{y}). \tag{8.38}$$

If strict inequality holds in (8.37), then the above reasoning can be used to show that the inequality (8.38) is also strict. Thus $u_{ij}(x, y)$ satisfies the single crossing property in (x, y), whence (u1) holds.   □

As a consequence of Propositions 8.54 and 8.56 we obtain the following proposition.

Proposition 8.57. *Assume that* $\Gamma = \{S_i, u_i\}_{i=1}^N$ *is a normal form game, where strategy spaces* S_i *satisfy condition (S0) of Proposition 8.54, and that the components* u_{ij} *of the utility functions* $u_i = (u_{i1}, \ldots, u_{im_i})$ *are given by (8.36), with*

$$v_{ij}(s_{ij}, s_{-i}) = f_{ij}(s_{ij}) + g_{ij}(s_{-i}), \quad s_{ij} \in S_{ij}, \; s_{-i} \in S_{-i} \tag{8.39}$$

for $i = 1, \ldots, N$, $j = 1, \ldots, m_i$, *where* $f_{ij} : S_{ij} \to \mathbb{R}_+$, $g_{ij} : S_{-i} \to \mathbb{R}_+$, *and every* g_{ij} *is increasing. If the functions* q_{ij} *in (8.36) are increasing, and if the hypothesis (v0) holds, then* Γ *has the smallest and greatest pure Nash equilibria. Moreover, the utilities of the greatest pure Nash equilibrium for* Γ *majorize the utilities of its all pure Nash equilibria.*

Proof: By the proof of Proposition 8.56, hypothesis (v0) implies that hypothesis (u0) of Proposition 8.54 is fulfilled for the functions u_{ij} defined by (8.36). To show that the hypothesis (u1) is valid, assume that $x < \hat{x}$ in S_{ij} and $y < \hat{y}$ in S_{-i}, and that

$$u_{ij}(x, y) < u_{ij}(\hat{x}, y).$$

The last inequality, (8.36), and (8.39) shows that

$$(f_{ij}(x) + g_{ij}(y))q_{ij}(x) < (f_{ij}(\hat{x}) + g_{ij}(y))q_{ij}(\hat{x}),$$

which is equivalent to

$$f_{ij}(x)q_{ij}(x) - f_{ij}(\hat{x})q_{ij}(\hat{x}) < g_{ij}(y)(q_{ij}(\hat{x}) - q_{ij}(x)).$$

Since each $g_{ij} : S_{-i} \to \mathbb{R}_+$ is increasing and $q_{ij}(\hat{x}) - q_{ij}(x) \geq 0$, from the last inequality we get

$$f_{ij}(x)q_{ij}(x) - f_{ij}(\hat{x})q_{ij}(\hat{x}) < g_{ij}(\hat{y})(q_{ij}(\hat{x}) - q_{ij}(x))$$

which is equivalent to

$$(f_{ij}(x) + g_{ij}(\hat{y}))q_{ij}(x) < (f_{ij}(\hat{x}) + g_{ij}(\hat{y}))q_{ij}(\hat{x}).$$

Multiplying both sides of the last inequality by $h_{ij}(\hat{y})$, and applying (8.36) and (8.39), we obtain

$$u_{ij}(x, \hat{y}) < u_{ij}(\hat{x}, \hat{y}).$$

The above reasoning can be applied to show that $u_{ij}(x, y) \leq u_{ij}(\hat{x}, y)$ implies $u_{ij}(x, \hat{y}) \leq u_{ij}(\hat{x}, \hat{y})$. Thus $u_{ij}(x, y)$ satisfies the single crossing property in (x, y), whence (u1) holds. The existence of the smallest and greatest Nash equilibria for Γ follows then from Proposition 8.54. Since the functions g_{ij} are increasing, then the functions $u_i(s_i, s_{-i})$ are increasing in s_{-i}. Thus the last conclusion follows from Proposition 8.51.   □

According to the proof of Theorem 8.47, the smallest Nash equilibrium of Γ is the smallest fixed point of G_- and the greatest Nash equilibrium of Γ is the greatest fixed point of G_+, where the mappings $G_\pm : S \to S$, $S = S_1 \times \cdots \times S_N$, are defined by (8.29). By Theorem 2.16 the smallest fixed point of G_- is the maximum of a well-ordered chain C of aG_--iterations. The smallest elements of C are the elements of the iteration sequence $(G_-^n(a))_{n=0}^\infty$, as long as this sequence is strictly increasing. In particular, if $G_-^n(a) = G_-^{n+1}(a)$ for some $n \in \mathbb{N}_0$, then $\underline{s}^* = G_-^n(a)$ is the smallest fixed point of G_-, and hence the smallest pure Nash equilibrium for Γ. Dually, the greatest fixed point of G_+ is the minimum of an inversely well-ordered chain D of bG_+-iterations. The greatest elements of D are the elements of the iteration sequences $(G_+^n(b))_{n=0}^\infty$, as long as this sequence is strictly decreasing. If $G_+^n(b) = G_+^{n+1}(b)$ for some $n \in \mathbb{N}_0$, then $\overline{s}^* = G_+^n(b)$ is the greatest fixed point of G_+, and thus the greatest pure Nash equilibrium for Γ. In particular, if chains of $G_\pm[S]$ are finite, the smallest and greatest pure Nash equilibria for Γ are obtained by finite number of iterations.

Example 8.58. Consider the normal form game $\Gamma = \{S_1, S_2, u_1, u_2\}$. Assume that the strategy spaces are $S_i = S_{i1} \times S_{i2}$, where $S_{ij} = [\frac{3}{2}, \frac{5}{2}]$, $i, j = 1, 2$, and that the utility functions are $u_i = (u_{i1}, u_{i2})$, where the components u_{ij}, $i, j = 1, 2$, are defined by ('*sgn*' denotes the sign function and $[\cdot]$ the greatest integer function)

$$u_{11}(s, s_{-1}) = \begin{cases} (52 - 21s + 10^{-30}[10^{30}s_{21}] + 4 \cdot 10^{-30}[10^{30}s_{22}])(s - 1) \\ +8sgn(s_{21}s_{22} - 4)(s - 1), \end{cases}$$

$$u_{12}(s, s_{-1}) = \begin{cases} (51 - 21s - sgn(s - \frac{11}{5}) + 2 \cdot 10^{-30}[10^{30}s_{21}])(s - \frac{11}{10}) \\ +(3 \cdot 10^{-30}[10^{30}s_{22}] + 4sgn(s_{21} + s_{22} - 4)(s - \frac{11}{10}), \end{cases}$$

$$u_{21}(s, s_{-2}) = \begin{cases} (50 - 20s - sgn(s - \frac{11}{5}) + 3 \cdot 10^{-30}[10^{30}s_{11}])(s - \frac{11}{10}) \\ +(2 \cdot 10^{-30}[10^{30}s_{12}] + 2sgn(s_{11} + s_{21} - 4)(s - \frac{11}{10}), \end{cases}$$

$$u_{22}(s, s_{-2}) = \begin{cases} (49 - 20s + 4 \cdot 10^{-30}[10^{30}s_{11}] + 10^{-30}[10^{30}s_{12}])(s - 1) \\ +sgn(s_{11}s_{12} - 4)(s - 1). \end{cases}$$

Show that the smallest and greatest Nash equilibria for Γ exist and calculate them along with the corresponding utilities.

Solution. The hypotheses of Proposition 8.57 are valid, which ensures the existence of the smallest and greatest Nash equilibria for Γ. The functions $u_{ij}(\cdot, s_{-i})$ have unique maximum points $f_{ij}(s_{-i})$, which are solutions of equations $\frac{d}{ds}u_{ij}(s, s_{-i}) = 0$:

$$f_{11}(s_{-1}) = \frac{73}{42} + \frac{1}{42}10^{-30}[10^{30}s_{21}] + \frac{2}{21}10^{-30}[10^{30}s_{22}] + \frac{4}{21}sgn(s_{21}s_{22} - 4),$$

$$f_{12}(s_{-1}) = \frac{247}{140} + \frac{1}{21}10^{-30}[10^{30}s_{21}] + \frac{1}{14}10^{-30}[10^{30}s_{22}]$$
$$+ \frac{2}{21}sgn(s_{21} + s_{22} - 4),$$

$$f_{21}(s_{-2}) = \frac{9}{5} + \frac{3}{40}10^{-30}[10^{30}s_{11}] + \frac{1}{20}10^{-30}[10^{30}s_{12}]$$
$$+ \frac{1}{20}sgn(s_{11} + s_{21} - 4),$$

$$f_{22}(s_{-2}) = \frac{69}{40} + \frac{1}{10}10^{-30}[10^{30}s_{11}] + \frac{1}{40}10^{-30}[10^{30}s_{12}] + \frac{1}{40}sgn(s_{11}s_{12} - 4).$$
$$(8.40)$$

These solutions can be used to define function $G = G_- = G_+$, where $G : [\frac{3}{2}, \frac{5}{2}]^2 \times [\frac{3}{2}, \frac{5}{2}]^2 \to \mathbb{R}^2 \times \mathbb{R}^2$, as follows:

$$G((s_{11}, s_{12}), (s_{21}, s_{22})) = ((f_{11}(s_{-1}), f_{12}(s_{-1})), (f_{21}(s_{-2}), f_{22}(s_{-2}))).$$

It follows from (8.40) that G is increasing, and that the range of G is finite. In particular, choosing $a = ((\frac{3}{2}, \frac{3}{2}), (\frac{3}{2}, \frac{3}{2}))$, then $G^n(a) = G^{n+1}(a)$ for some n, and the components of $G^n(a)$ form the smallest pure Nash equilibrium $(\underline{s}_{11}, \underline{s}_{12}), (\underline{s}_{21}, \underline{s}_{22})$ of Γ. The components of $G^n(a)$, $n = 1, 2, \ldots$, can be calculated by the following algorithm, written as a Maple program, where sij $= s_{ij}$ and fij $= f_{ij}(s_{-i})$:

s11 := 3/2: s12 := 3/2: s21 := 3/2: s22 := 3/2: for n while s11 < f11 or s12 < f12 or s21 < f21 or s22 < f22 do: if s11 < f11 then s11 := f11 end if: if s12 < f12 then s12 := f12 end if: if s21 < f21 then s21 := f21 end if: if s22 < f22 then s22 := f22 end if: end do:

It turns out that $G^{18}(a) = G^{19}(a)$. The components of the smallest pure Nash equilibrium of Γ are:

$$\underline{s}_{11} = \frac{49786472670216230708467311 68207}{2800000000000000000000000000000000}$$
$$= 1.7780883096505796681595468457882\overline{142857},$$

$$\underline{s}_{12} = \frac{285117058240320720325182771359}{150000000000000000000000000000000}$$
$$= 1.90078038826880480216788514239\overline{3},$$

$$\underline{s}_{21} = \frac{158271651410978697217628 8216443}{80000000000000000000000000000000}$$
$$= 1.97839564263723371522036027055375,$$

$$\underline{s}_{22} = \frac{15402626725374224694961214505109}{8000000000000000000000000000000}$$
$$= 1.925328340671778086870151813138625.$$

The components of the corresponding utilities are:

$$\underline{u}_{11} = \begin{cases} \dfrac{14239511742302362132227672084565915307692791426579572782784547}{1120000000000000000000000000000000000000000000000000000000000000} \\ \approx 12.71384976991282333235, \end{cases}$$

$$\underline{u}_{12} = \begin{cases} \dfrac{100996753762160199030902949634319394672470462241037694948167}{750000000000000000000000000000000000000000000000000000000000} \\ \approx 13.46623383495469320412, \end{cases}$$

$$\underline{u}_{21} = \begin{cases} \dfrac{4938104992026104327527256770958419624001926964486540155722249}{3200000000000000000000000000000000000000000000000000000000000} \\ \approx 15.43157810008157602352, \end{cases}$$

$$\underline{u}_{22} = \begin{cases} \dfrac{5479888243522471708128976279027387892105948724812393978710188 1}{3200000000000000000000000000000000000000000000000000000000000} \\ \approx 17.12465076100772408790. \end{cases}$$

Choosing $b = ((\frac{5}{2}, \frac{5}{2}), (\frac{5}{2}, \frac{5}{2}))$, the components of $G^n(b)$, $n = 1, 2, \ldots$, can be calculated by the above algorithm replacing $\frac{3}{2}$ by $\frac{5}{2}$ and reversing inequalities. Also in this case 18 iterations suffice, and the components of $G^{18}(b)$, which form the greatest pure Nash equilibrium $((\overline{s}_{11}, \overline{s}_{12}), (\overline{s}_{21}, \overline{s}_{22}))$ of Γ, are:

$$\overline{s}_{11} = \begin{cases} \dfrac{455985468259951884753728751164867}{21000000000000000000000000000000} \\ = 2.171359372666437546446327388803190\overline{4761}, \end{cases}$$

$$\overline{s}_{12} = \begin{cases} \dfrac{88395425992669352531448549615023}{4200000000000000000000000000000} \\ = 2.1046529998254607745582988003576\overline{904761}, \end{cases}$$

$$\overline{s}_{21} = \begin{cases} \dfrac{84723384117650234188455579767123}{40000000000000000000000000000000} \\ = 2.11808460294125585471138949417807 5, \end{cases}$$

$$\overline{s}_{22} = \begin{cases} \dfrac{80790090490491210960343608355569}{40000000000000000000000000000000} \\ = 2.0197522622622802740085902088892 25. \end{cases}$$

The components of the corresponding utilities are:

$$\overline{u}_{11} = \begin{cases} \dfrac{6050885059506779612483104582572359765530677625042245924311276 89}{2100000000000000000000000000000000000000000000000000000000000 0} \\ \approx 28.81373837860371244040, \end{cases}$$

$$\overline{u}_{12} = \begin{cases} \dfrac{17804539747028364144710382884842427573570226381035632815072905 29}{8400000000000000000000000000000000000000000000000000000000000 0} \\ \approx 21.19588065122424302942, \end{cases}$$

$$\overline{u}_{21} = \begin{cases} \dfrac{16583940139936873429343874881968031603948784589264598469116971 29}{8000000000000000000000000000000000000000000000000000000000000 0} \\ \approx 20.72992517492109178668, \end{cases}$$

$$\overline{u}_{22} = \begin{cases} \dfrac{16638314822224615191444324608411209008581442063744968323333137 61}{8000000000000000000000000000000000000000000000000000000000000 0} \\ \approx 20.79789352778076898931. \end{cases}$$

The functions $u_{ij}(s, s_{-i})$ are increasing with respect to s_{-i}, whence the utility functions u_i satisfy also the hypotheses of Proposition 8.57. Thus the utilities $\bar{u}_1 = (\bar{u}_{11}, \bar{u}_{12})$ and $\bar{u}_2 = (\bar{u}_{21}, \bar{u}_{22})$ majorize the utilities of all other pure Nash equilibria of Γ.

Remark 8.59. The functions $u_i(\cdot, s_{-i})$ in Examples 8.53 and 8.58 are neither upper semicontinuous nor upper closed because the functions u_{12} and u_{21} contain the term $-sgn(s - \frac{11}{5})$.

8.3.4 Applications to a Multiproduct Pricing Game

Consider a normal-form game, where players i, $i = 1, \ldots, N$, stand for firms that compete in prices. Assume that firm i sell products e_{ij}, $j = 1, \ldots, m_i$. Some of the products can be the same or substitutes in different firms. The feasible set S_{ij} of prices s_{ij} of e_{ij} per unit is assumed to be a finite subset of $\mathbb{R}$, bounded from below by c_{ij}, which stands for a purchase price of e_{ij} per unit for firm i. Denote $s_i = (s_{i1}, \ldots, s_{im_i})$ and $S_i = \times_{i=1}^{m_i} S_{ij}$, $i = 1, \ldots, N$, and let a function $d_{ij} : S_{ij} \times S_{-i} \to \mathbb{R}_+$ denote the demand of product e_{ij} (values $d_{ij}(s_{ij}, s_{-i})$ represent the amounts of products e_{ij} sold by firms during a fixed time period when the prices are s_{ij}). Since no d_{ij} depends on s_{in}, $n \neq j$, the demands of different products are assumed to be independent in every fixed firm i. The profit that the firm gets from the sale of e_{ij} is

$$u_{ij}(s_{ij}, s_{-i}) = d_{ij}(s_{ij}, s_{-i})(s_{ij} - c_{ij}). \tag{8.41}$$

The utility functions $u_i = (u_{i1}, \ldots, u_{im_i})$, whose components satisfy (8.41), are considered as profit functions of firms i. The above described pricing game and the one considered in Sect. 8.2.6 are generalizations to a Bertrand oligopoly model in the sense that, instead of a single product, they can include all the products that are for sale in every firm i.

The results of the next two propositions can be applied to the above described pricing game. The first proposition is a special case of Proposition 8.56.

Proposition 8.60. *Assume that* $S_i = \times_{i=1}^{m_i} S_{ij}$, $i = 1, \ldots, N$, *where every* S_{ij} *is a finite subset of* $[c_{ij}, \infty)$, *and that the components of utility functions* $u_i = (u_{i1}, \ldots, u_{im_i})$ *are given by*

$$u_{ij}(s_{ij}, s_{-i}) = v_{ij}(s_{ij}, s_{-i})h_{ij}(s_{-i})k_{ij}(s_{ij})(s_{ij} - c_{ij}), \tag{8.42}$$

where $v_{ij} : S_{ij} \times S_{-i} \to (0, \infty)$, $h_{ij} : S_{-i} \to (0, \infty)$, *and* $k_{ij} : S_{ij} \to \mathbb{R}_+$. *If for all* $i = 1, \ldots, N$, $j = 1, \ldots, m_i$,

(v) $\log v_{ij}(x, y)$ *has increasing differences in* (x, y);

then the normal-form game $\Gamma = \{S_i, u_i\}_{i=1}^N$ *has the smallest and greatest pure Nash equilibria.*

Proof: It suffices tho show that the hypotheses of Proposition 8.56 are valid when $q_{ij}(x) = k_{ij}(x)(x - c_{ij})$, since then (8.42) is of the form (8.36). Because every S_{ij} is a finite chain, hypotheses (S0) and (v0) of Proposition 8.56 are valid. To prove that hypothesis (v1) holds, suppose that $y < \hat{y}$ in S_{-i} and $x <_{ij} \hat{x}$ in S_{ij}, and that

$$v_{ij}(x, y)k_{ij}(x)(x - c_{ij}) \leq (<)v_{ij}(\hat{x}, y)k_{ij}(\hat{x})(\hat{x} - c_{ij}).$$

We may also suppose that $k_{ij}(x)(x - c_{ij}) > 0$. Thus both sides of the above inequality are positive. Taking logarithms, we obtain

$$\log v_{ij}(x, y) + \log(k_{ij}(x)(x - c_{ij})) \leq (<)\log v_{ij}(\hat{x}, y) + \log(k_{ij}(\hat{x})(\hat{x} - c_{ij})).$$

This inequality, the hypothesis (v), and the choices $y < \hat{y}$ and $x <_{ij} \hat{x}$ imply that

$$\log v_{ij}(x, \hat{y}) + \log(k_{ij}(x)(x - c_{ij})) \leq (<)\log v_{ij}(\hat{x}, \hat{y}) + \log(k_{ij}(\hat{x})(x - c_{ij})),$$

or

$$v_{ij}(x, \hat{y})k_{ij}(x)(x - c_{ij}) \leq (<)v_{ij}(\hat{x}, \hat{y})k_{ij}(\hat{x})(x - c_{ij}).$$

This proves that the hypothesis (v1) holds. The assertions follow then from Proposition 8.56. $\square$

The next result is a consequence of Proposition 8.57.

Proposition 8.61. *Assume that* $S_i = \times_{i=1}^{m_i} S_{ij}$, $i = 1, \ldots, N$, *where every* S_{ij} *is a finite subset of* $[c_{ij}, \infty)$, *and that the components of utility functions* $u_i = (u_{i1}, \ldots, u_{im_i})$ *are given by*

$$u_{ij}(s_{ij}, s_{-i}) = (f_{ij}(s_{ij}) + g_{ij}(s_{-i}))h_{ij}(s_{-i})k_{ij}(s_{ij})(s_{ij} - c_{ij}), \qquad (8.43)$$

where $f_{ij} : S_{ij} \to \mathbb{R}_+$, $g_{ij} : S_{-i} \to \mathbb{R}_+$, $h_{ij} : S_{-i} \to (0, \infty)$, *and* $k_{ij} : S_{ij} \to \mathbb{R}_+$. *If the functions* g_{ij} *and* $s_{ij} \mapsto k_{ij}(s_{ij})(s_{ij} - c_{ij})$ *are increasing, then the normal-form game* $\Gamma = \{S_i, u_i\}_{i=1}^N$ *has the smallest and greatest pure Nash equilibria. Moreover, the utilities of the greatest pure Nash equilibrium for* Γ *majorize the utilities of its all pure Nash equilibria.*

Proof: Because every S_{ij} is a finite chain and the functions v_{ij} are $\mathbb{R}_+$-valued, it is easy to see that the hypothesis (v0) of Proposition 8.57 holds when $v_{ij}(s_{ij}, s_{-i}) = (f_{ij}(s_{ij}) + g_{ij}(s_{-i}))$ and $q_{ij}(x) = k_{ij}(x)(x - c_{ij})$. Moreover, every g_{ij} is increasing and every q_{ij} is increasing. The conclusions follow then from Proposition 8.57. $\square$

Remark 8.62. In Proposition 8.61 the sets S_{ij} are finite. This implies that the smallest and greatest pure Nash equilibria for Γ are of the form $G_-^n(a)$ and $G_+^m(b)$, where $G_\pm$ are defined by (8.29), $a = \min S_1 \times \cdots \times S_N$, and $b = \max S_1 \times \cdots \times S_N$. This result and simple Maple programming are applied to solve the following concrete pricing problems.

Example 8.63. Assume that in the pricing game $\{S_1, S_2, u_1, u_2\}$, the utilities are given by

$$u_i = (d_{i1}(s_{i1}, s_{-i})(s_{i1} - c_{i1}), d_{i2}(s_{i2}, s_{-2})(s_{i2} - c_{i2})), \quad i = 1, 2,$$

with $c_{11} = c_{22} = 1$, $c_{12} = c_{21} = \frac{11}{10}$, and the strategy sets are $S_i = S_{i1} \times S_{i2}$, where

$$S_{11} = S_{22} = S_{12} = S_{21} = \{\frac{j}{20}, 24 \le j \le 48\}.$$

The demands are assumed to be of the form

$$d_{11}(s_{11}, s_{-1}) = \begin{cases} (53 - 22s_{11} + 4s_{21} + s_{22})(2 + \frac{1}{4}[4\sin(\frac{s_{11}}{2})]) \times \\ (8 + [4\sin(\frac{1}{22}(5s_{21} + 6s_{22}))]), \end{cases}$$

$$d_{12}(s_{12}, s_{-1}) = \begin{cases} (52 - 21s_{12} + 3s_{21} + 2s_{22})(2 + \frac{1}{5}[5\sin(\frac{s_{12}}{2})]) \times \\ (8 + [4\sin(\frac{1}{22}(6s_{21} + 5s_{22}))]), \end{cases}$$

$$d_{21}(s_{21}, s_{-2}) = \begin{cases} (48 - 19s_{21} + 2s_{11} + 3s_{12})(2 + \frac{1}{6}[6\sin(\frac{s_{21}}{2})]) \times \\ (8 + [4\sin(\frac{1}{22}(7s_{11} + 4s_{12}))]), \end{cases}$$

$$d_{22}(s_{22}, s_{-2}) = \begin{cases} (49 - 20s_{22} + s_{11} + 4s_{12})(2 + \frac{1}{7}[7\sin(\frac{s_{22}}{2})]) \times \\ (8 + [4\sin(\frac{1}{22}(8s_{11} + 3s_{12}))]). \end{cases}$$

Note, $s \mapsto [s]$ denotes the integer function. Show that the maximal and minimal pure Nash equilibria for prices exist. Calculate these equilibria and the corresponding profits.

Solution. From (8.44) it follows that $d_{ij} = v_{ij}h_{ij}k_{ij}$, where

$$v_{11}(s_{11}, s_{-1}) = 53 - 22s_{11} + 4s_{21} + s_{22},$$
$$v_{12}(s_{12}, s_{-1}) = 52 - 21s_{12} + 3s_{21} + 2s_{22},$$
$$v_{21}(s_{21}, s_{-2}) = 48 - 19s_{21} + 2s_{11} + 3s_{12},$$
$$v_{22}(s_{22}, s_{-2}) = 49 - 20s_{22} + s_{11} + 4s_{12},$$

$$h_{11}(s_{21}, s_{22}) = 8 + [4\sin(\frac{1}{22}(5s_{21} + 6s_{22}))],$$

$$h_{12}(s_{21}, s_{22}) = 8 + [4\sin(\frac{1}{22}(6s_{21} + 5s_{22}))],$$

$$h_{21}(s_{11}, s_{12}) = 8 + [4\sin(\frac{1}{22}(7s_{11} + 4s_{12}))],$$

$$h_{22}(s_{11}, s_{12}) = 8 + [4\sin(\frac{1}{22}(8s_{11} + 3s_{12}))],$$

$$k_{11}(s_{11}) = 2 + \frac{1}{4}[4\sin(s_{11}/2)], \quad k_{12}(s_{12}) = 2 + \frac{1}{5}[5\sin(s_{12}/2)],$$

$$k_{21}(s_{21}) = 2 + \frac{1}{6}[6\sin(s_{21}/2)], \quad k_{22}(s_{22}) = 2 + \frac{1}{7}[7\sin(s_{22}/2)].$$

It is easy to show that the hypotheses of Proposition 8.61 are valid, whence the prices have the smallest and greatest pure Nash equilibria. These equilibria are:

$$\overline{s}_1^* = (1.95, 2.05), \ \overline{s}_2^* = (2.1, 2.1), \ \underline{s}_1^* = (1.90, 2), \ \underline{s}_2^* = (2.05, 2.1).$$

Corresponding profits are

$$\overline{u}_1^* = (591, 569), \ \overline{u}_2^* = (565, 662), \ \underline{u}_1^* = (586, 564), \ \underline{u}_2^* = (561, 655).$$

By the last conclusion of Proposition 8.61, the profits $\overline{u}_1^*$ and $\overline{u}_2^*$ majorize the profits of all other Nash equilibria of prices.

Example 8.64. Assume that $N = 3$, and that the demands d_i, $i = 1, 2, 3$ are of the form

$$d_1(s_1, (s_2, s_3)) = v_1(s_1, (s_2, s_3))(\frac{1}{2} + \frac{1}{16}[4\sin(\frac{5s_2 + 6s_3}{22})]),$$

$$d_2(s_2, (s_1, s_3)) = v_2(s_2, (s_1, s_3))(\frac{1}{2} + \frac{1}{16}[4\sin(\frac{6s_1 + 5s_3}{22})]), \qquad (8.44)$$

$$d_3(s_3, (s_1, s_2)) = v_3(s_3, (s_1, s_2))(\frac{1}{2} + \frac{1}{16}[4\sin(\frac{7s_1 + 4s_2}{22})]),$$

where again $s \mapsto [s]$ denotes the integer function, and the functions v_i are defined by

$$v_1(s_1, (s_2, s_3)) = 370 + 213(s_2 + s_3) + 60s_1 - 230s_1^2,$$
$$v_2(s_2, (s_1, s_3)) = 360 + 233(s_1 + s_3) + 55s_2 - 220s_2^2, \qquad (8.45)$$
$$v_3(s_3, (s_1, s_2)) = 375 + 226(s_1 + s_2) + 50s_3 - 200s_3^2.$$

Assume that $c_1 := 1.10$, $c_2 := 1.2$, $c_3 := 1.25$, $a_i = 1.30$ and $b_i := 2.10$, $i = 1, 2, 3$, that the smallest price shift is five cents, and that profits are counted in euros. Show that the smallest and greatest pure Nash equilibria for prices exist. Calculate these equilibria and the corresponding profits.

Solution. The problem equals to the pricing game $\{S_1, S_2, S_3, u_1, u_2, u_3\}$, where

$$S_i = \{\frac{j}{20}, \ 26 \le j \le 42\} \ \text{and} \ u_i = d_i(s_i, s_{-i})(s_i - c_i), \ i = 1, 2, 3.$$

The demands are of the form $d_i = v_i h_i$, where the functions v_i are as in (8.45), and

$$h_1(s_2, s_3) = \frac{1}{2} + \frac{1}{16}[4\sin(\frac{5s_2 + 6s_3}{22}))],$$

$$h_2(s_1, s_3) = \frac{1}{2} + \frac{1}{16}[4\sin(\frac{6s_1 + 5s_3}{22})],$$

$$h_3(s_1, s_2) = \frac{1}{2} + \frac{1}{16}[4\sin(\frac{7s_1 + 4s_2}{22})].$$

The hypotheses of Proposition 8.61 are valid, whence we conclude that the prices have the smallest and greatest pure Nash equilibria. The components of the greatest pure Nash equilibrium are

$$\bar{s}_1^* = 1.80, \quad \bar{s}_2^* = 1.90, \quad \text{and} \quad \bar{s}_3^* = 2.00,$$

and the corresponding profits are:

$$\bar{u}_1^* = 271, \quad \bar{u}_2^* = 267, \quad \bar{u}_3^* = 263.$$

The components of the smallest pure Nash equilibrium result in

$$\underline{s}_1^* = 1.80, \quad \underline{s}_2^* = 1.90 \quad \text{and} \quad \underline{s}_3^* = 1.95,$$

and for its corresponding profits one gets

$$\underline{u}_1^* = 266, \quad \underline{u}_2^* = 261, \quad \underline{u}_3^* = 263.$$

The profits $\bar{u}_1^*$ and $\bar{u}_2^*$ majorize by the last conclusion of Proposition 8.61 the profits of all other Nash equilibria of prices.

8.3.5 Minimal and Maximal Pure Nash Equilibria

In this subsection we study the existence of minimal and maximal pure Nash equilibria for normal-form games. Recall that a nonempty subset A of a subset Y of a poset P is order compact upward in Y if for every chain C of Y that has a supremum in P the intersection $\bigcap\{[y) \cap A : y \in C\}$ is nonempty whenever $[y) \cap A$ is nonempty for every $y \in C$. If for every chain C of Y that has the infimum in P the intersection of all the sets $(y] \cap A$, $y \in C$ is nonempty whenever $(y] \cap A$ is nonempty for every $y \in C$, then A is order compact downward in Y. If both these properties hold, then A is order compact in Y. If $Y = A$, the phrase 'in Y' is omitted.

Lemma 8.65. *Let A be a nonempty subset of a poset P.*

(a) If A is order compact upward, and if chains of A have supremums in P, then A has a maximal element. If A is also directed upward, then $\max A$ exists.

(b) If A is order compact downward, and if chains of A have infimums in P, then A has a minimal element. If A is also directed downward, then $\min A$ exists.

Proof: Ad (a) Let C be a nonempty chain in A. $\sup C$ exists in P by a hypothesis. Every element y of C is in $[y) \cap A$. Since A is order compact upward, then $\bigcap\{[y) \cap A : y \in C\}$ contains at least one element x. Thus $x \in A$ and $y \leq x$ for every $y \in C$, so that x is an upper bound of C in A. This holds for every nonempty chain C of A, whence A has a maximal element z by Zorn's Lemma. If A is directed upward, then $z = \max A$.

Ad (b) The proof of (b) is similar.   □

Given posets X and P, recall that $\mathcal{F} : X \to 2^P \setminus \emptyset$ is increasing upward if $x \le y$ in X and $z \in \mathcal{F}(x)$ imply that $[z) \cap \mathcal{F}(y)$ is nonempty. $\mathcal{F}$ is increasing downward if $x \le y$ in X and $w \in \mathcal{F}(y)$ imply that $(w] \cap \mathcal{F}(x)$ is nonempty. If $\mathcal{F}$ is increasing upward and downward, then $\mathcal{F}$ is increasing.

In the next study, we apply the following fixed point theorem for set-valued functions, which is a consequence of Propositions 2.8 and 2.9 and Theorem 2.12.

Theorem 8.66. *Given a poset P, and assume that the values of $\mathcal{F} : P \to 2^P \setminus \emptyset$ are order compact in $\mathcal{F}[P]$.*

(a) If $\mathcal{F}$ is increasing upward, if the set $S_+ = \{x \in P : [x) \cap \mathcal{F}(x) \ne \emptyset\}$ is nonempty, and if well-ordered chains of $\mathcal{F}[S_+]$ have supremums in P, then $\mathcal{F}$ has a maximal fixed point, which is also a maximal element of S_+.
(b) If $\mathcal{F}$ is increasing downward, if the set $S_- = \{x \in P : (x] \cap \mathcal{F}(x) \ne \emptyset\}$ is nonempty, and if inversely well-ordered chains of $\mathcal{F}[S_-]$ have infimums in P, then $\mathcal{F}$ has a minimal fixed point, which is also a minimal element of S_-.
(c) If $\mathcal{F}$ is increasing, if chains of $\mathcal{F}[P]$ have supremums and infimums in P, and if the set of these supremums and infimums has a sup-center or an inf-center in P, then $\mathcal{F}$ has minimal and maximal fixed points.

Let $\Gamma = \{S_i, u_i\}_{i=1}^N$ be a normal-form game. Assume that every strategy set S_i is a nonempty subset of a poset $X_i = (X_i, \le_i)$, and that every u_i is a mapping from $S_1 \times \cdots \times S_N$ to a poset $Y_i = (Y_i, \preceq_i)$. Denote $P = S_1 \times \cdots \times S_N$ and $S_{-i} = S_1 \cdots \times S_{i-1} \times S_{i+1} \times \cdots \times S_N$, and assume that all these sets are ordered componentwise. If each mapping $u_i(\cdot, s_{-i})$, $s_{-i} \in S_{-i}$, $i = 1, \dots, N$, has the maximum value, one can define a mapping $\mathcal{F} : P \to 2^P \setminus \emptyset$ by

$$\mathcal{F}(s) := F_1(s_{-1}) \times \cdots \times F_N(s_{-N}), \ s = (s_1, \dots, s_N) \in P, \qquad (8.46)$$

where

$$F_i(s_{-i}) := \{t_i \in S_i : u_i(t_i, s_{-i}) = \max_{s_i \in S_i} u_i(s_i, s_{-i})\}, \ 1 \le i \le N.$$

The components of $s^* = (s_1^*, \dots, s_N^*)$ form a pure Nash equilibrium for Γ if and only if $s^* \in \mathcal{F}(s^*)$, i.e., s^* is a fixed point of $\mathcal{F}$.

As an application of Theorem 8.66 we obtain the following result.

Theorem 8.67. *Let $\Gamma = \{S_i, u_i\}_{i=1}^N$ be a normal-form game with the following property.*

(H0) For every $i = 1, \dots, N$ the sets $F_i(s_{-i})$, $s_{-i} \in S_{-i}$, are nonempty and order compact in $F_i[S_{-i}]$, and chains of $F_i[S_{-i}]$ have supremums and infimums in S_i.

Then the following assertions hold:

(a) *If for every $i = 1, \ldots, N$ the mapping $s_{-i} \mapsto F_i(s_{-i})$ is increasing upward, and if the sets $F_i[S_{-i}]$ have lower bounds in S_i, then Γ has a maximal pure Nash equilibrium.*

(b) *If for every $i = 1, \ldots, N$ the mapping $s_{-i} \mapsto F_i(s_{-i})$ is increasing downward, and if the sets $F_i[S_{-i}]$ have upper bounds in S_i, then Γ has a minimal pure Nash equilibrium.*

(c) *For every $i = 1, \ldots, N$ the mapping $s_{-i} \mapsto F_i(s_{-i})$ is increasing, and if every S_i has a sup-center (respectively an inf-center), then Γ has minimal and maximal pure Nash equilibria.*

Proof: Ad (a) We are going to show that the mapping $\mathcal{F} : P \to 2^P \setminus \emptyset$ defined by (8.46) satisfies the hypotheses of Theorem 8.66(a). Assume that $s = (s_1, \ldots, s_N) \le \bar{s} = (\bar{s}_1, \ldots, \bar{s}_N)$ in P, and let $y = (y_1, \ldots, y_N)$ be chosen from $F(s)$. Given $i \in \{1, \ldots, N\}$, we have $y_i \in F_i(s_{-i})$, and $s_{-i} \le \bar{s}_{-i}$ in S_{-i}. Since $s_{-i} \mapsto F_i(s_{-i})$ is increasing upward by hypothesis, there exists a $\bar{y}_i \in F_i(\bar{s}_{-i})$ such that $y_i \le \bar{y}_i$ in S_i. This holds for every $i = 1, \ldots, N$, whence $\bar{y} = (\bar{y}_1, \ldots, \bar{y}_N) \in \mathcal{F}(\bar{s})$, and $y \le \bar{y}$ in P. This proves that $\mathcal{F}$ is increasing upward. Because of the componentwise orderings, the hypotheses imposed on F_i, and the definition (8.46) of $\mathcal{F}$, it follows from (H0) that the values of $\mathcal{F}$ are order compact upward in $\mathcal{F}[P]$, and that chains of $\mathcal{F}[P]$ have supremums in P. Since every $F_i[S_{-i}]$ has a lower bound $s_i \in S_i$, then $[s) \cap \mathcal{F}[P] \ne \emptyset$ when $s = (s_1, \ldots, s_N)$. The above proof shows that $\mathcal{F}$ satisfies the hypotheses of Theorem 8.66(a), whence $\mathcal{F}$ has a maximal fixed point. Its components form a maximal pure Nash equilibrium for Γ.

Ad (b) If the hypotheses of (b) hold, one can show by dual reasoning that $\mathcal{F}$, defined by (8.46), satisfies the hypotheses of Theorem 8.66(b). Thus $\mathcal{F}$ has a minimal fixed point. Its components form a minimal pure Nash equilibrium for Γ.

Ad (c) If every S_i has a sup-center (respectively an inf-center) c_i, then $c = (c_1, \ldots, c_N)$ is a sup-center (respectively an inf-center) of P. This result and the above proof show that $\mathcal{F}$ satisfies the hypotheses of Theorem 8.66(c), whence $\mathcal{F}$ has a minimal and maximal fixed points. Their components form minimal and maximal pure Nash equilibria for Γ. $\square$

To obtain more concrete results, we impose the following hypotheses on the strategy spaces S_i and on the utilities u_i of a normal-form game $\Gamma = \{S_i, u_i\}_{i=1}^N$.

(S) Every strategy space S_i is a compact ordered topological space.

(u) For every $s_{-i} \in S_{-i}$, $u_i(\cdot, s_{-i})$ is a directedly upper closed function from S_i to a poset $Y_i = (Y_i, \preceq_i)$.

(h1) If $y < \hat{y}$ in S_{-i}, and $x \not\succ_i \hat{x}$ in S_i, then $u_i(x, y) \preceq_i u_i(\hat{x}, y)$ implies $u_i(x, \hat{y}) \preceq_i u_i(\hat{x}, \hat{y})$.

(h2) If $y < \hat{y}$ in S_{-i}, and $x \not\prec_i \hat{x}$ in S_i, then $u_i(x, \hat{y}) \preceq_i u_i(\hat{x}, \hat{y})$ implies $u_i(x, y) \preceq_i u_i(\hat{x}, y)$.

As a consequence of Theorem 8.67(a) and Theorem 8.37 we obtain the following result.

Theorem 8.68. *Let* $\Gamma = \{S_i, u_i\}_{i=1}^N$ *be a normal-form game.*

(a) *If (S), (u), and (h1) hold, and if every* S_i *is directed downward, then* Γ *has a maximal pure Nash equilibrium.*

(b) *If (S), (u), and (h2) hold, and if every* S_i *is directed upward, then* Γ *has a minimal pure Nash equilibrium.*

(c) *If (S), (u), (h1), and (h2) hold, and if every* S_i *has a sup-center or every* S_i *has an inf-center, then* Γ *has minimal and maximal pure Nash equilibria.*

Proof: Ad (a) Assume that (S), (u), and (h1) hold. Let $i \in \{1, \ldots, N\}$ and $s_{-i} \in S_{-i}$ be fixed. The hypotheses (S) and (u) imply by Theorem 8.37 that the set $F_i(s_{-i})$ of all the maximum points of $u_i(\cdot, s_{-i})$ is nonempty and compact, and hence also order compact in $F_i[S_{-i}]$. Moreover, it follows from (S) by [133, Proposition 1.1.4] and its dual that every chain of $F_i[S_{-i}]$ has supremums and infimums in S_i. Thus the hypothesis (H0) of Theorem 8.67 holds. To show that the set-valued mapping $s_{-i} \mapsto F_i(s_{-i})$ is increasing upward, assume that $s_{-i} < \hat{s}_{-i}$ in S_{-i}. If $\hat{s}_i \in F_i(s_{-i})$, then $\hat{s}_i$ maximizes $u_i(\cdot, s_{-i})$. Thus

$$u_i(s_i, s_{-i}) \preceq_i u_i(\hat{s}_i, s_{-i}) \quad \text{whenever } s_i \not\succ_i \hat{s}_i.$$

This result implies by hypothesis (h1) that

$$u_i(s_i, \hat{s}_{-i}) \preceq_i u_i(\hat{s}_i, \hat{s}_{-i}) \quad \text{whenever } s_i \not\succ_i \hat{s}_i. \tag{8.47}$$

If $\hat{s}_i$ maximizes $u_i(\cdot, \hat{s}_{-i})$, then $\hat{s}_i \in F_i(\hat{s}_{-i})$. Otherwise, it follows from (8.47) that $\hat{s}_i <_i \bar{s}_i$ for every $\bar{s}_i \in F_i(\hat{s}_{-i})$. Consequently, there is $\bar{s}_i \in F_i(\hat{s}_{-i})$ such that $\hat{s}_i \leq_i \bar{s}_i$. This proves that $s_{-i} \mapsto F_i(s_{-i})$ is increasing upward. Since the chains of S_i have lower bounds by (S), then S has a minimal element. If S_i is directed downward, then S_i has the smallest element s_i, which is a lower bound of $F_i[S_{-i}]$. The above proof shows that if the hypotheses of (a) hold, then the hypotheses of Theorem 8.67(a) are valid, whence Γ has a maximal pure Nash equilibrium.

Ad (b) Applying hypothesis (h2) one can show similarly that $s_{-i} \mapsto F_i(s_{-i})$ is increasing downward. Nonempty chains of S_i have upper bounds by (S), whence S_i has a maximal element s_i. It is the greatest element if S_i is directed upward, in which case s_i is an upper bound of $F_i[S_{-i}]$. Thus the hypotheses of (b) imply that the hypotheses of Theorem 8.67(b) are valid, whence Γ has a minimal pure Nash equilibrium.

Ad (c) Assume that the hypotheses (S), (u), (h1), and (h2) hold. The above proofs show that the hypothesis (H0) of Theorem 8.67 holds, and that every mapping $s_{-i} \mapsto F_i(s_{-i})$ is increasing. Hence, if every S_i has a sup-center or every S_i has an inf-center, then the hypotheses of Theorem 8.67(c) are valid, whence Γ has minimal and maximal pure Nash equilibria.   □

If all strictly increasing sequences of the values of every $u_i(\cdot, s_{-i})$ and all strictly monotone sequences of every S_i are finite, we can drop from the hypotheses (S) and (u) all the conditions that refer to topology or to order convergence. This leads, in particular, to the following proposition.

Proposition 8.69. *Assume that for all $i = 1, \ldots, N$ and $s_{-i} \in S_{-i}$ the range of $u_i(\cdot, s_{-i})$ is directed upward and its strictly increasing sequences are finite, that strictly monotone sequences of every S_i are finite, and that the hypotheses (h1) and (h2) of Theorem 8.68 hold. If every S_i has a sup-center or every S_i has an inf-center, then Γ has minimal and maximal pure Nash equilibria.*

Remark 8.70. (i) No lattice properties are imposed on the strategy sets S_i. However, if S_i is a lattice, then each point of S_i is both a sup-center and an inf-center of S_i. If each $F_i(s_{-i})$ is a lattice or directed, then maximal and minimal pure Nash equilibria for Γ are its smallest and greatest pure Nash equilibria.

(ii) In the above considerations the ranges of utility functions are posets, which generalizes the usual assumption that the utility functions are real-valued.

(iii) Results corresponding to those derived above can be obtained also for games of more general types, for instance, for those considered in [135, 222].

Example 8.71. Assume that for every $i = 1, \ldots, N$ the strategy spaces S_i are closed and bounded balls in lattice-ordered reflexive Banach spaces X_i equipped with weak topologies. Given ordered vector spaces Y_i, $i = 1, \ldots, N$, assume that the utilities are of the form

$$u_i(s_i, s_{-i}) = f_i(s_i)g_i(s_{-i}) + h_i(s_{-i}), \quad s_i \in S_i, \ s_{-i} \in S_{-i}, \qquad (8.48)$$

where $f_i : S_i \to \mathbb{R}_+$ is upper semicontinuous, $g_i, h_i : S_{-i} \to Y_i$, and $0 \prec_i g_i(s_{-i})$ for all $s_{-i} \in S_{-i}$. The hypotheses imposed on S_i imply that every S_i is compact. Every f_i is directedly upper closed by Corollary 8.34. Moreover, $0 \prec_i g_i(s_{-i})$ for all $s_{-i} \in S_{-i}$. From (8.48) then it follows that every $u_i(\cdot, s_{-i})$ satisfies the hypothesis (u). Since $u_i(\cdot, s_{-i})$ has same maximum points for every s_{-i}, the hypotheses (h1) and (h2) are valid.

Assume moreover that the spaces X_i have the property

(I0) $\| \sup\{0, x_i\} \| \leq \|x_i\|$ for all $x_i \in X_i$, $i = 1, \ldots, N$.

The geometrical center of every S_i is both a sup-center and an inf-center of S_i. Therefore, from Theorem 8.68 it follows that $\Gamma = \{S_i, u_i\}_{i=1}^N$ has minimal and maximal pure Nash equilibria.

Remark 8.72. (i) In Example 8.71 the balls S_i can be replaced by the following nonconvex sets:

$$S_i = \{(x_1, \ldots, x_{m_i}) \in \mathbb{R}^{m_i} : \sum_{j=1}^{m_i} |x_i - c_{ij}|^{p_i} \le r_i^{p_i}\},$$

where $c_{ij} \in \mathbb{R}$, $p_i \in (0,1)$ and $r_i > 0$, and the spaces $\mathbb{R}^{m_i}$ are ordered coordinatewise and equipped by any norm.

(ii) All reflexive Banach lattices are lattice-ordered and possess property (I0), which was required in Example 8.71. Each of the following spaces have also these properties when $1 < p < \infty$.

- Function spaces $L^p(\Omega)$, ordered a.e. pointwise, where $(\Omega, \mathcal{A}, \mu)$ is a σ-finite measure space.
- Sobolev spaces $W^{1,p}(\Omega)$, and $W_0^{1,p}(\Omega)$, ordered a.e. pointwise, where Ω is a domain in $\mathbb{R}^N$.
- Sequence spaces l^p, ordered coordinatewise and normed by the p-norm.
- $\mathbb{R}^N$, ordered coordinatewise and normed by the p-norm.

8.4 Pure and Mixed Nash Equilibria of Normal-Form Games

Here we extend the results of Sect. 8.2 to the normal-form game $\Gamma = \{S_i, u_i\}_{i=1}^N$ of N players whose strategy spaces S_i are compact sublattices of ordered Polish spaces, and the utility functions u_i defined on $S = S_1 \times \cdots \times S_N$ are vector-valued.

We first prove that if Γ has a quasisupermodular mixed extension, then among all possible Nash equilibria of Γ formed by mixed strategies, i.e., probability measures on S_i ordered by first order stochastic dominance, there exist the smallest and greatest Nash equilibrium. Moreover, their strategies are pure, that is, they are indicator functions of singletons $\{\underline{s}_1\}, \ldots, \{\underline{s}_N\}$ and $\{\overline{s}_1\}, \ldots, \{\overline{s}_N\}$. Conditions are provided under which one of the utilities $u_i(\underline{s}_1, \ldots, \underline{s}_N)$ and $u_i(\overline{s}_1, \ldots, \overline{s}_N)$ majorize the expected utilities at each mixed Nash equilibrium of Γ. In addition, we prove monotone comparative statics results for normal-form games whose utility functions depend on a parameter.

Next we present special cases of the above results. First we consider the case where the utility functions $u_i(s_i, s_{-i})$ are real-valued, supermodular in s_i, and have increasing differences in (s_i, s_{-i}). We show that the smallest and greatest Nash equilibria of such a game exist, and that they are pure. The dependence of Nash equilibria on a parameter is studied as well, and the above stated comparison results for expected utilities are shown in case that each $u_i(s_i, s_{-i})$ is increasing or decreasing in s_{-i}.

The results obtained here and in Sect. 8.2 justify the conclusion that dealing with extremal Nash equilibria and their utilities, there exist general classes of normal-form games with strategic complementarities for which randomization of strategies does not give any benefit. The main difficulty in the proofs of these results is that the space of probability measures on S_i ordered by first order stochastic dominance is not a lattice if S_i is not a chain. As stated in [86]: "This implies that we lack the mathematical structure needed for the theory of complementarities." The results of Theorems 2.20 and 2.21 and their duals provide tools to overcome this difficulty. These results present conditions under which fixed points of set-valued functions can be bounded by smallest and greatest fixed points of a single-valued increasing function.

8.4.1 Definitions and Auxiliary Results

We say that an ordered metric space $X = (X, d, \leq)$ is an **ordered Polish space** if X is complete and separable, and if the partial ordering $\leq$ is **closed** in the sense that $d(x_n, x) \to 0$, $d(y_n, y) \to 0$ and $x_n \leq y_n$ for each n imply $x \leq y$. For instance, nonempty and closed subsets of separable ordered Banach spaces are ordered Polish spaces.

Let $\Gamma = \{S_i, u_i\}_{i=1}^N$ be a normal-form game of N players whose strategy spaces S_i are nonempty closed subsets of ordered Polish spaces X_i, and the utility functions u_i defined on $S = S_1 \times \cdots \times S_N$ have values in ordered Banach spaces E_i. Let $\mathcal{B}_i$ denote the family of all Borel sets of S_i. According to [149, Theorem 2], the **first order stochastic dominance** $\preceq_i$, defined by

(SD) $\sigma_i \preceq_i \hat{\sigma}_i$ if and only if $\sigma_i(A) \leq \hat{\sigma}_i(A)$ for each $A \in \mathcal{B}_i$ which is increasing,
       i.e., $[x) \subset A$ whenever $x \in A$,

is a partial ordering on the space Σ_i of probability measures on S_i, i.e., on the space of all countably additive functions $\sigma_i : \mathcal{B}_i \to [0, 1]$ for which $\sigma_i(S_i) = 1$. Given $\sigma_i \in \Sigma_i$, $i = 1, \ldots, N$, denote by σ^{-i} the product measure of σ_j, $j \neq i$ on S_{-i}, $i = i, \ldots, N$ (recall notations $x_{-i} = (x_1, \ldots, x_{i-1}, x_{i+1}, \ldots, x_N)$ and $(x_1, \ldots, x_N) = (x_i, x_{-i})$ for N-tuples and $Y_{-i} = Y_1 \times \cdots \times Y_{i-1} \times Y_{i+1} \times \cdots \times Y_N$ for products of sets).

As for the integration theory needed in the sequel, see, e.g., Chap. 9 and [164, Chapter VI].

Definition 8.73. *We say that $\sigma_i \in \Sigma_i$, $i = 1, \ldots, N$ form a **mixed strategy** for a normal-form game $\Gamma = \{S_i, u_i\}_{i=1}^N$ if for each $i = 1, \ldots, N$ and for all $s_i \in S_i$ the integral*

$$U_i(s_i, \sigma_{-i}) := \int_{S_{-i}} u_i(s_i, s_{-i}) d\sigma^{-i}(s_{-i}) \tag{8.49}$$

exists, and if $s_i \mapsto U_i(s_i, \sigma_{-i})$ is σ_i-integrable on S_i. The integral

$$\mathcal{U}_i(\sigma_1, \sigma_{-i}) := \int_{S_i} \left(\int_{S_{-i}} u_i(s_i, s_{-i}) d\sigma^{-i}(s_{-i}) \right) d\sigma_i(s_i) \tag{8.50}$$

*is called the **expected utility** of player i. Γ is said to admit a **mixed extension** if all probability measures σ_i over S_i, $i = 1, \ldots, N$, form a mixed strategy for Γ.*

Strategies of the form $\delta_{s_i}(x_i) = \begin{cases} 1, & x_i = s_i \\ 0, & x_i \neq s_i \end{cases}$, $s_i \in S_i$, are called **pure strategies** for player i. The set of pure strategies for i is denoted by P_i.

Definition 8.74. *We say that mixed strategies* $\sigma_1, \ldots, \sigma_N$ *form a* **Nash equilibrium** *for* Γ *if for each* $i = 1, \ldots, N$, σ_i *belongs to the set*

$$\mathcal{F}(\sigma_{-i}) := \{\sigma_i \in \Sigma_i : \mathcal{U}_i(\sigma_i, \sigma_{-i}) = \max_{\tau_i \in \Sigma_i} \mathcal{U}_i(\tau_i, \sigma_{-i})\}. \tag{8.51}$$

A Nash equilibrium for Γ *is called* **pure** *if all its strategies are pure.*

Definition 8.75. *A normal-form game* $\Gamma = \{S_i, u_i\}_{i=1}^N$ *is said to have a* **quasisupermodular mixed extension** *if* Γ *admits a mixed extension, and if for every* $i = 1, \ldots, N$,

(I) S_i *is a compact sublattice of an ordered Polish space* X_i, *and* u_i *has values in an ordered Banach space* E_i.

(II) $U_i(\cdot, \sigma_{-i})$ *is for each* $\sigma_{-i} \in \Sigma_{-i}$ *directedly upper closed and* **quasisupermodular**, *i.e., for all* $x, \hat{x} \in S_i$, $U_i(x \wedge \hat{x}, \sigma_{-i}) \leq_i U_i(x, \sigma_{-i})$ *implies* $U_i(\hat{x}, \sigma_{-i}) \leq_i U_i(x \vee \hat{x}, \sigma_{-i})$, *and* $U_i(x \wedge \hat{x}, \sigma_{-i}) <_i U_i(x, \sigma_{-i})$ *implies* $U_i(\hat{x}, \sigma_{-i}) <_i U_i(x \vee \hat{x}, \sigma_{-i})$.

(III) $U_i(x, \sigma_{-i})$ *has a* **partial single crossing property** *in* (x, σ_{-i}), *i.e., if* $x <_i y$ *in* S_i, $\sigma_{-i} < \hat{\sigma}_{-i}$ *in* Σ_{-i}, *and if* σ_{-i} *or* $\hat{\sigma}_{-i}$ *belongs to* P_{-i}, *then* $U_i(x, \sigma_{-i}) \leq_i U_i(y, \sigma_{-i})$ *implies that* $U_i(x, \hat{\sigma}_{-i}) \leq_i U_i(y, \hat{\sigma}_{-i})$, *and* $U_i(x, \sigma_{-i}) <_i U_i(y, \sigma_{-i})$ *implies that* $U_i(x, \hat{\sigma}_{-i}) <_i U_i(y, \hat{\sigma}_{-i})$.

In what follows we assume that the set Σ_i of probability measures on S_i is ordered by the first order stochastic dominance $\preceq_i$, $i = 1, \ldots, N$. We assume also that products of posets are ordered by componentwise ordering.

The following auxiliary result plays an important role in the proof of our main results.

Lemma 8.76. *Let* $\Gamma = \{S_i, u_i\}_{i=1}^N$ *be a normal-form game that has a quasisupermodular mixed extension. Then for each* $i = 1, \ldots, N$, *the sets*

$$F_i(\sigma_{-i}) := \{s_i \in S_i : U_i(s_i, \sigma_{-i}) = \max_{t_i \in S_i} U_i(t_i, \sigma_{-i})\}, \quad \sigma_{-i} \in \Sigma_{-i}, \tag{8.52}$$

are nonempty and compact sublattices of S_i. *Moreover, for every* $\sigma_{-i} \in \Sigma_{-i}$,

$$\mathcal{F}_i(\sigma_{-i}) = \{\tau_i \in \Sigma_i : \tau_i(F_i(\sigma_{-i})) = 1\} \tag{8.53}$$

and

$$\min \mathcal{F}_i(\sigma_{-i}) = \delta_{\min F_i(\sigma_{-i})}, \quad \max \mathcal{F}_i(\sigma_{-i}) = \delta_{\max F_i(\sigma_{-i})}. \tag{8.54}$$

Proof: Let $i \in \{1, \ldots, N\}$ and $\sigma_{-i} \in \Sigma_{-i}$ be fixed. In view of Theorem 8.37, conditions (I) and (II) imply that $F_i(\sigma_{-i})$ is nonempty and compact. To prove that $F_i(\sigma_{-i})$ is a sublattice of S_i, let x and $\hat{x}$ be in $F_i(\sigma_{-i})$. Then $U_i(x \wedge \hat{x}, \sigma_{-i}) \leq_i U_i(x, \sigma_{-i})$. If $U_i(x \wedge \hat{x}, \sigma_{-i}) <_i U_i(x, \sigma_{-i})$, then $U_i(\hat{x}, \sigma_{-i}) <_i U_i(x \vee \hat{x}, \sigma_{-i})$ because $U_i(\cdot, \sigma_{-i})$ is quasisupermodular. But this is impossible, since $\hat{x}$ is in $F_i(\sigma_{-i})$. Thus $U_i(x \wedge \hat{x}, \sigma_{-i}) = U_i(x, \sigma_{-i})$, so that $x \wedge \hat{x}$ is in $F_i(\sigma_{-i})$. Because $U_i(x \wedge \hat{x}, \sigma_{-i}) = U_i(x, \sigma_{-i})$ and $U_i(\cdot, \sigma_{-i})$ is quasisupermodular, then $U_i(\hat{x}, \sigma_{-i}) \leq_i U_i(x \vee \hat{x}, \sigma_{-i})$, whence $x \vee \hat{x}$ is in $F_i(\sigma_{-i})$. Thus $F_i(\sigma_{-i})$ is a sublattice of S_i. From (8.49) and (8.50) it follows that

$$\mathcal{U}_i(\tau_i, \sigma_{-i}) = \int_{S_i} U_i(s_i, \sigma_{-i}) d\tau_i(s_i). \tag{8.55}$$

Denoting $c_i = \max_{s_i \in S_i} U_i(s_i, \sigma_{-i})$, we obtain

$$\mathcal{U}_i(\tau_i, \sigma_{-i}) = c_i \cdot \tau_i(F_i(\sigma_{-i})) + \int_{S_i \setminus F_i(\sigma_{-i})} U_i(s_i, \sigma_{-i}) d\tau_i(s_i),$$

which by the definition of c_i and Lemma 9.4 yields $\mathcal{U}_i(\tau_i, \sigma_{-i}) \leq_i c_i$. Equality holds if $\tau_i(F_i(\sigma_{-i})) = 1$. Thus $c_i = \max_{\tau_i \in \Sigma_i} \mathcal{U}_i(\tau_i, \sigma_{-i})$, and if $\tau_i(F_i(\sigma_{-i})) = 1$, then $\tau_i \in \mathcal{F}_i(\sigma_{-i})$. Conversely, if $\tau_i \in \mathcal{F}_i(\sigma_{-i})$, then

$$0 = c_i - \mathcal{U}_i(\tau_i, \sigma_{-i}) = \int_{S_i} (c_i - U_i(s_i, \sigma_{-i})) d\tau_i(s_i).$$

This result, the definition of c_i, and [133, Proposition 1.4.3] imply that if A is a Borel measurable subset of S_i, then

$$0 \leq_i \int_A (c_i - U_i(s_i, \sigma_{-i})) d\tau_i(s_i) \leq_i \int_{S_i} (c_i - U_i(s_i, \sigma_{-i})) d\tau_i(s_i) = 0.$$

In view of this result and [164, Chapter VI, Corollary 5.16], we have $U_i(\cdot, \sigma_{-i}) = c_i$, τ_i-a.e. on S_i, which proves that $\tau_i(F_i(\sigma_{-i})) = 1$.

The above proof shows that if $\tau_i \in \Sigma_i$, then $\tau_i \in \mathcal{F}_i(\sigma_{-i})$ if and only if $\tau_i(F_i(\sigma_{-i})) = 1$. Thus (8.53) holds. Because $F_i(\sigma_{-i})$ is a compact sublattice of S_i, then $\min F_i(\sigma_{-i})$ and $\max F_i(\sigma_{-i})$ exist. According to the definition (SD) we then have

$$\delta_{\min F_i(\sigma_{-i})} = \min\{\tau_i \in \Sigma_i : \tau_i(F_i(\sigma_{-i})) = 1\} \quad \text{and}$$
$$\delta_{\max F_i(\sigma_{-i})} = \max\{\tau_i \in \Sigma_i : \tau_i(F_i(\sigma_{-i})) = 1\}.$$

This result and (8.53) imply that (8.54) holds true.   □

Let $\Gamma = \{S_i, u_i\}_{i=1}^N$ be a normal-form game that admits a mixed extension. Denote $\Sigma = \Sigma_1 \times \cdots \times \Sigma_N$, and define a mapping $\mathcal{F} : \Sigma \to 2^\Sigma$ by

$$\mathcal{F}(p) := (\mathcal{F}_1(\sigma_{-1}), \ldots, \mathcal{F}_N(\sigma_{-N})), \quad p = (\sigma_1, \ldots, \sigma_N) \in \Sigma. \tag{8.56}$$

An element $p^* = (\sigma_1^*, \ldots, \sigma_N^*)$ of Σ is a fixed point of $\mathcal{F}$, i.e., $p^* \in \mathcal{F}(p^*) = (\mathcal{F}_1(\sigma_{-1}^*), \ldots, \mathcal{F}_N(\sigma_{-N}^*))$ if and only if σ_i^* belongs to the set

$$\mathcal{F}_i(\sigma_{-i}^*) = \{\sigma_i \in \Sigma_i : \mathcal{U}_i(\sigma_i, \sigma_{-i}^*) = \max_{\tau_i \in \Sigma_i} \mathcal{U}_i(\tau_i, \sigma_{-i}^*)\}$$

for every $i = 1, \ldots, N$. As a consequence of this result and Definition 8.74 we obtain the following result.

Lemma 8.77. *Mixed strategies $\sigma_1^*, \ldots, \sigma_N^*$ form a Nash equilibrium for a normal-form game Γ that admits a mixed extension if and only if $p^* = (\sigma_1^*, \ldots, \sigma_N^*)$ is a fixed point of $\mathcal{F}$ defined by (8.56).*

The following result is a consequence of Theorem 2.21 and its dual.

Theorem 8.78. *Given a nonempty subset P of a poset X, assume that $\mathcal{F} : X \to 2^X$ has the following properties:*

(H0) $\mathcal{F}[X]$ is an order-bounded subset of X.

(H1) If $p \in P$, then both $\max \mathcal{F}(p)$ and $\min \mathcal{F}(p)$ exist in X, they belong to P, $\max \mathcal{F}(p)$ is an upper bound of $\mathcal{F}[X \cap (p]]$, and $\min \mathcal{F}(p)$ is a lower bound of $\mathcal{F}[X \cap [p)]$.

(H2) Every well-ordered chain of the set $\{\min \mathcal{F}(p) : p \in P\}$ has a supremum in P, and every inversely well-ordered chain of the set $\{\max \mathcal{F}(p) : p \in P\}$ has an infimum in P.

Then $\mathcal{F}$ has the smallest and greatest fixed points, and they belong to P.

The next comparison result is a useful tool in monotone comparative statics studies.

Proposition 8.79. *Assume that $\mathcal{F}, \hat{\mathcal{F}} : X \to 2^X$ satisfy the hypotheses of Theorem 8.78.*

(a) If $\max \hat{\mathcal{F}}(p) \leq \max \mathcal{F}(p)$ for all $p \in P$, then the greatest fixed point of $\mathcal{F}$ is an upper bound for all the fixed points of $\hat{\mathcal{F}}$.

(b) If $\min \hat{\mathcal{F}}(p) \leq \min \mathcal{F}(p)$ for all $p \in P$, then the smallest fixed point of $\hat{\mathcal{F}}$ is a lower bound for all the fixed points of $\mathcal{F}$.

Proof: By the proof of Theorem 2.21, the greatest fixed points of $\mathcal{F}$ and $\hat{\mathcal{F}}$ are the greatest fixed points of the mappings $G, \hat{G} : P \to P$ defined by

$$G(p) = \max \mathcal{F}(p), \quad \hat{G}(p) = \max \hat{\mathcal{F}}(p), \quad p \in P.$$

If p^* denotes the greatest fixed point of G and $\hat{p}^*$ the greatest fixed point of $\hat{G}$, then

$$\hat{p}^* = \hat{G}(\hat{p}^*) = \max \hat{\mathcal{F}}(\hat{p}^*) \leq \max \mathcal{F}(\hat{p}^*) = G(\hat{p}^*).$$

This result implies by (8.31) that $\hat{p}^* \leq p^*$, which proves (a), since $\hat{p}^*$ is the greatest fixed point of $\hat{\mathcal{F}}$ and p^* is the greatest fixed point of $\mathcal{F}$.

The proof of (b) is done similarly by applying (8.30) and replacing maximums by minimums. $\square$

8.4.2 Existence and Comparison Results

Now we are ready to prove our main existence and comparison result.

Theorem 8.80. *Let* $\Gamma = \{S_i, u_i\}_{i=1}^N$ *be a normal-form game that has a quasisupermodular mixed extension. Then* Γ *has the smallest and greatest pure Nash equilibrium* $\delta_{\underline{s}_1}, \ldots, \delta_{\underline{s}_N}$ *and* $\delta_{\overline{s}_1}, \ldots, \delta_{\overline{s}_N}$, *respectively. Moreover, if* $\sigma_1^*, \ldots, \sigma_N^*$ *is any mixed Nash equilibrium for* Γ, *then* $\delta_{\underline{s}_i} \preceq_i \sigma_i^* \preceq_i \delta_{\overline{s}_i}$ *for each* $i = 1, \ldots, N$.

Proof: We are going to show that $\mathcal{F}$, defined by (8.56), satisfies the hypotheses (H0)–(H2) of Theorem 8.78. Since $a = (\delta_{a_1}, \ldots, \delta_{a_N})$, $a_i = \min S_i$, $i = 1, \ldots, N$ and $b = (\delta_{b_1}, \ldots, \delta_{b_N})$, $b_i = \max S_i$, $i = 1, \ldots, N$ belong to P and are lower and upper bounds of $\mathcal{F}[\Sigma]$, then the hypothesis (H0) is valid. To prove that (H1) holds, assume that $p = (\sigma_1, \ldots, \sigma_N) \in P = P_1 \times \cdots \times P_N$, $\hat{p} = (\hat{\sigma}_1, \ldots, \hat{\sigma}_N) \in \Sigma$, and that $p \preceq \hat{p}$ in Σ. Then $\sigma_{-i} \leq \hat{\sigma}_{-i}$ in Σ_{-i} for each $i = 1, \ldots, N$. From Lemma 8.76 it follows that $x_i = \min F_i(\sigma_{-i})$ and $y_i = \min F_i(\hat{\sigma}_{-i})$ exist. To prove that $x_i \wedge y_i \in F_i(\sigma_{-i})$, assume that this is not true. Then $\mathcal{U}_i(x_i \wedge y_i, \sigma_{-i}) <_i \mathcal{U}_i(x_i, \sigma_{-i})$ by (8.52), which implies by (II) that $\mathcal{U}_i(y_i, \sigma_{-i}) <_i \mathcal{U}_i(x_i \vee y_i, \sigma_{-i})$. Applying this inequality and (III) we obtain $\mathcal{U}_i(y_i, \hat{\sigma}_{-i}) <_i \mathcal{U}_i(x_i \vee y_i, \hat{\sigma}_{-i})$, which contradicts the choice of y_i. Thus $x_i \wedge y_i \in F_i(\hat{\sigma}_{-i})$, and hence $x_i = x_i \wedge y_i$, by the choice of x_i. This implies that $\min F_i(\sigma_{-i}) = x_i \leq_i y_i = \min F_i(\hat{\sigma}_{-i})$, so that $\min \mathcal{F}_i(\sigma_{-i}) = \delta_{x_i} \preceq_i \delta_{y_i} = \min \mathcal{F}_i(\hat{\sigma}_{-i})$ by (8.54) and (SD). This holds for each $i = 1, \ldots, N$, whence $\min \mathcal{F}(p) \preceq \min \mathcal{F}(\hat{p})$ if $p \preceq \hat{p}$ in Σ and $p \in P$. The latter shows that $\min \mathcal{F}(p)$, which belongs to P by (8.54) and (8.56), is for each $p \in P$ a lower bound of $\cup \{\mathcal{F}(\hat{p}) : \hat{p} \in \Sigma, p \preceq \hat{p}\}$. Assume next that $p \preceq \hat{p}$ in Σ, and that $\hat{p} \in P$. Then $\sigma_{-i} \leq \hat{\sigma}_{-i}$ in Σ_{-i} for each $i = 1, \ldots, N$. From Lemma 8.76 it follows that $x_i = \max F_i(\sigma_{-i})$ and $y_i = \max F_i(\hat{\sigma}_{-i})$ exist, and $\mathcal{U}_i(x_i \wedge y_i, \sigma_{-i}) \leq_i \mathcal{U}_i(x_i, \sigma_{-i})$ by (8.52). This implies by (II) that $\mathcal{U}_i(y_i, \sigma_{-i}) \leq_i \mathcal{U}_i(x_i \vee y_i, \sigma_{-i})$. Applying the last inequality and (III) we obtain $\mathcal{U}_i(y_i, \hat{\sigma}_{-i}) \leq_i \mathcal{U}_i(x_i \vee y_i, \hat{\sigma}_{-i})$. Thus $x_i \vee y_i \in F_i(\hat{\sigma}_{-i})$, and hence $y_i = x_i \vee y_i$, by the choice of y_i. Therefore we have $\max F_i(\sigma_{-i}) = x_i \leq_i y_i = \max F_i(\hat{\sigma}_{-i})$, so that $\max \mathcal{F}_i(\sigma_{-i}) = \delta_{x_i} \preceq_i \delta_{y_i} = \max \mathcal{F}_i(\hat{\sigma}_{-i})$ by (8.54) and (SD). This holds for each $i = 1, \ldots, N$, whence $\max \mathcal{F}(p) \preceq \max \mathcal{F}(\hat{p})$ if $p \preceq \hat{p}$ in Σ and $\hat{p} \in P$. Consequently, $\max \mathcal{F}(\hat{p})$, which belongs to P by (8.54) and (8.56), is for each $\hat{p} \in P$ an upper bound of $\cup \{\mathcal{F}(p) : p \in \Sigma, p \preceq \hat{p}\}$. Thus the hypothesis (H1) of Theorem 8.78 is valid.

If $\sigma_i = \delta_{s_i}$, $i = 1, \ldots, N$, it follows from (8.50) and (8.49) that

$$\mathcal{U}_i(\sigma_i, \sigma_{-i}) = U_i(s_i, \sigma_{-i}) = u_i(s_i, s_{-i}), \quad i = 1, \ldots, N.$$

Moreover, if $\sigma_i = \delta_{s_i}$ and $\hat{\sigma}_i = \delta_{\hat{s}_i}$, $i = 1, \ldots, N$, then $\sigma_i \preceq_i \hat{\sigma}_i$ in P_i if and only if $s_i \leq_i \hat{s}_i$ in S_i. Let $\mathcal{C}$ be a nonempty well-ordered chain in P. The elements of $\mathcal{C}$ are of the form $p = (\delta_{s_1}, \ldots, \delta_{s_N})$, where the elements $(s_1, \ldots, s_N)$ form a well-ordered chain C in S. Since $S = S_1 \times \cdots \times S_N$ is a finite product of compact subsets S_i of ordered metric spaces X_i, then S is a compact subset of the ordered metric space $X = X_1 \times \cdots \times X_N$ equipped with product metric and componentwise ordering. Thus all monotone sequences of C converge in S. This property along with [133, Proposition 1.1.5] prove that $(s_{*1}, \ldots, s_{*N}) = \sup C$ exists in S, whence $\sigma_* = (\delta_{s_{*1}}, \ldots, \delta_{s_{*N}}) = \sup C$ in Σ, and σ_* belongs to P.

Finally, if $\mathcal{D}$ is a nonempty inversely well-ordered chain in P, one can show similarly that $p^* = (\delta_{s_1^*}, \ldots, \delta_{s_N^*}) = \inf \mathcal{D}$ exists in Σ, and p^* belongs to P. This proves that the hypothesis (H2) of Theorem 8.78 is fulfilled. Then by applying Theorem 8.78 we see that $\mathcal{F}$, defined by (8.56), has the smallest fixed point $\underline{p} = (\delta_{\underline{s}_1}, \ldots, \delta_{\underline{s}_N})$ and the greatest fixed point $\overline{p} = (\delta_{\overline{s}_1}, \ldots, \delta_{\overline{s}_N})$, and they belong to P. According to Lemma 8.77 this result means that $\delta_{\underline{s}_1}, \ldots, \delta_{\underline{s}_N}$ is the smallest Nash equilibrium for Γ and $\delta_{\overline{s}_1}, \ldots, \delta_{\overline{s}_N}$ is the greatest Nash equilibrium for Γ in Σ, and both are pure. $\square$

The next result, dealing with monotone comparative statics, provides sufficient conditions for the monotone dependence of extremal pure Nash equilibria on a parameter that belongs to a poset. As a consequence of Proposition 8.79 we obtain the following result.

Proposition 8.81. *Let T be a poset, and assume that $\{\Gamma^t = \{S_i, u_i^t\}_{i=1}^N : t \in T\}$ is a family of normal-form games that have quasisupermodular mixed extensions. Then each Γ^t has the smallest pure Nash equilibrium $\delta_{\underline{s}_1^t}, \ldots, \delta_{\underline{s}_N^t}$ and the greatest pure Nash equilibrium $\delta_{\overline{s}_1^t}, \ldots, \delta_{\overline{s}_N^t}$. Assume, in addition, that each $u_i^t(s_i, s_{-i})$ has the* **single crossing property** *in (s_i, t):*

(IV) If $x_i <_i y_i$, $s_{-i} \in S_{-i}$ and $\hat{t} < t$ in T, then $u_i^{\hat{t}}(x_i, s_{-i}) \leq_i u_i^{\hat{t}}(y_i, s_{-i})$ implies that $u_i^t(x_i, s_{-i}) \leq_i u_i^t(y_i, s_{-i})$, and $u_i^{\hat{t}}(x_i, s_{-i}) <_i u_i^{\hat{t}}(y_i, s_{-i})$ implies that $u_i^t(x_i, s_{-i}) <_i u_i^t(y_i, s_{-i})$.

If $\hat{t} < t$ then $\delta_{\underline{s}_i^{\hat{t}}} \preceq_i \delta_{\underline{s}_i^t}$ and $\delta_{\overline{s}_i^{\hat{t}}} \preceq_i \delta_{\overline{s}_i^t}$ for each $i = 1, \ldots, N$.

Proof: The existence of the smallest and greatest Nash equilibria for each Γ^t follows from Theorem 8.80. Define for each $i = 1, \ldots, N$, for each $t \in T$ and for each $\sigma_{-i} \in \Sigma_{-i}$,

$$U_i^t(s_i, \sigma_{-i}) := \int_{S_{-i}} u_i^t(s_i, s_{-i}) d\sigma^{-i}(s_{-i}), \quad \mathcal{U}_i^t(\tau_i, \sigma_{-i}) = \int_{S_i} U_i^t(s_i, \sigma_{-i}) d\tau_i(s_i).$$
$$(8.57)$$

It follows from Lemma 8.77 that the smallest and greatest Nash equilibria of Γ^t are the components of the smallest and greatest fixed points of the set-valued function $\mathcal{F}^t : \Sigma \to 2^\Sigma$, defined by

$$\mathcal{F}^t(p) := (\mathcal{F}_1^t(\sigma_{-1}), \ldots, \mathcal{F}_N^t(\sigma_{-N})), \quad p = (\sigma_1, \ldots, \sigma_N) \in \Sigma,$$
$$\mathcal{F}_i^t(\sigma_{-i}) := \{\sigma_i \in \Sigma_i : \mathcal{U}_i^t(\sigma_i, \sigma_{-i}) = \max_{\tau_i \in \Sigma_i} \mathcal{U}_i^t(\tau_i, \sigma_{-i})\},$$
$$(8.58)$$

where $\sigma_{-i} \in \Sigma_{-i}$, $i = 1, \ldots, N$. Moreover, these extremal Nash equilibria are pure, i.e., they are of the form $\delta_{\underline{s}_1^t}, \ldots, \delta_{\underline{s}_N^t}$ and $\delta_{\overline{s}_1^t}, \ldots, \delta_{\overline{s}_N^t}$. Given $\hat{t} < t$ and $i \in \{1, \ldots, N\}$ and $\delta_{s_{-i}} \in P_{-i}$, we introduce

$$F_i^t(\delta_{s_{-i}}) = \{s_i \in S_i : U_i^t(s_i, \sigma_{-i}) = \max_{t_i \in S_i} U_i^t(t_i, \sigma_{-i})\},$$

$$F_i^{\hat{t}}(\delta_{s_{-i}}) = \{s_i \in S_i : U_i^{\hat{t}}(s_i, \sigma_{-i}) = \max_{t_i \in S_i} U_i^{\hat{t}}(t_i, \sigma_{-i})\}.$$

If $x_i = \min F_i^{\hat{t}}(\delta_{s_{-i}})$ and $y_i = \min F_i^t(\delta_{s_{-i}})$, then $U_i^{\hat{t}}(x_i \wedge y_i, \delta_{s_{-i}}) \leq_i U_i^{\hat{t}}(x_i, \delta_{s_{-i}})$. Equality must hold, for otherwise $U_i^{\hat{t}}(x_i \wedge y_i, \delta_{s_{-i}}) <_i U_i^{\hat{t}}(x_i, \delta_{s_{-i}})$. By condition (II), this implies that $U_i^{\hat{t}}(y_i, \delta_{s_{-i}}) <_i U_i^{\hat{t}}(x_i \vee y_i, \delta_{s_{-i}})$, or equivalently, $u_i^{\hat{t}}(y_i, s_{-i}) <_i u_i^{\hat{t}}(x_i \vee y_i, s_{-i})$. Applying condition (IV), we then obtain $u_i^t(y_i, s_{-i}) <_i u_i^t(x_i \vee y_i, s_{-i})$, or equivalently, $U_i^t(y_i, \delta_{s_{-i}}) <_i U_i^t(x_i \vee y_i, \delta_{s_{-i}})$. But this contradicts the choice of y_i. Thus $U_i^{\hat{t}}(x_i \wedge y_i, \delta_{s_{-i}}) = U_i^{\hat{t}}(x_i, \delta_{s_{-i}})$, so that $x_i \wedge y_i \in F^{\hat{t}}(\delta_{s_{-i}})$. Since $x_i = \min F_i^{\hat{t}}(\delta_{s_{-i}})$, then $x_i \wedge y_i = x_i$, whence $x_i \leq_i y_i$, i.e., $\min F_i^{\hat{t}}(\delta_{s_{-i}}) \leq_i \min F_i^t(\delta_{s_{-i}})$. This result, (8.54), and (SD) imply that $\min \mathcal{F}_i^{\hat{t}}(\delta_{s_{-i}}) \preceq_i \min \mathcal{F}_i^t(\delta_{s_{-i}})$. The latter holds for all $i = 1, \ldots, N$ and $\delta_{s_{-i}} \in P_{-i}$, whence $\min \mathcal{F}^{\hat{t}}(p) \preceq \min \mathcal{F}^t(p)$ for all $p \in P$. Consequently, the hypothesis of Proposition 8.79(b) holds true when $\mathcal{F} = \mathcal{F}^t$ and $\hat{\mathcal{F}} = \mathcal{F}^{\hat{t}}$. Thus $\underline{p}^{\hat{t}} \preceq \underline{p}^t$, where $\underline{p}^{\hat{t}} = (\delta_{\underline{s}_1^{\hat{t}}}, \ldots, \delta_{\underline{s}_N^{\hat{t}}})$ and $\underline{p}^t = (\delta_{\underline{s}_1^t}, \ldots, \delta_{\underline{s}_N^t})$ are the smallest fixed points of $\hat{\mathcal{F}}$ and $\mathcal{F}$. In other words, if $\hat{t} < t$, then for each $i = 1, \ldots, N$, $\delta_{\underline{s}_i^{\hat{t}}} \preceq_i \delta_{\underline{s}_i^t}$, where $\delta_{\underline{s}_1^t}, \ldots, \delta_{\underline{s}_N^t}$ is the smallest Nash equilibrium Γ^t and $\delta_{\underline{s}_1^{\hat{t}}}, \ldots, \delta_{\underline{s}_N^{\hat{t}}}$ is the smallest Nash equilibrium for $\Gamma^{\hat{t}}$.

To prove the similar comparison result for greatest Nash equilibria of Γ^t, choose $x_i = \max F_i^{\hat{t}}(\delta_{s_{-i}})$ and $y_i = \max F_i^t(\delta_{s_{-i}})$. Then $U_i^{\hat{t}}(x_i \wedge y_i, \delta_{s_{-i}}) \leq_i U_i^{\hat{t}}(x_i, \delta_{s_{-i}})$. This implies by condition (II) that $U_i^{\hat{t}}(y_i, \delta_{s_{-i}}) \leq_i U_i^{\hat{t}}(x_i \vee y_i, \delta_{s_{-i}})$, or equivalently, $u_i^{\hat{t}}(y_i, s_{-i}) \leq_i u_i^{\hat{t}}(x_i \vee y_i, s_{-i})$. Applying condition (IV) we obtain $u_i^t(y_i, s_{-i}) \leq_i u_i^t(x_i \vee y_i, s_{-i})$, or equivalently, $U_i^t(y_i, \delta_{s_{-i}}) \leq_i U_i^t(x_i \vee y_i, \delta_{s_{-i}})$. Equality must hold because of the choice of y_i, whence $x_i \vee y_i \in F^t(\delta_{s_{-i}})$. Since $y_i = \max F_i^t(\delta_{s_{-i}})$, then $x_i \vee y_i = y_i$, whence $x_i \leq_i y_i$, i.e., $\max F_i^{\hat{t}}(\delta_{s_{-i}}) \leq_i \max F_i^t(\delta_{s_{-i}})$. This result allows us to replace min by max in the above reasoning and apply Proposition 8.79(a) to obtain the asserted comparison result for greatest Nash equilibria of Γ^t.   □

Next we derive a comparison result for expected utilities of mixed Nash equilibria. The following hypotheses are used.

(V) $U_i(s_i, \hat{\sigma}_{-i}) \leq_i U_i(s_i, \sigma_{-i})$ if $\sigma_{-i} \leq \hat{\sigma}_{-i}$ in Σ_{-i}, $\sigma_{-i} \in P_{-i}$ and $s_i \in S_i$.
(VI) $U_i(s_i, \sigma_{-i}) \leq_i U_i(s_i, \hat{\sigma}_{-i})$ if $\sigma_{-i} \leq \hat{\sigma}_{-i}$ in Σ_{-i}, $\hat{\sigma}_{-i} \in P_{-i}$ and $s_i \in S_i$.

Proposition 8.82. *Assume that a normal-form game Γ has a quasisupermodular mixed extension. Denote by $\delta_{\underline{s}_1}, \ldots, \delta_{\underline{s}_N}$ and $\delta_{\overline{s}_1}, \ldots, \delta_{\overline{s}_N}$ the smallest and greatest pure Nash equilibria of Γ, respectively, and let $\sigma_1^*, \ldots, \sigma_N^*$ be a mixed Nash equilibrium for Γ.*

(a) If (V) holds for each $i = 1, \ldots, N$, then $\mathcal{U}_i(\sigma_i^*, \sigma_{-i}^*) \leq_i u_i(\underline{s}_1, \ldots, \underline{s}_N)$ for each $i = 1, \ldots, N$.

(b) If (VI) holds for each $i = 1, \ldots, N$, then $\mathcal{U}_i(\sigma_i^*, \sigma_{-i}^*) \leq_i u_i(\overline{s}_1, \ldots, \overline{s}_N)$ for each $i = 1, \ldots, N$.

Proof: Ad (a) Since $\delta_{\underline{s}_{-i}} \preceq \sigma_{-i}^*$ for each $i = 1, \ldots, N$, from (V) and (8.51) and by applying Lemma 9.4 it follows that for all $i = 1, \ldots, N$,

$$\mathcal{U}_i(\sigma_i^*, \sigma_{-i}^*) = \int_{S_i} U_i(s_i, \sigma_{-i}^*) d\sigma_i^*(s_i) \leq_i \int_{S_i} U_i(s_i, \delta_{\underline{s}_{-i}}) d\sigma_i^*(s_i)$$
$$= \mathcal{U}_i(\sigma_i^*, \delta_{\underline{s}_{-i}}) \leq_i \mathcal{U}_i(\delta_{\underline{s}_i}, \delta_{\underline{s}_{-i}}) = u_i(\underline{s}_i, \underline{s}_{-i}).$$

This implies the assertion of (a).

Ad (b) Because $\sigma_{-i}^* \leq \delta_{\overline{s}_{-i}}$ in Σ_{-i} for all $i = 1, \ldots, N$, from (VI) and from the equilibrium condition (8.51) and by Lemma 9.4 it follows that for all $i = 1, \ldots, N$,

$$\mathcal{U}_i(\sigma_i^*, \sigma_{-i}^*) = \int_{S_i} U_i(s_i, \sigma_{-i}^*) d\sigma_i^*(s_i) \leq_i \int_{S_i} U_i(s_i, \delta_{\overline{s}_{-i}}) d\sigma_i^*(s_i)$$
$$= \mathcal{U}_i(\sigma_i^*, \delta_{\overline{s}_{-i}}) \leq_i \mathcal{U}_i(\delta_{\overline{s}_i}, \delta_{\overline{s}_{-i}}) = u_i(\overline{s}_i, \overline{s}_{-i}).$$

This proves the assertion of (b). □

Remark 8.83. (i) The results of this section cannot be obtained by the methods used, e.g., in [86, 180, 217, 218, 222, 223, 232], because the first order stochastic dominance is not a lattice ordering for Σ_i if S_i is not a chain, and since the utility mappings u_i need not be chain-valued.

(ii) If each S_i is only a join sublattice of X_i, i.e., if $x \vee y := \sup\{x, y\}$ exist in X_i and belongs to S_i for all $x, y \in S_i$ then quasisupermodularity condition (II) is not available. The results derived above for greatest Nash equilibria can be obtained when conditions (II) and (III) are replaced by the following conditions:

(II') $U_i(\cdot, \sigma_{-i})$ is for each $\sigma_{-i} \in \Sigma_{-i}$ directedly upper closed, and if $x_i, y_i \in S_i$ and $\sigma_{-i} \in \Sigma_{-i}$, then $U_i(y_i, \sigma_{-i}) \leq_i U_i(x_i, \sigma_{-i})$ implies that $U_i(y_i, \sigma_{-i}) \leq_i U_i(x_i \vee y_i, \sigma_{-i})$.

(III') If $x_i <_i y_i$ in S_i, $\sigma_{-i} < \hat{\sigma}_{-i}$ in Σ_{-i} and $\hat{\sigma}_{-i}$ belongs to P_{-i} then $U_i(x_i, \sigma_{-i}) \leq_i U_i(y_i, \sigma_{-i})$ implies that $U_i(x_i, \hat{\sigma}_{-i}) \leq_i U_i(y_i, \hat{\sigma}_{-i})$.

These conditions correspond to concepts of weak quasisupermodularity and weak single crossing property defined in [197, Section 4.1] for player i's interim payoff function V_i in a Bayesian game. Notice however the restriction $\hat{\sigma}_{-i} \in P_{-i}$ in condition (III').

(iii) Because of the partial monotonicity hypothesis (H1) on $\mathcal{F}$ in Theorem 8.78 and in Proposition 8.79, the partial single crossing property (III) is sufficient.

8.4.3 Applications to Supermodular Games

We apply the results of Sect. 8.4.2 to supermodular normal-form games. As for examples of such games, see, e.g., [218, Section 4.4].

Definition 8.84. *A normal-form game $\Gamma = \{S_i, u_i\}_{i=1}^N$ is called **supermodular**, if for every $i = 1, \ldots, N$ the following conditions are satisfied:*

(1) S_i is a compact sublattice of an ordered Polish space, u_i is real-valued and bounded, $u_i(\cdot, s_{-i})$ is upper semicontinuous in S_i, uniformly over $s_{-i} \in S_{-i}$, and $u_i(s_i, \cdot)$ is Borel measurable in S_{-i} for each $s_i \in S_i$;

*(2) $u_i(\cdot, s_{-i})$ is **supermodular**, i.e., if $x_i, y_i \in S_i$ and $s_{-i} \in S_{-i}$, then*
$$u_i(x_i, s_{-i}) + u_i(y_i, s_{-i}) \le u_i(x_i \wedge y_i, s_{-i}) + u_i(x_i \vee y_i, s_{-i});$$

*(3) u_i has **increasing differences** in (s_i, s_{-i}), i.e., if $x_i <_i y_i$ in S_i and $s_{-i} < \hat{s}_{-i}$ in S_{-i}, then $u_i(y_i, s_{-i}) - u_i(x_i, s_{-i}) \le u_i(y_i, \hat{s}_{-i}) - u_i(x_i, \hat{s}_{-i})$.*

In the proof of the main theorem of this subsection we make use of the following auxiliary result.

Lemma 8.85. *A supermodular game $\Gamma = \{S_i, u_i\}_{i=1}^N$ admits a mixed extension.*

Proof: Let $i \in \{1, \ldots, N\}$ and $(\sigma_1, \ldots, \sigma_N) \in \Sigma$ be fixed. Since $u_i(s_i, \cdot)$ is bounded and Borel measurable in S_{-i} for each $s_i \in S_i$, then (8.49) defines a function $U_i(\cdot, \sigma_{-i})$ on S_i. Because $u_i(\cdot, s_{-i})$ is upper semicontinuous in S_i, uniformly over $s_{-i} \in S_{-i}$, then for any $\epsilon > 0$, each $x \in S_i$ has a neighborhood $V(x)$ in S_i such that $u_i(y, s_{-i}) \le u_i(x, s_{-i}) + \epsilon$ for all $y \in V(x)$ and $s_{-i} \in S_{-i}$. Thus

$$U_i(y, \sigma_{-i}) \le \int_{S_{-i}} (u_i(x, s_{-i}) + \epsilon) d\sigma^{-i}(s_{-i}) = U_i(x, \sigma_{-i}) + \epsilon$$

for all $y \in V(x)$, which proves that $U_i(\cdot, \sigma_{-i})$ is upper semicontinuous, and hence also Borel measurable. This ensures that $U_i(\cdot, \sigma_{-i})$ is σ_i-integrable on S_i. The assertion of the lemma follows now from the above results and from the definition 8.73. $\qquad \square$

The following theorem can be derived from Theorem 8.80 as a special case.

Theorem 8.86. *Let $\Gamma = \{S_i, u_i\}_{i=1}^N$ be a supermodular normal-form game. Then Γ has the smallest pure Nash equilibrium $\delta_{\underline{s}_1}, \ldots, \delta_{\underline{s}_N}$ and the greatest pure Nash equilibrium $\delta_{\bar{s}_1}, \ldots, \delta_{\bar{s}_N}$, and, if $\sigma_1^*, \ldots, \sigma_N^*$ is any mixed Nash equilibrium for Γ, then $\delta_{\underline{s}_i} \preceq \sigma_i^* \preceq_i \delta_{\bar{s}_i}$ for each $i = 1, \ldots, N$.*

Proof: It suffices to show that the conditions (I), (II), and (III) of Definition 8.75 are valid. Condition (I) is a consequence of condition (1) of Definition 8.84. Because the functions $U_i(\cdot, \sigma_{-i})$ are upper semicontinuous by the proof of Lemma 8.85, they are also directedly upper closed due to Corollary 8.34.

To prove quasisupermodularity of $U_i(\cdot, \sigma_{-i})$, assume that x_i, $y_i \in S_i$ and $\sigma_{-i} \in \Sigma_i$. Applying condition (2) and Lemma 9.4 we get

$$U_i(x_i, \sigma_{-i}) - U_i(x_i \wedge y_i, \sigma_{-i}) = \int_{S_{-i}} (u_i(x_i, s_{-i}) - u_i(x_i \wedge y_i, s_{-i})) d\sigma^{-i}(s_{-i})$$

$$\leq \int_{S_{-i}} (u_i(x_i \vee y_i, s_{-i}) - u_i(y_i, s_{-i})) d\sigma^{-i}(s_{-i})$$

$$= U_i(x_i \vee y_i, \sigma_{-i}) - U_i(y_i, \sigma_{-i}).$$

This result implies quasisupermodularity of $U_i(\cdot, \sigma_{-i})$, so that the condition (II) is satisfied.

Next we shall show that condition (III) holds. Given $i \in \{1, \ldots, N\}$, assume first that $\sigma_{-i} \leq \hat{\sigma}_{-i}$ in Σ_{-i} and that

$$\sigma_{-i} = (\delta_{z_1}, \ldots, \delta_{z_{i-1}}, \delta_{z_{i+1}}, \ldots, \delta_{z_N}) \in P_{-i}.$$

If $j \neq i$, then $\delta_{z_j} \preceq_j \hat{\sigma}_j$. The order interval $[z_j)$ is an increasing Borel set that contains z_j. Thus $1 = \delta_{z_j}([z_j)) \leq \hat{\sigma}_j([z_j))$ by the definition (SD). This result holds for each $j \neq i$, whence $\hat{\sigma}_{-i}([z_{-i})) = 1$. Hence, if $x_i <_i y_i$ in S_i, then applying condition (3) and Lemma 9.4 we obtain

$$U_i(y_i, \sigma_{-i}) - U_i(x_i, \sigma_{-i}) = \int_{S_{-i}} (u_i(y_i, s_{-i}) - u_i(x_i, s_{-i})) d\sigma^{-i}(s_{-i})$$

$$= u_i(y_i, z_{-i}) - u_i(x_i, z_{-i}) = \int_{[z_{-i})} (u_i(y_i, z_{-i}) - u_i(x_i, z_{-i})) d\hat{\sigma}^{-i}(s_{-i})$$

$$\leq \int_{[z_{-i})} (u_i(y_i, s_{-i}) - u_i(x_i, s_{-i})) d\hat{\sigma}^{-i}(s_{-i})$$

$$= \int_{S_{-i}} u_i(y_i, s_{-i}) - u_i(x_i, s_{-i})) d\hat{\sigma}^{-i}(s_{-i}) = U_i(y_i, \hat{\sigma}_{-i}) - U_i(x_i, \hat{\sigma}_{-i}).$$

Assume next that $\sigma_{-i} \leq \hat{\sigma}_{-i}$ in Σ_{-i}, and that

$$\hat{\sigma}_{-i} = (\delta_{\hat{s}_1}, \ldots, \delta_{\hat{s}_{i-1}}, \delta_{\hat{s}_{i+1}}, \ldots, \delta_{\hat{s}_N}) \in P_{-i}.$$

If $j \neq i$, then $\sigma_j \preceq_j \delta_{\hat{s}_j}$. Since the set $S_j \setminus (\hat{s}_j]$ is an increasing Borel set, then $\sigma_j(S_j \setminus (\hat{s}_j]) \leq \delta_{\hat{s}_j}(S_j \setminus (\hat{s}_j]) = 0$ by the definition (SD). This holds for each $j \neq i$, whence $\sigma_{-i}((\hat{s}_{-i}]) = 1$. If $x_i <_i y_i$ in S_i, then applying condition (3), the above result, and Lemma 9.4, we obtain

$$U_i(y_i, \sigma_{-i}) - U_i(x_i, \sigma_{-i}) = \int_{S_{-i}} (u_i(y_i, s_{-i}) - u_i(x_i, s_{-i})) d\sigma^{-i}(s_{-i})$$

$$= \int_{(\hat{s}_{-i}]} (u_i(y_i, s_{-i}) - u_i(x_i, s_{-i})) d\sigma^{-i}(s_{-i})$$

$$\leq \int_{(\hat{s}_{-i}]} (u_i(y_i, \hat{s}_{-i}) - u_i(x_i, \hat{s}_{-i})) d\sigma^{-i}(s_{-i}) = u_i(y_i, \hat{s}_{-i}) - u_i(x_i, \hat{s}_{-i})$$

$$= \int_{S_{-i}} (u_i(y_i, s_{-i}) - u_i(x_i, s_{-i})) d\hat{\sigma}_{-i}(s_{-i}) = U_i(y_i, \hat{\sigma}_{-i}) - U_i(x_i, \hat{\sigma}_{-i}).$$

The above proof shows that if $x_i <_i y_i$ in S_i, $\sigma_{-i} \leq \hat{\sigma}_{-i}$ in Σ_{-i}, and σ_{-i} or $\hat{\sigma}_{-i}$ is in P_{-i}, then $U_i(y_i, \sigma_{-i}) - U_i(x_i, \sigma_{-i}) \leq U_i(y_i, \hat{\sigma}_{-i}) - U_i(x_i, \hat{\sigma}_{-i})$. This result implies that condition (III) is valid. Thus all the conditions (I), (II), and (III) of Definition 8.75 hold true. The above proof and Lemma 8.85 imply that Γ is a normal-form game that has a quasisupermodular mixed extension. Thus the assertions follow from Theorem 8.80. $\square$

The proof of Theorem 8.86 shows that if Γ is a supermodular normal-form game, then it has a quasisupermodular mixed extension. Therefore, the following comparison result is a special case of Proposition 8.81.

Proposition 8.87. *Let T be a poset, and let $\{\Gamma^t = \{S_i, u_i^t\}_{i=1}^N : t \in T\}$ be a family of supermodular normal-form games. Then each Γ^t has the smallest pure Nash equilibrium $\delta_{\underline{s}_1^t}, \ldots, \delta_{\underline{s}_N^t}$ and the greatest pure Nash equilibrium $\delta_{\overline{s}_1^t}, \ldots, \delta_{\overline{s}_N^t}$. Moreover, if each $u_i^t(s_i, s_{-i})$ has the single crossing property (IV) in (s_i, t), then $\delta_{\underline{s}_i^{\hat{t}}} \preceq_i \delta_{\underline{s}_i^t}$ and $\delta_{\overline{s}_i^{\hat{t}}} \preceq_i \delta_{\overline{s}_i^t}$ whenever $\hat{t} < t$ and $i \in \{1, \ldots, N\}$.*

As a consequence of Proposition 8.82 we obtain a comparison result for the expected utilities of mixed Nash equilibria of Γ.

Proposition 8.88. *Let $\Gamma = \{S_i, u_i\}_{i=1}^N$ be a supermodular normal-form game, let $\delta_{\underline{s}_1}, \ldots, \delta_{\underline{s}_N}$ and $\delta_{\overline{s}_1}, \ldots, \delta_{\overline{s}_N}$ denote the smallest and greatest Nash equilibria of Γ in Σ, and let $\sigma_1^*, \ldots, \sigma_N^*$ be any mixed Nash equilibrium for Γ. Then we have:*

(a) If $u_i(s_i, \cdot)$ is decreasing in S_{-i} for all $s_i \in S_i$, then $U_i(\sigma_i^*, \sigma_{-i}^*) \leq u_i(\underline{s}_i, \ldots, \underline{s}_N)$.

(b) If $u_i(s_i, \cdot)$ is increasing in S_{-i} for all $s_i \in S_i$, then $U_i(\sigma_i^*, \sigma_{-i}^*) \leq u_i(\overline{s}_1, \ldots, \overline{s}_N)$.

Proof: By the proof of Theorem 8.86, Γ has a quasisupermodular mixed extension. Therefore, in view of Proposition 8.82 it suffices to prove that the hypotheses (V) and (VI) of Proposition 8.82 hold true.

Ad (a) To prove the validity of (V), let $i \in \{1, \ldots, N\}$ and $s_i \in S_i$ be given, and assume that $\sigma_{-i} \leq \hat{\sigma}_{-i}$ in Σ_{-i}, and that $\sigma_{-i} = \delta_{x_{-i}} \in P_{-i}$. As in the proof of Theorem 8.86, one can show that $\hat{\sigma}_{-i}([x_{-i})) = 1$. This result and taking into account Lemma 9.4 as well as the hypothesis that $u_i(s_i, \cdot)$ is decreasing in S_{-i} imply that

$$U_i(s_i, \sigma_{-i}) = \int_{S_{-i}} u_i(s_i, s_{-i}) d\sigma^{-i}(s_{-i}) = u_i(s_i, x_{-i})$$

$$= \int_{[x_{-i})} u_i(s_i, x_{-i}) d\hat{\sigma}^{-i}(s_{-i}) \geq \int_{[x_{-i})} u_i(s_i, s_{-i}) d\hat{\sigma}_{-i}(s_{-i})$$

$$= \int_{S_{-i}} u_i(s_i, s_{-i}) d\hat{\sigma}^{-i}(s_{-i}) = U_i(s_i, \hat{\sigma}_{-i}).$$

Thus condition (V) of Proposition 8.82 is fulfilled.

Ad (b) Next we shall show that condition (VI) of Proposition 8.82 holds true. Assume that $s_i \in S_i$, $\sigma_{-i} \leq \hat{\sigma}_{-i}$ in Σ_{-i} and $\hat{\sigma}_{-i} = \delta_{\hat{s}_{-i}} \in P_{-i}$ for each $i = 1, \ldots, N$. Let $i \in \{1, \ldots, N\}$ be given. One can show as in the proof of Theorem 8.86 that $\sigma_{-i}((\hat{s}_{-i}]) = 1$. Applying this result, the hypothesis that $u_i(s_i, \cdot)$ is increasing in S_{-i}, and Lemma 9.4, we get

$$U_i(s_i, \sigma_{-i}) = \int_{S_{-i}} u_i(s_i, s_{-i}) d\sigma^{-i}(s_{-i}) = \int_{(\hat{s}_{-i}]} u_i(s_i, s_{-i}) d\sigma^{-i}(s_{-i})$$

$$\leq \int_{(\hat{s}_{-i}]} u_i(s_i, \hat{s}_{-i}) d\sigma^{-i}(s_{-i}) = u_i(s_i, \hat{s}_{-i}) = U_i(s_i, \hat{\sigma}_{-i}).$$

This implies that condition (VI) of Proposition 8.82 is fulfilled.   □

The following special case includes also monotone comparative statics results.

Corollary 8.89. *Let T be a poset, and let $\{\Gamma^t = \{S_i, u_i^t\}_{i=1}^N : t \in T\}$ be a family of normal-form games with the following properties:*

(A) Strategy spaces are products $S_i = \times_{k=1}^{m_i} S_{ik}$, $1 \leq m_i \leq \infty$, of compact chains S_{ik} of ordered Polish spaces (X_{ik}, d_{ik}).

(B) Each $u_i^t(s_1, \ldots, s_N)$ is bounded, real-valued, upper semicontinuous in s_i, uniformly over s_j, $j \neq i$, and continuous in each s_j, $j \neq i$ with respect to product topologies of $X_i = \times_{k=1}^{m_i} X_{ik}$.

(C) Each $u_i^t(s_i, s_{-i})$ is supermodular in s_i and has increasing differences in (s_i, s_{-i}) and in (s_i, t) with respect to componentwise orderings of X_i and X_{-i}.

Then the following assertions hold:

(a) If each $u_i^t(s_1, \ldots, s_N)$ is decreasing in each s_j, $j \neq i$, then each Γ^t has the smallest pure Nash equilibrium $\delta_{\underline{s}_1^t}, \ldots, \delta_{\underline{s}_N^t}$, it is a lower bound of all mixed Nash equilibria for $\Gamma^{\hat{t}}$, $\hat{t} \leq t$, and the utilities $u_i^t(\underline{s}_1^t, \ldots, \underline{s}_N^t)$ majorize the expected utilities of all mixed Nash equilibria for Γ^t.

(b) If each $u_i^t(s_1, \ldots, s_N)$ is increasing in each s_j, $j \neq i$, then each Γ^t has the greatest pure Nash equilibrium $\delta_{\bar{s}_1^t}, \ldots, \delta_{\bar{s}_N^t}$, it majorizes all mixed Nash equilibria for $\Gamma^{\hat{t}}$, $\hat{t} \leq t$, and the utilities $u_i^t(\bar{s}_1^t, \ldots, \bar{s}_N^t)$ majorize the expected utilities of all mixed Nash equilibria for Γ^t.

Proof: Since each S_{ik} is compact, it follows from Tychonoff's Theorem ([85, Theorem 2.2.8]) that S_i is a compact subset of the product $X_i = \times_{k=1}^{m_i} X_{ik}$. According to [85, Proposition 2.4.4], the product topology of X_i is metrizable by the metric

$$d_i(s_i, \hat{s}_i) := \sum_{k=1}^{m_i} \frac{d_{ik}(s_{ik}, \hat{s}_{ik})}{2^k(1 + d_{ik}(s_{ik}, \hat{s}_{ik}))}, \quad s_i = (s_{ik})_{k=1}^{m_i}, \hat{s}_i = (\hat{s}_{ik})_{k=1}^{m_i} \in S_i.$$

$$(8.59)$$

As noticed in [85, p. 388], the countable product X_i of Polish spaces X_{ik} is a Polish space with respect to the metric d_i. Moreover, the products S_i of chains S_{ik} of X_{ik}, $k = 1, \ldots, m_i$ are sublattices of X_i with respect to the componentwise ordering, and each u_i^t is bounded and real-valued by (B). Since each $u_i^t(s_1, \ldots, s_N)$ is by (B) upper semicontinuous in s_i, uniformly over s_j, $j \neq i$, and continuous in each s_j, $j \neq i$, then each $u_i^t(s_i, s_{-i})$ is Borel measurable in s_{-i} by [35] and upper semicontinuous in s_i, uniformly over s_{-i}. Because each $u_i^t(s_i, s_{-i})$ is by (C) supermodular in s_i and has increasing differences in (s_i, s_{-i}) and in (s_i, t), then conditions of Definition 8.84 are valid for each $t \in T$. Thus the assertions follow from Theorem 8.86, Propositions 8.87 and 8.88.   □

As a consequence of Theorem 8.86 and Proposition 8.88, we obtain the following corollary.

Corollary 8.90. *Let* $\Gamma = \{S_i, u_i\}_{i=1}^N$ *be a normal-form game in which strategy spaces* S_i *are products of compact subsets* S_{ij} *of* $[c_{ij}, \infty)$, $j = 1, \ldots, m_i$, *ordered coordinatewise, and utilities are of the form*

$$u_i(s_i, s_{-i}) = \sum_{j=1}^{m_j} (f_{ij}(s_{ij}) + g_{ij}(s_{-i}))(s_{ij} - c_{ij}), \quad i = 1, \ldots, N. \tag{8.60}$$

If each $f_{ij} : S_{ij} \to \mathbb{R}_+$ *is bounded from below and upper semicontinuous, and each* $g_{ij} : S_{-i} \to \mathbb{R}_+$ *is increasing and upper semicontinuous, then* Γ *has the smallest pure Nash equilibrium* $\delta_{\underline{s}_1}, \ldots, \delta_{\underline{s}_N}$ *and the greatest pure Nash equilibrium* $\delta_{\overline{s}_1}, \ldots, \delta_{\overline{s}_N}$. *If* $\sigma_1^*, \ldots, \sigma_N^*$ *is any mixed Nash equilibrium for* Γ, *then* $\delta_{\underline{s}_i} \preceq \sigma_i^* \preceq_i \delta_{\overline{s}_i}$ *for each* $i = 1, \ldots, N$. *The utilities* $u_i(\overline{s}_1, \ldots, \overline{s}_N)$ *majorize the expected utilities of all mixed Nash equilibria for* Γ.

Proof: The given hypotheses imply that condition (1) of Definition 8.84 is valid. Since every S_{ij} is a compact chain in $\mathbb{R}$, then every function f_{ij} is supermodular. This implies that condition (2) of Definition 8.84 holds. Assume next that $s_i <_i \hat{s}_i$ in S_i and $s_{-i} < \hat{s}_{-i}$ in S_{-i}. Since every g_{ij} is increasing, then every $u_i(s_i, s_{-i})$ is increasing in s_{-i}, and

$$u_i(\hat{s}_i, \hat{s}_{-i}) - u_i(s_i, \hat{s}_{-i}) - (u_i(\hat{s}_i, s_{-i}) - u_i(s_i, s_{-i}))$$
$$= \sum_{j=1}^{m_j} (g_{ij}(\hat{s}_{-i}) - g_{ij}(s_{-i}))(\hat{s}_{ij} - s_{ij}) \geq 0.$$

Thus every $u_i(s_i, s_{-i})$ has increasing differences in (s_i, s_{-i}). The above proof shows that Γ is a supermodular normal-form game, and that the functions $u_i(s_i, s_{-i})$ are increasing in s_{-i}. Thus the assertions follow from Theorem 8.86 and Proposition 8.88(b).   □

Example 8.91. Consider a two-person normal form game $\Gamma = \{S_1, S_2, u_1, u_2\}$, where the strategy spaces are $S_i = [\frac{3}{2}, \frac{5}{2}] \times [\frac{3}{2}, \frac{5}{2}]$. The utility functions are of the form $u_i((s_{i1}, s_{12}), s_{-i}) = u_{i1}(s_{i1}, s_{-i}) + u_{i2}(s_{i2}, s_{-i})$, where the functions $u_{ij} : [\frac{3}{2}, \frac{5}{2}] \to \mathbb{R}$, $i, j = 1, 2$, are defined as follows:

$$u_{11}(s, s_{-1}) = (52 - 21s + s_{21} + 4s_{22})(s - 1),$$

$$u_{12}(s, s_{-1}) = (51 - 21s + 2s_{21} + 3s_{22})(s - \frac{11}{10}),$$

$$u_{21}(s, s_{-2}) = (50 - 20s + 3s_{11} + 2s_{12})(s - \frac{11}{10}),$$

$$u_{22}(s, s_{-2}) = (49 - 20s + 4s_{11} + s_{12})(s - 1).$$

Show that Γ has exactly one Nash equilibrium. Calculate it and the corresponding utilities.

Solution. The hypotheses of Corollary 8.90 are valid, whence Γ has the smallest and the greatest pure Nash equilibria. Moreover, the functions $u_i(\cdot, s_{-i})$ have unique maximum points $(f_{i1}(s_{-i}), f_{i2}(s_{-i}))$, whose coordinates $f_{ij}(s_{-i})$ are solutions of equations $\frac{d}{ds} u_{ij}(s, s_{-i}) = 0$:

$$
\begin{aligned}
f_{11}(s_{-1}) &= \frac{73}{42} + \frac{1}{42}s_{21} + \frac{2}{21}s_{22}, \\
f_{12}(s_{-1}) &= \frac{247}{140} + \frac{1}{21}s_{21} + \frac{1}{14}s_{22}, \\
f_{21}(s_{-2}) &= \frac{9}{5} + \frac{3}{40}s_{11} + \frac{1}{20}s_{12}, \\
f_{22}(s_{-2}) &= \frac{69}{40} + \frac{1}{10}s_{11} + \frac{1}{40}s_{12}.
\end{aligned}
\tag{8.61}
$$

Equalizing points s_i of S_i and corresponding pure strategies δ_{s_i}, it follows that $((s_{11}, s_{12}), (s_{21}, s_{22}))$ is a pure Nash equilibrium of Γ if and only if $(s_{11}, s_{12}, s_{21}, s_{22})$ is a solution of the following system of linear equations:

$$
\begin{aligned}
s_{11} &= \frac{73}{42} + \frac{1}{42}s_{21} + \frac{2}{21}s_{22}, \\
s_{12} &= \frac{247}{140} + \frac{1}{21}s_{21} + \frac{1}{14}s_{22}, \\
s_{21} &= \frac{9}{5} + \frac{3}{40}s_{11} + \frac{1}{20}s_{12}, \\
s_{22} &= \frac{69}{40} + \frac{1}{10}s_{11} + \frac{1}{40}s_{12}.
\end{aligned}
\tag{8.62}
$$

The obtained system has the unique solution:

$$s_{11}^* = \frac{54992649}{27757240} \approx 1.98120018416816657563936472070,$$

$$s_{12}^* = \frac{28539271}{13878620} \approx 2.02708082647986615383950277477,$$

$$s_{21}^* = \frac{29174031}{13878620} \approx 2.0563478933784482895273449375,$$

$$s_{22}^* = \frac{56266169}{27757240} \approx 2.1020844291435315614953071031.$$

Thus the only pure Nash equilibrium for Γ is $((s_{11}^*, s_{12}^*), (s_{21}^*, s_{22}^*))$. The values of the corresponding utilities $\underline{u}_i = \underline{u}_{i1} + \underline{u}_{i2}$ are

$$\underline{u}_1 = \frac{867861128850819}{22013267783360} \approx 39.4244569862018379774764101105,$$

$$\underline{u}_2 = \frac{317287741895329}{7704643724176} \approx 41.1813645450896490849943708262.$$

Since Γ has only one pure Nash equilibrium, and since every Nash equilibrium of Γ is bounded from above and from below by its smallest and greatest pure Nash equilibria, then Γ has no properly mixed Nash equilibria.

8.5 Undominated and Weakly Dominating Strategies and Weakly Dominating Pure Nash Equilibria for Normal-Form Games

The main goal of this section is to present necessary and sufficient conditions for the existence of undominated strategies, weakly dominating strategies, and weakly dominating pure Nash equilibria for a normal-form game $\Gamma = \{S_i, u_i\}_{i=1}^N$. The strategy spaces S_i are assumed to be nonempty sets, topological spaces, or pseudometric spaces. The utilities u_i are functions from $S_1 \times \cdots \times S_N$ to nonempty posets $Y_i = (Y_i, \preceq_i)$, $i \in \{1, \ldots, N\}$. The obtained results are illustrated by examples and remarks. Unless otherwise stated, we assume that each strategy set S_i is partially ordered by the weak dominance relation $<_i$, defined as follows.

(wd) $s_i <_i t_i$ if $u_i(s_i, s_{-i}) \preceq_i u_i(t_i, s_{-i})$ for all $s_{-i} \in S_{-i}$, and $u_i(s_i, t_{-i}) \prec_i u_i(t_i, t_{-i})$ for some $t_{-i} \in S_{-i}$.

8.5.1 Existence of Undominated Strategies

Strategy $s_i \in S_i$ is said to be **undominated** if $s_i \not<_i t_i$ for all $t_i \in S_i$. We say that $t_i \in S_i$ is a **weak majorant** of a subset W of S_i if $u_i(s_i, s_{-i}) \preceq_i u_i(t_i, s_{-i})$ for all $s_{-i} \in S_{-i}$ and $s_i \in W$. Given $t_i \in S_i$, denote by $D(t_i)$ the set of all **weak dominants** of t_i, i.e.,

$$D(t_i) = \{s_i \in S_i : t_i <_i s_i\}. \tag{8.63}$$

We are going to derive necessary and sufficient conditions for the existence of undominated strategies for player i. The proof of our main result is based on the Chain Generating Recursion Principle presented in Lemma 2.1.

Proposition 8.92. *Player i in a game $\Gamma = \{S_i, u_i\}_{i=1}^{N}$ has undominated strategies if and only if there is a $t_i \in S_i$ such that each nonempty well-ordered set in $D(t_i)$ has a weak majorant.*

Proof: Assume first that each nonempty well-ordered set in $D(t_i)$ has a weak majorant for some $t_i \in S$. If $D(t_i)$ is empty, then t_i is an undominated strategy of player i. Otherwise, denote by $\mathcal{D}$ the set of all those well-ordered subsets of $D(t_i)$ which have strict upper bounds in S_i, or equivalently, in $D(t_i)$. Let $f : \mathcal{D} \to D(t_i)$ be a function that assigns to each $W \in \mathcal{D}$ a strict upper bound $f(W)$ of W in $D(t_i)$. By Lemma 2.1 there is a unique well-ordered set C in $D(t_i)$ so that $x \in C$ if and only if $x = f(\{y \in C : y <_i x\})$. By hypothesis, C has a weak majorant, say z. To show that z is undominated, assume there is $y \in S_i$ such that $z <_i y$. Let $x \in C$ be given. Because z is a weak majorant of C, and $z <_i y$, then $u_i(x, s_{-i}) \preceq_i u_i(z, s_{-i}) \preceq_i u_i(y, s_{-i})$ for each $s_{-i} \in S_{-i}$, and there is $t_{-i} \in S_{-i}$ such that $u_i(z, t_{-i}) \prec u_i(y, t_{-i})$. Thus $x <_i y$. This holds for each $x \in C$, whence y is a strict upper bound of C in $D(t_i)$. In particular, $C \in \mathcal{D}$, so that $f(C)$ exists and is, by definition, a strict upper bound of C. But this contradicts the last conclusion of Lemma 2.1, and proves that z is undominated.

If player i has an undominated strategy t_i, it is not weakly dominated by any other strategy in S_i, whence $D(t_i) = \emptyset$. Thus each nonempty well-ordered set in $D(t_i)$ has a weak majorant. □

Consider next the case when S_i is a topological space. Recall that a function $f : S_i \to Y_i$ is upper closed if the set $f^{-1}[[y)]$ is closed for all $y \in f[S_i]$.

Proposition 8.93. *Assume that S_i is a topological space, and that $u_i(\cdot, s_{-i})$ is upper closed for each $s_{-i} \in S_{-i}$. Then player i has undominated strategies if and only if there is a t_i in S_i such that the closure of each nonempty well-ordered subset of $D(t_i)$ is compact.*

Proof: Assume that the closure of each nonempty well-ordered subset of $D(t_i)$ is compact. Let C be a nonempty well-ordered set in $D(t_i)$. By Proposition 8.92 it suffices to show that C has a weak majorant. By hypothesis, the closure $\overline{C}$ of C is a compact subset of S_i. For each $x \in C$ and $s_{-i} \in S_{-i}$ denote

$$R(x, s_{-i}) = \{z \in S_i : u_i(x, s_{-i}) \preceq_i u_i(z, s_{-i})\}.$$

The sets $R(x, s_{-i})$ are closed because the functions $u_i(\cdot, s_{-i})$ are upper closed. Thus the set $R(x) = \bigcap\{R(x, s_{-i}) : s_{-i} \in S_{-i}\}$ is closed for every $x \in C$. Since $x \in R(x)$, then $R(x)$ is nonempty. Moreover, $R(z) \subset R(x)$ whenever $x <_i z$. Because C is well-ordered, then $\{R(x) \cap \overline{C} : x \in C\}$ is a family of nested nonempty closed subsets of $\overline{C}$. Clearly, every finite subfamily of these sets has a nonempty intersection. Since $\overline{C}$ is compact, then $\bigcap\{R(x) \cap \overline{C} : x \in C\}$ is nonempty. It is easy to see that each element of this set is a weak majorant of C. This result implies that the sufficiency part follows from Proposition 8.92.

If player i has an undominated strategy t_i, then $D(t_i) = \emptyset$, whence there is no nonempty well-ordered subset in $D(t_i)$. This proves the necessity part.   □

Corollary 8.94. *If S_i is a compact topological space, and if $u_i(\cdot, s_{-i})$ is upper closed for every $s_{-i} \in S_{-i}$, then player i has undominated strategies.*

In the special case that $Y_i = \mathbb{R}$ and the functions $u_i(\cdot, s_{-i})$ are upper semicontinuous, Corollary 8.94 is reduced to [199, Proposition 0].

Assume next that S_i is a pseudometric space. For $y \in S_i$ and $r > 0$, denote $B(y, r) = \{x \in S_i : d(x, y) \le r\}$, where d is the pseudometric of S_i.

Lemma 8.95. *Let S_i be a pseudometric space, $t_i \in S_i$, and assume that each increasing sequence of $D(t_i)$ has a convergent subsequence. If C is a nonempty well-ordered subset of $D(t_i)$, then for each $n \in \mathbb{N}$ there is a finite number of points $x_0^n, \ldots, x_{m_n}^n \in C$ such that $\overline{C} \subseteq \bigcup_{k=0}^{m_n} B(x_k^n, \frac{1}{2^n})$.*

Proof: Assume that C is a nonempty well-ordered subset of $D(t_i)$. Given $n \in \mathbb{N}$, denote $x_0^n = \min C$, and when $x_0^n, \ldots, x_m^n$ are chosen, let x_{m+1}^n be the smallest element in C, if such an element exists, such that $d(x_k^n, x_{m+1}^n) > \frac{1}{2^n}$ for each $k = 0, \ldots, m$. The so constructed sequence (x_k^n) is finite, for otherwise it is an increasing sequence of C that has no convergent subsequence. The points x_k^n form a finite subset $\{x_0^n, \ldots, x_{m_n}^n\}$ of C, and $\overline{C} \subseteq \bigcup_{k=0}^{m_n} B(x_k^n, \frac{1}{2^n})$.
    □

Proposition 8.96. *Let S_i be a pseudometric space, and let $u_i(\cdot, s_{-i})$ be upper closed for each $s_{-i} \in S_{-i}$. Then player i has undominated strategies if and only if there is a $t_i \in S_i$ such that each increasing sequence of $D(t_i)$ has a convergent subsequence.*

Proof: Assume there is a $t_i \in S_i$ such that each increasing sequence of $D(t_i)$ has a convergent subsequence. To prove that the closure of each well-ordered subset of $D(t_i)$ is compact, make a counter-hypothesis: There is a well-ordered subset C of $D(t_i)$ and an open covering $\{U_\alpha\}_{\alpha \in \Gamma}$ of $\overline{C}$ such that no finite subfamily of $\{U_\alpha\}_{\alpha \in \Gamma}$ covers $\overline{C}$. For each $n \in \mathbb{N}$, let B_n be the first of the balls $B(x_0^n, \frac{1}{2^n}), \ldots, B(x_{m_n}^n, \frac{1}{2^n})$, constructed in Lemma 8.95, such that no finite subfamily of $\{U_\alpha\}_{\alpha \in \Gamma}$ covers $B_n \cap \overline{C}$. Let y_n be the center of B_n. Defining inductively $n_0 = \min\{j : y_j = \min\{y_n : n \in \mathbb{N}\}\}$ and $n_{k+1} = \min\{j : y_j = \min\{y_n : n > n_k\}\}$ when $k \in \mathbb{N}_0$, we obtain an increasing subsequence $(y_{n_k})_{k=0}^\infty$ of $(y_n)_{n=1}^\infty$. By hypothesis the sequence $(y_{n_k})_{k=0}^\infty$ has a convergent subsequence $(y_{n_{k_i}})_{i=0}^\infty$. Denote $y = \lim_i y_{n_{k_i}}$. Since $y \in \overline{C}$, there is U_α such that $y \in U_\alpha$. Because U_α is open, there is $r > 0$ such that $B(y, r) \subset U_\alpha$. Since $\lim_i d(y_{n_{k_i}}, y) = 0$, there is n_{k_i} such that $d(y_{n_{k_i}}, y) < \frac{r}{2}$ and $\frac{1}{2^{n_{k_i}}} < \frac{r}{2}$. But then $B_{n_{k_i}} \subset B(y, r) \subset U_\alpha$, which is a contradiction, since no finite subfamily of $\{U_\alpha\}_{\alpha \in \Gamma}$ covers $B_{n_{k_i}}$. The above proof shows that the closure of each well-ordered subset of $D(t_i)$ is compact, whence Proposition 8.93 implies that player i has undominated strategies. The proof of the necessity part is obvious. $\square$

Remark 8.97. (i) The function f defined in the proof of Proposition 8.92 is a choice function over a set of subsets of S_i. If the set $D(t_i)$ has a subset B equipped with a well-ordering relation $\prec$ such that each element of $D(t_i)$ has a weak majorant in B, one can choose $f(W)$, $W \in \mathcal{D}$, to be that element of B that is the smallest with respect to $\prec$ of the strict upper bounds (with respect to $<_i$) of W. In particular, if the set B is countable, it has a natural well-ordering through its sequence representation. In these cases all the proofs of this subsection are independent on the axiom of choice. This is the case also when $D(t_i) = \emptyset$ for all $t_i \in S_i$. In such a case each strategy of S_i is undominated.

(ii) The smallest elements of well-ordered subset C of $D(t_i)$, constructed in the proof of Proposition 8.92 are of the form: $x_0 = f(\emptyset)$, $x_1 = f(\{x_0\})$, ..., $x_n = f(\{x_0, \ldots, x_{n-1}\})$, as long as x_n is defined. If $x_\omega = f(\{x_n\}_{n=0}^\infty)$ exists, then x_ω is the next element of C, and so on.

(iii) The existence of smallest and greatest undominated Nash equilibria in games of strategic complementarities is studied in [159, 160].

8.5.2 Existence of Weakly Dominating Strategies and Pure Nash Equilibria

Strategy $t_i \in S_i$ is called **weakly dominating** if $s_i <_i t_i$ for every $s_i \in S_i \setminus \{t_i\}$, where $<_i$ is defined by (wd). In this subsection we first derive necessary and sufficient conditions for the existence of the weakly dominating strategy of a player in a normal-form game.

Proposition 8.98. *Player i in a game $\Gamma = \{S_i, u_i\}_{i=1}^N$ has the weakly dominating strategy if and only if S_i is directed upward (with respect to '$<_i$'), and there is a $t_i \in S_i$ such that each nonempty well-ordered subset of $D(t_i)$ has a weak majorant.*

Proof: Assume first that S_i is directed upward and there is a $t_i \in S_i$ such that each nonempty well-ordered subset of $D(t_i)$ has a weak majorant. By Proposition 8.92 these hypotheses imply that player i has at least one undominated strategy, say z, in S_i. To prove that z is the weakly dominating strategy in S_i, assume there is $y \in S_i \setminus \{z\}$ such that $y \not<_i z$. Because S_i is directed upward, there is $x \in S_i$ such that $y \leq_i x$ and $z \leq_i x$. Since $y \not\leq_i z$, it follows that $x \neq z$, so that $z <_i x$. However, the latter is impossible because z is undominated. Hence, z is the weakly dominating strategy in S_i. If player i has the weakly dominating strategy x in S_i, then for all y, $z \in S_i$ we have $y \leq_i x$ and $z \leq_i x$, whence S_i is directed upward. Moreover, since x is undominated, by the proof of Proposition 8.92 it follows that each nonempty well-ordered subset of $D(t_i)$ has a weak majorant when $t_i = x$. $\square$

The following proposition is a consequence of Propositions 8.93 and 8.98.

Proposition 8.99. *Assume that S_i is a topological space, and that $u_i(\cdot, s_{-i})$ is upper closed for each $s_{-i} \in S_{-i}$. Then player i has the weakly dominating strategy if and only if S_i is directed upward and there is a $t_i \in S_i$ such that the closure of each well-ordered subset of $D(t_i)$ is compact.*

Proof: In view of the proof of Proposition 8.93, the hypotheses imply that each nonempty well-ordered subset of $D(t_i)$ has a weak majorant. Thus the sufficiency part follows from Proposition 8.98. If player i has the weakly dominating strategy t_i, then the upward directness of S_i can be proved as in Proposition 8.98. Since t_i is undominated, then $D(t_i) = \emptyset$, whence the upper closeness and compactness hypotheses are trivially satisfied. This proves the necessity part of the assertion.   □

As an easy consequence of Propositions 8.96 and 8.99 we obtain the following result.

Proposition 8.100. *Let S_i be a pseudometric space, and let $u_i(\cdot, s_{-i})$ be upper closed for each $s_{-i} \in S_{-i}$. Then player i has the weakly dominating strategy if and only if S_i is directed upward, and there is a $t_i \in S_i$ such that each increasing sequence of $D(t_i)$ has a convergent subsequence.*

Recall that $s_1^*, \ldots, s_N^*$ is a pure Nash equilibrium for Γ if $u_i(s_i, s_{-i}^*) \preceq_i u_i(s_i^*, s_{-i}^*)$ for all $s_i \in S_i$, and $i = 1, \ldots, N$. We say that a pure Nash equilibrium $s_1^*, \ldots, s_N^*$ for Γ is weakly dominating if each s_i^* is weakly dominating. Proposition 8.98 and the definition of pure Nash equilibrium imply the following theorem.

Theorem 8.101. *A normal-form game $\Gamma = \{S_i, u_i\}_{i=1}^N$ has the weakly dominating pure Nash equilibrium if and only if for each $i = 1, \ldots, N$, S_i is directed upward, and there is a $t_i \in S_i$ such that each nonempty well-ordered subset of $D(t_i)$ has a weak majorant.*

Proof: By Proposition 8.98, the given hypotheses are necessary and sufficient for the existence of the weakly dominating strategy s_i^* for each $i = 1, \ldots, N$. Obviously, $(s_1^*, \ldots, s_N^*)$ is the only weakly dominating pure Nash equilibrium for game Γ.   □

By similar reasoning, one can derive the following consequences of Propositions 8.99 and 8.100 and Corollary 8.94.

Theorem 8.102. *Let $\Gamma = \{S_i, u_i\}_{i=1}^N$ be a normal-form game, and assume that each S_i is a topological (respectively pseudometric) space, and that $u_i(\cdot, s_{-i})$ is upper closed for each $s_{-i} \in S_{-i}$, and for each $i = 1, \ldots, N$. Then Γ has the weakly dominating pure Nash equilibrium if and only if for each $i = 1, \ldots, N$, S_i is directed upward, and there is a t_i in S_i such that each nonempty well-ordered subset of $D(t_i)$ has compact closure (respectively each increasing sequence of $D(t_i)$ has a convergent subsequence).*

Corollary 8.103. *Let $\Gamma = \{S_i, u_i\}_{i=1}^{N}$ be a normal-form game, and assume that the strategy spaces S_i are compact and directed upward, and that $u_i(\cdot, s_{-i})$ is upper closed for all $s_{-i} \in S_{-i}$ and $i = 1, \ldots, N$. Then Γ has the weakly dominating pure Nash equilibrium.*

Remark 8.104. (i) The above results have obvious extensions to the game of the form $\Gamma = (\{S_i\}_{i \in \Lambda}, \{u_i\}_{i \in \Lambda})$, where Λ is any nonempty index set. Thus no quantitative limitations are required for the set of players or their strategy sets.

(ii) In Propositions 8.92–8.100 no hypotheses are imposed on the strategy sets S_j or on the utility functions u_j for $j \neq i$. In particular, Propositions 8.98–8.100 give necessary and sufficient conditions for the existence of exactly one s_i^* that maximizes $u_i(\cdot, s_{-i})$, no matter what is the 'state of nature' s_{-i}. Existence of such a choice s_i^* would be especially useful in situations where the decision maker is uncertain about the true state of the world, and cannot formulate even a probability distribution over the possible states (so that the maximization of expected utility is out of question).

(iii) In the case when the utility functions u_i are real-valued, the above results, as well as the examples in the next subsection, are introduced in [136]. Our assumption that the values of u_i are in posets allows us to evaluate the utilities of different players in different ordinal scales.

8.5.3 Examples

The purpose of the following examples is to illustrate the applicability of the results derived in the above subsections. The first four examples deal with two-person games.

Example 8.105. Let $\Gamma = \{S_1, S_2, u_1, u_2\}$ be a two-person game with $S_1 = S_2 = \mathbb{N}_0 = \{0, 1, 2, \ldots\}$, and u_i, $i = 1, 2$, given by

$$u_i(s_i, s_{-i}) = \begin{cases} s_{-i} & \text{if } s_{-i} \leq s_i, \\ 0 & \text{otherwise,} \end{cases}$$

where $\leq$ is the standard ordering of $\mathbb{N}_0$. The partial ordering $<_i$ of the sets S_i defined by the weak dominance relation equals to the standard strict ordering $<$ of $\mathbb{N}_0$ for both $i = 1, 2$. In particular, the strategy sets S_i are well-ordered, and hence also directed upward. However, $D(t_i)$ does not have a weak majorant for any $t_i \in S_i$. Therefore neither undominated nor weakly dominating strategies exist by Propositions 8.92 and 8.98. The set of pure Nash equilibria is an infinite and well-ordered set $\text{NE}(\Gamma) = \{(s_1, s_2) : s_1 = s_2\}$. The utility functions are not bounded above in $\text{NE}(\Gamma)$.

Example 8.106. Let $\Gamma = \{S_1, S_2, u_1, u_2\}$ be a two-person game, where $S_1 = S_2 = \mathbb{Z} = \{0, \pm 1, \pm 2, \ldots\}$, and $u_i(s_i, s_{-i}) = -\max\{|s_{-i}|, |s_i|\}$, $i = 1, 2$, with $|x|$ denoting the absolute value of x. It is easy to see that s_i weakly dominates

t_i, i.e., $t_i <_i s_i$ iff $|s_i| < |t_i|$, where $<$ is as in Example 8.105. Therefore S_i is directed upward and the strategy $s_i^* = 0$ is the weak majorant of S_i, $i = 1, 2$. The set of pure Nash equilibria is $NE(\Gamma) = \{(s_1, s_2) \in S : |s_1| = |s_2|\}$, so there is an infinite number of pure Nash equilibria. Utility functions are not bounded from below in $NE(\Gamma)$.

Example 8.107. Let $\Gamma = \{S_1, S_2, u_1, u_2\}$ be a two-person game with

$$S_i = \{\tfrac{n}{n+1} : n \in \mathbb{N}_0\} \cup \{1\}, \text{ and } u_i(s_i, s_{-i}) = \begin{cases} 1 - s_i & \text{if } s_{-i} < s_i \\ 0 & \text{otherwise,} \end{cases}$$

$i = 1, 2$, where $<$ is the standard strict ordering of $\mathbb{R}$. In this case $t_i <_i s_i$ iff $t_i \in \{0, 1\}$ and $s_i \in S_i \setminus \{0, 1\}$. Thus

$$D(t_i) = \begin{cases} \emptyset & \text{if } t_i \in S_i \setminus \{0, 1\}, \\ S_i \setminus \{0, 1\} & \text{if } t_i = 0 \text{ or } t_i = 1. \end{cases}$$

In particular, for any given $t_i \in S_i$, each well-ordered subset C of $D(t_i)$ contains at most one element. Thus Proposition 8.92 implies that S_i contains undominated strategies. In fact, for each $i = 1, 2$, the set of all the undominated strategies is $S_i \setminus \{0, 1\}$, which contains more than one element, and therefore S_i cannot be directed upward. Thus no weakly dominating strategies exist (Proposition 8.98). Note that the unique pure Nash equilibrium is $(1, 1)$, so there is no equilibria in undominated strategies.

Example 8.108. Let $\Gamma = \{S_1, S_2, u_1, u_2\}$ be a two-person game with $S_i = [0, 1]$, and

$$u_i(s_i, s_{-i}) = \begin{cases} s_{-i} & \text{if } s_i \le s_{-i}, \\ 1 & \text{if } s_{-i} < s_i < 1, \\ s_{-i}^2 & \text{if } s_i = 1, \end{cases}$$

$i = 1, 2$, where $<$ is the standard strict ordering of $\mathbb{R}$. In this example $t_i <_i s_i$ iff $t_i < s_i$ or $t_i = 1 \neq s_i$, $i = 1, 2$. There is no $t_i \in S_i$ such that $u_i(\cdot, s_{-i})$ is upper semicontinuous for each $s_{-i} \in S_{-i}$. Therefore both Proposition 8.93 and Proposition 8.96 as well as Corollary 8.94 imply that no undominated strategies exist. The unique pure Nash equilibrium is $(1, 1)$.

Given a vector $p = (\sigma_1, \ldots, \sigma_m) \in \mathbb{R}^m$ and a positive number α, define an α-norm of p by $\|p\|_\alpha = (\sum_{i=1}^m |\sigma_i|^\alpha)^{\frac{1}{\alpha}}$.

Example 8.109. Assume that $\Gamma = \{S_i, u_i\}_{i=1}^N$ is a normal form game, where to each $i = 1, \ldots, N$ there corresponds a positive integer m_i, numbers $\alpha_i, \beta_i \in (0, 1)$, and a function $\varphi : S_{-i} \to \mathbb{R}$ such that

$$S_i = \{s_i \in \mathbb{R}_+^{m_i} : \tfrac{1}{2} < \|s_i\|_{\alpha_i} \le 1\}, \quad u_i(s_i, s_{-i}) = \|s_i\|_{\beta_i} + \varphi(s_{-i}).$$

If $\beta_i = \alpha_i$, $i = 1, \ldots, N$, it is easy to show that the hypotheses of Proposition 8.96 are satisfied for each $t_i \in S_i$, and that each strategy of the set $T_i = \{s_i \in$

$\mathbb{R}^{m_i}_+ : \|s_i\|_{\alpha_i} = 1\}$ is undominated. Thus player i has no weakly dominating strategies if $m_i > 1$. Obviously, $\mathrm{NE}(\Gamma) = T_1 \times \cdots \times T_N$. The strategy sets S_i are neither compact nor convex in $\mathbb{R}^{m_i}$.

8.6 Pursuit and Evasion Game

In this section the *Chain Generating Recursion Principle* and generalized iteration methods presented in Chap. 2 are applied to prove the existence of winning strategies for a pursuit and evasion game. The obtained results are used to study the solvability of equations and inclusions in ordered spaces.

In Sect. 8.6.2 we generalize the results stated in Chap. 1 for finite pursuit and evasion games to games of ordinal length. These generalizations are then applied in Sect. 8.6.3 to prove fixed point results for a set-valued mapping $\mathcal{F} : P \to 2^P \setminus \emptyset$ and for a single-valued mapping $G : P \to P$. Monotonicity hypotheses imposed on $\mathcal{F}$ and G are weaker than those assumed in Chap. 2. We study also the solvability of the equation $Lu = Nu$ as well as the inclusion $Lu \in Nu$, where L and N are mappings from a set V to a poset P and $\mathcal{N} : V \to 2^P \setminus \emptyset$, respectively. Moreover, special cases and examples are considered.

8.6.1 Preliminaries

Let $P = (P, \leq)$ be a nonempty poset. We say that a mapping $\beta \mapsto z_\beta$ from an ordinal ν to P is a (transfinite) sequence in P, and denote it by $(z_\beta)_{\beta<\nu}$. The sequence is called increasing if $z_\alpha \leq z_\beta$ whenever $\alpha < \beta < \nu$, decreasing if $z_\beta \leq z_\alpha$ whenever $\alpha < \beta < \nu$, and monotone if it is increasing or decreasing. If the above inequalities are strict, the sequence $(z_\beta)_{\beta<\nu}$ is called strictly increasing, strictly decreasing, or strictly monotone, respectively. If $(z_\beta)_{\beta<\nu}$ is increasing and $z = \sup\{z_\beta\}_{\beta<\nu}$ exists, or if $(z_\beta)_{\beta<\nu}$ is decreasing and $z = \inf\{z_\beta\}_{\beta<\nu}$ exists, we say that $(z_\beta)_{\beta<\nu}$ **order converges**, and that z is its **order limit**. A sequence $(x_\beta)_{\beta<\mu}$ is called an **initial sequence** of $(z_\beta)_{\beta<\nu}$ if $x_\beta = z_\beta$ for every $\beta < \mu$, and if $\mu \leq \nu$. If $\mu < \nu$, we say that the initial sequence is **proper**. The next result follows from [169, Chapter II, Theorem 3.23].

Lemma 8.110. *A nonempty subset W of P is well-ordered if and only if there is a unique strictly increasing sequence $(z_\beta)_{\beta<\mu}$ whose range is W.*

Applying Lemma 8.110, we can reformulate Lemma 2.2 as follows.

Lemma 8.111. *Given a poset P, a function $G : P \to P$, and $c \in P$, then there exists a unique strictly increasing sequence $(x_\beta^G)_{\beta<\mu(G)}$ in P, called* **an increasing sequence of cG-iterations**, *satisfying*

$$x_\alpha^G = \sup\{c, G[\{x_\beta^G\}_{\beta<\alpha}]\} \text{ for every } \alpha < \mu(G). \tag{8.64}$$

The following result is dual to that of Lemma 8.111.

Lemma 8.112. *Given a mapping $G : P \to P$ and $b \in P$, then there exists exactly one strictly decreasing sequence $(z_\beta)_{\beta < \nu(G)}$ in P, called* **a decreasing sequence of bG-iterations,** *such that*

$$z_\alpha = \inf\{b, G[\{z_\beta\}_{\beta < \alpha}]\} \text{ for every } \alpha < \nu(G). \tag{8.65}$$

8.6.2 Winning Strategy

In this subsection we study in a more general setting the pursuit and evasion game considered in Chap. 1. The game board is now a nonempty poset $P = (P, \leq)$. Assume that to every position $x \in P$ of player p there corresponds in P a nonempty subset $\mathcal{F}(x)$ of possible positions of player q. Instead of a finite game, we assume now that the game is of **ordinal length,** i.e., the positions of the players in a play of the game form sequences $(x_\beta)_{\beta < \mu}$ and $(y_\beta)_{\beta < \mu}$ in P, where μ is an ordinal. Next we present rules for an **ordered pursuit and evasion (o.p-e) game.**

Definition 8.113. *A sequence $((x_\beta, y_\beta))_{\beta < \mu}$ of $P \times P$ is a play of the o.p-e game if the following conditions hold:*

(r1) If $\gamma < \alpha < \mu$, and if $(x_\beta)_{\gamma \leq \beta \leq \alpha}$ is increasing (respectively decreasing), then $(y_\beta)_{\gamma \leq \beta \leq \alpha}$ is increasing (respectively decreasing).
(r2) μ is the smallest ordinal for which μ-th position does not exist for p or for q, or μ is a successor and p does not move further from $x_{\mu-1}$.

Our purpose is to derive conditions under which p has a winning strategy in the o.p-e game defined as follows.

Definition 8.114. *Player p wins a play $((x_\beta, y_\beta))_{\beta < \mu}$ of the o.p-e game if μ is a successor, and $x_{\mu-1} = y_{\mu-1}$. A* **strategy** *for player p is a rule that tells p what move to make at each of its turns depending on the moves played by player q previously. If a strategy of player p has the property that p always wins, it is called a* **winning strategy.**

Consider a play $((x_\beta, y_\beta))_{\beta < \mu}$ of the o.p-e game where the positions of p are determined by the following strategy, $c \in P$ being fixed:

$$x_\alpha = \sup\{c, \{y_\beta\}_{\beta < \alpha}\} \text{ for every } \alpha < \mu. \tag{8.66}$$

Consider first moves of such a play. Player p starts from $x_0 = \sup\{c, \emptyset\} = c$, and player q starts from a point y_0 of $\mathcal{F}(x_0)$. If $y_0 \leq c$, then $x_1 = \sup\{c, y_0\} = c = x_0$, so that by rule (r2), (x_0, y_0) is the only position pair of the play. After a finite number of move pairs, say, $((x_n, y_n))_{n=0}^m$, $m \in \mathbb{N}_0$, p can proceed to apply strategy (8.66) if and only if $\sup\{c, \{y_n\}_{n=0}^m\} = \sup\{c, y_m\}$ exists and $x_m < \sup\{c, y_m\}$, in which case p moves to $x_{m+1} = \sup\{c, y_m\}$. Otherwise the play stops. According to rule (r1), all the possible responds y_{m+1} of q belong to the set $[y_m) \cap \mathcal{F}(x_{m+1})$. If this set is empty, then the play stops. If play has proceeded to state $((x_n, y_n))_{n<\omega}$, where ω is the smallest infinite ordinal, p can further apply strategy (8.66) if and only if $\sup\{c, \{y_n\}_{n=0}^{\infty}\}$ exists. In such a case p moves to $x_\omega = \sup\{c, \{y_n\}_{n=0}^{\infty}\}$. According to (r1), q responds by an element y_ω that belongs to the set $\{y \in \mathcal{F}(x_\omega) : y_n \le y \text{ for all } n < \omega\}$ if such an element exists. If (x_ω, y_ω) exists, proceed as above when (x_0, y_0) is replaced by (x_ω, y_ω), and so on.

The next lemma deals with monotonicity properties of a play $((x_\beta, y_\beta))_{\beta<\mu}$ where p follows strategy (8.66).

Lemma 8.115. *Given $c \in P$. Let $((x_\beta, y_\beta))_{\beta<\mu}$ be a play of the o.p-e game where p uses strategy (8.66). Then $(x_\beta)_{\beta<\mu}$ is strictly increasing, and $(y_\beta)_{\beta<\mu}$ is strictly increasing, except in the case when μ and $\alpha = \mu - 1$ are successors, in which case $(y_\beta)_{\beta<\alpha}$ is strictly increasing and equality $y_\alpha = y_{\alpha-1}$ may hold.*

Proof: If $\mu = 1$, then (x_0, y_0) is the only element of the sequence $((x_\beta, y_\beta))_{\beta<\mu}$, whence the assertions hold trivially. Assume that $1 < \mu$. By (8.66) the sequence $(x_\beta)_{\beta<\mu}$ is increasing. Rule (r1) implies that also $(y_\beta)_{\beta<\mu}$ is increasing. To prove that $(x_\beta)_{\beta<\mu}$ is strictly increasing, assume first that $\alpha < \mu$, and that α is a successor. Then $x_{\alpha-1} < x_\alpha$, because otherwise $x_\alpha = x_{\alpha-1}$, and then $\mu = \alpha$ by rule (r2), which is impossible because $\alpha < \mu$. If α is a limit ordinal, then $x_\beta < x_\alpha$ for every $\beta < \alpha$, for otherwise $x_\beta = x_\alpha$ for some $\beta < \alpha$. Since $\beta + 1$ is a successor and $\beta + 1 < \alpha$, then $x_\alpha = x_\beta \le x_{\beta+1} \le x_\alpha$. Thus $x_\beta = x_{\beta+1}$, whence rule (r2) implies that $\mu = \beta + 1$, a contradiction. Consequently, x_α is a strict upper bound of $(x_\beta)_{\beta<\alpha}$. The above proof implies that $(x_\beta)_{\beta<\mu}$ is strictly increasing.

Next we prove the asserted strict monotonicity properties for $(y_\beta)_{\beta<\mu}$. Assume first that $0 < \alpha < \mu$, and that α is a successor. Because $(y_\beta)_{\beta<\mu}$ is increasing, then $y_{\alpha-1} = \max\{y_\beta\}_{\beta<\alpha}$. If $y_\alpha = y_{\alpha-1}$, then

$$x_\alpha = \sup\{c, \{y_\beta\}_{\beta<\alpha}\} = \sup\{c, y_{\alpha-1}\} = \sup\{c, y_\alpha\} = x_{\alpha+1}.$$

Applying rule (r2), this implies that $\mu = \alpha + 1$. Hence, μ is also a successor. Assume next that α is a limit ordinal. To show that $y_\beta < y_\alpha$ for every $\beta < \alpha$, assume on the contrary that $y_\beta = y_\alpha$ for some $\beta < \alpha$. Since $\beta + 1$ is a successor and $\beta + 1 < \alpha$, the above proof implies that $y_\beta < y_{\beta+1} = y_\alpha$, which is impossible. Thus y_α is a strict upper bound of $(y_\beta)_{\beta<\alpha}$. The above proof shows that $(y_\beta)_{\beta<\mu}$ is strictly increasing, except in the case when μ and $\alpha = \mu - 1$ are successors, in which case $(y_\beta)_{\beta<\alpha}$ is strictly increasing and equality $y_\alpha = y_{\alpha-1}$ may hold. $\square$

The axiom of choice is needed to formulate and prove our next results. Denote by

$$\mathcal{G} := \{G : P \to P : G(x) \in \mathcal{F}(x) \text{ for all } x \in P\} \qquad (8.67)$$

the set of all single-valued selection mappings $G : P \to P$ of the multi-function $x \mapsto \mathcal{F}(x)$.

Lemma 8.116. *Given $c \in P$, then $((x_\beta, y_\beta))_{\beta < \mu}$ is a play of the o.p-e game, where p uses strategy (8.66), if and only if there is a selection $G \in \mathcal{G}$ such that $x_\beta = x_\beta^G$ and $y_\beta = G(x_\beta^G)$ for every $\beta < \mu$, and $(G(x_\beta^G))_{\beta < \mu}$ is increasing. Here, $(x_\beta^G)_{\beta < \mu(G)}$ is the sequence of cG-iterations defined by (8.64).*

Proof: Let $((x_\beta, y_\beta))_{\beta < \mu}$ be a play of the o.p-e game where p uses strategy (8.66). Equation

$$g(x_\beta) := y_\beta, \quad \beta < \mu,$$

defines a function g from $W = \{x_\beta\}_{\beta < \mu}$ to P, and $g(x) \in \mathcal{F}(x)$ for each $x \in W$. Because $(x_\beta)_{\beta < \mu}$ is strictly increasing, rule (r1) implies that g is increasing. Let $G : P \to P$ be any selection from $\mathcal{F}$ such that $G|W = g$. From (8.66) it follows that

$$x_\alpha = \sup\{c, G[\{x_\beta\}_{\beta < \alpha}]\} \quad \text{for every } \alpha < \mu,$$

which in view of Lemma 8.111 implies that $x_\beta = x_\beta^G$ for every $\beta < \mu$. Moreover, $(G(x_\beta^G))_{\beta < \mu}$ is increasing and $y_\beta = G(x_\beta^G)$, $\beta < \mu$.

Conversely, let $G : P \to P$ be any selection from $\mathcal{F}$, and let $(x_\beta^G)_{\beta < \mu}$ be such an initial sequence of the sequence $(x_\beta^G)_{\beta < \mu(G)}$ of cG-iterations that $(G(x_\beta^G))_{\beta < \mu}$ is increasing. Denoting $x_\beta = x_\beta^G$, and $y_\beta = G(x_\beta^G)$, $\beta < \mu$, from (8.64) and (8.66) it follows that $((x_\beta, y_\beta))_{\beta < \mu}$ is a play of the o.-p-e game where p uses strategy (8.66). $\square$

We are going to prove that among the plays of the o.p-e game, where p uses strategy (8.66), there exists at least one play that cannot be extended. In that proof we apply Lemma 2.1 in $\mathcal{G}$ equipped with the following partial ordering: For every $G \in \mathcal{G}$ denote by $(x_\beta^G)_{\beta < \mu_G}$ the longest initial sequence of the sequence $(x_\beta^G)_{\beta < \mu(G)}$ of cG-iterations such that $(G(x_\beta^G))_{\beta < \mu_G}$ is increasing. Define a partial ordering $\prec$ on $\mathcal{G}$ as follows.

(O) $\overline{G} \prec G$ if and only if $((x_\beta^{\overline{G}}, \overline{G}(x_\beta^{\overline{G}})))_{\beta < \mu_{\overline{G}}}$ is a proper initial sequence of $((x_\beta^G, G(x_\beta^G)))_{\beta < \mu_G}$.

Lemma 8.117. *Given $c \in P$, there is a play $((x_\beta, y_\beta))_{\beta < \mu}$ of the o.p-e game where p uses strategy (8.66), and which is not a proper initial sequence of any such a play.*

Proof: Choosing $x_0 = c$ and $y_0 \in \mathcal{F}(x_0)$, then $((x_\beta, y_\beta))_{\beta < 1}$ is a play of the o.p-e game where p uses strategy (8.66). By Lemma 8.116 there exists a selection $G_0 \in \mathcal{G}$ such that $((x_\beta, y_\beta))_{\beta < 1}$ is an initial sequence of $((x_\beta^{G_0}, G_0(x_\beta^{G_0})))_{\beta < \mu_{G_0}}$. Let $\mathcal{D}$ contain the empty set and all those well-ordered subsets of $\mathcal{W}$ of $(\mathcal{G}, \preceq)$ for which $G_0 = \min \mathcal{W}$, and which have strict upper bounds in $(\mathcal{G}, \preceq)$. Let $f : \mathcal{D} \to \mathcal{G}$ be a mapping that assigns to each nonempty element $\mathcal{W}$ of $\mathcal{D}$ one of its strict upper bounds, and $f(\emptyset) = G_0$. By Lemma 2.1 there exists a unique well-ordered chain $\mathcal{C}$ in $\mathcal{G}$ such that $G \in \mathcal{C}$ if and only if $G = f(\mathcal{C}^{\prec G})$. Denote $\mu := \cup\{\mu_G : G \in \mathcal{C}\}$, and define

$$(x_\beta, y_\beta) = (x_\beta^G, G(x_\beta^G)), \quad \beta < \mu, \quad \text{where} \tag{8.68}$$

G is the smallest element of $\mathcal{C}$ such that $\beta < \mu_G$.

The definition (O) of $\prec$ implies that the sequence $((x_\beta, y_\beta))_{\beta<\mu}$ is well-defined, and that sequences $((x_\beta^G, G(x_\beta^G)))_{\beta<\mu_G}$, $G \in \mathcal{C}$, are its initial sequences. Moreover, $(x_\beta)_{\beta<\mu}$ is strictly increasing, $(y_\beta)_{\beta<\mu}$ is increasing, and $y_\beta \in \mathcal{F}(x_\beta)$ for every $\beta < \mu$. If $\alpha < \mu$, then $x_\alpha = x_\alpha^G$ for some $G \in \mathcal{C}$. The definitions of $\mathcal{C}$ and the partial ordering $\prec$ imply that $(x_\beta, y_\beta) = (x_\beta^G, G(x_\beta^G))$ for every $\beta \le \alpha$. Thus we obtain

$$x_\alpha = x_\alpha^G = \sup\{c, G[\{x_\beta^G\}_{\beta<\alpha}]\} = \sup\{c, \{y_\beta\}_{\beta<\alpha}\}. \tag{8.69}$$

Consequently, $((x_\beta, y_\beta))_{\beta<\mu}$ is a play of the o.p-e game where p uses strategy (8.66). To show that it is not a proper initial sequence of any such a play, assume on the contrary that it can be extended by (x_μ, y_μ), where

$$x_\mu = \sup\{c, \{y_\beta\}_{\beta<\mu}\}, \tag{8.70}$$

is a strict upper bound of $(x_\beta)_{\beta<\mu}$ and y_μ is an upper bound of $(y_\beta)_{\beta<\mu}$. By Lemma 8.116 there is a selection G from $\mathcal{F}$ such that $((x_\beta, y_\beta))_{\beta<\mu+1}$ is an initial sequence of $((x_\beta^G, G(x_\beta^G)))_{\beta<\mu_G}$. But then G would be a strict upper bound of $\mathcal{C}$, so that $f(\mathcal{C})$ would exist and would be a strict upper bound of $\mathcal{C}$, contradicting the last conclusion of Lemma 2.1. Consequently, $((x_\beta, y_\beta))_{\beta<\mu}$ is not a proper initial sequence of any play of the o.p-e game where p uses strategy (8.66). $\qquad\square$

Recall that a subset A of P has a **sup-center** c in P if $c \in P$ and $\sup\{c, x\}$ exists in P for every $x \in A$. If $\inf\{c, x\}$ exists in P for every $x \in A$, we say that c is an **inf-center** of A in P.

In the proof of our main result of this subsection we make use of the following auxiliary results.

Proposition 8.118. *Assume that the following conditions hold whenever $(x_\beta)_{\beta<\mu}$ and $(y_\beta)_{\beta<\mu}$ are increasing in P and $y_\beta \in \mathcal{F}(x_\beta)$ for every $\beta < \mu$.*

(Fa) $(y_\beta)_{\beta<\mu}$ has an order limit in P.
(Fb) If x is an upper bound of both $(x_\beta)_{\beta<\mu}$ and $(y_\beta)_{\beta<\mu}$, then $(y_\beta)_{\beta<\mu}$ has an upper bound in $\mathcal{F}(x)$.

Let $((x_\beta, y_\beta))_{\beta<\mu}$ be a maximally extended play of the o.p-e game where p uses strategy (8.66). Then the following assertions hold.

(a) If c in (8.66) is a sup-center of the set of order limits of increasing and order convergent sequences of $\mathcal{F}[P]$ in P, then $x_{\mu-1} = \sup\{c, y_{\mu-1}\}$.
(b) If $x_0 \le y_0$, if $c = x_0$ in (8.66), and if conditions (Fa) and (Fb) hold whenever $(x_\beta)_{\beta<\mu}$ and $(y_\beta)_{\beta<\mu}$ are increasing in P and $x_\beta \le y_\beta \in \mathcal{F}(x_\beta)$ for every $\beta < \mu$, then $x_{\mu-1} = y_{\mu-1}$, whence p wins the play.

Proof: Ad (a) In the play $((x_\beta, y_\beta))_{\beta<\mu}$, the sequence $(x_\beta)_{\beta<\mu}$ is strictly increasing, and $(y_\beta)_{\beta<\mu}$ is an increasing sequence of $\mathcal{F}[P]$. The order limit $y = \sup\{y_\beta\}_{\beta<\mu}$ exists by condition (Fa), and $x_\mu := \sup\{c, y\}$ exists by assumption. It is easy to see that (8.70) holds. Since $(y_\beta)_{\beta<\mu}$ is increasing, it follows from (8.66) and (8.70) that x_μ is an upper bound of $(x_\beta)_{\beta<\mu}$. To show that x_μ is not a strict upper bound of $(x_\beta)_{\beta<\mu}$, assume on the contrary that $x_\beta < x_\mu$ for each $\beta < \mu$. Now x_μ is an upper bound of both increasing sequences $(x_\beta)_{\beta<\mu}$ and $(y_\beta)_{\beta<\mu}$, and each y_β belongs to $\mathcal{F}(x_\beta)$. It then follows from condition (Fb) that $(y_\beta)_{\beta<\mu}$ has an upper bound in $\mathcal{F}(x_\mu)$. When q chooses such an upper bound y_μ as the response to x_μ, then $((x_\beta, y_\beta))_{\beta<\mu}$ has a proper extension to a play $((x_\beta, y_\beta))_{\beta\leq\mu}$ of the o.p-e game where p uses strategy (8.66). This contradicts the choice of $((x_\beta, y_\beta))_{\beta<\mu}$, whence $x_\mu = x_\beta$ for some $\beta < \mu$. If μ would be a limit ordinal, then $\beta + 1 < \mu$ and $x_\mu = x_\beta < x_{\beta+1} \leq x_\mu$, a contradiction. Thus μ is a successor, and $x_{\mu-1} = x_\mu$. Because $(y_\beta)_{\beta<\mu}$ is increasing, then $y_{\mu-1} = \max\{y_\beta\}_{\beta<\mu}$, and hence it follows

$$x_{\mu-1} = x_\mu = \sup\{c, \{y_\beta\}_{\beta<\mu}\} = \sup\{c, y_{\mu-1}\}.$$

This concludes the proof of (a).

Ad (b) Assume next that $x_0 \leq y_0$. Choosing $c = x_0$, then $c \leq y_0$. We show that under the hypotheses of (b) the positions of the play in question satisfy $x_\alpha \leq y_\alpha$ for every $\alpha < \mu$. Since $(y_\beta)_{\beta<\mu}$ is increasing, then $c \leq y_\beta$ for every $\beta < \mu$. It then follows from (8.66) that

$$x_\alpha = \sup\{y_\beta : \beta < \alpha\} \text{ for every } \alpha < \mu. \tag{8.71}$$

Because $(y_\beta)_{\beta<\mu}$ is increasing, then y_α is an upper bound of $(y_\beta)_{\beta<\alpha}$ for every $\alpha < \mu$. This result along with (8.71) imply that $x_\alpha \leq y_\alpha$ for every $\alpha < \mu$. As in the proof of part (a), one can show that μ is a successor, and that $x_{\mu-1} = \sup\{c, y_{\mu-1}\}$. Since $c \leq y_{\mu-1}$, then $x_{\mu-1} = y_{\mu-1}$, whence p wins the play. $\qquad\square$

The next results are duals to those proved in Proposition 8.118.

Proposition 8.119. *Assume that the following conditions hold whenever* $(z_\beta)_{\beta<\mu}$ *and* $(w_\beta)_{\beta<\mu}$ *are decreasing in* P *and* $w_\beta \in \mathcal{F}(z_\beta)$ *for every* $\beta < \mu$.

(Fc) $(w_\beta)_{\beta<\mu}$ *has an order limit in* P.
(Fd) If z *is a lower bound of both* $(z_\beta)_{\beta<\mu}$ *and* $(w_\beta)_{\beta<\mu}$, *then* $(w_\beta)_{\beta<\mu}$ *has a lower bound in* $\mathcal{F}(z)$.

Let $((z_\beta, w_\beta))_{\beta<\nu}$ *be a maximally extended play of the o.p-e game where* p *uses the following strategy,* $b \in P$ *being fixed:*

$$z_\alpha = \inf\{b, \{w_\beta\}_{\beta<\alpha}\} \text{ for every } \alpha < \nu. \tag{8.72}$$

Then the following assertions hold true:

(a) *If b is an inf-center of the set of order limits of decreasing and order convergent sequences of $\mathcal{F}[P]$ in P, then $z_{\nu-1} = \inf\{b, w_{\nu-1}\}$.*

(b) *If $w_0 \leq z_0$, $b = z_0$ in (8.72), and conditions (Fc) and (Fd) are fulfilled whenever $(z_\beta)_{\beta<\mu}$ and $(w_\beta)_{\beta<\mu}$ are decreasing in P and $\mathcal{F}(z_\beta) \ni w_\beta \leq z_\beta$ for every $\beta < \mu$, then $z_{\nu-1} = w_{\nu-1}$, so that p wins the play.*

By means of Propositions 8.118 and 8.119, we are now in the position to prove our main result, which reads as follows.

Theorem 8.120. *Assume that the following conditions hold for sequences $(x_\beta)_{\beta<\mu}$ and $(y_\beta)_{\beta<\mu}$ where $y_\beta \in \mathcal{F}(x_\beta)$ for every $\beta < \mu$.*

(F0) $(y_\beta)_{\beta<\mu}$ has an order limit in P if both $(x_\beta)_{\beta<\mu}$ and $(y_\beta)_{\beta<\mu}$ are increasing or decreasing.

(F1) If $(x_\beta)_{\beta<\mu}$ and $(y_\beta)_{\beta<\mu}$ are increasing and have a common upper bound x in P, then $(y_\beta)_{\beta<\mu}$ has an upper bound in $\mathcal{F}(x)$.

(F2) If $(x_\beta)_{\beta<\mu}$ and $(y_\beta)_{\beta<\mu}$ are decreasing and have a common lower bound x in P, then $(y_\beta)_{\beta<\mu}$ has a lower bound in $\mathcal{F}(x)$.

Then player p has a winning strategy for maximally extended plays of the o.p-e game if the set of order limits of increasing and order convergent sequences of $\mathcal{F}[P]$ has a sup-center in P, or if the set of order limits of decreasing and order convergent sequences of $\mathcal{F}[P]$ has an inf-center in P.

Proof: We prove the assertion in the case when the set of order limits of increasing and order convergent sequences of $\mathcal{F}[P]$ has a sup-center in P. The hypotheses of Proposition 8.118(a) are valid, whence every maximally extended play of the o.p-e game, where p uses strategy (8.66), stops at position $(x_{\mu-1}, y_{\mu-1})$, where $x_{\mu-1} = \sup\{c, y_{\mu-1}\}$. In particular, $y_{\mu-1} \leq x_{\mu-1}$. If equality holds, then p has won.

If $y_{\mu-1} < x_{\mu-1}$, choose $z_0 = x_{\mu-1}$ and $w_0 = y_{\mu-1}$, and consider maximally extended plays $((z_\beta, w_\beta))_{\beta<\nu}$ of the o.p-e game where the positions of p are determined by strategy (8.72) with $b = z_0$. The given hypotheses imply that the hypotheses of Proposition 8.119(b) are valid. Thus p wins. Consequently, the above described combination of strategies (8.66) and (8.72) is a winning strategy for p.

Applying first Proposition 8.119(a), and then Proposition 8.118(b), one can prove the assertion when the set of order limits of decreasing and order convergent sequences of $\mathcal{F}[P]$ has an inf-center in P.　　　□

The next consequence of Theorem 8.120 contains the result stated in Chap. 1.

Corollary 8.121. *Let P be a poset that has a sup-center or an inf-center, and whose monotone sequences are order convergent. Then p has a winning strategy in the o.p-e game where $\mathcal{F}(x) = P$ for all $x \in P$.*

Proof: The hypotheses (F1) and (F2) of Theorem 8.120 are trivially valid, since $\mathcal{F}(x) = P$, and (F0) follows from a hypothesis. Because P has a sup-center or an inf-center, it follows that p has the asserted winning strategy due to Theorem 8.120.   □

8.6.3 Applications and Special Cases

In this subsection we apply the results of Sect. 8.6.2 to solve fixed point problems, operator equations, and inclusion problems. In what follows, $P = (P, \leq)$ is a nonempty poset and $\mathcal{F} : P \to 2^P \setminus \emptyset$ is a set-valued mapping. When we refer to results derived in Sect. 8.6.2 for o.p-e games, the values $\mathcal{F}(x)$ of $\mathcal{F}$ are assumed to be sets of possible response positions y of q to positions x of p. If p wins such a play, it ends at a position pair (x, y) where $x = y$. Thus $x \in \mathcal{F}(x)$, i.e., x is a fixed point of $\mathcal{F}$.

Definition 8.122. *We say that a mapping* $\mathcal{F} : P \to 2^P \setminus \emptyset$ *is* **semi-increasing upward** *if* $x \leq y$, $z \in \mathcal{F}(x)$, *and* $z \leq y$ *imply that* $[z) \cap \mathcal{F}(y)$ *is nonempty, and* **semi-increasing downward** *if* $x \leq y$, $w \in \mathcal{F}(y)$, *and* $x \leq w$ *imply that* $(w] \cap \mathcal{F}(x)$ *is nonempty.* $\mathcal{F}$ *is* **semi-increasing** *if* $\mathcal{F}$ *is semi-increasing upward and downward.*

Recall that $\mathcal{F}$ is said to be **increasing upward** if $x \leq y$ in P and $z \in \mathcal{F}(x)$ imply that $[z) \cap \mathcal{F}(y)$ is nonempty, **increasing downward** if $x \leq y$ in P and $w \in \mathcal{F}(y)$ imply that $(w] \cap \mathcal{F}(x)$ is nonempty, and **increasing** if $\mathcal{F}$ is increasing upward and downward.

According to Definition 2.5 and Lemma 8.110, a nonempty subset A of a subset Y of P is **order compact upward** in Y if for every increasing and order convergent sequence $(y_\beta)_{\beta < \mu}$ of Y, the intersection $\cap \{[y_\beta) \cap A : \beta < \mu\}$ is nonempty whenever $[y_\beta) \cap A$ is nonempty for every $\beta < \mu$. If for every decreasing and order convergent sequence $(y_\beta)_{\beta < \mu}$ of Y the intersection of the sets $(y_\beta] \cap A$, $\beta < \mu$ is nonempty whenever all these sets are nonempty, then A is **order compact downward** in Y. If both these properties hold, then A is **order compact** in Y. If $Y = A$, then A is **order compact**.

If A has the greatest element (respectively the smallest element), then A is order compact upward (respectively downward) in any subset of P that contains A. Thus an order compact set is not necessarily topologically compact, not even closed. On the other hand, every compact subset A of an ordered topological space P (the sets $[a)$ and $(a]$ are closed for each $a \in P$) is order compact in every subset of P containing A. Every poset is order compact in itself.

The following lemma is an easy consequence of above definitions.

Lemma 8.123. *(a) If $\mathcal{F}$ is increasing (upward and/or downward), then it is semi-increasing (upward and/or downward).*
(b) If $[x) \cap \mathcal{F}(x) \neq \emptyset$ (respectively $(x] \cap \mathcal{F}(x) \neq \emptyset$) for every $x \in P$, then $\mathcal{F}$ is semi-increasing upward (respectively semi-increasing downward).

(c) If $\max \mathcal{F}(x)$ *(respectively* $\min \mathcal{F}(x)$*) exists for every* $x \in P$*, or if increasing (respectively decreasing) sequences of* $\mathcal{F}[P]$ *are finite, then every* $\mathcal{F}(x)$ *is order compact upward (respectively downward) in* $\mathcal{F}[P]$*.*

Our first fixed point result is a consequence of Proposition 8.118(b).

Proposition 8.124. *Assume that a mapping* $\mathcal{F} : P \rightarrow 2^P \setminus \emptyset$ *is semi-increasing upward, that its values are order compact upward in* $\mathcal{F}[P]$*, that* $[c) \cap \mathcal{F}(c) \neq \emptyset$ *for some* $c \in P$*, and that* (y_β) *has an order limit whenever* $x_\beta \leq y_\beta \in \mathcal{F}(x_\beta)$ *for every* β*, and both* (x_β) *and* (y_β) *are increasing. Then* $\mathcal{F}$ *has a fixed point.*

Proof: Assume that $x_\beta \leq y_\beta \in \mathcal{F}(x_\beta)$ for every β, and that (x_β) and (y_β) are increasing. By assumption, hypothesis (Fa) of Proposition 8.118 holds. To show that the hypothesis (Fb) of Proposition 8.118 is satisfied, assume that $(x_\beta)_{\beta < \mu}$ and $(y_\beta)_{\beta < \mu}$ have a common upper bound x in P. To show that $(y_\beta)_{\beta < \mu}$ has an upper bound in $\mathcal{F}(x)$, let $\beta < \mu$ be given. Since $y_\beta \in \mathcal{F}(x_\beta)$, $x_\beta \leq x$ and $y_\beta \leq x$, it follows that $[y_\beta) \cap \mathcal{F}(x)$ is nonempty due to the fact that $\mathcal{F}$ is semi-increasing upward. This holds for every $\beta < \mu$, and since $\mathcal{F}(x)$ is order compact upward in $\mathcal{F}[P]$, then the intersection $\cap \{ [y_\beta) \cap \mathcal{F}(x) : \beta < \mu \}$ is nonempty. Any element of that intersection is an upper bound of $(y_\beta)_{\beta < \mu}$ in $\mathcal{F}(x)$. By hypothesis, $[c) \cap \mathcal{F}(c) \neq \emptyset$ for some $c \in P$. Choosing $x_0 = c$ and $y_0 \in [c) \cap \mathcal{F}(c)$, then $x_0 \leq y_0$. By Proposition 8.118(b), a maximally extended play of the o.p-e game having (x_0, y_0) as its initial position ends at a position $x_{\mu-1} = y_{\mu-1} \in \mathcal{F}(x_{\mu-1})$. Thus $x_{\mu-1}$ is a fixed point of $\mathcal{F}$. □

As a consequence of Proposition 8.119 we obtain the following result.

Proposition 8.125. *Assume that a mapping* $\mathcal{F} : P \rightarrow 2^P \setminus \emptyset$ *is semi-increasing downward, that its values are order compact downward in* $\mathcal{F}[P]$*, that* $(b] \cap \mathcal{F}(b) \neq \emptyset$ *for some* $b \in P$*, and that* (y_β) *has an order limit whenever* $\mathcal{F}(x_\beta) \ni y_\beta \leq x_\beta$ *for every* β *and both* (x_β) *and* (y_β) *are decreasing. Then* $\mathcal{F}$ *has a fixed point.*

The following result is a consequence of Theorem 8.120.

Theorem 8.126. *A semi-increasing mapping* $\mathcal{F} : P \rightarrow 2^P \setminus \emptyset$ *whose values are order compact in* $\mathcal{F}[P]$ *has a fixed point if* (y_β) *has an order limit whenever* $y_\beta \in \mathcal{F}(x_\beta)$ *for every* β*, and both* (x_β) *and* (y_β) *are increasing or decreasing, and if the set of order limits of such sequences* (y_β) *has a sup-center or an inf-center in* P*.*

As an application of Theorem 8.126 we obtain an existence result for the inclusion problem $Lu \in \mathcal{N}u$, where L is a mapping from a set V to a poset P and $\mathcal{N} : V \rightarrow 2^P \setminus \emptyset$.

Proposition 8.127. *Assume that* $L : V \rightarrow P$ *is a bijection, that the values of* $\mathcal{N} : V \rightarrow 2^P \setminus \emptyset$ *are order compact in* $\mathcal{N}[V]$*, that monotone sequences of* $\mathcal{N}[V]$ *have order limits in* P*, that* $\mathcal{N} \circ L^{-1}$ *is semi-increasing, and that* P *has a sup-center or an inf-center. Then the inclusion problem* $Lu \in \mathcal{N}u$ *is solvable.*

Proof: We shall show that the set-valued mapping $\mathcal{F} := \mathcal{N} \circ L^{-1}$ from P to $2^P \setminus \emptyset$ satisfies the hypotheses of Theorem 8.126. Notice first that $\mathcal{F}[P] = \mathcal{N}L^{-1}[P] = \mathcal{N}[V]$. Hence, if $x \in P$ and $u = L^{-1}x$, then $\mathcal{F}(x) = \mathcal{N}L^{-1}x = \mathcal{N}u$. Thus $\mathcal{F}(x)$ is order compact in $\mathcal{F}[P] = \mathcal{N}[V]$ because $\mathcal{N}u$ is order compact. Since by hypothesis monotone sequences of $\mathcal{N}[V] = \mathcal{F}[P]$ have order limits in P, and since P has a sup-center or an inf-center, then $\mathcal{F}$ satisfies the hypotheses of Theorem 8.126. Thus $\mathcal{F}$ has a fixed point x. With $u = L^{-1}x$ we get $Lu = x \in \mathcal{F}(x) = \mathcal{N}L^{-1}x = \mathcal{N}u$, whence u is a solution of the inclusion problem $Lu \in \mathcal{N}u$. $\square$

In case that P is an ordered topological space we obtain the following results.

Corollary 8.128. *Let P be a second countable or metrizable ordered topological space. Then a mapping $\mathcal{F} : P \to 2^P \setminus \emptyset$ whose values are compact has a fixed point in the following cases.*

(a) $\mathcal{F}$ is semi-increasing upward, (y_n) converges whenever both (x_n) and (y_n) are increasing ordinary sequences, $x_n \leq y_n \in \mathcal{F}(x_n)$ for every n, and $[c) \cap \mathcal{F}(c) \neq \emptyset$ for some $c \in P$.

(b) $\mathcal{F}$ is semi-increasing downward, (y_n) converges whenever (x_n) and (y_n) are decreasing ordinary sequences, $y_n \in \mathcal{F}(x_n)$ for every n, and $(b] \cap \mathcal{F}(b) \neq \emptyset$ for some $b \in P$.

(c) $\mathcal{F}$ is semi-increasing, (y_n) converges whenever $y_n \in \mathcal{F}(x_n)$ for every n, (x_n) and (y_n) are increasing or decreasing ordinary sequences, and the set of limits of such sequences (y_n) has a sup-center or an inf-center in P.

Proof: Ad (a) Let $x \in P$ be given. To prove that $\mathcal{F}(x)$ is order compact upward, assume that $(y_\beta)_{\beta < \mu}$ is an increasing sequence in $\mathcal{F}[P]$, and that the sets $[y_\beta) \cap \mathcal{F}(x)$, $\beta < \mu$, are nonempty. Because $(y_\beta)_{\beta < \mu}$ is increasing, then the sets $[y_\beta) \cap \mathcal{F}(x)$, $\beta < \mu$, satisfy the finite intersection property. Thus their intersection is nonempty if $\mathcal{F}(x)$ is compact, and every element from that intersection is an upper bound of $(y_\beta)_{\beta < \mu}$ in $\mathcal{F}(x)$. This proves that $\mathcal{F}(x)$ is order compact upward.

Assume next that $y_\beta \in \mathcal{F}(x_\beta)$ for every β, and both (x_β) and (y_β) are increasing. Thus $C = \{x_\beta\}$ is well-ordered. If (y_n) is an increasing sequence in $W = \{y_\beta\}$, then $y_n \in \mathcal{F}(x_n)$, where $x_n = \min\{x_\beta \in C : y_\beta = y_n\}$ for every n, and (x_n) is increasing. Hence it follows that (y_n) converges by a hypothesis (b). By means of [133, Lemma 1.1.7 and Proposition 1.1.5] we infer that W contains an increasing sequence that converges to $\sup W$. This result and Lemma 8.110 imply that (y_β) has an order limit. The above proof shows that the hypotheses of Proposition 8.124 hold true, which completes the proof (a).

Ad (b) and (c) The assertions (b) and (c) are consequences of Proposition 8.125 and Theorem 8.126, respectively. $\square$

A single-valued mapping $G : P \to P$ is semi-increasing upward if $x \leq y$ and $G(x) \leq y$ imply $G(x) \leq G(y)$, and semi-increasing downward if $x \leq y$ and $x \leq G(y)$ imply $G(x) \leq G(y)$. G is semi-increasing if it is semi-increasing upward and downward. If G is increasing, i.e., if $G(x) \leq G(y)$ whenever $x \leq y$, then G is semi-increasing. If $x \leq G(x)$ for every $x \in P$, then G is semi-increasing upward by Lemma 8.123. Thus, for example, the absolute-value function $G(x) = |x|$, $x \in [0,1] \subset \mathbb{R}$, is semi-increasing upward, but not increasing.

The following results are special cases of Propositions 8.124 and 8.125 and Theorem 8.126.

Proposition 8.129. *A mapping $G : P \to P$ has a fixed point in the following cases.*

(a) *The set $S_+ = \{x \in P : x \leq G(x)\}$ is nonempty, G is semi-increasing upward, and $(G(x_\beta))$ has an order limit whenever both (x_β) and $(G(x_\beta))$ are increasing sequences in S_+.*

(b) *The set $S_- = \{x \in P : G(x) \leq x\}$ is nonempty, G is semi-increasing downward, and $(G(x_\beta))$ has an order limit whenever both (x_β) and $(G(x_\beta))$ are decreasing sequences in S_-.*

(c) *G is semi-increasing, (Gx_β) has an order limit whenever both (x_β) and (Gx_β) are increasing or decreasing sequences, and the set of order limits of such sequences (Gx_β) has a sup-center or an inf-center in P.*

The following examples are adopted from [124].

Example 8.130. Assume that $P = \{x, y, z\}$ is equipped with a partial ordering $<$ such that $x < y$, $x < z$, and y, z are unordered. There exist 27 different self-mappings of P; 11 of them are increasing. The mappings

$$\{(x,y),(y,y),(z,x)\}, \quad \{(x,z),(y,x),(z,z)\}, \quad \{(x,y),(y,y),(z,z)\}, \text{ and}$$
$$\{(x,z),(y,y),(z,z)\}$$

are semi-increasing upward, but not increasing. The first two of them are neither maximalizing nor minimalizing. All these mappings satisfy the hypotheses of Proposition 8.129(a).

Example 8.131. Let $[x]$ denote the greatest integer $\leq x$. The real function

$$G(x) = \begin{cases} D(x) + x, & x < 0, \\ \frac{[x]+1}{2}, & 0 \leq x, \end{cases} \text{ where } D(x) = \begin{cases} 1, & x \text{ is irrational}, \\ 0, & x \text{ is rational}, \end{cases} \quad (8.73)$$

satisfies the hypotheses of Proposition 8.129(a) when $P = \mathbb{R}$. The mapping G, defined by (8.73), is non-increasing, non-extensive, non-ascending, non-maximalizing, non-minimalizing, non-bounded, and non-continuous, and P is not strictly inductive. The greatest fixed point of G is $x = 1$. Every negative rational number and $x = \frac{1}{2}$ are also fixed points of G, whence G does not have the smallest fixed point. The real function

$$G(x) = \begin{cases} \frac{[x]-1}{2} & x \le 0, \\ D(x) + x - 1, & 0 < x, \end{cases}$$

where D is defined in (8.73), satisfies the hypotheses of Proposition 8.129(b) when $P = \mathbb{R}$. The function G is neither semi-increasing upward nor bounded. The smallest fixed point of G is $x = -\frac{3}{2}$. Every positive irrational number is also a fixed point of G, whence G does not have the greatest fixed point.

As a consequence of Proposition 8.129 and Theorem 8.126 we get existence results for the equation $Lu = Nu$.

Proposition 8.132. *Given a nonempty set V, a poset P and mappings L, N : $V \to P$, assume that L is a bijection, and that monotone sequences of $N[V]$ have order limits in P. Then the equation $Lu = Nu$ is solvable in the following cases.*

(a) $Lu \le Nu$ for some $u \in V$, and if $Lu \le Lv$ and $Nu \le Lv$, then $Nu \le Nv$.
(b) $Nu \le Lu$ for some $u \in V$, and if $Lu \le Lv$ and $Lu \le Nv$, then $Nu \le Nv$.
(c) $N \circ L^{-1}$ is semi-increasing, and P has a sup-center or an inf-center.

Proof: Ad (a) We shall show that the mapping $G = N \circ L^{-1}$ satisfies the hypotheses of Proposition 8.129(a). Let $u \in V$ satisfy $Lu \le Nu$. Denoting $u = L^{-1}x$ we have $x = Lu \le Nu = NL^{-1}x = G(x)$. Thus the set $S_+ = \{x \in P : x \le G(x)\}$ is nonempty. To prove that G is semi-increasing upward, assume that $x \le y$ and $G(x) \le y$. Denoting $u = L^{-1}x$ and $v = L^{-1}y$, we have $Lu = x \le y = Lv$ and $Nu = G(x) \le y = Nv$. Thus, by hypothesis, $Nu \le Nv$, or equivalently, $G(x) \le G(y)$. This proves that G is semi-increasing upward. Moreover, increasing sequences of $G[P]$ have order limits in P since $G[P] = NL^{-1}[P] = N[V]$, and since monotone sequences of $N[V]$ have order limits in P. Thus G satisfies the hypotheses of of Proposition 8.129(a), so that it has a fixed point x. Denoting $u = L^{-1}x$, then $Lu = x \in G(x) = NL^{-1}x = Nu$, whence u is a solution of the operator equation $Lu = Nu$.

Ad (b) The conclusion of (b), a consequence of Proposition 8.129(b), and the proof is similar to that of (a).

Ad (c) The hypotheses of Theorem 8.126 are valid when $\mathcal{N} = N$.   □

The following corollary is a consequence of Corollary 8.121 and Proposition 8.129.

Corollary 8.133. *Let P be a subset of an ordered normed space with the properties:*

(P0) Every monotone ordinary sequence $(x_n)_{n=0}^{\infty}$ of P has a weak or a strong limit in P.

(P1) P has a sup-center or an inf-center.

Then the following results are valid.

(a) p has a winning strategy in the o.p-e game if $\mathcal{F}(x) = P$ for every $x \in P$.
(b) Every semi-increasing mapping $G : P \to P$ has a fixed point.

Proof: By Lemma 9.31, property (P0) implies that well-ordered and inversely well-ordered chains of P have a supremums and infimums in P. Thus monotone transfinite sequences have order limits in P, since their ranges are well-ordered or inversely well-ordered chains.   □

Next we present subsets P of ordered normed spaces that satisfy the hypotheses (P0) and (P1) of Corollary 8.133.

Example 8.134. The ball $P = \{x \in E : \|x - c\| \le r\}$ has for every $r > 0$ and $c \in E$ properties (P0) and (P1) assumed in Corollary 8.133 if E is any of the following spaces (cf. [119, Remark 1.1], [152] and the references given therein):

- $\mathbb{R}^N$, ordered coordinatewise and normed by any norm.
- l^p, $1 \le p < \infty$, normed by p-norm and ordered coordinatewise.
- A weakly complete Banach lattice or a UMB-lattice.
- $L^p(\Omega, Y)$, $1 \le p < \infty$, normed by p-norm and ordered a.e. pointwise, where Ω is a σ-finite measure space and Y is any of the spaces given above.
- A Sobolev space $W^{1,p}(\Omega)$ or $W_0^{1,p}(\Omega)$, $1 < p < \infty$, ordered a.e. pointwise, where Ω is a bounded domain in $\mathbb{R}^N$.
- An Orlicz space $L_M(\Omega)$, ordered a.e. pointwise and normed by the Luxemburg norm, where M satisfies the Δ_2-condition.
- An Orlicz–Sobolev space $W^1 L_M(\Omega)$, ordered a.e. pointwise and normed by the Luxemburg norm, where M and its conjugate $\overline{M}$ satisfy the Δ_2-condition.
- A Newtonian space $N^{1,p}(Y)$, $1 < p < \infty$, ordered a.e. pointwise, where Y is a metric space with the doubling measure that supports a $(1, Y)$-Poincaré inequality.

From Corollary 8.133, we deduce that if such a ball P is a game board for the o.p-e game, and if $\mathcal{F}(x) = P$ for every $x \in P$, then p has a winning strategy. Moreover, each semi-increasing mapping $G : P \to P$ has a fixed point. These results hold also when P is the subset of $\mathbb{R}^2$ illustrated by Figure 1.1, or if P is a subset of $\mathbb{R}^N$, defined by

$$P = \{(x_1, \ldots, x_N) \in \mathbb{R}^N : \sum_{j=1}^{N} |x_j - c_j|^p \le r^p\},$$

where $p \in (0, 1)$ and $r > 0$ and $c = (c_1, \ldots, c_N) \in \mathbb{R}^N$.

As noticed in Chap. 1, the existence of the assumed sup-center or inf-center is needed, in general, to ensure the validity of Theorems 8.120 and 8.126. The following counterexample shows that neither the hypothesis (F0) nor (F1) can be dropped, in general.

Example 8.135. Let P be a compact real interval $[a, b]$, $a < b$, ordered by the usual ordering of reals. Define

$$\mathcal{F}(x) = [a, b) \setminus \{x\}, \quad x \in P. \tag{8.74}$$

It is easy to see that the hypotheses of Theorem 8.120 except (F1) are valid, and that a is a sup-center of P. Since $x \notin \mathcal{F}(x)$ for all $x \in P$, then $\mathcal{F}$ has no fixed point. Consequently, if (8.74) defines the set of possible positions of q for each $x \in P$, then p has no winning strategy.

If $P = [a, b)$, and $\mathcal{F}$ is defined by (8.74), then all the other hypotheses of Theorem 8.120 except (F0) hold, and a is a sup-center of P. Also in this case the results of Theorems 8.120 and 8.126 are not valid.

Remark 8.136. (i) As for other pursuit and evasion games in $\mathbb{R}^N$, see, e.g., [154] and the references therein. In [203] an evasion problem is considered on the unit disc of $\mathbb{R}^2$.

(ii) The proofs of classical fixed point theorems in ordered spaces (see, for instance, [1, 145, 176, 209, 214, 215, 228]) do not provide tools to prove Theorem 8.126.

(iii) Semi-monotonicity hypotheses imposed on $\mathcal{F}$ are weaker than the monotonicity hypotheses used so far to prove fixed point results for set-valued mapping in posets. As for a more detailed study of the single-valued case, see [124].

8.7 Notes and Comments

Sections 8.1 and 8.2 were devoted to finite normal-form games. The material was adopted from [218] and from lecture notes of the second author. The main purpose of Sects. 8.2 and 8.4 was to derive conditions under which the extremal Nash equilibria are pure also in case of randomized normal-form games. The major part of the material of Sect. 8.4 is based on [122]. In Sect. 8.3 we studied the existence of extremal pure Nash equilibria for a normal-form game of N players. Some of the results are based on lecture notes of the second author and on the papers [121] and [123]. The extreme value results proved in Sect. 8.3.1 are new. The results of Sect. 8.5 for the existence of undominated and weakly dominating strategies of normal-form games are presented in [136] in the special case when the utility functions are real-valued. In Sect. 8.5 the values of the utility functions are allowed to be in posets. The results of Sect. 8.6 are new. As by-products of the results dealing with the existence of winning strategies in a pursuit and evasion game, we obtain new fixed point results for set-valued functions in posets, as well as new existence results for operator equations and inclusions in posets.

9

Appendix

In this chapter we provide basic facts of the theory of operators of monotone type, as well as the calculus of Clarke's generalized gradient as it is used in Chaps. 3, 4, and 5. The focus, however, is on the basic analysis of vector-valued, HL integrable functions used in the theory of differential and integral equations and inclusions presented in Chaps. 6–7, which represents in itself a new development that is of interest in its own. With the tools provided by that theory we are able to convert the problems under consideration into operator equations and inclusions that then can be solved by means of the results derived in Chap. 2, see Chaps. 6–7. The application of the order-theoretic results of Chap. 2 to the problems studied in Chaps. 6–7 requires a detailed analysis about the existence of supremums and infimums of chains, as well as the existence of order centers of sets in ordered function spaces, which is provided in Sect. 9.2.

9.1 Analysis of Vector-Valued Functions

9.1.1 μ-Measurability and μ-Integrability of Banach-Valued Functions

Our first goal is the foundation of μ-measurability and μ-integrability of Banach-valued functions.

Let $\Omega = (\Omega, \mathcal{A}, \mu)$ be a measure space, i.e., Ω is a nonempty set, $\mathcal{A}$ is a σ-algebra of so-called measurable subsets of Ω, and $\mu : \mathcal{A} \to [0, \infty]$, called (positive) *measure*, is countably additive and $\mu(\emptyset) = 0$. A measurable subset Z of Ω is called μ-*null set* if $\mu(Z) = 0$. We say that a property $\mathbf{P}$ holds for almost every (a.e.) $t \in \Omega$, or a.e. on Ω, if there is a μ-null set Z in Ω such that $\mathbf{P}$ holds for all $t \in \Omega \setminus Z$. A measure μ is called *complete* if each subset of a μ-null set is μ-null set. μ is said to be σ-*finite* if Ω can be represented as a countable union of sets with finite measure.

S. Carl and S. Heikkilä, *Fixed Point Theory in Ordered Sets and Applications: From Differential and Integral Equations to Game Theory*, DOI 10.1007/978-1-4419-7585-0_9, © Springer Science+Business Media, LLC 2011

Let $\Omega = (\Omega, \mathcal{A}, \mu)$ be a measure space and $E = (E, \| \cdot \|)$ a Banach space. We say that a function $u : \Omega \to E$ is μ-*measurable* if u is a.e. pointwise limit of a sequence of *step functions* of the form $t \mapsto \sum_{i=1}^{n} \chi_{A_i}(t) u_i$, where χ_{A_i} denotes the characteristic function of $A_i \in \mathcal{A}$ with $\mu(A_i) < \infty$, and $u_i \in E$, $i = 1, \ldots, n$. By a result due to Pettis (cf., e.g., [227]), a function $u : \Omega \to E$ is μ-measurable if and only if it is weakly μ-measurable and a.e. separably-valued. Applying this property we get the following result.

Lemma 9.1. *If (u_n) is a sequence of μ-measurable functions u_n from Ω to E, and if $u_n(t)$ converges weakly to $u(t)$ for a.e. $t \in \Omega$, then u is μ-measurable.*

Proof: Since each u_n is weakly μ-measurable, then the limit u is also weakly μ-measurable. Moreover, for each $n \in \mathbb{N}_0$ there is a μ-null set Z_n in Ω such that a set $\{u_n(t) : t \in \Omega \setminus Z_n\}$ is separable. Denoting $Z = \bigcup_{n=0}^{\infty} Z_n$, then Z is a μ-null set and the set $D = \overline{co} \bigcup_{n=0}^{\infty} \{u_n(t) : t \in \Omega \setminus Z\}$ is separable. As a closed and convex set D is weakly closed, so that $u(t) \in D$ for a.e. $t \in \Omega$. Thus u is also a.e. separably-valued, and hence μ-measurable by Pettis Theorem. $\square$

Recall that a μ-measurable function $u : \Omega \to E$ is μ-*integrable* if and only if the function $t \mapsto \|u(t)\|$ is μ-integrable. The integral of a step function $u = \sum_{i=1}^{n} \chi_{A_i}(t) u_i$ over Ω is defined by

$$\int_{\Omega} u(s)\, d\mu(s) = \sum_{i=1}^{n} \mu(A_i) u_i.$$

If u is μ-integrable, define

$$\int_{\Omega} u(s)\, d\mu(s) = \lim_{n \to \infty} \int_{\Omega} u_n(s)\, d\mu(s), \qquad (9.1)$$

where $(u_n)_{n=1}^{\infty}$ is any sequence step functions satisfying $\lim_{n \to \infty} u_n(t) = u(t)$ for a.e. $t \in \Omega$. It is easy to see that

$$\left\| \int_{\Omega} u(s)\, d\mu(s) \right\| \leq \int_{\Omega} \|u(s)\|\, d\mu(s).$$

If $A \in \mathcal{A}$, define

$$\int_{A} u(s)\, d\mu(s) = \int_{\Omega} \chi_A(s) u(s)\, d\mu(s).$$

Let E and V be Banach spaces. We say that a function $g : V \to E$ is *demi-continuous* if $x_n \to x$ in V implies $g(x_n) \rightharpoonup g(x)$ in E.

Proposition 9.2. *Let E and V be Banach spaces. If $g : V \to E$ is demi-continuous, and if $u : \Omega \to V$ is μ-measurable, then $g \circ u : \Omega \to E$ is μ-measurable.*

Proof: Let $u : \Omega \to V$ be μ-measurable, and let $(u_n)_{n=0}^{\infty}$ be a sequence of step functions that converges pointwise a.e. on Ω to u. If $g : V \to E$, then the function $g \circ u_n : \Omega \to E$ is for each $n \in \mathbb{N}_0$ a step function, and hence μ-measurable. Assuming that g is demicontinuous, then $g(u_n(t)) \rightharpoonup g(u(t))$ for a.e. $t \in \Omega$. This implies by Lemma 9.1 that $g \circ u$ is μ-measurable.   □

Recall that a closed subset E_+ of a normed space E is an *order cone* if $E_+ + E_+ \subseteq E_+$, $E_+ \cap (-E_+) = \{0\}$ and $cE_+ \subseteq E_+$ for each $c \geq 0$. It is easy to see that the order relation $\leq$, defined by

$$x \leq y \text{ if and only if } y - x \in E_+,$$

is a partial ordering in E, and that $E_+ = \{y \in E : 0 \leq y\}$. The space E, equipped with this partial ordering, is called an *ordered normed space*. The order interval $[y, z] = \{x \in E : y \leq x \leq z\}$ is a closed subset of E for all $y, z \in E$. A sequence (subset) of E is called *order bounded* if it is contained in an order interval $[y, z]$ of E.

We say that an order cone E_+ of a normed space is *normal* if there is such a constant $\lambda \geq 1$ that

$$0 \leq x \leq y \text{ in } E \text{ implies } \|x\| \leq \lambda\|y\|. \tag{9.2}$$

The order cone E_+ is called *regular* if all increasing and order bounded sequences of E_+ converge. If all norm-bounded and increasing sequences of E_+ converge, we say that E_+ is *fully regular*. As for the proof of the following result, see, e.g., [106, Theorems 2.2.1 and 2.4.5].

Lemma 9.3. *Let E_+ be an order cone of a Banach space E. If E_+ is fully regular, it is also regular, and if E_+ is regular, it is also normal. The converse holds if E is weakly sequentially complete.*

The following lemma has been proved in [133].

Lemma 9.4. *Let E be an ordered Banach space. If $u, v : \Omega \to E$ are μ-integrable and $u(t) \leq v(t)$ for a.e. $t \in \Omega$, then*

$$\int_{\Omega} u(s)\, d\mu(s) \leq \int_{\Omega} v(s)\, d\mu(s). \tag{9.3}$$

The set $\mathcal{L}^1(\Omega, E)$ of all the μ-integrable functions from Ω to E is a vector space with respect to usual additions and scalar multiplications of functions. Moreover, $\mathcal{L}^1(\Omega, E)$ is complete with respect to the seminorm $\|\cdot\|_1$ defined by

$$\|u\|_1 = \int_{\Omega} \|u(s)\|\, d\mu(s).$$

Because $\|u\|_1 = 0$ if and only if $u(t) = 0$ for a.e. $t \in \Omega$, the factor space $L^1(\Omega, E) = \mathcal{L}^1(\Omega, E)/\mathcal{N}^1(\Omega, E)$, where $\mathcal{N}^1(\Omega, E) = \{u \in \mathcal{L}^1(\Omega, E) : u(t) =$

0 for a.e. $t \in \Omega\}$, is a Banach space. Unless otherwise stated we identify from now on the a.e. equal functions. Denote by $L^p(\Omega, E)$, $1 \le p \le \infty$, the space of all μ-measurable functions $u : \Omega \to E$ for which $t \mapsto \|u(t)\|^p$ belongs to $L^1(\Omega, \mathbb{R})$. If E is an ordered Banach space, then $L^p(\Omega, E)$ is an ordered Banach space with respect to the p-norm:

$$\|u\|_p = \left(\int_\Omega \|u(t)\|^p \, d\mu(t) \right)^{\frac{1}{p}}, \ 1 \le p < \infty, \quad \|u\|_\infty = \operatorname{essup}\{\|u(t)\| : t \in \Omega\},$$

and the partial ordering

$$u \le v \ \text{ if and only if } u(t) \le v(t) \text{ for a.e. } t \in \Omega. \tag{9.4}$$

Recall (cf., e.g., [227]) that a sequence (x_n) of E that converges weakly to x is bounded, i.e., $\sup_n \|x_n\| < \infty$, and

$$\|x\| \le \liminf_{n \to \infty} \|x_n\|. \tag{9.5}$$

The next result is a kind of weak monotone convergence theorem.

Proposition 9.5. *Given an ordered Banach space E and $1 \le p < \infty$, assume that a monotone and bounded sequence $(u_n)_{n=0}^\infty$ of $L^p(\Omega, E)$ converges weakly a.e. pointwise to $u : \Omega \to E$. Then $u \in L^p(\Omega, E)$, and $u = \sup_n u_n$ if $(u_n)_{n=0}^\infty$ is increasing, and $u = \inf_n u_n$ if $(u_n)_{n=0}^\infty$ is decreasing.*

Proof: Let $(u_n)_{n=0}^\infty$ be an increasing and bounded sequence in $L^p(\Omega, E)$, and assume that it converges weakly a.e. pointwise to $u : \Omega \to E$. Then

$$u(t) = \sup_n u_n(t) \ \text{ for a.e. } t \in \Omega. \tag{9.6}$$

The function u is measurable by Lemma 9.1. In view of (9.5) we obtain

$$\|u(t)\|^p \le \liminf_{n \to \infty} \|u_n(t)\|^p < \infty \ \text{ for a.e. } t \in \Omega.$$

The above inequality, Fatou's Lemma, and the boundedness of (u_n) in $L^p(\Omega, E)$ imply that

$$\int_\Omega \|u(t)\|^p d\mu(t) \le \int_\Omega \liminf_{n \to \infty} \|u_n(t)\|^p d\mu(t) \le \liminf_{n \to \infty} \int_\Omega \|u_n(t)\|^p d\mu(t) < \infty.$$

This proves that $t \mapsto \|u(t)\| \in L^p(\Omega, \mathbb{R})$, so that $u \in L^p(\Omega, E)$. Moreover, it follows from (9.6) that $u = \sup_n u_n$. $\qquad\square$

Definition 9.6. *Let E be a Banach space, and let μ be the Lebesgue measure on $\mathbb{R}^m$. A μ-measurable mapping u from a Lebesgue measurable subset Ω to a Banach space E is called* **strongly measurable** *on Ω. If u is also μ-integrable it is called* **Bochner integrable on** *on Ω. The μ-integral of u over Ω is called the* **Bochner integral**, *and is denoted by $\int_\Omega u(s) \, ds$.*

If Ω is a closed real interval $[a, b]$, denote

$$\int_\Omega u(s)\,ds = \int_a^b u(s)ds = -\int_b^a u(s)ds.$$

The next result has been proved in [133].

Lemma 9.7. *Let E be an ordered Banach space, let $u, v : [a, b] \to E$ be Bochner integrable, and assume $0 \le u(s) \le v(s)$ for a.e. $s \in [a, b]$. If A, B are Lebesgue measurable subsets of $[a, b]$ and $A \subseteq B$, then*

$$0 \le \int_A u(s)\,ds \le \int_B v(s)\,ds. \tag{9.7}$$

9.1.2 HL Integrability

In this subsection we study properties of the so-called Henstock–Lebesgue (shortly HL) integrable functions $u : [a, b] \to E$, where E is a Banach space and $[a, b]$ a nonempty closed interval of $\mathbb{R}$.

We say that $D = \{(\xi_i, I_i)\}$ is a *K-partition of $[a, b]$* if $\{I_i\}$ is a finite collection of closed subintervals I_i of $[a, b]$ that are non-overlapping, i.e., their interiors are pairwise disjoint, whose union is $[a, b]$, and if every tag ξ_i belongs to I_i. A partition D is called a *partial K-partition* if $\cup_i I_i$ is a proper subset of $[a, b]$. Given a function $\delta : [a, b] \to (0, \infty)$ (called a *gauge* of $[a, b]$), we say that a K-partition $D = \{(\xi_i, I_i)\}$ is *δ-fine* if $I_i \subset (\xi_i - \delta(\xi_i), \xi_i + \delta(\xi_i))$ for every i. Such a partition exists due to Cousin's Lemma (cf., e.g., [168, Lemma 6.2.6]). The length of I_i is denoted by $|I_i|$.

A function $u : [a, b] \to E$ is *HL integrable* if there is an E-valued function $I \mapsto F(I)$ of closed subintervals I of $[a, b]$ that is additive on non-overlapping intervals and has the following property: If $\epsilon > 0$, there is a gauge δ of $[a, b]$ such that for every δ-fine K-partition $D = \{(\xi_i, I_i)\}$,

$$\sum_i \|u(\xi_i)|I_i| - F(I_i)\| < \epsilon. \tag{9.8}$$

If u is HL integrable on $[a, b]$, it is HL integrable on every closed subinterval $I = [c, d]$ of $[a, b]$, and $F(I)$ is the *Henstock–Kurzweil* integral of u over I, i.e.,

$$F(I) = {}^K\!\!\int_I u(s)\,ds = {}^K\!\!\int_c^d u(s)\,ds. \tag{9.9}$$

If $u : [a, b] \to E$ is HL integrable and $c \in E$, then the relation

$$f(t) = c + {}^K\!\!\int_a^t u(s)\,ds, \quad t \in [a, b], \tag{9.10}$$

defines a function $f \in C([a, b], E)$, which is called a *primitive* of u. Denoting in (9.8) $I_i = [t_{i-1}, t_i]$, then $|I_i| = t_i - t_{i-1}$. Moreover, (9.9) and (9.10) imply that $F(I_i) = f(t_i) - f(t_{i-1})$. Thus (9.8) can be rewritten as

$$\sum_i \|f(t_i) - f(t_{i-1}) - u(\xi_i)(t_i - t_{i-1})\| < \epsilon. \tag{9.11}$$

This shows that the above definition and that given in Chap. 1 for HL integrability are equivalent.

The proofs for the results of the next lemma can be found, e.g., in [207].

Lemma 9.8. *(a) Every HL integrable function is strongly measurable.*
(b) A Bochner integrable function $u : [a, b] \to E$ is HL integrable, and

$$\int_I u(s)\, ds = {}^K\!\!\int_I u(s)\, ds$$

whenever I is a closed subinterval of $[a, b]$.

The following variant of Saks–Henstock Lemma is a consequence of [207, Lemma 3.4.1 and Lemma 3.6.15].

Lemma 9.9. *If $u : [a, b] \to E$ is HL integrable, then for every $\epsilon > 0$ there exists a gauge $\delta : [a, b] \to (0, \infty)$ such that if $D = \{(\xi_i, I_i)\}$ is any δ-fine K-partition or partial K-partition of $[a, b]$, then*

$$\sum_i \left\| u(\xi_i)|I_i| - {}^K\!\!\int_{I_i} u(s)\, ds \right\| \le \epsilon.$$

The following result is adapted from [192] and from [201].

Proposition 9.10. *If $v : I \to E$ is HL integrable and $\varphi : I \to \mathbb{R}$ is of bounded variation, then $\varphi \cdot v$ is HL integrable.*

The following lemma plays a central role in both the theory and application of HL integrability in ordered Banach spaces.

Lemma 9.11. *Let E be an ordered Banach space, and let $u_\pm : [a, b] \to E$ be HL integrable. If $u_-(s) \le u_+(s)$ for a.e. $s \in [a, b]$, and if I is a closed subinterval of $[a, b]$, then*

$${}^K\!\!\int_I u_-(s)\, ds \le {}^K\!\!\int_I u_+(s)\, ds. \tag{9.12}$$

Proof: It suffices to prove the assertion when $I = [a, b]$. In view of [207, Theorems 3.3.7 and 3.6.4] we may assume that $u_-(s) \le u_+(s)$ for all $s \in [a, b]$. Denoting $u = u_+ - u_-$, then $u(s)$ belongs to the order cone E_+ of E for all $s \in [a, b]$. Let $I \mapsto F(I)$ be as in the definition of HL integrability. Choose for each $n \in \mathbb{N}$ a gauge δ_n of $[a, b]$ and a δ_n-fine K-partition $D_n = \{(\xi_i^n, I_i^n)\}$ so that

$$\sum_i \|u(\xi_i^n)|I_i^n| - F(I_i^n)\| < \frac{1}{n}.$$

Denoting $y_n = \sum_i u(\xi_i^n)|I_i^n|$, we obtain for all $n \in \mathbb{N}$

$$\|y_n - F[a,b]\| = \|\sum_i u(\xi_i^n)|I_i^n| - \sum_i F(I_i^n)\| \le \sum_i \|u(\xi_i^n)|I_i^n| - F(I_i^n)\| < \frac{1}{n}.$$

Thus $F[a,b] = \lim_{n\to\infty} y_n \in E_+$, since E_+ is closed, and since $y_n \in E_+$ for every $n \in \mathbb{N}$. Consequently,

$$0 \le F[a,b] = {}^K\!\!\int_a^b u(s)\,ds = {}^K\!\!\int_a^b u_+(s)\,ds - {}^K\!\!\int_a^b u_-(s)\,ds.$$

This proves (9.12) when $I = [a,b]$. $\qquad\qquad\qquad\qquad\qquad\qquad\qquad$ □

Denote by $\mathcal{I}$ the set of all closed subintervals of $[a,b]$. The proof of the next lemma exploits ideas of the proof of [191, Theorem 4.1], and the normality hypothesis of E_+.

Lemma 9.12. *Let E be a Banach space ordered by a normal order cone E_+, and let $u : [a,b] \to E_+$ be HL integrable. Then there exist Lebesgue measurable sets B_j, $j \in \mathbb{N}$ such that $B_j \subseteq B_{j+1}$ for every $j \in \mathbb{N}$, $\cup_j B_j = [a,b]$, and u is Bochner integrable on every B_j satisfying*

$$\lim_{j\to\infty} \int_I \chi_{B_j}(s)u(s)\,ds = {}^K\!\!\int_I u(s)\,ds \quad \text{uniformly with respect to } I \in \mathcal{I}.$$

$$(9.13)$$

Proof: Since u is strongly measurable by Lemma 9.8, it can be shown (cf. [207, Proposition 1.1.3] and Remark after it) that the sets

$$B_j := \{t \in [a,b] : \|u(t)\| \le j\}, \quad j \in \mathbb{N} \qquad\qquad (9.14)$$

are Lebesgue measurable. Obviously, $B_j \subseteq B_{j+1}$ for every $j \in \mathbb{N}$, and $\cup_j B_j = [a,b]$. Because every restriction $u|B_j$ is bounded and strongly measurable, then u is Bochner integrable on every B_j. Thus the functions $u_j : [a,b] \to E_+$, $j \in \mathbb{N}$, defined by

$$u_j(t) := \chi_{B_j}(t)u(t), \quad t \in [a,b], \qquad\qquad (9.15)$$

are Bochner integrable, and hence also HL integrable on $[a,b]$. The definitions (9.14) and (9.15) imply that

$$0 \le u_j(t) \le u_{j+1}(t) \le u(t), \quad t \in [a,b],\ j \in \mathbb{N}, \qquad\qquad (9.16)$$

Define

$$F_{B_j}(I) := \int_I \chi_{B_j}(s)u(s)\,ds,\ j \in \mathbb{N},\ I \in \mathcal{I}, \quad F(I) := {}^K\!\!\int_I u(s)\,ds,\ I \in \mathcal{I}.$$

$$(9.17)$$

From Lemma 9.11, (9.16) and (9.17) it follows that if $0 < j < k$, then for all $I \in \mathcal{I}$,

$0 \leq F_{B_j}(I) \leq F_{B_k}(I) \leq F(I)$, and hence $0 \leq F(I) - F_{B_k}(I) \leq F(I) - F_{B_j}(I)$.

Since E_+ is normal, we obtain

$$\|F(I) - F_{B_k}(I)\| \leq \lambda \|F(I) - F_{B_j}(I)\|, \quad 0 < j < k, \ I \in \mathcal{I}. \tag{9.18}$$

Let $\epsilon > 0$ be given. By Lemma 9.9 there exist $\delta_j : [a, b] \to (0, \infty)$, $j \in \mathbb{N}$, such that if $D_j = \{(\xi_i, I_i)\}$ is a δ_j-fine partial K-partition of $[a, b]$ with $\{\xi_i\} \subset B_j$, then

$$\sum_i \|F_{B_j}(I_i) - u_j(\xi_i)|I_i|\| \leq 2^{-j-1}\epsilon, \quad \text{and} \quad \sum_i \|F(I_i) - u(\xi_i)|I_i|\| \leq 2^{-j-1}\epsilon.$$

These results imply, since $u_j(\xi) = u(\xi)$ for every $\xi \in B_j$, that

$$\sum_i \|F_{B_j}(I_i) - F(I_i)\| \leq 2^{-j}\epsilon \tag{9.19}$$

for every δ_j-fine partial K-partition $D_j = \{(\xi_i, I_i)\}$ of $[a, b]$ with $\{\xi_i\} \subset B_j$. Define $\delta : [a, b] \to (0, \infty)$ by

$$\delta(\xi) := \delta_{j(\xi)}(\xi), \ \xi \in [a, b], \quad \text{where } j(\xi) = \min\{j \in \mathbb{N} : \|u(\xi)\| \leq j\}.$$

Let $D = \{(\xi_i, I_i)\}$ be a δ-fine K-partition of $[a, b]$. Denoting

$$m = \min\{j(\xi_i) : \xi_i \in D\}, \quad n = \max\{j(\xi_i) : \xi_i \in D\}, \quad \text{and } B_0 = \emptyset,$$

it follows that partitions

$$D_j = \{(\xi_i, I_i) \in D : \xi_i \in B_j \setminus B_{j-1}\}, \quad m \leq j \leq n,$$

are δ_j-fine partial K-partitions of $[a, b]$ with $\{\xi_i\} \subset B_j$. Thus the results (9.18) and (9.19) imply that

$$\|F([a, b]) - F_{B_n}([a, b])\| \leq \sum_i \|F(I_i) - F_{B_n}(I_i)\|$$

$$= \sum_{j=m}^{n} \sum_{I_i \in D_j} \|F(I_i) - F_{B_n}(I_i)\|$$

$$\leq \sum_{j=m}^{n} \sum_{I_i \in D_j} \lambda \|F(I_i) - F_{B_j}(I_i)\| < \lambda\epsilon.$$

In view of the above result and (9.18) we then have

$$\|F([a, b]) - F_{B_k}([a, b])\| < \lambda^2 \epsilon \quad \text{whenever} \ n < k. \tag{9.20}$$

For every $j \in \mathbb{N}$, $u - u_j$ is E_+-valued and HL integrable on $[a, b]$. Applying Lemma 9.11, (9.15), (9.16), (9.17), and the additivity property of interval functions, it is easy to show that

$$0 \le F(I) - F_{B_j}(I) \le F([a,b]) - F_{B_j}([a,b]), \quad j \in \mathbb{N},$$

for every $I \in \mathcal{I}$. Thus

$$\|F(I) - F_{B_j}(I)\| \le \lambda \|F([a,b]) - F_{B_j}([a,b])\| \text{ whenever } j \in \mathbb{N} \text{ and } I \in \mathcal{I}.$$

This result and (9.20) imply that

$$\|F(I) - F_{B_j}(I)\| \le \lambda^3 \epsilon \text{ whenever } n < j \text{ and } I \in \mathcal{I}.$$

The above proof shows that $\lim_{j\to\infty} F_{B_j}(I) = F(I)$ uniformly with respect to $I \in \mathcal{I}$. In view of notations (9.17), this result is equivalent to (9.13). $\quad\square$

The next lemma is used in the proof of the Dominated Convergence Theorem for HL integrable functions.

Lemma 9.13. *Let E be an ordered Banach space with a normal order cone, let v, $v_+ : [a.b] \to E$ be strongly measurable, and assume that $0 \le v(s) \le v_+(s)$ for a.e. $s \in [a,b]$. If v_+ is HL integrable, then v is HL integrable.*

Proof: Assume that v_+ is HL integrable. By Lemma 9.12, there exist Lebesgue measurable sets B_j, $j \in \mathbb{N}$ such that $B_j \subseteq B_{j+1}$ for every $j \in \mathbb{N}$, $\cup_j B_j = [a,b]$, v_+ is Bochner integrable on every B_j, and

$$\lim_{j\to\infty} \int_I \chi_{B_j}(s)v_+(s)\,ds = {}^K\!\!\int_I v_+(s)\,ds \tag{9.21}$$

for every closed subinterval I of $[a,b]$. Since the order cone of E is normal, it follows

$$\|v(s\| \le \lambda \|v_+(s)\| \text{ for a.e. } s \in [a,b]. \tag{9.22}$$

This result implies that v is Bochner integrable on every B_j because v_+ is Bochner integrable. For every closed subinterval I of $[a,b]$ denote

$$F_+(I) = {}^K\!\!\int_I v_+(s)\,ds, \ F_j^+(I) = \int_I \chi_{B_j}(s)v_+(s)\,ds, \ F_j(I) = \int_I \chi_{B_j}(s)v(s)\,ds. \tag{9.23}$$

From Lemma 9.7 it follows that; if $j < i$, then

$$0 \le F_i(I) - F_j(I) = \int_I \chi_{B_i\backslash B_j}(s)v(s)\,ds \le \int_I \chi_{B_i\backslash B_j}(s)v_+(s)\,ds$$
$$= F_i^+(I) - F_j^+(I). \tag{9.24}$$

From (9.24) and (9.2) we obtain

$$\|F_i(I) - F_j(I)\| \le \lambda \|F_i^+(I) - F_j^+(I)\| \text{ whenever } j < i. \tag{9.25}$$

In view of (9.21) and (9.23) the sequence $(F_j^+(I))_{j\in\mathbb{N}}$ converges, and hence it is a Cauchy sequence. Thus $(F_j(I))_{j\in\mathbb{N}}$ is, by (9.25), a Cauchy sequence, whence it converges. It is easy to see that the limit relation

$$F(I) := \lim_{j \to \infty} F_j(I) = \sup_j F_j(I) \qquad (9.26)$$

defines an additive E-valued interval function. Passing to the limit in (9.25) as $i \to \infty$, it follows that

$$\|F(I) - F_j(I)\| \le \lambda \|F_+(I) - F_j^+(I)\| \quad \text{for all } j \in \mathbb{N}. \qquad (9.27)$$

Denote

$$u_j(s) = \chi_{B_j}(s)v(s), \quad u_j^+(s) = \chi_{B_j}(s)v_+(s), \quad s \in [a, b], \ j \in \mathbb{N}. \qquad (9.28)$$

For each $\xi \in [a, b]$, there exists a $j(\xi) \in \mathbb{N}$ such that $\xi \in B_j$ for every $j \ge j(\xi)$. Thus

$$u_j(\xi) = v(\xi) \ \text{ and } u_j^+(\xi) = v_+(\xi) \ \text{ whenever } \xi \in [a, b] \text{ and } \ j \ge j(\xi). \qquad (9.29)$$

Given any point ξ_i of $[a, b]$, any closed subinterval I_i of $[a, b]$, and any integer $j \ge j(\xi_i)$, then from (9.27), (9.28), and (9.29) it follows that

$$
\begin{aligned}
\|v(\xi_i)|I_i| - F(I_i)\| &= \|u_j(\xi_i)|I_i| - F(I_i)\| \\
&\le \|u_j(\xi_i)|I_i| - F_j(I_i)\| + \|F_j(I_i) - F(I_i)\| \\
&\le \|u_j(\xi_i)|I_i| - F_j(I_i)\| + \lambda \|F_j^+(I_i) - F_+(I_i)\| \\
&\le \|u_j(\xi_i)|I_i| - F_j(I_i)\| + \lambda \|v_+(\xi_i)|I_i| - F_j^+(I_i)\| + \lambda \|v_+(\xi_i)|I_i| - F_+(I_i)\| \\
&= \|u_j(\xi_i)|I_i| - F_j(I_i)\| + \lambda \|u_j^+(\xi_i)|I_i| - F_j^+(I_i)\| + \lambda \|v_+(\xi_i)|I_i| - F_+(I_i)\|.
\end{aligned}
$$
$$(9.30)$$

Let $\epsilon > 0$ be given. Since the functions u_j and u_j^+ are Bochner integrable, they are HL integrable in view of Lemma 9.8. Thus, by Lemma 9.9, there exist gauges $\delta_j : [a, b] \to (0, \infty)$, $j \in \mathbb{N}$, for which

$$\sum_i \|u_j(\xi_i)|I_i| - F_j(I_i)\| < \frac{\epsilon}{2^j} \ \text{ and } \ \sum_i \|u_j^+(\xi_i)|I_i| - F_j^+(I_i)\| < \frac{\epsilon}{2^j} \qquad (9.31)$$

whenever $D = \{(\xi_i, I_i)\}$ is a δ_j-fine K-partition or partial K-partition of $[a, b]$. Choose also a gauge $\delta_+ : [a, b] \to (0, \infty)$ such that

$$\sum_i \|v_+(\xi_i)|I_i| - F_+(I_i)\| < \epsilon \qquad (9.32)$$

whenever $D = \{(\xi_i, I_i)\}$ is a δ_+-fine K-partition of $[a, b]$, and define a gauge $\delta : [a, b] \to (0, \infty)$ by

$$\delta(\xi) = \min\{\delta_+(\xi), \delta_{j(\xi)}(\xi)\}, \quad \xi \in [a, b].$$

Let $D = \{(\xi_i, I_i)\}$ be a δ-fine K-partition of $[a, b]$. By (9.30) we have, for each i,

$$\|v(\xi_i)|I_i| - F(I_i)\| \leq \|u_{j(\xi_i)}(\xi_i)|I_i| - F_{j(\xi_i)}(I_i)\| + \lambda\|u_{j(\xi_i)}^+(\xi_i)|I_i| - F_{j(\xi_i)}^+(I_i)\|$$
$$+ \lambda\|v_+(\xi_i)|I_i| - F_+(I_i)\|. \tag{9.33}$$

Summing both sides of (9.33) over i, and denoting $m = \min\{j(\xi_i)\}$ and $n = \max\{j(\xi_i)\}$, we obtain

$$\sum_i \|v(\xi_i)|I_i| - F(I_i)\| \leq \sum_{j=m}^{n} \sum \{\|u_j(\xi_i)|I_i| - F_j(I_i)\| : j(\xi_i) = j\}$$
$$+ \lambda \sum_{j=m}^{n} \sum \{\|u_j^+(\xi_i)|I_i| - F_j^+(I_i)\| : j(\xi_i) = j\}$$
$$+ \lambda \sum_i \|v_+(\xi_i)|I_i| - F_+(I_i)\|. \tag{9.34}$$

The inner sums on the right-hand side of (9.34) correspond to δ_j-fine partial K-partitions of $[a, b]$, whence they are $\leq \frac{\epsilon}{2^j}$, since (9.31) is valid also for partial K-partitions of $[a, b]$. The last sum on the right-hand side of (9.34) is $\leq \epsilon$ by (9.32). It then follows from (9.34) that

$$\sum_i \|v(\xi_i)|I_i| - F(I_i)\| \leq \sum_{j=m}^{n} \frac{\epsilon}{2^j} + \lambda \sum_{j=m}^{n} \frac{\epsilon}{2^j} + \lambda\epsilon < (3\lambda + 2)\epsilon.$$

This proves that v is HL integrable over $[a, b]$, and $F(I) = {}^K\!\!\int_I v(s)\,ds$ for each closed subinterval I of $[a, b]$. □

As a consequence of Lemma 9.13 we obtain the following result.

Proposition 9.14. *Suppose E is an ordered Banach space with a normal order cone, and $u : [a, b] \to E$ is a function that is a.e. pointwise order bounded by HL integrable functions. Then u is HL integrable if and only if u is strongly measurable.*

Proof: If u is HL integrable, it is strongly measurable by Lemma 9.8. Conversely, assume that u is strongly measurable, and let $u_\pm : [a, b] \to E$ be such HL integrable functions that $u_-(s) \leq u(s) \leq u_+(s)$ for a.e. $s \in [a, b]$. Then $u_+ - u_-$ is HL integrable, $u - u_-$ is strongly measurable, and

$$0 \leq u(s) - u_-(s) \leq u_+(s) - u_-(s) \quad \text{for a.e.} \quad s \in [a, b]. \tag{9.35}$$

Then $u - u_-$ is HL integrable by Lemma 9.13, so that $u = u_- + (u - u_-)$ is HL integrable. □

Remark 9.15. Henstok–Kurzweil integral is considered, e.g., in the books [18, 24, 95, 143, 162, 166, 168, 178, 194, 207].

9.1.3 Integrals of Derivatives of Vector-Valued Functions

The results of this subsection are needed in the theory of differential equations in Banach spaces.

Let $E = (E, \| \cdot \|)$ be a Banach space and $J = [a, b]$, $a < b$. We say that a mapping $u : J \to E$ is *absolutely continuous (shortly AC)*, if for each $\epsilon > 0$ there exists a $\delta > 0$ such that for any sequence $[a_j, b_j]$, $j = 1, \dots, n$ of disjoint subintervals of J with $\sum_{j=1}^n (b_j - a_j) < \delta$ we have $\sum_{j=1}^n \|u(b_j) - u(a_j)\| < \epsilon$. If $u : J \to E$ is absolutely continuous, it is also of *bounded variation*, i.e.,

$$V_J(x) = \sup_P \left\{ \sum \|u(t_k) - u(t_{k-1})\| : P = \{t_k\} \text{ is a partition of } J \right\} < \infty.$$

Definition 9.16. *We say that a function* $u : J \to E$ *satisfies the* **Strong Lusin Condition** *if for each null set* Z *of* J *and* $\epsilon > 0$ *there is a gauge* $\delta : J \to (0, \infty)$ *such that* $\sum_i \|u(s_{2i}) - u(s_{2i-1})\| < \epsilon$ *for every* δ-*fine partial* K-*partition* $D = \{(\xi_i, [s_{2i-1}, s_{2i}])\}$ *of* J *with* $\{\xi_i\} \subseteq Z$. *A function* $u : J \to E$ *is said to be* **a.e. differentiable,** *if the derivative*

$$u'(t) = \lim_{\Delta t \to 0} \frac{u(t + \Delta t) - u(t)}{\Delta t}$$

exists for a.e. $t \in J$.

The following result is proved, e.g., in [133].

Theorem 9.17. *If* $u, v : J \to E$ *and* $(t_0, x_0) \in J \times E$, *then the following conditions are equivalent.*

(a) u is absolutely continuous, $u'(t) = v(t)$ for a.e. $t \in J$ and $u(t_0) = x_0$.
(b) v is Bochner integrable and $u(t) = x_0 + \int_{t_0}^t v(s)ds$ for all $t \in J$.

Another version of the *Fundamental Theorem of Calculus* that plays an important role in the application of HL integrability theory to differential equations is provided by the following theorem, cf. [92, Theorem 2.4].

Theorem 9.18. *Given* $u, v : J \to E$ *and* $(t_0, x_0) \in J \times E$, *then the following conditions are equivalent.*

(a) u satisfies the Strong Lusin Condition, $u'(t) = v(t)$ for a.e. $t \in J$ and $u(t_0) = x_0$.
(b) v is HL integrable and $u(t) = x_0 + {}^K\!\!\int_{t_0}^t v(s)ds$ for all $t \in J$ (i.e., u is a primitive of v).

Proof: Assume that (a) holds. Denote by Z the set of those $t \in J$ for which $u'(t)$ does not exist or $u'(t) \neq v(t)$. Given $\epsilon > 0$, there exists for every $\xi \in J \backslash Z$ a $\delta_1(\xi) > 0$ such that

$$\|u(t) - u(\bar{t}) - u'(\xi)(t - \bar{t})\| \leq \epsilon(t - \bar{t})$$

whenever $\xi - \delta_1(\xi) < \bar{t} \leq \xi \leq t < \xi + \delta_1(\xi)$. Since Z is a null set, the Strong Lusin Condition implies the existence of a $\delta_2 : J \to (0, \infty)$ such that if $D = \{(\xi_i, [s_{2i-1}, s_{2i}])\}$ is a partial K-partition of J with $\{\xi_i\} \subseteq Z$, then

$$\sum_i \|u(s_{2i}) - u(s_{2i-1})\| \leq \epsilon.$$

Define $\delta : J \to (0, \infty)$ and $w : J \to E$ by

$$\delta(\xi) = \begin{cases} \min\{\delta_1(\xi), \delta_2(\xi)\}, & \xi \in J \setminus Z, \\ \delta_2(\xi), & \xi \in Z, \end{cases} \qquad w(t) = \begin{cases} u'(t), & t \in J \setminus Z, \\ 0, & t \in Z. \end{cases}$$

Then, for every δ-fine K-partition $D = \{(\xi_i, [t_{i-1}, t_i])\}$ of J,

$$\sum_i \|u(t_i) - u(t_{i-1}) - w(\xi_i)(t_i - t_{i-1})\| = \sum_{\xi_i \in Z} \|u(t_i) - u(t_{i-1})\|$$

$$+ \sum_{\xi_i \in J \setminus Z} \|u(t_i) - u(t_{i-1}) - u'(\xi_i)(t_i - t_{i-1})\| \leq \epsilon(1 + |J|).$$

Thus w is HL integrable and u is its primitive. Moreover, $w(t) = v(t)$ for a.e. $t \in J$, whence v is HL integrable by [207, Theorem 3.6.4], and u is a primitive of v, i.e.,

$$u(t) = c + {}^K\!\!\int_a^t v(s)ds = c + {}^K\!\!\int_a^{t_0} v(s)\,ds + {}^K\!\!\int_{t_0}^t v(s)\,ds.$$

Since $u(t_0) = x_0$, then $c + {}^K\!\!\int_a^{t_0} v(s)\,ds = x_0$, so that $u(t) = x_0 + {}^K\!\!\int_{t_0}^t v(s)ds$, whence (b) holds.

Conversely, assume that (b) holds. Then $u(t_0) = x_0 + {}^K\!\!\int_{t_0}^{t_0} v(s)ds = x_0$. Since u is a primitive of v, from [207, Theorem 7.4.2] it follows that u is a.e. differentiable, and that $u'(t) = v(t)$ for a.e. $t \in J$. Moreover, the proof of [207, Theorem 7.5.1] implies that u satisfies the Strong Lusin Condition. Thus (a) holds. $\qquad\qquad\square$

If $u : J \to E$ is a.e. differentiable, define $u'(t) = 0$ at those points $t \in J$ where u is not differentiable. In view of Theorem 9.17 we then obtain the following result.

Corollary 9.19. *Let* $u : J \to E$ *be a.e. differentiable. Then* u *is absolutely continuous if and only if* u' *is Bochner integrable, and*

$$u(t) - u(t_0) = \int_{t_0}^t u'(s)ds \quad \text{for all } t_0, t \in J.$$

Lemma 9.20. *If* $u : J \to \mathbb{R}$ *is absolutely continuous and* $v : J \to E$ *is both absolutely continuous and a.e. differentiable, then*

$$u(t)v(t) - u(t_0)v(t_0) = \int_{t_0}^t (u(s)v'(s) + u'(s)v(s))ds \quad \text{for all } t_0, t \in J.$$

Proof: For $t, t + \Delta t \in J$, $\Delta t \neq 0$ we have

$$u(t+\Delta t)v(t+\Delta t) - u(t)v(t) = (u(t+\Delta t)-u(t))v(t+\Delta t)+u(t)(v(t+\Delta t)-v(t)). \tag{9.36}$$

Since u and v are absolutely continuous, they are also bounded, whence it follows from (9.36) that

$$\|u(t + \Delta t)v(t + \Delta t) - u(t)v(t)\| \leq M\,|u(t + \Delta t) - u(t)| + m\,\|v(t + \Delta t) - v(t)\|, \tag{9.37}$$

where $M = \max\{\|v(t)\| : t \in J\}$ and $m = \max\{|u(t)| : t \in J\}$. Thus $u \cdot v$ is also absolutely continuous. Because u and v are a.e. differentiable, an easy application of (9.36) implies that

$$(u \cdot v)'(t) = u(t)v'(t) + u'(t)v(t) \quad \text{for a.e. } t \in J. \tag{9.38}$$

(Note, any real-valued absolutely continuous function $u : J \to \mathbb{R}$ is a.e. differentiable.) The assertion follows then from Corollary 9.19.   □

The next result is a consequence of Theorem 9.18.

Corollary 9.21. *The function $u : J \to E$ is a.e. differentiable and satisfies the Strong Lusin Condition if and only if u' is HL integrable, and*

$$u(t) - u(t_0) = {}^K\!\!\int_{t_0}^{t} u'(s)ds \quad \text{for all } t_0, t \in J.$$

Lemma 9.22. *Assume that $u : J \to \mathbb{R}$ is absolutely continuous, and that $v : J \to E$ is a.e. differentiable and satisfies the Strong Lusin Condition. Then*

$$u(t)v(t) - u(t_0)v(t_0) = {}^K\!\!\int_{t_0}^{t} (u(s)v'(s) + u'(s)v(s))ds \quad \text{for all } t_0, t \in J.$$

Proof: In view of Corollary 9.21, v' is HL integrable on J. Since u is absolutely continuous, then $t \mapsto u(t)v'(t)$ is HL integrable on J by Proposition 9.10. The given hypotheses imply also that u' is Lebesgue integrable and v is continuous, whence $t \mapsto u'(t)v(t)$ is strongly measurable by [133, Theorem 1.4.3]. Moreover, $\|u'(t)v(t)\| \leq M|u'(t)|$, where $M = \max\{\|v(t)\| : t \in J\}$. Thus $t \mapsto u'(t)v(t)$ is Bochner integrable, and hence also HL integrable. The above results along with (9.38) imply that $(u \cdot v)'$ is HL integrable. The assertion follows then from Corollary 9.21.   □

As an application of Theorem 9.18 we obtain the following result.

Lemma 9.23. *Let E be a Banach space ordered by an order cone E_+. Assume that $v : [a, b] \to E$ is HL integrable, and that ${}^K\!\!\int_c^d v(s)\,ds \in E_+$ for every closed subinterval $[c, d]$ of $[a, b]$. Then $v(t) \in E_+$ for a.e. $t \in [a, b]$.*

Proof: It follows from Theorem 9.18, that $v(t_0)$ is the derivative of the function $t \mapsto {}^K\!\int_a^t v(s)\, ds$ at t_0 for a.e. $t_0 \in [a, b]$. If $t_0 \in (a, b)$ is such a point, and $(t_n)_{n=1}^\infty$ is a decreasing sequence in (t_0, b) converging to t_0, then

$$\frac{1}{t_n - t_0} {}^K\!\int_{t_0}^{t_n} v(s)\, ds \to v(t_0) \quad \text{when} \quad n \to \infty. \tag{9.39}$$

The given hypothesis and the positivity of $t_n - t_0$ imply that

$$\frac{1}{t_n - t_0} {}^K\!\int_t^{t_n} v(s)\, ds \in E_+, \ \forall\, n.$$

Because E_+ is a closed subset of E, then $v(t_0) \in E_+$. This holds for a.e. $t_0 \in [a, b]$, which concludes the proof.   □

The next result is consequence of Lemma 9.23 and [207, Theorems 3.3.7 and 3.6.4].

Lemma 9.24. *Let E be an ordered Banach space. If f, $g : [a, b] \to E$ are HL integrable, then*

$$ {}^K\!\int_c^d f(s)\, ds = {}^K\!\int_c^d g(s)\, ds $$

for every closed subinterval $[c, d]$ of $[a, b]$ if and only if $f(t) = g(t)$ for a.e. $t \in [a, b]$.

Proof: The necessity part follows from [207, Theorem 3.6.4]. To prove sufficiency, assume that ${}^K\!\int_c^d f(s)\, ds = {}^K\!\int_c^d g(s)\, ds$ for every closed subinterval $[c, d]$ of $[a, b]$. Denoting $v = f - g$, then ${}^K\!\int_c^d v(s)\, ds = 0 \in E_+$ and ${}^K\!\int_c^d (-v(s)),\, ds = 0 \in E_+$ for every closed subinterval $[c, d]$ of $[a, b]$. This result implies by Lemma 9.23 that $v(t) \in -E_+ \cap E_+ = \{0\}$ for a.e. $t \in [a, b]$. Thus $v(t) = f(t) - g(t) = 0$, for a.e. $t \in [a, b]$.   □

Remark 9.25. A function $u : [a, b] \to E$ is called *Henstock–Kurzweil (shortly HK) integrable* if there is a function $f : [a, b] \to E$, called a *primitive of u*, which has the following property: For every $\epsilon > 0$ there is a function $\delta : [a, b] \to (0, \infty)$ such that

$$\left\| \sum_i (u(\xi_i)(t_i - t_{i-1}) - (f(t_i) - f(t_{i-1}))) \right\| < \epsilon$$

whenever $D = \{(\xi_i, [t_{i-i}, t_i])\}$ is a δ-fine K-partition of $[a, b]$.

    Some of the characteristic features of HL and HK integrability are as follows:

- If $u : [a, b] \to E$ is HL integrable, then it is HK integrable (cf. [207, Proposition 3.6.5]).

- If E is finite-dimensional, then $u : [a, b] \to E$ is HK integrable if and only if it is HL integrable (cf. [207, Proposition 3.6.6]).
- Both HK and HL integrability encompasses improper integrals on finite intervals. For instance, if u is HK (respectively HL) integrable on every closed subinterval $[a, c]$ of $[a, b)$, and if $\alpha = \lim\limits_{c \to b-} {}^K\!\!\int_a^c u(s)\,ds$ exists, then u is

  HK (respectively HL) integrable on $[a, b]$, and ${}^K\!\!\int_a^b u(s)\,ds = \alpha$. This result, called *Hake's Theorem*, is proved in [207, Theorem 3.4.5] for HK integrable functions. The proof for HL integrable functions is similar when Lemma 9.9 is used instead of [207, Lemma 3.4.1]. In particular, HL integrability encompasses improper integrals of Bochner integrable functions on finite intervals.
- In [90, Example 2.1], a HK integrable function $f : [a, b] \to l_2[a, b]$ is constructed whose values are $\neq 0$ for a.e. $t \in [a, b]$, and then calculated that ${}^K\!\!\int_a^b f(s)\,ds = 0$. Thus the result of Lemma 9.24 is not valid for all Banach-valued HK integrable functions.

Further, the functions $u_n(t) = \chi_{[\frac{1}{n},1]}(t)/t$, $t \in [0, 1]$, $n \in \mathbb{N}$, form an increasing sequence of Lebesgue integrable functions from $[0, 1]$ to $\mathbb{R}_+$. It converges pointwise on $[0, 1]$ to the function $u(t) = \chi_{(0,1]}(t)/t$, $t \in [0, 1]$, which is Lebesgue measurable but not HL integrable. Note, for $u(t) = \chi_{(0,1]}(t)/t$, $t \in [0, 1]$, the sets B_j defined by formula (9.14) have the explicit form $B_j = \{0\} \cup [\frac{1}{j}, 1]$, $j \in \mathbb{N}$. Therefore, the following hypotheses are indispensable:
  - The HL integrability of u in Lemma 9.12.
  - The existence of a HL integrable upper bound v_+ in Lemma 9.13.

Notice that $f(t) = \chi_{(0,1]}(t) \ln t$ is differentiable on $(0, 1]$, and $f'(t) = u(t) = \chi_{(0,1]}(t)/t$, $t \in (0, 1]$. However, no extension of f on the interval $[0, 1]$ satisfies the Strong Lusin Condition, because otherwise u would be HL integrable on $[0, 1]$.

The Cantor function f (cf. [85, Proposition 4.2.1]) is increasing and continuous, and hence bounded function from $[0, 1]$ onto itself, and $f'(t) = 0$ for a.e. $t \in [0, 1]$. Because f is not constant function, it is not a primitive of f'. Thus f does not satisfy the Strong Lusin Condition.

9.1.4 Convergence Theorems for HL Integrable Functions

Given a Banach space E and a closed interval $[a, b]$ of $\mathbb{R}$, we denote by $HL([a, b], E)$ the space of HL integrable functions from $[a, b]$ to E (a.e. equal functions are identified).

A dominated convergence theorem for real-valued Henstock integrable functions has been proved in [166, Theorem 4.3]. The following result whose proof is adopted from [132] can be considered as a generalization of [166, Theorem 4.3], since real-valued Henstock–Kurzweil integrable functions are also HL integrable (cf. [207, Proposition 3.6.6]).

Theorem 9.26 (Dominated Convergence Theorem for HL integrable functions). *Let E be an ordered Banach space having a normal order cone, let $u_\pm$ and u_n, $n \in \mathbb{N}$, be strongly measurable functions from $[a, b]$ to E, and assume that $u_-(s) \leq u_n(s) \leq u_+(s)$ for each $n \in \mathbb{N}$ and for a.e. $s \in [a, b]$, and that $u_n(s) \to u(s)$ for a.e. $s \in [a, b]$. If $u_\pm \in HL([a, b], E)$, then $u, u_n \in HL([a, b], E)$, $n \in \mathbb{N}$, and*

$$K\!\int_a^b u_n(s)\,ds \to {}^K\!\!\int_a^b u(s)\,ds.$$

Proof: The a.e. pointwise limit function u of the sequence (u_n) of strongly measurable functions is strongly measurable. Since $u_n(s) \in [u_-(s), u_+(s)]$ for a.e. $s \in [a, b]$ and the order intervals $[x, y]$ of E are closed, then $u(s) \in [u_-(s), u_+(s)]$ for a.e. $s \in [a, b]$. Then from Proposition 9.14 it follows that functions u and u_n, $n \in \mathbb{N}$, are HL integrable. Moreover,

$$u_-(s) - u_+(s) \leq u(s) - u_n(s) \leq u_+(s) - u_-(s) \quad \text{for a.e. } s \in [a, b],$$

so that

$$0 \leq u(s) - u_n(s) + u_+(s) - u_-(s) \leq 2(u_+(s) - u_-(s)) \quad \text{for a.e. } s \in [a, b]. \tag{9.40}$$

Since the order cone of E is normal, from (9.2) and (9.40) it follows that

$$\|u(s) - u_n(s) + u_+(s) - u_-(s)\| \leq 2\lambda\|u_+(s) - u_-(s)\| \quad \text{for a.e. } s \in [a, b]. \tag{9.41}$$

By Lemma 9.12, there exists a sequence of Lebesgue measurable sets $(B_j)_{j \in \mathbb{N}}$ such that $B_j \subset B_{j+1}$ for every $j \in \mathbb{N}$, $\cup_j B_j = [a, b]$, $u_+ - u_-$ is Bochner integrable on every B_j, and

$$\lim_{j \to \infty} \int_{B_j} (u_+(s) - u_-(s))\,ds = {}^K\!\!\int_a^b (u_+(s) - u_-(s))\,ds. \tag{9.42}$$

Setting $v = u - u_n + u_+ - u_-$ and $v_+ = 2(u_+ - u_-)$, then the hypotheses of Lemma 9.13 are satisfied, and thus by means of Lemma 9.13 along with (9.27) and (9.40) it follows that for every $j \in \mathbb{N}$,

$$\left\| {}^K\!\!\int_a^b (u(s) - u_n(s) + u_+(s) - u_-(s))\,ds \right.$$
$$\left. - \int_{B_j} (u(s) - u_n(s) + u_+(s) - u_-(s))\,ds \right\|$$
$$\leq 2\lambda \left\| {}^K\!\!\int_a^b (u_+(s) - u_-(s))\,ds - \int_{B_j} (u_+(s) - u_-(s))\,ds \right\|. \tag{9.43}$$

This result implies that, for all $j, n \in \mathbb{N}$,

$$\left\| {}^{K}\!\!\int_a^b (u(s) - u_n(s))\, ds - \int_{B_j} (u(s) - u_n(s))\, ds \right\|$$

$$\leq (2\lambda + 1) \left\| {}^{K}\!\!\int_a^b (u_+(s) - u_-(s))\, ds - \int_{B_j} (u_+(s) - u_-(s))\, ds \right\|. \qquad (9.44)$$

Let $\epsilon > 0$ be given. Because of (9.42) there exists a $j_\epsilon \in \mathbb{N}$ such that

$$\left\| {}^{K}\!\!\int_a^b (u_+(s) - u_-(s))\, ds - \int_{B_{j_\epsilon}} (u_+(s) - u_-(s))\, ds \right\| \leq \frac{\epsilon}{4\lambda + 2}. \qquad (9.45)$$

It follows from (9.44) and (9.45) that

$$\left\| {}^{K}\!\!\int_a^b (u(s) - u_n(s))\, ds - \int_{B_{j_\epsilon}} (u(s) - u_n(s))\, ds \right\| \leq \frac{\epsilon}{2}. \qquad (9.46)$$

In view of (9.41) we have

$$\|u(s) - u_n(s)\| \leq (2\lambda + 1)\|u_+(s) - u_-(s)\| \quad \text{for a.e. } s \in B_{j_\epsilon}.$$

Thus the dominated convergence theorem in $L^1(B_{j_\epsilon}, E)$ (cf., e.g., [164]) implies the existence of an $n_\epsilon \in \mathbb{N}$ such that

$$\left\| \int_{B_{j_\epsilon}} (u(s) - u_n(s))\, ds \right\| \leq \frac{\epsilon}{2} \quad \text{for } n \geq n_\epsilon. \qquad (9.47)$$

Then from (9.46) and (9.47) it follows that

$$\left\| {}^{K}\!\!\int_a^b (u(s) - u_n(s))\, ds \right\| \leq \epsilon \quad \text{for } n \geq n_\epsilon.$$

The above proof shows that ${}^{K}\!\!\int_a^b u_n(s)\, ds \to {}^{K}\!\!\int_a^b u(s)\, ds.$ □

The normality of the order cone E_+ of E was a sufficient assumption in Proposition 9.14 and in Theorem 9.26. In the proof of the next result, which is a generalization of the monotone convergence theorem proved in [166, Theorem 4.1] for real-valued Henstock integrable functions, we need a stronger condition; the regularity of E_+.

Theorem 9.27 (Monotone Convergence Theorem for HL integrable functions). *Let E be an ordered Banach space having a regular order cone. Given a monotone sequence $(u_n)_{n=1}^\infty$ of strongly measurable functions from $[a, b]$ to E, assume there exist HL integrable functions $u_\pm : [a, b] \to E$ such that $u_-(s) \leq u_n(s) \leq u_+(s)$ for each $n \in \mathbb{N}$ and for a.e. $s \in [a, b]$. Then $u(s) := \lim_{n \to \infty} u_n(s)$ exists for a.e. $s \in [a, b]$, and*

$$ {}^{K}\!\!\int_a^b u_n(s)\, ds \to {}^{K}\!\!\int_a^b u(s)\, ds.$$

Proof: The given hypotheses imply that for a.e. $s \in [a,b]$, the sequence $(u_n(s))_{n=1}^{\infty}$ is monotone and order bounded in E. Because the order cone of E is regular, then $u(s) = \lim_{n \to \infty} u_n(s)$ exists for a.e. $s \in [a,b]$. Consequently, the hypotheses of the dominated convergence theorem hold for $(u_n)_{n=1}^{\infty}$ and u, whence ${}^K\!\int_a^b u_n(s)\,ds \to {}^K\!\int_a^b u(s)\,ds$. □

Remark 9.28. (i) Recalling the example of Remark 9.25, i.e., the increasing sequence $u_n(t) = \chi_{[\frac{1}{n},1]}(t)/t$ of Lebesgue integrable functions from $[0,1]$ to $\mathbb{R}_+$ that converges pointwise on $[0,1]$ to the function $u(t) = \chi_{(0,1]}(t)/t$, $t \in [0,1]$, which is Lebesgue measurable but not HL integrable. This shows that the existence of a HL integrable upper bound u_+ in the Dominated and Monotone Convergence Theorems (Theorems 9.26 and 9.27) is indispensable.

(ii) Further, the Dominated Convergence Theorem for HL integrable functions is reduced to that for Bochner integrable functions if order boundedness of the sequence (u_n) is replaced by the norm-boundedness: $\|u_n(t)\| \leq g(t)$ for a.e. $t \in [a,b]$, where $g \in L^1([a,b], \mathbb{R}_+)$.

9.1.5 Ordered Normed Spaces of HL Integrable Functions

Given a Banach space E and a closed interval $[a,b]$ of $\mathbb{R}$, define the Alexiewicz norm on $HL([a,b], E)$ by

$$\|u\|_A = \sup\left\{\left\|{}^K\!\int_c^d u(s)\,ds\right\| : [c,d] \subseteq [a,b]\right\}. \tag{9.48}$$

Lemma 9.29. *Let E be a Banach space ordered by an order cone E_+. Then the set $HL([a,b], E_+)$ of a.e. E_+-valued functions of $HL([a,b], E)$ equipped with the norm $\|\cdot\|_A$ is an order cone of $HL([a,b], E)$, which induces the a.e. pointwise ordering to $HL([a,b], E)$.*

Proof: Obviously, $HL([a,b], E_+)$ is a cone in $HL([a,b], E)$. To prove that $HL([a,b], E_+)$ is closed, assume that (u_n) is a sequence in $HL([a,b], E_+)$ that converges to $u \in HL([a,b], E)$, i.e., $\|u_n - u\|_A \to 0$ as $n \to \infty$. This implies by (9.48) that

$${}^K\!\int_c^d u_n(s)\,ds \to {}^K\!\int_c^d u(s)\,ds \text{ for all } [c,d] \subseteq [a,b] \text{ as } n \to \infty. \tag{9.49}$$

Since $0 \leq u_n(s)$ for every n and for a.e. $s \in [a,b]$, it follows from Lemma 9.11 that $0 \leq {}^K\!\int_c^d u_n(s)\,ds$, or equivalently, ${}^K\!\int_c^d u_n(s)\,ds \in E_+$, for all n and $[c,d] \subseteq [a,b]$. Because E_+ is closed, from (9.49) we get ${}^K\!\int_c^d u(s)\,ds \in E_+$, for all n and $[c,d] \subseteq [a,b]$. Thus $u(t) \in E_+$ for a.e. $t \in [a,b]$ by Lemma 9.23, so that $u \in HL([a,b], E_+)$. This proves that $HL([a,b], E_+)$ is closed. Moreover, if $u, v \in HL([a,b], E)$, then

$u(t) \leq v(t)$ for a.e. $t \in [a, b]$ if and only if $v - u \in HL([a, b], E_+)$.

Thus $HL([a, b], E_+)$ induces the a.e. pointwise ordering to $HL([a, b], E)$.   □

The next result provides properties for the order cone $HL([a, b], E_+)$.

Theorem 9.30. *Let E be an ordered Banach space and E_+ its order cone.*
(a) If E_+ is normal, then $HL([a, b], E_+)$ is normal.
(b) If E_+ is regular, then $HL([a, b], E_+)$ is regular.

Proof: Ad (a) Assume that E_+ is normal, that $u, v \in HL([a, b], E_+)$, and that $u \leq v$. Then

$$0 \leq u(s) \leq v(s) \text{ for a.e. } s \in [a, b].$$

This result and Lemma 9.11 imply that if I is a closed subinterval of $[a, b]$, then

$$0 \leq {}^K\!\!\int_I u(s)\, ds \leq {}^K\!\!\int_I v(s)\, ds.$$

Since E_+ is normal, there exists a $\lambda \geq 1$ such that

$$\left\| {}^K\!\!\int_I u(s)\, ds \right\| \leq \lambda \left\| {}^K\!\!\int_I v(s)\, ds \right\|.$$

Since this result holds for every closed subinterval I of $[a, b]$, we then have

$$\|u\|_A = \sup\left\{ \left\| {}^K\!\!\int_c^d u(s)\, ds \right\| : [c, d] \subseteq [a, b] \right\}$$

$$\leq \lambda \sup\left\{ \left\| {}^K\!\!\int_c^d v(s)\, ds \right\| : [c, d] \subseteq [a, b] \right\} = \lambda \|v\|_A.$$

This shows that $HL([a, b], E_+)$ is normal.

Ad (b) Assume that $(u_n)_{n=1}^\infty$ is an increasing sequence in $HL([a, b], E_+)$, and that it has an upper bound u_+ in $HL([a, b], E_+)$. Thus

$$0 \leq u_n(s) \leq u_{n+1}(s) \leq u_+(s) \quad \text{for a.e. } s \in [a, b]. \tag{9.50}$$

According to Theorem 9.27 there exists an HL integrable function $u : [a, b] \to E$ such that $u(s) = \lim_n u_n(s)$ for a.e. $s \in [a, b]$. Moreover, it follows from (9.50) and [133, Proposition 1.1.3] that

$$0 \leq u_n(s) \leq u_{n+1}(s) \leq u(s) \quad \text{for a.e. } s \in [a, b]. \tag{9.51}$$

In particular, $u - u_n \in HL([a, b], E_+)$ for every $n \in \mathbb{N}$. This result and Lemma 9.11 imply that if I is a closed subinterval of $[a, b]$, then

$$^K\!\!\int_I (u(s) - u_n(s))\, ds \in E_+.$$

Consequently, if $[c, d]$ is a closed subinterval of $[a, b]$, then

$$0 \le {}^K\!\!\int_c^d (u(s) - u_n(s))\, ds \le \left({}^K\!\!\int_a^c + {}^K\!\!\int_c^d + {}^K\!\!\int_d^b \right) (u(s) - u_n(s))\, ds$$

$$= {}^K\!\!\int_a^b (u(s) - u_n(s))\, ds.$$

The last inequality and the normality of E_+ result in

$$\left\| {}^K\!\!\int_c^d (u(s) - u_n(s))\, ds \right\| \le \lambda \left\| {}^K\!\!\int_a^b (u(s) - u_n(s))\, ds \right\|.$$

This inequality is valid for every $n \in \mathbb{N}$ and for every closed subinterval $[c, d]$ of $[a, b]$, whence we conclude

$$\|u - u_n\|_A = \sup\left\{ \left\| {}^K\!\!\int_c^d (u(s) - u_n(s))\, ds \right\| : [c, d] \subseteq [a, b] \right\}$$

$$\le \lambda \left\| {}^K\!\!\int_a^b (u(s) - u_n(s))\, ds \right\|.$$

Since ${}^K\!\!\int_a^b (u(s) - u_n)(s)\, ds \to 0$ by Theorem 9.27, we see that $\|u - u_n\|_A \to 0$ as $n \to \infty$. This proves that $HL([a, b], E_+)$ is regular.   □

9.2 Chains in Ordered Function Spaces

In this section we study the existence of supremums of chains in ordered function spaces. In this study we apply the next result, which follows, e.g., from [44, Lemma A.3.1] and [133, Proposition 1.1.5].

Lemma 9.31. *Let C be a well-ordered subset of an ordered normed space E, and assume that each increasing sequence of C has a weak (respectively strong) limit in E. Then C contains an increasing sequence that converges weakly (respectively strongly) to $\sup C$.*

9.2.1 Chains in L^p-Spaces

Our goal in this subsection is to study chains of the function space $L^p(\Omega, E)$, $1 \le p \le \infty$, ordered a.e. pointwise, where $\Omega = (\Omega, \mathcal{A}, \mu)$ is a σ-finite measure space and E is an ordered Banach space.

As an application of Proposition 9.5 and Lemma 9.31 we prove the following result.

Lemma 9.32. *Given an ordered Banach space E whose bounded and increasing sequences have weak limits, and $p \in [1, \infty)$, assume that C is a bounded and well-ordered chain of $L^p(\Omega, E)$. If $\mu(\Omega) < \infty$, then C contains an increasing sequence that converges weakly a.e. pointwise to $\sup C$.*

Proof: Let C be a bounded and well-ordered chain of $L^p(\Omega, E)$. Because $\mu(\Omega) < \infty$, then $L^p(\Omega, E)$ is continuously embedded in $L^1(\Omega, E)$, so that C is a bounded and well-ordered chain in $L^1(\Omega, E)$. If $u, v \in C$ and $u \le v$, then $\int_\Omega u \le \int_\Omega v$ by Lemma 9.4. Thus $\{\int_\Omega v\}_{v \in C}$ is a bounded and well-ordered subset of E whose bounded and increasing sequences have weak limits. By Lemma 9.31 there is an increasing sequence $(u_n)_{n=1}^\infty$ in C such that

$$ {}^w\lim_{n\to\infty} \int_\Omega u_n = \sup_n \int_\Omega u_n = \sup_{v \in C} \int_\Omega v. \tag{9.52}$$

Moreover, for a.e. $t \in \Omega$ the sequence $(u_n(t))_{n=1}^\infty$ is increasing and bounded, so that

$$ u(t) = {}^w\lim_{n\to\infty} u_n(t) = \sup_n u_n(t) \tag{9.53}$$

exists for a.e. $t \in \Omega$ by a hypothesis, and $u \in L^p(\Omega, E)$ due to Proposition 9.5.

To show that u is an upper bound of C, let $w \in C$ be given. Assume first that $u_n \le w$ for each $n \in \mathbb{N}$. Then $u \le w$ by (9.53). It follows from (9.52) and (9.53) that

$$ \int_\Omega w \le \sup_{v \in C} \int_\Omega v = \sup_n \int_\Omega u_n \le \int_\Omega u. \tag{9.54}$$

If A is a measurable subset of Ω, then $\int_A u \le \int_A w$ and $\int_{\Omega \setminus A} u \le \int_{\Omega \setminus A} w$ by Lemma 9.4. If $\int_A u < \int_A w$, then

$$ \int_\Omega u = \int_A u + \int_{\Omega \setminus A} u < \int_A w + \int_{\Omega \setminus A} w = \int_\Omega w, $$

which contradicts with (9.54). Thus $\int_A u = \int_A w$ for each measurable subset A of Ω, so that $w = u$ by [164, VI, Corollary 5.16]. The above proof shows that

$$ w = u, \quad \text{whenever } w \in C \text{ and } u_n \le w \text{ for all } n \in \mathbb{N}. $$

If $w \le u_n$ for some $n \in \mathbb{N}$, then $w \le u$ by (9.53). Thus $w \le u$ for each $w \in C$.

To show that $u = \sup C$, let $v \in L^p(\Omega, E)$ be an upper bound of C. Then $u_n(t) \le v(t)$ for a.e. $t \in \Omega$ and for all $n \in \mathbb{N}$. This result and (9.53) imply that $u(t) = \sup_n u_n(t) \le v(t)$ for a.e. $t \in \Omega$, i.e., $u \le v$. Thus $u = \sup C$.   □

Consider next the case when $p = \infty$.

Lemma 9.33. *Let E be an ordered Banach space whose bounded and increasing sequences have weak limits. If $\mu(\Omega) < \infty$, then each bounded and well-ordered chain C of $L^\infty(\Omega, E)$ contains an increasing sequence that converges weakly a.e. pointwise to $\sup C$.*

Proof: If $u \in L^\infty(\Omega, E)$, then $\|u(t)\| \le \|u\|_\infty$ for a.e. $t \in \Omega$, which implies that $\|u\|_1 \le \mu(\Omega)\|u\|_\infty$. This shows that $L^\infty(\Omega, E)$ is continuously embedded in $L^1(\Omega, E)$. Let C be a bounded and well-ordered chain in $L^\infty(\Omega, E)$. Then C is a bounded and well-ordered chain also in $L^1(\Omega, E)$. If $(u_n)_{n=1}^\infty$ is an increasing sequence in C, there is $M > 0$ such that for each $n \in \mathbb{N}$,

$$\|u_n(t)\| \le \|u_n\|_\infty \le M \quad \text{for a.e. } t \in \Omega.$$

Thus $(u_n)_{n=1}^\infty$ is also a.e. pointwise bounded. Hence, C has by the proof of Lemma 9.32 a supremum u in $L^1(\Omega, E)$, and there is an increasing sequence $(u_n)_{n=1}^\infty$ in C that converges weakly a.e. pointwise in Ω to u. Moreover, (9.5) and the above inequality imply that

$$\|u(t)\| \le \liminf_{n\to\infty} \|u_n(t)\| \le M \quad \text{for a.e. } t \in \Omega.$$

Therefore, u belongs to $L^\infty(\Omega, E)$, and it is easy to see that $u = \sup C$ is also in $L^\infty(\Omega, E)$. $\qquad\square$

The next result extends the results of Lemma 9.32 to the case when μ is σ-finite.

Proposition 9.34. *Let $(\Omega, \mathcal{A}, \mu)$ be a σ-finite measure space and E an ordered Banach space whose bounded and increasing sequences have weak limits. Assume that C is a bounded well-ordered chain in $L^p(\Omega, E)$, $1 \le p < \infty$. Then C contains an increasing sequence that converges weakly a.e. pointwise to $\sup C$.*

Proof: Since Ω is σ-finite, then $\Omega = \bigcup_{n=0}^\infty \Omega_n$, where $\Omega_n \subseteq \Omega_{n+1}$ and $\mu(\Omega_n) < \infty$ for each $n \in \mathbb{N}_0$. Let C be a well-ordered and bounded chain in $L^p(\Omega, E)$, $1 \le p < \infty$. The restriction $C|\Omega_n = \{u|\Omega_n : u \in C\}$ is for each $n \in \mathbb{N}_0$ a well-ordered and bounded chain in $L^p(\Omega_n, E)$. It follows from Lemma 9.32 that

$$v_n = \sup(C|\Omega_n)$$

exists in $L^p(\Omega_n, E)$. Denoting $\underline{u} = \min C$, and defining $v_n(t) = \underline{u}(t)$ for $t \in \Omega \setminus \Omega_n$, we obtain a sequence of μ-measurable functions $v_n : \Omega \to E$. This sequence is increasing, since $\Omega_n \subseteq \Omega_{n+1}$. It is also a.e. pointwise bounded, whence

$$u^*(t) = {}^w\lim_{n\to\infty} v_n(t) = \sup_{n\in\mathbb{N}_0} v_n(t)$$

exists for a.e. $t \in \Omega$. Defining $u^*(t) = 0$ for the remaining $t \in \Omega$, we obtain by Proposition 9.5 a function $u^* \in L^p(\Omega, E)$. By the proof of Lemma 9.32, for each $n \in \mathbb{N}_0$ there exists an increasing sequence $(u_k^n)_{k=0}^\infty$ of C and a μ-null set $Z_n \subset \Omega_n$ such that

$$v_n(t) = {}^w\lim_{k\to\infty} u_k^n(t) = \sup_{k\in\mathbb{N}_0} u_k^n(t) \quad \text{for each } t \in \Omega_n \setminus Z_n.$$

Denoting

$$u_n = \max\{u_n^j : 0 \le j \le n\}, \quad n \in \mathbb{N}_0,$$

we obtain an increasing sequence (u_n) of C, which satisfies

$$u_k^n(t) \le u_n(t) \le u^*(t) \quad \text{for each } k = 0, \ldots, n \text{ and } t \in \Omega_n \setminus Z_n.$$

In particular $(u_n(t))_{n=0}^\infty$ is increasing and bounded for each $t \in \Omega \setminus Z$, where $Z = \cup_{n=0}^\infty Z_n$, so that

$$u(t) = {}^w\lim_{k \to \infty} u_n(t) = \sup_{n \in \mathbb{N}_0} u_n(t)$$

exists for each $t \in \Omega \setminus Z$. Moreover, the definitions of u and v_n imply that

$$v_n(t) \le u(t) \le u^*(t) \quad \text{for each } t \in \Omega_n \setminus Z_n.$$

Thus

$$u^*(t) = {}^w\lim_{n \to \infty} v_n(t) \le u(t) \le u^*(t)$$

for a.e. $t \in \Omega$. This result shows that $u = u^*$, whence $u_n(t) \rightharpoonup u^*(t)$ for a.e. $t \in \Omega$.

It remains to prove that $u^* = \sup C$. If $w \in C$, then $w|\Omega_n \le v_n$, so that

$$w(t) \le v_n(t) \le u^*(t) \quad \text{for a.e. } t \in \Omega_n \text{ and for each } n \in \mathbb{N}_0.$$

Thus $w \le u^*$ for each $w \in C$, so that u^* is an upper bound of C. If $v \in L^p(\Omega, E)$ is another upper bound of C, then $w(t) \le v(t)$ for a.e. $t \in \Omega$ and for each $w \in C$. Thus $w|\Omega_n \le v|\Omega_n$ for all $n \in \mathbb{N}_0$ and $w \in C$, whence $v_n(t) \le v(t)$ for a.e. $t \in \Omega$ and for each $n \in \mathbb{N}_0$. This result and the definition of u^* imply that $u^* \le v$, so that $u^* = \sup C$ in $L^p(\Omega, E)$.   □

9.2.2 Chains of Locally Bochner Integrable Functions

We say that a nonempty subset Ω of the space $\mathbb{R}^m$ is **hemicompact** if Ω is a countable union of compact subsets of $\mathbb{R}^m$. A strongly measurable mapping u from a hemicompact set to a Banach space is called **locally Bochner integrable on** Ω, and denote $u \in L^1_{loc}(\Omega, E)$, if u is Bochner integrable on each compact subset K of Ω.

Next we study chains in the a.e. pointwise ordered space $L^1_{loc}(\Omega, E)$, where E is an ordered Banach space. In this study we use the following result, which is an immediate consequence of Proposition 1.3.2, Lemma 5.8.2, and Proposition 5.8.7 of [133].

Lemma 9.35. *Assume that C is a nonempty subset of $L^p(\Omega, E)$, $1 \le p < \infty$, where Ω is a measure space and E is an ordered Banach space with regular order cone. If C is well-ordered with respect to a.e. pointwise ordering, and if there exist functions $u_\pm \in L^p(\Omega, E)$ such that $u_-(t) \le u(t) \le u_+(t)$ for all $u \in C$ and for a.e. $t \in \Omega$, then C contains an increasing sequence that converges a.e. pointwise to $\sup C$.*

As an application of Lemma 9.35 we obtain the following proposition.

Proposition 9.36. *Let Ω be a hemicompact subset of $\mathbb{R}^m$. Assume that C is a nonempty subset of $L^1_{loc}(\Omega, E)$, E is an ordered Banach space with regular order cone, and that there exist functions $u_\pm \in L^1_{loc}(\Omega, E)$ such that $C \subseteq [u_-, u_+]$, i.e.,*

$$u_-(t) \le u(t) \le u_+(t) \quad \text{for all } u \in C \text{ and for a.e. } t \in \Omega. \tag{9.55}$$

If C is well-ordered with respect to the a.e. pointwise ordering, it contains an increasing sequence that converges a.e. pointwise to $\sup C$.

Proof: Assume that C is well-ordered and (9.55) holds. Choose a sequence of compact subsets Ω_n, $n \in \mathbb{N}_0$, of Ω such that $\Omega = \cup_{n=0}^\infty \Omega_n$, and that $\Omega_n \subset \Omega_{n+1}$ for each $n \in \mathbb{N}_0$. The given assumptions ensure that for each $n \in \mathbb{N}_0$ the restrictions $u|\Omega_n$, $u \in C$, form a well-ordered and order-bounded chain C_n in $L^1(\Omega_n, E)$, ordered a.e. pointwise. It follows from Lemma 9.35 that for each $n \in \mathbb{N}_0$

$$v_n = \sup C_n$$

exists in $L^1(\Omega_n, E)$, and there exists an increasing sequence $(u_n^k)_{k=0}^\infty$ of C and a null-set $Z_n \subset \Omega_n$ such that

$$v_n(t) = \lim_{k \to \infty} u_n^k(t) = \sup_{k \in \mathbb{N}_0} u_n^k(t) \quad \text{for each } t \in \Omega_n \setminus Z_n. \tag{9.56}$$

Denoting $\underline{u} = \min C$, and defining $v_n(t) = \underline{u}(t)$ for $t \in \Omega \setminus \Omega_n$, we obtain a sequence of strongly measurable functions $v_n : \Omega \to E$. The sequence (v_n) is also increasing since $\Omega_n \subset \Omega_{n+1}$, $n \in \mathbb{N}_0$. It is also a.e. pointwise order bounded by (9.55) and (9.56), whence

$$u^*(t) = \lim_{n \to \infty} v_n(t) = \sup_{n \in \mathbb{N}_0} v_n(t) \tag{9.57}$$

exists for a.e. $t \in \Omega$. Defining $u^*(t) = 0$ for the remaining $t \in \Omega$ we get a strongly measurable function $u^* : \Omega \to E$. Denoting

$$u_n = \max\{u_j^n : 0 \le j \le n\}, \quad n \in \mathbb{N}_0,$$

we obtain an increasing sequence (u_n) of C that satisfies

$$u_n^k(t) \le u_n(t) \le u^*(t)$$

for each $k = 0, \ldots, n$ and $t \in \Omega_n \setminus Z_n$. Moreover, by (9.55) the sets Z_n can be chosen in such a way that $(u_n(t))_{n=0}^\infty$ is order bounded and increasing for each $t \in \Omega \setminus Z$, where $Z = \cup_{n=0}^\infty Z_n$. Thus

$$u(t) = \lim_{n \to \infty} u_n(t) = \sup_{n \in \mathbb{N}_0} u_n(t)$$

exists for each $t \in \Omega \setminus Z$. The definitions of v_n and u imply that

$$v_n(t) \leq u(t) \leq u^*(t) \quad \text{for each } t \in \Omega_n \setminus Z_n.$$

Thus

$$u^*(t) = \lim_{n \to \infty} v_n(t) \leq u(t) \leq u^*(t)$$

for a.e. $t \in \Omega$. This result implies that $u = u^*$, whence $u_n(t) \to u^*(t)$ for a.e. $t \in \Omega$. Since $(u_n)_{n=0}^{\infty}$ is a sequence of C, it follows from (9.55) that

$$u_-(t) \leq u^*(t) \leq u_+(t) \quad \text{for a.e } t \in \Omega.$$

This result and the strong measurability of u^* imply that $u^* \in L^1_{loc}(\Omega, E)$.

The proof that $u^* = \sup C$ is similar to that presented in the proof of Proposition 9.34. □

Since each increasing sequence of $L^1_{loc}(\Omega, E)$ is well-ordered and each decreasing sequence of $L^1_{loc}(\Omega, E)$ is inversely well-ordered, as a consequence of Proposition 9.36 and its dual, as well as [133, Proposition 1.1.3 and Corollary 1.1.3] we obtain the following results.

Corollary 9.37. *Assume that (u_n) is a sequence of $L^1_{loc}(\Omega, E)$, that the order cone of E is regular, and that there exist functions $w_\pm \in L^1_{loc}(\Omega, E)$ such that $u_n \in [w_-, w_+]$ for each n.*

(a) If (u_n) is increasing, it converges a.e. pointwise to $u_* = \sup_n u_n$ in $L^1_{loc}(\Omega, E)$, and u_* belongs to $[w_-, w_+]$.
(b) If (u_n) is decreasing, it converges a.e. pointwise to $u_* = \inf_n u_n$ in $L^1_{loc}(\Omega, E)$, and u_* belongs to $[w_-, w_+]$.

9.2.3 Chains of HL Integrable and Locally HL Integrable Functions

First we study the a.e. pointwise ordered spaces $HL([a, b], E)$ of HL integrable functions from $[a, b]$ to an ordered Banach space E that has a regular order cone.

Proposition 9.38. *Let C be a nonempty set of strongly measurable functions from $[a, b]$ to an ordered Banach space E with a regular order cone. Assume that there exist $u_\pm \in HL([a, b], E)$ such that $u_-(s) \leq u(s) \leq u_+(s)$ for all $u \in C$ and for a.e. $s \in [a, b]$. Then the following results hold.*

(a) $u \in HL([a, b], E)$, for each $u \in C$.
(b) If C is well-ordered with respect to a.e. pointwise ordering, then C contains an increasing sequence that converges a.e. pointwise to $\sup C$ in $HL([a, b], E)$.
(c) If C is inversely well-ordered with respect to a.e. pointwise ordering, then C contains a decreasing sequence that converges a.e. pointwise to $\inf C$ in $HL([a, b], E)$.

Proof: Ad (a) is a consequence of Proposition 9.14.

Ad (b) By hypothesis, C is an order bounded and well-ordered chain in the normed space $(HL([a,b], E), \|\cdot\|_A)$, ordered by the cone $HL([a,b], E_+)$, which is regular by Theorem 9.30. Thus every increasing sequence of C converges in $(HL([a,b], E), \|\cdot\|_A)$. It then follows from Lemma 9.31 that C contains an increasing sequence (u_n) that converges to $\sup C$ in $(HL([a,b], E), \|\cdot\|_A)$. Since (u_n) is order bounded, it satisfies also the hypotheses of Theorem 9.27. Thus (u_n) converges also a.e. pointwise to $\sup C$.

Ad (c) The proof of (c) is similar to that of (b). $\square$

We say that a function u from a real interval J to a Banach space E is **locally HL integrable on** J, if u is HL integrable on every compact subinterval of J. Denote by $HL_{loc}(J, E)$ the space of such functions. In the next proposition the interval J may be unbounded, open, or half-open.

Proposition 9.39. *Let C be a nonempty set of strongly measurable functions from a real interval J to an ordered Banach space E with a regular order cone. Assume that there exist $u_\pm \in HL_{loc}(J, E)$ such that $u_-(s) \le u(s) \le u_+(s)$ for all $u \in C$ and for a.e. $s \in J$. Then the following results hold.*

(a) $u \in HL_{loc}(J, E)$, for each $u \in C$.

(b) If C is well-ordered with respect to a.e. pointwise ordering, then C contains an increasing sequence that converges a.e. pointwise to $\sup C$ in $HL_{loc}(J, E)$.

(c) If C is inversely well-ordered with respect to a.e. pointwise ordering, then C contains a decreasing sequence that converges a.e. pointwise to $\inf C$ in $HL_{loc}(J, E)$.

Proof: Ad (a) The conclusion follows from the definition of $HL_{loc}(J, E)$ and from Proposition 9.14.

Ad (b) Assume next that C is well-ordered. Choose a sequence of compact subintervals J_n, $n \in \mathbb{N}_0$, of J such that $J = \cup_{n=0}^\infty J_n$, and that $J_n \subset J_{n+1}$ for each $n \in \mathbb{N}_0$. The given assumptions ensure that for each $n \in \mathbb{N}_0$, the restrictions $u_{|J_n}$, $u \in C$ form a well-ordered and order-bounded chain C_n in $HL(J_n, E)$ ordered a.e. pointwise. It follows from Proposition 9.38 that for each $n \in \mathbb{N}_0$,

$$v_n = \sup C_n$$

exists in $HL(J_n, E)$, and there exist an increasing sequence $(u_n^k)_{k=0}^\infty$ of C and a null-set $Z_n \subset J_n$ such that

$$v_n(t) = \lim_{k \to \infty} u_n^k(t) = \sup_{k \in \mathbb{N}_0} u_n^k(t) \quad \text{for each } t \in J_n \setminus Z_n.$$

Defining $v_n(t) = u_-(t)$ for $t \in J \setminus J_n$, we obtain a sequence of strongly measurable functions $v_n : J \to E$. The sequence (v_n) is also increasing since $J_n \subset J_{n+1}$, $n \in \mathbb{N}_0$. By assumption it is also a.e. pointwise order bounded, whence

$$u^*(t) = \lim_{n \to \infty} v_n(t) = \sup_{n \in \mathbb{N}_0} v_n(t)$$

exists for a.e. $t \in [a, b]$. Defining $u^*(t) = 0$ for the remaining $t \in J$ we get a strongly measurable function $u^* : J \to E$. Denoting

$$u_n = \max\{u_k^k : 0 \le k \le n\}, \quad n \in \mathbb{N}_0,$$

we obtain an increasing sequence (u_n) of C that satisfies

$$u_n^k(t) \le u_n(t) \le u^*(t)$$

for each $k = 0, \ldots, n$ and $t \in J_n \setminus Z_n$. Moreover, the sets Z_n can be chosen in such a way that $(u_n(t))_{n=0}^{\infty}$ is order bounded and increasing for each $t \in [a, b) \setminus Z$, where $Z = \cup_{n=0}^{\infty} Z_n$. Thus

$$u(t) = \lim_{n \to \infty} u_n(t) = \sup_{n \in \mathbb{N}_0} u_n(t)$$

exists for each $t \in [a, b) \setminus Z$. The definitions of v_n and u imply that

$$v_n(t) \le u(t) \le u^*(t) \quad \text{for each } t \in J_n \setminus Z_n.$$

Thus

$$u^*(t) = \lim_{n \to \infty} v_n(t) \le u(t) \le u^*(t)$$

for a.e. $t \in [a, b)$. This result implies that $u = u^*$, whence $u_n(t) \to u^*(t)$ for a.e. $t \in J$. Since $(u_n)_{n=0}^{\infty}$ is a sequence of C, it belongs to the order interval $[u_-, u_+]$ of $HL_{loc}(J, X)$. The latter result and Theorem 9.27 imply that $u^* \in HL_{loc}([a, b), E)$.

It remains to prove that $u^* = \sup C$. If $w \in C$, then $w|J_n \in C_n$, whence $w|J_n \le \sup C_n = v_n$, so that

$$w(t) \le v_n(t) \le u^*(t) \quad \text{for a.e. } t \in J_n \text{ and for each } n \in \mathbb{N}_0.$$

Thus $w \le u^*$ for each $w \in C$, so that u^* is an upper bound of C. If $v \in HL_{loc}([a, b), E)$ is another upper bound of C, then

$$w(t) \le v(t) \quad \text{for a.e. } t \in J_n \text{ and for each } n \in \mathbb{N}_0.$$

Thus $w|J_n \le v|J_n$ for all $n \in \mathbb{N}_0$ and $w \in C$, whence

$$v_n(t) \le v(t) \quad \text{for a.e. } t \in J_n \text{ and for each } n \in \mathbb{N}_0.$$

This result and the definition of u^* imply that $u^* \le v$. Consequently, $u^* = \sup C$.

Ad (c) The proof of (c) is similar to that of (b).   □

9.2.4 Chains of Continuous Functions

Consider first the existence of supremums of well-ordered chains in the space $C(X, E)$ of continuous functions $u : X \to E$, where X is a topological space and E is an ordered normed space. Define a partial ordering in $C(X, E)$ by

$$u \leq v \text{ if and only if } u(t) \leq v(t) \text{ for each } t \in X. \qquad (9.58)$$

We say that a subset C of $C(X, E)$ is *equicontinuous* if for each $t \in X$ and for each $\epsilon > 0$ there exists a neighborhood U of t such that

$$\|u(s) - u(t)\| \leq \epsilon \text{ for all } u \in C \text{ and } s \in U. \qquad (9.59)$$

Proposition 9.40. *Let E be an ordered normed space, X a topological space, and let C be an equicontinuous and well-ordered subset of $C(X, E)$, whose increasing sequences have weak pointwise limits. Then $v = \sup C$ exists in $C(X, E)$.*

Proof: The hypotheses given for C imply that for each $t \in E$ the set $\{u(t)\}_{u \in C}$ is a well-ordered subset of E whose increasing sequences have weak limits, which in view of Lemma 9.31 implies that

$$v(t) = \sup\{u(t)\}_{u \in C} \qquad (9.60)$$

exists in E for each $t \in \Omega$. To prove that the so obtained function $v : X \to E$ is the supremum of C in $C(X, E)$, it suffices to show its continuity. Let $t \in X$ and $\epsilon > 0$ be given. By the equicontinuity hypothesis there is such a neighborhood U of t that

$$\|u(s) - u(t)\| \leq \epsilon \text{ whenever } s \in U \text{ and } u \in C.$$

Let $s \in U$ be fixed. By Lemma 9.31 there exists an increasing sequence $(v_n)_{n=0}^{\infty}$ in C such that $(v_n(s))_{n=0}^{\infty}$ converges weakly in E to $v(s)$, and an increasing sequence $(u_n)_{n=0}^{\infty}$ in C such that $(u_n(t))_{n=0}^{\infty}$ converges weakly in E to $v(t)$. Denoting $z_n = \max\{v_n, u_n\}$, $n \in \mathbb{N}_0$, we obtain an increasing sequence $(z_n)_{n=0}^{\infty}$ in C. Taking into account the hypotheses it converges weakly pointwise, whence $(z_n(s))_{k=0}^{\infty}$ converges weakly in E to $v(s)$, and $(z_n(t))_{k=0}^{\infty}$ converges weakly in E to $v(t)$. Thus $(z_n(s) - z_n(t))_{k=0}^{\infty}$ converges weakly in E to $v(s) - v(t)$, so that

$$\|v(s) - v(t)\| \leq \liminf_{n \to \infty} \|z_n(s) - z_n(t)\| \leq \epsilon.$$

This holds for each $s \in U$, which shows that v is continuous at t. Thus $v \in C(X, E)$, whence v is the supremum of C in $C(X, E)$. $\qquad \square$

If X is a separable topological space we have the following result.

Proposition 9.41. *Let E be an ordered normed space, X a separable topological space, and let C be an equicontinuous well-ordered subset of $C(X, E)$ whose increasing sequences have weak pointwise limits. Then $v = \sup C$ exists, and there is an increasing sequence (u_n) of C such that $u_n(t) \rightharpoonup v(t)$ for each $t \in X$.*

Proof: Let $D = \{t_j\}_{j\in\mathbb{N}_0}$ be a dense subset of X. It follows from Proposition 9.40 that $v = \sup C$ exists in $C(X, E)$ and satisfies (9.60). Moreover, the proof of Proposition 9.40 implies that the set $C \cup \{v\}$ is equicontinuous, and that for each $j \in \mathbb{N}_0$, there is a sequence $(u_k^j)_{k=0}^\infty$ in C that converges weakly to $v(t_j)$. Denote

$$u_n = \max\{u_k^j : 0 \leq j,\, k \leq n\}, \quad n \in \mathbb{N}_0.$$

The so obtained sequence (u_n) is increasing, is contained in C, and $u_n(t) \rightharpoonup v(t)$ for each $t \in D$. To prove this convergence also when t belongs to the complement of D, let $t \in X \setminus D$, $\epsilon > 0$, and $f \in E'$ be given. Choose a neighborhood U of t such that

$$\|u(s) - u(t)\| \leq \frac{\epsilon}{1 + 4\|f\|} \quad \text{for all } u \in C \cup \{x\} \text{ and } s \in U. \tag{9.61}$$

Since D is a dense subset of E, we can choose s in (9.61) so that it belongs to $U \cap D$. The sequence $(u_n(t))$ has by hypothesis a weak limit z. We have to prove that $z = v(t)$. Since $u_n(t) \rightharpoonup z$ and $u_n(s) \rightharpoonup v(s)$, there is $n \in \mathbb{N}_0$ such that

$$|f(u_n(t)) - f(z)| \leq \frac{\epsilon}{4} \quad \text{and} \quad |f(u_n(s)) - f(v(s))| \leq \frac{\epsilon}{4}. \tag{9.62}$$

Applying (9.61) and (9.62) we get

$$|f(z - v(t))| = |f(z) - f(v(t))| \leq |f(z) - f(u_n(t))| + \|f\|\,\|u_n(t) - (u_n(s)\|$$
$$+ |f(u_n(s)) - f(v(s))| + \|f\|\,\|v(s)) - v(t)\| \leq \epsilon.$$

This holds for each $f \in E'$ and for each $\epsilon > 0$, whence $z = v(t)$. Thus $u_n(t) \rightharpoonup v(t)$ also when $t \in X \setminus D$, which concludes the proof. $\quad\square$

Remark 9.42. The results of Propositions 9.40 and 9.41 are valid also when weak convergence is replaced by strong convergence.

Next we consider ordered topological spaces X that have the following property.

(C) Each nonempty well-ordered chain C of X whose increasing sequences converge contains an increasing sequence that converges to $\sup C$, and each nonempty inversely well-ordered chain D of X whose decreasing sequences converge contains a decreasing sequence that converges to $\inf D$.

For instance, the following ordered topological spaces possess property (C).

- If X satisfies the second countability axiom, then each chain of X is separable, whence property (C) follows from [133, Lemma 1.1.7] and its dual.
- Each ordered metric space has property (C) by [133, Proposition 1.1.5] and its dual.
- Each nonempty subset X of an ordered normed space E has property (C) with respect to the norm topology, and with respect to the weak topology as well due to [44, Lemma A.3.1] and its dual.

Consider next the case when X is a subset of the space $C(Y,Z)$ of continuous functions $x : Y \to Z$, where Y is a topological space and Z is an ordered topological space. *In what follows, we assume that $C(Y,Z)$, and all its subsets are equipped with the pointwise ordering and the topology of pointwise convergence.*

Lemma 9.43. *Let Y be a separable topological space and Z an ordered Hausdorff space that has property (C). Then each nonempty subset X of $C(Y,Z)$ is an ordered topological space that has property (C).*

Proof: Let W be a well-ordered chain in a nonempty subset X of $C(Y,Z)$, and assume that each increasing sequence of W converges pointwise to a mapping of X. For each $s \in Y$ the set $W(s) = \{x(s)\}_{x \in W}$ is a well-ordered chain of Z. For if A is a nonempty subset of $W(s)$, then the set $B = \{x \in W : x(s) \in A\}$ is nonempty, and thus it has the smallest element y. Therefore, $y(s) = \min A$ because of the pointwise ordering of $C(Y,Z)$. Since Y is separable, it contains a countable and dense subset $D = \{s_j\}$, $0 \le j < m \le \infty$. Let $s \in D$ be fixed, and let $(z_{n_k})_{k=0}^{\infty}$ be a subsequence of an increasing sequence $(z_n)_{n=0}^{\infty}$ of $W(s)$. If $(z_{n_k})_{k=k_0}^{\infty}$ is a constant sequence for some $k_0 \in \mathbb{N}_0$, then $z_{n_{k_0}}$ is the limit of $(z_{n_k})_{k=0}^{\infty}$. Otherwise $(z_{n_k})_{k=0}^{\infty}$ has a strictly increasing subsequence $(z_{n_{k_i}})_{i=0}^{\infty}$. Since the members of this subsequence belong to $W(s)$, and since W is well-ordered with respect to the pointwise ordering of $C(Y,Z)$, there exists an increasing sequence $(x_i)_{i=0}^{\infty}$ in W such that $x_i(s) = z_{n_{k_i}}$ for each $i \in \mathbb{N}_0$ (take $x_i = \min\{x \in W : x(s) = z_{n_{k_i}}\}$). Because $(x_i)_{i=0}^{\infty}$ converges pointwise, then $(x_i(s))_{i=0}^{\infty} = (z_{n_{k_i}})_{i=0}^{\infty}$ converges. Consequently, each subsequence of $(z_n)_{n=0}^{\infty}$ has a convergent subsequence, whence $(z_n)_{n=0}^{\infty}$ converges due to [133, Corollary 1.1.3].

The above proof shows that each increasing sequence $(z_n)_{n=0}^{\infty}$ of $W(s)$ converges when $s \in D$. Since $W(s)$ is a well-ordered chain in Z that has property (C), then for each $s_j \in D$, there exists an increasing sequence $(x_k^j(s_j))_{k=0}^{\infty}$ in $W(s_j)$ such that

$$\lim_{k \to \infty} x_k^j(s_j) = \sup W(s_j). \tag{9.63}$$

Denote

$$x_n = \max\{x_k^j : 0 \le j, \, k \le n\}, \quad n \in \mathbb{N}_0. \tag{9.64}$$

The so obtained sequence $(x_n)_{n=0}^{\infty}$ is an increasing sequence of W, whence it converges pointwise to a mapping of X by hypothesis. Denoting

$$x(s) = \lim_{n \to \infty} x_n(s), \quad s \in Y, \tag{9.65}$$

it follows from (9.63), (9.64), and (9.65) that

$$x(s_j) = \sup W(s_j) \quad \text{for each } s_j \in D. \tag{9.66}$$

To show that $x = \sup W$, let $s \in Y \setminus D$ be given. The above proof shows that there exists an increasing sequence $(y_n)_{n=0}^{\infty}$ in W such that

$$\lim_{n \to \infty} y_n(s) = \sup W(s). \qquad (9.67)$$

With z_n given by $z_n = \max\{x_n, y_n\}$, $n \in \mathbb{N}_0$, we obtain an increasing sequence $(z_n)_{n=0}^{\infty}$ of W. Denoting by z its limit function, which by hypothesis is continuous, it follows from (9.66) and (9.67) that

$$z(s_j) = \sup W(s_j), \; s_j \in D \text{ and } z(s) = \sup W(s). \qquad (9.68)$$

Both x and z are continuous, their restrictions to the dense subset D of Y are equal by (9.66) and (9.68). Since Z is a Hausdorff space, then $z = x$. In particular,

$$x(s) = \lim_{n \to \infty} x_n(s) = \sup W(s).$$

This result holds for all $s \in Y \setminus D$. It holds by (9.65) and (9.66) also for all $s \in D$, and hence it follows that x is the pointwise supremum of W. Obviously, $x = \sup W$ with respect to the pointwise ordering of X. Moreover, x is the pointwise limit of an increasing sequence $(x_n)_{n=0}^{\infty}$ of W.

The proof that each inversely well-ordered chain W of X whose decreasing sequences converge pointwise in X contains a decreasing sequence that converges pointwise to $\inf W$ in X is dual to the above proof. □

Assume next that Y is a topological space and $Z = (Z, d)$ is a metric space. We say that a subset W of $C(Y, Z)$ is *equicontinuous* if for each $t \in Y$ and for each $\epsilon > 0$ there exists a neighborhood U of t such that

$$d(x(s), x(t)) \le \epsilon \text{ for all } x \in W \text{ and } s \in U.$$

The next result is an easy consequence of the proofs of [133, Proposition 1.3.8], and [84, (7.5.6)].

Lemma 9.44. *Let Y be a topological space and $Z = (Z, d)$ an ordered metric space. If a pointwise monotone and equicontinuous sequence of functions from Y to Z has a pointwise limit, this limit function is continuous. If Y is a compact metric space, then the convergence is uniform.*

9.2.5 Chains of Random Variables

Let B be a closed and bounded ball of a separable and weakly sequentially complete ordered Banach space E whose order cone is normal. Denote by $\mathcal{B}$ the σ-algebra of Borel sets of B. Let (Ω, P) denote a probability space and X the space of all B-valued random variables on (Ω, P), i.e., measurable mappings $x : (\Omega, P) \to (B, \mathcal{B})$. Define a *Ky Fan* metric α and a partial ordering $\le_r$ in X by

$$\alpha(x, y) = \inf\{\epsilon > 0 : P\{\omega \in \Omega : (\|x(\omega) - y(\omega)\| > \epsilon\} \le \epsilon\},$$
$$x \le_r y \text{ if and only if } P\{\omega \in \Omega : x(\omega) \le y(\omega)\} = 1.$$

It can be shown that $(X, \le_r, \alpha)$ is an ordered metric space.

Lemma 9.45. *Each monotone sequence of* $(X, \leq_r)$ *converges in* (X, α).

Proof: Let $(x_n)_{n=1}^\infty$ be a monotone sequence in $(X, \leq_r)$. The definition of $\leq_r$ implies that for a.e. $\omega \in \Omega$ the sequence $(x_n(\omega))_{n=1}^\infty$ of E is monotone. Since B is bounded, it follows from Lemma 9.3 that the limit $x(\omega) = \lim_{n \to \infty} x_n(\omega)$ exists for a.e. $\omega \in \Omega$. Thus x is (equal a.e. to) a B-valued random variable by [85, Theorem 4.2.2], and $x_n \to x$ a.e., and hence also in probability. Because the Ky Fan metric α metrizes the convergence in probability by [85, Theorem 9.2.2], then $\alpha(x_n, x) \to 0$. □

Because every ordered metric space has property (C), the following result is a consequence of Lemma 9.45.

Proposition 9.46. *Let C be a chain in* $(X, \leq_r)$. *Then the following properties hold.*

(a) C contains an increasing sequence that converges in probability a.e. pointwise to $\sup C$ in X.

(b) C contains a decreasing sequence that converges in probability a.e. pointwise to $\inf C$ in X.

9.2.6 Properties of Order Intervals and Balls in Ordered Function Spaces

First we consider the existence of supremums and infimums of chains in ordered spaces of functions whose values are in a normed space that is ordered by a regular order cone.

Proposition 9.47. *Let E be an ordered Banach space whose order cone is regular, and assume that $[\underline{u}, \overline{u}]$ is an order interval in any of the following a.e. pointwise ordered function spaces.*

(a) $L^p(\Omega, E)$, $1 \leq p < \infty$, where Ω is a σ-finite measure space.
(b) $L^1_{loc}(\Omega, E)$, where Ω is a hemicompact subset of $\mathbb{R}^m$.
(c) $HL(J, E)$, where J is a compact interval in $\mathbb{R}$.
(d) $HL_{loc}(J, E)$, where J is an open interval in $\mathbb{R}$.

Then $\sup W$ and $\inf W$ exist and belong to $[\underline{u}, \overline{u}]$ whenever W is a chain in $[\underline{u}, \overline{u}]$. Moreover, there exists an increasing sequence in W that converges a.e. pointwise to $\sup W$ and a decreasing sequence in W that converges a.e. pointwise to $\inf W$.

Proof: Let W be a chain in $[\underline{u}, \overline{u}]$, and let C be a well-ordered cofinal subchain W. Since C is contained in $[\underline{u}, \overline{u}]$, it follows from Lemma 9.35 and Propositions 9.36, 9.38, and 9.39 that in each of the cases (a)–(d), C contains an increasing sequence (u_n) that converges a.e. pointwise to $u = \sup C = \sup W$, and u belongs to $[\underline{u}, \overline{u}]$. The proof that $\inf W$ exists, and belongs to $[\underline{u}, \overline{u}]$, and that it is the a.e. pointwise limit of a decreasing sequence of W is similar. □

The reasoning of the above proof, Remark 9.42, and the results of Propositions 9.40 and 9.41 lead to the following result.

Proposition 9.48. *Given an ordered normed space E whose order cone is regular, and an ordered topological space X. Assume that $[\underline{u}, \overline{u}]$ is an order interval in the pointwise ordered space $C(X, E)$ of continuous functions from X to E. Then $\sup W$ and $\inf W$ exist and belong to $[\underline{u}, \overline{u}]$ whenever W is an equicontinuous chain in $[\underline{u}, \overline{u}]$. Moreover, if X is separable, there exists an increasing sequence in W that converges pointwise to $\sup W$ and a decreasing sequence in W that converges pointwise to $\inf W$.*

Next we study the existence of supremums and infimums in bounded balls and in order intervals of ordered function spaces. In what follows, E is an ordered Banach space having the following properties.

(E0) Bounded and monotone sequences of E have weak or strong limits.
(E1) E is lattice-ordered and $\|x^+\| \le \|x\|$ for all $x \in E$, where $x^+ = \sup\{0, x\}$.
(E2) The mapping $E \ni x \to x^+$ is continuous.

Proposition 9.49. *Let Ω be a σ-finite measure space, E an ordered Banach space that has properties (E0)–(E2), and $1 \le p < \infty$. Assume that $L^p(\Omega, E)$ is ordered a.e. pointwise. Given $c \in L^p(\Omega, E)$ and $R \in [0, \infty)$, we denote*

$$B^p(c, R) := \{u \in L^p(\Omega, E) : \|u - c\|_p \le R\}.$$

(a) $\sup\{c, u\} \in B^p(c, R)$ and $\inf\{c, u\} \in B^p(c, R)$ for every $u \in B^p(c, R)$.
(b) $\sup W \in B^p(c, R)$ and $\inf W \in B^p(c, R)$ for every chain W of $B^p(c, R)$.

Proof: Ad (a) Property (E2) and Proposition 9.2 imply that for each $v \in L^p(\Omega, E)$ the mapping $v^+ = t \mapsto v(t)^+$ is μ-measurable. Moreover, in view of property (E1) and the definition of the p-norm we have $\|v^+\|_p \le \|v\|_p$, so that $v^+ \in L^p(\Omega, E)$. Thus for each $u \in B^p(c, R)$ we obtain

$$\| \sup\{c, u\} - c\|_p = \| \inf\{c, u\} - c\|_p = \|(u - c)^+\|_p \le \|u - c\|_p \le R,$$

which proves (a).

Ad (b) Let W be a chain in $B^p(c, R)$, and let C be a well-ordered cofinal subchain W. Since C is a chain in $B^p(c, R)$, it is bounded, whence there is by Proposition 9.34 and property (E0) an increasing sequence (u_n) in C that converges weakly a.e. pointwise to $u = \sup C$. Moreover, it follows from (9.5) that

$$\|u(t) - c(t)\| \le \liminf_{n \to \infty} \|u_n(t) - c(t)\| \quad \text{for a.e. } t \in \Omega. \tag{9.69}$$

The above inequality, Fatou's Lemma, and the fact that (u_n) is a sequence in $B^p(c, R)$ imply that if $p \in [1, \infty)$, then

$$\int_{\Omega} \|u(t) - c(t)\|^p d\mu(t) \leq \int_{\Omega} \liminf_{n \to \infty} \|u_n(t) - c(t)\|^p d\mu(t)$$

$$\leq \liminf_{n \to \infty} \int_{\Omega} \|u_n(t) - c(t)\|^p d\mu(t) \leq R.$$

The above proof shows that $u = \sup C = \sup W \in B^p(c, R)$. Similarly one can show that $\inf W$ exists and belongs to $B^p(c, R)$. This proves the conclusion (b). $\qquad \square$

Next we prove corresponding results in the space of continuous functions.

Proposition 9.50. *Let X be a topological space, and E be an ordered normed space with properties (E0)–(E2), and assume that $C(X, E)$ is ordered pointwise. Given $c \in C(X, E)$ and $h : X \to \mathbb{R}_+$, denote*

$$B(c, h) := \{u \in C(X, E) : \|u(t) - c(t)\| \leq h(t) \ \text{for all} \ t \in X\}.$$

Then the following assertions hold.

(a) $\sup\{c, u\} \in B(c, h)$ and $\inf\{c, u\} \in B(c, h)$ for every $u \in B(c, h)$.
(b) If X is separable, then $\sup W \in B(c, h)$ and $\inf W \in B(c, h)$ for every equicontinuous chain W of $B(c, h)$.

Proof: Ad (a) Property (E2) implies that for each $v \in C(X, E)$ the mapping $v^+ = t \mapsto v(t)^+$ is continuous. Moreover, in view of property (E1) we have $\|v^+(t)\| \leq \|v(t)\|$ for each $t \in X$, so that

$$\|\sup\{c, u\}(t) - c(t)\| = \|\inf\{c, u\}(t) - c(t)\|_p = \|(u - c)^+(t)\|$$
$$\leq \|u(t) - c(t)\| \leq h(t)$$

for all $u \in B(c, h)$ and $t \in X$, which proves conclusion (a).

Ad (b) Let W be an equicontinuous chain in $B(c, h)$, and let C be a well-ordered cofinal subchain of W. By Proposition 9.41 there is an increasing sequence (u_n) in C that converges weakly pointwise to $u = \sup C = \sup W$. Since (u_n) is a sequence in $B(c, h)$, it follows from (9.5) that

$$\|u(t) - c(t)\| \leq \liminf_{n \to \infty} \|u_n(t) - c(t)\| \leq h(t) \quad \text{for all} \ t \in U. \tag{9.70}$$

This proves that $u = \sup W \in B(c, h)$. The proof that $\inf W \in B(c, h)$ is similar. $\qquad \square$

Remark 9.51. In the considered function spaces, excluding the spaces of continuous functions with nonseparable domain, the well-ordered chains and their subchains whose supremums are proved to exist contain cofinal sequences. This implies by [133, Lemma 1.1.4] and its dual that all those well-ordered chains are countable. Similarly, all those inversely well-ordered chains of these function spaces whose infimums are proved to exist are countable.

9.3 Sobolev Spaces

In this section, we summarize the main properties of Sobolev spaces. These properties include, e.g., the approximation of Sobolev functions by smooth functions (density theorems), continuity properties and compactness conditions (embedding theorems), the definition of the boundary values of Sobolev functions (trace theorem), and calculus for Sobolev functions (chain rule).

9.3.1 Definition of Sobolev Spaces

Let $\alpha = (\alpha_1, \ldots, \alpha_N)$ with nonnegative integers $\alpha_1, \ldots, \alpha_N$ be a multi-index, and denote its order by $|\alpha| = \alpha_1 + \cdots + \alpha_N$. Set $D_i = \partial/\partial x_i$, $i = 1, \ldots, N$, and $D^\alpha u = D_1^{\alpha_1} \cdots D_N^{\alpha_N} u$, with $D^0 u = u$. Let Ω be a domain in $\mathbb{R}^N$ with $N \geq 1$. Then $w \in L^1_{\text{loc}}(\Omega)$ is called the αth weak or generalized derivative of $u \in L^1_{\text{loc}}(\Omega)$ if and only if

$$\int_\Omega u D^\alpha \varphi \, dx = (-1)^{|\alpha|} \int_\Omega w\varphi \, dx, \quad \text{for all } \varphi \in C_0^\infty(\Omega),$$

holds, where $C_0^\infty(\Omega)$ denotes the space of infinitely differentiable functions with compact support in Ω. The generalized derivative w denoted by $w = D^\alpha u$ is unique up to a change of the values of w on a set of Lebesgue measure zero.

Definition 9.52. Let $1 \leq p \leq \infty$ and $m = 0, 1, 2, \ldots$. The Sobolev space $W^{m,p}(\Omega)$ is the space of all functions $u \in L^p(\Omega)$, which have generalized derivatives up to order m such that $D^\alpha u \in L^p(\Omega)$ for all α: $|\alpha| \leq m$. For $m = 0$, we set $W^{0,p}(\Omega) = L^p(\Omega)$.

With the corresponding norms given by

$$\|u\|_{W^{m,p}(\Omega)} = \left(\sum_{|\alpha| \leq m} \|D^\alpha u\|_{L^p(\Omega)}^p \right)^{1/p}, \quad 1 \leq p < \infty,$$

$$\|u\|_{W^{m,\infty}(\Omega)} = \max_{|\alpha| \leq m} \|D^\alpha u\|_{L^\infty(\Omega)},$$

$W^{m,p}(\Omega)$ becomes a Banach space.

Definition 9.53. $W_0^{m,p}(\Omega)$ is the closure of $C_0^\infty(\Omega)$ in $W^{m,p}(\Omega)$.

$W_0^{m,p}(\Omega)$ is a Banach space with the norm $\| \cdot \|_{W^{m,p}(\Omega)}$.

Theorem 9.54. Let $\Omega \subset \mathbb{R}^N$ be a bounded domain, $N \geq 1$. Then we have the following:

(i) $W^{m,p}(\Omega)$ is separable for $1 \leq p < \infty$.
(ii) $W^{m,p}(\Omega)$ is reflexive for $1 < p < \infty$.

(iii) Let $1 \leq p < \infty$. Then $C^\infty(\Omega) \cap W^{m,p}(\Omega)$ is dense in $W^{m,p}(\Omega)$, and if $\partial\Omega$ is a Lipschitz boundary then $C^\infty(\overline{\Omega})$ is dense in $W^{m,p}(\Omega)$, where $C^\infty(\Omega)$ and $C^\infty(\overline{\Omega})$ are the spaces of infinitely differentiable functions in Ω and $\overline{\Omega}$, respectively (cf., e.g., [99]).

As for the proofs of these properties we refer to, e.g., [99].

Now we state some Sobolev embedding theorems. Let X, Y be two normed linear spaces with $X \subseteq Y$. We recall, the operator $i : X \to Y$ defined by $i(u) = u$ for all $u \in X$ is called the embedding operator of X into Y. We say X is continuously (compactly) embedded in Y if $X \subseteq Y$ and the embedding operator $i : X \to Y$ is continuous (compact).

Theorem 9.55 (Sobolev Embedding Theorem). *Let $\Omega \subset \mathbb{R}^N$, $N \geq 1$, be a bounded domain with Lipschitz boundary $\partial\Omega$. Then the following holds.*

(i) If $mp < N$, then the space $W^{m,p}(\Omega)$ is continuously embedded in $L^{p^*}(\Omega)$, $p^* = Np/(N - mp)$, and compactly embedded in $L^q(\Omega)$ for any q with $1 \leq q < p^*$.

(ii) If $0 \leq k < m - \frac{N}{p} < k + 1$, then the space $W^{m,p}(\Omega)$ is continuously embedded in $C^{k,\lambda}(\overline{\Omega})$, $\lambda = m - \frac{N}{p} - k$, and compactly embedded in $C^{k,\lambda'}(\overline{\Omega})$ for any $\lambda' < \lambda$.

(iii) Let $1 \leq p < \infty$, then the embeddings

$$L^p(\Omega) \supset W^{1,p}(\Omega) \supset W^{2,p}(\Omega) \supset \cdots$$

are compact.

Here $C^{k,\lambda}(\overline{\Omega})$ denotes the *Hölder space*; cf. [99]. As for the proofs we refer to, e.g., [99, 229].

The proper definition of boundary values for Sobolev functions is based on the following theorem.

Theorem 9.56 (Trace Theorem). *Let $\Omega \subset \mathbb{R}^N$ be a bounded domain with Lipschitz ($C^{0,1}$) boundary $\partial\Omega$, $N \geq 1$, and $1 \leq p < \infty$. Then there exists exactly one continuous linear operator*

$$\gamma : W^{1,p}(\Omega) \to L^p(\partial\Omega)$$

such that:

(i) $\gamma(u) = u|_{\partial\Omega}$ if $u \in C^1(\overline{\Omega})$.
(ii) $\|\gamma(u)\|_{L^p(\partial\Omega)} \leq C \|u\|_{W^{1,p}(\Omega)}$ with C depending only on p and Ω.
(iii) If $u \in W^{1,p}(\Omega)$, then $\gamma(u) = 0$ in $L^p(\partial\Omega)$ if and only if $u \in W_0^{1,p}(\Omega)$.

Definition 9.57 (Trace). *We call $\gamma(u)$ the trace (or generalized boundary function) of u on $\partial\Omega$.*

The following compactness result of the trace operator holds, see [157].

Theorem 9.58. *Let* $\Omega \subset \mathbb{R}^N$ *be a bounded domain with Lipschitz boundary* $\partial\Omega$, $N \geq 1$.

(i) If $1 < p < N$, *then*

$$\gamma : W^{1,p}(\Omega) \to L^q(\partial\Omega)$$

is completely continuous for any q *with* $1 \leq q < (Np - p)/(N - p)$.
(ii) If $p \geq N$, *then for any* $q \geq 1$

$$\gamma : W^{1,p}(\Omega) \to L^q(\partial\Omega)$$

is completely continuous.

9.3.2 Chain Rule and Lattice Structure

In this section we assume that $\Omega \subset \mathbb{R}^N$ is a bounded domain with Lipschitz boundary $\partial\Omega$.

Lemma 9.59 (Chain Rule). *Let* $f \in C^1(\mathbb{R})$ *and* $\sup_{s \in \mathbb{R}} |f'(s)| < \infty$. *Let* $1 \leq p < \infty$ *and* $u \in W^{1,p}(\Omega)$. *Then the composite function* $f \circ u \in W^{1,p}(\Omega)$, *and its generalized derivatives are given by*

$$D_i(f \circ u) = (f' \circ u)D_i u, \quad i = 1, \dots, N.$$

Lemma 9.60 (Generalized Chain Rule). *Let* $f : \mathbb{R} \to \mathbb{R}$ *be continuous and piecewise continuously differentiable with* $\sup_{s \in \mathbb{R}} |f'(s)| < \infty$, *and* $u \in W^{1,p}(\Omega)$, $1 \leq p < \infty$. *Then* $f \circ u \in W^{1,p}(\Omega)$, *and its generalized derivative is given by*

$$D_i(f \circ u)(x) = \begin{cases} f'(u(x))D_i u(x) & \text{if } f \text{ is differentiable at } u(x), \\ 0 & \text{otherwise.} \end{cases}$$

The chain rule may further be extended to Lipschitz continuous f; see, e.g., [99, 229].

Lemma 9.61 (Generalized Chain Rule). *Let* $f : \mathbb{R} \to \mathbb{R}$ *be a Lipschitz continuous function and* $u \in W^{1,p}(\Omega)$, $1 \leq p < \infty$. *Then* $f \circ u \in W^{1,p}(\Omega)$, *and its generalized derivative is given by*

$$D_i(f \circ u)(x) = f_B(u(x))D_i u(x) \quad \text{for a.e. } x \in \Omega,$$

where $f_B : \mathbb{R} \to \mathbb{R}$ *is a Borel-measurable function such that* $f_B = f'$ *a.e. in* $\mathbb{R}$.

The generalized derivative of the following special functions are frequently used.

Example 9.62. Let $1 \leq p < \infty$ and $u \in W^{1,p}(\Omega)$. Then $u^+ = \max\{u, 0\}$, $u^- = \max\{-u, 0\}$, and $|u|$ are in $W^{1,p}(\Omega)$, and their generalized derivatives are given by

$$(D_i u^+)(x) = \begin{cases} D_i u(x) & \text{if} \quad u(x) > 0, \\ 0 & \text{if} \quad u(x) \leq 0; \end{cases}$$

$$(D_i u^-)(x) = \begin{cases} 0 & \text{if} \quad u(x) \geq 0, \\ -D_i u(x) & \text{if} \quad u(x) < 0; \end{cases}$$

$$(D_i |u|)(x) = \begin{cases} D_i u(x) & \text{if} \quad u(x) > 0, \\ 0 & \text{if} \quad u(x) = 0, \\ -D_i u(x) & \text{if} \quad u(x) < 0. \end{cases}$$

As for the traces of u^+ and u^- we have (cf., e.g., [77])

$$\gamma(u^+) = (\gamma(u))^+, \quad \gamma(u^-) = (\gamma(u))^-.$$

Lemma 9.63 (Lattice Structure). *Let u, $v \in W^{1,p}(\Omega)$, $1 \leq p < \infty$. Then $\max\{u, v\}$ and $\min\{u, v\}$ are in $W^{1,p}(\Omega)$ with generalized derivatives*

$$D_i \max\{u, v\}(x) = \begin{cases} D_i u(x) & \text{if} \quad u(x) > v(x), \\ D_i v(x) & \text{if} \quad v(x) \geq u(x); \end{cases}$$

$$D_i \min\{u, v\}(x) = \begin{cases} D_i u(x) & \text{if} \quad u(x) < v(x), \\ D_i v(x) & \text{if} \quad v(x) \leq u(x). \end{cases}$$

Proof: The assertion follows easily from the above examples and the generalized chain rule by using $\max\{u, v\} = (u-v)^+ + v$ and $\min\{u, v\} = u - (u-v)^+$; see, e.g., [142, Theorem 1.20]. □

Lemma 9.64. *If $(u_j), (v_j) \subset W^{1,p}(\Omega)$ $(1 \leq p < \infty)$ are such that $u_j \to u$ and $v_j \to v$ in $W^{1,p}(\Omega)$, then $\min\{u_j, v_j\} \to min\{u, v\}$ and $\max\{u_j, v_j\} \to \max\{u, v\}$ in $W^{1,p}(\Omega)$ as $j \to \infty$.*

For the proof see, e.g., [142, Lemma 1.22]. By means of Lemma 9.64 we readily obtain the following result.

Lemma 9.65. *Let $\underline{u}$, $\bar{u} \in W^{1,p}(\Omega)$ satisfy $\underline{u} \leq \bar{u}$, and let T be the truncation operator defined by*

$$Tu(x) = \begin{cases} \bar{u}(x) & \text{if} \quad u(x) > \bar{u}(x), \\ u(x) & \text{if} \quad \underline{u}(x) \leq u(x) \leq \bar{u}(x), \\ \underline{u}(x) & \text{if} \quad u(x) < \underline{u}(x). \end{cases}$$

Then T is a bounded continuous mapping from $W^{1,p}(\Omega)$ (respectively, $L^p(\Omega)$) into itself.

Proof: The truncation operator T can be represented in the form

$$Tu = \max\{u, \underline{u}\} + \min\{u, \bar{u}\} - u.$$

Thus the assertion easily follows from Lemma 9.64. $\square$

Lemma 9.66 (Lattice Structure). *If* u, $v \in W_0^{1,p}(\Omega)$, *then* $\max\{u, v\}$ *and* $\min\{u, v\}$ *are in* $W_0^{1,p}(\Omega)$.

Lemma 9.66 implies that $W_0^{1,p}(\Omega)$ has a lattice structure as well; see, e.g., [142].

A partial ordering of traces on $\partial\Omega$ is given as follows.

Definition 9.67. *Let* $u \in W^{1,p}(\Omega)$, $1 \le p < \infty$. *Then* $u \le 0$ *on* $\partial\Omega$ *if* $u^+ \in W_0^{1,p}(\Omega)$.

9.4 Operators of Monotone Type

In this section we provide the basic results on pseudomonotone operators from a Banach space X into its dual space X^*.

9.4.1 Main Theorem on Pseudomonotone Operators

Let X be a real, reflexive Banach space with norm $\|\cdot\|$, X^* its dual space, and denote by $\langle\cdot,\cdot\rangle$ the duality pairing between them. The norm convergence in X and X^* is denoted by "$\to$" and the weak convergence by "$\rightharpoonup$".

Definition 9.68. *Let* $A : X \to X^*$; *then* A *is called*

(i)   *continuous (resp. weakly continuous) iff* $u_n \to u$ *implies* $Au_n \to Au$ *(resp.* $u_n \rightharpoonup u$ *implies* $Au_n \rightharpoonup Au$);

(ii) *demicontinuous iff* $u_n \to u$ *implies* $Au_n \rightharpoonup Au$;

(iii) *hemicontinuous iff the real function* $t \to \langle A(u + tv), w\rangle$ *is continuous on* $[0, 1]$ *for all* $u, v, w \in X$;

(iv) *strongly continuous or completely continuous iff* $u_n \rightharpoonup u$ *implies* $Au_n \to Au$;

(v)   *bounded iff* A *maps bounded sets into bounded sets*;

(vi) *coercive iff* $\lim_{\|u\| \to \infty} \frac{\langle Au, u\rangle}{\|u\|} = +\infty$.

Definition 9.69 (Operators of Monotone Type). *Let* $A : X \to X^*$; *then* A *is called*

(i)   *monotone (resp. strictly monotone) iff* $\langle Au - Av, u - v\rangle \ge$ *(resp.* $>$)0 *for all* $u, v \in X$ *with* $u \ne v$;

(ii) *strongly monotone iff there is a constant* $c > 0$ *such that*
      $\langle Au - Av, u - v\rangle \ge c\|u - v\|^2$ *for all* $u, v \in X$;

(iii) uniformly monotone iff $\langle Au - Av, u - v \rangle \geq a(\|u - v\|)\|u - v\|$ *for all*
   $u, v \in X$ *where* $a : [0, \infty) \to [0, \infty)$ *is strictly increasing with* $a(0) = 0$
   and $a(s) \to +\infty$ *as* $s \to \infty$;

(iv) pseudomonotone iff $u_n \rightharpoonup u$ *and* $\limsup_{n \to \infty} \langle Au_n, u_n - u \rangle \leq 0$ *implies*
   $\langle Au, u - w \rangle \leq \liminf_{n \to \infty} \langle Au_n, u_n - w \rangle$ *for all* $w \in X$;

(v) to satisfy (S_+)-*condition iff* $u_n \rightharpoonup u$ *and* $\limsup_{n \to \infty} \langle Au_n, u_n - u \rangle \leq 0$
   imply $u_n \to u$.

One can show (cf., e.g., [19]) that the pseudomonotonicity according to (iv) of Definition 9.69 is equivalent to the following definition.

Definition 9.70. *The operator* $A : X \to X^*$ *is pseudomonotone iff* $u_n \rightharpoonup u$ *and* $\limsup_{n \to \infty} \langle Au_n, u_n - u \rangle \leq 0$ *implies* $Au_n \rightharpoonup Au$ *and* $\langle Au_n, u_n \rangle \to \langle Au, u \rangle$.

For the following result see, e.g., [229, Proposition 27.6].

Lemma 9.71. *Let* $A, B : X \to X^*$ *be operators on the real reflexive Banach space* X. *Then the following implications hold:*

(i) If A is monotone and hemicontinuous, then A is pseudomonotone.
(ii) If A is strongly continuous, then A is pseudomonotone.
(iii) If A and B are pseudomonotone, then $A + B$ is pseudomonotone.

The main theorem on pseudomonotone operators due to Brézis is given by the next theorem (see [229, Theorem 27.A]).

Theorem 9.72 (Main Theorem on Pseudomonotone Operators). *Let* X *be a real, reflexive Banach space, and let* $A : X \to X^*$ *be a pseudomonotone, bounded, and coercive operator, and* $b \in X^*$. *Then there exists a solution of the equation* $Au = b$.

Remark 9.73. Theorem 9.72 contains several important surjectivity results as special cases, such as Lax–Milgram's theorem and the Main Theorem on Monotone Operators. The latter will be formulated in the following corollary.

Corollary 9.74 (Main Theorem on Monotone Operators). *Let* X *be a real, reflexive Banach space, and let* $A : X \to X^*$ *be a monotone, hemicontinuous, bounded, and coercive operator, and* $b \in X^*$. *Then there exists a solution of the equation* $Au = b$.

For the proof of Corollary 9.74 we have only to mention that in view of Lemma 9.71, a monotone and hemicontinuous operator is pseudomonotone.

9.4.2 Leray–Lions Operators

An important class of operators of monotone type are the so-called Leray–Lions operators (see, e.g., [219] or [172]). These kinds of operators occur in the functional analytical treatment of nonlinear elliptic and parabolic problems.

Definition 9.75 (Leray–Lions Operator). *Let X be a real, reflexive Banach space. We say that $A : X \to X^*$ is a Leray–Lions operator if it is bounded and satisfies*

$$Au = \mathcal{A}(u, u), \quad \text{for } u \in X,$$

where $\mathcal{A} : X \times X \to X^*$ has the following properties:

(i) *For any $u \in X$, the mapping $v \mapsto \mathcal{A}(u, v)$ is bounded and hemicontinuous from X to its dual X^*, with*

$$\langle \mathcal{A}(u, u) - \mathcal{A}(u, v), u - v \rangle \geq 0 \quad \text{for } v \in X;$$

(ii) *For any $v \in X$, the mapping $u \mapsto \mathcal{A}(u, v)$ is bounded and hemicontinuous from X to its dual X^*;*

(iii) *For any $v \in X$, $\mathcal{A}(u_n, v)$ converges weakly to $\mathcal{A}(u, v)$ in X^* if $(u_n) \subset X$ is such that $u_n \rightharpoonup u$ in X and*

$$\langle \mathcal{A}(u_n, u_n) - \mathcal{A}(u_n, u), u_n - u \rangle \to 0;$$

(iv) *For any $v \in X$, $\langle \mathcal{A}(u_n, v), u_n \rangle$ converges to $\langle F, u \rangle$ if $(u_n) \subset V$ is such that $u_n \rightharpoonup u$ in X, and $\mathcal{A}(u_n, v) \rightharpoonup F$ in X^*.*

As for the proof of the next theorem, see, e.g., [219].

Theorem 9.76. *Every Leray–Lions operator $A : X \to X^*$ is pseudomonotone.*

Next we will see that quasilinear elliptic operators satisfying certain structure and growth conditions represent Leray–Lions operators. To this end we need to study first the mapping properties of superposition operators, which are also called Nemytskij operators.

Definition 9.77 (Nemytskij Operator). *Let $\Omega \subseteq \mathbb{R}^N$, $N \geq 1$, be a nonempty measurable set and let $f : \Omega \times \mathbb{R}^m \to \mathbb{R}$, $m \geq 1$, and $u : \Omega \to \mathbb{R}^m$ be a given function. Then the superposition or Nemytskij operator F assigns $u \mapsto f \circ u$, i.e., F is given by*

$$F(u)(x) = (f \circ u)(x) = f(u(x)) \quad \text{for } x \in \Omega.$$

Definition 9.78 (Carathéodory Function). *Let $\Omega \subseteq \mathbb{R}^N$, $N \geq 1$, be a nonempty measurable set and let $f : \Omega \times \mathbb{R}^m \to \mathbb{R}$, $m \geq 1$. The function f is called a Carathéodory function if the following two conditions are satisfied:*

(i) $x \mapsto f(x, s)$ *is measurable in* Ω *for all* $s \in \mathbb{R}^m$;
(ii) $s \mapsto f(x, s)$ *is continuous on* $\mathbb{R}^m$ *for a.e.* $x \in \Omega$.

Lemma 9.79. *Let* $f : \Omega \times \mathbb{R}^m \to \mathbb{R}$, $m \geq 1$, *be a Carathéodory function that satisfies a growth condition of the form*

$$|f(x, s)| \leq k(x) + c \sum_{i=1}^m |s_i|^{p_i/q}, \quad \forall\, s = (s_1, \ldots, s_m) \in \mathbb{R}^m, \quad a.e.\ x \in \Omega,$$

for some positive constant c *and some* $k \in L^q(\Omega)$, *and* $1 \leq q, p_i < \infty$ *for all* $i = 1, \ldots, m$. *Then the Nemytskij operator* F *defined by*

$$F(u)(x) = f(x, u_1(x), \ldots, u_m(x))$$

is continuous and bounded from $L^{p_1}(\Omega) \times \cdots \times L^{p_m}(\Omega)$ *into* $L^q(\Omega)$. *Here* u *denotes the vector function* $u = (u_1, \ldots, u_m)$. *Furthermore,*

$$\|F(u)\|_{L^q(\Omega)} \leq c \left(\|k\|_{L^q(\Omega)} + \sum_{i=1}^m \|u_i\|_{L^{p_i}(\Omega)}^{p_i/q} \right).$$

Definition 9.80. *Let* $\Omega \subseteq \mathbb{R}^N$, $N \geq 1$, *be a nonempty measurable set. A function* $f : \Omega \times \mathbb{R}^m \to \mathbb{R}$, $m \geq 1$, *is called superpositionally measurable (or sup-measurable) if the function* $x \mapsto F(u)(x)$ *is measurable in* Ω *whenever the component functions* $u_i : \Omega \to \mathbb{R}$ *of* $u = (u_1, \ldots, u_m)$ *are measurable.*

9.4.3 Multi-Valued Pseudomonotone Operators

In this section we briefly recall the main results of the theory of pseudomonotone multi-valued operators developed by Browder and Hess to an extent as it will be needed in the study of variational and hemivariational inequalities. For the proofs and a more detailed presentation we refer, e.g., to the monographs [229, 184].

First we present basic results about the continuity of multi-valued functions (multi-functions) and provide useful equivalent descriptions of these notions. Even though these notions can be defined in a much more general context, we confine ourselves to mappings between Banach spaces, which is sufficient for our purpose.

Definition 9.81 (Semicontinuous Multi-Functions). *Let* X, Y *be Banach spaces, and* $A : X \to 2^Y$ *be a multi-function.*

(i) *A is called upper semicontinuous at x_0, if for every open subset $V \subseteq Y$ with $A(x_0) \subseteq V$, there exists a neighborhood $U(x_0)$ such that $A(U(x_0)) \subseteq V$. If A is upper semicontinuous at every $x_0 \in X$, we call A upper semicontinuous in X.*

(ii) *A is called lower semicontinuous at x_0 if for every neighborhood $V(y)$ of every $y \in A(x_0)$, there exists a neighborhood $U(x_0)$ such that*

$$A(u) \cap V(y) \neq \emptyset \quad \text{for all} \quad u \in U(x_0).$$

If A is lower semicontinuous at every $x_0 \in X$, we call A lower semicontinuous in X.

(iii) *A is called continuous at x_0 if A is both upper and lower semicontinuous at x_0. If A is continuous at every $x_0 \in X$, we call A continuous in X.*

Alternative equivalent continuity criteria are given in the following propositions. To this end we introduce the *preimage* of a multi-function.

Definition 9.82 (Preimage). *Let $M \subseteq Y$ and $A : X \to 2^Y$ be a multifunction. The preimage $A^{-1}(M)$ is defined by*

$$A^{-1}(M) = \{x \in X : A(x) \cap M \neq \emptyset\}.$$

Proposition 9.83. *Let X, Y be Banach spaces, and $A : X \to 2^Y$ be a multifunction. Then the following statements are equivalent.*

(i) *A is upper semicontinuous.*
(ii) *For all closed sets $C \subseteq Y$, the preimage $A^{-1}(C)$ is closed.*
(iii) *If $x \in X$, (x_n) is a sequence in X with $x_n \to x$ as $n \to \infty$, and V is an open set in Y such that $A(x) \subseteq V$, then there exists $n_0 \in \mathbb{N}$ depending on V such that for all $n \geq n_0$ we have $A(x_n) \subseteq V$.*

Proposition 9.84. *Let X, Y be Banach spaces, and $A : X \to 2^Y$ be a multifunction. Then the following statements are equivalent.*

(i) *A is lower semicontinuous.*
(ii) *For all open sets $O \subseteq Y$, the preimage $A^{-1}(O)$ is open.*
(iii) *If $x \in X$, (x_n) is a sequence in X with $x_n \to x$ as $n \to \infty$, and $y \in A(x)$, then for every $n \in \mathbb{N}$ one can find a $y_n \in A(x_n)$, such that $y_n \to y$, as $n \to \infty$.*

Remark 9.85. For a single-valued operator $A : X \to Y$, upper semicontinuous and lower semicontinuous in the multi-valued setting is identical with continuous. For $A : M \to 2^N$ having the same corresponding properties, where M and N are subsets of the Banach spaces X and Y, respectively, then M and N have to be equipped with the induced topology.

Next we introduce the notion of multi-valued monotone and pseudomonotone operators from a real, reflexive Banach space X into its dual space, and formulate the main surjectivity result for these kinds of operators.

Definition 9.86 (Graph). *Let X be a real Banach space and let $A : X \to 2^{X^*}$ be a multi-valued mapping, i.e., to each $u \in X$ there is assigned a subset*

$A(u)$ of X^*, which may be empty if $u \notin D(A)$, where $D(A)$ is the domain of A given by

$$D(A) = \{u \in X : A(u) \neq \emptyset\}.$$

The graph of A denoted by $\mathrm{Gr}(A)$ is given by

$$\mathrm{Gr}(A) = \{(u, u^*) \in X \times X^* : u^* \in A(u)\}.$$

Definition 9.87 (Monotone Operator). *The mapping* $A : X \to 2^{X^*}$ *is called*

(i)   monotone iff

$$\langle u^* - v^*, u - v \rangle \geq 0 \quad \textit{for all} \ \ (u, u^*), \ (v, v^*) \in \mathrm{Gr}(A);$$

(ii) strictly monotone iff

$$\langle u^* - v^*, u - v \rangle > 0 \quad \textit{for all} \ \ (u, u^*), \ (v, v^*) \in \mathrm{Gr}(A), \ u \neq v;$$

(iii) maximal monotone iff A *is monotone and there is no monotone mapping* $\tilde{A} : X \to 2^{X^*}$ *such that* $\mathrm{Gr}(A)$ *is a proper subset of* $\mathrm{Gr}(\tilde{A})$, *which is equivalent to the following implication*

$$(u, u^*) \in X \times X^* : \quad \langle u^* - v^*, u - v \rangle \geq 0 \ \ \textit{for all} \ \ (v, v^*) \in \mathrm{Gr}(A)$$

implies $(u, u^*) \in \mathrm{Gr}(A)$.

The notions of strongly and uniformly monotone multi-valued operators are defined in a similar way as for single-valued operators.

A single-valued operator

$$A : D(A) \subseteq X \to X^*$$

is to be understood as a multi-valued operator $A : X \to X^*$ by setting $Au = \{Au\}$ if $u \in D(A)$ and $Au = \emptyset$ otherwise. Thus, A is monotone iff

$$\langle Au - Av, u - v \rangle \geq 0 \quad \text{for all} \ \ u, v \in D(A),$$

and $A : D(A) \subseteq X \to X^*$ is maximal monotone iff A is monotone and the condition

$$(u, u^*) \in X \times X^* : \quad \langle u^* - Av, u - v \rangle \geq 0 \ \ \text{for all} \ \ v \in D(A)$$

implies $u \in D(A)$ and $u^* = Au$.

Definition 9.88 (Pseudomonotone Operator). *Let* X *be a real reflexive Banach space. The operator* $A : X \to 2^{X^*}$ *is called pseudomonotone if the following conditions hold.*

(i)   The set $A(u)$ *is nonempty, bounded, closed, and convex for all* $u \in X$.

(ii) A is upper semicontinuous from each finite dimensional subspace of X to the weak topology on X^*.

(iii) If $(u_n) \subset X$ with $u_n \rightharpoonup u$, and if $u_n^* \in A(u_n)$ is such that

$$\limsup \langle u_n^*, u_n - u \rangle \leq 0,$$

then to each element $v \in X$ there exists $u^*(v) \in A(u)$ with

$$\liminf \langle u_n^*, u_n - v \rangle \geq \langle u^*(v), u - v \rangle.$$

Definition 9.89 (Generalized Pseudomonotone Operator). *Let X be a real reflexive Banach space. The operator $A : X \to 2^{X^*}$ is called generalized pseudomonotone if the following holds:*
Let $(u_n) \subset X$ and $(u_n^*) \subset X^*$ with $u_n^* \in A(u_n)$. If $u_n \rightharpoonup u$ in X and $u_n^* \rightharpoonup u^*$ in X^* and if $\limsup \langle u_n^*, u_n - u \rangle \leq 0$, then the element u^* lies in A(u) and

$$\langle u_n^*, u_n \rangle \to \langle u^*, u \rangle.$$

The next two propositions provide the relation between pseudomonotone and generalized pseudomontone operators.

Proposition 9.90. *Let X be a real reflexive Banach space. If the operator $A : X \to 2^{X^*}$ is pseudomonotone, then A is generalized pseudomonotone.*

Under the additional assumption of boundedness, the following converse of Proposition 9.90 is true.

Proposition 9.91. *Let X be a real reflexive Banach space, and assume that $A : X \to 2^{X^*}$ satisfies the following conditions.*

(i) *For each $u \in X$ we have that A(u) is a nonempty, closed, and convex subset of X^*.*
(ii) *$A : X \to 2^{X^*}$ is bounded.*
(iii) *If $u_n \rightharpoonup u$ in X and $u_n^* \rightharpoonup u^*$ in X^* with $u_n^* \in A(u_n)$ and if $\limsup \langle u_n^*, u_n - u \rangle \leq 0$, then $u^* \in A(u)$ and $\langle u_n^*, u_n \rangle \to \langle u^*, u \rangle$.*

Then the operator $A : X \to 2^{X^*}$ is pseudomonotone.

As for the proof of Proposition 9.91 we refer, e.g., to [184, Chap. 2]. Note that the notion of boundedness of a multi-valued operator is exactly the same as for single-valued operators, i.e., the image of a bounded set is again bounded.

The main theorem on pseudomonotone multi-valued operators is formulated in the next theorem.

Theorem 9.92. *Let X be a real reflexive Banach space, and let $A : X \to 2^{X^*}$ be a pseudomonotone and bounded operator that is coercive in the sense that there exists a real-valued function $c : \mathbb{R}_+ \to \mathbb{R}$ with*

$$c(r) \to +\infty, \quad as \ r \to +\infty$$

such that for all $(u, u^*) \in Gr(A)$ one has

$$\langle u^*, u - u_0 \rangle \geq c(\|u\|_X)\|u\|_X$$

for some $u_0 \in X$. Then A is surjective, i.e., range(A) = X.

9.5 First Order Evolution Equations

In this section we present the basic functional analytic tools needed in the study of first order single- and multi-valued evolution equations in the form

$$u \in X, \ u' \in X^* : \quad u' + Au \ni f \quad \text{in } X^*, \quad u(0) = u_0, \qquad (9.71)$$

where $X = L^p(0, \tau; V)$, $1 < p < \infty$, with $\tau > 0$ is the L^p-space of vector-valued functions $u : (0, \tau) \to V$ defined on the interval $(0, \tau)$ with values in some Banach space V, and u' is the generalized or distributional derivative of the function $t \mapsto u(t)$ with respect to $t \in (0, \tau)$. The right-hand side $f \in X^*$ is given and $A : X \to 2^{X^*}$ is some (in general) multi-valued operator. The initial values u_0 are taken from some Hilbert space H such that the embedding $V \subseteq H$ is continuous and dense. Problem (9.71) provides an abstract framework for the functional analytic treatment of initial-boundary value problems for parabolic differential equations and inclusions.

9.5.1 Evolution Triple and Generalized Derivative

The material of this subsection is mainly taken from [202, 229].

Definition 9.93 (Evolution Triple). *A triple (V, H, V^*) is called an evolution triple if the following properties hold.*

(i) V is a real, separable, and reflexive Banach space, and H is a real, separable Hilbert space endowed with the scalar product $(\cdot, \cdot)$.
(ii) The embedding $V \subseteq H$ is continuous, and V is dense in H.
(iii) Identifying H with its dual H^* by the Riesz map, we then have $H \subseteq V^*$ with the equation

$$\langle h, v \rangle_V = (h, v) \quad \text{for } h \in H \subseteq V^*, \ v \in V,$$

where $\langle \cdot, \cdot \rangle_V$ denotes the duality pairing between V and V^*.

Example 9.94. Let $\Omega \subset \mathbb{R}^N$ be a bounded domain with Lipschitz boundary $\partial\Omega$, and let V be a closed subspace of $W^{1,p}(\Omega)$ with $2 \le p < \infty$ such that $W_0^{1,p}(\Omega) \subseteq V \subseteq W^{1,p}(\Omega)$. Then (V, H, V^*) with $H = L^2(\Omega)$ is an evolution triple with all the embeddings being, in addition, compact.

Definition 9.95. *Let Y, Z be Banach spaces, and $u \in L^1(0, \tau; Y)$ and $w \in L^1(0, \tau; Z)$. Then, the function w is called the generalized derivative of the function u in $(0, \tau)$ iff the following relation holds*

$$\int_0^\tau \varphi'(t) u(t) \, dt = - \int_0^\tau \varphi(t) w(t) \, dt \quad \text{for all } \varphi \in C_0^\infty(0, \tau).$$

We write $w = u'$.

Theorem 9.96. *Let $V \subseteq H \subseteq V^*$ be an evolution triple and let $1 \leq p, q \leq \infty$, $0 < \tau < \infty$. Let $u \in L^p(0, \tau; V)$, then there exists the generalized derivative $u' \in L^q(0, \tau; V^*)$ iff there is a function $w \in L^q(0, \tau; V^*)$ such that*

$$\int_0^\tau (u(t), v)_H \varphi'(t) \, dt = - \int_0^\tau \langle w(t), v \rangle_V \varphi(t) \, dt$$

for all $v \in V$ and all $\varphi \in C_0^\infty(0, \tau)$. The generalized derivative u' is uniquely defined and $u' = w$.

Definition 9.97. *Let V be a real, separable, and reflexive Banach space, and let $X = L^p(0, \tau; V)$, $1 < p < \infty$. A space W is defined by*

$$W = \{u \in X : u' \in X^*\},$$

where u' is the generalized derivative, and $X^* = L^q(0, \tau; V^*)$, $1/p + 1/q = 1$.

Theorem 9.98 (Lions–Aubin). *Let B_0, B, B_1 be reflexive Banach spaces with $B_0 \subseteq B \subseteq B_1$, and assume $B_0 \hookrightarrow B$ is compactly and $B \hookrightarrow B_1$ is continuously embedded. Let $1 < p < \infty$, $1 < q < \infty$ and define $\mathcal{W}$ by*

$$\mathcal{W} = \{u \in L^p(0, \tau; B_0) : u' \in L^q(0, \tau; B_1)\}.$$

Then $\mathcal{W} \hookrightarrow L^p(0, \tau; B)$ is compactly embedded.

Example 9.99. Let $\Omega \subset \mathbb{R}^N$ be a bounded domain with Lipschitz boundary $\partial\Omega$. Since $W^{1,p}(\Omega) \subset L^p(\Omega)$ is compactly embedded, and $L^p(\Omega) \subset W^{1,p}(\Omega)^*$ is continuously embedded for $2 \leq p < \infty$, Theorem 9.98 can be applied by setting $B_0 = W^{1,p}(\Omega)$, $B = L^p(\Omega)$ and $B_1 = W^{1,p}(\Omega)^*$, $2 \leq p < \infty$. Thus W defined in Definition 9.97, i.e.,

$$W = \{u \in L^p(0, \tau; W^{1,p}(\Omega)) : u' \in L^q(0, \tau; W^{1,p}(\Omega)^*)\}$$

is compactly embedded in $L^p(0, \tau; L^p(\Omega)) \equiv L^p(Q)$, where $Q = \Omega \times (0, \tau)$.

Proposition 9.100. *Let $\Omega \subset \mathbb{R}^N$ be a bounded domain with Lipschitz boundary $\partial\Omega$, and let $X = L^p(0, \tau; W^{1,p}(\Omega))$ with $2 \leq p < \infty$. Then the trace operator $\gamma : W \to L^p(\Gamma)$ is compact.*

Proof: We apply Theorem 9.98. To this end let $B_0 = W^{1,p}(\Omega)$, $B = W^{1-\varepsilon,p}(\Omega)$, and $B_1 = B_0^*$. Since $B_0 \subseteq B$ is compactly embedded for any $\varepsilon \in (0, 1)$, and $B \subseteq B_1$ is continuously embedded, from Theorem 9.98 it follows that $W \subseteq L^p(0, \tau; W^{1-\varepsilon,p}(\Omega))$ is compactly embedded. If we select ε such that $0 < \varepsilon < 1 - 1/p$, then $\gamma : W^{1-\varepsilon,p}(\Omega) \to W^{1-\varepsilon-1/p,p}(\partial\Omega)$ is linear and continuous, and thus $\gamma : L^p(0, \tau; W^{1-\varepsilon,p}(\Omega)) \to L^p(0, \tau; W^{1-\varepsilon-1/p,p}(\partial\Omega)) \subset L^p(\Gamma)$ is linear and continuous, which due to the compact embedding of $W \hookrightarrow L^p(0, \tau; W^{1-\varepsilon,p}(\Omega))$ completes the proof. $\square$

Theorem 9.101. *Let $V \subseteq H \subseteq V^*$ be an evolution triple, and let $1 < p < \infty$, $1/p + 1/q = 1$, $0 < \tau < \infty$. Then the following hold.*

(i) *The space W defined in Definition 9.97 is a real, separable, and reflexive Banach space with the norm*

$$\|u\|_W = \|u\|_X + \|u'\|_{X^*}.$$

(ii) *The embedding $W \hookrightarrow C([0,\tau]; H)$ is continuous.*

(iii) *For all $u, v \in W$ and arbitrary t, s with $0 \le s \le t \le \tau$, the following generalized integration by parts formula holds:*

$$(u(t), v(t))_H - (u(s), v(s))_H = \int_s^t \langle u'(\zeta), v(\zeta) \rangle_V + \langle v'(\zeta), u(\zeta) \rangle_V \, d\zeta.$$

Remark 9.102. The integration by parts formula is equivalent to

$$\frac{d}{dt}(u(t), v(t))_H = \langle u'(t), v(t) \rangle_V + \langle v'(t), u(t) \rangle_V \quad \text{for a.e. } t \in (0, \tau).$$

In particular, for $u = v$ we obtain

$$\frac{d}{dt}\|u(t)\|_H^2 = 2\langle u'(t), u(t) \rangle_V,$$

which implies

$$\int_s^t \langle u'(\zeta), u(\zeta) \rangle_V \, d\zeta = \frac{1}{2}(\|u(t)\|_H^2 - \|u(s)\|_H^2). \tag{9.72}$$

In case that $V = W^{1,p}(\Omega)$, $2 \le p < \infty$, and $H = L^2(\Omega)$, we obtain the following generalization of formula (9.72), which will be useful for obtaining comparison principles in evolutionary problems.

Lemma 9.103. *Let $X = L^p(0, \tau; W^{1,p}(\Omega))$ with $2 \le p < \infty$ and $W = \{u \in X : u' \in X^*\}$, where $\Omega \subset \mathbb{R}^N$ is a bounded domain with Lipschitz boundary $\partial\Omega$. Let $\theta : \mathbb{R} \to \mathbb{R}$ be continuous and piecewise continuously differentiable with $\theta' \in L^\infty(\mathbb{R})$, and $\theta(0) = 0$, and let Θ denote the primitive of θ defined by*

$$\Theta(r) = \int_0^r \theta(s) \, ds.$$

Then for $w \in W$ the following formula holds.

$$\int_r^s \langle w'(t), \theta(w(t)) \rangle \, dt = \int_\Omega \Theta(w(s)) \, dx - \int_\Omega \Theta(w(r)) \, dx, \tag{9.73}$$

for a.e. $0 \le r < s \le \tau$.

Proof: The proof makes use of density arguments and the generalized chain rule for Sobolev functions, see Lemma 9.60. Note first that in view of the assumptions on θ and Lemma 9.60, the composed function $\theta(w)$ is in X for $w \in W$. The space $C^1([0,\tau]; C^1(\overline{\Omega}))$ of smooth functions is dense in W, cf., e.g., [229, Chap. 23]. Let $w \in W$ be given. Then there is a sequence $(w_n) \subset C^1([0,\tau]; C^1(\overline{\Omega}))$ with $w_n \to w$ as $n \to \infty$. For the smooth functions w_n we have

$$\int_r^s \langle w_n'(t), \theta(w_n(t)) \rangle \, dt = \int_r^s \int_\Omega w_n'(x,t)\theta(w_n(x,t)) \, dxdt$$
$$= \int_r^s \int_\Omega \frac{\partial}{\partial t}\Big(\Theta(w_n(x,t))\Big) \, dxdt$$
$$= \int_\Omega \Big(\Theta(w_n(x,s)) - \Theta(w_n(x,r))\Big) \, dx. \quad (9.74)$$

The assumptions on θ imply that θ is Lipschitz continuous, and thus it follows that for some subsequence of (w_n) (again denoted by (w_n))

$$\theta(w_n) \to \theta(w) \quad \text{in } X, \quad (9.75)$$

and due to the continuous embedding $W \hookrightarrow C([0,\tau]; L^2(\Omega))$ one gets for all $t \in [0,\tau]$

$$\Theta(w_n(t)) \to \Theta(w(t)) \quad \text{in } L^2(\Omega). \quad (9.76)$$

By using (9.75), (9.76) we may pass to the limit in (9.74) for some subsequence, which completes the proof. $\square$

Example 9.104. Let $\theta(s) = s$. Then θ trivially satisfies all the assumptions of Lemma 9.103, and the primitive Θ is given by $\Theta(s) = (1/2)s^2$, and thus formula (9.73) becomes

$$\int_r^s \langle w'(t), w(t) \rangle \, dt = \frac{1}{2}\int_\Omega (w(s))^2 \, dx - \frac{1}{2}\int_\Omega (w(r))^2 \, dx$$
$$= \frac{1}{2}\Big(\|w(s)\|_H^2 - \|w(r)\|_H^2\Big), \quad (9.77)$$

for all $0 \le r < s \le \tau$, where $H = L^2(\Omega)$, which is formula (9.72.)

The following example will play a crucial rule in obtaining comparison results.

Example 9.105. If $\theta(s) = s^+ = \max\{s,0\}$, then its primitive can easily be seen to be $\Theta(s) = (1/2)(s^+)^2$, and thus for $w \in W$ we get the formula

$$\int_r^s \langle w'(t), (w(t))^+ \rangle \, dt = \frac{1}{2}\Big(\|(w(s))^+\|_H^2 - \|(w(r))^+\|_H^2\Big). \quad (9.78)$$

9.5.2 Existence Results for Evolution Equations

The material of this subsection is mainly based on results taken from [20, 21, 202]; see also [172, 229].

Let $V \subseteq H \subseteq V^*$ be an evolution triple, and let $X = L^p(0, \tau; V)$, X^* and W be the spaces of vector-valued functions with $1 < p < \infty$, $1/p + 1/q = 1$, and $0 < \tau < \infty$. We provide an existence result for the evolution equation

$$u \in W : \quad u'(t) + A(t)u(t) = f(t), \ 0 < t < \tau, \quad u(0) = 0, \qquad (9.79)$$

where $f \in X^*$ is given, and $A(t) : V \to V^*$ is some operator specified later. Without loss of generality, homogeneous initial values have been assumed, since inhomogeneous initial values can be transformed to homogeneous ones by translation. The generalized derivative $Lu = u'$ restricted to the subset

$$D(L) = \{u \in X : u' \in X^* \text{ and } u(0) = 0\} = \{u \in W : u(0) = 0\}$$

defines a linear operator $L : D(L) \to X^*$ given by

$$\langle Lu, v \rangle = \int_0^\tau \langle u'(t), v(t) \rangle \, dt \quad \text{for all } v \in X.$$

The operator L has the following properties.

Lemma 9.106. *Let $V \subseteq H \subseteq V^*$ be an evolution triple, and let $X = L^p(0, \tau; V)$, where $1 < p < \infty$. Then the operator $L : D(L) \subseteq X \to X^*$ is densely defined, closed, and maximal monotone.*

Let us state the following conditions on the time-dependent operators $A(t) : V \to V^*$.

(H1) $\|A(t)u\|_{V^*} \leq c_0 \left(\|u\|_V^{p-1} + k_0(t) \right)$ for all $u \in V$ and $t \in [0, \tau]$ with some positive constant c_0 and $k_0 \in L^q(0, \tau)$.

(H2) $A(t) : V \to V^*$ is demicontinuous for each $t \in [0, \tau]$.

(H3) The function $t \to \langle A(t)u, v \rangle$ is measurable on $(0, \tau)$ for all $u, v \in V$.

(H4) $\langle A(t)u, u \rangle \geq c_1(\|u\|_V^p - k_1(t))$ for all $u \in V$ and $t \in [0, \tau]$ with some constant $c_1 > 0$ and some function $k_1 \in L^1(0, \tau)$.

Define an operator $\hat{A}$ related with $A(t)$ by

$$\hat{A}(u)(t) = A(t)u(t), \quad t \in [0, \tau], \qquad (9.80)$$

which may be considered as the associated Nemytskij operator generated by the operator-valued function $t \mapsto A(t)$. Thus problem (9.79) corresponds to the following one.

$$u \in D(L) : Lu + \hat{A}(u) = f \quad \text{in } X^*. \qquad (9.81)$$

Definition 9.107. *Let $D(L)$ be equipped with the graph norm; that is,*

$$\|u\|_L = \|u\|_X + \|Lu\|_{X^*}.$$

The operator $\hat{A} : X \to X^*$ is called pseudomonotone with respect to the graph norm topology of $D(L)$ (or pseudomonotone w.r.t. $D(L)$ for short), if for any sequence $(u_n) \in D(L)$ satisfying

$$u_n \rightharpoonup u \text{ in } X, \quad Lu_n \rightharpoonup Lu \text{ in } X^*, \quad \text{and} \quad \limsup_{n\to\infty}\langle\hat{A}(u_n), u_n - u\rangle \le 0,$$

it follows that

$$\hat{A}(u_n) \rightharpoonup \hat{A}(u) \text{ in } X^* \text{ and } \langle\hat{A}(u_n), u_n\rangle \to \langle\hat{A}(u), u\rangle.$$

In an obvious similar way the (S_+)-condition with respect to $D(L)$ is defined.

For the following surjectivity result, which yields the existence for problem (9.81), we refer to [20, 172].

Theorem 9.108. *Let $L : D(L) \subseteq X \to X^*$ be as given above, and let $\hat{A} : X \to X^*$ defined by (9.80) be bounded, demicontinuous, and pseudomonotone w.r.t. $D(L)$. If $\hat{A}$ is coercive, then $(L + \hat{A})(D(L)) = X^*$, that is, $L + \hat{A}$ is surjective.*

The next result shows that certain properties of the operators $A(t)$ are transfered to its Nemytskij operator $\hat{A}$; cf. [21].

Theorem 9.109. *(a) Let hypotheses (H1)–(H4) be satisfied. Then we have the following results.*
  (i) If $A(t) : V \to V^*$ is pseudomonotone for all $t \in [0, \tau]$, then $\hat{A} : X \to X^*$ is pseudomonotone with respect to $D(L)$ according to Definition 9.107
  (ii) If $A(t) : V \to V^*$ has the (S_+)-property for all $t \in [0, \tau]$, then $\hat{A} : X \to X^*$ has the (S_+)-property with respect to $D(L)$.
(b) (iii) Hypotheses (H1) and (H3) imply that $\hat{A} : X \to X^*$ is bounded.
  (iv) Hypotheses (H1)–(H3) imply that $\hat{A} : X \to X^*$ is demicontinuous.
  (v) Hypothesis (H4) implies that $\hat{A} : X \to X^*$ is coercive.

9.6 Calculus of Clarke's Generalized Gradient

The material of this section is mainly based on [80]. Throughout this section X stands for a real Banach space endowed with the norm $\|\cdot\|$. The dual space of X is denoted X^*, and the notation $\langle\cdot, \cdot\rangle$ means the duality pairing between X^* and X.

We recall the following well-known definition.

Definition 9.110. *A functional $f : X \to \mathbb{R}$ is said to be locally Lipschitz if for every point $x \in X$ there exist a neighborhood V of x in X and a constant $K > 0$ such that*

$$|f(y) - f(z)| \le K \|y - z\|, \quad \forall y, z \in V.$$

Example 9.111. A convex and continuous function $f : X \to \mathbb{R}$ is locally Lipschitz. More generally, a convex function $f : X \to \mathbb{R}$, that is bounded above on a neighborhood of some point is locally Lipschitz (see, e.g., [80, p. 34]).

The classical theory of differentiability does not work in the case of locally Lipschitz functions. However, a suitable subdifferential calculus approach has been developed by Clarke [80]. Here we give a brief introduction.

Definition 9.112 (Generalized Directional Derivative). *Let $f : X \to \mathbb{R}$ be a locally Lipschitz function and fix two points $u, v \in X$. The generalized directional derivative of f at u in the direction v is defined as follows*

$$f^o(u; v) = \limsup_{\substack{x \to u \\ t \downarrow 0}} \frac{f(x + tv) - f(x)}{t}.$$

Since f is locally Lipschitz, it is clear that $f^o(u; v) \in \mathbb{R}$.

Proposition 9.113. *If $f : X \to \mathbb{R}$ is a locally Lipschitz function, then the following holds.*

(i)   *The function $f^o(u; \cdot) : X \to \mathbb{R}$ is subadditive, positively homogeneous, and satisfies the inequality*

$$|f^o(u; v)| \le K \|v\|, \quad \forall v \in X,$$

   where $K > 0$ is the Lipschitz constant of f near the point $u \in X$.
(ii) *$f^o(u; -v) = (-f)^o(u; v), \quad \forall v \in X$.*
(iii) *The function $(u, v) \in X \times X \mapsto f^o(u; v) \in \mathbb{R}$ is upper semicontinuous.*

Proof: The result follows directly from Definition 9.112.   □

The next definition focuses on the case where $f^o(u; v)$ reduces to the usual directional derivative

$$f'(u; v) = \lim_{t \downarrow 0} \frac{f(u + tv) - f(u)}{t}.$$

Definition 9.114. *A locally Lipschitz function $f : X \to \mathbb{R}$ is said to be regular at a point $u \in X$ if*

(i) *there exists the directional derivative $f'(u; v)$, for every $v \in X$;*
(ii) *$f^o(u; v) = f'(u; v), \quad \forall v \in X$.*

Significant classes of regular functions are given in the following examples.

Example 9.115. If the function $f : X \to \mathbb{R}$ is strictly differentiable, that is for all $u \in X$ there exists $f'(u) \in X^*$ such that

$$\lim_{\substack{w \to u \\ t \downarrow 0}} \frac{f(w + tv) - f(w)}{t} = \langle f'(u), v \rangle, \quad \forall v \in X,$$

where the convergence is uniform for v in compact sets, then f is locally Lipschitz and regular in the sense of Definition 9.114. In particular, if $f : X \to \mathbb{R}$ is a continuously differentiable function, then f is strictly differentiable, so it is locally Lipschitz and regular.

Example 9.116. A convex and continuous function $f : X \to \mathbb{R}$ is regular.

Definition 9.117 (Clarke's Generalized Gradient). *The generalized gradient of a locally Lipschitz functional $f : X \to \mathbb{R}$ at a point $u \in X$ is the subset of X^* defined by*

$$\partial f(u) = \{\zeta \in X^* : f^\circ(u; v) \geq \langle \zeta, v \rangle, \ \forall v \in X\}.$$

By using the Hahn–Banach theorem (see, e.g., [30, p. 1]), it follows $\partial f(u) \neq \emptyset$.

Example 9.118. If $f : X \to \mathbb{R}$ is a locally Lipschitz function that is Gâteaux differentiable and regular at the point $u \in X$, then one has $\partial f(u) = \{D_G f(u)\}$, where $D_G f(u)$ denotes the Gâteaux differential of f at u. Indeed, since f is Gâteaux differentiable and regular at u, we may write

$$\langle D_G f(u), v \rangle = f'(u; v) = f^\circ(u; v), \quad \forall v \in X,$$

which implies $D_G f(u) \in \partial f(u)$. Conversely, if $\zeta \in \partial f(u)$, from Definitions 9.117 and 9.114 in conjunction with the assumption that f is Gâteaux differentiable at u, it turns out that

$$\langle \zeta, v \rangle \leq f^\circ(u; v) = f'(u; v) = \langle D_G f(u), v \rangle, \quad \forall v \in X,$$

so $\zeta = D_G f(u)$.

Example 9.119. If $f : X \to \mathbb{R}$ is continuously differentiable, then $\partial f(u) = \{f'(u)\}$ for all $u \in X$, where $f'(u)$ denotes the Fréchet differential of f at u. This is a direct consequence of Example 9.118.

Example 9.120. If $f : X \to \mathbb{R}$ is convex and continuous, then the generalized gradient $\partial f(u)$ coincides with the subdifferential of f at u in the sense of Convex Analysis. This follows from Examples 9.111 and 9.116.

Remark 9.121. It is seen from Definition 9.117, Example 9.120, and Proposition 9.113 (i) that the generalized gradient of a locally Lipschitz functional $f : X \to \mathbb{R}$ at a point $u \in X$ is given by

$$\partial f(u) = \partial(f^{\circ}(u; \cdot))(0),$$

where in the right-hand side the subdifferential in the sense of convex analysis is written.

The next proposition presents some important properties of the generalized gradient.

Proposition 9.122. *Let $f : X \to \mathbb{R}$ be a locally Lipschitz function. Then for any $u \in X$ the below properties hold:*

(i) $\partial f(u)$ is a convex, weak*-compact subset of X^* and

$$\|\zeta\|_{X^*} \leq K, \quad \forall \zeta \in \partial f(u),$$

where $K > 0$ is the Lipschitz constant of f near u;

(ii) $f^{\circ}(u; v) = \max\{\langle \zeta, v \rangle : \zeta \in \partial f(u)\}, \quad \forall v \in X$.

(iii) The mapping $u \mapsto \partial f(u)$ is weak*-closed from X into X^*.

(iv) The mapping $u \mapsto \partial f(u)$ is upper semicontinuous from X into X^*, where X^* is equipped with the weak*-topology.

Proof: As for (i) and (ii), one applies Definitions 9.112 and 9.117, and as for (iv) see, e.g., [80]. To see (iii), let $(u_n) \subset X$ satisfy $u_n \to u$ in X, and let $\zeta_n \in \partial f(u_n)$ with $\zeta_n \rightharpoonup^* \zeta$ in X^*. We need to show that $\zeta \in \partial f(u)$. By Definition 9.117 one has $\langle \zeta_n, v \rangle \leq f^{\circ}(u_n; v)$ for all $v \in X$, which due to the weak*-convergence of (ζ_n) and the upper semicontinuity of the function $x \mapsto f^{\circ}(x; v)$ according to Proposition 9.113 (iii) implies

$$\langle \zeta, v \rangle \leq \limsup_{n \to \infty} f^{\circ}(u_n; v) \leq f^{\circ}(u; v) \text{ for all } v \in X,$$

and thus $\zeta \in \partial f(u)$. $\square$

Remark 9.123. The definitions and results given here are applicable to a locally Lipschitz function $f : U \to \mathbb{R}$ on a nonempty, open subset U of the Banach space X.

In what follows we recall basic calculus rules for the generalized gradient.

Proposition 9.124. *Let $f : X \to \mathbb{R}$ be a locally Lipschitz function, let $\lambda \in \mathbb{R}$, and let $u \in X$. Then the following formula holds*

$$\partial(\lambda f)(u) = \lambda \partial f(u).$$

In particular, one has

$$\partial(-f)(u) = -\partial f(u).$$

Proposition 9.125. *Let $f, g : X \to \mathbb{R}$ be locally Lipschitz functions. Then for every $u \in X$ the following inclusion holds*

$$\partial(f + g)(u) \subseteq \partial f(u) + \partial g(u).$$

If, in addition, the functions f and g are regular at the point $u \in X$, then the above inclusion becomes an equality, and $f + g$ is regular at u.

Remark 9.126. The inclusion of Proposition 9.125 becomes an equality also in case that at least one of the two locally Lipschitz functions is strictly differentiable, because, in general, one has the following rule.

Proposition 9.127 (Finite Sums). *Let $f_i : X \to \mathbb{R}, i = 1, \ldots, m$, be locally Lipschitz functions. Then for every $u \in X$ the following inclusion holds*

$$\partial \left(\sum_{i=1}^{m} f_i \right)(u) \subseteq \sum_{i=1}^{m} \partial f_i(u).$$

If all but at most one of the locally Lipschitz functions f_i are strictly differentiable, then the inclusion above becomes an equality.

The result below presents the mean value property for locally Lipschitz functionals due to Lebourg [165].

Theorem 9.128. *Let $f : X \to \mathbb{R}$ be a locally Lipschitz function. Then for all $x, y \in X$, there exist $u = x + t_0(y - x)$, with $0 < t_0 < 1$, and $\zeta \in \partial f(u)$ such that*

$$f(y) - f(x) = \langle \zeta, y - x \rangle.$$

Another important result in the calculus with generalized gradients is the chain rule.

Theorem 9.129. *Let $F : X \to Y$ be a continuously differentiable mapping between the Banach spaces X, Y, and let $g : Y \to \mathbb{R}$ be a locally Lipschitz function. Then the function $g \circ F : X \to \mathbb{R}$ is locally Lipschitz, and for any point $u \in X$ the following formula holds:*

$$\partial(g \circ F)(u) \subseteq \partial g(F(u)) \circ DF(u) \tag{9.82}$$

in the sense that every element $z \in \partial(g \circ F)(u)$ can be expressed as

$$z = DF(u)^* \zeta, \quad \text{for some } \zeta \in \partial g(F(u)),$$

where $DF(u)^*$ denotes the adjoint operator associated with the Fréchet differential $DF(u)$ of F at u. If, in addition, F maps every neighborhood of u onto a dense subset of a neighborhood of $F(u)$, then (9.82) becomes an equality.

Corollary 9.130. *Under the assumptions of the first part of Theorem 9.129, if g (or $-g$) is regular at $F(u)$, then $g \circ F$ (or $-g \circ F$) is regular at u and equality holds in (9.82).*

Corollary 9.131. *If there exists a linear continuous embedding $i : X \to Y$ of the Banach space X into a Banach space Y, then for every locally Lipschitz function $g : Y \to \mathbb{R}$ one has*

$$\partial(g \circ i)(u) \subseteq i^* \partial g(i(u)), \quad \forall u \in X.$$

If, in addition, $i(X)$ is dense in Y, then

$$\partial(g \circ i)(u) = i^* \partial g(i(u)), \quad \forall u \in X.$$

Finally, we give Aubin–Clarke's Theorem [14] of subdifferentiation under the integral sign.

Let numbers $m \geq 1$, $1 < p < +\infty$, and let T be a complete measure space with $|T| < \infty$, where $|T|$ stands for the measure of T. Let $j : T \times \mathbb{R}^m \to \mathbb{R}$ be a function such that $j(\cdot, y) : T \to \mathbb{R}$ is measurable whenever $y \in \mathbb{R}^m$, and satisfies either

$$|j(x, y_1) - j(x, y_2)| \leq k(x)\|y_1 - y_2\|, \quad \text{for a.a. } x \in T, \ \forall y_1, y_2 \in \mathbb{R}^m, \quad (9.83)$$

with a function $k \in L^q_+(T)$ and $1/p + 1/q = 1$, or, $j(x, \cdot) : \mathbb{R}^m \to \mathbb{R}$ is locally Lipschitz for almost all $x \in T$ and there are a constant $c > 0$ and a function $h \in L^q_+(T)$ such that

$$\|z\| \leq h(x) + c\|y\|^{p-1}, \quad \text{a.a. } x \in T, \quad \forall y \in \mathbb{R}^m, \ \forall z \in \partial_y j(x, y). \quad (9.84)$$

The notation $\partial_y j(x, y)$ in (9.84) means the generalized gradient of j with respect to the second variable $y \in \mathbb{R}^m$, i.e., $\partial_y j(x, y) = \partial j(x, \cdot)(y)$. We introduce the functional $J : L^p(T; \mathbb{R}^m) \to \mathbb{R}$ by

$$J(v) = \int_T j(x, v(x))dx, \quad \forall v \in L^p(T; \mathbb{R}^m). \quad (9.85)$$

Theorem 9.132 (Aubin–Clarke's Theorem). *Under assumption (9.83) or (9.84), one has that the functional $J : L^p(T; \mathbb{R}^m) \to \mathbb{R}$ in (9.85) is Lipschitz continuous on bounded subsets of $L^p(T; \mathbb{R}^m)$ and its generalized gradient satisfies*

$$\partial J(u) \subseteq \{w \in L^q(T; \mathbb{R}^m) : \ w(x) \in \partial_y j(x, u(x)) \ \text{for a.e. } x \in T\}. \quad (9.86)$$

Moreover, if $j(x, \cdot)$ is regular at $u(x)$ for almost all $x \in T$, then J is regular at u and (9.86) holds with equality.

List of Symbols

$\mathbb{N}$	natural numbers
$\mathbb{N}_0$	$\mathbb{N} \cup \{0\}$
$\mathbb{Z}$	integer numbers
$\mathbb{R}$	real numbers
$\mathbb{R}_+$	nonnegative real numbers
$\mathbb{R}^N$	N-dimensional Euclidean space
$|E|$	Lebesgue-measure of a subset $E \subset \mathbb{R}^N$
X	real normed linear space
X^* or X'	dual space of X
$\langle \cdot, \cdot \rangle$	duality pairing
X_+	positive (or order) cone of X
$x \wedge y$	$\min\{x, y\}$
$x \vee y$	$\max\{x, y\}$
x^+	$\max\{x, 0\}$
x^-	$\max\{-x, 0\}$
$\sup A$	least upper bound (or supremum) of A
$\inf A$	greatest lower bound (or infimum) of A
$\max A$	greatest element (or maximum) of A
$\min A$	smallest element (or minimum) of A
$\mathrm{ocl}(K)$	order closure of a subset K of a poset X
$\mathrm{cl}(K)$ or $\overline{K}$	topological closure of a subset K of X
$\mathrm{int}(K)$ or $\overset{\circ}{K}$	interior of K
$X \subseteq Y$	X is a subset of Y including Y
$X \subset Y$	X is a proper subset of Y not including Y
2^X	power set of the set X, i.e., the set of all subsets of X
$\emptyset$	empty set
$C^{<x}$	the set of those elements of $C \subset X$ that are $< x \in X$
$[x)$	the set of those elements of X that are $\geq x \in X$
$(y]$	the set of those elements of X that are $\leq y \in X$
$[x, y]$	$[x) \cap (y]$

S. Carl and S. Heikkilä, *Fixed Point Theory in Ordered Sets and Applications:
From Differential and Integral Equations to Game Theory*,
DOI 10.1007/978-1-4419-7585-0, © Springer Science+Business Media, LLC 2011

$[s]$	greatest integer $\leq s \in \mathbb{R}$
$\|x\|$	absolute value of x $(= x^+ + x^-)$
$\|x\|$	norm of a vector x
$\|f\|_A$	Alexiewicz norm of an HL integrable function f
a.a.	"almost all"
a.e.	"almost every"
iff	"if and only if"
$D(A)$	domain of the operator A
$\text{dom}(A)$	effective domain of the mapping A
I_K	indicator function, i.e., $I_K(x) = 0$ if $x \in K$, $+\infty$ otherwise
χ_E	characteristic function of the set E
$\text{Gr}(A)$	graph of the mapping A
A^*	adjoint or dual operator to A
$M(\Omega, E)$	space of measurable functions from Ω to E
$L^p(\Omega, E)$	space of p integrable functions (whose L^p norm is finite)
$\|f\|_{L^p(\Omega)}$	$\left(\int_\Omega \|f\|^p d\mu\right)^{1/p}$, the L^p norm
$L^p_{\text{loc}}(\Omega)$	space of locally p integrable functions
$\int$	Bochner (or Lebesgue) integral
$^K\!\int$	Henstock–Kurzweil integral
$W^1_{SL}(J, E)$	space of a.e. differentiable functions from J to E satisfying the Strong Lusin Condition
$HL(J, E)$	space of HL integrable functions from J to E
$HL_{loc}(J, E)$	space of locally HL integrable functions from J to E
c_0	space of those sequences of reals that converge to 0
l^p	space of such sequences $(x_n)_{n=1}^\infty$ of reals that $\sum \|x_n\|^p < \infty$
γ	Euler constant $= \lim_{n\to\infty} \left(\sum_{i=1}^n \frac{1}{i} - ln(n)\right)$
$Si(x)$	sine integral $= \int_0^x \frac{\sin t}{t} dt$
$Ci(x)$	cosine integral $= \int_0^x \frac{\cos t - 1}{t} dt + \gamma + \ln(x)$
$FrS(x)$	Fresnel sine integral $= \int_0^x \sin(\frac{\pi}{2}t^2) dt$
$Ci(x)$	Fresnel cosine integral $= \int_0^x \cos(\frac{\pi}{2}t^2) dt$
x_{-i}	$(x_1, \ldots, x_{i-1}, x_{i+1}, \ldots, x_N)$
(x_i, x_{-i})	$(x_1, \ldots, x_N)$
X_{-i}	$X_1 \times \cdots \times X_{i-1} \times X_{i+1} \times \cdots \times X_N$
$\rightharpoonup$	weak convergence
$f^\circ(u; h)$	generalized directional derivative
∂f	subdifferential of f or Clarke's generalized gradient of f

∇f $\qquad$ $(\partial f/\partial x_1, \partial f/\partial x_2, \ldots, \partial f/\partial x_N)$, the gradient of f

Δf $\qquad$ $\partial^2 f/\partial x_1^2 + \partial^2 f/\partial x_2^2 + \cdots + \partial^2 f/\partial x_N^2$, the Laplacian of f

$\Delta_p f$ $\qquad$ the p-Laplacian of f

$C_0^\infty(\Omega)$ $\qquad$ space of infinitely differentiable functions
with compact support in Ω

$\|f\|_{W^{m,p}(\Omega)}$ $\left(\sum_{|\beta| \le m} \int_\Omega |D^\beta f|^p dx \right)^{1/p}$, the Sobolev norm

$W^{m,p}(\Omega)$ $\qquad$ space of functions with bounded $W^{m,p}(\Omega)$ Sobolev norm–Sobolev space

$W_0^{m,p}(\Omega)$ $\qquad$ $W^{m,p}(\Omega)$-functions with generalized homogeneous boundary values

$\gamma(u)$ or γu $\qquad$ trace of u or generalized boundary values of u

$L^p(0, \tau; B)$ $\qquad$ space of p integrable vector-valued functions
$u : (0, \tau) \to B$

$C([0, \tau]; B)$ $\qquad$ space of continuous vector-valued functions
$u : [0, \tau] \to B$

$C^1([0, \tau]; B)$ $\qquad$ space of continuously differentiable vector-valued
functions $u : [0, \tau] \to B$

References

1. Abian, S., Brown, A.B.: A theorem on partially ordered sets, with applications to fixed point theorems. Canad. J. Math. **13**, 78–82 (1961)
2. Agarwal, R.P., O'Regan, D., Sahu, D.R.: Fixed Point Theory for Lipschitzian-type Mappings with Applications. Springer, New York (2009)
3. Agarwal, R.P., O'Regan, D.: Infinite Interval Problems for Differential, Difference and Integral Equations. Kluwer Acad. Publ., New York (2001)
4. Agarwal, R.P., Meehan, M., O'Regan, D.: Fixed Point Theory and Applications. Cambridge University Press, Cambridge (2001)
5. Agarwal, R.P., O'Regan, D., Sikorska-Nowak, A.: The set of solutions of integrodifferential equations in Banach spaces. Bull. Aust. Math. Soc. **78**, 507–522 (2008)
6. Anane, A.: Simplicity and isolation of the first eigenvalue of the p-Laplacian with weight. C. R. Acad. Sci. Paris Sér. I Math., **305** (16), 725–728 (1987)
7. Amann, H.: Fixed point equations and nonlinear eigenvalue problems in ordered Banach spaces. SIAM Rev. **18**, 620–709 (1976)
8. Amann, H.: Order structures and fixed points. ATTI 2° Sem. Anal. Funz. Appl. (1977)
9. Amir, R.: Supermodularity and complementarity in economics: An elementary survey. Southern Economical J. **71** (3), 636–660 (2005)
10. Amir, R., Bloch, F.: Comparative statics in a simple class of strategic market games. Games and Economic Behavior **65** (1), 7–24 (2009)
11. Ashworth, S., de Mesquita, E.B.: Monotone comparative statics for models of politics. Amer. J. of Political Sci. **50**, 214–231 (2006)
12. Athey, S.: Single crossing properties and the existence of pure strategy equilibria in games of incomplete information. Econometrica **69** (4), 861–890 (2007)
13. Athey, S., Milgrom, P., Roberts, J.: Robust Comparative Statics. Draft Monograph (1998)
14. Aubin, J.P., Clarke, F.H.: Shadow prices and duality for a class of optimal control problems. SIAM J. Control Optim. **17**, 567–586 (1979)
15. Baclawski, K., Björner, A.: Fixed points in partially ordered sets. Adv. in Math. **31** (3), 263–287 (1979)
16. Bai Chuanzhi, Fang Jinxuan: On positive solutions of boundary value problems for second order functional differential equations on infinite intervals. J. Math. Anal. Appl. **282**, 711–731 (2003)

17. Baoqiang Yan: Boundary value problems on the half-line with impulse and infinite delay. J. Math. Anal. Appl. **259**, 94–114 (2001)
18. Bartle, R.G.: A Modern Theory of Integration. Amer. Math. Soc., Graduate Studies in Math., Providence, RI (2001)
19. Berkovits, J., Mustonen, V.: Nonlinear mappings of monotone type. Report, University of Oulu, Oulu (1988)
20. Berkovits, J., Mustonen, V.: Topological degree for perturbations of linear maximal monotone mappings and applications to a class of parabolic problems. Rend. Mat. Appl., Serie VII **12**, 597–621 (1992)
21. Berkovits, J., Mustonen, V.: Monotonicity methods for nonlinear evolution equations. Nonlinear Anal., **27**, 1397–1405 (1996)
22. Birkhoff, G.: Lattice Theory. Amer. Math. Soc. Publ. **XXV**, Providence, RI (1940)
23. Birkhoff, G.: Lattice Theory. 3rd edition, Amer. Math. Soc. Publ. Volume 3, Providence, RI (1967)
24. Boccuto, A., Riecan, B., Vrábelová, M.: Kurzweil–Henstock Integral in Riesz Spaces. Bentham eBooks (2009)
25. Bohl, E.: Monotonie, Lösbarkeit und Numerik bei Operatorgleichungen. Springer-Verlag, Berlin (1974)
26. Bongiorno, B.: Relatively weakly compact sets in the Denjoy Space. J. Math. Study **27**, 37–43 (1994)
27. Bonafede, S., Marano, S. A.: Implicit parabolic differential equations. Bull. Austral. Math. Soc. **51**, 501–509 (1995)
28. Bourbaki, N.: Eléments de mathématique: I. Théorie des ensembles. Fascicule de Résultats, Actualités Scientifiques et Industrielles, no. 846, Hermann, Paris (1939)
29. Bourbaki, N.: General Topology. Hermann, Paris (1948)
30. Brézis, H.: Analyse Fonctionnelle - Théorie et Applications. Masson, Paris (1983)
31. Browder, F.E.: Fixed point theory and nonlinear problems. Bull. AMS **9**, 1–39 (1983)
32. Brown, R.F., Furi, M., Górniewicz, L., Jiang, B. (eds): Handbook of Topological Fixed Point Theory. Springer, Dordrecht (2005)
33. Büber, T., Kirk, W.A.: A constructive proof of a fixed point theorem of Soardi. Math Japonica **41**, 233–237 (1995)
34. Büber, T., Kirk, W.A.: Constructive aspects of fixed point theory for nonexpansive mappings. Proc. First World Congress of Nonlinear Analysts, Tampa 1992, Walter de Gruyter, 2115–2125 (1996)
35. Burke, M.: Borel measurability of separately continuous functions. Topology and its Applications **129**, 29–65 (2003)
36. Cantor, G.: Beiträge zur Begründung der transfiniten Mengenlehre II. Math. Ann. **49** (1897)
37. Carathéodory, C.: Vorlesuncen über Reelle Funktionen. Chelsea Publishing Company, New York, (1948)
38. Carl, S.: Existence and comparison results for noncoercive and nonmonotone multivalued elliptic problems. Nonlinear Anal. **65** (8), 1532–1546 (2006)
39. Carl, S.: The sup-supersolution method for variational-hemivariational inequalities. Nonlinear Anal. **69**, 816–822 (2008)
40. Carl, S.: Multiple solutions of quasilinear periodic-parabolic inclusions. Nonlinear Anal. **72**, 2909–2922 (2010)

41. Carl, S.: Equivalence of some multi-valued elliptic variational inequalities and variational-hemivariational inequalities. Adv. Nonlinear Stud. (2010) (in press)
42. Carl, S., Gilbert, R.P.: Extremal solutions of a class of dynamic boundary hemivariational inequalities. J. Inequal. Appl. **7**, 479–502 (2002)
43. Carl, S., Heikkilä, S.: On discontinuous implicit evolution equation. J. Math Anal. Appl. **219**, 455–471 (1998)
44. Carl, S., Heikkilä, S.: Nonlinear Differential Equations in Ordered Spaces. Chapman & Hall/CRC, Boca Raton (2000)
45. Carl, S., Heikkilä, S.: Discontinuous reaction-diffusion equations under discontinuous and nonlocal flux condition. Math. Comput. Modelling **32**, 1333–1344 (in Special issue: "Advanced Topics in Nonlinear Operator Theory") (2000)
46. Carl, S., Heikkilä, S.: On discontinuous implicit and explicit abstract impulsive boundary value problems. Nonlinear Anal. **41**, 701–723 (2000)
47. Carl, S., Heikkilä, S.: Operator equations in ordered sets and discontinuous implicit parabolic equations. Nonlinear Anal. **43**, 605–622 (2001)
48. Carl, S., Heikkilä, S.: Elliptic problems with lack of compactness via a new fixed point theorem. J. Differential Equations **186**, 122–140 (2002)
49. Carl, S., Heikkilä, S.: Existence results for equations in ordered Hilbert spaces and elliptic BVP's. Nonlinear Funct. Anal. & Appl. **7** (4), 531–546 (2002)
50. Carl, S., Heikkilä, S.: Existence results for implicit functional evolution equations and parabolic IBVP's. Nonlinear Studies **10**, 237–246 (2003)
51. Carl, S., Heikkilä, S.: Fixed point theorems for multivalued operators and application to discontinuous quasilinear BVP's. Appl. Anal. **82**, 1017–1028 (2003)
52. Carl, S., Heikkilä, S.: Fixed point theorems for multivalued functions with applications to discontinuous operator and differential equations. J. Math. Anal. Appl. **297** (1), 56–69 (2004)
53. Carl, S., Heikkilä, S.: Nonsmooth and nonlocal differential equations in lattice-ordered Banach spaces. Boundary Value Problems **2005** (2), 165–179 (2005)
54. Carl, S., Heikkilä, S.: Existence results for nonlocal and nonsmooth hemivariational inequalities. J. Inequal. Appl. **2006**, Art. ID 79532, 13 pp. (2006)
55. Carl, S., Heikkilä, S.: Existence results for discontinuous functional evolution equations in abstract spaces. Int. J. Math. Game Theory Algebra **16**(6), 495–506 (2007)
56. Carl, S., Heikkilä, S.: Nonsmooth and nonlocal implicit differential equations in lattice-ordered Banach spaces. Nonlinear Stud. **15**, 11–28 (2008)
57. Carl, S., Heikkilä, S.: p-Laplacian inclusions via fixed points for multifunctions in posets. Set-Valued Analysis **16**, 637–649 (2008)
58. Carl, S., Heikkilä, S.: Existence and multiplicity for quasilinear elliptic inclusions with nonmonotone discontinuous multifunction. Nonlinear Analysis Series B: Real World Applications **10**, 2326–2334 (2009)
59. Carl, S., Heikkilä, S., Jerome J.W.: Trapping regions for discontinuously coupled systems of evolution variational inequalities and applications. J. Math. Anal. Appl. **282**, 421–435 (2003)
60. Carl, S., Heikkilä, S., Guoju Ye: Order properties of spaces of non-absolutely integrable vector-valued functions and applications to differential equations. Differential and Integral Equations, **22** (1–2), 135–156 (2009)
61. Carl, S., Le, V.K.: Sub-supersolution method for quasilinear parabolic variational inequalities. J. Math. Anal. Appl. **293**, 269–284 (2004)
62. Carl, S., Le, V.K., Motreanu, D.: Nonsmooth Variational Problems and Their Inequalities. Springer, New York (2007)

63. Carl, S., Le, V.K., Motreanu, D.: Evolutionary variational-hemivariational inequalities: existence and comparison results. J. Math. Anal. Appl. **345**, 545–558 (2008)

64. Carl, S., Motreanu, D.: Extremal solutions of quasilinear parabolic inclusions with generalized Clarke's gradient. J. Differential Equations **191**, 206–233 (2003)

65. Carl, S., Motreanu, D.: Extremality in solving general quasilinear parabolic inclusions. J. Optim. Theory Appl. **123**, 463–477 (2004)

66. Carl, S., Motreanu, D.: Quasilinear elliptic inclusions of Clarke's gradient type under local growth conditions. Appl. Anal. **85** (12), 1527–1540 (2006)

67. Carl, S., Motreanu, D.: Constant-sign and sign-changing solutions for nonlinear eigenvalue problems. Nonlinear Anal. **68**, 2668–2676 (2008)

68. Carl, S., Motreanu, D.: Multiple solutions of nonlinear elliptic hemivariational problems. Pacific Journal of Applied Mathematics **1** (4), 39–59 (2008)

69. Carl, S., Motreanu, D.: General comparison principle for quasilinear elliptic inclusions. Nonlinear Anal. **70**, 1105–1112 (2009)

70. Carl, S., Motreanu, D.: Comparison principle for quasilinear parabolic inclusions with Clarke's gradient. Adv. Nonlinear Stud. **9**, 69–80 (2009)

71. Carl, S., Motreanu, D.: Sign-changing solutions for nonlinear elliptic problems depending on parameters. International Journal of Differential Equations **2010**, Article ID 536236, doi: 10.1155/2010/536236 (2010)

72. Carl, S., Motreanu, D.: Sub-supersolution method for multi-valued elliptic and evolution problems. In: Handbook of Nonconvex Analysis and Applications (Editors: David Yang Gao and Dumitru Motreanu), International Press, Inc., of Somerville, Massachusetts, pp. 45–98, (in press)

73. Carl, S., Winkert, P.: General comparison principle for variational-hemivariational inequalities. J. Inequal. Appl. **2009**, Article ID 184348, 29 pages (2009)

74. Chambers, C.P., Echenique, P.: Supermodularity and preferences. J. Econ. Theory **144**, 1004–1014 (2009)

75. Chen, Y., Gazzale, R.: When does learning in games generate convergence to Nash equilibria? The role of supermodularity in an experimental selling. Amer. Econ. Rev. **94** (5), 1505–1535 (2004)

76. Chew, T.S., Flordeliza, F.: On $x' = f(t, x)$ and Henstock-Kurzweil integrals. Differential Integral Equations **4** (4), 861–868 (1991)

77. Chipot, M.: Nonlinear Analysis. Birkhäuser Verlag, Basel (2000)

78. Cichoń, M.: Solutions of differential equations in Banach spaces. Nonlinear Anal. **60**, 651–667 (2005)

79. Cichoń, M., Kubiazyk, I., Sikorska, A.: The Henstock-Kurzweil-Pettis integrals and existence theorems for the Cauchy problem. Czechoslovak Math. J. **54** (129), 279–289 (2004)

80. Clarke, F.H.: Optimization and Nonsmooth Analysis. Society for Industrial and Applied Mathematics (SIAM) Philadelphia, PA (1990)

81. Davis, A.C.: A characterization of complete lattices. Pacific J. of Math. **5**, 311–319 (1955)

82. Denkowski, Z., Migórski, S., Papageorgiou, N.S.: An Introduction to Nonlinear Analysis: Theory. Kluwer Academic Publishers, Boston, Dordrecht, London (2003)

83. Dessy, S., Djebbari, H.: Career choice, marriage-timing and the attraction of unequals. IZA Discussion Paper Ser. No. 1561, 1–32 (2005)

84. Dieudonné, J.: Foundations of Modern Analysis. Academic Press, New York and London (1960)
85. Dudley, R.M.: Real Analysis and Probability. Chapman & Hall, New York (1989)
86. Echenique, F.: Mixed equilibria in games of strategic complementarities. Economic Theory **22**, 33–44 (2003)
87. Echenique, F.: A characterization of strategic complementarities. Games and Economic Behavior **46** (2), 325–347 (2004)
88. Echenique, P., Edlin, A.: Mixed strategy equilibria are unstable in games of strategic complements. J. Econ. Theory **118**, 161–179 (2004)
89. Evans, L. C.: Partial Differential Equations. **19**, AMS, Providence, RI (1998)
90. Federson, M.: Some peculiarities of the Henstock and Kurzweil integrals of Banach space-valued functions. Real Analysis Exchange **29**(1), 1–22 (2003/2004)
91. Federson, M., Bianconi, M.: Linear Fredholm integral equations and the integral of Kurzweil. J. Appl. Analysis **8** (1), 81–108 (2002)
92. Federson, M., Táboas, P.: Impulsive retarded differential equations in Banach spaces via Bochner-Lebesgue and Henstock integrals. Nonlinear Anal. **50**, 389–407 (2002)
93. Federson, M., Schwabik, Š.: Generalized ODE approach to impulsive retarded functional differential equations. Differential and Integral Equations **19**, 1201–1234 (2006)
94. Fraenkel, A., Bar-Hillel, Y., Levy, A.: Foundations of Set Theory. North-Holland, Amsterdam, London (1973)
95. Fremlin, D.H.: Measure Theory. Vol. 4: Topological Measure Spaces. Torres Fremlin, Colchester (2003)
96. Fuchssteiner, B.: Iterations and fixpoints. Pacific J. Math. **68**, 73–80 (1977)
97. Gilioli, A.: Natural ultrabornological non-complete normed function spaces. Arch. Math. **61**, 465–477 (1993)
98. Gill, T.L., Zachary, W.W.: Constructive representation theory for the Feynman operator calculus. arXiv: math-ph/0701039 (2007)
99. Gilbarg, D., Trudinger, N.S.: Elliptic Partial Differential Equations of Second Order. Springer-Verlag, Berlin (1983)
100. Goebel, K., Kirk, W.A.: Topics in Metric Fixed Point Theory. Cambridge University Press, Cambridge (1990)
101. Goebel, K., Reich, S.: Uniform Convexity, Hyperbolic Geometry and Nonexpansive Mappings. Marcel Dekker, New York and Basel (1984)
102. Gordon, R.A.: The Integrals of Lebesgue, Denjoy, Perron and Henstock. Grad. Stud. Math. Vol. 4, Amer. Math. Soc., Providence, RI (1994)
103. Górniewicz, L.: Topological Fixed Point Theory of Multivalued Mappings. Springer, Dordrecht (2006)
104. Granas, A., Dugundji, J.: Fixed Point Theory. Springer-Verlag, New York (2003)
105. Guo, Da Jun: Impulsive integral equations in Banach spaces with applications. J. Math. Appl. Stochastic Anal. **5** (2), 111–122 (1992)
106. Guo, D., Cho, Y.J., Zhu, J.: Partial Ordering Methods in Nonlinear Problems. Nova Science Publishers, Inc., New York (2004)
107. Guo, Dajun, Lakshmikantham, V., Liu, Xinzhi: Nonlinear Integral Equations in Abstract Spaces. Kluwer Academic Publishers, Dordrecht (1996)
108. Guo, D., Lakshmikantham, V.: Nonlinear Problems in Abstract Cones. Academic Press, New York (1988)

109. Heal, G., Kunreuter, H.: The vaccination game. Columbia Business School and The Wharton School, manuscript, 18 pp. (2005)
110. Heikkilä, S.: On recursions, iterations and well-orderings. Nonlinear Times and Digest **2**, (1995)
111. Heikkilä, S.: Applications of a recursion method to maximization problems in partially ordered spaces. Nonlinear Studies **3** (2), 249–260 (1996)
112. Heikkilä, S.: Monotone methods with applications to nonlinear analysis. Proceedings of the First Congress of Nonlinear Analysts (1992), Walter de Gruyter, Berlin-New York, 2147–2158 (1996)
113. Heikkilä, S.: Existence results for first order discontinuous differential equations of general form. Pitman Research Notes in Mathematics Ser. **374**, 79–83, Longman, Harlow (1997)
114. Heikkilä, S.: First order discontinuous differential equations with discontinuous boundary conditions. Proc. 2nd World Congress of Nonlinear Analysts, Nonlinear Anal. **30** (3), 1753–1761 (1997)
115. Heikkilä, S.: On chain methods used in fixed point theory. Nonlinear Studies **6**, 171–180 (1999)
116. Heikkilä, S.: A method to solve discontinuous boundary value problems. Nonlinear Anal. **47**, 2387–2394 (2001)
117. Heikkilä, S.: Existence and comparison results for operator and differential equations in abstract spaces. J. Math. Anal. Appl. **274**, 586–607 (2002)
118. Heikkilä, S.: Operator equations in ordered function spaces. In: R.P. Agarwal and D. O'Regan (Eds.), Nonlinear Analysis and Applications: To V. Lakshmikantham on his 80th Birthday, Kluwer Acad. Publ., Dordrecht, ISBN 1-4020-1688-3, 595–616 (2003)
119. Heikkilä, S.: Existence results for operator and differential equations and inclusions. Nonlinear Anal. **63**, e267–e276 (2005)
120. Heikkilä, S.: Discontinuous functional integral equations in abstract spaces. in R. P. Agarwal and D. O'Regan (Eds.), Advances in Integral Equations, Dynamic Systems and Applications **14** (1), 39–56 (2005)
121. Heikkilä, S.: Fixed point results for multifunctions in ordered topological spaces with applications to inclusion problems and to game theory. Dynamic Syst. Appl. **16** (1), 105–120 (2007)
122. Heikkilä, S.: On extremal pure Nash equilibria for mixed extensions of normal form games. Nonlinear Anal. **66**, 1645–1659 (2007)
123. Heikkilä, S.: On extremal solutions of inclusion problems with applications to game theory. Nonlinear Anal. **69**, 3060–3069 (2008)
124. Heikkilä, S.: Fixed point results for semi-increasing mappings. Nonlinear Studies, to appear
125. Heikkilä, S.: Fixed point results for set-valued and single-valued mappings in ordered spaces. CUBO **10**, 119–135 (2008)
126. Heikkilä, S.: Well-posedness results for Cauchy problems containing non-absolutely integrable vector-valued functions. Nonlinear Studies, to appear
127. Heikkilä, S.: On fixed points of maximalizing mappings in posets. Fixed Point Theory and Appl. **2010**, 8 pp, Art. ID 634109
128. Heikkilä, S., Kumpulainen, M.: Differential equations with non-absolutely integrable functions in ordered Banach spaces. Nonlinear Anal. **72**, 4082–4090 (2010)
129. Heikkilä, S., Kumpulainen, M.: On improper integrals and differential equations in ordered Banach spaces. J. Math. Anal Appl. **319**, 579–603 (2006)

130. Heikkilä, S., Kumpulainen, M., Seikkala, S.: Uniqueness, comparison and existence results for discontinuous implicit differential equations. Dynam. Systems Appl. **7**, 237–244 (1998)
131. Heikkilä, S., Kumpulainen, M., Seikkala, S.: On functional improper Volterra integral equations and impulsive differential equations in ordered Banach spaces. J. Math. Anal. Appl. **341**, 433–444 (2008)
132. Heikkilä, S., Kumpulainen, M., Seikkala, S.: Convergence theorems for HL integrable vector-valued functions with applications. Nonlinear Anal. **70** (5), 1939-1955 (2009)
133. Heikkilä, S., Lakshmikantham, V.: Monotone Iterative Techniques for Discontinuous Nonlinear Differential Equations. Marcel Dekker Inc., New York (1994)
134. Heikkilä, S., O'Regan, D.: On discontinuous functional Volterra integral equations and impulsive differential equations in abstract spaces. Glasgow Math. J. **46**, 529–536 (2004)
135. Heikkilä, S. Reffett, K.: Fixed point theorems and their applications to theory of Nash equilibria. Nonlinear Anal. **64**, 1415–1436 (2006)
136. Heikkilä, S., Salonen, H.: Applications of a recursion method to game theory. Nonlinear Studies **4** (1), 81–87 (1997)
137. Heikkilä, S., Seikkala, S.: On the existence of extremal solutions of phi-Laplacian initial and boundary value problems. Int. J. Pure Appl. Math. **17** (1), 119–138 (2004)
138. Heikkilä, S., Seikkala, S.: On singular, functional, nonsmooth and implicit phi-Laplacian initial and boundary value problems. J. Math. Anal. Appl. **308** (2), 513–531 (2005)
139. Heikkilä, S., Seikkala, S.: On non-absolute functional Volterra integral equations and impulsive differential equations in ordered Banach spaces. EJDE, **2008**, No. 103, 1–11 (2008)
140. Heikkilä, S., Seikkala, S.: On singular, functional nonsmooth and implicit phi-Laplacian initial and boundary value problems. J. Math. Anal. Appl. **308**, 513–531 (2005)
141. Heikkilä, S., Seikkala, S.: On non-absolute functional Volterra integral equations and impulsive differential equations in ordered Banach spaces. EJDE **2008**, No. 103, 1–11 (2008)
142. Heinonen, J., Kilpeläinen, T., Martio, O.: Nonlinear Potential Theory of Degenerate Elliptic Equations. Clarendon Press, Oxford (1993)
143. Henstock, R.: The General Theory of Integration. Clarendon Press, Oxford (1991)
144. Holt, C.A., Roth, A.E.: The Nash equilibrium: A perspective. Proc. Natl. Acad. Sci. USA **101** (12), 3999–4002 (2004)
145. Huy, N. B.: Fixed points of increasing multivalued operators and an application to discontinuous elliptic equations. Nonlinear Anal. **51** (4), 673–678 (2002)
146. Iwasaki, N., Kudo, Y., Tremblay, C., Tremblay, V.: The advertising-price relationship: Theory and evidence. Int. J. Economics of Business **15** (2), 149–167 (2008)
    Supermodularity and advertizing-price relationship: Theory and evidence. Int. J. Economics of Business, to appear
147. Kanamori, A.: The mathematical import of Zermelo's well-ordering theorem. The Bulletin of Symbolic Logic **3** (1), 281–311 (1997)
148. Kakutani, S.: A generalization of Brouwer's fixed point theorem. Duke Math. J. **8** (3), 457–459 (1941)

149. Kamae, T., Krengel, U.: Stochastic partial ordering. The Annals of Probability **6**, 1044–1049 (1978)

150. Kamae, T., Krengel, U., O'Brien, G.L.: Stochastic inequalities in partially ordered spaces. The Annals of Probability **5**, 899–912 (1977)

151. Kinderlehrer, D., Stampacchia, G.: An Introduction to Variational Inequalities and Their Applications. Academic Press, New York (1980)

152. Kinnunen, J., Martio, O.: Nonlinear potential theory on metric spaces. Illinois J. Math. **46** (3), 857–883 (2002)

153. Knaster, B.: Un théorème sur les fonctions d'ensembles. Ann. Soc. Polon. Math. **6**, 133–134 (1927)

154. Kopparty, S., Ravishankar, C.V.: A framework for pursuit evasion games in $\mathbb{R}^n$. Information Processing Letters **96**, 114–122 (2005)

155. Krasnoselskiĭ, M.A., Zabreĭko, P.P.: Geometric Methods of Nonlinear Analysis. Izdat. "Nauka," Moscow (1975)

156. Krzyśka, S.: On the existence of continuous solutions of Urysohn and Volterra integral equations in Banach spaces. Demonstratio Math. **28** (2), 353–360 (1995)

157. Kufner, A., John, O., Fučik, S.: Function Spaces. ACADEMIA, Praque (1977)

158. Kukushkin, N.: On the existence of monotone selections. MPRA paper 15845, 17 pp. (2009)

159. Kultti, K., Salonen, H.: Undominated equilibria in games with strategic complements. Games and Economic Behavior **18**(1), 98–115 (1997)

160. Kultti, K., Salonen, H.: Iterated dominance in supermodular games with strict single crossing property. Int. J. Game Theory **27**(1), 305–309 (1998)

161. Kumpulainen, M.: On extremal and unique solutions of discontinuous ordinary differential equations and finite systems of differential equations. Dissertation, Oulu University Press, ISBN 952-90-8275-4, 98 pp. (1996)

162. Kurzweil, J.: Nichtabsolut Konvergente Integrale. Teubner-Verlag, Leipzig (1980)

163. Ladas, G., Lakshmikantham, V.: Differential Equations in Abstract Spaces. Academic Press, New York–London (1972)

164. Lang, S.: Real and Functional Analysis. Springer-Verlag, Berlin (1993)

165. Lebourg, G.: Valeur moyenne pour gradient généralisé. C. R. Acad. Sci. Paris, **281**, 795–797 (1975)

166. Lee, P.-Y.: Landzhou Lectures on Henstock Integration. World Scientific, Singapore (1989)

167. Lee, P.-Y., Lu, J.: On singularity of Henstock integrable functions. Real Anal. Exchange **25** (2), 795–798 (1999)

168. Lee, P-.Y., Výborný, R.: The Integral: An Easy Approach after Kurzweil and Henstock. Cambridge University Press (2000)

169. Levy, A.: Basic Set Theory. Springer Verlag, Berlin Heidelberg New York (1979)

170. Lindenstraus, J., Tzafriri, L.: Classical Banach Spaces II, Function Spaces. Springer-Verlag, Berlin (1979)

171. Lindqvist, P.: On the equation $\operatorname{div}|\nabla u|^{p-2}\nabla u) + \lambda |u|^{p-2}u = 0$. Proc. Amer. Math. Soc., **109**, 157–164 (1990), Addendum: Proc. Amer. Math. Soc., **116**, 583–584 (1992)

172. Lions, J.L.: Quelques Méthodes de Résolutions des Problèmes aux Limites Nonlinéaires. Dunod, Paris (1969)

173. Liu, Weian: Cones in integrable vector valued function spaces and abstract impulsive integro-differential equations. Northeast. Math. J. **12** (4), 441-446 (1996)

174. Ma, Ruyun: Existence of positive solutions for second-order boundary value problems on infinite intervals. Appl. Math. Letters **16**, 33–39 (2003)

175. Manka, R: On generalized methods of successive approximations. Nonlinear Anal. **72**, 1438–1444 (2010)

176. Markowsky, G.: Chain-complete posets and directed sets with applications. Algebra Univ. **6**, 53–68 (1976)

177. Mathevet, T.: Selection, learning and nomination: Essays on supermodular games, design and political theory. Thesis, California Institute of Technology, 146 pp. (2008)

178. McLeod, R.: The Generalized Riemann Integral. Math. Assoc. America, Washington, D.C., (1980)

179. Milgrom, P., Roberts, J.: Rationalizability, learning and equilibrium in games of strategic complementarities. Econometrica **58**(6), 1255–1277 (1990)

180. Milgrom, P., Shannon, C.: Monotone comparative statics. Econometrica **62** (1), 157–180 (1994)

181. Motreanu, D., Panagiotopoulos, P.D.: Minimax Theorems and Qualitative Properties of the Solutions of Hemivariational Inequalities and Applications. Kluwer Academic Publishers, Boston, Dordrecht, London (1999)

182. Muldowney, P.: A General Theory of Integration in Function Spaces. Pitman Research Notes in Mathematics, John Wiley & Sons, New York (1987)

183. Myerson, R.B.: Nash equilibrium and the history of economic theory. J. Econ. Lit **36**, 1067–1082 (1999)

184. Naniewicz, Z., Panagiotopoulos, P.D.: Mathematical Theory of Hemivariational Inequalities and Applications. Marcel Dekker, New York (1995)

185. Nash, J.F.: Equilibrium points in n-person games. Proc. Nat. Acad. Sci. USA **36**, 48–49 (1950)

186. Nash, J.F.: Non-cooperative games. Ann. Math., 2nd Ser., **54** (2), 286–295 (1951)

187. O'Regan, D.: Theory of Singular Boundary Value Problems. World Scientific, Singapore (1994)

188. O'Regan, D.: Integral equations in reflexive Banach spaces and weak topologies. Proc. Amer. Math. Soc. **124** (2), 607–614 (1996)

189. O'Regan, D.: Volterra and Urysohn integral equations in Banach spaces. J. Appl. Math. Stochastic Anal. **11** (4), 449–464 (1998)

190. Panagiotopoulos, P.D.: Hemivariational Inequalities and Applications in Mechanics and Engineering. Springer-Verlag, New York (1993)

191. Paredes, L.I., Seng, Chew Tuan, and Yee, Lee Peng: Controlled convergence theorems for strong variational Banach-valued multiple integrals. Real Anal. Exchange **28**, 579–591 (2002/2003)

192. Di Piazza, L., Marraffa, V.: The McShane, PU and Henstock integrals of Banach valued functions. Czechoslovak Math. J. **52** (127), 609–633 (2002)

193. Peng Sun, Liu Yang, de Véricomt, F.: Selfish drug allocation for containing an international influenza pandemic at the onset. Duke University of Durham, manuscript (2009)

194. Rao, M. M.: Measure Theory and Integration. Marcel Dekker, Inc., New York (2004)

195. Reem, D., Reich, S.: Zone and douple zone diagrams in abstract spaces. Colloquium Math. **115**, 129–145 (2009)
196. Reich, S., Shoikhet, S.: Nonlinear Semigroups, Fixed Points, and Geometry of Domains in Banach Spaces. Imperial College Press, London (2005)
197. Reny, P.J.: On the existence of monotone pure strategy equilibria in Bayesian games. Manuscript (2006).
198. Royden, H.L.: Real Analysis. The MacMillan Company, London (1968)
199. Salonen, H.: On the existence of undominated Nash equilibria in normal form games. Games Econom. Behav. **14**, 208–219 (1996)
200. Satco, B.: Second order three boundary value problem in Banach spaces via Henstock and Henstock-Kurzweil-Pettis integral. J. Math. Anal. Appl. **332**, 919–933 (2007)
201. Satco, B.: Nonlinear Volterra integral equations in Henstock integrability setting. Electron. J. Differential Equations **2008** (39), 9 pp. (2008)
202. Showalter, R.E.: Monotone Operators in Banach Space and Nonlinear Partial Differential Equations. **49** AMS, Providence, RI (1997)
203. Satimov, N. Yu, Muminov, G.M.: The evasion problem on the unit disc. Problemy Upravlen. Inform. **3**, 70–75 (2001)
204. Schaefer, H.H.: Topological Vector Spaces. The Macmillan Company, New York (1966)
205. Scharb, H.E.: The allocation of resources in the presence of invisibilities. J. Econ. Perspectives **8**(4), 111–128 (1994)
206. Schwabik, Š.: The Perron integral in ordinary differential equations. Differential Integral Equations **6**, 836–882 (1993)
207. Schwabik, Š., Ye, Guoju: Topics in Banach Space Integration. World Scientific, Singapore (2005)
208. Seikkala, S., Heikkilä, S.: Uniqueness, comparison and existence results for discontinuous implicit differential equations. Proc. 2nd World Congress of Nonlinear Analysts, Nonlinear Anal. **30** (3) 1771–1780 (1997)
209. Shuhong, F.: Semilattice structure of fixed point set for increasing operators. Nonlinear Anal. **27** (7) 793–796, (1996)
210. Siegel, D., Shapiro, J.: Underfunding in terrorist organizations. Int. Stud. Org. **51**(2), 402–429 (2007)
211. Sikorska-Nowak, A.: On the existence of solutions of nonlinear integral equations in Banach spaces and Henstock-Kurzweil integrals. Ann. Polon. Math. **83** (3), 257–267 (2004)
212. Sikorska-Nowak, A.: The existence theory for the differential equation $x^{(m)} = f(t, x)$ in Banach spaces and Henstock-Kurzweil integral. Demonstratio Mathematicae **40** (1), 115–124 (2007)
213. Sikorska-Nowak, A.: Existence of solutions of nonlinear integral equations and Henstock-Kurzweil integrals. Commentationes Mathematicae **47** (2), 227–238 (2007)
214. Smithson, R.E.: Fixed points of order preserving multifunctions. Proc. Amer. Math. Soc. **28** (1), 304–310 (1971)
215. Tarski, A.: A lattice-theoretical fixpoint theorem and its applications. Pacific J. Math. **5**, 285–309 (1955)
216. Topkis, D.: Minimizing a submodular function on a lattice. Operations Research **26**, 305–321 (1978)
217. Topkis, D.: Equilibrium points in nonzero-sum n-person submodular game. SIAM J. of Optimization and Control **17**, 773–787 (1989)

218. Topkis, D.: Supermodularity and Complementarity. Princeton University Press, New Jersey (1998)
219. Troianiello, G.M.: Elliptic Differential Equations and Obstacle Problems. Plenum Press, New York (1987)
220. Van Zandt, T., Vives, X.: Monotone equilibria in Bayesian games of strategic complementarities. J. Econ. Theory, Elsevier **134** (1), 339–260 (2007)
221. Vázquez, J.L.: A strong maximum principle for some quasilinear elliptic equations. Appl. Math. Optim., **12**(3), 191–202 (1984)
222. Veinott, A.F. Jr: Lattice Programming. Draft for OR 375 Seminar, The Johns Hopkins University Press, Baltimore and London (1992)
223. Vives, X.: Nash equilibrium with strategic complementarities. J. Math. Econom. **19**, 305–321 (1990)
224. Vives, X.: Games of strategic complementarities: New applications to industrial organization. Industrial Organization **23** (7), 625–637 (2005)
225. Yan, Baoqiang, Liu, Yansheng: Unbounded solutions for the singular boundary value problems for second order differential equations on the half-line. Appl. Math. Comput. **147**, 629–644 (2004)
226. Ye, Guoju: On Kurzweil and McShane integrals of Banach space-valued functions. J. Math. Anal. Appl. **330**, 753–765 (2007)
227. Yoshida, K.: Functional Analysis. Springer-Verlag, Berlin Heidelberg New York (1974)
228. Zeidler, E.: Nonlinear Functional Analysis and its Applications, Vol. I. Springer-Verlag, Berlin (1985)
229. Zeidler, E.: Nonlinear Functional Analysis and Its Applications, Vols. II A/B. Springer-Verlag, Berlin (1990)
230. Zeidler, E.: Nonlinear Functional Analysis and Its Applications, Vol. III. Springer-Verlag, New York (1985)
231. Zermelo, E.: Beweis dass jede Menge wohlgeordnet werden kann. Math. Ann. **59**, 514–516 (1904)
232. Zhou, L.: The set of Nash equilibria of a supermodular game is a complete lattice. Games Econom. Behav. **7**, 295–300 (1994)

Index